ALGEBRAIC GEOMETRY
ARCATA 1974

Redwoods at Arcata

PROCEEDINGS OF SYMPOSIA
IN PURE MATHEMATICS
Volume 29

ALGEBRAIC GEOMETRY
ARCATA 1974

AMERICAN MATHEMATICAL SOCIETY
PROVIDENCE, RHODE ISLAND
1975

PROCEEDINGS OF THE SYMPOSIUM IN PURE MATHEMATICS
OF THE AMERICAN MATHEMATICAL SOCIETY

HELD AT HUMBOLDT STATE UNIVERSITY

ARCATA, CALIFORNIA

JULY 29–AUGUST 16, 1974

EDITED BY

ROBIN HARTSHORNE

Prepared by the American Mathematical Society
with partial support from National Science Foundation grant GP-43762

Library of Congress Cataloging in Publication Data

Symposium in Pure Mathematics, Humboldt State University, 1974.
Algebraic geometry, Arcata 1974.

(Proceedings of symposia in pure mathematics; v. 29)
"Twenty-first Summer Research Institute."
1. Geometry, Algebraic–Congresses. I. Hartshorne, Robin. II. American Mathematical Society. III. Title. IV. Series.
QA564.S96 1974 516'.35 75-9530
ISBN 0-8218-1429-X

AMS (MOS) subject classifications.
Primary 13-02, 13D05, 13F15, 14-02, 14B05, 14C10, 14C30, 14D20, 14E15, 14F20, 14F99, 14G10, 14G15, 14J10, 14L15, 20G05, 32C10, 32J25, 55A25, 57C15
Secondary 12A70, 13C10, 14C15, 14C25
14F10, 14J15, 18F25, 18G05, 32B20, 53C55, 55E99

PREFACE

This volume contains the proceedings of the Summer Institute in Algebraic Geometry held in July and August 1974 in Arcata, California. Some details of the organization and activities of the institute will be found in the report following.

In preparing this volume, we have attempted to adhere to the same principles which guided the organization of the summer institute: we hope to make algebraic geometry accessible to a wider audience, and we hope to dispel some of the confusion and mystery which such a rapidly changing and complex field has inspired in the uninitiate.

To this end, we have included the texts of almost all the expository lectures series, and those seminar talks which were of a sufficiently broad nature to serve as introductions to their respective areas. This volume should therefore provide orientation to the newcomer or the specialist exploring new fields, by surveying the present state of the art, and giving references for further study.

Two papers originally intended for publication here have outgrown this volume. E. Brieskorn's much appreciated lectures on *Special singularities–resolution, deformation and monodromy* will be published later as a separate monograph. P. Deligne's lectures *Inputs of étale cohomology* are being written up jointly with J.-F. Boutot, and will appear in a companion volume to SGA5. While we have lost these two papers, we are fortunate to be able to include a survey article, not presented at Arcata, *Classification and embedding of surfaces* by E. Bombieri and D. Husemoller.

ROBIN HARTSHORNE
BERKELEY, CALIFORNIA
FEBRUARY 21, 1975

Preface

This volume contains the proceedings of the [illegible] Institute in Algebraic Geometry held [illegible] 1974 [illegible] Arcata [illegible]. Some details on the organization and activities of the institute will be found in the report following.

In preparing this volume, we have [illegible] to the [illegible] which [illegible] of the [illegible] we [illegible] of the [illegible] and [illegible] which [illegible] complex [illegible] in [illegible].

[illegible] we have [illegible] and [illegible] papers [illegible] to [illegible] in the [illegible] of the [illegible] and [illegible] for further [illegible].

[illegible] papers [illegible] will [illegible] and [illegible] with [illegible] and will [illegible] volume [illegible] papers [illegible] for [illegible].

Robin Hartshorne

Berkeley, California

February [illegible], 1975

CONTENTS

REPORT ON THE SUMMER INSTITUTE

The American Mathematical Society held its twenty-first Summer Research Institute at Humboldt State University, Arcata, California, from July 29 to August 16, 1974. "Algebraic Geometry" was selected as the topic. Members of the Committee on Summer Institutes at the time were Louis Auslander, Richard E. Bellman, William Browder (chairman), Louis Nirenberg, Walter Rudin, and John T. Tate. The institute was supported by a grant from the National Science Foundation. The Organizing Committee for the institute consisted of Michael Artin, Phillip A. Griffiths, Robin Hartshorne, Heisuke Hironaka, Nicholas Katz, and David Mumford (chairman).

The Institute marked the 10th anniversary of the first Summer Institute to be devoted exclusively to algebraic geometry: the 1964 Woods Hole Institute; and the 20th anniversary of the 1954 Boulder Summer Institute in several complex variables and algebraic geometry. The subject has grown immensely in this period, and thanks to the efforts of the fathers of this modern period of growth—Zariski, Weil and Grothendieck—seems to have attained a certain maturity. All the basic foundational work needed to put the subject on a modern axiomatic footing with sufficient generality to encompass characteristic p and mixed characteristic cases as well as the traditional complex case seems to have been done. A large proportion of the work of the Italian school has been relearned and assimilated. Our hope is that with such ample preparation, our generation will be fortunate enough to penetrate deeper into the nature and structure of algebraic varieties.

Program

Although the central subject of the institute was algebraic geometry, because of the broad overlap of this field with many neighboring fields, the program expanded to express the interests of those present in arithmetic problems, analytic geometry, commutative algebra, K-theory (and to a lesser extent, algebraic groups) as well. The committee attempted to plan about half the program in advance, by soliciting specific lectures on specific topics, and to let the remaining half develop spontaneously at the meeting. In particular, they invited ten mathematicians to present series of expository talks on particular areas of algebraic geometry and encouraged them to put together a survey of these areas including not only their own particular specialty but the area as a whole. As far as the committee could tell, all the speakers involved seemed to enjoy the challenge of communicating the central ideas of these areas in a relatively short space of time and did an excellent job. The committee also planned five seminars in advance, appointed chairmen for these and asked them to approach the first few speakers in advance. At Arcata, ten more seminars were organized spontaneously, and although many seminars had to meet simultaneously, most people were able to attend all of the seminars in which they had a very strong interest.

Lecture Series

Special singularities–resolution, deformation, and monodromy by E. Brieskorn (5 lectures)

**Some transcendental aspects of algebraic geometry* by M. Cornalba and P. Griffiths (3 lectures each)

Inputs of étale cohomology by P. Deligne (6 lectures)

**Serre problem and homological methods in commutative algebra* by D. Eisenbud (3 lectures)

**Equivalence relations on algebraic cycles, and subvarieties of small codimension* by R. Hartshorne (4 lectures)

**Triangulation of algebraic sets* by H. Hironaka (1 lecture)

**Introduction to resolution of singularities* by J. Lipman (3 lectures)

**Eigenvalues of Frobenius acting on algebraic varieties over finite fields* by B. Mazur (2 lectures)

Problems on l-adic representations by J.-P. Serre (3 lectures)

**Theory of moduli* by C. S. Seshadri (4 lectures)

Seminar Series

(1) Proof of the Weil conjectures and the hard Lefschetz theorem (Chairman, M. Artin)

Beginning of the proof of Weil conjectures by J. Milne

Continuation of proof: Lefschetz pencils by M. Artin

Kazdan and Margulis theorem by M. Artin

Chebotarev density theorem and proof of Deligne's Main Lemma by S. Bloch

* {
Deligne's proof of hard Lefschetz theorem. I by W. Messing
Hard Lefschetz theorem. II: *group theoretic reductions* by W. Messing
Hard Lefschetz theorem. III: *the Hadamard-de la Vallée Poussin argument* by W. Messing

(2) Varieties of low codimension (Chairman, R. Hartshorne)

Local cohomological dimension of algebraic varieties (*Ogus thesis*) by L. Szpiro

Conditions for embedding varieties in projective space (*work of Holme*) by R. Speiser

* denotes a paper in this volume

**Homotopy groups of projective varieties* (*work of Larsen*) by W. Barth
Gorenstein ideals in codimension 3 by G. Evans
Rings of invariants are Cohen-Macaulay (*work of Hochster and Roberts*) by A. Ogus
**Unique factorization in complete local rings, etc.* by J. Lipman
Vector bundles on projective spaces by W. Barth

(3) Classification questions and special varieties (Chairman, D. Mumford)
Introduction to Enriques' classification by D. Mumford
Hilbert modular surfaces. I by F. Hirzebruch
Hilbert modular surfaces. II by F. Hirzebruch
Kodaira dimension and classification in higher dimensions by K. Ueno
**Matsusaka's big theorem* by D. Lieberman
A finiteness theorem for curves over function fields (*Parshin and Arakelov*) by F. Oort
Surfaces of general type by A. Van de Ven

(4) Topics in analytic algebraic geometry (Chairman, P. Griffiths)
Toledo-Tong proof of Riemann-Roch by J. King (3 lectures)
Schmid's difficult theorem by J. Carlson (4 lectures)

(5) Toroidal embeddings (Chairman, P. Wagreich)
Varieties with torus actions by T. Oda
**Varieties with commutative group action* by P. Wagreich
Varieties with algebraic vector fields by J. Carrell

(6) deRham and crystalline cohomology (Chairman, N. Katz)
**Differentials of the first, second and third kinds* by W. Messing
Gauss-Manin connection by D. Lieberman (2 hours)
Differential equations on algebraic varieties ... by A. Ogus (2 hours)
**Report on crystalline cohomology: What's true!* by L. Illusie (2 hours)
**The slopes of Frobenius.* I, II by P. Berthelot (2 hours)
Report on flat duality by J. Milne
Report on Mazur-Messing by S. Bloch

(7) Singularities, equisingularity (Chairman, H. Hironaka)
**Introduction to equisingularity problems* by B. Teissier
Polar curves of plane curve singularities by M. Merle
**Topological use of polar curves* by Lê Dũng Tráng
Singularities with group actions by P. Wagreich
**A survey of knot theoretic invariants of singularities* by A. Durfee
Equisingular deformations by J. Wahl
Subanalytic chains, integration and intersections by P. Dolbeault

Rigid singularities and nonsmoothable singularities by H. Pinkham
More on resolution of singularities by H. Hironaka

(8) Recent work on compactification of moduli (Chairman, M. Rapoport)
Compactification of tube domains by D. Mumford
Regularity of automorphic forms at the cusps by Y.-S. Tai
A 'good' compactification of the Siegel moduli space by Y. Namikawa
Seshadri and Oda's compactification of the Picard variety of a singular curve by M. Rapoport
Stable vector bundles on a degenerating family of curves by D. Gieseker

(9) Arithmetic (Chairmen, J.-P. Serre and J. Tate)
Modular forms of weight 1 by J. Tate
**p-adic l-functions via moduli* (*Hurwitz case*) by N. Katz
**p-adic l-functions via moduli* (*Siegel case*) by K. Ribet
The Mordell conjecture for the modular curves $X_0(N)$ *and* $X_1(N)$ *over* **Q** by B. Mazur
Problems on l-adic representations by J.-P. Serre
Representations and nonabelian class field theory by P. Deligne
Formal groups and their division points by J.-M. Fontaine

(10) *K*-theory (Chairmen, H. Bass and S. Gersten)
Report on SGA6 by L. Illusie
**Riemann-Roch theorem for varieties with singularities* by W. Fulton
Finite generation of K_i *of curves over finite fields* (*Quillen*) by H. Bass
Algebraic cycles by S. Bloch
Higher K-groups of finite fields by E. Friedlander
Vector bundles on affine surfaces by M. P. Murthy
Higher regulators and zeta functions by A. Borel

(11) Special examples of intermediate Jacobians, etc. (Chairmen, H. Clemens and A. Landman)
Prym varieties associated to plane curves of odd degree by H. Clemens
Intersection of two quadrics in $\mathbf{P}_{2n+1}$ by A. Landman
Cubics containing a linear space of codimension two by H. Clemens
Prym varieties associated to cubic threefolds (*algebraic theory*) by J. Murre
Intersection of two quadrics as a moduli space of vector bundles by M. S. Narasimhan
A survey of moduli of vector bundles on curves by M. S. Narasimhan

(12) Algebraic groups (Chairman, A. Fauntleroy)
**Survey of representation theory.* I, II by A. Borel

Mumford's conjecture by C. S. Seshadri
Picard groups of algebraic groups by B. Iversen

(13) Weierstrass points (Chairman, R. Lax)
The number of Weierstrass points of a line bundle by J. Hubbard
Existence of Weierstrass points and deformations of curves by H. Pinkham
Arbarello's thesis (*Weierstrass points and moduli of curves*) by R. Lax
Weierstrass points of forms by A. Iarrobino

(14) Complex manifolds (Chairman, K. Ueno)
Surfaces of class VII_0 by M. Inoue
On deformations of quintic surfaces by E. Horikawa

(15) Deformation of complex analytic spaces (Chairman, A. Douady)
Deformation of complex analytic spaces by A. Douady (3 lectures)

Summary

The broad areas emphasized by this summer institute can be summarized as follows:

Étale cohomology and the Weil conjectures. Because of Deligne's recent and spectacular proof of the Weil conjectures for the absolute values of the Frobenius acting on the étale cohomology of a variety over a finite field, a great deal of interest focused on seeing not only the details of his proof, but on learning thoroughly the techniques of étale cohomology needed for this proof. Deligne talked for the first seven days, each day on the étale cohomology in general; this was followed by Artin's seminar which discussed his proof both of the Weil conjectures and the hard Lefschetz theorem.

Singularities. Many lectures concerned singularities from various points of view. Lipman discussed various approaches to proving resolution of singularities; Brieskorn discussed the topology of singularities and the tie-in with the work of the Thom school on unfoldings as well as many special types of singularities. Hironaka gave one morning talk explaining his beautiful proof of the triangulability of varieties; Hironaka led a seminar covering many more topics such as equisingularity, rigidity of singularities, and singularities with group action.

Cycles, commutative algebra, K-theory. This is a broad area united to some extent by the theme of wanting to understand the geometry of cycles of codimension greater than one, but very diverse in its methods. Hartshorne explained recent progress on the structure of the Chow ring and desingularizing cycles mod rational equivalence; Eisenbud lectured on developments in commutative algebra, e.g. on Serre's conjecture, on the intersection problem and on the structure of resolutions. Hartshorne led a seminar centered on varieties of small codimension but covering related topics too; Bass led a seminar in *K*-theory proper, which at several points tied in with the Chow ring.

Analytic geometry. Analytic and algebraic geometry are, of course, inseparable as was amply demonstrated by the survey talks of Cornalba and Griffiths. Griffiths led a seminar exploring particularly the recent Toledo-Tong proof of Riemann-Roch and Schmid's deep work on degeneration of Hodge structures. Ueno and Douady both ran seminars in the last week concerning classification of complex manifolds and deformations of complex analytic spaces respectively.

Moduli. Although closely tied at points to the theory of deformations of singularities and to the theory of variation of Hodge structure, the theory of moduli proper was the topic of Seshadri's survey talks. Mumford ran a seminar on classification questions related to moduli. Clemens ran a seminar on cubic three-folds, their intermediate Jacobians and rationality, and on the moduli space of vector bundles. Rapoport ran a seminar on the compactification of various moduli spaces, which tied in with the seminar of Wagreich on torus actions and toroidal embeddings. Lax ran a seminar on Weierstrass points of curves which ties in very closely with moduli problems for curves.

Number theory. As with analytic geometry, recent developments have tied number theory and algebraic geometry very intimately together. Serre gave general lectures on l-adic representations. Katz led a seminar on de Rham and crystalline cohomology which tied together the analytic de Rham approach to cohomology with the p-adic absolute values of Frobenius with crystalline sheaves (characterized by their ability to grow in a suitable medium and their rigidity!) acting as go between.

Location

The institute was held at Humboldt State University in Arcata, California; the university administration provided excellent support services. The participants lived almost entirely in residence halls overlooking the Jolly Giant Commons and immediately next to a beautiful redwood forest. Except for a bit of fog, the site was perfect and its remoteness from the distractions of civilization is believed to have promoted concentration on mathematics.

Two-hundred seventy mathematicians registered for the institute. Forty-nine were accompanied by their families and eighty-four participants were from foreign countries.

Saturday and Sunday excursions were organized to the Redwood National Park and Trinity Alps, where participants and their families had the opportunity to hike their choice of well-marked trails. Wine and cheese tasting parties were held on two evenings. On one occasion the group was transported to a picnic site on the banks of the Mad River. Some hardy individuals sharpened their appetites by a swim in the river before enjoying the steak cookout that was the highlight of the evening.

DAVID MUMFORD

PART I: LECTURE SERIES

Proceedings of Symposia in Pure Mathematics
Volume 29, 1975

SOME TRANSCENDENTAL ASPECTS OF ALGEBRAIC GEOMETRY

Maurizio Cornalba and Phillip A. Griffiths[1]

TABLE OF CONTENTS

0. INTRODUCTORY REMARKS

The purpose of these seven lectures is to discuss some transcendental aspects of algebraic geometry. Historically, a great deal of the subject was initially developed by analytical and topological methods. This was probably due to the origins of much of algebraic geometry as a branch of complex function theory (Gauss, Abel, Jacobi, Riemann, Weierstrass,

AMS (MOS) subject classifications (1970). Primary 14C30, 14D05, 22E15, 30A42, 32C10, 32C25, 32C30, 32C35, 32G05, 32G20, 32H20, 32H25, 32J25, 32L10, 32M10, 32M15, 32G13, 35D05, 35D10, 35N15, 53B35, 53C30, 53C55, 53C65, 58C10, 58G05, 58G10; Secondary 14F05, 14F10, 14F25, 14H15, 14M15, 30A70, 14F35.

[1] Research partially supported by National Science Foundation Grant GP 38886.

Poincaré, Picard, etc.). Another possible reason is that the intimate relationship between algebraic geometry and topology is more visible in the complex case (Lefschetz, and more recently the Hirzebruch-Riemann-Roch formula). Finally, the beautiful local and global methods of differential geometry are available in the complex case (Hodge, de Rham, Chern, Kodaira, etc.).

During the last 100 years, and especially in the last 40 years, much of the theory which was initially discovered by analytical methods has, quite properly, been put in a purely algebraic setting (the foundations, Torelli for curves, Riemann-Roch, etc.). Moreover, longstanding problems have been proved by algebraic methods (resolution of singularities). Finally, due primarily to the input from arithmetic, new and striking results over the complex numbers have been suggested and sometimes proved by algebra (Deligne's theory of mixed Hodge structures, the Tate conjectures, etc.). However, because of the fundamental nature of the complex numbers, there remains a transcendental aspect of algebraic geometry which is both essential to an understanding of the subject and quite beautiful for its own sake. In these lectures, we hope to focus on some facets of this theory.

Specifically, we shall concentrate on the following topics as illustrating transcendental algebraic geometry:

a) The _Hirzebruch-Riemann-Roch formula_ for compact, complex manifolds;

b) _Hodge theory_ for a single compact Kähler manifold, and the related _vanishing theorems_ for cohomology, theory of _mixed Hodge structures_, and _homotopy type_ of Kähler manifolds;

c) _variation of Hodge structure_ culminating in the recent work of Schmid; and

d) the global theory of transcendental holomorphic mappings (_Nevanlinna theory_) viewed as non-compact algebraic geometry.

The Hirzebruch-Riemann-Roch formula is of course well-known and has a purely algebraic proof in much stronger form. However, it is a basic result first proved by transcendental methods and has inspired a great deal of mathematics over the last 20 years. Moreover, there has recently been given an "elementary" proof by Toledo and Tong, one in which the local and global properties of the $\bar{\partial}$-operator are brought nicely into focus, and so this proof will be discussed in the second lecture. The complete argument will be presented in the analysis seminar.

Hodge theory for a compact Kähler manifold is again a subject which has been around for quite some time. However, we have deemed it worthwhile to sketch the theory in some detail, emphasizing Chern's conceptual explanation of the plethora of Kähler identities. As applications, we have given Le Potier's recent extension of the Kodaira-Nakano vanishing theorem to vector bundles, and a brief account of Deligne's theory of mixed Hodge structures and the homotopy type of Kähler varieties. Finally, in the belief that many, if not most, algebraic geometers are aware of the formal aspect of Hodge theory but have not been through the grubby analysis, we have (in the appendix to lecture one) given an account of the underlying analysis of the Laplacian which hopefully may appeal to tastes of the algebraists.

The main thrust of these lectures will be to discuss variation of Hodge structure leading up to the recent work of Schmid. Here the methods of complex analysis, Hermitian differential geometry, and Lie group theory blend together to maximally illustrate the flavor of transcendental algebraic geometry. The complete proofs of most of the main results will be covered between the lectures and the accompanying analysis seminar.

We shall also discuss some "non-compact" algebraic geometry. For example, <u>Picard's theorem</u> and its subsequent refinement, the beautiful <u>value distribution theory</u> of R. Nevanlinna, appear naturally as transcendental analogues of the fundamental theorem of algebra. The extension of this theory to transcendental <u>holomorphic curves</u> in $\mathbb{P}^n$ is based on the <u>non-compact Plücker formulae</u> of H. and J. Weyl. Although it is not yet established, one may adopt the philosophy that a global result concerning complex algebraic varieties is not properly understood unless one has analogous results for non-compact manifolds, and this is to some extent the viewpoint we shall take. It is on non-compact varieties that the full richness of the larger class of generally transcendental analytic functions and holomorphic mappings can be perhaps best exploited. An example of this is the consequence of a theorem of Grauert that all of the rational, even dimensional homology on a smooth, affine algebraic variety is representable by analytic subvarieties, although very little of it is so by algebraic cycles. Conversely, the existing theory in the compact case frequently points the way to profound analytical results in the non-compact case, as illustrated by the L^2-methods for studying the $\bar{\partial}$-operator on open manifolds which has developed so fruitfully in the last decade, and whose basic estimate is just the identity Kodaira initially used in his vanishing theorem for a compact Kähler manifold.

GENERAL BIBLIOGRAPHICAL REFERENCES

We shall give a few references to articles and books which were turning points in the development of the subject.

N. Abel, Démonstration d'une propriété générale d'une certaine classe de fonctions transcendentes, J. reine angew. Math., vol. 4 (1829), 200-201.

B. Riemann, Theorie der Abelschen Funktionen, J. reine angew. Math., vol. 54 (1857), 115-155.

E. Picard and G. Simart, Théorie des fonctions algébriques de deux variables independantes (tomes I et II), Gauthier-Villars, Paris, 1892-1906.

H. Weyl, Die Idee der Riemannschen Fläche, Teubner, Berlin (1913); third edition Teubner, Stuttgart (1955).

S. Lefschetz, L'analysis situs et la géométrie algébrique, Gauthier-Villars, Paris (1924).

W. V. D. Hodge, The theory and applications of harmonic integrals, Cambridge Univ. Press, New York (second edition 1952).

S. S. Chern, Characteristic classes of hermitian manifolds, Annals of Math., vol. 47 (1946), 85-121.

K. Kodaira, On Kähler varieties of restricted type, Annals of Math., vol. 60 (1954), 43-76.

F. Hirzebruch, Topological methods in algebraic geometry, Springer-Verlag, New York (1966).

1. GENERALITIES ON COMPLEX MANIFOLDS

A <u>complex manifold</u> M is a manifold M provided with a distinguished open covering $\{U, V, \ldots\}$ and coordinate charts $f_U : U \to C^n$, $f_V : V \to C^n, \ldots$ such that $f_{UV} = f_U \circ f_V^{-1}$ is biholomorphic where defined. Such manifolds are even-dimensional, oriented, and we shall assume them to be connected. Some standard examples are:

a) $\mathbb{C}^n$ with global coordinates $z = (z_1, \ldots, z_n)$;

b) $\mathbb{P}^n$ with homogeneous coordinates $Z = [z_0, \ldots, z_n]$,

c) the Grassmannian $G(k, n)$ of k-planes through the origin in $\mathbb{C}^n$,

d) complex tori $T_\Lambda = \mathbb{C}^n/\Lambda$ where Λ is a lattice in $\mathbb{C}^n$;

e) among the spheres, only $S^2 - \mathbb{P}^1$ and possibly S^6 have complex structures; however, the product $S^{2p+1} \times S^{2q+1}$ of odd spheres has a continuous family of complex structures (Calabi-Eckmann);

f) let Q be the skew-form with matrix $\begin{pmatrix} 0 & I_n \\ -I_n & 0 \end{pmatrix}$, and define $\mathcal{H}_n \subset G(n, 2n)$ to be all subspaces Λ satisfying the <u>Riemann bilinear relations</u>

$$Q(e, e') = 0 \qquad (e, e' \in \Lambda)$$

$$\sqrt{-1}\, Q(e, \bar{e}) > 0 \qquad (0 \neq e \in \Lambda)$$

then $\mathcal{H}_n$ is a complex manifold biholomorphic to the <u>Siegel-upper-half-plane</u> of all $n \times n$ matrices $Z = X + \sqrt{-1}\, Y$ satisfying $Z = {}^tZ$, $Y > 0$;

g) a one-dimensional complex manifold is a <u>Riemann surface</u>; and

h) the most important examples are the <u>projective algebraic manifolds</u>, which are the complex submanifolds of $\mathbb{P}^n$ given as the zeroes of homogeneous polynomials. This class includes the Grassmannians, compact Riemann surfaces, some but not all complex tori, any quotient D/Γ of a bounded domain D in $\mathbb{C}^n$ by a discrete group of $\mathrm{Aut}(D)$ acting without fixed points and with compact quotient; except for $p = q = 0$, the product of odd spheres is not projective.

An important tool for linearizing the study of complex manifolds are the holomorphic vector bundles, defined as in the real case but with holomorphic transition functions. We denote such bundles by $E \to M$; the fibres $E_p = \pi^{-1}(p)$ $(p \in M)$ are (non-canonically) isomorphic to $\mathbb{C}^r$ where r is the <u>rank</u>. Aside from the trivial bundle $M \times \mathbb{C}^r$, some examples are:

a) the <u>holomorphic tangent bundle</u> $T(M) \to M$, whose local holomorphic sections are the holomorphic vector fields $\sum_i \theta_i(z) \frac{\partial}{\partial z_i}$ on M ;

b) the <u>normal bundle</u> to a complex submanifold;

c) the line bundle $(r = 1)$ $[D] \to M$ determined by a <u>divisor</u> D on

M ; here D has local defining equations $f_U = 0$ $(f_U \in \mathcal{O}(U))$, and the transition functions of $[D]$ are $f_{UV} = f_U/f_V \in \mathcal{O}^*(U \cap V)$;

d) the universal sub-bundle S and quotient bundle Q over the Grassmannian $G(k, n)$; for a k-plane $\Lambda \in G(k, n)$

$$S_\Lambda = \Lambda , \qquad Q_\Lambda = C^n/\Lambda ;$$

e) over a projective variety, there are generally "very few" holomorphic (= algebraic) vector bundles, since the Chern classes (defined below) of such a bundle must have Hodge type (p, p) . On the other hand, it is a theorem of Grauert that on an affine algebraic variety (= algebraic subvariety of C^N), every topological bundle has a unique analytic structure. A nice problem is to determine the "growth" of such bundles.

The standard constructions of linear algebra $(\otimes, \oplus, \Lambda^p$, Hom, duality, ...) apply fibre-wise to vector bundles. We follow the usual notational conventions: $\mathcal{O}(E)$ is the sheaf of holomorphic sections of $E \to M$; E^* is the dual of E , $\Omega^p = \mathcal{O}(\Lambda^p T(M)^*)$, etc. Noteworthy are the universal exact sequence

$$0 \to S \to G(k, n) \times C^n \to Q \to 0 ,$$

and natural identification

$$T(G(k, n)) \cong \mathrm{Hom}(S, Q) . \tag{1.1}$$

(PROOF: A variation Λ_t $(|t| < \varepsilon)$ of a k-plane Λ_0 is measured infinitesimally by choosing $e_t \in \Lambda_t$ and then projecting $\left.\frac{de_t}{dt}\right|_{t=0}$ into C^n/Λ_0 .)

We also denote by $A^{p,q}(M, E)$ the C^∞ , E-valued (p, q) forms on M . The operator

$$\bar\partial : A^{p,q}(M, E) \to A^{p,q+1}(M, E)$$

is well-defined since the transition functions of E are holomorphic.

In some sense, transcendental algebraic geometry is distinguished by the use of metrics. A holomorphic vector bundle with Hermitian metric in the fibres will be called a Hermitian vector bundle. Given such, the fundamental invariant is the curvature matrix $\Theta \in A^{1,1}(M, \mathrm{Hom}(E, E))$.

The curvature is of twofold importance: First, it gives rise to Chern forms and Chern classes, as in the real case. Secondly, peculiar to the complex case are the notions of positivity and negativity, which tie in with the analytic concept of pluri-subharmonicity.

Given a Hermitian vector bundle $E \to M$, there exists a canonical <u>metric connection</u>

$$D : A^0(M, E) \to A^1(M, E)$$

uniquely characterized by the conditions

$$(1.2)\qquad \begin{aligned} d(\xi, \eta) &= (D\xi, \eta) + (\xi, D\eta) \qquad (\xi, \eta \in A^0(M, E)) \\ D'' &= \bar\partial \end{aligned}$$

where $D = D' + D''$ is the type decomposition of D. A <u>unitary frame field</u> $\xi_1, \ldots, \xi_r$ is given by smooth sections ξ_ν of E over an open set $U \subset M$ which give a unitary basis in each fibre E_p $(p \in U)$. For such a frame field, the connection matrix θ and curvature Θ are defined by

$$(1.3)\qquad \begin{aligned} D\xi_\nu &= \sum_\mu \theta_{\nu\mu} \xi_\mu \\ d\theta + \theta \wedge \theta &= \Theta . \end{aligned}$$

The conditions (1.2) characterizing D imply that

$$(1.4)\qquad \begin{aligned} &\Theta + {}^t\bar\Theta = 0 \\ &\Theta \text{ is a matrix of } (1, 1) \text{ forms.} \end{aligned}$$

As examples of Hermitian vector bundles, observe that the inclusions

$$S \subset C^n$$

induce a Hermitian metric on the universal sub-bundle. Similarly, for the universal sub-bundle $S|H_n$ restricted to the Siegel-upper-half-plane, we may set

$$(\xi, \eta) = \sqrt{-1}\, Q(\xi, \bar\eta)$$

to obtain a canonical Hermitian metric of the sort encountered in variation of Hodge structure.

From the curvature, one constructs the <u>Chern forms</u> $c_q(\Theta)$ by taking elementary symmetric functions:

$$\sum_{q=0}^{r} c_q(\Theta) t^{r-q} = \det\left(\frac{\sqrt{-1}\,\Theta}{2\pi} + tI\right) .$$

Using (1.4), the Chern forms are global real forms of type (q, q), and the Bianchi identity $D\Theta = 0$ implies that

$$dc_q(\Theta) = 0 .$$

Thus, the Chern forms define classes

$$c_q \in H^{2q}_{DR}(M) ,$$

the de Rham cohomology of M. A basic result is that these Chern classes c_q are independent of the metric. In fact, this is true for real manifolds. In the complex case Bott and Chern proved more; namely that the Chern forms associated to two metrics $(\ ,\ )$ and $(\ ,\ )'$ satisfy

$$c_q(\Theta) - c_q(\Theta') = dd^c \eta_{q-1}$$

where $d^c = \sqrt{-1}\,(\bar{\partial} - \partial)$.

To define positivity, we use the curvature form

$$\text{(1.5)} \qquad \Theta(\eta) = \frac{\sqrt{-1}}{2\pi}(\Theta\eta, \eta) = \frac{\sqrt{-1}}{2\pi}\sum_{\mu,\nu} \Theta_{\mu\nu}\eta^{\mu}\bar{\eta}^{\nu}$$

where $\eta = \sum_\mu \eta^\mu \xi_\mu$ is a vector. Each $\Theta(\eta)$ is a real $(1, 1)$ form,

$$\Theta(\eta) = \frac{\sqrt{-1}}{2\pi}\sum_{i,j} a_{ij}\, dz_i \wedge d\bar{z}_j \qquad (a_{ij} = \bar{a}_{ji}) ,$$

and the Hermitian bundle is positive in case $\Theta(\eta) > 0$ for every non-zero η, in the sense that $(a_{ij}) > 0$. For line bundles, we note that

$$\Theta(\eta) = c_1(\Theta)\cdot(\eta, \eta) .$$

Here is an illustration of the use of the curvature form.

(1.6) Suppose that M is compact and that $E \to M$ is a Hermitian vector bundle whose curvature form has everywhere at least one negative eigenvalue. Then $H^0(M, \mathcal{O}(E)) = 0$.

PROOF: Suppose that $0 \neq \xi \in H^0(M, \mathcal{O}(E))$, and let $p_0 \in M$ be a point where the length $|\xi(p)|^2$ has a maximum. Then, at p_0,

$$(1.7)\qquad \frac{\sqrt{-1}}{2\pi}\,\partial\bar{\partial}|\xi(p)|^2 \leqq 0 .$$

On the other hand, using (1.2), $\bar{\partial}\xi = 0$, and $D^2 = \Theta$, we obtain

$$\begin{aligned} \partial\bar{\partial}(\xi, \xi) &= \partial(\xi, D'\xi) \\ &= (D'\xi, D'\xi) + (\xi, D''D'\xi) \\ &= (D'\xi, D'\xi) + (\xi, \Theta\xi) ; \quad \text{i.e.} \end{aligned}$$

$$(1.8)\qquad \begin{aligned} \frac{\sqrt{-1}}{2\pi}\,\partial\bar{\partial}(\xi, \xi) &= \frac{\sqrt{-1}}{2\pi}(D'\xi, D'\xi) - \frac{\sqrt{-1}}{2\pi}(\Theta\xi, \xi) \\ &\geqq -\Theta(\xi) , \end{aligned}$$

which contradicts (1.7). Q.E.D.

The relationship between positivity and Chern forms is not yet sufficiently understood. For example, aside from vector bundles of rank $\leqq 2$, it is not known if the Chern forms of a positive bundle are positive. In general, the relationship between positive (differential-geometric), ample (algebro-geometric), and numerically positive (topological and algebro-geometric) vector bundles has not been explained.

Before discussing harmonic forms, we want to give a device for computing curvatures. Suppose given an exact sequence

$$0 \to E' \to E \to E'' \to 0$$

of holomorphic vector bundles. A Hermitian metric in E induces one in E' and E'', and the differential geometry is analogous to that of submanifolds in $\mathbb{R}^n$. More precisely, the metric connection D in $E \to M$ induces a map

$$\sigma : E' \to E'' \otimes T(M)^*$$

in the obvious way (σ is of type $(1, 0)$ since $D'' = \bar{\partial}$ and E' is a holomorphic sub-bundle). We call σ the $2^{\underline{nd}}$ fundamental form of E' in E. As an example, in the universal exact sequence over the Grassmannian, $\sigma \in \mathrm{Hom}(T(M), \mathrm{Hom}(S, Q))$ gives the isomorphism (1.1). The curvature form of E' is given by

$$(1.9)\qquad \Theta_{E'}(\xi) = \Theta_E(\xi) - (\sigma(\xi), \sigma(\xi)) \qquad (\xi \in E') .$$

In particular,

$$\Theta_{E'} \leqq \Theta_E . \tag{1.10}$$

The principle that <u>curvatures decrease in sub-bundles</u> is of fundamental importance in Hermitian differential geometry.

A second major use of Hermitian metrics is the study of cohomology by using harmonic forms. Given a holomorphic vector bundle $E \to M$ over a compact complex manifold M, we consider the Dolbeault complex

$$\cdots \to A^{p,q-1}(M, E) \xrightarrow{\bar\partial} A^{p,q}(M, E) \to \cdots .$$

The holomorphic analogue of de Rham's theorem is the isomorphism

$$H^q(M, \Omega^p(E)) = H^{p,q}_{\bar\partial}(M, E) \tag{1.11}$$

between the Čech cohomology of the sheaf $\Omega^p(E) = \Omega^p \otimes \mathcal{O}(E)$ and the cohomology of the Dolbeault complex. If we introduce metrics in E and in the tangent bundle, then the Dolbeault cohomology $H^{p,q}_{\bar\partial}(M, E)$ is represented by harmonic forms as follows: The spaces $A^{p,q}(M, E)$ are pre-Hilbert spaces using the L^2 inner product

$$(\phi, \psi) = \int_M \phi \wedge *\psi$$

where

$$* : A^{p,q}(M, E) \to A^{n-p,n-q}(M, E^*) \tag{1.12}$$

is the pointwise duality operator. Next, the adjoint operator

$$\bar\partial^* : A^{p,q}(M, E) \to A^{p,q-1}(M, E)$$

and Laplacian

$$\Box = \bar\partial\bar\partial^* + \bar\partial^*\bar\partial$$

are defined as usual, as is the harmonic space

$$\begin{aligned} \mathcal{H}^{p,q}(M, E) &= \{\phi \in A^{p,q}(M, E) : \Box\phi = 0\} \\ &= \{\phi \in A^{p,q}(M, E) : \bar\partial\phi = 0 = \bar\partial^*\phi\} . \end{aligned}$$

What is by now standard P.D.E. gives, among other things, the isomorphism

$$H^{p,q}_{\bar\partial}(M, E) \cong \mathcal{H}^{p,q}(M, E) . \tag{1.13}$$

There are three immediate applications of this representation of cohomology by harmonic forms:

a) The cohomology is finite dimensional;

$$\dim H^q(M, \Omega^p(E)) < \infty .$$

b) Next, we consider two compact, complex manifolds M, N over which we are given holomorphic vector bundles E, F. Denoting by $E \otimes F \to M \times N$ the bundle with fibres $(E \otimes F)_{(x \times y)} = E_x \otimes F_y$ $(x \in M, y \in N)$, if we choose the obvious product metrics throughout and make straightforward considerations of the various Laplacians, then

$$\mathcal{H}^{p,q}(M \times N, E \otimes F) = \sum_{r,s} \mathcal{H}^{r,s}(M, E) \otimes \mathcal{H}^{p-r,q-s}(N, F) .$$

This leads to the Künneth formula

$$H^*(M \times N, \Omega^*_{M \times N}(E \otimes F)) \cong H^*(M, \Omega^*_M(E)) \otimes H^*(N, \Omega^*_N(F)) . \tag{1.14}$$

c) Since $\bar\partial^* = \pm * \bar\partial *$ where $*$ is the operator (1.12), it follows that $* \Box = \Box *$ and

$$* : \mathcal{H}^{p,q}(M, E) \to \mathcal{H}^{m-p,m-q}(M, E^*)$$

is an isometry. This implies that the pairing

$$H^q(M, \Omega^p(E)) \otimes H^{n-q}(M, \Omega^{m-p}(E^*)) \to C \tag{1.15}$$

given by using (1.11) and

$$\phi \otimes \psi \to \int_M \phi \wedge \psi$$

is a pairing of dual vector spaces (Kodaira-Serre duality).

The relationship between harmonic forms and positivity will be discussed in lecture 3 when we talk about vanishing theorems.

REFERENCES FOR CHAPTER ONE

As general references for complex manifolds and Hermitian differential geometry, we suggest:

S. S. Chern, Complex manifolds without potential theory, Van Nostrand, New York (1967).

R. O. Wells, Differential analysis on complex manifolds, Prentice-Hall, Englewood Cliffs (1973).

J. Morrow and K. Kodaira, Complex manifolds, Holt, Rinehart and Winston, New York (1971).

The curvature form, 2nd fundamental form of a holomorphic sub-bundle, etc. are discussed in

P. Griffiths, Hermitian differential geometry, Chern classes, and positive vector bundles, in Global Analysis, Princeton University Press, Princeton, N. J. (1969).

APPENDIX TO LECTURE 1: PROOF OF THE HODGE THEOREM*

Let M be a compact Hermitian manifold. We shall prove the Hodge theorem for the Laplace-Beltrami operator $\Box = \bar\partial\bar\partial^* + \bar\partial^*\bar\partial$ acting on the Dolbeault complex $A^{0,*}(M)$.

a) FOURIER SERIES. Let $T = \mathbb{R}^n/(2\pi\,\mathbb{Z})^n$ be a standard torus with coordinate $x = (x_1,\ldots,x_n)$. We consider the space F of formal Fourier series

$$u \sim \sum_{\xi\in\mathbb{Z}^n} u_\xi e^{\sqrt{-1}\,<\xi,x>}$$

with complex coefficients. For any integer s, the Sobolev s-norm is

$$\|u\|_s^2 = \sum_\xi (1 + \|\xi\|^2)^s |u_\xi|^2 ,$$

and we denote by H_s the Hilbert space of all $u \in F$ with finite s-norm. Note that

$$H_s \subset H_r \qquad (s \geq r) ,$$

and we set

$$H_\infty = \bigcap_s H_s$$

$$H_{-\infty} = \bigcup_s H_s .$$

* We shall assume only elementary Hilbert space theory up to the spectral theorem for compact operators.

The mapping

$$u_\xi \rightarrow (1 + \|\xi\|^2)^s u_\xi$$

is an isometry from H_s to H_{-s} , so that we may identify H_{-s} with the dual of H_s by the pairing

$$(u, v) \rightarrow \sum_\xi u_\xi \bar{v}_\xi .$$

A basic tool is the

RELLICH LEMMA: For $s > r$, the inclusion $H_s \subset H_r$ is a compact operator.

PROOF: We must show that a bounded sequence $\{u_k\}$ in H_s has a convergent subsequence in H_r . Since $\|u_k\|_s^2 \leqq C < \infty$, the sequences

$$(1 + \|\xi\|^2)^{r/2} u_{k,\xi}$$

are bounded. By diagonalization and passing to a subsquence of the u_k , we may assume that

$$(1 + \|\xi\|^2)^{r/2} u_{k,\xi}$$

is Cauchy for each fixed ξ . Given $\varepsilon > 0$, we may then choose R and m such that

$$\frac{4C}{(1 + \|\xi\|^2)^{s-r}} < \varepsilon/2 \qquad \text{for } \|\xi\| > R$$

$$\sum_{\|\xi\| \leqq R} (1 + \|\xi\|^2)^r |u_{k,\xi} - u_{\ell,\xi}|^2 < \varepsilon/2 \qquad \text{for } k, \ell \geqq m .$$

Then, for $k, \ell \geqq m$,

$$\|u_k - u_\ell\|_r^2 = \sum_{\|\xi\| \leqq R} (1 + \|\xi\|^2)^r |u_{k,\xi} - u_{\ell,\xi}|^2$$

$$+ \sum_{\|\xi\| > R} (1 + \|\xi\|^2)^r |u_{k,\xi} - u_{\ell,\xi}|^2 < \varepsilon/2 + \varepsilon/2 = \varepsilon .$$

Denote by $C^s(T)$ the functions of differentiability class s on the torus. Each continuous function $\phi \in C^0(T)$ generates a formal Fourier series $\sum_\xi \phi_\xi e^{\sqrt{-1} < \xi, x >}$ whose <u>Fourier</u> <u>coefficients</u> are given by

$$\phi_\xi = \int_T \phi(x) e^{-\sqrt{-1} < \xi, x >} .$$

Set $D_j = \frac{1}{\sqrt{-1}} \frac{\partial}{\partial x_j}$ and use the standard multi-index notations:

$$D^\alpha = D_1^{\alpha_1} \cdots D_n^{\alpha_n} \qquad \alpha = (\alpha_1, \ldots, \alpha_n)$$

$$[\alpha] = \alpha_1 + \cdots + \alpha_n$$

$$\xi^\alpha = \xi_1^{\alpha_1} \cdots \xi_n^{\alpha_n} .$$

Then the formulae

$$\text{(A.1)} \qquad \begin{aligned} D^\alpha e^{\sqrt{-1} < \xi, \alpha >} &= \xi^\alpha e^{\sqrt{-1} < \xi, \alpha >} \\ \int_T (D^\alpha \phi)(\bar{\psi}) &= \int_T (\phi)\overline{(D^\alpha \psi)} \qquad (\phi, \psi \in C^\infty(T)) \end{aligned}$$

are valid. If $\phi \in C^s(T)$ and $[\alpha] \leqq s$, the Fourier coefficients of $D^\alpha \phi$ are

$$(D^\alpha \phi)_\xi = \xi^\alpha \phi_\xi .$$

<u>Parseval's identity</u>

$$\int_T |\phi|^2 = \sum_\xi |\phi_\xi|^2$$

shows that the Fourier series mapping $C^0(T) \to H_0$ is injective, and that $C^s(T)$ maps to H_s. A partial converse is given by our second basic tool, the important

> SOBOLEV LEMMA: $H_{s+[n/2]+1} \subset C^s(T)$, in the sense that each $u \in H_{s+[n/2]+1}$ is the Fourier series of a unique function $u \in C^s(T)$, and moreover this Fourier series converges uniformly to the function.

PROOF: In case $s = 0$, we let $\sum_\xi u_\xi e^{\sqrt{-1} < \xi, x >}$ satisfy $\sum_\xi (1 + \|\xi\|^2)^{[n/2]+1} |u_\xi|^2 < \infty$. The partial sums $S_R = \sum_{\|\xi\| \leqq R} u_\xi e^{\sqrt{-1} < \xi, x >}$ are continuous functions, and for $R \leqq R'$

$$\begin{aligned} |S_R(x) - S_{R'}(x)| &\leqq \sum_{\|\xi\| > R} |u_\xi| \\ &= \sum_{\|\xi\| > R} \frac{((1 + \|\xi\|^2)^{[n/2]+1} |u_\xi|^2)^{1/2}}{((1 + \|\xi\|^2)^{[n/2]+1})^{1/2}} \\ &\leqq C \left(\sum_{\|\xi\| > R} (1 + \|\xi\|^2)^{[n/2]+1} |u_\xi|^2 \right)^{1/2} \end{aligned}$$

by the Schwarz inequality and $\sum_\xi \frac{1}{(1 + \|\xi\|^2)^{[n/2]+1}} \leqq C < \infty$. Consequently, the $S_R(x)$ converge uniformly to a continuous function $u(x)$, whose Fourier coefficients are in turn just the u_ξ by virtue of the orthogonality relations

$$\int_T e^{\sqrt{-1}<\xi,x>} e^{-\sqrt{-1}<\xi',x>} = \begin{cases} 1 & \xi = \xi' \\ 0 & \xi \neq \xi' \end{cases}$$

We shall do the case $s = n = 1$ to illustrate the general situation for $s > 0$. Thus, let $u(x) = \sum_{\xi\in\mathbb{Z}} u_\xi e^{\sqrt{-1}<\xi,x>}$ be a continuous function where $\sum_\xi |\xi|^4 |u_\xi|^2 < \infty$. Using the previous case, the partial sums of $v(x) = \sum_\xi \sqrt{-1}\,\xi u_\xi e^{\sqrt{-1}<\xi,x>}$ converge uniformly to a continuous function. Integrating term by term gives

$$\int_0^x v(t)dt = \sum_{\xi\neq 0} u_\xi e^{\sqrt{-1}<\xi,x>} = u(x) - u_0 .$$

It follows that $u(x)$ is of class C^1 and $u'(x) = v(x)$.

In summary, we have proved:

<u>The Fourier series mapping</u> $C^0(T) \to F$ <u>leads to inclusions</u>

$$C^s(T) \subset H_s$$
$$H_{s+[n/2]+1} \subset C^s(T)$$

In particular, we shall make the identification

$$C^\infty(T) = H_\infty .$$

We remark that the Fourier series of a function $\phi \in C^\infty(T)$ may be differentiated term by term. Indeed, $D^\alpha\phi$ is given by a Fourier series whose Fourier coefficients are $(D^\alpha\phi)_\xi = \xi^\alpha\phi$ by (A.1).

Another useful comment is that the proof of the Sobolev lemma gives an estimate

$$\sup_{x\in T} |\phi(x)| \leqq C\|\phi\|_{[n/2]+1} . \tag{A.2}$$

The norm $\| \|_0$ is just the usual L^2 norm on $C^\infty(T)$. Similarly, the Sobolev s-norm is equivalent to the norm

(A.3) $$|||\phi|||_s^2 = \sum_{[\alpha]\leq s} \|D^\alpha\phi\|_0^2 = \sum_{\substack{[\alpha]\leq s \\ \xi}} |\xi^{2\alpha}||u_\xi|^2 .$$

This follows from the inequalities

$$\sum_{[\alpha]\leq s} |\xi^{2\alpha}| \leq (1 + \|\xi\|^2)^s \leq C_s \sum_{[\alpha]\leq 2s} |\xi^{2\alpha}| .$$

Thus, H_s <u>is the L^2-completion of</u> $C^\infty(T)$ <u>with respect to the norm (A.3)</u>. Combining these remarks with (A.2) gives

(A.4) $$\sup_{x\in T} |D^\alpha\phi(x)| \leq C_\alpha\|\phi\|_{[\alpha]+[n/2]+1} .$$

We shall conclude this discussion of Fourier series with some remarks concerning the <u>distributions</u>, defined as the linear functionals

$$\lambda : C^\infty(T) \to C$$

which are continuous in the sense that

(A.5) $$|\lambda(\phi)| \leq C \sup_{\left\{\substack{[\alpha]\leq k \\ x\in T}\right.} |D^\alpha\phi(x)| .$$

Each distribution generates a formal Fourier series $\sum_\xi \lambda_\xi e^{\sqrt{-1}\,<\xi,x>}$ where

$$\lambda_\xi = \lambda(e^{-\sqrt{-1}\,<\xi,x>}) .$$

It follows from (A.5) that λ is continuous in the norm $\|\ \|_k$, so that $\lambda \in H_{-k}$ by previous remarks. Indeed, the formal Fourier series of λ in H_{-k} is just $\sum_\xi \lambda_\xi e^{\sqrt{-1}\,<\xi,x>}$. On the other hand, by (A.4) any $\lambda \in H_{-k}$ gives a distribution by setting

$$\lambda(\phi) = \sum_\xi \lambda_\xi\phi_\xi .$$

Thus, we may identify the distributions with $H_{-\infty}$.

The derivatives of any distribution λ are defined by

$$(D^\alpha\lambda)\phi = \lambda(D^\alpha\phi) .$$

A distribution λ is said to <u>be in</u> L^2 in case $\lambda \in H_0 \subset H_{-\infty}$. Putting our remarks together, we may give the following description of the Sobolev spaces H_s for $s \geq 0$:

H_s consists of the distributions λ such that all derivatives $D^\alpha \lambda$ are in L^2 for $[\alpha] \leqq s$.

Referring to lecture 2, the delta-function distribution

$$\delta(\phi) = \psi(0)$$

has formal Fourier series $\sum_\xi e^{\sqrt{-1}<\xi,x>}$. Letting $dV = dx_1 \wedge \cdots \wedge dx_n$, the equation of currents

$$dk = \delta \cdot dV - dV$$

has the solution $k = \sum_{j=1}^n (-1)^j k_j dx_1 \wedge \cdots \wedge d\hat{x}_j \wedge \cdots \wedge dx_n$ where $k_j = \sum_\xi \sqrt{-1} \frac{C_j(\xi)}{\xi_j} e^{\sqrt{-1}<\xi,x>}$ and $C_j(\xi)$ is the number of non-zero ξ_j in $\xi = (\xi_1, \ldots, \xi_n)$.

b) GLOBALIZATION TO MANIFOLDS; THE HODGE DECOMPOSITION. Any compactly supported function on $\mathbb{R}^n$ may be regarded as a function on the torus. Using a partition of unity, we may thus globalize the above discussion to a compact Hermitian manifold. Let ∇ be the metric connection on the complexified tangent bundle $T_C = T' \oplus T''$. ∇ induces a connection on all associated tensor bundles. Thus

$$\nabla : C^\infty(\Lambda^q T''^*) \to C^\infty(\Lambda^q T''^* \otimes T_C^*) ,$$

$$\nabla : C^\infty(\Lambda^q T''^* \otimes T_C^*) \to C^\infty(\Lambda^q T''^* \otimes T_C^* \otimes T_C^*) ,$$

etc. are defined. If $\phi \in C^\infty(\Lambda^q T''^*) = A^{0,q}(M)$, we set $\nabla^k = \underbrace{\nabla(\nabla(\cdots(\nabla\phi)\cdots)}_{k\text{-times}}$. In any local coordinate system,

$$\sum_{[\alpha]\leqq s} |D^\alpha\phi|^2 \leqq C \sum_{k\leqq s} (\nabla^k\phi, \nabla^k\phi) \leqq C' \sum_{[\alpha]\leqq s} |D^\alpha\phi|^2$$

since, for any tensor τ , in local coordinates

$$\nabla\tau = \sum_{i=1}^m \frac{\partial\tau}{\partial z_i} dz_i + \sum_{i=1}^m \frac{\partial\tau}{\partial \bar{z}_i} d\bar{z}_i + \chi\cdot\tau$$

where χ is an algebraic operator. Thus, if we let $H_s^{0,q}(M)$ be the completion of $A^{0,q}(M)$ in the norm

$$\|\phi\|_s^2 = \sum_{k\leqq s} \int_M (\nabla^k\phi, \nabla^k\phi) dV ,$$

then $H_s^{0,q}(M)$ localizes to the Sobolev spaces H_s for C^∞ forms compactly supported in a fixed coordinate patch. In particular,

$$\bigcap_s H_s^{0,q}(M) = A^{0,q}(M) ;$$

$$\bigcup_s H_s^{0,q}(M) = D^{0,q}(M) \quad \text{are the currents of type } (0,q) ;^*$$

$$H_0^{0,q}(M) = L_2^{0,q}(M) \quad \text{are the } L^2\text{-forms; and}$$

the inclusion

$$H_s^{0,q}(M) \to H_r^{0,q}(M)$$

is compact for $s > r$.

In addition to the local Fourier analysis, our basic tool in the study of harmonic theory is

GÅRDING'S INEQUALITY: For $\phi \in A^{0,q}(M)$,

$$\|\phi\|_1^2 \leq C\mathcal{D}(\phi, \phi) \tag{A.6}$$

where $\mathcal{D}(\phi, \phi)$ is the Dirichlet norm

$$\mathcal{D}(\phi, \phi) = (\phi, \phi) + (\bar\partial\phi, \bar\partial\phi) + (\bar\partial^*\phi, \bar\partial^*\phi) = (\phi, (I + \Box)\phi) . \tag{A.7}$$

Assuming (A.6) we shall prove the basic results in Hodge theory, and then derive (A.6) in the next section. The Gårding inequality says that the norm $\mathcal{D}(\phi, \phi)^{1/2}$ is equivalent to $\|\phi\|_1$ on $H_1^{0,q}$. Moreover, for $\psi \in H_1^{0,q}(M)$ and $\eta \in A^{0,q}(M)$,

$$\mathcal{D}(\psi, \eta) = (\psi, (I + \Box)\eta) . \tag{A.8}$$

(A.9) LEMMA: Given $\phi \in L_2^{0,q}$, there exists a unique $\psi \in H_1^{0,q}$ such that

$$(\phi, \eta) = \mathcal{D}(\psi, \eta) \qquad (\eta \in A^{0,q}(M)) . \tag{A.10}$$

The map $T(\phi) = \psi$ is a compact self-adjoint operator on $L_2^{0,q}(M)$ whose range is contained in $H_1^{0,q}(M)$. As a mapping of $L_2^{0,q}(M)$ into $H_1^{0,q}(M)$, T is continuous.

---

* Actually, we have only defined the $H_s^{0,q}(M)$ for $s \geq 0$, but this may be extended to all s.

PROOF: The linear functional

$$\eta \to (\phi, \eta) \qquad (\eta \in A^{0,q}(M))$$

extends to a bounded linear form on $H_1^{0,q}(M)$, by virtue of $|(\phi, \eta)| \leq \|\phi\|_0 \|\eta\|_0 \leq \|\phi\|_0 \mathcal{D}(\eta)$. Thus the equation (A.10) has a unique solution $\psi \in H_1^{0,q}(M)$. The mapping $T(\phi) = \psi$ is characterized by

$$(\phi, \eta) = \mathcal{D}(T\phi, \eta) = (T\phi, (I + \Box)\eta) \qquad (\eta \in A^{0,q}(M)) .$$

T is self-adjoint since

$$(\phi, T\gamma) = \mathcal{D}(T\phi, T\gamma) = \overline{\mathcal{D}(T\gamma, T\phi)} = \overline{(\gamma, T\phi)}$$

for smooth forms ϕ, γ. From

$$\|T\phi\|_1^2 \leq \mathcal{D}(T\phi, T\phi) = (\phi, T\phi) \leq \|\phi\|_0 \cdot \|T\phi\|_0$$

and inequalities of the form

$$2\alpha\beta \leq \varepsilon\alpha^2 + \frac{1}{\varepsilon}\beta^2 ,$$

it follows that T is bounded, and Rellich's theorem implies that T is compact. Q.E.D.

We may now prove the regularity theorem. Given $\phi \in H_s^{0,q}(M)$ and $\psi \in L_2^{0,q}(M)$, we say ψ <u>is a weak solution of the equation</u>

(A.11) $$\Box\, \psi = \phi$$

in case

$$(\psi, \Box\, \eta) = (\phi, \eta) \qquad (\eta \in A^{0,q}(M)) .^*$$

REGULARITY THEOREM: If ψ is a weak solution of (A.11) where $\phi \in H_s^{0,q}(M)$, then $\psi \in H_{s+2}^{0,q}(M)$.

PROOF: We write

$$\Box = P^2$$
$$P = \bar{\partial} + \bar{\partial}^*$$

---

* A <u>strong solution</u> is a solution in the usual sense.

and notice that in proving the regularity theorem we may assume that ϕ, ψ are both supported in a fixed coordinate patch. Regularity then follows from

(A.12) PROPOSITION: Let P be a first order differential operator on $\mathbb{R}^n$:

$$Pu = Qu + Ru$$
$$(Qu)_i = \sum a^h_{ij} \frac{\partial u_j}{\partial x_h}$$
$$(Ru)_i = \sum b_{ij} u_j$$

where a^h_{ij}, b_{ij} are smooth functions. Suppose that the Gårding inequality

$$\alpha \|Pu\|_0 + \beta \|u\|_0 \geqq \|u\|_1 \tag{A.13}$$

holds for any compactly supported smooth u. If $u, v \in H_s$ and

$$Pu = v$$

in the weak sense (as distributions), then $u \in H_{s+1}$.

PROOF: We will assume that $s \geqq 0$, although the proof works for any s. Before actually proving (A.12) we recall some standard facts about mollifiers. Let χ be a positive, compactly supported smooth function on $\mathbb{R}^n$ such that

$$\chi(-x) = \chi(x)$$
$$\int \chi(x)dx = 1 .$$

We define

$$\chi_\varepsilon(x) = \frac{1}{\varepsilon^n} \chi\left(\frac{x}{\varepsilon}\right)$$
$$(v * u)(x) = \int v(y)u(x-y)dy = \int v(x-y)u(y)dy$$

and notice that, no matter what the differentiability properties of u are, $\chi_\varepsilon * u$ is C^∞ and

$$\frac{\partial}{\partial x_h}(\chi_\varepsilon * u) = \frac{\partial}{\partial x_h}\chi_\varepsilon * u = \chi_\varepsilon * \frac{\partial u}{\partial x_h} .$$

Moreover it is a standard and easily proved fact that, if u is compactly supported and belongs to L_2 (resp. H_s), then $\chi_\varepsilon * u$ converges to u in L_2-norm (H_s-norm) as $\varepsilon \to 0$. To prove (A.12) it is sufficient to uniformly bound the H_{s+1}-norm of $\chi_\varepsilon * u$, since then a sequence $\chi_{\varepsilon_n} * u$, $\varepsilon_n \to 0$ converges weakly to an element of H_{s+1} which can only be u.

It is a consequence of (A.13) that we can bound the H_{s+1}-norm of $\chi_\varepsilon * u$ in terms of the H_s-norms of $Q(\chi_\varepsilon * u)$ and $\chi_\varepsilon * u$. The latter, in turn, is bounded by a constant times the H_s-norm of u. We know how to bound the H_s-norm of $\chi_\varepsilon * Qu$, so we must bound the H_s-norm of the difference

$$\chi_\varepsilon * Qu - Q(\chi_\varepsilon * u) . \tag{A.14}$$

For simplicity we will do this when $s = 0$, the argument being the same in general. The i-th component of (A.14) is

$$\left[\frac{\partial}{\partial x_h} \chi_\varepsilon * \Big(\sum_{j,h} a^h_{ij} u_j\Big) - \sum_{j,h} a^h_{ij} \frac{\partial}{\partial x_h} \chi_\varepsilon * u_j\right] - \chi_\varepsilon * \Big(\sum_{j,h} u_j \frac{\partial}{\partial x_h} a^h_{ij}\Big) .$$

The second term is bounded by a constant times the L_2-norm of u. The other term is

$$\frac{1}{\varepsilon^{n+1}} \sum_{j,h} \int \frac{\partial}{\partial x_h} \chi(y/\varepsilon)\,[a^h_{ij}(x-y) - a^h_{ij}(x)]u_j(x-y)dy$$

and the Minkowski inequality implies that its L_2-norm is less than

$$\frac{1}{\varepsilon^{n+1}} \int_{|y| \leq \varepsilon} \left|\frac{\partial}{\partial x_h} \chi\left(\frac{y}{\varepsilon}\right)\right| K|y| \, \|u_j\|_0 \, dy \leq K' \|u\|_0$$

for suitable constants K, K'. Q.E.D.

As an application, we call any weak solution of the equation

$$\Box \phi = \lambda\phi \qquad (\lambda \in C)$$

an <u>eigenfunction</u> for the Laplacian.

COROLLARY: Any eigenfunction of $\Box$ is smooth.

PROOF: From $\Box\phi = \lambda\phi$, it follows inductively on s that $\phi \in \bigcap_s H^{0,q}_s(M)$, which is just $A^{0,q}(M)$ by the Sobolev lemma. Q.E.D.

The eigenspace $A^{0,q}_\lambda(M) = \{\phi \in L^{0,q}_2(M) : \Box\phi = \lambda\phi\}$ is just the eigenspace with eigenvalue $1/(1+\lambda)$ for T. Since T is compact and self-adjoint these eigenspaces are all finite dimensional. In particular, the harmonic space $\mathcal{H}^{0,q}(M)$ corresponding to the eigenvalue $\lambda = 0$ is finite dimensional. Moreover, the spectral theorem for such operators gives the discrete decomposition

$$L_2^{0,q}(M) = \bigoplus_{\lambda_m} A_{\lambda_m}^{0,q}(M)$$

where $A_{\lambda_m}^{0,q}$ is the eigenspace for $1/(1+\lambda_m)$ for T. In particular, $0 \leqq \lambda_0 < \lambda_1$ and $\lambda_m \to \infty$ as $m \to \infty$. On the orthogonal complement $\mathcal{H}^{0,q}(M)^{\perp} = \bigoplus_{m \geqq 1} A_{\lambda_m}^{0,q}(M)$ of the harmonic space, the estimate

$$\|\Box\eta\|_0 \geqq \lambda_1\|\eta\|_0 \qquad (\lambda_1 > 0)$$

is valid for all smooth forms. This is the essential estimate needed to prove the

HODGE DECOMPOSITION: The equation

$$\Box\psi = \phi \qquad (\phi \in A^{0,q}(M))$$

has a unique solution $\psi \in A^{0,q}(M) \cap \mathcal{H}^{0,q}(M)^{\perp}$ if and only if $\phi \in \mathcal{H}^{0,q}(M)^{\perp}$. The map $\phi \to \psi$ is a compact self-adjoint operator G (the Green's operator) which commutes with $\bar\partial$ and $\bar\partial^*$. For any $\phi \in A^{0,q}(M)$, one has the Hodge decomposition

$$\phi = \mathcal{H}(\phi) + \Box G(\phi) \tag{A.15}$$

where $\mathcal{H}(\phi)$ is the projection of ϕ in $\mathcal{H}^{0,q}(M)$ and G has been extended to all of $L_2^{0,q}(M)$ by setting $G = 0$ on $\mathcal{H}^{0,q}(M)$.

PROOF: The necessity condition $\phi \in \mathcal{H}^{0,q}(M)$ and the uniqueness statement are obvious. An explicit formula for G is as follows. Write

$$\phi = \sum \phi_i$$

where $\phi_i \in A_{\lambda_i}^{0,q}$, $\lambda_i \neq 0$. Then

$$G\phi = \sum \frac{1}{\lambda_i}\phi_i .$$

It follows from the regularity theorem that $\psi = G\phi$ is smooth and is a solution of $\Box\psi = \phi$ in the usual sense. Moreover the Hodge decomposition (A.15) is valid, and the operator properties of G follow from those of T. The commutation relations $[G, \bar{\partial}] = 0 = [G, \bar{\partial}^*]$ result from $[\Box, \bar{\partial}] = 0 = [\Box, \bar{\partial}^*]$. Q.E.D.

In particular, any $\phi \in A^{0,q}(M)$ may be written in the chain homotopy form

$$\phi = H(\phi) + \bar{\partial}(G\bar{\partial}^*)\phi + (G\bar{\partial}^*)\bar{\partial}\phi, \tag{A.16}$$

and this, together with what has been previously said, proves the results on cohomology which were stated in lecture 1.

The proof of Hodge theory we have given depends only on the Gårding inequality (A.6), which is the basic ellipticity estimate for $\Box$. It works equally well for Riemannian manifolds and vector bundle cohomology $H^*_{\bar{\partial}}(M, E)$. The basic defect is that the more subtle properties of the Green's operator as a kernel on $M \times M$ with certain precise singularities along the diagonal are not readily visible by the Hilbert space method. We remark that the total operator

$$G\bar{\partial}^* : A^{0,*}(M) \to A^{0,*-1}(M)$$

comes from a very beautiful kernel $k(x, y)$ on $M \times M$, which in particular solves the equation of currents

$$\bar{\partial}k = T_{0,\Delta} - s$$

with $s(x, y)$ being the smooth form $\Psi(x, y)$ in lecture 2 corresponding to taking the ψ^q_μ and ψ^{*q}_μ to be an orthonormal basis for the harmonic forms.

c) THE GÅRDING INEQUALITY. For any differential operator D of order k on a compact manifold M, one trivially has estimates

$$\|D\phi\|^2_s \leqq C_s\|\phi\|^2_{k+s}.$$

Very roughly speaking, ellipticity is the converse. For compactly supported functions $\phi(x)$ on $\mathbb{R}^n$, the Euclidean Laplacian $\Delta = -\sum_{i=1}^n \partial^2/\partial x_i^2$

satisfies

$$\int_{\mathbb{R}^n} (\Delta\phi\cdot\phi)dV = \int_{\mathbb{R}^n} \sum_{i=1}^{n} \left|\frac{\partial\phi}{x_i}\right|^2 dV$$

by an obvious integration by parts. Since

$$\Delta\Big(\sum_I \phi_I\, dx^I\Big) = \sum_I (\Delta\phi_I)\, dx^I ,$$

the same is true for compactly supported p-forms. Thus, one has the equality

$$\|\phi\|_1^2 = \mathcal{D}(\phi, \phi)$$

in the flat Euclidean case. The same is true for the Laplacian $\Box$ on $\mathbb{C}^m$, since $\Box = \frac{1}{2}\Delta$ by the Kähler property (Lecture 3).

In the general case, one proves (2.6) by writing out the formula for $\Box$ (Weitzenböck identity) and doing an integration by parts (Stokes' theorem). The principal part of $\Box$ has the same form as on $\mathbb{C}^m$, but there are first order terms coming from the torsion and zero order terms coming from the curvature. These may be estimated out by repeatedly using the inequality

$$2\alpha\beta \leqq \varepsilon\alpha^2 + \frac{1}{\varepsilon}\beta^2 .$$

Here is the argument in more detail. Write locally $ds^2 = \sum_{i=1}^m \omega_i\bar{\omega}_i$ where $\omega_1, \ldots, \omega_m$ is an orthonormal coframe for the $(1, 0)$ forms. On any tensor τ, $\nabla'\tau = \sum_{i=1}^m \nabla_i\tau \otimes \omega_i$ where ∇_i is covariant differentiation along the $(1, 0)$ vector field dual to ω_i. A $(0, q)$ form is written as $\phi = \sum_I \phi_I\bar{\omega}_I$ where $\bar{\omega}_I = \bar{\omega}_{i_1} \wedge \cdots \wedge \bar{\omega}_{i_q}$, etc. The Weitzenböck identity (in crude form) is

$$(\Box\phi)_I = -\sum_j \nabla_j\nabla_{\bar{j}}\phi_I + (\delta\phi)_I \tag{A.17}$$

where $\delta\phi$ is a first order operator. In the Kähler case, $\delta\phi$ is the zero order operator given by

$$(\delta\phi)_I = \sum_j R_{i_p\bar{j}}\,\phi_{\bar{i}_1\cdots\underset{p}{\bar{j}}\cdots\bar{i}_q}$$

where $R_{i\bar{j}} = c_1(K_M)$ is the Ricci tensor of the metric. To prove (A.17), one first derives the formulae

$$(A.18)\qquad \begin{aligned} (\bar\partial\phi)_{\bar i_1\cdots\bar i_{q+1}} &= \sum_{k=1}^{q+1} (-1)^k \nabla_{\bar i_k} \phi_{\bar i_1\cdots\hat{\bar i}_k\cdots\bar i_{q+1}} + (\tau\phi)_{\bar i_1\cdots\bar i_{q+1}} \\ (\bar\partial^*\phi)_{\bar i_1\cdots\bar i_{q-1}} &= \sum_{k,i} (-1)^k \nabla_i \phi_{\bar i_1\cdots\underset{k}{\bar i}\cdots\bar i_{q-1}} + (\tau^*\phi)_{\bar i_1\cdots\bar i_{q-1}} \end{aligned}$$

where τ, τ^* are algebraic operators involving the torsion. These equations follow in turn from the basic structure equations

$$d\omega_i = \sum_j \theta_{ij} \wedge \omega_j + \tau_i$$

$$\theta_{ij} + \bar\theta_{ji} = 0\ , \quad \tau_i \text{ has type } (2,\ 0)$$

for the Hermitian connection in $T'(M)$, θ being the connection matrix.* Now then the Weitzenböck formula (A.17) results from (A.18) and $\square = \bar\partial\bar\partial^* + \bar\partial^*\bar\partial$ together with $[\nabla_i, \nabla_{\bar j}] = \chi_{i\bar j}$, an operator involving the curvature and torsion.

A little reflection should convince the reader that whatever computation works for C^m will carry over to a general manifold in the crude form (A.17).

Assuming (A.17), we proceed as follows: The $(m-1,\ m)$ form

$$\Psi = C_1 \sum_j (-1)^j \nabla_{\bar j}\phi_I\bar\phi_I\ \omega_1 \wedge \cdots \wedge \hat\omega_j \wedge \cdots \wedge \omega_m \wedge \bar\omega_1 \wedge \cdots \wedge \bar\omega_m$$

is intrinsically defined on M . Choosing the constant C_1 properly,

$$d\Psi = \sum_{I,j} |\nabla_{\bar j}\phi_I|^2 dV + C_2 \sum_{j,I} (\nabla_j\nabla_{\bar j}\phi_I\bar\phi_I)dV$$

for some constant C_2 . Applying Stokes' theorem and keeping track of constants (including signs) gives

$$\int_M \sum_{I,j} |\nabla_{\bar j}\phi_I|^2 dV = (\square\phi,\ \phi) + (\delta_1\phi,\ \phi)$$

where δ_1 is first order. Similarly,

$$\int_M \sum_{I,j} |\nabla_j\phi_I|^2 = (\square\phi,\ \phi) + (\delta_2\phi,\ \phi)\ .$$

---

* A proof of (A.18) in the Kähler case $\tau \equiv 0$ runs as follows: Both sides are intrinsically defined first order operators. Hence, using osculation of the metric to C^m to second order, it will suffice to verify (A.18) in C^m , which is easy.

Adding $\|\phi\|_0^2$ to these equations gives

$$\|\phi\|_1^2 \leqq C(\mathcal{D}(\phi, \phi) + 2(\delta_3\phi, \phi)) .$$

Then, by the Schwarz inequality,

$$2|(\delta_3\phi, \phi)| \leqq \varepsilon\|\delta_3\phi\|_0^2 + \frac{1}{\varepsilon}\|\phi\|_0^2 ,$$

and taking $\varepsilon\|\delta_3\phi\|_0^2 \leqq C'\varepsilon\|\phi\|_1^2$ to the left gives

$$\|\phi\|_1^2 \leqq C\mathcal{D}(\phi, \phi)$$

for a suitable large constant C . Q.E.D.

REMARK: In the Kähler case, the precise Weitzenböck identity plus integration by parts as above gives the <u>Kodaira identity</u>

$$(\Box\phi, \phi) = \|\bar{\nabla}\phi\|_0^2 + (R\phi, \phi) .$$

In particular, if the canonical bundle is negative (as on $\mathbf{P}^m$), then $H_{\bar\partial}^{0,q}(M) = 0$ for $q > 0$. Applying a similar method to the groups $H_{\bar\partial}^{0,q}(M, E)$ gives Kodaira's original proof of his vanishing theorem.

REFERENCES FOR THE APPENDIX TO LECTURE ONE

A proof of the Hodge theorem along the same general lines as that above was given by J. J. Kohn in a course at Princeton University in 1961-62. A different argument, based on Fourier transforms and pseudo-differential operators as opposed to Fourier series, is in the book by R. O. Wells listed in the references to lecture one.

2. RIEMANN-ROCH AND FIXED POINT FORMULAE

a) FORMALISM. Let $E \to M$ be a holomorphic vector bundle over a compact, complex manifold. Since the cohomology $H^*(M, \mathcal{O}(E)) = \sum_{q=0}^m H^q(M, \mathcal{O}(E))$ is finite dimensional, the <u>Euler characteristic</u>

$$\chi(M, E) = \sum_{q=0}^{m} (-1)^q \dim H^q(M, \mathcal{O}(E))$$

is defined. The <u>Hirzebruch-Riemann-Roch formula</u>

$$\chi(M, E) = \int_M T(c_1, \ldots, c_m; d_1, \ldots, d_r) \tag{2.1}$$

expresses the Euler characteristic as a universal polynomial (the Todd polynomial) in the Chern classes c_i of M and d_μ of E evaluated on M. We shall outline Toledo and Tong's proof of (2.1), giving as an application the Atiyah-Bott fixed point formula. For notational simplicity, we shall take E to be the trivial line bundle, and shall write $\chi(M, \mathcal{O})$ for the Euler characteristic in this case.

The basic formal step is to use Kodaira-Serre duality and the Künneth formula to express the Euler characteristic as the value of a cohomology class on $M \times M$ on the diagonal Δ. This is the holomorphic analogue of the usual formula for the topological Euler characteristic as the self-intersection number of the diagonal, and goes as follows: Let $\psi_\mu^q \in A^{0,q}(M)$ be $\bar\partial$-closed forms giving a basis for $H_{\bar\partial}^{0,q}(M)$, and $\psi_\mu^{*q} \in A^{m,m-q}(M)$ forms yielding a dual basis for $H_{\bar\partial}^{m,m-q}(M)$. Using (x, y) to denote points on $M \times M$ and $A^{(p,q)(r,s)}(M \times M)$ to denote forms on the product which are of type (p, q) in x and (r, s) in y (bi-type for short), we set

$$\Psi_q(x, y) = (-1)^q \sum_\mu \psi_\mu^q(x) \wedge \psi_\mu^{*q}(y)$$

$$\Psi(x, y) = \sum_{q=0}^{m} \Psi_q(x, y) .$$

Then clearly

$$\chi(M, \mathcal{O}) = \int_\Delta \Psi(x, x) .$$

In general, any smooth form $s_q(x, y) \in A^{(0,q)(m,m-q)}(M \times M)$ gives an operator

$$S_q : A^{0,q}(M) \to A^{0,q}(M)$$

defined by

$$(S_q\phi)(x) = \int_M s_q(x, y) \wedge \phi(y) ,$$

which we call a smoothing operator with kernel $s_q(x, y)$. Suppose that $S = \sum_{q=0}^m S_q$ is any smoothing operator with kernel $s(x, y) = \sum_{q=0}^m s_q(x, y)$ having the properties

$s_q(x, y) \in A^{(0,q)(m,m-q)}(M \times M)$;

S_q commutes with $\bar{\partial}$; and

S_q induces the identity on $H^{0,q}_{\bar{\partial}}(M)$.

Then, from the Künneth formula on $M \times M$, $s(x, y)$ is $\bar{\partial}$-cohomologous to $\Psi(x, y)$, and consequently

$$\chi(M, \mathcal{O}) = \int_{\Delta} s(x, x) . \tag{2.2}$$

Toledo and Tong give a universal procedure for finding such a smoothing operator.

To explain in other terms what (2.2) means, it is convenient to use the language of currents. On a (possibly non-compact) complex manifold N , the <u>currents of type</u> (p, q) , denoted $D^{p,q}(N)$, are the continuous linear functionals

$$T : A^{n-p,n-q}_c(N) \to \mathbf{C}$$

on the compactly supported forms $(n = \dim N)$. The usual formula for distributional derivatives

$$\bar{\partial}T(\phi) = (-1)^{p+q+1}T(\bar{\partial}\phi)$$

defines $\bar{\partial} : D^{p,q}(N) \to D^{p,q+1}(N)$ with $\bar{\partial}^2 = 0$.

Currents are introduced to have a formalism including both subvarieties and smooth forms in one large complex. For example, a codimension-k analytic subvariety $Z \subset N$ defines a current $T_Z \in D^{q,q}(N)$ by integration over the regular points of Z :

$$T_Z(\phi) = \int_{Z_{reg}} \phi .$$

It can be proved that $\bar{\partial}T_Z = 0$ (intuitively, this is because the boundary $\bar{Z} - Z$ has real codimension <u>two</u> in $\bar{Z}$). As another example, any form $\eta \in A^{p,q}(N)$ defines a current $T_\eta \in D^{p,q}(N)$ by the formula

$$T_\eta(\phi) = \int_M \eta \wedge \phi .$$

Stokes' theorem implies that $\bar{\partial}T_\eta = T_{\bar{\partial}\eta}$.

A basic result (<u>smoothing of cohomology</u>) is that the inclusion

$$A^{p,q}(N) \to D^{p,q}(N)$$

induces an isomorphism on cohomology. In particular, given a subvariety $Z \subset N$ as above, the equation of currents

$$\bar{\partial}k = T_Z - \phi \tag{2.3}$$

can be solved for a smooth (k, k)-form ϕ. We may take k to be a locally integrable $(k, k-1)$-form which is smooth outside Z and has "residue type" singularities along Z, i.e. singularities generalizing the Cauchy formula

$$\bar{\partial}\left(\frac{1}{2\pi\sqrt{-1}}\frac{dz}{z}\right) = T_{\{0\}}$$

in the complex plane.

Returning to $M \times M$, the current T_Δ defined by the diagonal has an expansion into bi-type

$$T_\Delta = \sum_{p=0}^{m} T_{p,\Delta} \quad \text{where}$$

$$T_{p,\Delta} = \sum_{q=0}^{m} (T_\Delta)^{(p,q)(m-p,m-q)} .$$

The equation $\bar{\partial}T_\Delta = 0$ implies that $\bar{\partial}T_{p,\Delta} = 0$ for each p. In particular, $\bar{\partial}T_{0,\Delta} = 0$, and as in (2.3) we may solve the equation of currents

$$\bar{\partial}k(x, y) = T_{0,\Delta} - s(x, y) \tag{2.4}$$

where $s(x, y) \in \sum_{q=0}^{m} A^{(0,q)(m,m-q)}(M \times M)$ is a smooth form and $k(x, y)$ is a residue form. $s(x, y)$ is the kernel of a smoothing operator with the properties listed above (2.2), and by that equation we may write the Euler characteristic

$$\chi(M, \mathcal{O}) = T_{0,\Delta} \cdot T_{m,\Delta} \tag{2.5}$$

as the <u>holomorphic</u> <u>self-intersection</u> number of the diagonal (note that $T_{0,\Delta} \cdot T_{m,\Delta} = T_{0,\Delta} \cdot T_\Delta$ by type considerations).

A final remark concerning (2.4) is that, if we have $k_0(x, y)$ defined only in a neighborhood of the diagonal in $M \times M$ and satisfying

$$\bar{\partial}k_0 = T_{0,\Delta} - s_0$$

there, then we may find a global k satisfying (2.4) and with $s_0(x, x) = s(x, x)$. Indeed, let ρ be a bump function which is one near Δ and set $k = \rho k_0$. Then (2.4) is satisfied with $s = \rho s_0 + \bar{\partial}\rho \wedge k_0$.

b) THE ČECH PARAMETRIX. Let M be a complex manifold with coordinate covering $\mathcal{U} = \{U_\alpha; \phi_\alpha : U_\alpha \to \mathbb{C}^m ; \phi_{\alpha\beta} = \phi_\alpha \circ \phi_\beta^{-1}$ where defined$\}$. We may consider $\mathcal{U}$ as the raw data of the manifold. The cohomology groups $H^*(M, \mathcal{O})$ are given in terms of this raw data by the Čech method (for a suitable such covering). The Dolbeault cohomology $H_{\bar{\partial}}^{0,*}(M)$ is also defined intrinsically by the $\bar{\partial}$-operator, but the explicit formulae in the isomorphism $H^*(M, \mathcal{O}) \cong H_{\bar{\partial}}^{0,*}(M)$ depends on a choice of partition of unity (or something similar). On the other side of the Hirzebruch-Riemann-Roch formula, the Chern classes have been defined using a metric on M. However, there is a procedure due to Atiyah and reviewed below for defining the Chern classes in terms of the raw data. Thus, both sides of (2.1) are defined purely in terms of $\mathcal{U}$, and it makes formal sense to look for a means of evaluating the holomorphic intersection number $T_{0,\Delta} \cdot T_\Delta$ without introducing the extraneous data of a partition of unity and metric. In other words, we should try and solve the Čech analogue of (2.4) purely in terms of the raw data of M. Furthermore, since currents are an essential part of the problem, we should use a Čech complex containing currents as well as holomorphic material. Finally, by the remark at the end of (a) it will suffice to solve the Čech analogue of (2.4) in a neighborhood of the diagonal.

Putting all of this together leads to the following formulation of the problem: On a complex manifold N, we denote by $\mathcal{D}^{p,q}$ and $A^{p,q}$ the respective sheaves of currents and smooth forms of type (p, q). Given a coordinate covering $\mathcal{V} = \{V_\mu\}$ of N, we denote by

$$C^*(\mathcal{V}, \mathcal{D}^{0,*})$$

the Čech bi-complex where $C^p(\mathcal{V}, \mathcal{D}^{0,q})$ are the p-cochains with coefficients in $\mathcal{D}^{0,q}$. The total differential $D = \delta \pm \bar{\partial}$ satisfies $D^2 = 0$, and thus the Čech hypercohomology groups

$$\mathbb{H}^*(\mathcal{V}, \mathcal{D}^{0,*})$$

are defined. Assuming that the covering $\mathcal{V}$ is $\bar{\partial}$-acyclic, the inclusion $\mathcal{O} \to \mathcal{D}^{0,*}$ induces an isomorphism on cohomology

$$H^*(\mathcal{V}, \mathcal{O}) \cong \mathbb{H}^*(\mathcal{V}, \mathcal{D}^{0,*}) . \tag{2.6}$$

In fact, there are two spectral sequences E' and E'' abutting to $\mathbb{H}^*(\mathcal{V}, \mathcal{D}^{0,*})$ and with

$$E_2' = H_\delta^*(\mathcal{V}, \mathcal{H}_{\bar{\partial}}^*(\mathcal{D}^{0,*}))$$

$$E_2'' = H_{\bar{\partial}}^*(H_\delta^*(\mathcal{V}, \mathcal{D}^{0,*})) .$$

By the $\bar{\partial}$-acyclicity of $\mathcal{V}$, $\mathcal{H}_{\bar{\partial}}^0(\mathcal{D}^{0,*}) = \mathcal{O}$ and $\mathcal{H}_{\bar{\partial}}^q(\mathcal{D}^{0,*}) = 0$ for $q > 0$. Thus, the first spectral sequence degenerates and gives (2.6). Using a partition of unity, $H_\delta^q(\mathcal{V}, \mathcal{D}^{0,*}) = 0$ for $q > 0$ while $H_\delta^0(\mathcal{V}, \mathcal{D}^{0,*}) = D^{0,*}(N)$. Consequently the second spectral sequence degenerates also and gives

$$H_{\bar{\partial}}^*(N) \cong \mathbb{H}^*(\mathcal{V}, \mathcal{D}^{0,*}) .$$

The composition of this isomorphism and (2.6) gives the Dolbeault isomorphism, whose explicit formulae depend on the partition of unity.

On $M \times M$ we consider the covering $\mathcal{U} \times \mathcal{U} = \{U_\alpha \times U_\alpha, \phi_\alpha \times \phi_\alpha, \ldots\}$ of a neighborhood of the diagonal. The current $T_{0,\Delta} \in C^0(\mathcal{U} \times \mathcal{U}, \mathcal{D}^{(0,*)(m,m-*)})$ satisfies

$$\bar{\partial} T_{0,\Delta} = 0 = \delta T_{0,\Delta} ,$$

and the Čech analogue of (2.4) is the relation

$$Dk = T_{0,\Delta} - s \tag{2.7}$$

where s is to be smooth. When written out, what this amounts to are the equations

$$k = k_0 + \cdots + k_{m-1} ;$$

$$\begin{aligned} \bar{\partial} k_0 &= T_{0,\Delta} \\ \bar{\partial} k_j &= \delta k_{j-1} \qquad (j = 1, \ldots, m-1) \\ \delta k_{m-1} &= s ; \end{aligned}$$

and

$$k_j \in C^j(\mathcal{U} \times \mathcal{U}, \mathcal{D}^{(0,*)(m,m-j-1-*)})$$
$$s \in C^m(\mathcal{U} \times \mathcal{U}, \mathcal{D}^{(0,0)(m,0)}) .$$

It follows that

$$\bar{\partial} s = \pm \delta^2 k_{m-2} = 0 , \quad \delta s = 0$$

so that, by Hartog's theorem for $m > 1$ and by elementary reasons in case $m = 1$,

$$s \in H^m(\mathcal{U} \times \mathcal{U}, \Omega^m_{M \times M}) .$$

A solution of (2.7) is called a <u>Čech parametrix</u>, and for any such we have by (2.2)

$$\chi(M, \mathcal{O}) = s[\Delta] . \tag{2.8}$$

The main result of Toledo and Tong may be informally stated as follows:

(2.9) Given a complex manifold M and raw data $\{U_\alpha, \phi_\alpha, \phi_{\alpha\beta}\}$, there exists a Čech parametrix $s = s\{U_\alpha, \phi_\alpha, \phi_{\alpha\beta}\}$ which is functorial in the raw data. In particular, the formulae for the Čech cocycle $s \in H^m(\mathcal{U}, \Omega^m_M)$ are given by universal expressions in terms of the $\phi_\alpha, \phi_{\alpha\beta}$ and finitely many of their derivatives.

c) ČECH CHERN CLASSES. To explain how (2.9) gives a Riemann-Roch formula, we recall the following construction of Atiyah:

Given a complex manifold M , an open covering $\{U_\alpha\}$, and transition functions $g_{\alpha\beta} : U_\alpha \cap U_\beta \to GL(r, C)$ for a holomorphic vector bundle $E \to M$, the <u>Atiyah curvature cocycle</u> $\Theta \in Z^1(\{U_\alpha\}, \Omega^1_M(\mathrm{Hom}(E, E)))$ is defined by

$$\Theta_{\alpha\beta} = g^{-1}_{\alpha\beta} \cdot dg_{\alpha\beta} .$$

Note that Θ is functorial in the raw data $\{U_\alpha, g_{\alpha\beta}\}$. For any invariant polynomial $P(X)$ $(X \in \mathfrak{gl}(r, C))$ with corresponding multilinear form $P(X_1,\ldots,X_q)$ the <u>Atiyah Chern polynomial</u> is defined by

$$P(\Theta) \in Z^q(\{U_\alpha\}, \Omega^q_M)$$

where $q = \deg P$ and where

$$P(\Theta)_{\alpha_0 \cdots \alpha_q} = P(\Theta_{\alpha_0\alpha_1}, \mathrm{Ad} g_{\alpha_0\alpha_1}\Theta_{\alpha_1\alpha_2}, \ldots, \mathrm{Ad} g_{\alpha_{q-2}\alpha_{q-1}}\Theta_{\alpha_{q-1}\alpha_q}) .$$

The Atiyah Chern polynomials are functorial in the raw data, and a fairly straightforward argument using invariant theory shows that:

(2.10) Any method of functorially assigning to the raw data $\{U_\alpha, g_{\alpha\beta}\}$ a cocycle $c = c(\{U_\alpha\}, g_{\alpha\beta}) \in Z^m(\{U_\alpha\}, \Omega^m_M(\mathrm{Hom}(E, E))$ is necessarily an Atiyah Chern polynomial.

Combining (2.8), (2.9), and (2.10) gives a formula

$$\chi(M, \mathcal{O}) = \int_M T(c_1, \ldots, c_m)$$

for some universal polynomial T. We shall give the explicit formula for T after discussing the first step in the construction of a Čech parametrix and using this to deduce the Atiyah-Bott fixed point formula.

d) THE FIXED POINT FORMULA. We shall give the first step in the construction of a universal Čech parametrix, and shall use this to deduce the Atiyah-Bott-Lefschetz-Woods Hole fixed point formula (this conference being on the tenth anniversary of Woods Hole).

For this we use the <u>Bochner-Martinelli kernel</u> on $C^m \times C^m$ defined by

$$k_0(z, \zeta) = C_m \frac{\sum_{i=1}^m (-1)^i \overline{(z_i - \zeta_i)\Phi_i(z-\zeta)} \wedge \Phi(\zeta)}{\|z - \zeta\|^{2m}} \tag{2.11}$$

where

$$\Phi(w) = dw_1 \wedge \cdots \wedge dw_m \quad \text{and} \quad \Phi_i(w) = dw_1 \wedge \cdots \wedge \widehat{dw_i} \wedge \cdots \wedge dw_m .$$

Note that $k_0(z, \zeta)$ has bi-type $(0, *)(m, m-1-*)$, and is formally $\bar\partial$-closed. As distributions,

$$\bar\partial k_0(z, \zeta) = T_{0,\Delta}$$

for a suitable normalizing constant C_m.

Using the coordinate charts $\phi_\alpha \times \phi_\alpha : U_\alpha \times U_\alpha \to C^m \times C^m$ to pull back the Bochner-Martinelli kernels gives $k_0 \in C^0(\mathcal{U} \times \mathcal{U}, \mathcal{D}^{(0,*)(m,m-1-*)})$ satisfying

$$\bar{\partial} k_0 = T_{0,\Delta} .$$

It is usually not the case that $k_{0,\alpha} = k_{0,\beta}$ in $(U_\alpha \times U_\alpha) \cap (U_\beta \times U_\beta)$. However, from general principles, the equation

$$\bar{\partial} k_{1,\alpha\beta} = k_{0,\alpha} - k_{0,\beta}$$

has a solution, as does the succeeding equation

$$\bar{\partial} k_{2,\alpha\beta\gamma} = k_{1,\alpha\beta} - k_{1,\beta\gamma} + k_{1,\alpha\gamma} ,$$

and so forth. The thrust of (2.9) is that all of these equations may be solved in a suitable universal manner.

Now to the fixed point formula! Let $f : M \to M$ be a biholomorphic automorphism having isolated transversal fixed points. There are induced maps on cohomology

$$f^{*,q} : H^q(M, \mathcal{O}) \to H^q(M, \mathcal{O}) ,$$

and the <u>holomorphic Lefschetz number</u> is defined by

$$L(f, \mathcal{O}) = \sum_{q=0}^{m} (-1)^q \text{ Trace } f^{*,q} .$$

What we wish to find is a formula for $L(f, \mathcal{O})$ in terms of the eigenvalues of the induced linear map

$$f_{*,p} : T_p(M) \to T_p(M)$$

at the fixed points of f , and the result is:

$$(2.12) \qquad L(f, \mathcal{O}) = \sum_{f(p)=p} \frac{1}{\det(I - f_{*,p})} .$$

For the proof, we denote by $G_f = \{(p, f(p)\}$ the <u>graph</u> of f in $M \times M$. As in the formal part of the Riemann-Roch, one may prove the intersection relation

$$(2.13) \qquad L(f, \mathcal{O}) = T_{0,\Delta} \cdot G_f .$$

Choose a coordinate covering $\{U_\alpha\}$ of M such that each fixed point p is in exactly one open set U_α , and let $\{\rho_\alpha\}$ be a partition of unity subordinate to this covering and such that $\rho_\alpha \equiv 1$ near any p . Let $k_{0,\alpha}$ be the Bochner-Martinelli kernel in $U_\alpha \times U_\alpha$ and consider the form

$$k = \sum_\alpha \rho_\alpha k_{0,\alpha}$$

on $M \times M$. The equation of currents

$$\bar{\partial} k = T_{0,\Delta} - s_0 , \tag{2.14}$$

where s_0 is the integrable form $\sum_\alpha \bar{\partial}\rho_\alpha \wedge k_{0,\alpha}$, is valid. This form s_0 is not smooth on $M \times M$, but it is smooth on the graph of f, and consequently

$$T_{0,\Delta}\cdot G_f = \int_{G_f} s_0 .$$

Letting $B_\varepsilon(p)$ be a ball of radius ε around p, Stokes' theorem together with (2.13) give

$$L(f, \mathcal{O}) = \sum_{f(p)=p} \lim_{\varepsilon\to 0} \int_{\partial B_\varepsilon(p)} k . \tag{2.15}$$

To evaluate the integrals, we choose local holomorphic coordinates $z = (z_1, \ldots, z_m)$ in a neighborhood U_α around p such that $f(z) = A\cdot z + (\cdot\cdot)$ where A is a non-singular matrix and $(\cdot\cdot)$ denotes higher order terms. These latter will disappear in the limit, and so we shall assume for simplicity that $f(z) = Az$. Letting (z, ζ) denote coordinates in $U_\alpha \times U_\alpha$, the form

$$\sigma = C_m \frac{\sum_{i=1}^m (-1)^i \overline{(z_i - \zeta_i)\Phi_i(z - \zeta) \wedge \Phi(z - \zeta)}}{\|z - \zeta\|^{2m}}$$

gives the standard d-<u>closed</u> volume form on the normal spheres to the diagonal having constant value one on any such sphere. On the graph $\zeta = f(z)$,

$$k_0 = \frac{\sigma}{\det(I - A)}$$

so that $\lim_{\varepsilon\to 0} \int_{\partial B_\varepsilon(p)} k_0 = \frac{1}{\det(I - A)}$.

We shall conclude by explaining how (2.12) leads to the expression for the Todd polynomials in the Riemann-Roch formula. For this, we first need to discuss <u>Bott's residue theorem</u>. In simplest form, this result states that, if M has a nowhere vanishing holomorphic vector field v and if $P(c_1, \ldots, c_m)$ is any polynomial of degree m in the Chern classes, then

(2.16) $$\int_M P(c_1, \ldots, c_m) = 0 .$$

To prove (2.16), we shall choose a special connection for the tangent bundle such that the Chern polynomial $P(c_1(\Theta), \ldots, c_m(\Theta))$ is identically zero. For this we take a coordinate covering $\{U_\alpha\}$ of M with coordinates $z_\alpha = (z_\alpha^1, \ldots, z_\alpha^m)$ in U_α such that $v = \partial/\partial z_\alpha^1$. Let D_α be the flat Euclidean connection in U_α and $D = \sum_\alpha \rho_\alpha D_\alpha$. The curvature Θ of D satisfies the contraction relation

$$< \Theta, v > \equiv 0 ;$$

i.e., in U_α the curvature does not involve dz_α^1. To see this, we need only observe that in $U_\alpha \cap U_\beta$ the connection matrix θ_β for D_β <u>written in terms of the coordinates</u> z_α satisfies $< \theta_\beta, v > \equiv 0$. Now it is clear that $P(c_1(\Theta), \ldots, c_m(\Theta)) \equiv 0$ for this curvature.

Bott's general formula deals with a holomorphic vector field v having isolated non-degenerate zeroes. Near such a zero p, we choose holomorphic coordinates $(z_1, \ldots, z_m)$ such that

$$v = \sum_{i,j} a_{ij} z_i \frac{\partial}{\partial z_j} + \text{(higher order terms)} .$$

Denote by $\lambda_1, \ldots, \lambda_m$ the eigenvalues of the matrix (a_{ij}) and by σ_q the qth elementary symmetric function of the λ_i's. Bott's formula is the relation

(2.17) $$\int_M P(c_1, \ldots, c_m) = \sum_{v(p)=0} \frac{P(\sigma_1, \ldots, \sigma_m)}{\lambda_1 \cdots \lambda_m} .$$

To prove (2.17), we choose a connection D_1 on M which is flat near each zero of v, and a connection D_2 on $M - \{\text{zeroes of } v\}$ satisfying $< \Theta_2, v > \equiv 0$ as before. Now then the proof that the de Rham Chern classes are independent of the choice of connection gives a relation

$$P(c_1(\Theta_1), \ldots, c_m(\Theta_1)) - P(c_1(\Theta_2), \ldots, c_m(\Theta_2)) = d\eta$$

on $M - \{\text{zeroes of } v\}$ for an explicit η (note that $P(c_1(\Theta_2), \ldots, c_m(\Theta_2)) \equiv 0$). Stokes' theorem then gives

$$\int_M P(c_1(\Theta_1), \ldots, c_m(\Theta_1)) = \sum_{v(p)=0} \lim_{\varepsilon \to 0} \int_{\partial B_\varepsilon(p)} \eta ,$$

which when computed out yields (2.17).

Finally, the explicit form of the Todd genus arises as follows: Suppose that v is a holomorphic vector field with isolated zeroes and set $f_t = \exp(tv)$. From the homotopy formula $\mathrm{Lie}_v = v \lrcorner \bar\partial + \bar\partial v \lrcorner$ on the Dolbeault complex, it follows that $f_t^{*,q}$ is the identity on $H^q(M, \mathcal{O})$. Thus, for all t,

$$\chi(M, \mathcal{O}) = L(f_t, \mathcal{O}) = \sum_{v(p)=0} \prod_{i=1}^{m} \frac{1}{1 - e^{t\lambda_i}} .$$

Taking $t = -1$ gives

$$\chi(M, \mathcal{O}) = \sum_{v(p)=0} \frac{1}{\lambda_1 \cdots \lambda_m} \prod_{i=1}^{m} \frac{\lambda_i}{1 - e^{-\lambda_i}} .$$

Combining this with (2.1) and (2.17) yields

$$\chi(M, \mathcal{O}) = \int_M T(c_1, \ldots, c_m)$$

where the Todd polynomial T is obtained by writing formally $1 + c_1 + \cdots + c_m = \prod_{i=1}^{m} (1 + \gamma_i)$ and then setting

$$T(c_1, \ldots, c_m) = \Big\{ \prod_{i=1}^{m} \frac{\gamma_i}{1 - e^{-\gamma_i}} \Big\}_m$$

where $\{\cdots\}_m$ is the component of degree m.

REFERENCES FOR LECTURE 2

The Hirzebruch-Riemann-Roch formula first appeared in the book by Hirzebruch listed as a general reference. Since then the subject has mushroomed into index theory, for which the most recent account is

M. Atiyah, R. Bott, and V. K. Patodi, On the heat equation and the index theorem, Inventiones Math., vol. 19 (1973), 279-330.

The proof of the Hirzebruch formula we are following is from

D. Toledo and Y. Tong, A parametrix for $\bar\partial$ and Riemann-Roch in Čech theory, to appear.

The general theory of currents and smoothing of cohomology is given in

G. de Rham, Variétés differentiables, Hermann, Paris (1955).

A discussion of the currents which appear in complex analysis appears in

J. King, The currents defined by analytic varieties, Acta Math., vol. 127 (1971), 185-220.

The Čech Chern class construction is given in

M. F. Atiyah, Complex analytic connections in fibre bundles, Trans. Amer. Math. Soc., 85 (1957), 181-207.

The various integral formulae which interpolate between the Bochner-Martinelli kernel and Cauchy kernel are discussed in

R. Harvey, Integral formulae connected by Dolbeault's isomorphism, Rice Univ. Studies, vol. 56 (1970), 77-97.

The Atiyah-Bott fixed point formula is given in

M. F. Atiyah and R. Bott, A Lefschetz fixed point formula for elliptic complexes (II), Annals of Math., vol. 88 (1968), 451-491.

Proofs of Bott's residue theorem may be found in

R. Bott, Vector fields and characteristic numbers, Michigan Math. J., vol. 14 (1967), 231-244;

S. S. Chern, Geometry of characteristic classes, Proc. 13th Biennial Seminar (1972), 1-40.

3. KÄHLER MANIFOLDS; HODGE THEORY

Among complex manifolds some are singled out by the fact that their holomorphic tangent bundle can be endowed with a special kind of hermitian metric, a _Kähler metric_. The presence of such a metric has important global implications and we shall discuss some of them.

There are several possible definitions of what a Kähler metric is. To a hermitian metric

$$h = \sum_{i,j} h_{ij}\, dz_i d\bar{z}_j$$

on a complex manifold M there is naturally attached a distinguished connection D_h on the holomorphic tangent bundle $T_h(M)$, the _metric_ connection. $D = D_h \oplus \bar{D}_h$ is a connection on the complexification $T_h \oplus \bar{T}_h$ of the real tangent bundle of M.

i) The hermitian metric h is said to be Kähler if D is the riemannian connection on M. In other terms, set

$$\tau(X, Y) = D_X Y - D_Y X - [X, Y]$$

for any couple X, Y of vector fields on M. τ is easily seen to be a tensor, the so-called torsion tensor of the metric h. Saying that h is Kähler means that τ vanishes identically.

ii) To give a second, equivalent definition of what a Kähler metric is, let

$$\omega = \frac{\sqrt{-1}}{2} \sum h_{ij}\, dz_i \wedge d\bar{z}_j$$

be the exterior form associated to h. A necessary and sufficient condition for h to be Kähler is that ω be d-closed.

iii) A third definition is the following: a hermitian metric h is said to be Kähler if it can be written locally as

$$\sum_i dz_i d\bar{z}_i + [2]$$

where $[2]$ stands for terms which vanish of order at least two at $z = 0$, relative to suitable holomorphic coordinate systems.

Each of the above definitions has some advantage over the others. Definition ii), for example, makes it clear that the property of having a Kähler metric is inherited by subvarieties and that a Kähler metric gives a distinguished cohomology class in $H^2(M, \mathbb{R})$. Definition iii) is most useful in computation as it allows us to prove the basic Kähler identities by verifying them for the flat metric on $\mathbb{C}^m$.

As for the proof of the equivalence of i), ii), iii), ii) follows from i) by noticing that for the metric connection D_h

$$\partial h_{ij} = (D_h' \frac{\partial}{\partial z_i}, \frac{\partial}{\partial z_j})$$

holds and that i) means that:

$$D_{\frac{\partial}{\partial z_i}} \frac{\partial}{\partial z_j} = D_{\frac{\partial}{\partial z_j}} \frac{\partial}{\partial z_i} .$$

The converse is done in the same way. iii) clearly implies ii), and one obtains iii) from i) by choosing the coordinate systems to be geodesic coordinate systems up to second order terms.

We now want to discuss the basic <u>Kähler identities</u> satisfied by the operators acting on the algebra

$$A^*(M) = \sum A^{p,q}(M) .$$

We define

$$L : A^{p,q}(M) \to A^{p+1,q+1}(M)$$

to be multiplication by ω, and Λ to be the adjoint of L relative to the given metric. Also we denote by $\pi_{p,q}$ the projection onto $A^{p,q}(M)$ and set $\pi_k = \sum_{p+q=k} \pi_{p,q}$. It is a consequence of the theory of unitary invariants that on a "general" Kähler manifold the operators L, Λ, $\pi_{p,q}$ generate the algebra of all invariant <u>algebraic</u> operators on $A^*(M)$.

Probably the neatest and most compact way of presenting the identities satisfied by L, Λ and $\pi_{p,q}$ is the following: Write

$$H = \sum_k (m - k)\pi_k$$

where m is the complex dimension of M. An easy computation then gives the commutation relations:

$$(3.1) \qquad \begin{aligned} [H, \Lambda] &= 2\Lambda \\ [H, L] &= -2L \\ [\Lambda, L] &= H . \end{aligned}$$

This means that the assignments:

$$\begin{pmatrix} 0 & 1 \\ 0 & 0 \end{pmatrix} \to \Lambda$$

$$\begin{pmatrix} 1 & 0 \\ 0 & -1 \end{pmatrix} \to H$$

$$\begin{pmatrix} 0 & 0 \\ 1 & 0 \end{pmatrix} \to L$$

give a <u>representation</u>

$$\rho : s\ell_2 \to \mathrm{End}(A^*(M))$$

which can also be viewed as a continuously varying family of finite dimensional representations. The standard representation theory for $s\ell_2$ then implies that:

$$L^k : A^{m-k}(M) \to A^{m+k}(M) \tag{3.2}$$

is an isomorphism and that

$$A^k(M) = \sum L^h P^{k-2h} A(M) \qquad \text{(direct sum)} \tag{3.3}$$

where $PA^\ell(M)$ (the <u>primitive part</u> of A^ℓ) stands for the kernel of $L^{m-\ell+1} \mid A^\ell(M)$. It is also clear that ρ is compatible with decomposition into (p, q)-type, hence $PA^\ell(M)$ decomposes into a direct sum:

$$PA^\ell(M) = \sum_{p+q=\ell} PA^{p,q}(M) .$$

What we have done so far does not require that the metric be Kähler. The crucial fact, however, is that, <u>when the metric is Kähler</u>, (3.2) <u>and</u> (3.3), <u>together with the decomposition into type, pass over to cohomology</u>.

Proving this involves exploring the commutation relation between ρ and exterior differentiation. The basic such relation, from which all others can be deduced, is:

$$[\Lambda, d] = (d^C)^* \tag{3.4}$$

where $(d^C)^*$ is the adjoint of

$$d^C = C^{-1}dC = \sqrt{-1}(\bar\partial - \partial)$$

and $C = \sum \sqrt{-1}^{\,p-q}\pi_{p,q}$ is the <u>Weil operator</u>. (3.4) is actually equivalent to the metric being Kähler. The two (easy) consequences of (3.4) that will be important for us are:

$$[\Delta, \rho] = 0 \tag{3.5}$$

where $\Delta = dd^* + d^*d$ is the Laplace operator, and

$$\Delta = 2\,\Box \tag{3.6}$$

where $\Box = \bar\partial\bar\partial^* + \bar\partial^*\bar\partial$ is the complex Laplacian. (3.6) in turn implies that

$$[\Delta, \pi_{p,q}] = 0 \tag{3.7}$$

since $\square$ is of pure $(0, 0)$ type.

As for the proof of (3.4), it suffices to do the case of the flat metric on C^m, which is straightforward.

The main consequence of (3.5) is that, on a compact Kähler manifold, ρ passes over to cohomology, and therefore, by standard representation theory,

$$L^k : H^{m-k}(M, \mathbb{R}) \to H^{m+k}(M, \mathbb{R}) \tag{3.8}$$

is an isomorphism (Hard Lefschetz Theorem) and one has the Lefschetz decomposition

$$H^k(M, \mathbb{R}) = \bigoplus_h L^h P^{k-2h}(M, \mathbb{R}) \tag{3.9}$$

where $P^\ell(M, \mathbb{R})$ (primitive cohomology) is the kernel of $L^{m-\ell+1}|H^\ell(M, \mathbb{R})$.

On the other hand, (3.7) implies that $H^k(M, C)$ has a direct sum decomposition (Hodge decomposition)

$$H^k(M, C) = \sum_{p+q=k} H^{p,q}(M) \tag{3.10}$$

such that

$$H^{p,q}(M) = \overline{H^{q,p}(M)} .$$

Here $H^{p,q}(M)$ stands for the subspace of $H^k(M, C)$ generated by d-closed (p, q)-forms. Moreover the Hodge decomposition is obviously compatible with the Lefschetz decomposition. It should also be noticed that the equality (3.6) gives an isomorphism (which does not depend on the metric) between $H^{p,q}(M)$ and $H^q(M, \Omega^p)$.

As a final ingredient, if we define a bilinear form

$$Q_k(\xi, \eta) = \int_M \omega^{m-k} \wedge \xi \wedge \eta$$

on $P^k(M)$, then the Hodge-Riemann bilinear relations hold:

$$\text{(3.11)} \quad \begin{aligned} &\text{I)} \quad Q_k(P^{p,q}, P^{q',p'}) = 0 \quad \text{unless } p = p',\ q = q' \\ &\text{II)} \quad Q_k(C\xi, \bar{\xi}) > 0 \qquad \xi \in P^k(M),\ \xi \neq 0 \end{aligned}$$

In case the cohomology class of ω is rational — which, by a theorem of Kodaira, is equivalent to saying that an integral multiple of the class of ω is induced by the generator of $H^2(\mathbb{P}^n, \mathbb{Z})$ via an embedding into $\mathbb{P}^n$ — the representation ρ, and hence the Lefschetz decomposition, are defined over $\mathbb{Q}$.

To conclude, we may remark that, while the Lefschetz decomposition is of a topological nature, the Hodge decomposition depends on the complex structure of M, and in fact is a very significant invariant for this.

As another application of the Kähler identities we will prove the following result, due to Kodaira, Nakano and Le Potier.

(3.12) Let $E \to M$ be a rank r positive vector bundle on a compact complex manifold M of dimension m. Then

$$H^q(M, \Omega^p(E)) = 0$$

if $p + q \geq m + r$.

PROOF: The proof is done by reducing to the rank 1 case. We shall deal with this first. It follows from the hypotheses that M is Kähler, and we shall choose (the hermitian form associated to) $\sqrt{-1}\,\Theta$ as a metric on M, where Θ is the curvature form of E. We shall show that, for any E-valued harmonic (p, q)-form ϕ, the Nakano inequalities

$$(3.13) \qquad \begin{aligned} (\Lambda L\phi, \phi) &\geq 0 \\ (L\Lambda\phi, \phi) &\leq 0 \end{aligned}$$

hold. Combining them gives

$$0 \leq ([\Lambda, L]\phi, \phi) = (m - p - q)(\phi, \phi)$$

which implies that ϕ is zero if $p + q > m$, as desired. Now to the proof of (3.13)! A slight generalization of (3.4) gives the commutation relation:

$$[\Lambda, D''] = -\sqrt{-1}\, D'^*$$

where $D = D' + D''$, $D'' = \bar{\partial}$, is the metric connection of E. Therefore:

$$\begin{aligned} (\Lambda L\phi, \phi) &= \sqrt{-1}\,(\Lambda\Theta\phi, \phi) = \\ &= \sqrt{-1}\,(\Lambda\bar{\partial}D'\phi, \phi) = \\ &= \sqrt{-1}\,(\bar{\partial}\Lambda D'\phi, \phi) + (D'^*D'\phi, \phi) = \end{aligned}$$

$$= (D'\phi, D'\phi) \geq 0$$

since ϕ, being harmonic, is $\bar\partial$-closed and $\bar\partial^*$-closed. This is the first of the Nakano inequalities and the proof of the other one is similar.

When the rank of E is larger than one, we may argue as follows. Let $\mathbb{P}(E^*)$ be the complex manifold whose points are the hyperplanes lying in the fibres of E. $\mathbb{P}(E^*)$ has dimension $m + r - 1$ and is obviously a bundle

$$\mathbb{P}(E^*) \xrightarrow{\pi} M$$

with projective $(r-1)$-spaces as fibres. Let H be the standard tautological line bundle over $\mathbb{P}(E^*)$. Direct computation shows that:

H is positive

$$R^i\pi_*\Omega^j(H) = \begin{cases} 0 & \text{if } i > 0 \\ \Omega^j(E) & \text{if } i = 0 . \end{cases}$$

In particular $\pi_*\mathcal{O}(H) = \mathcal{O}(E)$ and the E_2-term of the Leray spectral sequence abutting to $H^*(\mathbb{P}(E^*), \Omega^p(H))$ is

$$E_2^{ij} = H^j(M, R^i\pi_*\Omega^p(H)) = \begin{cases} 0 & \text{if } i > 0 \\ H^j(M, \Omega^p(E)) & \text{if } i = 0 . \end{cases}$$

Therefore the above spectral sequence degenerates at the E_2-term and

$$H^q(M, \Omega^p(E)) = H^q(\mathbb{P}(E^*), \Omega^p(H)) = 0 \quad \text{if } p + q \geq m + r$$

by applying to H the line bundle case of the theorem.

A classical application of the Kodaira-Nakano vanishing theorem is an analytic proof of the Lefschetz theorem on hyperplane sections with $\mathbb{R}$-coefficients. Let D be a smooth, ample divisor on M (this is the same as saying that the line bundle $[D]$ is positive). $[D]|_D$ can be identified with the normal bundle to D in M. It follows that there are exact sequences

$$0 \to \Omega^p_M[D]^* \to \Omega^p_M \to \Omega^p_{M|D} \to 0$$

$$0 \to \Omega^{p-1}_D[D]^* \to \Omega^p_{M|D} \to \Omega^p_D \to 0 .$$

Applying the vanishing theorem and Serre duality to $H^q(M, \Omega^p_M[D]^*)$ and

$H^q(D, \Omega_D^p[D]^*)$ gives that <u>the restriction map</u>

$$H^{p,q}(M) \to H^{p,q}(D)$$

<u>is an isomorphism if</u> $p + q \leq m - 2$ <u>and is injective if</u> $p + q = m - 1$. It follows from (3.10) that this is just the Lefschetz theorem on hyperplane sections. Conversely, Mumford and Ramanujan have proved that the topological Lefschetz theorem implies the Kodaira-Nakano vanishing theorem.

It is perhaps worth noticing that the integral formulae underlying the vanishing theorems are <u>exactly</u> the same as those used to establish the basic estimates used in the study of the $\bar{\partial}$-operator on non-compact manifolds by L^2-methods. This is an illustration of our contention that the compact case has been properly understood only after there are results for general possibly non-compact complex manifolds which specialize to the existing ones in the compact case.

Another implication of the fundamental Kähler identity (3.4) and the subsequent equality (3.6) among the various Laplacians is the

PRINCIPLE OF TWO TYPES: Given a (p, q) form ϕ which is also exact, $\phi = d\eta$, then we may write either

$$\text{(3.14)} \qquad \begin{aligned} \phi &= d\eta', \quad \eta' \text{ has type } (p-1, q), \text{ or} \\ \phi &= d\eta'', \quad \eta'' \text{ has type } (p, q-1). \end{aligned}$$

An application of this is that:

(3.15) All Massey products on a compact Kähler manifold are zero.

Recall that, given closed differential forms α, β, γ of degrees p, q, and r,

$$\alpha \wedge \beta = d\rho$$
$$\beta \wedge \gamma = d\sigma$$

the Massey triple product is the closed $p + q + r - 1$ form

$$\alpha \wedge \sigma + (-1)^{(\cdots)} \gamma \wedge \rho .$$

Its cohomology class $[\alpha, \beta, \gamma]$ is well-defined in

$$\text{(3.16)} \qquad H^{p+q+r-1}(M)/H^p(M) \cup H^{q+r-1}(M) + H^r(M) \cup H^{p+q-1}(M) .$$

To see that $[\alpha, \beta, \gamma] = 0$, we decompose α, β, γ under the Hodge decomposition and consider the case where they all three have pure Hodge type. Using (3.14), we may write

$$\alpha \wedge \beta = d\rho' \quad \text{or} \quad \alpha \wedge \beta = d\rho''$$
$$\beta \wedge \gamma = d\sigma' \quad \text{or} \quad \beta \wedge \gamma = d\sigma''$$

where ρ', ρ'' and σ', σ'' have different Hodge types. But then $[\alpha, \beta, \gamma]$ has the two representatives $\alpha \wedge \sigma' + (-1)^{(\cdots)}\gamma \wedge \rho'$ and $\alpha \wedge \sigma'' + (-1)^{(\cdots)}\gamma \wedge \rho''$ *of different Hodge types*, and so must be zero since the quotient space (3.16) has a direct sum Hodge decomposition.

Now it is well-known to topologists that Massey triple products are homotopy invariants of a space which are not contained in its cohomology. What is suggested by the above argument, together with the vanishing of the higher Massey products which is proved similarly, is that: *among all spaces with a given cohomology ring, the Kähler manifolds (if there are any) have the "simplest" homotopy type*. In particular, for simply-connected M the rational homotopy groups and rational Whitehead products should be determined from the cohomology alone. Likewise, the rational nilpotent completion of $\pi_1(M)$ should be determined by $H^1(M, Q)$ together with the cup product $H^1(M, Q) \otimes H^1(M, Q) \to H^2(M, Q)$. This can all be easily proved using (3.15) together with Sullivan's recent de Rham homotopy theory.

The philosophy here, as well as in Deligne's degeneration of spectral sequence arguments used in proving the existence of mixed Hodge structures on general varieties (to be discussed in the next talk), is that any sort of naturally defined higher cohomology operation on a compact Kähler manifold must, because of the principle of two types, be zero.

REFERENCES FOR LECTURE 3

The basic Kähler identities are proved in

A. Weil, Introduction à l'étude des variétés kähleriennes, Hermann, Paris (1958).

The conceptual explanation of the Kähler identities via $s\ell_2$ appears in

S. S. Chern, On a generalization of Kähler geometry, in A Symposium in Honor of S. Lefschetz, Princeton Univ. Press, Princeton, N. J. (1957), 103-121.

The theorem of Kodaira referred to just below (3.11) is given in his paper listed in the general references.

The vanishing theorem (3.12) appears in

S. Nakano, On complex analytic vector bundles, J. Math. Soc. Japan, vol. 7 (1955), 1-12; and

J. Le Potier, Annulation de la cohomologie à valeurs dans certains fibrés vectoriels holomorphes et applications, Thèse présentée à l'Université de Poitiers,(1974).

The equivalence of the vanishing theorem and Lefschetz theorem on hyperplane sections is given in

C. P. Ramanujam, Remarks on the Kodaira vanishing theorem, Jour. Indian Math. Soc. Vol. 36 (1972), 41-50.

The result on the homotopy type of Kähler manifolds will appear in

P. Deligne, P. Griffiths, J. Morgan, and D. Sullivan, On the homotopy type of compact Kähler manifolds, to appear.

4. GENERALIZATIONS OF HODGE STRUCTURES

a) The first topic we shall discuss today is Hodge structure and mixed Hodge structure. The definition of a Hodge structure is obtained by extracting the essential features of the Hodge decomposition on the primitive part of the cohomology of a projective variety. Formally, a *Hodge structure of weight* M consists of a real vector space $H_{\mathbb{R}}$, a lattice $H_{\mathbb{Z}}$ and a decreasing filtration (the Hodge filtration) on the complexification $H_{\mathbb{C}}$ of $H_{\mathbb{R}}$

$$0 \subset F^m \subset \cdots \subset F^0 = H_{\mathbb{C}}$$

such that, for any p,

$$H_{\mathbb{C}} = F^p \oplus \overline{F^{m-p+1}} \quad (4.1)$$

An equivalent way of defining a Hodge structure is saying that H_C should have a direct sum decomposition

$$(4.2)\qquad \begin{aligned} H_C &= \sum_{p+q=m} H^{p,q} \\ H^{p,q} &= \overline{H^{q,p}} . \end{aligned}$$

From the filtration one obtains (4.2) by setting

$$H^{p,m-p} = F^p \cap \overline{F^{m-p}}$$

and conversely, given (4.2) one can recover the Hodge filtration by setting

$$F^p = \sum_{i \geq p} H^{i,m-i} .$$

The <u>Weil operator</u> of the Hodge structure H is defined, as usual, to be

$$C \sum \phi^{pq} = \sum \sqrt{-1}^{\,p-q} \phi^{p,q} .$$

A <u>polarization</u> on H is a rational (non-degenerate) bilinear form Q, symmetric if m is even, skew if m is odd, such that the Hodge-Riemann bilinear relation

$$(4.3)\qquad \begin{aligned} Q(F^p, F^{m-p+1}) &= 0 \\ Q(C\xi, \bar{\xi}) &> 0 \quad \text{if } \xi \neq 0 \end{aligned}$$

hold. In lecture 3 we have sketched a proof of the fact that the primitive cohomology

$$P^m(M)$$

of a projective variety M carries a natural polarized Hodge structure of weight m.

A morphism of type (r, r) between two Hodge structures H and H' is a rationally defined homomorphism

$$\phi : H_C \to H'_C$$

such that

$$\phi(F^p) \subset F'^{p+r}$$

or, equivalently:

$$\phi(H^{p,q}) \subset H^{p+r,q+r} .$$

Most linear algebra constructions, like taking Hom's, tensor products, etc. of Hodge structures of arbitrary weights, or direct sums of Hodge structures of the same weight, can be performed within the category of Hodge structures.

The structure of the cohomology of general open or singular algebraic varieties is more complex than a plain Hodge structure. However it is a fundamental theorem of Deligne that:

The cohomology groups of a general algebraic variety carry natural, functorial mixed Hodge structures.

We shall now define what such objects are and prove a very special case of Deligne's theorem.

A _mixed Hodge structure_ consists of a real vector space $H_{\mathbb{R}}$, a lattice $H_{\mathbb{Z}}$ and two finite filtrations of $H_C = H_{\mathbb{R}} \otimes C$:

$$0 \subset \cdots \subset W_m \subset W_{m+1} \subset \cdots \subset H_C \quad \text{(weight filtration)}$$
$$0 \subset \cdots \subset F^p \subset F^{p-1} \subset \cdots \subset H_C \quad \text{(Hodge filtration)}$$

such that:

i) $\{W_m\}$ is rationally defined

ii) $\{F^p\}$ induces a Hodge structure of weight m on each of the quotients W_m/W_{m-1}.

"Regular" Hodge structures of weight m can be viewed as mixed Hodge structures with a trivial weight filtration:

$$0 = W_{m-1} \subset W_m = H_C .$$

A _morphism of type_ (r, r) of mixed Hodge structures is a rationally defined homomorphism

$$\phi : H_C \to H'_C$$

such that

$$\phi(W_m) \subset W'_{m+2r}$$

$$\phi(F^p) \subset F'^{p+r} .$$

Such morphisms turn out to be <u>strict</u> relative to both the weight and Hodge filtration, i.e.

$$\mathrm{Im}\,\phi \cap W'_{m+2r} = \phi(W_m)$$

$$\mathrm{Im}\,\phi \cap F'^{p+r} = \phi(F^p) .$$

All the standard linear algebra constructions can be performed within the category of mixed Hodge structure, which is, moreover, <u>abelian</u>.

We shall now prove, using differential forms, Deligne's theorem for varieties of the very special form

$$X = \bigcup_i X_i$$

where the X_i's are smooth, compact Kähler subvarieties of X of the same dimension meeting <u>transversally</u>.

To do this we first have to give an analogue of de Rham's theorem for X. For every multiindex $I = (i_0, \ldots, i_q)$ we set:

$$|I| = q + 1$$

$$X_I = X_{i_0} \cap \cdots \cap X_{i_q} .$$

We also set:

$$X^{[q]} = \coprod_{|I|=q+1} X_I$$

$$A^{r,s}(X) = r\text{-forms on } X^{[s]} .$$

The differentials d (exterior differentiation) and

$$\delta : A^{r,s-1}(X) \to A^{r,s}(X)$$

defined by the formula:

$$\delta\phi(i_0 \cdots i_s) = \sum (-1)^j \phi(i_0 \cdots \hat{i}_j \cdots i_s)|_{X_{(i_0 \cdots i_s)}}$$

make

$$A^{*,*}(X) = \bigoplus A^{r,s}(X)$$

into a double complex.

(4.4) LEMMA: The (total) cohomology of $A^{**}(X)$ is canonically isomorphic to the cohomology of X.

PROOF: We may define <u>sheaves</u> $A^{r,s}(X)$ whose sections over an open set U are $A^{r,s}(U)$. These sheaves are obviously acyclic and their direct sum is made into a double complex by d and δ. All we have to do is then to show that $A^{**}(X)$ is a resolution of the constant sheaf $\mathbb{C}$ (or $\mathbb{R}$, depending on the coefficients we are using). It is clear that a section ϕ of $A^{0,0}(X)$ such that

$$d\phi = \delta\phi = 0$$

must be constant. It remains to prove the Poincaré lemma for $A^{**}(X)$. The E_2-term of one of the two spectral sequences of the double complex $A^{**}(X)$ is:

$$E_2^{pq} = H_\delta^q H_d^p (A^{**}(X)) .$$

By the usual Poincaré lemma this is zero whenever $p > 0$, therefore $E_2 = E_\infty$ and

$$H^q(A^{**}(X)) = E_2^{0q} .$$

On the other hand E_2^{0q} is the q-th cohomology sheaf of the complex of sheaves

$$(4.5) \qquad \cdots \to \mathbb{C}_{X^{[q]}} \xrightarrow{\delta} \mathbb{C}_{X^{[q+1]}} \to \cdots$$

where $\mathbb{C}_Y$ stands for the constant sheaf $\mathbb{C}$ on Y. Now the formula for δ is the same as the one for the coboundary operator on a simplex, therefore

$$E_2^{0q} = \begin{cases} 0 & \text{if } q > 0 \\ \mathbb{C}_X & \text{if } q = 0 \end{cases}$$

which proves the Lemma. Q.E.D.

We now define the weight and Hodge filtrations on $A^{**}(X)$ as follows:

$$W_m A^{**}(X) = \sum_{r \leq m} A^{r,*}(X)$$

$$F^p A^{**}(X) = \sum_{r,s} F^p A^{r,s}(X) .$$

These two filtrations induce filtration $\{W_m\}$ and $\{F^p\}$ on the cohomology of X .

We are going to show that:

$\{W_m\}$ and $\{F^p\}$ give a mixed Hodge structure on the cohomology of X .

PROOF: The first step is to replace the filtration $\{W_m\}$ with a decreasing filtration $\{\tilde{W}^h\}$ which will induce the same filtration on each of the cohomology groups of X (up to indices, of course) and will allow us to use a spectral sequence argument. We set:

$$\tilde{W}^h A^{**} = \sum_{s \geq h} A^{*s} .$$

We shall show that the spectral sequence associated with the filtration $\{\tilde{W}^h\}$ degenerates at the E_2-term and that E_2^{pq} has a Hodge structure of pure weight p . When coupled with the remark that

$$\tilde{W}^h H^m(X, C) = W_{m-h} H^m(X, C)$$

this will prove our contention.

We notice that the E_1-term of the above spectral sequence is

$$E_1^{pq} = H^p(X^{[q]}, C)$$

and that

$$d_1 : H^p(X^{[q]}, C) \to H^p(X^{[q+1]}, C)$$

is obviously a morphism of Hodge structures, therefore E_2^{pq} has a Hodge structure of pure weight p . We shall show that $d_2 = 0$, the proof that $d_3 = d_4 = \cdots = 0$ being similar. We choose a differential form α representing a class ξ in E_2^{pq} . We may assume that α has pure type (r, s). Then

$$d\alpha = 0$$

$$\delta\alpha = d\beta$$

and $\delta\beta$ is a representative of d_2 . By the principle of two types we may choose β to have type $(r, s-1)$ or $(r-1, s)$, therefore $d_2\xi$ has two different types and must be zero. Q.E.D.

A by-product of the above proof is that the weight filtration on $H^m(X, C)$ is of the form:

$$0 \subset W_0 \subset \cdots \subset W_m = H^m(X, C) .$$

The same limitations hold for the mixed Hodge structure on the cohomology of a general complete variety.

The main reason for discussing mixed Hodge structures in these lectures is that, given a one-parameter family $\{V_t\}_{0<|t|<1}$ of projective varieties degenerating into a singular one, V_0 , the Hodge structure on the cohomology of V_t tends in a precise manner to a mixed Hodge structure which is related to the mixed Hodge structure on the cohomology on V_0 . The comparison of these two structures yields important geometric information.

b) VARIATION OF HODGE STRUCTURE. As mentioned in the preceding lecture, the Hodge decomposition

$$H^*(M, C) = \sum_{p,q} H^{pq}(M, C) \tag{4.6}$$

of the cohomology of a compact Kähler manifold reflects the particular complex structure on M and it is therefore natural to study how (4.6) behaves as the complex structure on M varies.

Geometrically, the basic situation is a family

$$f : X \to S \tag{4.7}$$

of compact Kähler manifolds. Here X and S are generally non-compact complex manifolds, X Kähler, and f is smooth and proper. The fibres $X_s = f^{-1}(s)$ constitute an analytic family of Kähler manifolds. The most important case is when everything is algebraic, X and S are quasi-projective and S is a curve. Thus S is obtained from a compact Riemann surface $\bar{S}$ by deleting finitely many points and, by resolution of singularities, we may embed (4.7) in a <u>smooth compactification</u>

$$\begin{array}{ccc} X & \hookrightarrow & \bar{X} \\ f \downarrow & & \downarrow \bar{f} \\ S & \hookrightarrow & \bar{S} \end{array} \tag{4.8}$$

where $\bar{X}$, $\bar{S}$ are smooth, but where $\bar{f}$ may fail to be smooth on $\bar{f}^{-1}(\bar{S} - S)$. Moreover it may be assumed that the fibre of $\bar{f}$ over each point of $\bar{S} - S$ is a divisor with normal crossings.

Returning to the situation (4.7), we will denote by E^n the holomorphic vector bundle on S whose sheaf of sections is

$$R^n f_*(C_X) \otimes \mathcal{O}_S .$$

E^n comes to us naturally equipped with a distinguished sheaf of locally constant sections, namely $R^n f_*(C_X)$, and hence with a flat holomorphic connection ∇' (the so-called Gauss-Manin connection). The solutions of

$$\nabla' s = 0$$

are precisely the sections of $R^n f_*(C)$. In the following we will denote by $\nabla = \nabla' + \nabla''$ the flat connection on E^n whose (0, 1)-part ∇'' is the $\bar{\partial}$ operator.

The Hodge numbers $h^{pq}(s) = \dim H^{pq}(X_s, C)$, $p + q = n$ are upper semi-continuous functions of s, as follows from general elliptic principles; on the other hand they add up to $\dim H^n(X_s, C)$, which is locally constant, hence they are locally constant, too. It follows, again from general principles, that the groups $H^{pq}(X_s, C)$ fit together to give a smooth vector subbundle E^{pq} of E^n. E^{pq} has a natural holomorphic structure, which comes to it from being the vector bundle associated to the holomorphic sheaf $R^q f_* \Omega^p_{X/S}$, where $\Omega^p_{X/S}$ stands for holomorphic p-forms along the fibres of f. However, <u>in general</u> E^{pq} <u>is not a holomorphic subbundle of</u> E^n. We also set:

$$F^p = \sum_{r \geq p} E^{p,n-p} .$$

To see how the $E^{p,q}$ behave relative to the complex structure of E^n, we need an explicit formula for the connection ∇. The argument we shall give is due, in an algebraic setting, to Katz-Oda.

Let $\{\phi_s\}$ be a smooth family of closed n-forms along the fibres of

(4.7) giving a smooth section e of E^n. Let u be a vector field on S, v a C^∞ lifting of u to X. Then it makes sense to consider the Lie derivative of $\{\phi_s\}$ along v, $\mathrm{Lie}_v(\{\phi_s\})$. Since the Lie derivative commutes with exterior differentiation along the fibres, this yields a family of closed n-forms whose cohomology classes depend only on the cohomology classes of the ϕ_s. Therefore $\mathrm{Lie}_v(\{\phi_s\})$ determines a section of E^n which we may denote by $\mathrm{Lie}_v(e)$. We want to show that:

$$\nabla_u e = \mathrm{Lie}_v e . \tag{4.9}$$

It is quite obvious from the definition of Lie derivative that if e is flat $\mathrm{Lie}_v(e)$ is zero. One thing that has to be proved is that $\mathrm{Lie}_v(e)$ does not depend on the particular lifting v of u. Granting this, Lie_v behaves like a connection except for the fact that it might not depend linearly, over the C^∞ functions, on u.

Now to take the Lie derivative of $\{\phi_s\}$ we may proceed as follows: we lift $\{\phi_s\}$ to a form ϕ on X, take the Lie derivative of ϕ and restrict the result to the fibres of (4.7). But then we may use the <u>homotopy formula</u>:

$$\mathrm{Lie}_v \phi = v \,\bar{\wedge}\, d\phi + d(v \,\bar{\wedge}\, \phi) \tag{4.10}$$

where $\bar{\wedge}$ stands for contraction. $d(v \,\bar{\wedge}\, \phi)$ restricts to a family of exact form, therefore Lie_v has the required linearity properties. It is also clear from (4.10) that, when v projects to zero on S, $\mathrm{Lie}_v(e)$ is zero: in fact, in this case, $v \,\bar{\wedge}\, d\phi$ vanishes. This proves (4.9).

It follows immediately from (4.9) and (4.10) that, if u is a vector field of type $(1, 0)$, then

$$\nabla_u C^\infty(F^p) \subset C^\infty(F^{p-1}) \tag{4.11}$$

and that if v has type $(0, 1)$, then

$$\nabla_u C^\infty(F^p) \subset C^\infty(F^p) . \tag{4.12}$$

Formula (4.12) says that F^p <u>is a holomorphic subbundle</u> of E^n, whereas (4.11) is the <u>horizontality property</u> of variations of Hodge structure which is a crucial ingredient in all applications. One should notice that there is a natural <u>isomorphism of holomorphic vector bundles</u>:

$$F^p/F^{p+1} \xrightarrow{\sim} E^{p,n-p} .$$

To apply methods of differential geometry and analysis to study the variation of the Hodge structures of the X_s, one needs to have metrics in the various vector bundles involved. Such metrics are given, quite naturally, by the Hodge-Riemann bilinear relations, provided of course that we pass to the primitive cohomology. This is permissible, since the Lefschetz decomposition is of a topological nature and hence is locally constant in $E \to S$.

The resulting data may be codified in the following, somewhat lengthy

DEFINITION: A <u>variation of Hodge structure of weight</u> n, $\mathcal{V} = \{S, E, E_{\mathbb{Z}}, \nabla, Q, \{F^p\}\}$ is given by a holomorphic vector bundle $E \to S$ over a complex manifold S having the following structure:

i) E has a flat holomorphic connection ∇ and contains a flat bundle of lattices $E_{\mathbb{Z}}$ (integral cohomology, in the geometric case).

ii) The F^p, $p = 0, \ldots, n$, are a decreasing filtration of E by holomorphic subbundles which satisfy the <u>horizontality condition</u>

$$\nabla\mathcal{O}(F^p) \subset \Omega^1_S(F^{p-1}) .$$

iii) $Q : E_{\mathbb{Z}} \otimes E_{\mathbb{Z}} \to \mathbb{Z}$ is a non degenerate, flat bilinear form such that

$$Q(e, e') = (-1)^n Q(e', e) .$$

iv) For each $s \in S$ the filtration $\{F^p_s\}$ and Q_s give a polarized Hodge structure on E_s. Thus if we set:

$$E^{p,n-p} = F^p \cap \overline{F^{n-p}}$$

then $E = \sum_{p+q=n} E^{pq}$ and the Hodge-Riemann bilinear relations (4.3) are satisfied.

All we have done here is to abstract the data arising from the variation of the Hodge decomposition in a family of compact Kähler manifolds whose metric form is rational. It is, however, important to note that $\mathcal{V}$ is not assumed to arise from geometry in this way. Moreover, even when the base space S is algebraic, one should not assume that an algebraic structure is given on $\mathcal{V}$.

We will discuss some foundational results on variation of Hodge structure. To state the first of these we assume for simplicity that S is an algebraic curve, i.e. that $S = \bar{S} - \{s_1, \ldots, s_N\}$ is a compact Riemann surface minus N points, and we set

$$S[r] = \{s \in \bar{S} \mid |s - s_\mu| \geqq \frac{1}{r} \text{ for all } \mu\}.$$

The result is the following:

(4.13) Let $\mathcal{V} = \{S, E, E_{\mathbb{Z}}, \nabla, Q, \{F^p\}\}$ be a variation of Hodge structure. Then E has an <u>intrinsic algebraic structure</u> where the algebraic sections of E are those holomorphic sections ξ which satisfy the growth estimate

$$\max_{s \in S[r]} \log (Q(C_s\xi(s), \overline{\xi(s)})) = O(\log r) \tag{4.14}$$

where C_s is the <u>Weil operator</u>. Moreover, the F^p are algebraic subbundles of E and ∇ is algebraic.

Intuitively, (4.14) means that ξ has at most "poles at infinity" using the intrinsic Hodge norm to measure size.

Actually the proof of the above result will yield more, namely that <u>the Gauss-Manin connection</u> ∇ <u>has regular singular points</u>; this means that E has an algebraic extension $\bar{E}$ to $\bar{S}$ such that for every holomorphic section e of $\bar{E}$, ∇e has <u>at most simple poles</u>. Another, equivalent way of defining regular singular points is that the length of a (multivalued) flat section of E, measured using a C^∞ metric on $\bar{E}$, should grow at most like a polynomial when approaching a point in $\bar{S} - S$.

When $\mathcal{V}$ comes from algebraic geometry there is another, possibly different, algebraic structure on E, arising from Grothendieck's algebraic de Rham theorem in relative form:

$$R^q f_* C_X \otimes \mathcal{O}_S \simeq \mathbb{R}^q f_*(\Omega^*_{X/S})$$

where $\Omega^*_{X/S}$ is the complex of <u>algebraic</u> forms along the fibres of $X \to S$. One can prove, however, that this algebraic structure is the same as the one given by (4.13), thus proving the usual regularity theorem in algebraic geometry.

The next results center around the local monodromy group. Given a variation of Hodge structure $\mathcal{V} = \{S, E, E_{\mathbb{Z}}, \nabla, Q, \{F^p\}\}$, there is a monodromy representation

$$\rho : \pi_1(S, s_0) \to \mathrm{Aut}(E_{s_0})$$

gotten by displacing elements of E_{s_0} around closed paths by parallel translation. $\Gamma = \rho\pi_1(S, s_0)$ is called the <u>monodromy group</u> of $\mathcal{V}$. When S is a punctured disk Δ^*, the image of the generator of $\pi_1(\Delta^*)$ will be denoted by T and called the <u>Picard-Lefschetz transformation</u>. This is suggested by looking at the geometric situation $f : X \to S$, where $S = \bar{S} - \{s_1, \ldots, s_N\}$ is a curve and then localizing around one of the s_i. The basic fact in this localized situation is the

MONODROMY THEOREM: The Picard-Lefschetz transformation T is quasi-unipotent of index of unipotency n, where n is the weight of the variation of Hodge structure $\mathcal{V}$. In other terms,

$$(T^\mu - I)^{n+1} = 0$$

for some positive integer μ.

The simplest situation occurs when T is of finite order. Then one may go to a finite covering and prove the

REMOVABLE SINGULARITY THEOREM: If the Picard-Lefschetz transformation T is the identity, the variation of Hodge structure $\mathcal{V}$ extends across the puncture of Δ^*.

In the general case, it was conjectured by Deligne and proved by Schmid that, as $\zeta \in \Delta^*$ tends to zero, the Hodge decomposition of E_ζ tends to a mixed Hodge structure whose weight filtration is constructed from T.

The final results we wish to mention center around the global monodromy representation

$$\rho : \pi_1(S, s_0) \to \mathrm{Aut}(E_{s_0})$$

when the base space S is an algebraic variety. One of these results is the

RIGIDITY THEOREM: Let V, V' be two variations of Hodge structure of weight n and assume that there is an isomorphism σ between them at one point s_0. If σ is equivariant with respect to ρ, ρ', then it extends to a global isomorphism between V and V'.

Finally, there is Deligne's:

SEMI-SIMPLICITY THEOREM: The global monodromy representation ρ is completely reducible and the variation of Hodge structure V decomposes accordingly.

In classical terms, one has known for a long time that the position of the singular points and global monodromy determine a wide class of ordinary differential equations (e.g. the hypergeometric equations) on $\mathbb{P}^1$ having regular singular points. The above results assert the overwhelming influence of monodromy in variation of Hodge structures.

REFERENCES FOR LECTURE 4

Deligne's theory of mixed Hodge structures is given in

P. Deligne, Théorie de Hodge II, Publ. Math. I.H.E.S., vol. 40 (1972), 5-57.

An alternate more analytic account of his results may be found in

P. Griffiths and W. Schmid, Variation of Hodge structure (a discussion of recent results and methods of proof), to appear in Proc. Tata Institute Conf. on Discrete Groups and Moduli.

This paper also contains an exposition of the theory of variation of Hodge structure, together with additional references on the subject.

The original proof of the monodromy theorem has finally appeared in

A. Landman, On the Picard-Lefschetz transformations, Trans. Amer. Math. Soc. vol. 181 (1973), 89-126.

5. HERMITIAN DIFFERENTIAL GEOMETRY

In this talk we shall use the E. Cartan method of moving frames to discuss the theory of holomorphic curves, including local versions of the Wirtinger theorem and Plücker formulae as illustrations of non-compact algebraic geometry, and classifying spaces for variations of Hodge structure.

We begin with a homogeneous space G/H of a Lie group G by a closed subgroup H. In practice, G may frequently be identified with a set of "frames" on G/H, and when this is done the left-invariant Maurer-Cartan forms on G appear in the structure equations of a moving frame. Furthermore, when mapping a manifold M into G/H, there will frequently appear natural "Frênet frames" or liftings of the map to G. Restricting the Maurer-Cartan forms on G to these natural frames gives a complete set of invariants for the map, by virtue of the following general principle:

(5.1) Let M be a connected manifold and G a Lie group with basis $\{\omega_i\}$ for the Maurer-Cartan forms. Two maps $f, \tilde{f} : M \to G$ differ by a left translation in G if, and only if, $f^*\omega_i = \tilde{f}^*\omega_i$ for all i.

(PROOF IN CASE G IS A MATRIX GROUP: In this case the ω_i are the matrix entries in $\omega = g^{-1}dg$. Writing $f(m) = g(m)\cdot\tilde{f}(m)$ $(m \in M)$, $f(m)^{-1}df(m) = \tilde{f}(m)^{-1}d\tilde{f}(m) + \tilde{f}(m)^{-1}[g(m)^{-1}dg(m)]\tilde{f}(m)$. Thus $f^*\omega = \tilde{f}^*\omega \Longleftrightarrow dg(m) = 0$. Q.E.D.)

Here are some examples.

a) COMPLEX PROJECTIVE SPACE. Points in $\mathbb{P}^n$ will be written as homogeneous coordinate vectors $Z = [z_0, \ldots, z_n]$. The frame manifold $F(\mathbb{P}^n)$ consists of all unitary bases $\mathbb{F} = \{Z_0, Z_1, \ldots, Z_n\}$ for C^{n+1}.

Choosing a reference frame $\mathbb{F}_0$, any frame $\mathbb{F}$ is uniquely of the form

$$\mathbb{F} = T \cdot \mathbb{F}_0$$

for some unitary transformation $T \in U_{n+1}$. The correspondence $\mathbb{F} \longleftrightarrow T$ gives an identification $F(\mathbb{P}^n) \cong U_{n+1}$.

The vectors Z_i may be considered as smooth maps $Z_i : F(\mathbb{P}^n) \to \mathbb{C}^{n+1}$. Expanding the differential $dZ_i(\mathbb{F})$ <u>in terms of the basis vectors in the</u> frames $\mathbb{F}$ leads to the <u>structure equations of a moving frame</u>:

$$(5.2) \qquad \begin{aligned} dZ_i &= \sum_j \theta_{ij} Z_j \\ \theta_{ij} + \bar{\theta}_{ji} &= 0 , \end{aligned}$$

which should be read: "Under infinitesimal displacement, the frame $\mathbb{F}$ undergoes an infinitesimal unitary transformation with coefficient matrix θ_{ij}." Since $Z_i(T \cdot \mathbb{F}) = TZ_i(\mathbb{F})$ for any fixed T, the θ_{ij} give a basis for the left-invariant Maurer-Cartan forms on U_{n+1}. The <u>Maurer-Cartan equations</u>

$$(5.3) \qquad d\theta_{ij} = \sum_k \theta_{ik} \wedge \theta_{kj}$$

follow from $d(dZ_i) = 0$ in (5.2).

A <u>holomorphic curve</u> is a holomorphic mapping $Z : S \to \mathbb{P}^n$ from a Riemann surface into $\mathbb{P}^n$. In case S is compact, $Z(S)$ is an algebraic curve and hence has a <u>degree</u>, satisfies various <u>Plücker formulae</u>, etc. We shall eventually discuss certain non-compact analogues of these. In terms of a local coordinate ζ on S, Z is given by $Z(\zeta) = [z_0(\zeta), \ldots, z_n(\zeta)]$ where the $z_i(\zeta)$ are holomorphic. A <u>frame field</u> is given by a C^∞ lifting of Z to $F(\mathbb{P}^n)$; i.e. by a C^∞ frame $\mathbb{F}(\zeta) = \{Z_0(\zeta), \ldots, Z_n(\zeta)\}$ where $Z_0(\zeta) \wedge Z(\zeta) \equiv 0$. For such a frame field, the Maurer-Cartan forms $\theta_{ij} = \theta'_{ij} + \theta''_{ij}$ are linear combinations of $d\zeta$ and $d\bar{\zeta}$, and we claim that

$$(5.4) \qquad \theta''_{0\alpha} = 0 \qquad (\alpha = 1, \ldots, n) .$$

(PROOF: $0 = \bar{\partial}(Z_0(\zeta) \wedge Z(\zeta)) = \bar{\partial} Z_0(\zeta) \wedge Z(\zeta) = \sum_{\alpha=1}^{n} \theta''_{0\alpha} Z_\alpha(\zeta) \wedge Z(\zeta)$.)

Similarly, one may prove that

(5.5) $$\Omega_0 = \frac{\sqrt{-1}}{2\pi}\left\{\sum_{\alpha=1}^{n} \theta_{0\alpha} \wedge \bar{\theta}_{0\alpha}\right\}$$

is independent of the frame field, and is the pull-back to S of the standard Kähler form on $\mathbb{P}^n$ (Fubini-Study metric).

b) CLASSIFYING SPACES FOR VARIATION OF HODGE STRUCTURE. Let E be a complex vector space with integral lattice $E_{\mathbb{Z}}$ and non-degenerate bilinear form

$$Q : E_{\mathbb{Z}} \otimes E_{\mathbb{Z}} \to \mathbb{Z}$$

$$Q(e, e') = (-1)^n Q(e', e) .$$

Given a set of Hodge numbers $h^{p,q}$ with $\sum_{p+q=n} h^{p,q} = \dim E$, $h^{p,q} = h^{q,p}$, the set of all polarized Hodge structures

(5.6) $$E = \bigoplus_{p+q=n} E^{p,q}$$

with $\dim E^{p,q} = h^{p,q}$ * forms a classifying space D for polarized Hodge structures of weight n.

A Hodge frame associated to a Hodge decomposition (5.6) is a collection $\mathbb{F} = \{\underline{f}_n, \underline{f}_{n-1}, \ldots, \underline{f}_0\}^{\dagger}$ where each $\underline{f}_p$ is a set $\{f_{p_1}, \ldots, f_{p_k}\}$ $(k = h^{p,q})$ of vectors giving an orthonormal basis for $E^{p,q}$, and where $\underline{f}_{n-p} = \bar{\underline{f}}_p$. Upon choosing a reference frame $\mathbb{F}_0$, the relation

$$\mathbb{F} = T\cdot\mathbb{F}_0$$

gives an identification $F(D) \cong G_{\mathbb{R}}$ of the manifold $F(D)$ of all Hodge frames with the Lie group $G_{\mathbb{R}}$ of real automorphisms of E which preserve Q. In particular, the classifying space is a homogeneous manifold

$$D = G_{\mathbb{R}}/H$$

with compact isotropy group H.

A $G_{\mathbb{R}}$-invariant complex structure on D may be given by the require-

---

* This means that $E^{p,q} = \bar{E}^{q,p}$, and that the Hodge-Riemann bilinear relations (I) and (II) from lecture 3 are satisfied.

† Throughout our discussions of Hodge theory, indices will appear in decreasing order.

ment that a C^∞ curve $\{E^{p,q}(\zeta)\}$ $(\zeta \in U \subset C)$ in D varies holomorphically if, and only if,

$$\frac{\partial F^p(\zeta)}{\partial \bar{\zeta}} \subseteq F^p(\zeta) \tag{5.7}$$

where $F^p(\zeta) = E^{n,0}(\zeta) + \cdots + E^{p,n-p}(\zeta)$ is the associated <u>Hodge filtration</u>.

A variation of Hodge structure $V = \{S, \mathbb{E}, \mathbb{F}^p, \nabla, Q\}$, as defined in the third lecture, gives rise to:

a) A classifying space D as above, where $E = \mathbb{E}_{s_0}$, $Q = Q_{s_0}$, $h^{p,q} = \dim \mathbb{E}^{p,q}_{s_0}$, etc.;

b) the monodromy group Γ, which is the subgroup of the arithmetic group $G_{\mathbb{Z}}$ of all linear automorphisms of $E_{\mathbb{Z}}$ which preserve Q obtained by displacing flat frames around closed paths $\gamma \in \pi_1(S, s_0)$; and

c) a holomorphic <u>period mapping</u>

$$\Phi : S \to \Gamma\backslash D \tag{5.8}$$

satisfying the infinitesimal period relation

$$\frac{\partial F^p(\zeta)}{\partial \zeta} \subseteq F^{p-1}(\zeta), \tag{5.9}$$

where by definition $\Phi(s)$ is the Hodge decomposition $\mathbb{E}_s = \bigoplus_{p+q=n} \mathbb{E}^{p,q}(s)$ combined with an isomorphism $\mathbb{E}_s \cong \mathbb{E}_{s_0}$ depending on a homotopy class of paths from s to s_0, and $F^p(\zeta)$ is the image of $\mathbb{E}^{n,0}(s) + \cdots + \mathbb{E}^{p,n-p}(s)$ with ζ being the coordinate of s.

Conversely, such a period mapping Φ gives rise to a variation of Hodge structure by pulling back the universal family over $\Gamma\backslash D$. <u>Henceforth, a variation of Hodge structure shall mean either the bundle data</u> $V = \{S, \mathbb{E}, \mathbb{F}^p, \nabla, Q\}$, <u>or a period mapping</u> (5.8) <u>satisfying</u> (5.9).

Suppose now that we are either working locally or on a universal covering so that Γ may be taken to be trivial. Then we have $\Phi : S \to D$ satisfying (5.9). In terms of a local coordinate ζ on S, a <u>Hodge frame field</u> $\mathbb{F}(\zeta) = \{\underline{f}_n(\zeta), \underline{f}_{n-1}(\zeta), \ldots, \underline{f}_0(\zeta)\}$ is defined to be a smooth lifting of $\Phi(\zeta)$ to $F(D)$. Given such a frame field, we set

$$d\underline{f}_p = \sum_{q=n}^{0} \omega_{p,q}\underline{f}_q$$

where the $\omega_{p,q}$ are matrices of 1-forms on S which are the pull-backs of the Maurer-Cartan forms on $F(D) \cong G_{\mathbb{R}}$. The symmetry relations

$$\begin{aligned} &\omega_{p,q} + (-1)^{q-p}\,{}^t\bar{\omega}_{q,p} = 0\,, \quad \omega_{p,q} = \bar{\omega}_{n-p,n-q}\,, \\ (5.10) \qquad &\omega''_{p,q} = 0 \quad \text{for} \quad q < p \\ &\omega_{p,q} = 0 \quad \text{for} \quad |p-q| \geq 2 \end{aligned}$$

result from the orthonormal symmetry relations on $\mathbb{F}(\zeta)$, the Cauchy-Riemann equations (5.7) (as in the case of holomorphic curves in $\mathbb{P}^n$, cf. (5.4)), and the infinitesimal period relation (5.9). Setting $\psi_p = \omega_{p,p-1}$* and $\phi_p = \omega_{p,p}$, the Maurer-Cartan matrix for a variation of Hodge structure has the form

$$(5.11) \qquad \omega = \begin{pmatrix} \phi_n & \psi_n & & 0 & \cdots & 0 \\ {}^t\bar{\psi}_n & \phi_{n-1} & & & & \vdots \\ 0 & & \ddots & & & 0 \\ \vdots & & & & \phi_1 & \psi_1 \\ 0 & \cdots & 0 & & {}^t\bar{\psi}_1 & \phi_0 \end{pmatrix}$$

and satisfies the Maurer-Cartan equation

$$(5.12) \qquad d\omega = \omega \wedge \omega$$

as in the case of curves in $\mathbb{P}^n$.

The idea of how one proves the global results on variation of Hodge structure is: (a) To apply curvature arguments as in lecture one to the Hodge bundles $\mathbb{E}^p \longrightarrow S$ in the case when S is compact, the ϕ_p in (5.11) are the connection matrices in the Hodge bundles, and the curvature is computed by (5.12); and (b) in case S is non-compact, the Ahlfors lemma (lecture 6) gives an estimate on ψ_p, and judiciously choosing our frame field then allows one to estimate ϕ_p using (5.12), the upshot being that the arguments in the compact case carry over to the general situation. This will all be explained in more detail in lecture 7 and in the seminar.

---

* The ψ_p are (1, 0) forms with values in $\mathrm{Hom}(\mathbb{E}^p, \mathbb{E}^{p-1})$ which measure the variation of Hodge structure, and may be identified with the Kodaira-Spencer class (cf. lecture 7 below).

For the remainder of this lecture, we shall return to the study of holomorphic curves in projective space.

A holomorphic curve $Z : S \to \mathbb{P}^n$ is <u>non-degenerate</u> in case the image does not lie in a proper linear subspace. Analytically, this is expressed by the Wronskian condition

$$W(\zeta) = Z(\zeta) \wedge Z'(\zeta) \wedge \cdots \wedge Z^{(n)}(\zeta) \not\equiv 0 .$$

Near a <u>regular point</u> ζ_0 where $W(\zeta_0) \neq 0$, we define <u>Frênet frames</u> by the conditions:

$$\text{Span}\,\{Z_0, \ldots, Z_k\} = \text{Span}\,\{Z, Z', \ldots, Z^{(k)}\} = P^k(\zeta) ,$$

where $P^k(\zeta)$ is the $k^{\underline{th}}$ <u>osculating space</u>. For a Frênet frame field, dZ_k is clearly a linear combination of $Z_0, \ldots, Z_{k+1}$ and $\bar{\partial} Z_k$ is a linear combination of $Z_0, \ldots, Z_k$. This implies that $\theta_{k,\ell} = 0$ for $|k - \ell| \geqq 2$ and $\theta''_{k,k+1} = 0$, and thus the structure equations (5.2) for a Frênet frame field reduce to the Frênet equations

$$(5.13) \qquad \begin{aligned} dZ_k &= \theta_{k,k-1}Z_{k-1} + \theta_{k,k}Z_k + \theta_{k,k+1}Z_{k+1} \\ \theta_{k,k} + \bar{\theta}_{k,k} &= 0 , \quad \theta''_{k,k+1} = 0 , \quad \theta_{k,k-1} + \bar{\theta}_{k-1,k} = 0.^* \end{aligned}$$

The $k^{\underline{th}}$ <u>associated curve</u> is the locus of the osculating spaces $P^k(\zeta)$. Analytically, this curve is given by the holomorphic mapping

$$\Lambda_k(\zeta) = Z(\zeta) \wedge Z'(\zeta) \wedge \cdots \wedge Z^{(k)}(\zeta)$$

from S into the Grassmannian $\mathbb{P}G(k, n)$ of projective k-planes in $\mathbb{P}^n$. Here, we are tacitly using the Plücker coordinates on $\mathbb{P}G(k, n)$. Since first $\Lambda_k(\zeta) = \lambda_k Z_0(\zeta) \wedge \cdots \wedge Z_k(\zeta)$ is a multiple of $Z_0 \wedge \cdots \wedge Z_k$ as vectors in $\Lambda^{k+1}C^{n+1}$, and secondly by the Frênet equations (5.13)

$$d(Z_0 \wedge \cdots \wedge Z_k) = \mu_k Z_0 \wedge \cdots \wedge Z_k \pm \theta_{k,k+1} Z_0 \wedge \cdots \wedge Z_{k-1} \wedge Z_{k+1} ,$$

it follows from (5.5) that

$$(5.14) \qquad \Omega_k = \frac{\sqrt{-1}}{2\pi}\, \theta_{k,k+1} \wedge \bar{\theta}_{k,k+1}$$

---

* We note the similarity between the Maurer-Cartan matrices for a variation of Hodge structure and for holomorphic curves.

is the pull-back under Λ_k of the standard Kähler form on $\mathbb{P}G(k, n)$. The quantities $\Omega_0, \Omega_1, \Omega_2, \ldots$ are the complex analogues of arclength, curvature, torsion, ... for ordinary curves in $\mathbb{R}^n$. As in that situation, one may show:

(5.15) Two holomorphic curves $Z, \tilde{Z} : S \to \mathbb{P}^n$ differ by a rigid unitary motion if, and only if, $\Omega_k = \tilde{\Omega}_k$ for $k = 0, \ldots, n-1$.

(PROOF: The result is local, and we choose Frênet frame fields $\{Z_i\}$ and $\{\tilde{Z}_i\}$ for Z and $\tilde{Z}$. Fixing the $\tilde{Z}_i$, we seek to rotate the Z_i by

$$(5.16) \qquad Z_i \to e^{\sqrt{-1}\,\psi_i} Z_i \qquad (i = 0, \ldots, n)$$

such that the Maurer-Cartan matrices agree on the new frame fields. The result then follows from (5.1). Writing $\theta_{k,k+1} = h_k d\zeta$ and $\tilde{\theta}_{k,k+1} = \tilde{h}_k d\zeta$, the assumption $\Omega_k = \tilde{\Omega}_k$ gives $|h_k| = |\tilde{h}_k|$ or

$$h_k = e^{\sqrt{-1}\,\gamma_k}\, \tilde{h}_k .$$

Under a rotation (5.16),

$$\theta_{k,k+1} \to e^{\sqrt{-1}\,(\psi_k - \psi_{k+1})}\, \theta_{k,k+1} ,$$

so that choosing $\psi_k - \psi_{k+1} = \gamma_k$ for $k = 0, \ldots, n-1$ gives

$$(5.17) \qquad \begin{aligned} \theta_{k,k+1} &= \tilde{\theta}_{k,k+1}, \quad \text{and by (5.13)} \\ \theta_{k,k-1} &= \tilde{\theta}_{k,k-1} . \end{aligned}$$

We are still free to rotate all Z_i through the same angle ψ.

By the structure equations (5.3) and (5.13),

$$(5.18) \qquad \begin{aligned} d\theta_{k,k+1} &= (\theta_{k,k} - \theta_{k+1,k+1}) \wedge \theta_{k,k+1} \\ d\theta_{k,k} &= \theta_{k,k-1} \wedge \theta_{k-1,k} + \theta_{k,k+1} \wedge \theta_{k+1,k} . \end{aligned}$$

Using (5.17) in the second equation gives $d(\theta_{k,k} - \tilde{\theta}_{k,k}) = 0$. By the Poincaré lemma, we may rotate all Z_i through angle ψ where $\theta_{n,n} - \tilde{\theta}_{n,n} = \sqrt{-1}\, d\psi$ to have $\theta_{n,n} = \tilde{\theta}_{n,n}$. Now, using (5.17) and the characterization of the Hermitian connection given in lecture 3 in the first equation in (5.18) gives $\theta_{k,k} - \theta_{k+1,k+1} = \tilde{\theta}_{k,k} - \tilde{\theta}_{k+1,k+1}$ for

$k = 0, \ldots, n-1$. Combining, we obtain $\theta_{k,k} = \tilde{\theta}_{k,k}$ for all k , and then $\theta = \tilde{\theta}$. Q.E.D.)

At this point, we make the following notational convention: For a positive (1, 1) form

$$\Omega = \frac{\sqrt{-1}}{2\pi} h d\zeta \wedge d\bar{\zeta} = \frac{\sqrt{-1}}{2\pi} \theta \wedge \bar{\theta}$$

where $\theta = e^{\sqrt{-1}\psi}\sqrt{h}\,d\zeta$, the connection form ϕ is characterized by

$$d\theta = \phi \wedge \theta$$
$$\phi + \bar{\phi} = 0 ,$$

and we define the <u>Ricci form</u>*

$$\operatorname{Ric} \Omega = \frac{\sqrt{-1}}{2\pi} d\phi .$$

Since $\operatorname{Ric} \Omega$ is a constant times the curvature form, it depends only on Ω.

From the first equation in (5.18), the connection form for Ω_k is $\theta_{k,k} - \theta_{k+1,k+1}$. By the second equation there,

$$\operatorname{Ric} \Omega_k = -2\Omega_k + \Omega_{k-1} + \Omega_{k+1} . \tag{5.19}$$

This beautiful relation, which is due originally to H. and J. Weyl, has many applications, especially to "non-compact algebraic geometry." A first one is:

(5.20) Ω_0 uniquely determines $\Omega_1, \ldots, \Omega_{n-1}$.

This follows from $\operatorname{Ric} \Omega_0 = -2\Omega_0 + \Omega_1$, $\operatorname{Ric} \Omega_1 = -2\Omega_1 + \Omega_0 + \Omega_2$, etc. In geometric terms, (5.20) states:

> For holomorphic curves, the curvature, torsion, ... are all functions of arclength alone. In particular, if two such curves osculate to first order, they are congruent (theorem of Calabi).

Before giving the second application we need some preliminary remarks on the Wirtinger theorem. Suppose that S is a relatively compact open set in a larger Riemann surface S' on which Z is defined (an extreme

---

* This terminology will be justified in the next lecture.

case is when $S = S'$ is compact). For each hyperplane $H \in \mathbb{P}^{n*}$, the projective space dual to $\mathbb{P}^n$, the number of points of intersection $n(S, H)$ of $Z(S)$ with H is finite. Crofton's formula from integral geometry is the relation

$$\int_S \Omega_0 = \int_{\mathbb{P}^{n*}} n(S, H)\, dH \tag{5.21}$$

expressing the area of $Z(S)$ as the average number of intersections of $Z(S)$ with a hyperplane. A proof of (5.21) using frames is given in the appendix to this lecture.

In case $S = S'$ is compact, $n(S, H)$ is independent of the hyperplane H since, by the Cauchy integral formula, a meromorphic function has the same number of zeroes as poles. The integer $n(S, H)$ is called the degree of the algebraic curve, and (5.21) is the Wirtinger theorem

$$\text{area }(S) = \text{degree }(S)\ . \tag{5.22}$$

For non-compact S, say for definiteness that $S' = \mathbb{C}$ and $S = \Delta_R = \{\zeta \in \mathbb{C} : |\zeta| < R\}$,* the proof that $n(S, H)$ is independent of H leads to the following significant analytical generalization of the Wirtinger theorem (5.22), illustrating quite well, we think, the principle of non-compact algebraic geometry: Setting $n(r, H) = n(\Delta_r, H)$ and $n(0, H) = \lim_{r\to 0} n(r, H)$, the Nevanlinna inequality, also to be proved in the appendix

$$\int_0^R \{n(r, H) - n(o, H)\} \frac{dr}{r} \leqq \int_0^R \Big\{ \int_{\Delta_r} \Omega_0 \Big\} \frac{dr}{r} + O(1)\ , \tag{5.23}$$

bounding the growth of $n(r, H)$ by the growth of the area, is valid.

To give our second application of (5.19), we define the mean degree of the $k^{\underline{th}}$ osculating curve by

$$\delta_k(S) = \int_S \Omega_k\ .$$

For plane curves $(n = 2)$, $\delta_0(S)$ is the average number of intersections of $Z(S)$ with a line, and $\delta_1(S)$ is the average number of tangent lines to $Z(S)$ passing through a point in $\mathbb{P}^2$ (mean class).

---

* $Z : \mathbb{C} \to \mathbb{P}^n$ may be called an entire holomorphic curve.

When $S = S'$ is compact, we may apply the Gauss-Bonnet theorem for singular metrics to (5.19) to obtain a formula

$$\chi(S) + N_k(S) + 2\delta_k(S) = \delta_{k-1}(S) + \delta_{k+1}(S) . \tag{5.24}$$

Here, $\chi(S) = 2 - 2g$ is the Euler characteristic of S, and $N_k(S)$ measures the number and type of singular points on the $k^{\underline{th}}$ osculating curve. The relations (5.24) are the general Plücker formulae of an algebraic curve.

When S is non-compact, e.g. in the study of an entire holomorphic curve, the same Gauss-Bonnet method applies to give the Plücker estimates, which roughly state that

$$\int_0^R \{\int_{\Delta_r} \Omega_{k-1} + \Omega_{k+1}\} \frac{dr}{r} \leqq (2+\varepsilon) \int_0^R \{\int_{\Delta_r} \Omega_k\} \frac{dr}{r} ,$$

and which serve to relate the orders of growth of the various associated curves. These inequalities are of fundamental importance in the transcendental theory of holomorphic curves, and serve to illustrate once again the principle of non-compact algebraic geometry.

APPENDIX TO LECTURE 5:
PROOF OF CROFTON'S FORMULA AND THE NEVANLINNA INEQUALITY

A hyperplane H in $\mathbb{P}^n$ is spanned by n orthonormal vectors $W_0, \ldots, W_{n-1}$; i.e.

$$H = W_0 \wedge \cdots \wedge W_{n-1} . \tag{A.5.1}$$

We consider the manifold $F(\mathbb{P}^n)$ of unitary frames $\{W_0, \ldots, W_n\}$ in C^{n+1} and fibering $F(\mathbb{P}^n) \to \mathbb{P}^{n*}$ given by $\{W_0, \ldots, W_n\} \to W_0 \wedge \cdots \wedge W_{n-1}$. Writing, as in (5.2),

$$dW_i = \sum_{j=0}^{n} \phi_{ij} W_j$$

$$\phi_{ij} + \bar{\phi}_{ji} = 0 ,$$

we have that

$$d(W_0 \wedge \cdots \wedge W_{n-1}) = (\sum_{i=0}^{n-1} \phi_{ii}) W_0 \wedge \cdots \wedge W_{n-1} + \sum_{\mu=0}^{n-1} \phi_{\mu n} W_0 \wedge \cdots \wedge W_n \wedge \cdots \wedge W_{n-1} .$$

It follows from (5.5) that the Kähler form Ω^* on $\mathbb{P}^{n*}$, pulled back to $F(\mathbb{P}^n)$, is given by

$$\Omega^* = \frac{\sqrt{-1}}{2}\{\sum_{\mu=0}^{n-1} \phi_{\mu n} \wedge \bar{\phi}_{\mu n}\} .$$

The invariant measure on $\mathbb{P}^{n*}$ is thus

$$\text{(A.5.2)} \qquad dH = (\Omega^*)^n = n!\left(\frac{\sqrt{-1}}{2}\right)^n \{\sum_{\mu=0}^{n-1} \phi_{\mu n} \wedge \bar{\phi}_{\mu n}\}^n .$$

In $S \times \mathbb{P}^{n*}$ we consider the incidence divisor I of all points (ζ, H) such that $Z(\zeta) \in H$. Clearly,

$$\int_{\mathbb{P}^{n*}} n(S, H) = \int_I dH .$$

On I, we shall rewrite dH as

$$dH(\zeta, H) = \Omega_0(\zeta) \times \Psi(\zeta, H)$$

where $\Psi(\zeta, H)$ is the measure on the set of hyperplanes passing through $Z(\zeta)$, and then apply the Fubini theorem to conclude (5.21).

Choose a Frênet frame field $\{Z_0(\zeta), \ldots, Z_n(\zeta)\}$. Then all hyperplanes (A.5.1) passing through $Z(\zeta)$ are given by frames $\{W_0, W_1, \ldots, W_n\}$ where

$$W_0 = Z_0(\zeta)$$

$$W_\alpha = \sum_{\beta=1}^{n} A_{\alpha\beta} Z_\beta(\zeta)$$

and $A = (A_{\alpha\beta})$ is an arbitrary unitary matrix. Using ${}^t\bar{A} = A^{-1}$ and the Frênet equation (5.13),

$$dW_0 = \theta_{00} Z_0 + \theta_{01}\left(\sum_{\alpha=1}^{n} \bar{A}_{\alpha 1} W_\alpha\right)$$

$$dW_\alpha \equiv \sum_{\beta,\gamma=1}^{n} dA_{\alpha\beta}\bar{A}_{\alpha\beta} W_\gamma \qquad (d\zeta, d\bar{\zeta}) ,$$

which implies that, on the incidence divisor I,

$$\phi_{0n} = \bar{A}_{n1}\theta_{01}$$

$$\phi_{\alpha n} \equiv \sum_{\beta=1}^{n} dA_{\alpha\beta}\bar{A}_{n\beta} \qquad (d\zeta, d\bar{\zeta}) .$$

From (A.5.2) it follows that

$$dH = \Omega_0 \wedge \Psi$$

where Ψ is a differential form involving the matrix entries in A and dA. Integrating Ψ over all hyperplanes containing $Z(\zeta)$ gives a constant C independent of $Z(\zeta)$, and thus, by the Fubini theorem,

$$\int_I dH = C \int_S \Omega_0 .$$

Taking S to be a line in $\mathbb{P}^n$ gives $C = 1$. Q.E.D.

Now to the proof of (5.23). Let $A \in (\mathbb{C}^{n+1})^*$ be a unit vector such that the hyperplane H is defined by

$$< A, Z > = 0 .$$

On $\mathbb{C}$ we define the potential function

$$u(\zeta) = \log \frac{|< A, Z(\zeta) >|^2}{\|Z(\zeta)\|^2} .$$

This function has logarithmic singularities on the divisor D_H where the holomorphic curve meets the hyperplane H, and

$$dd^c u = - \Omega_0$$

on $\mathbb{C} - D_H$. Applying Stokes' theorem to $d^c u$ and taking into account the singularities of u gives

$$n(r, H) = \int_{|\zeta|=r} d^c u + \int_{\Delta_r} \Omega_0 .$$

In polar coordinates, $d^c = \frac{1}{2\pi} r \frac{d}{dr} \otimes d\theta + \frac{1}{2\pi} \frac{d}{d\theta} \otimes dr$, and thus

$$n(r, H) = r \frac{d}{dr} \left(\frac{1}{2\pi} \int_{|\zeta|=r} u \, d\theta \right) + \int_{\Delta_r} \Omega_0 .$$

Assuming for simplicity that $\{0\} \notin D_H$, we may integrate this equation and obtain the First Main Theorem

$$\text{(A.5.3)} \qquad \int_0^R n(r, H) \frac{dr}{r} = \frac{1}{2\pi} \int_{|\zeta|=R} u \, d\theta + \int_0^R \left(\int_{\Delta_r} \Omega_0 \right) \frac{dr}{r} .$$

In particular, since $u \leq 0$, we find the Nevanlinna inequality (5.33).

REFERENCES FOR LECTURE 5

The classic reference for moving frames is

E. Cartan, Groupes finis et continus et la géométrie différentielle, Gauthier-Villars, Paris (1935).

A recent account is given in

P. Griffiths, On Cartan's method of Lie groups and moving frames as applied to uniqueness and existence questions in differential geometry, to appear in Duke Math. J.

The theory of holomorphic curves is discussed in

M. Cowen and P. Griffiths, Holomorphic curves and metrics of negative curvature, to appear in Jour. d'analyse.

Integral geometry is treated in

L. Santalo, Introduction to integral geometry, Hermann, Paris (1953).

6. CURVATURE AND HOLOMORPHIC MAPPINGS

The goal of this lecture is to provide an introduction to *hyperbolic complex analysis*, which can be described as the study of how negative curvature conditions influence holomorphic mappings. Together with Lie group theory, this provides the basic tools for dealing with general variations of Hodge structure: examples of applications will be given in the next lecture. On the other hand, hyperbolic complex analysis applies to Picard-type theorems and their beautiful quantitative refinement, the value distribution theory of R. Nevanlinna. It is mainly this aspect that we will be discussing today.

In particular, we will be giving applications of the *Ahlfors lemma*, a simple but extremely powerful generalization of the classical Schwarz lemma. This result alone, for example, provides the estimates leading to the results on variations of Hodge structure discussed in the fourth lecture. Before formally stating the Ahlfors lemma, we will give some examples of hermitian manifolds to which it applies.

A <u>volume form</u> Ψ on a complex manifold M of dimension m is a smooth, positive (m, m) form. In local coordinates we may write

$$\Psi = h\Phi$$

where

$$\Phi = \prod_{j=1}^{m} \left(\frac{\sqrt{-1}}{2} dz_j \wedge d\bar{z}_j \right)$$

is the euclidean volume form and h is a positive function. We will also be using <u>pseudo-volume forms</u>, these being non-negative, smooth (m, m) forms Ψ that can be written locally as

$$\Psi = |f|^2 h\Phi$$

where h is a positive function and f is a holomorphic function that is not identically zero. Pseudo-volume forms naturally arise by pulling back volume forms under non-degenerate holomorphic mappings. The Ricci form of Ψ is the global, smooth, real $(1, 1)$-form given locally by

$$\text{Ric } \Psi = \frac{\sqrt{-1}}{2\pi} \partial\bar{\partial} \log h .$$

Volume forms which satisfy the curvature estimates:

$$\begin{aligned} &\text{Ric } \Psi \geqq 0 \\ &(\text{Ric } \Psi)^m \geqq \Psi \end{aligned} \tag{6.1}$$

play a crucial role in the theory. The estimates (6.1) are obviously invariant under non-degenerate holomorphic mappings. Here are a few examples.

a) When $m = 1$, there is a natural correspondence between hermitian metrics and volume forms, given by:

$$ds^2 = h\, dz d\bar{z} \longleftrightarrow \frac{\sqrt{-1}}{2} h\, dz \wedge d\bar{z} = \Psi .$$

For such a Ψ ,

$$\text{Ric } \Psi = -K\Psi$$

where

$$K = -\frac{1}{\pi}\frac{1}{h} \frac{\partial^2 \log h}{\partial z \partial \bar{z}}$$

is the <u>Gaussian curvature</u> of ds^2. Hence, in this case (6.1) just says that ds^2 has Gaussian curvature bounded above by -1. From this point of view (6.1) appears to be a generalization to higher dimension of the condition of having Gaussian curvature bounded above by a negative constant.

In particular, on the disk:

$$\Delta(R) = \{z \in C : |z| < R\}$$

the Poincaré metric

$$\eta(R) = \frac{\sqrt{-1}}{\pi} \frac{R^2 dz \wedge d\bar{z}}{(R^2 - |z|^2)^2}$$

is the unique invariant metric such that

$$\mathrm{Ric}\, \eta(R) = \eta(R) .$$

We write $\eta(1) = \eta$, $\Delta(1) = \Delta$. The holomorphic mapping

$$w \to (\sqrt{-1}\, w + 1)/(-\sqrt{-1}\, w + 1)$$

gives a conformal equivalence between Δ and the upper half-plane:

$$H = \{w \in C \mid \mathrm{Im}\, w > 0\} .$$

Under this equivalence, η pulls back to

$$\frac{\sqrt{-1}}{4\pi} \frac{dw \wedge d\bar{w}}{(\mathrm{Im}\, w)^2} .$$

This metric will also be denoted by η and referred to as the Poincaré metric. The same will apply to the metric

$$\frac{\sqrt{-1}}{\pi} \frac{d\zeta \wedge d\bar{\zeta}}{|\zeta|^2(\log |\zeta|^2)^2}$$

on the punctured disk

$$\Delta^* = \{\zeta \in C \mid 0 < |\zeta| < 1\}$$

which corresponds to the Poincaré metric on H via the covering map

$$\zeta = e^{2\pi\sqrt{-1}\, w} .$$

b) On the polycylinder

$$\Delta^m(R) = \{(z_1, \ldots, z_m) \in C^m \mid |z_i| < R, \quad i = 1, \ldots, m\}$$

the product of the Poincaré metrics induces a volume form $\eta_m(R)$ such that

$$\text{(6.2)} \qquad \begin{aligned} \text{Ric } \eta_m(R) &> 0 \\ (\text{Ric } \eta_m(R))^m &= \eta_m(R) . \end{aligned}$$

As above, η_m induces volume forms, which will be denoted by the same symbol, on the punctured polycylinders

$$\Delta^*_{k,m} \doteq \Delta^{*k} \times \Delta^{m-k} .$$

The Bergmann volume form on the ball $\{z \in C^m \mid \|z\| < R\}$ also satisfies the inequalities (6.1).

c) A volume form Ψ on M is the same thing as a metric for the dual of the canonical bundle K_M, and $\text{Ric } \Psi$ is the same as the first Chern form of K_M relative to this metric. In case M is compact it follows that, after adjusting constants, we may find a volume form on M satisfying (6.1) exactly when K_M is _positive_.

This suggests that, for a general compact M, we look for such "negatively curved" volume forms on $M - D$, where D is an effective divisor such that $K_M \otimes [D]$ is positive. Some restrictions on the singularities of D are necessary, and we will assume that D has _simple normal crossings_, i.e. that

$$D = D_1 + \cdots + D_N$$

where the D_i are distinct smooth divisors meeting transversally. In a suitable neighborhood of each point $p \in D$, $M - D$ looks like a punctured polycylinder and this suggests the following global version of the Poincaré volume form. Choose a smooth volume form Ψ_M on M and metrics in the bundles $[D_i]$ such that the inequality of _Chern forms_:

$$\text{(6.3)} \qquad c_1([D]) + c_1(K_M) > 0$$

holds, and for each i let σ_i be a section of $[D_i]$ which defines D_i. Then for a suitable choice of the constants α_i, the volume form:

$$\text{(6.4)} \qquad \Psi = \Psi_M \Big/ \prod_{i=1}^{N} |\sigma_i|^2 (\log(\alpha_i|\sigma_i|^2))^2$$

on $M - D$ satisfies the curvature conditions (6.1).

The simplest special case is when M is projective m-space and the D_i are hyperplanes. Recalling that the canonical bundle of $\mathbb{P}^m$ is H^{-m-1}, where H is the hyperplane bundle, (6.3) translates into $N > m + 1$. When $m = 1$, this means that D must consist of at least three points.

We now state and prove the ubiquitous

AHLFORS LEMMA: If Ψ is a pseudo-volume form on $\Delta^m(R)$ such that (6.1) holds then $\Psi \leqq \eta_m(R)$.

PROOF: It obviously suffices to prove that $\Psi \leqq \eta_m(r)$ for $r < R$. On $\Delta^m(r)$ Ψ is bounded, whereas $\eta_m(r)$ goes to infinity at the boundary. Therefore, if we write

$$\Psi = u\eta_m(r)$$

u has an interior maximum at some z_0. It follows that, at z_0,

$$0 \geqq \frac{\sqrt{-1}}{2\pi} \partial\bar{\partial} \log u = \text{Ric}\,\Psi - \text{Ric}\,\eta_m(r) .$$

Taking m^{th} exterior powers and using (6.1) and (6.2) gives

$$\Psi \leqq (\text{Ric}\,\Psi)^m \leqq (\text{Ric}\,\eta_m(r))^m = \eta_m(r)$$

i.e. $u(z_0) \leqq 1$, and therefore $u \leqq 1$ everywhere as was to be proved.

A COROLLARY to the Ahlfors lemma is that if we write

$$\Psi = h\Phi$$

where Φ is the euclidean metric and Ψ is a metric on $\Delta^m(R)$ satisfying (6.1), then

$$R \leqq C_m h(0)^{-1/2m} \tag{6.5}$$

for some universal constant C_m.

We now give an immediate application of the Ahlfors lemma. Let

$$f : \Delta^m(R) \to N$$

be a holomorphic mapping into an m-dimensional complex manifold N with a volume form Ψ. Write

$$f^*\Psi = |Jf|^2\Phi$$

where Φ is the euclidean metric. If Ψ satisfies (6.1) then (6.5) implies that

$$R \leqq C_m |Jf(0)|^{-1/m} . \tag{6.6}$$

Applying this to the volume form Ψ constructed in example c) gives the

GENERALIZED PICARD THEOREM IN FINITE FORM: Let M be an m-dimensional compact, complex manifold, and let D be a divisor with simple normal crossings whose Chern class satisfies (6.3). Then for any non-degenerate holomorphic mapping:

$$f : \Delta^m(R) \to M - D$$

the estimate (6.6) holds. In particular, an entire holomorphic mapping:

$$f : C^m \to M - D$$

is degenerate.

When $M = \mathbb{P}^1$ and D consists of three distinct points, this implies the usual Picard theorem. However, the above result gives more: restricting to the case of an entire meromorphic mapping

$$f : C \to \mathbb{P}^1$$

it says that, for any three points z_1, z_2, z_3 in $\mathbb{P}^1$ and any point $\zeta_0 \in C$ such that $f'(\zeta_0) \neq 0$, any disc around ζ_0 of radius $\geqq R(f'(\zeta_0))$ will meet $f^{-1}(\{z_1, z_2, z_3\})$. In principle this gives a lower bound on the "size" of $f^{-1}(\{z_1, z_2, z_3\})$ which, when made precise, leads to the beautiful <u>defect relations</u> of R. Nevanlinna.

We will now describe another, closely related, application of the Ahlfors lemma, which leads to the basic estimates for studying variations of Hodge structure. To give this we must first discuss holomorphic sectional curvature.

Let M be a complex manifold, and suppose a hermitian metric with associated exterior form ω is given on its tangent bundle. In the first lecture we have defined the curvature form $\Theta(\xi)$ attached to such a metric. For any holomorphic tangent vector ξ, the <u>holomorphic sectional curvature in the ξ-direction</u>, $K(\xi)$, is defined as follows:

$$K(\xi) = \frac{2\sqrt{-1} < \Theta(\xi), \xi \wedge \bar{\xi} >}{\|\xi\|^4} .$$

It is clear from the definition that $K(\xi)$ depends only on the direction of ξ. When $m = 1$, $K(\xi)$ is just the Gaussian curvature of M. We say that M <u>is negatively curved</u> when

$$K(\xi) \leqq -A < 0$$

for some positive A and every ξ. Multiplying the metric by $1/A$, we may assume that $A = 1$.

PROPOSITION (GENERALIZED SCHWARZ LEMMA): Let $f : \Delta \to M$ be a holomorphic mapping of the unit disc into a negatively curved complex manifold. Then f is <u>distance decreasing</u>, in the sense that

$$f^*\omega \leqq \eta$$

where η is the Poincaré metric on Δ.

PROOF: To apply the Ahlfors lemma it suffices to show that $f^*\omega$ satisfies the inequality:

$$\text{(6.7)} \qquad \operatorname{Ric}(f^*\omega) \geqq f^*\omega$$

at points where $f' \neq 0$. If $\mathcal{U}$ is a sufficiently small neighborhood of such a point, $f(\mathcal{U})$ is a submanifold of M, and has an induced hermitian metric $\omega|f(\mathcal{U})$. (6.7) now follows from the curvature assumptions on M and from the principle that curvature decreases on submanifolds (cf. lecture 1).

REMARK: The same proof as given above applies to the case when we only know that $K(\xi) \leqq -1$ for all vectors ξ which are <u>tangent to</u> $f(\Delta)$. This will be the case in applications to variation of Hodge structure.

When applied to M = unit disc and ω = Poincaré metric, the above proposition gives the invariant form of the Schwarz lemma due to Pick

$$\frac{|f'(z)|^2}{(1 - |f(z)|^2)^2} \leqq \frac{1}{(1 - |z|^2)^2} .$$

The usual statement

$$|f'(0)| \leqq 1$$

follows by assuming that $f(0) = 0$ and setting $z = 0$ in the above inequality.

REFERENCES FOR LECTURE 6

The basic reference for hyperbolic complex analysis is

S. Kobayashi, Hyperbolic manifolds and holomorphic mappings, Marcel Dekker, New York (1970).

For value distribution theory, the classic is

R. Nevanlinna, Le théorème de Picard-Borel et la théorie des fonctions méromorphes, Gauthier-Villars, Paris (1929).

A modern treatment is given in the monograph

P. Griffiths, Entire holomorphic mappings in one and several complex variables, to appear as an Annals of Math. Studies, Princeton Univ. Press.

7. PROOF OF SOME RESULTS ON VARIATIONS OF HODGE STRUCTURE

We first establish some notational conventions, which will be valid for the rest of this lecture. Let $V = \{S, E, E_{\mathbb{Z}}, \nabla, Q, \{F^p\}\}$ be a variation of Hodge structure of weight m. The second Hodge-Riemann bilinear relation says that, on $E^{p,m-p}$, the hermitian form

$$(-1)^p \sqrt{-1}^{\,-m} Q(\xi, \bar{\eta})$$

is positive definite. To simplify notations in the following we shall write Q for $\sqrt{-1}^{\,-m} Q$, so that the Hodge hermitian metric $(\ ,\ )$ is related to $Q(\xi, \bar{\eta})$ by

$$(7.1) \qquad (\alpha, \beta) = \sum (\alpha_p, \beta_p) = \sum (-1)^p Q(\alpha_p, \bar{\beta}_p)$$

where α_p and β_p are the $(p, m-p)$ components of α and β, respectively. It will be convenient to consider E^{pq} and E/F^p as C^∞ subbundles of E ; for example E/F^p will be identified to the orthogonal complement of F^p in E , and so on.

In lecture one the metric connection and curvature of a hermitian metric were defined. That discussion carries over verbatim to non-degenerate, but <u>possibly indefinite</u>, hermitian metrics. From this point of view the Gauss-Manin connection ∇ appears as the $(1, 0)$ part of the metric connection for E relative to the indefinite metric $Q(\xi, \bar{\eta})$. The metric connection on F^p is related to ∇ by:

$$D'_{F^p} + \psi_p = \nabla$$

where ψ_p is the second fundamental form of F^p in E (Kodaira-Spencer map). The $(1, 0)$ part of the metric connection D_{E/F^p} on E/F^p , on the other hand, agrees with ∇ , whereas the $(0, 1)$ part is

$$\bar{\partial} \pm {}^t\psi_p$$

where $\bar{\partial}$ is the Cauchy-Riemann operator for the complex structure of E and ${}^t\psi_p$ is the adjoint of ψ_p . With these notations we may now prove the

a) CURVATURE PROPERTIES OF HODGE BUNDLES. We will show that the following formula holds for the curvature Θ_p of $E^{p,m-p}$:

$$(7.2) \qquad (\Theta_p e, e') = (\psi_p e, \psi_p e') + ({}^t\psi_{p+1} e, {}^t\psi_{p+1} e') .$$

To prove (7.2) we have to describe how curvature behaves when one goes to subbundles or quotient bundles. To some extent, this has already been done in the first of these lectures. Suppose we are given an exact sequence of holomorphic vector bundles

$$0 \to H' \to H \to H'' \to 0$$

and that H has a (possibly indefinite) hermitian metric which induces non-degenerate metrics on H' and H'' . As usual, identify H'' , as a C^∞ bundle, with the orthogonal complement of H' in H . Let σ be the second fundamental form of H' in H . Then the curvatures $\Theta_{H'}$, $\Theta_{H''}$ are given by:

$$(7.3)\qquad \begin{aligned}(\Theta_{H'}e, e') &= (\Theta_H e, e') - (\sigma e, \sigma e')\\ (\Theta_{H''}e, e') &= (\Theta_H e, e') - ({}^t\sigma e, {}^t\sigma e')\ .\end{aligned}$$

The two equations (7.3) express the principle that <u>curvature</u> <u>decreases</u> <u>on</u> <u>subbundles</u> <u>and</u> <u>increases</u> <u>on</u> <u>quotient</u> <u>bundles</u> (notice that ${}^t\sigma$ has type $(0, 1)$). Applying (7.3) to the exact sequences:

$$0 \to F_p \to E \to E/F^p \to 0$$

$$0 \to F_{p+1} \to F_p \to E^{p,m-p} \to 0$$

and taking into account the <u>alternation</u> <u>of</u> <u>signs</u> (7.1) gives formula (7.2).

b) APPLICATIONS TO RIGIDITY THEOREMS AND RELATED MATTERS WHEN THE BASE IS COMPACT. Let $V = \{S, E, E_{\mathbb{Z}}, Q, \nabla, \{F^p\}\}$ be a variation of Hodge structure with <u>compact</u> base S . Let ξ be a global holomorphic section of F^p . Suppose that ξ is <u>quasi-horizontal</u>, i.e. suppose that $\nabla\xi$ is a section of $\Omega^1(F^p)$. In other terms this means that $\psi_p\xi = 0$. Let ξ_p be the $(p, m - p)$-component of ξ . Then, as follows from the results of the first lecture,

$$\partial\bar\partial\|\xi_p\|^2 = (D'_p\xi_p, D'_p\xi_p) - (\Theta_p\xi_p, \xi_p)$$

where D_p is the metric connection of $E^{p,m-p}$. Taking into account formula (7.2) and the quasi-horizontality of ξ , it follows that $\|\xi_p\|^2$ is plurisubharmonic, hence <u>constant</u>, due to the compactness of the base, therefore:

$$D'_p\xi_p = {}^t\psi_{p+1}\xi_p = \psi_p\xi_p = 0\ .$$

This means that ξ_p is <u>holomorphic</u>, <u>as</u> <u>a</u> <u>section</u> <u>of</u> E , and <u>horizontal</u>. If we assume that ξ is horizontal, we may apply the same procedure to $\xi - \xi_p$ and inductively obtain the following statement:

THEOREM OF THE FIXED PART: Let $V = \{S, E, E_{\mathbb{Z}}, \nabla, Q, \{F^p\}\}$ be a variation of Hodge structure with compact base S . Let ξ be a flat global holomorphic section of E . Then each of the (p, q)-components of ξ is holomorphic and flat. In particular, if ξ has pure (p, q)-type at one point, then it has pure (p, q)-type everywhere.

A straightforward application of the above theorem is a proof of the <u>rigidity theorem</u> when the base is compact. Let $\mathcal{V}$ and $\mathcal{V}'$ be variations of Hodge structure of the same weight m. A $\pi_1(S, s_0)$-equivariant isomorphism of Hodge structures

$$\phi_{s_0} : E_{s_0} \to E'_{s_0}$$

extends by parallel translation to a flat section ϕ of $\mathrm{Hom}(E, E')$, which has type $(0, 0)$ at s_0. Applying the theorem of the fixed part to the variation of Hodge structure $\mathrm{Hom}(\mathcal{V}, \mathcal{V}')$ gives that ϕ has type $(0, 0)$ everywhere, which is just the rigidity theorem.

Another application is the <u>complete reducibility theorem</u> of Deligne, always when the base S is compact. The full argument is too long to be given here and we will content ourselves with the weaker statement:

Let $\mathcal{V}$ be a variation of Hodge structure with compact base S and monodromy group Γ. Then Q is non-degenerate on the space E^Γ of Γ-invariants. In particular,

$$E = E^\Gamma \oplus (E^\Gamma)^\perp .$$

To conclude we would like to remark that, since a bounded plurisubharmonic function on an algebraic (possibly non-compact) variety is constant, the above proofs of the theorem of the fixed part and its corollaries would go through for an arbitrary algebraic base S if we knew that a flat section of $E \to S$ has <u>bounded length</u>. This, and much more, follows from the results of Schmid that we will discuss in a short while.

c) THE DISTANCE-DECREASING PROPERTY AND SOME APPLICATIONS. When studying variations of Hodge structure on a non-complete algebraic curve, one is naturally led, by localizing near points at infinity, to study variations of Hodge structure on a punctured disk Δ^*. In the language of classifying spaces for Hodge structures we have a diagram

$$\begin{array}{ccc} H & \xrightarrow{\tilde{\Phi}} & D \\ {\scriptstyle e^{2\pi\sqrt{-1}\,w}}\downarrow & & \downarrow \\ \Delta^* & \xrightarrow{\Phi} & \Gamma\backslash D \end{array}$$

The monodromy group Γ consists of multiples of the Picard-Lefschetz transformation T and $\tilde{\Phi}$ satisfies

$$T\tilde{\Phi}(w) = \tilde{\Phi}(w+1) .$$

Moreover $\tilde{\Phi}$ is <u>horizontal</u> in the following sense. The tangent bundle of D has a distinguished holomorphic subbundle, the horizontal subbundle $TH(D)$, consisting of all tangent vectors X such that

$$XF^p \subset F^{p-1}/F^p$$

(recall that a tangent vector at a point $\{F^p\}$ of D can be identified with a collection of homomorphisms from F^p to E/F^p, for each p). Saying that a mapping of a complex manifold into D is horizontal means that it is tangent to $TH(D)$. Mappings arising from variations of Hodge structure are horizontal by definition (cf. (5.9)).

A consequence of the curvature properties of Hodge bundles is that, for any suitably normalized $G_{\mathbb{R}}$-invariant metric, the holomorphic sectional curvature of D satisfies an inequality

$$K(\xi) \leqq -1$$

<u>whenever ξ lies in the horizontal subbundle</u> $TH(D)$. Hence the Ahlfors lemma applies to $\tilde{\Phi}$ and says that $\tilde{\Phi}$ is <u>distance decreasing</u>, in other terms:

$$\rho_D(\tilde{\Phi}(w), \tilde{\Phi}(w')) \leqq \rho_H(w, w') \tag{7.4}$$

where ρ_D is the invariant distance on D and ρ_H is the Poincaré distance on H.

A very elegant application of the distance decreasing property in this form is <u>Borel's proof of the quasi-unipotency of the Picard-Lefschetz transformation</u>. Since T is integral, by a theorem of Kronecker (an algebraic integer all of whose conjugates have absolute value one is a root of unity) it will suffice to prove that the eigenvalues of T have absolute value one. Choose a sequence $\{w_n\}$ of points of H whose imaginary part goes to infinity with n and write

$$\tilde{\Phi}(w_n) = g_n p_0$$

where p_0 is a reference Hodge filtration and g_n belongs to $G_{\mathbb{R}}$. Then

$$\begin{aligned}\rho_D(\tilde{\Phi}(w_n+1), \tilde{\Phi}(w_n)) &= \rho_D(Tg_n p_0, g_n p_0) \\ &= \rho_D(g_n^{-1}Tg_n p_0, p_0) .\end{aligned}$$

On the other hand

$$\rho_H(w_n+1, w_n) = O\left(\frac{1}{\operatorname{Im} w_n}\right)$$

so that $\rho_D(g_n^{-1}Tg_n p_0, p_0)$ geos to zero as n tends to infinity. Since D is the quotient of $G_{\mathbb{R}}$ by a <u>compact</u> subgroup H, this says that a sequence of conjugates of T tends to an element of H, and therefore all eigenvalues of T must have absolute value one.

Another way of looking at the distance decreasing property for a variation of Hodge structure over Δ, H or Δ^* is the following:

For every p and every section e of F^p

(7.5) $$\frac{\sqrt{-1}}{2\pi}(\psi_p e, \psi_p e) \leqq C(e, e)\eta$$

where η is the Poincaré metric and C a suitable positive constant.

This is seen to be equivalent to (7.4) by recalling the explicit description of the tangent bundle to a Grassmannian (lecture 1).

Formula (7.5) can be viewed as giving an estimate

(7.6) $$-C\eta \leqq \frac{(\Theta_p e, e)}{\|e\|^2} \leqq C\eta$$

on the curvature of the Hodge bundles $E^{p,m-p}$. The inequality (7.6) has important consequences. The first one is the <u>algebrization theorem</u> for E.

The basic tool is the following general result.

(7.7) Let $V \to A$ be a holomorphic, hermitian vector bundle over an affine variety A . Choose a smooth compactification $\bar{A}$ of A . If the curvature form of V satisfies estimates

$$-C\eta \leqq \frac{(\Theta e, e)}{\|e\|^2} \leqq C\eta$$

where η is the Poincaré metric in the punctured polycylinders at infinity, then V has an algebraic structure whose global sections e satisfy estimates

$$\max_{\bar{A}-T_\varepsilon} (\log \|e\|) = O(\log (\tfrac{1}{\varepsilon}))$$

where T_ε is an ε-tube around $\bar{A} - A$.

This result immediately gives an algebraic structure on $E^{p,m-p}$, for each p . In addition to (7.7) one has a description of the <u>algebraic cohomology</u> of E , this being given by closed modulo exact forms with appropriate growth conditions at infinity.

To give the algebraic structure on E , we proceed by steps. Suppose F^{p+1} has been algebraicized. Then it follows by the above description of cohomology and (again!) the distance decreasing property that the extension class of $E^{p,m-p}$ by F^{p+1} is algebraic, which allows to algebraicize F^p , and so on.

The following local version of (7.7) also holds.

(7.8) Let $V = \{\Delta^*, E, E_{\mathbb{Z}}, \nabla, Q, \{F^p\}\}$ be a variation of Hodge structures over the punctured disk $\Delta^* = \{\zeta \in C \mid 0 < |\zeta| < 1\}$. Then each of the F^p is generated by a finite number of sections e such that

$$\max_{|\zeta| \geqq r} (\log \|e\|) = O(\log (\tfrac{1}{r})) . \tag{7.9}$$

Rather than going into the details of the proof of (7.8) or of its global counterpart we wish to show how one can deduce from the distance decreasing property that the Gauss-Manin connection is algebraic and has regular singular points, when the base is an algebraic curve $S = \bar{S} - \{s_1, \ldots, s_N\}$, $\bar{S}$ being smooth and complete.

By localizing at infinity we may work on a punctured disk $\Delta^* = \{\zeta \mid 0 < |\zeta| < 1\}$. Let e be a (multi-valued) holomorphic flat section of E and write $\zeta = re^{i\theta}$. We decompose e into $(p, m-p)$-components

$$e = \sum e_p \qquad e_p \in E^{p,m-p} .$$

The condition that e be flat means that, for any p

$$D'_p e_p = -\psi_{p+1} e_{p+1}$$

and to say that e is holomorphic means that

$$D''_p e_p = \pm\, {}^t\psi_{p-1} e_{p-1} .$$

Now we take the radial derivative of the Hodge length of e.

$$\begin{aligned} \frac{\partial}{\partial r}(e, e) &= \sum \frac{\partial}{\partial r}(e_p, e_p) \\ &= \sum 2\,\mathrm{Re}\left(\frac{\partial}{\partial r} \lrcorner\, (D_p e_p, e_p)\right) \\ &= -\sum 2\,\mathrm{Re}\left(\frac{\partial}{\partial r} \lrcorner\, (\psi_{p-1} e_{p-1}, e_p)\right) \pm \sum 2\,\mathrm{Re}\left(\frac{\partial}{\partial r} \lrcorner\, ({}^t\psi_{p+1} e_{p+1}, e_p)\right) \end{aligned}$$

Taking absolute values and using the Schwarz inequality plus the distance decreasing property and the explicit form of the Poincaré metric on Δ^* one gets

$$\left| \frac{\partial}{\partial r}(e, e) \right| \leqq \frac{2(e, e)}{r \log \frac{1}{r}} . \tag{7.10}$$

By integrating the differential inequality (7.10) we obtain inequalities

$$C_1\left(\log \frac{1}{|\zeta|}\right)^{-k} \leqq \|e\|^2 \leqq C_2\left(\log \frac{1}{|\zeta|}\right)^k \tag{7.11}$$

which hold uniformly on any angular sector, for suitable C_1, C_2, k.

Now let T be the Picard-Lefschetz transformation. For the sake of simplicity, here, and in the following, we will always assume that the Picard-Lefschetz transformations involved are <u>unipotent</u>. Write

$$N = \log T = \sum_{q=0}^{\infty} \frac{(I-T)^{q+1}}{q+1} .$$

Due to the unipotency of T, the sum on the right hand side is a finite sum. Then for every flat section e of E,

$$\exp\left(-\frac{\log N}{2\pi\sqrt{-1}}\right) e \tag{7.12}$$

is a single-valued never vanishing holomorphic section of E . These sections give a privileged extension of E across the puncture of Δ^* , so that, returning to the global situation, we have a privileged extension of E to a vector bundle E' over $\bar{S}$. By Serre's G.A.G.A. this has an algebraic structure: the upper bound in (7.11) together with (7.12) tells us that this agrees with the intrinsic algebraic structure of E . Also

$$\nabla \exp\left(-\frac{\log N}{2\pi\sqrt{-1}}\right)e = -\frac{N}{2\pi\sqrt{-1}}\exp\left(-\frac{\log N}{2\pi\sqrt{-1}}\right)e\,\frac{d\zeta}{\zeta}$$

which shows that ∇ is algebraic and has regular singular points.

d) THE NILPOTENT ORBIT THEOREM. Let V be a variation of Hodge structure of weight m on Δ^* . Another straightforward consequence of (7.11) is the following metric comparison lemma. We denote by E' the privileged extension of E to Δ constructed at the end of c) and let $<\,,\,>$ be any C^∞ inner product on E' . Then, on every compact subset of Δ and for every section e of E ,

$$A\left(\log\left(\frac{1}{|\zeta|}\right)\right)^{-k}\|e\|^2 \leq \langle e, e\rangle \leq B\left(\log\left(\frac{1}{|\zeta|}\right)\right)^{k}\|e\|^2 \tag{7.13}$$

where A, B, k are suitable constants.

Now let f be a section of F^p which satisfies the estimate (7.9). The metric comparison lemma tells us that f is a linear combination with meromorphic coefficients of sections of the form (7.12). Since the base is one-dimensional, this and (7.8) imply that the filtration $\{F^p\}$ extends to a filtration $\{F'^p\}$ on E' .

What we have done here is essentially deducing from (7.8) the first part of W. Schmid's nilpotent orbit theorem. To formally state this we have to define the compact dual $\check{D}$ of D . Recall that D can be viewed as the set of all filtrations (with appropriate Hodge numbers) on a fixed vector space which satisfy the two Hodge-Riemann bilinear relations. $\check{D}$ is defined to be the set of all filtrations satisfying the first Hodge-Riemann bilinear relation, but not necessarily the second. $\check{D}$ is easily seen to be a projective homogeneous algebraic variety: D is an open subset of $\check{D}$. With these notations what we have shown, in the language of classifying spaces for Hodge filtrations, reads as follows.

EXTENSION LEMMA. Let

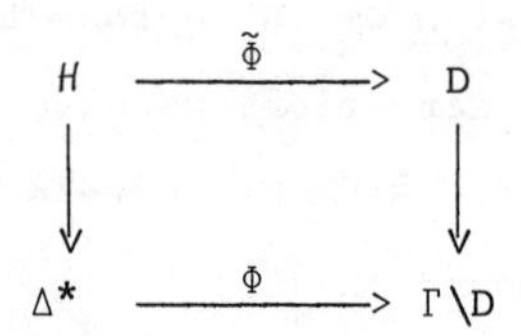

be a variation of Hodge structure over Δ^*. Let T be the Picard-Lefschetz transformation (which, for simplicity, we will assume to be unipotent) and set $N = \log T$. Then the mapping

$$w \to \exp(-wN)\tilde{\Phi}(w)$$

descends to a mapping

$$\Psi : \Delta^* \to D$$

which extends across the puncture.

The remaining part of the nilpotent orbit theorem, which follows fairly easily from (7.13), says that, if we write $p_0 = \Psi(0)$,

(i) $\exp(wN)p_0$ is horizontal;

(ii) there is a non-negative number α such that, if $\mathrm{Im}(w) > \alpha$, then $\exp(wN)p_0$ belongs to D;

(iii) $\exp(wN)p_0$ is strongly asymptotic to $\tilde{\Phi}(w)$ in the sense that

$$\rho_D(\tilde{\Phi}(w), \exp(wN)p_0) \leqq (\mathrm{Im}\, w)^{\beta} e^{-2\pi\, \mathrm{Im}\, w}$$

for some $\beta \geqq 0$ and large enough $\mathrm{Im}\, w$.

We now give a sketch of the proof of (i), (ii), (iii). In the Hodge bundle framework, the nilpotent orbit $\exp(wN)p_0$ corresponds to a filtration $\{\tilde{F}^p\}$ of E' which agrees with $\{F'^p\}$ at 0 and is <u>constant</u>, relative to the sections of the type:

$$\exp\left(-\frac{\log \zeta N}{2\pi\sqrt{-1}}\right)e$$

where e is flat. N can be viewed as a flat endomorphism of E which extends across the puncture. (i) just says that $\{\tilde{F}^p\}$ satisfies the infinitesimal period relations (4.11). This follows from the fact that

$$NF'^{p} \subset F'^{p-1}$$

at the origin, which in turn is a consequence of the infinitesimal period relations for $\{F^p\}$.

To prove (ii) we must show that, for any non vanishing section e of E'

$$\sum (-1)^p Q(e'_p, \overline{e'_p}) > 0$$

in a neighborhood of the origin, where $e = \sum e'_p$ is the decomposition of e into $(p, m-p)$ type relative to the filtration $\{\tilde{F}^p\}$. Assume, inductively, that this has already been proved for sections of $\tilde{F}^{p+1}$ and let e be a section of $\tilde{F}^p$ whose projection into $\tilde{F}^p/\tilde{F}^{p+1}$ does not vanish at the origin. What we have to show is that

$$(-1)^p Q(e'_p, \overline{e'_p}) > 0$$

near the origin. Notice that the length of e'_p, $< e'_p, e'_p >^{1/2}$, relative to any C^∞ metric on E' , is bounded both above and below. We may write

$$e'_p = \alpha + \beta + \gamma$$

where α belongs to $E^{p,m-p}$, β belongs to F^{p+1} , γ belongs to $\overline{F^{m-p+1}}$. The metric comparison lemma plus the fact that $\{F'^p\}$ and $\{\tilde{F}^p\}$ agree at the origin imply the estimates

$$(7.14)\qquad \begin{aligned} (\alpha, \alpha) &\geqq C_1\left(\log \frac{1}{|\zeta|}\right)^{-k} \\ (\beta, \beta) &\leqq C_2|\zeta|^2\left(\log \frac{1}{|\zeta|}\right)^k \geqq (\gamma, \gamma) \end{aligned}$$

in a neighborhood of the origin. It follows that

$$(-1)^p Q(e'_p, \overline{e'_p}) \geqq (\alpha, \alpha) - (\beta, \beta) - (\gamma, \gamma) > 0$$

near 0 , as was to be proved.

When suitably interpreted, the estimates (7.14) also give part (iii) of the nilpotent orbit theorem.

In a way the nilpotent orbit theorem enables one to reduce most questions about a general variation of Hodge structure on Δ^* to questions about the approximating nilpotent orbit. However, studying nilpotent orbits seems to be a deeper matter than the nilpotent orbit theorem itself. Very detailed information about nilpotent orbits is given by W. Schmid's SL_2-<u>orbit theorem</u> which will be discussed in the appendix to this lecture. This theorem will not be formally stated here. We will just say that, roughly, it enables us to construct a mapping

$$\Xi : H \to D$$

which lifts to a homomorphism of Lie groups

$$\tilde{\Xi} : SL_2(\mathbb{R}) \to G_{\mathbb{R}}$$

and is asymptotic to the nilpotent orbit $\exp(wN)p_0$ (in a much weaker sense than above). Moreover the theorem gives very detailed information on the way Ξ is related to $\exp(wN)p_0$, and this is crucial in most applications.

Instead of insisting on the SL_2-orbit theorem we will give some of its consequences. To do so we have to go back to the original situation, when we have a variation of Hodge structure of weight m over Δ^*: as usual the Picard-Lefschetz transformation T will be assumed to be unipotent. The monodromy theorem in strong form, as follows from the SL_2-orbit theorem, says that the index of unipotency of T is at most m, provided the Hodge numbers h^{pq} are zero if $p < 0$ or $q > m$. We have constructed extensions of E and F^p to Δ. We will denote these with the same symbols as their restrictions to Δ^*. N may be viewed as a flat endomorphism of E on all of Δ. The <u>weight filtration</u> of E is the unique filtration

$$0 \subset W_0 \subset \cdots \subset W_{2m} = E$$

which satisfies the following conditions:

$$NW_i \subset W_{i-2}$$

$$N^\ell : W_{m+\ell}/W_{m+\ell-1} \to W_{m-\ell}/W_{m-\ell-1}$$
is an isomorphism for every $\ell \geq 0$

$\{W_i\}$ can be constructed as follows. It is clear that $W_{2m-1} = \ker N^m$ and $W_0 = N^m(E)$. Then we consider the linear mapping induced by N on

W_{2m-1}/W_0 and work our way down by induction on m.

Now denote by $\{\tilde{F}^p\}$ an approximating nilpotent orbit to $\{F^p\}$. One of the consequences of the SL_2-orbit theorem is that

(7.15) $\{W_i\}$, $\{\tilde{F}^p\}$, $E_{\mathbb{Z}}$ induce a mixed Hodge structure on each of the fibres of $E|\Lambda^*$. Relative to this, N is a morphism of type $(-1, -1)$.

Let us remark that the last assertion follows immediately from the infinitesimal period relation for the nilpotent orbit $\{\tilde{F}^p\}$ and from the definition of $\{W_i\}$.

What (7.15) tells us is that _the Hodge structures on the fibres of_ E _asymptotically approach a mixed Hodge structure as we go into the puncture_.

Another consequence is a _concrete description of the weight filtration_. This is a refinement of the estimates (7.11). Let e be a _flat_ section of E. _Saying that_ e _belongs to_ W_ℓ _(at each point) means that the Hodge length_ $\|e\|$ _satisfies an estimate_

$$\|e\| = O\left(\log\left(\frac{1}{|\zeta|}\right)^{\frac{\ell-m}{2}}\right)$$

uniformly on each angular sector. In particular, since obviously

$$\ker N \subset W_m$$

then every _invariant_ e ($Te = e$, or, which is the same, $Ne = 0$) is bounded near the puncture. This is the statement needed to prove the theorem of the fixed part and its corollaries without completeness assumptions on the base.

REFERENCES FOR LECTURE 7

The curvature properties of Hodge bundles and subsequent applications to variation of Hodge structure over a compact base were given in

P. Griffiths, Periods of integrals on algebraic manifolds (III), Publ. Math. I.H.E.S., vol. 38 (1970), 125-180.

The algebraization theorem 7.7 is proved in

M. Cornalba and P. Griffiths, Analytic cycles and vector bundles on non-compact algebraic varieties. To appear in Inventiones Math.

The theory of regular singular points is discussed in

P. Deligne, Equations différentielles à points singuliers réguliers, Lecture notes in math. no. 163, Springer-Verlag (1970).

Finally, the basic results of Schmid appear in

W. Schmid, The singularities of the period mapping, Inventiones Math., vol. 22 (1973), 211-319.

APPENDIX TO LECTURE 7: ON SCHMID'S SECOND THEOREM

This appendix is intended as an aid in understanding Schmid's proof of his basic technical theorem, the $\mathfrak{sl}_2$-orbit theorem, which appears in his paper listed in the references to lecture 7. It is not our purpose to give either the precise statement or complete proof, but rather to extract the essential aspects of the argument and put them in a more differential-geometric and less Lie-theoretic setting.

a) HEURISTIC REASONING. We consider a variation of Hodge structure

$$\Phi : \Delta^* \to \{T^n\}\backslash D$$

over the punctured disc. By passing to a finite covering of Δ^*, we may assume that the Picard-Lefschetz transformation T is unipotent with logarithm

$$N = \log T = \sum_q (-1)^q \frac{(T-I)^q}{q} .$$

Lifting up to the universal covering $H \to \Delta^*$, there is an induced variation of Hodge structure, still denoted by Φ,

$$\Phi : H \to D \qquad \text{satisfying}$$
$$\Phi(z+1) = T\Phi(z) .$$

Schmid's results allow us to approximate any such Φ by an equivariant variation of Hodge structure

$$\Sigma : H \to D$$

satisfying the asymptotic estimate $(z = x + \sqrt{-1}\, y)$

$$\rho_D(\Phi(z), \Sigma(z)) = O(y^{-1})$$

for $y \geq C$. Σ is induced by a representation

$$\eta : SL_2(\mathbb{R}) \to G_{\mathbb{R}}$$

having very special properties, and most aspects of the behavior of $\Phi(z)$ as $\operatorname{Im} z \to \infty$ are the same as for $\Sigma(z)$ and can therefore be deduced by Lie algebra calculations.

Here is the heuristic reasoning behind Schmid's theorem. Let $F(z)$ be a Hodge frame field for $\Phi(z)$ with Maurer-Cartan matrix ω defined by

$$\text{(A.7.1)} \qquad \begin{aligned} dF &= \omega\cdot F, \quad \text{where} \\ d\omega &= \omega \wedge \omega . \end{aligned}$$

We adopt the viewpoint that Φ is completely determined by ω, and thus we should try to select the Hodge frame field $F(z)$ such that the asymptotic behavior of $\omega(z)$ as $\operatorname{Im} z \to \infty$ becomes most transparent. As in lecture 5 we write

$$\omega = \phi + \psi + {}^t\bar{\psi}$$

where

$$\psi = \begin{pmatrix} 0 & \psi_n & \cdot & 0 \\ \cdot & & & \cdot \\ \cdot & & & \psi_1 \\ 0 & \cdot & \cdot & 0 \end{pmatrix}$$

is a matrix of $(1, 0)$ forms giving the Kodaira-Spencer class for the variation of Hodge structure, and where

$$\phi = \begin{pmatrix} \phi_n & \cdot & \cdot & 0 \\ \cdot & & & \cdot \\ \cdot & & & \cdot \\ 0 & \cdot & \cdot & \phi_0 \end{pmatrix}$$

gives the connection matrices in the various Hodge bundles. The pull-back of the $G_{\mathbb{R}}$-invariant metric on D is

$$\sum_q \text{Trace}\,(\psi_p \wedge {}^t\bar{\psi}_p)\,,$$

so that if we write $\psi = Adz$, the Ahlfors lemma from lecture 6 gives the basic estimate

$$A(z) = O(y^{-1}) \qquad (z = x + iy) \tag{A.7.2}$$

which underlies the whole development. What is suggested is that we try to choose $\underline{F}(z)$ such that ϕ can also be estimated.

The integrability condition (A.7.1) is now

$$\begin{aligned} d\phi - \phi \wedge \phi &= \psi \wedge {}^t\bar{\psi} + {}^t\bar{\psi} \wedge \psi \\ d\psi &= \phi \wedge \psi + \psi \wedge \phi\,. \end{aligned} \tag{A.7.3}$$

Writing $\phi = Bdx + Cdy$, the first equation gives

$$\frac{\partial C}{\partial x} - \frac{\partial B}{\partial y} - [B, C] = -2\sqrt{-1}\,[A, {}^t\bar{A}] = O(y^{-2})\,. \tag{A.7.4}$$

This fails to yield an estimate on B or C, unless one or the other is zero. The form of (A.7.4) suggests that we try to select our frame field such that $C = 0$. Then $\frac{\partial B}{\partial y} = O(y^{-2})$ so that

$$B = O(y^{-1})\,. \tag{A.7.5}$$

Frame fields with the property that $\phi \equiv O(dx)$ are characterized geometrically by the condition that $\underline{F}(z)$ remains parallel to itself along the vertical lines $x = \text{constant}$.* We shall call these geodesic frame fields; they are easy to construct by prescribing the initial values $\underline{F}(x + \sqrt{-1})$ and then making parallel displacement up and down vertical lines. Geodesic frame fields are the analogue of normal coordinates in Riemannian geometry, and allow one to most easily recover properties of the connection matrix from the curvature.

EXAMPLE: Over the upper half plane H we consider the universal family of Hodge structures of weight one and genus one. The vector space E is C^2, the quadratic form $Q = \begin{pmatrix} 0 & -1 \\ 1 & 0 \end{pmatrix}$, and each point $z \in H$ gives the polarized Hodge decomposition

---

* Here, parallel displacement is relative to the Hodge connection ϕ and not the Gauss-Manin connection.

$$C^2 = [C\cdot(z, 1)] \oplus [C\cdot(\bar{z}, 1)] .$$

Since $\sqrt{-1}\, Q((z, 1), \overline{(z, 1)}) = 2y$, the manifold $F(H)$ of Hodge frames consists of all pairs

$$\{e, \bar{e}\} \quad \text{where}$$

(A.7.6)
$$e = \frac{\exp \sqrt{-1}\,\sigma}{\sqrt{2y}}(z, 1) .$$

The Maurer-Cartan matrix

(A.7.7)
$$\omega_H = \begin{pmatrix} \sqrt{-1}\, d\sigma + \dfrac{\sqrt{-1}\, dx}{2y} & \exp 2\sqrt{-1}\,\sigma \dfrac{dz}{2y} \\ \exp(-2\sqrt{-1}\,\sigma) \dfrac{d\bar{z}}{2y} & -\sqrt{-1}\, d\sigma - \dfrac{\sqrt{-1}\, dx}{2y} \end{pmatrix}$$

Geodesic frame fields are those frame fields $\{e(z), \overline{e(z)}\} \in F(H)$ for which $\sigma = \sigma(x)$ is a function of x alone. In particular, $\sigma = \text{constant}$ defines one such. This example shows that the estimates (A.7.2) and (A.7.5) are sharp.

Returning to the general case, we let $\underline{F}(z)$ be a geodesic frame field lying over a variation of Hodge structure $\Phi(z)$. On the basis of (A.7.2), (A.7.5) and the above example, we might hope to have an expansion

$$\omega = \omega_{-1} + \tilde{\omega}$$

where ω_{-1} is a linear combination of $\dfrac{dz}{y}$ and $\dfrac{d\bar{z}}{y}$ whose coefficient matrices are constant, and where $\tilde{\omega}(z)$ is of lower order as $\operatorname{Im} z \to \infty$. The second equation in (A.7.1) should imply that

$$d\omega_{-1} = \omega_{-1} \wedge \omega_{-1} .$$

In this case, we may define a mapping

$$\Sigma : H \to D$$

by solving the differential equation

(A.7.9)
$$d\underline{G}(z) = \omega_{-1}(z)\underline{G}(z)$$

to find a Hodge frame field $\underline{G}$ for Σ. By comparing the form $\omega = \phi + \psi + {}^t\bar{\psi}$ of ω with the form (A.7.7) of the Maurer-Cartan matrix for $SL_2(\mathbb{R}) \cong F(H)$, and taking into account that ω_{-1} is a constant linear combination of $\frac{dz}{y}$ and $\frac{d\bar{z}}{y}$, Σ <u>as defined by</u> (A.7.9) <u>is an equivariant variation of Hodge structure induced from a homomorphism</u> $\eta : SL_2(\mathbb{R}) \to G_{\mathbb{R}}$ <u>whose induced Lie algebra mapping is just</u> ω_{-1} . Moreover, it is plausible that Σ , viewed as the "principal part" of Φ as $\mathrm{Im}\, z \to \infty$, should share its essential properties. Making this argument precise constitutes Schmid's theorem.

To carry all this out, one essentially needs to know that for the original variation of Hodge structure $\Phi(x + \sqrt{-1}\, y)$ looked at along lines parallel to the imaginary axis, the Maurer-Cartan matrix $\omega(x + \sqrt{-1}\, y)$ for a geodesic frame field has a Laurent series expansion in powers of y^{-1} as $y \to \infty$. In this case, the decomposition (A.7.6) can be analyzed by using series expansions in the structure equations (A.7.3). However, it is not at all clear that, for our given variation of Hodge structure $\Phi(z)$, there is some Hodge frame field $\underline{F}(z)$ (much less a geodesic one) whose Maurer-Cartan matrix has such a Laurent series. Thus, the proof consists of first replacing $\Phi(z)$ by a nilpotent orbit $\Psi(z)$ where it is easy to find a (non-geodesic) frame field with this property, and then proving that rotating into a geodesic frame field still yields a Maurer-Cartan matrix having the desired Laurent series.

b) FRAMING THE NILPOTENT ORBIT. Given a variation of Hodge structure $\Phi : H \to D$ satisfying $\Phi(z + 1) = T\Phi(z)$ where T is unipotent with logarithm N , we consider nilpotent orbits

$$\Psi(z) = \exp(zN)\cdot\Psi_0 \qquad (\Psi_0 \in D) .$$

Any such nilpotent orbit is the restriction to $H \in \mathbb{P}^1$ of a polynomial mapping into the dual classifying space $\check{D}$ of all filtrations $\{F^p\}$ on E satisfying the first bilinear relation $Q(F^p, F^{n-p-1}) = 0$. In general, however, $\Psi(z)$ is neither horizontal nor maps H into D . Schmid's first theorem asserts that, for a suitable choice of reference point Ψ_0 , the nilpotent orbit gives a variation of Hodge structure which is strongly asymptotic to $\Phi(z)$ in the sense that

$$\rho_D(\Phi(z), \Psi(z)) = O(y^{\alpha}e^{-y})$$

for $z = x + \sqrt{-1}\, y \in H$, $y \geqq C$. Henceforth, we shall restrict our attention to $\Psi(z)$.

We first note that

$$\exp(zN) = \exp(xN) \exp(\sqrt{-1}\, yN)$$

where $\exp(xN) \subset C_{\mathbb{R}}$. Thus, the $G_{\mathbb{R}}$-invariant properties of $\Psi(z)$ are all described by looking at $\Psi(\sqrt{-1}\, y)$.

In order to frame $\Psi(z)$, we fix a reference Hodge frame $\underline{F}_0$ lying over Ψ_0 . Then the frames

$$\underline{\tilde{F}}(z) = \exp(zN)\underline{F}_0$$

give a field of frames lying over $\Psi(z)$ which depend in a polynomial fashion on z . However, these frames are not Hodge frames, but are only frames adapted to the filtration $\Psi(z)$. To obtain a Hodge frame field, it is natural to apply the Gram-Schmidt process, which we now review.

Given a vector space V with Hermitian form $(\, , \,)$, a subspace W such that $(\, , \,)$ is definite on W and on $W^{\perp}$, and a basis $\{w_1, \ldots, w_d; v_1, \ldots, v_e\}$ for V such that $w_1, \ldots, w_d$ is a basis for W , the _Gram-Schmidt process without normalization_ consists of the following three steps:

(i) a transformation

$$\begin{aligned} w_1 &\to w_1 \\ w_2 &\to w_2 - \frac{(w_1, w_2)}{(w_1, w_1)} w_1 \\ &\vdots \end{aligned}$$

converts $w_1, \ldots, w_d$ into an orthogonal basis for W ;

(ii) assuming (i), the transformation

$$v_\alpha \to v_\alpha - \sum_i \frac{(w_i, v_\alpha)}{(w_i, w_i)} w_i$$

where the v_α's orthogonal to the w_i's ; and

(iii) assuming (i) and (ii), we apply (i) to the v_α's .

In an obvious way, we may apply this process to the vectors in $\underline{F}(z)$ relative to the Hodge filtration $\Psi(z)$ and obtain an orthogonal frame $\underline{F}^{\#}(z)$ in which the vectors are rational functions of z and $\bar{z}$. Finally, we may normalize to make the vectors in $\underline{F}^{\#}(z)$ have unit length to obtain a Hodge frame field $\underline{F}(z)$. Since, if $a(y)$ is a rational function of y with $a(y) > 0$ for $y \geq C$ then $\sqrt{a(y)}$ has a convergent Laurent series expansion in $y^{-1/2}$ for $y \geq C'$, our <u>rational frame field</u> $\underline{F}(z)$ has the properties:

$\underline{F}(z)$ is a Hodge frame field for $\Psi(z)$;

$\underline{F}(z + x) = \exp(xN)\underline{F}(z) \qquad (\exp(xN) \in G_{\mathbb{R}})$; and

$\underline{F}(\sqrt{-1}\, y)$ has a convergent Laurent series expansion in $y^{-1/2}$ for $y \geq C$.

Moreover, if the Maurer-Cartan matrix $\omega_{\underline{F}}(z)$ is defined by $d\underline{F} = \omega_{\underline{F}}\underline{F}$, then

$$\omega_{\underline{F}}(z + x) = \omega_{\underline{F}}(z)$$

since ω is $G_{\mathbb{R}}$-invariant. Writing as usual

$$\begin{aligned} \omega_{\underline{F}} &= \phi_{\underline{F}} + \psi_{\underline{F}} + {}^{t}\bar{\psi}_{\underline{F}} \\ \psi_{\underline{F}} &= a(y)\,dz \\ \phi_{\underline{F}} &= b(y)\,dx + c(y)\,dy , \end{aligned}$$

the Ahlfors lemma gives

$$a(y) = \sum_{n=0}^{\infty} a_n y^{-(n+2)/2} .$$

As discussed above, this does not seem to easily yield information on the Laurent series expansions of $b(y)$ and $c(y)$.

Therefore, for the reasons discussed previously, we are led to consider a rotation

$$\underline{G}(z) = h(z)\underline{F}(z) \qquad (h(z) \in H)$$

taking $\underline{F}(z)$ into a geodesic frame field $\underline{G}(z)$. We may obviously assume that $h(x + \sqrt{-1}\, y) = h(y)$ is independent of x . The relation between Maurer-Cartan matrices is

$$\mathrm{Adh}\,\omega_{\underline{G}} = h^{-1}dh + \omega_{\underline{F}} .$$

Writing $\omega_{\underline{G}} = \phi_{\underline{G}} + \psi_{\underline{G}} + {}^t\bar{\psi}_{\underline{G}}$ where

$$\text{(A.7.10)} \qquad \begin{aligned} \psi_{\underline{G}}(z) &= A(y)dz \\ \phi_{\underline{G}}(z) &= B(y)dx \qquad (\text{since } \underline{G} \text{ is geodesic}), \end{aligned}$$

it follows that $h(y)$ is defined by the O.D.E.

$$\text{(A.7.11)} \qquad h'(y) + h(y)c(y) = 0 .$$

Our major step will be to prove that (A.7.11) has a regular singular point at $y = \infty$, which, when taken together with certain addition properties of the solution matrix $h(y)$, will lead to the desired Laurent series expansion. Writing

$$c(y) = \sum_{n=-N}^{\infty} c_n y^{-(n+2)/2} = c_{-N}y^{N/2-1} + (\cdots) \qquad (c_{-N} \neq 0) ,$$

we want to show that $N = 0$. The idea is to use the O.D.E. (A.7.11) to obtain certain growth estimates on the derivatives $h^{(k)}(y)$, and then to compare these with estimates coming from the Ahlfors lemma and the structure equations (A.7.3).

LEMMA: $h^{(k)}(y) = O(y^{(kN/2)-k})$, and no better estimate is possible.

PROOF: Differentiating (A.7.11) leads to

$$\begin{aligned} h'' &= hc^2 + hc' \\ &\vdots \\ h^{(k)} &= (-1)^k hc^k + hd_k \end{aligned}$$

where d_k has a Laurent series beginning with $y^{(kN/2)-(k+1)}$. This follows inductively from

$$d_{k+1} = (-1)^k c'c^k - cd_k + d_{k+1} .$$

Since H is a compact matrix group, $h(y) = O(1)$ and $c_{-N} + {}^t\bar{c}_{-N} = 0$. Thus c_{-N} can be diagonalized, and in particular $(c_{-N})^k \neq 0$ for all k. Consequently

$$h^{(k)} = O((c_{-N})^k y^{(kN/2)-k}) = O(y^{(kN/2)-k})$$

and no better estimate is possible. Q.E.D.

Next, referring to (A.7.10), we shall prove the

LEMMA: $A^{(k)}(y)$ and $B^{(k)}(y)$ are $O(y^{-k-1})$.

PROOF: The structure equations (A.7.3) give

$$B'(y) = [A(y), {}^t\bar{A}(y)]$$
$$A'(y) = [B(y), A(y)] .$$

The Ahlfors lemma gives the estimate

$$A(y) = O(y^{-1}) .$$

It follows from the first equation that $B'(y) = O(y^{-2})$ so that $B(y) = O(y^{-1})$. The second equation then implies that $A'(y) = O(y^{-2})$. Differentiating the first equation then yields $B''(y) = O(y^{-3})$, and doing the same in the second equation gives $A''(y) = O(y^{-3})$. Continuing this process gives the lemma.

LEMMA: $h^{(k)}(y) = O(y^{\gamma-k})$ for some fixed γ .

PROOF: We write $\underline{G}(y) = \underline{G}(\sqrt{-1}\, y)$, $\underline{F}(y) = \underline{F}(\sqrt{-1}\, y)$ and identify the Hodge frame manifold with the matrix group $G_{\mathbb{R}}$. Then

$$h(y) = \underline{G}(y)\underline{F}^{-1}(y) . \tag{A.7.12}$$

Moreover, $\underline{F}^{-1}(y)$ has a Laurent series and so

$$\underline{F}^{-1}(y)^{(k)} = O(y^{\alpha-k})$$

for some fixed α . If we can prove that

$$\underline{G}(y)^{(k)} = O(y^{\beta-k}) , \tag{A.7.13}$$

then the lemma will follow by differentiating (A.7.12) and taking $\gamma = \alpha + \beta$.

Since $h(y) = O(1)$, $\underline{G}(y) = h(y)\underline{F}(y)$ is $O(y^{\beta+1})$. Moreover, the structure equation $d\underline{G} = \omega_{\underline{G}}\cdot\underline{G}$ and previous lemma give

$$\underline{G}'(y) = \omega(y)\underline{G}(y)$$

where $\omega(y)^{(k)} = O(y^{-k-1})$. (A.7.13) now results inductively by differentiating this equation. Q.E.D.

Comparing the first and third lemmas gives $N \leq 0$; i.e.

The O.D.E. (A.7.11) has a regular singular point at $y = \infty$.

The solution matrix to any such equation has an expansion in powers of $y^{-1/2}$ and $(\log y)^{\alpha} y^{-\sigma/2}$ where σ is an eigenvalue of c_0 . Moreover, the log terms can occur only if two eigenvalues of c_0 differ by an integer. Since $c_0 + {}^t\bar{c}_0 = 0$, the eigenvalues of c_0 are purely imaginary, and thus no log terms can appear. To insure that $h(y)$ has a Laurent series in $y^{-1/2}$, some further argument is necessary. What must be proved is that $c_0 = 0$.

For this, we consider the rational frame field $\underline{F}(z)$, which we know has a Laurent series in $y^{-1/2}$ along the imaginary axis. Since $h^{-1}h'$ and $Adh \cdot \omega_{\underline{G}}$ are both $O(y^{-1})$, the Maurer-Cartan matrix

$$\omega_{\underline{F}} = Adh\omega_{\underline{G}} - h^{-1}dh$$

is $O(y^{-1})$. Thus $b(y) = \sum_{n \geq 0} b_n y^{-(n+2)/2}$. The integrability conditions (A.7.3) give

$$\begin{aligned} b_0 + [c_0, b_0] &= -2\sqrt{-1}\,[a_0, {}^t\bar{a}_0] \\ a_0 &= \sqrt{-1}\,[b_0, a_0] - \sqrt{-1}\,[c_0, a_0] \\ {}^t\bar{a}_0 &= -\sqrt{-1}\,[b_0, {}^t\bar{a}_0] + \sqrt{-1}\,[c_0, {}^t\bar{a}_0] . \end{aligned}$$

Under these conditions, a lemma on Lie algebras due to Deligne gives that

$$[c_0, a_0] = 0 = [c_0, {}^t\bar{a}_0] .$$

Along the imaginary axis,

$$\underline{F}'(y) = [(c_0 + a_0 + {}^t\bar{a}_0)y^{-1} + \cdots]\underline{F}(y) .$$

The matrix $a_0 + {}^t\bar{a}_0$ has real eigenvalues while those of c_0 are purely imaginary. By the commutation relation, the eigenvalues of $c_0 + a_0 + {}^t\bar{a}_0$ are real if, and only if, $c_0 = 0$. But $\underline{F}(y)$ has a Laurent series in $y^{-1/2}$, which is possible only if $c_0 + a_0 + {}^t\bar{a}_0$ has only real eigenvalues. In conclusion:

$h(y)$ has a convergent Laurent series $\sum_{n\geq 0} h_n y^{-n/2}$.

At this point, we have proved that the geodesic frame field $\underline{G}(z)$ for the nilpotent orbit has the following properties:

$\underline{G}(z + x) = \exp(xN)\underline{G}(z) \qquad (\exp(xN) \in G_{\mathbb{R}})$;

$\underline{G}(\sqrt{-1}\, y)$ has a Laurent series in $y^{-1/2}$;

the Maurer-Cartan matrix $\omega_{\underline{G}} = \phi + \psi + {}^t\bar{\psi}$ where $\phi = b(y)dx$ and $\psi = a(y)dz$ with $a(y)$, $b(y)$ having Laurent series in $y^{-1/2}$ beginning with y^{-1} .

We are now in a position to pursue further the heuristic reasoning given in (a).

c) USE OF THE REPRESENTATION THEORY OF $\mathfrak{sl}_2$. Let $\Psi(z)$ be the nilpotent orbit and $\underline{G}(z)$ the geodesic frame with Maurer-Cartan matrix $\omega_{\underline{G}}$. According to what was proved in (b), we may write

$$\omega_{\underline{G}} = \omega_{\underline{H}} + \tilde{\omega}$$

where $\omega_{\underline{H}}$ contains the terms in $\omega_{\underline{G}}$ involving y^{-1} and $\tilde{\omega}$ is a Laurent series in $y^{-1/2}$ beginning with $y^{-3/2}$. The integrability condition $d\omega_{\underline{G}} = \omega_{\underline{G}} \wedge \omega_{\underline{G}}$ obviously implies that

$$d\omega_{\underline{H}} = \omega_{\underline{H}} \wedge \omega_{\underline{H}} \, .$$

According to the discussion in (a), $\omega_{\underline{H}}$ is thus the Maurer-Cartan matrix for a geodesic frame $\underline{H}(z)$ associated to an equivariant variation of Hodge structure

$$\Sigma : H \rightarrow D \, .$$

To see better what is going on, we write as usual (cf. (A.7.7) for motivation of constants)

$$\omega_{\underline{G}} = \phi + \psi + {}^t\bar{\psi}$$

$$\phi = -A(y)\,\frac{\sqrt{-1}\,dx}{2} \, , \qquad \psi = B(y)\,\frac{dz}{2}$$

where

$$A(y) = \sum_{n \geq 0} A_n y^{-(n+2)/2} , \qquad B(y) = \sum_{m \geq 0} B_m y^{-(m+2)/2} .$$

The integrability relations (A.7.3) give

(A.7.14)
$$\begin{aligned} - A'(y) &= [B(y), {}^t\overline{B(y)}] \\ - 2B'(y) &= [A(y), B(y)] \\ 2\,{}^t\overline{B'(y)} &= [A(y), {}^t\overline{B(y)}] . \end{aligned}$$

When expanded out, these become

(A.7.15)
$$\begin{aligned} \frac{n+2}{2} A_n &= \sum_{p+q=n} [B_p, {}^t\bar{B}_q] \\ (n+2) B_n &= \sum_{p+q=n} [A_p, B_q] \\ (n+2)\,{}^t\bar{B}_n &= \sum_{p+q=n} [A_p, {}^t\bar{B}_q] . \end{aligned}$$

For $n = 0$, we obtain

(A.7.16)
$$\begin{aligned} A_0 &= [B_0, {}^t\bar{B}_0] \\ 2B_0 &= [A_0, B_0] \\ - 2\,{}^t\bar{B}_0 &= [A_0, {}^t\bar{B}_0] . \end{aligned}$$

This says exactly that the assignment

$$h = \begin{pmatrix} 1 & 0 \\ 0 & -1 \end{pmatrix} \rightarrow A_0$$

$$e_+ = \begin{pmatrix} 0 & 1 \\ 0 & 0 \end{pmatrix} \rightarrow B_0$$

$$e_- = \begin{pmatrix} 0 & 0 \\ 1 & 0 \end{pmatrix} \rightarrow {}^t\bar{B}_0$$

gives a representation of $\mathfrak{sl}_2$ on $\mathfrak{gl}(E)$. Since

$$\omega_{\underline{H}} = \phi_0 + \psi_0 + {}^t\bar{\psi}_0 \qquad \text{where}$$

$$\phi_0 = - A_0\sqrt{-1}\,\frac{dx}{2y} , \qquad \psi_0 = B_0\,\frac{dz}{2y}$$

this is certainly consistent with Σ being induced by a representation $\eta : SL_2(\mathbb{R}) \rightarrow G_{\mathbb{R}}$.

To use this, we identify the manifold $F(D)$ of Hodge frames with the group $G_{\mathbb{R}}$ by choosing a reference frame $\underline{F}_0$. Thus $\underline{G}(z) = G(z)\cdot\underline{F}_0$ where $G(z), F(z) \in G_{\mathbb{R}}$, and the Maurer-Cartan forms are given by the matrices

$$\omega_{\underline{G}} = dG\cdot G^{-1}$$
$$\omega_{\underline{H}} = dH\cdot H^{-1} .$$

Along the imaginary axis we may explicitly describe $H(y) = H(\sqrt{-1}\, y)$ by

$$H(y) = \exp(\tfrac{1}{2} \log y\, C_0)$$
$$C_0 = \sqrt{-1}\,(B_0 - {}^t\bar{B}_0) .$$

(PROOF: The vector field $2y \frac{\partial}{\partial y}$ is the infinitesimal generator of the 1-parameter subgroup $\begin{pmatrix} e^t & 0 \\ 0 & e^{-t} \end{pmatrix}$ of $SL_2(\mathbb{R})$ acting by linear fractional transformations on $\mathcal{H}$. Setting $\Sigma_0 = \Sigma(\sqrt{-1})$,

$$\sqrt{-1}\, y = \begin{pmatrix} y^{1/2} & 0 \\ 0 & y^{-1/2} \end{pmatrix}\cdot\sqrt{-1}$$
$$= \exp\left(\frac{1}{2} \log y \begin{pmatrix} 1 & 0 \\ 0 & -1 \end{pmatrix}\right)\cdot\sqrt{-1} ,$$

which implies that

$$\Sigma(\sqrt{-1}\, y) = \exp\left(\frac{1}{2} \log y\, \eta_*\begin{pmatrix} 1 & 0 \\ 0 & -1 \end{pmatrix}\right)\cdot\Sigma_0$$
$$= \exp[\, \frac{1}{2} \log y < \omega_{\underline{H}}, 2y \frac{\partial}{\partial y} >]\cdot\Sigma_0$$
$$= \exp(\, \frac{1}{2} \log y\, C_0)\cdot\Sigma_0 .)$$

We now set $J(y) = H(y)^{-1}G(y)$ so that

$$dJ\cdot J^{-1} = -H^{-1}dH + AdH^{-1}(\omega_{\underline{G}}) . \tag{A.7.17}$$

By what was just proved, along the imaginary axis

$$-H^{-1}dH = -C_0 \frac{dy}{2} . \tag{A.7.18}$$

The strategy is to first prove that the right hand side of (A.7.17) is a Laurent series in y^{-1} beginning with y^{-2}. If this has been done, $J(y)$ is regular at $y = \infty$. Then

$$\begin{aligned}\rho_D(\Sigma(\sqrt{-1}\,y),\ \Psi(\sqrt{-1}\,y)) &= \rho_D(H(y)\Sigma_0,\ G(y)\Psi_0)\\ &= \rho_D(\Sigma_0,\ H(y)^{-1}G(y)\Psi_0)\\ &= O(1)\end{aligned}$$

since $J(y) \in G_{\mathbb{R}}$ is regular at $y = \infty$. Choosing $\Sigma_0 = \Psi_0$ gives

$$\rho_D(\Sigma(\sqrt{-1}\,y),\ \Psi(\sqrt{-1}\,y)) = O(y^{-1}),$$

from which it follows that

$$\begin{aligned}\rho_D(\Sigma(z),\ \Psi(z)) &= \rho_D(\exp(xN)\Sigma(\sqrt{-1}\,y),\ \exp(xN)\Psi(\sqrt{-1}\,y))\\ &= \rho_D(\Sigma(\sqrt{-1}\,y),\ \Psi(\sqrt{-1}\,y))\\ &= O(y^{-1})\end{aligned}$$

proving that the orbits are asymptotic.

To carry this out, we write $\omega_{\underline{G}} = \omega_{\underline{H}} + \tilde{\omega}$ and observe that by (A.7.18)

$$- H^{-1}dH + \mathrm{Ad}(\exp - \tfrac{1}{2}\log y\ C_0)\omega_{\underline{H}} = 0 .$$

It follows that

$$\begin{aligned}dJ\cdot J^{-1} &= \mathrm{Ad}(\exp - \tfrac{1}{2}\log y\ C_0)\tilde{\omega}\\ \text{(A.7.19)}\qquad &= \sum_{n>0} \mathrm{Ad}(\exp - \tfrac{1}{2}\log y\ C_0)C_n y^{-(n+2)/2} \quad \text{where}\\ C_n &= \sqrt{-1}\,(B_n - {}^t\bar{B}_n) .\end{aligned}$$

The matrix $C_0 = {}^t\bar{C}_0$ is Hermitian and has integral eigenvalues, since $H(y) = \exp(\tfrac{1}{2}\log y\ C_0)$ is a Laurent series in $y^{-1/2}$. Thus we may write $C_n = \sum_s C_{n,s}$ where

$$[C_0,\ C_{n,s}] = sC_{n,s} .$$

Expanding (A.7.19) out gives

$$\text{(A.7.20)}\qquad dJ\cdot J^{-1} = \sum_{n>0,s} C_{n,s}y^{-(n+s+2)/2} .$$

What we must show is that $C_{n,s} = 0$ unless $n + s$ is even and positive.

The idea for proving this is to use (A.7.16) to have an action of $\mathfrak{sl}_2$

on $g\ell(E)$. We may decompose $g\ell(E)$ into the well-known irreducible representations, and decompose the coefficients A_n, B_n, ${}^t\bar{B}_n$ accordingly. When this information is fed into (A.7.15), many of the pieces in the resulting decomposition are forced to be zero.

There is, however, one further hitch. Referring to (A.7.19) and (A.7.20), it is desirable to have an $s\ell_2$ action in which C_0 , rather than A_0 , plays the role of h . For this we set

$$\sqrt{-1}\,(B(y) - {}^t\overline{B(y)}) = C(y) = \sum_{n\geq 0} C_n y^{-(n+2)/2}$$

$$\frac{\sqrt{-1}\,A(y)}{2} - \frac{B(y)}{2} - \frac{{}^t\bar{B}(y)}{2} = E(y) = \sum_{n\geq 0} E_n y^{-(n+2)/2}$$

$$-\frac{\sqrt{-1}\,A(y)}{2} - \frac{2B(y)}{3} - \frac{2\,{}^t\overline{B(y)}}{3} = D(y) = \sum_{n\geq 0} D_n y^{-(n+2)/2} .$$

The equations (A.7.14) then give

$$\text{(A.7.21)}\qquad \begin{aligned} -C'(y) &= [D(y), E(y)] \\ -2D'(y) &= [C(y), D(y)] \\ 2E'(y) &= [C(y), E(y)] . \end{aligned}$$

In particular, $\{C_0, D_0, E_0\}$ give an $s\ell_2$ action on $g\ell(E)$ with C_0 corresponding to $h = \begin{pmatrix} 1 & 0 \\ 0 & -1 \end{pmatrix}$. Moreover, the relations (A.7.15) are now satisfied with C_n, D_n, E_n replacing A_n, B_n, ${}^t\bar{B}_n$.

It is well known that the irreducible $s\ell_2$-modules V_r are indexed by non-negative integers r . Moreover, V_r decomposes under h into 1-dimensional eigenspaces $V_{r,s}$ for $s = r, r-2, \ldots, -r$ on which h has eigenvalue s . Write

$$\text{(A.7.22)}\qquad g\ell(E) = \bigoplus_{r,s} g\ell(r, s)$$

where $g\ell(r) = \bigoplus_{s=-r}^{r} g\ell(r, s)$ are the copies of V_r appearing in the above $s\ell_2$ action and $g\ell(r, s)$ is the sth eigenspace. (A.7.22) induces

$$C_n = \sum_{r,s} C_n^{r,s}$$

$$D_n = \sum_{r,s} D_n^{r,s}$$

$$E_n = \sum_{r,s} E_n^{r,s} .$$

Under these conditions, the representation theory of $s\ell_2$ applied to the relations (A.7.15) involving the C_n, D_n, E_n's gives the following (lemma 9.48) in Schmid's paper):

(i) If $n < r$ or $n - r$ is not even, $C_n^{r,s} = D_n^{r,s} = E_n^{r,s} = 0$;

(ii) $C_n^{n,n} = C_n^{n,-n} = D_n^{n,-n} = D_n^{n,2-n} = E_n^{n,n} = E_n^{n,n-2} = 0$;

(iii) $C_{n-2}^{n-2,n-2} = C_{n-2}^{n-2,2-n} = D_{n-2}^{n-2,n-2} = C_{n-2}^{n-2,4-n} = E_{n-2}^{n-2,n-2} = E_{n-2}^{n-2,n-4} = 0$.

Referring to (A.7.20),

$$dJ\cdot J^{-1} = \sum_{\substack{n>0\\ r,s}} C_n^{r,s} y^{-(n+s+2)/2} .$$

Now $C_n^{r,s} = 0$ unless $r \equiv s$ (2), so that by (i) the only non-zero $C_n^{r,s}$ are of the form

$$C_n^{n+2t,2p} \qquad (-2t \leq 2p \leq 2t) .$$

Thus $dJ\cdot J^{-1}$ has a Laurent series in y^{-1} beginning with y^{-2} as desired.

Schmid's theorems on variation of Hodge structure now follow from the preceding analysis together with further discussion about the representation theory of $\mathfrak{sl}_2$. Here is the idea.

Fixing a reference Hodge structure $E = \bigoplus_{p+q=n} E_0^{p,q}$, the Lie algebra $\mathfrak{g} = \mathfrak{g}_{\mathbb{R}} \otimes \mathbb{C}$ has a Hodge structure of weight zero:

$$\mathfrak{g} = \bigoplus_r \mathfrak{g}^{r,-r} ,$$

where $\mathfrak{g}^{r,-r}$ are those linear transformations taking $E_0^{p,q}$ into $E_0^{p+r,q-r}$. It is visibly the case that

$$\eta_* : \mathfrak{sl}_2(\mathbb{C}) \to \mathfrak{g}$$

is a morphism of Hodge structures whose image lies in $\mathfrak{g}^{0,0} + \mathfrak{g}^{1,-1} + \mathfrak{g}^{-1,1}$.

To give an application of this, we fix the reference frame $e_0 = \frac{1}{\sqrt{2}}(\sqrt{-1}, 1)$, $\bar{e}_0 = \frac{1}{\sqrt{2}}(-\sqrt{-1}, 1)$ in $F(H)$. Setting $e(t) = \frac{1}{\sqrt{2}}(\sqrt{-1} + t, 1)$, $\overline{e(t)} = \frac{1}{\sqrt{2}}(-\sqrt{-1} + t, 1)$, the logarithm L of the monodromy matrix is obviously defined by

$$\begin{pmatrix} e'(t) \\ \overline{e'(t)} \end{pmatrix}_{t=0} = L \begin{pmatrix} e_0 \\ \bar{e}_0 \end{pmatrix}$$

relative to the fixed frame. Explicitly,

$$L = \frac{1}{2}\begin{pmatrix} -\sqrt{-1} & \sqrt{-1} \\ -\sqrt{-1} & \sqrt{-1} \end{pmatrix} .$$

In $\mathfrak{sl}_2(\mathbb{C})$, L is conjugate by some element $g \in SL_2(\mathbb{C})$ to the standard nilpotent matrix

$$e_+ = \begin{pmatrix} 0 & 1 \\ 0 & 0 \end{pmatrix} ,$$

which lies in $\mathfrak{sl}_2^{1,-1}$. Then

$$N = \eta_*(L) = \eta_*(ge_+g^{-1}) = \eta(g)\eta_*(e_+)\eta(g)^{-1}$$

where $\eta_*(e_+) \in \mathfrak{g}^{-1,1}$ by the previous remarks. This implies the strong monodromy theorem

$$N^{n+1} = 0 .$$

Pursuing the matter further, one may completely analyze any representation

$$\eta_* : \mathfrak{sl}_2(\mathbb{R}) \to \mathfrak{g}_{\mathbb{R}}$$

with the above properties, and this leads to proofs of Schmid's theorems <u>for the</u> SL_2-<u>orbit</u> $\Sigma : H \to D$. If the SL_2 and nilpotent orbits were asymptotic like $y^{\alpha}e^{-y}$ as $y \to \infty$, then the general results on mixed Hodge structures and monodromy weight filtrations would easily follow. However, since the approximation is only of the order y^{-1} , one must go back to the vanishing coefficient relations (i)-(iii) above to carry out the proof. For the details together with further applications we refer to Schmid's paper, listed in the references to lecture 7.

NOTE: The approach to Schmid's theorems using Hodge frames and their structure equations was worked out with J. Carlson. Schmid's proofs are heavily based on Lie theory using Iwasawa decompositions and the like. Although the flavor of this approach is perhaps different, the three essential steps (the lemma on the regular singular points of the equation $h' + ch = 0$, the use of Deligne's lemma to show that $c_0 = 0$, and the coefficient relations (i)-(iii)) are the same.

HARVARD UNIVERSITY

Proceedings of Symposia in Pure Mathematics
Volume 29, 1975

SOME DIRECTIONS OF RECENT PROGRESS IN COMMUTATIVE ALGEBRA

David Eisenbud*

ABSTRACT

Three recently active areas of commutative algebra are discussed, and some results from each are presented. The areas are

1) Projective modules over polynomial rings.
2) The recent work on the existence of Cohen-Macaulay modules, and its relation to the conjectures on the rigidity of Tor, and on multiplicities.
3) Ideals of low codimension; two applications of the structure theorem for perfect ideals of codimension 2.

This article contains the write-ups of three independent talks on areas of commutative algebra which have shown what seems to me striking recent progress. They are also areas which ought to go on developing -- nearly all the main problems are still unsolved.

I have not tried to merge the three talks; each even retains its own references.

I. THE SERRE PROBLEM ON PROJECTIVE MODULES

The problem, posed by Serre in 1954 [4], is: Let k be a field; is every projective $k[x_1,\ldots,x_n]$-module free? Equivalently, is every algebraic vector-bundle on affine n-space over k free?

Progress on this question was smooth, if slow, until the early sixties, thanks to the work of Serre, Seshadri, and Bass. By that time the answer to the question itself was known only for $n \leq 2$ ("Seshadri's theorem" - the answer is "yes" in this case); but there was a wealth of subsidiary information on stable freeness and cancellation, which showed, for instance, that projective $k[x_1,\ldots,x_n]$-modules of rank $\geq n+1$ are free.

Though many people continued to work on Serre's problem, little direct progress was made for the next ten years. Then, in 1973, Murthy-Towber, Swan, Roitman, and

* Partially supported by grants from the NSF, and by a Sloan Foundation Fellowship during preparation of this work.

Suslin-Vaserstein all announced important progress. The detailed history is very well portrayed in Bass' paper [2]; the reader may also find there a very complete exposition of the new results. (The material exposed here comes, in fact, from Bass' paper. I am very grateful to Bass for giving me an early copy to study.) Suffice it now to say that we now know that if k is a field, then projective $k[x_1 \dots x_n]$-modules are free if either

1) $n \leq 3$,

2) $n \leq 4$, char $k \neq 2$,

or 3) $n = 5$, k finite, char $k \neq 2$,

and that the best results of this sort, which apply to much more general rings, are due to Suslin and Vaserstein.

Part of the Suslin-Vaserstein proof is so simple and elegant that it deserves to be seen by everyone. We will give it in full, and sketch what remains to be done for case 2) above.

THEOREM 1 (Roitman-Swan, Suslin): Let $A = k[x_1 \dots x_n]$.

a) Projective A-modules of rank $\geq n/2 + 1$ are free.

b) Any unimodular row $(f_1, \dots, f_{r+1})$ of elements of A can be reduced by elementary transformations to $(0, \dots, 0, 1)$ if $r \geq n/2 + 1$. Here a "unimodular row" is a sequence of elements of A which generate the unit ideal.

Part a) follows from part b) because every projective A-module P is stably free: that is, $P \oplus A \oplus \dots \oplus A \cong A^m$ for some m, and part b) allows one to "cancel" the copies of A.

For the proof of b) we need one prerequisite:

Stable Range Theorem (Bass [1]): If B is a noetherian ring of dimension d, and $(c_1, \dots, c_s)$ is a unimodular row of length $s > d + 1$, then there is a unimodular row $(c_1', \dots, c_{s-1}')$ with

$$c_i \equiv c_i' \qquad (\mathrm{mod}\ c_s) \ .$$

Proof of Theorem 1.b): After a change of variables, we may assume that f_{r+1} is monic of degree $p \geq 1$ in x_n. We can now forget all but the last variable and write

$$B = k[x_1 \dots x_{n-1}] \ , \ t = x_n \ , \ A = B[t] \ .$$

Adding multiples of f_{r+1} to the other f_i, we may assume that they have degree $\leq p-1$.

Case 1 $p = 1$: In this case we may write $f_{r+1} = f_{r+1}' + t$, so that $f_1, \dots, f_r$,

$f'_{r+1} \in B$. The substitution $t = -f'_{r+1}$ shows that $(f_1,\ldots,f_r)$ is a unimodular row, so we can make elementary transformations

$$(f_1,\ldots,f_{r+1}) \longrightarrow (f_1,\ldots,f_r, 1) \longrightarrow (0,\ldots,0,1) .$$

Case 2 $p \geq 2$: We will show how to lower p by 1, thus eventually reducing to case 1. Let $\underline{c} = (c_1,\ldots)$ be the vector of coefficients of degree $\leq p-1$ of the polynomials $f_1,\ldots,f_r$. Write (c) for the ideal generated by the c_i.

If $(c) = B$, then the ideal generated by $f_1,\ldots,f_{r-1}, f_{r+1}$ contains a monic polynomial of degree $p-1$; for, every c_i occurs as the leading coefficient of a polynomial of degree $p-1$. (Hint: use f_{r+1} to kill off terms of degree $\geq p$ in $t^i f_j$.) By elementary transformations we can replace f_r by a polynomial monic of degree $p-1$, and interchange f_{r+1} and f_r, lowering p as required.

If $(c) \neq B$, then, since $f_1,\ldots,f_{r-1}$ are congruent to 0 in $B/_{(c)}[t]$, I assert that the ideal (f_r, f_{r+1}) contains an element $g \equiv 1 \pmod{(c)}$, and $g \in B$.

For, since f_{r+1} is monic, $B[t]/(f_r,f_{r+1})$ is a finitely generated B-module. Also, we have $(c)B[t]/(f_r,f_{r+1}) = B[t]/(f_r,f_{r+1})$, so for some element $h \in (c)$, $(1+h)B[t]/(f_r,f_{r+1}) = 0$. Take $g = 1+h \in (f_r,f_{r+1})$. (One could also use a resultant). Thus the row $(\underline{c},g)$ is unimodular, and it has $p(r-1) + 1 \geq 2(r-1) + 1 \geq 2(\frac{n}{2}) + 1 = n + 1$ elements. By the stable range theorem applied to B, there exists a unimodular row $\underline{c}' = (c'_1 \ldots)$, congruent to $\underline{c}$ mod (g). Clearly $\underline{c}'$ will be the row of coefficients of some row of polynomials of degree $\leq p-1$

$$f'_1,\ldots,f'_{r-1}, \text{ with } f'_i \equiv f_i \pmod{f_r, f_{r+1}} .$$

Thus we can reduce by elementary transformations to the case $(c) = B$. //

Note that we could replace n, in the proof, by the "stable range" of B, and that hypotheses on B much weaker than that it is a polynomial ring over a field will allow us to make a change of variables making f_{r+1} monic.

To finish the proof that projective A-modules are free if $n \leq 4$, we need only deal with modules of rank 2, and it suffices to show that the special linear group $SL_3(A)$ is transitive on $Un_3(A)$, the set of unimodular rows of length 3. Now given $f = (f_1,f_2,f_3)$, unimodular, we may regard it as the matrix of a map

$$A^3 \xrightarrow{f} A .$$

Let P_f be the kernel. From the exactness of the Koszul complex

$$0 \longrightarrow \overset{3}{\Lambda} A^3 \longrightarrow \overset{2}{\Lambda} A^3 \xrightarrow{\partial} A^3 \xrightarrow{f} A$$

and the easy computation $\partial(\overset{2}{\Lambda} P_f) = 0$ we get a map

$$\overset{2}{\Lambda} P_f \longrightarrow \overset{3}{\Lambda} A^3 \cong A \,. \qquad \text{(This is in fact an isomorphism).}$$

Such a map may be regarded as an alternating form on P. We thus get a map $Un_3(A) \longrightarrow KSp_0(A)$, the Grothendieck group of projective modules with alternating forms. It turns out that the orbits of $SL_3(A)$ are identified under this map, so we get a map

$$Un_3(A)/_{SL_3 A} \xrightarrow{\varphi} KSp_0 A \ .$$

Recall that for any r, $E_r(A)$ is the subgroup of $SL_r(A)$ generated by the elementary transformations. Theorem 1b asserts that $E_{r+1}(A)$ is transitive on $Un_{r+1}(A)$ for all $r \geq n/2 + 1$. Thus we can apply the following theorem of Vaserstein to $A = k[x_1,\ldots,x_n]$ in the case $n \leq 4$:

Theorem (Vaserstein): For any commutative ring A, if $E_{r+1}(A)$ is transitive on $Un_{r+1}(A)$ for all $r \geq 3$, then φ is injective.

Note that P_f is not only projective, it is stably free; thus the image of φ is in the kernel $W(A)$ of the composite map

$$\begin{array}{ccc} KSp_0(A) & \xrightarrow{\text{forgetful map}} & K_0(A) \\ & & \downarrow \\ & & K_0(A)/\{\text{stably free modules}\} \\ & & \| \\ & & \tilde{K}_0(A) \ . \end{array}$$

Now we want to compute $W(A)$:

Theorem (Karoubi [3]): If $A = B[t]$, and $\frac{1}{2} \in B$, then $W(A) = W(B)$.

Thus, in our case, $A = k[x_1,\ldots,x_4]$, $SL_3(A)$ is transitive on $Un_3(A)$, so $W(A) = 0$ and projectives of rank 2 are free. Thus all projectives over $k[x_1,\ldots,x_4]$ are free.

REFERENCES

1) Bass, H.; K-theory and stable algebra. Publ. I.H.E.S. 22 (1964), 5-60.

2) Bass, H.; Liberation des modules projectifs sur certains anneaux de polynômes. Sem. Bourb. 1974.

3) Karoubi, M.; Periodicité de la K-theorie hermitienne, Proc. Batelle Conf. on Algebraic K-theory, vol. III, Springer Lecture Notes 343, Springer-Varlag, N.Y. 1973.

4) Serre, J.-P.; Faisceaux algebriques coherents. Ann. Math. 61 (1955), 197-278.

II. SOME RECENT PROGRESS ON THE "HOMOLOGICAL CONJECTURES" IN COMMUTATIVE RING THEORY

I. Survey of the situation

Probably the biggest advance in commutative rings in the 1950's was the introduction of homological methods, mostly by Auslander, Buchsbaum, and Serre, leading to the solution to a number of old problems (regular local rings are factorial, etc.). A few of the problems as to the nature of the new homological machines have never been completely resolved, though there has been quite a bit of work and progress. Here are two "central" examples:

a) The Multiplicity Problem: In his 1957 notes "Algebre Locale-Multiplicités", which consolidated and explained the new homological process, Serre gave a homological treatment of intersection multiplicities. He proved, for regular local rings R containing a field (and more generally for "unramified" regular local rings), that if M and N are two modules such that $M \otimes N$ has finite length, then, writing $\dim M$ for $\dim \operatorname{supp} M$, and $\ell(T)$ for the length of an R-module of finite length,

i) $\dim M + \dim N \leq \dim R$

ii) $\chi(M,N) = \sum_{i \geq 0} (-1)^i \ell(\operatorname{Tor}_i(M,N) \geq 0$

iii) $\chi(M,N) = 0$ if and only if $\dim M + \dim N < \dim R$.

(The central case, and that corresponding to classical intersection theory, is that in which $M = R/P$, $N = R/Q$, and P and Q are prime ideals).

The problem that remains is to extend this theorem to all regular local rings, or, most optimistically, to prove that it holds for any local ring R, provided only that p.d. M , the projective dimension of M, is finite.

b) The Rigidity of Tor: Auslander conjectured that, if R is a regular local ring, and if M and N are finitely generated R-modules, such that $\operatorname{Tor}_1(M,N) = 0$, then $\operatorname{Tor}_i(M,N) = 0$ for every $i > 0$; this was subsequently proved by Lichtenbaum. Again, the remaining question is: is this rigidity property true when R is any local ring, assuming p.d. $M < \infty$?

There has been some progress on problem a) since 1957, for example by Malliavin and Hochster and others, and an important step toward the "optimistic" version has recently been taken by Peskine-Szpiro [3]. The biggest progress, however, has been made on our understanding of the consequences and relations of problem b). It is this work that I want to tell you about, especially some new work of Hochster on a conjecture which has among its consequences most of the well-known consequences of b).

The "modern" development of the problem seems to me to begin with the Thesis of

Peskine-Szpiro [2], in which they exhibited a consequence of b), and verified it in a large number of cases. It is as follows:

c) The Intersection property for a local ring R: If M and N are finitely generated R-modules and $M \otimes N$ has finite length, then

$$\text{p.d. } M \geq \dim N .$$

Conjecture c) is actually a consequence of conjecture ai) in the regular, or even Cohen-Macaulay case. To see this, note that c) is vacuous if $\text{p.d. } M = \infty$, while if $\text{p.d. } M < \infty$, then $\text{p.d. } M = \text{depth } R - \text{depth } M$, so c) becomes

$$\text{depth } R \geq \dim N + \text{depth } M .$$

(Here, depth M is the maximal length of an M-sequence. Note that $\text{depth } M \leq \dim M$ always holds -- in fact depth M is bounded by the dimension of any associated prime of M -- and that $\text{depth } R = \dim R$ if R is Cohen-Macaulay.)

One reason for the significance of the intersection property becomes clear if, with Hochster, we note that it is equivalent to an extension of the Krull principal ideal theorem:

d) Homological height conjecture: Let R be a local ring, S an R-algebra, and M a finitely generated R-module. Then $\text{p.d. } M \geq \text{height } (\text{annihilator}_S(S \otimes M))$.

(To get the usual version of the principal ideal theorem, take R to be a localization of $Z[t]$ at a prime containing t, and take $M = R/(t)$.)

Another reason why the intersection property is important is that though it seems to be much easier to check than the rigidity conjecture, its truth would imply most of the known consequences of the rigidity conjecture, such as the following conjectures:

i) (Auslander) If M is a finitely generated module over the local ring R, with

$$\text{p.d. } M < \infty ,$$

and if $x \in R$ is a nonzerodivisor on M, then x is a nonzerodivisor on R.

ii) (Bass) If R is a local ring which possesses a finitely generated module of finite injective dimension, then R is Cohen-Macaulay.

Within the last year, a new step in the direction of these conjectures has been taken by Hochster, and it is his work that I wish to describe here. A further definition is necessary: If M is a (possibly not finitely generated) R-module, then a sequence of elements $(x_1,\ldots,x_n)$ of R is an M-sequence if

i) $(x_1,\ldots,x_n)M \neq M$

ii) x_i is a nonzerodivisor on $M/(x_1,\dots,x_{i-1})M$ for $i = 1,\dots,n$.

Recall that if R is a local ring of dimension n with maximal ideal J, then a system of parameters is a sequence of n elements $x_1,\dots,x_n \in R$ such that the ideal $(x_1,\dots,x_n)$ is J-primary.

Theorem (Hochster): If R is a local ring containing a field, and if $x_1,\dots,x_n$ is a system of parameters in R, then there exists an R-module M such that $x_1,\dots,x_n$ is an M-sequence.

(More generally, this works even if only R_{red} contains a field.) The interest for us of this theorem is clear from the following:

Proposition (Hochster): If R is a local ring, and if for each prime ideal P of R, every system of parameters in R/P is an $\mathfrak{n}$-sequence for some R/P-module $\mathfrak{n}$ then the intersection property holds for R.

Proof: Assume that M and N are finitely generated modules such that $M \otimes N$ has finite length, and let $P \in \operatorname{supp} N$ be such that $\dim R/P = \dim N$; it is easy to see that $M \otimes R/P$ has finite length. Since $P + \operatorname{ann} M$ is primary to the maximal ideal of R, we may choose a system of parameters $(x_1,\dots,x_d)$ for R/P contained in $\operatorname{ann} M$, and let $\mathfrak{n}$ be an R/P module such that $(x_1,\dots,x_d)$ is an $\mathfrak{n}$-sequence.

Note that

$$\begin{aligned}\operatorname{Tor}_0^R(M,\mathfrak{n}) &= M \otimes_R \mathfrak{n} \\ &= M/PM \otimes_{R/P+(x_1,\dots,x_d)} \mathfrak{n}/(x_1,\dots,x_d)\mathfrak{n} \\ &\neq 0 ,\end{aligned}$$

since both tensor factors are $\neq 0$ and the maximal ideal of $R/P + (x_1,\dots,x_d)$ is nilpotent. Thus we may choose $i \geq 0$ maximal with respect to the condition

$$\operatorname{Tor}_i^R(M,\mathfrak{n}) \neq 0 .$$

Since $\operatorname{Tor}_*^R(M,-)$ is annihilated by each x_i, the long exact sequences associated to the sequences

$$0 \longrightarrow \mathfrak{n}(x_1,\dots,x_i)\mathfrak{n} \xrightarrow{x_{i+1}} \mathfrak{n}(x_1,\dots,x_i)\mathfrak{n} \longrightarrow \mathfrak{n}/(x_1,\dots,x_{i+1})\mathfrak{n} \longrightarrow 0$$

show that

$$\operatorname{Tor}_{i+d}^R(M,\mathfrak{n}/(x_1,\dots,x_d)\mathfrak{n} \cong \operatorname{Tor}_i(M,\mathfrak{n}) \neq 0 .$$

Thus p.d. $M \geq i + d \geq d = \dim N$. //

The existence of Macaulay modules can be used for other things besides the intersection property. One nice example comes from the following:

Theorem (Hochster): Let R be a regular local ring, and let $S \supseteq R$ be a finite R-algebra. Suppose that there is an S-module M such that some regular system of parameters for R is an M sequence. Then R is a direct summand of S (as R-modules).

Corollary: If R is a regular local ring containing a field, and $S \supseteq R$ is a finite algebra, then R is a direct summand of S.

(In fact Hochster has given a proof of this corollary which is independent of the existence of Macaulay modules.)

II. Sketch of the construction of Cohen-Macaulay modules

Hochster's construction of Cohen-Macaulay modules is childishly simple. Furthermore, it is easy to show that if Cohen-Macaulay modules exist at all, then the construction does lead to them. Nevertheless, Hochster's proof that the construction does work is quite intricate, and I'll confine myself to describing the more canonical looking (and therefore easier) portion.

A.) The construction: Fix a local ring R and a system of parameters $x_1, \ldots, x_n \in R$. If M is any R-module, and if $m \in M$ is such that

$$x_{k+1} m \in (x_1, \ldots, x_k)M \quad \text{for some } k < n ,$$

we may define a new module M', called a modification of M, by

$$M' = \frac{M \oplus R^k}{R(m, -x_1, -x_2, \ldots, -x_k)} .$$

The natural map $M \longrightarrow M'$ is clearly universal for maps $M \xrightarrow{\varphi} N$ such that $\varphi(m) \in (x_1, \ldots, x_k)N$. Now start with $M = R$, and consider, for each k, a finite set of elements in R which generates

$$((x_1, \ldots, x_k)M : x_{k+1})/(x_1, \ldots, x_k)M .$$

Taking finitely many successive modifications, we may construct a module M_1 and a map $\varphi_1 : R \longrightarrow M_1$ so that

$$\varphi((x_1, \ldots, x_k)M : x_{k+1}) = (x_1, \ldots, x_k)M_1 .$$

Repeating this procedure countably many times, and taking the limit, we will obtain a module M_∞ with the property that for each $k < n$,

$$(x_1, \ldots, x_k)M_\infty : x_{k+1} = (x_1, \ldots, x_k)M_\infty .$$

The elements $x_1, \ldots, x_n$ will be an M_∞-sequence if and only if we have in addition

$$(x_1,\ldots,x_n)M_\infty \neq M_\infty .$$

We can actually improve this assertion considerably:

Proposition: With the notation above, let $\varphi_\infty : R \longrightarrow M$ be the natural map. There exists an R-module N such that $x_1,\ldots,x_n$ is an N-sequence if and only if $\varphi_\infty(1) \notin (x_1,\ldots,x_n)M_\infty$ (in which case $x_1,\ldots,x_n$ is an M_∞- sequence).

Proof: We need only show that if $(x_1,\ldots,x_n)$ is an N-sequence, then $\varphi_\infty(1) \notin (x_1,\ldots,x_n)M_\infty$. Let $e \in N - (x_1,\ldots,x_n)N$, and let $f: R \longrightarrow N$ be defined by $f(1) = e$. By the universal property of modifications, there exists an extension $\psi: M_\infty \longrightarrow N$.

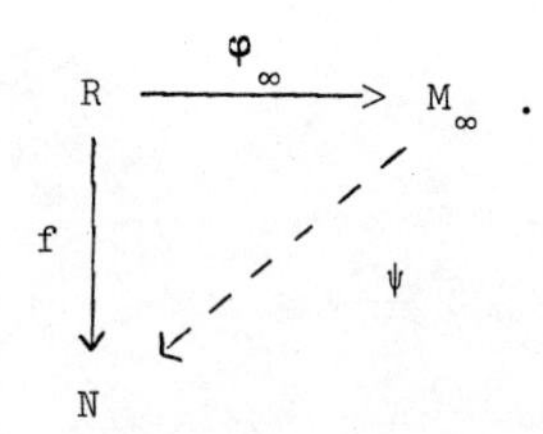

Thus if $\varphi_\infty(1) \in (x_1,\ldots,x_n)M_\infty$, we would have

$$\psi\varphi_\infty(1) = f(1) = e \in (x_1,\ldots,x_n)N ,$$

a contradiction. //

B.) The proof: We now see that to prove the existence of a module M on which a given system of parameters is an M-sequence, it suffices (and is also necessary) to show:

(*) No sequence of modifications

$$R \longrightarrow M_1 \longrightarrow \cdots \longrightarrow M_i \quad (\text{composite } \varphi)$$

has the property that $\varphi(1) \in (x_1,\ldots,x_n)M_i$.

Hochster's proof that this is so for a local ring R containing a field k has two parts: a proof when $\operatorname{char} k = p > 0$, which uses the Frobenius homomorphism, and a reduction to this case. Hochster's paper (see the references) contains a very illuminating and fairly full account of these processes, the second of which involves Artin's Approximation Theorem. Here we will content ourselves with giving the "punch line" -- the case in which R is a complete local domain of characteristic $p > 0$.

The usual proof that any variety maps birationally to a hypersurface can be applied to show that, at least after extending R slightly, we may assume $R \supseteq R_0 \ni x_1,\ldots,x_n$, where R_0 is a Cohen-Macaulay local ring, $x_1,\ldots,x_n$ are an R_0-sequence, and R is a finitely generated R_0-submodule of the quotient field of R_0. Thus there exists an element $c \in R_0$ such that $cR \subseteq R_0$.

We claim that for any integer $t \geq 1$, and any $k < n$,

$$c[(x_1^t,\ldots,x_k^t) : x_{k+1}^t] \subseteq (x_1^t,\ldots,x_k^t) .$$

(Here all ideals are to be thought of as ideals of R.) For, if $ax_{k+1}^t \in (x_1^t,\ldots,x_k^t)$, then $cax_{k+1}^t \in \Sigma_1^k R_0 x_i^t$, so $ca \in \Sigma R_0 x_i^t$, since $x_1^t,\ldots,x_{k+1}^t$ is an R_0-sequence.

Now suppose that

$$*) \qquad R \xrightarrow{\varphi_1} M_1 \xrightarrow{\varphi_2} M_2 \xrightarrow{\varphi_3} \cdots \xrightarrow{\varphi_n} M_n$$

is a sequence of modifications with respect to the system of parameters $x_1^t,\ldots,x_n^t$, for any t. We will show that there are maps $\psi_i : M_i \longrightarrow R$ making the following diagram commute:

$$(**) \qquad \begin{array}{ccccccccc} R & \xrightarrow{\varphi_1} & M_1 & \xrightarrow{\varphi_2} & M_2 & \longrightarrow & \cdots & \xrightarrow{\varphi_n} & M_n \\ \| & & \downarrow{\scriptstyle\psi_1} & & \downarrow{\scriptstyle\psi_2} & & & & \downarrow{\scriptstyle\psi_n} \\ R & \xrightarrow[c]{} & R & \xrightarrow[c]{} & R & \xrightarrow[c]{} & \cdots & \xrightarrow[c]{} & R \end{array}$$

In fact, given $M \xrightarrow{\psi} R$ and an element $m \in M$ such that $x_{k+1}^t m \in (x_1^t,\ldots,x_k^t)$, we have $x_{k+1}^t \psi(m) \in (x_1^t,\ldots,x_k^t)$, and thus $c\psi(m) \in (x_1^t,\ldots,x_k^t)$. If $\varphi : M \longrightarrow M'$ is a modification corresponding to the equation $x_{k+1}^t m \in (x_1^t,\ldots,x_k^t)$, then by the universal property of φ, there exists a map ψ' making the diagram

$$\begin{array}{ccc} M & \xrightarrow{\varphi} & N \\ \downarrow{\scriptstyle\psi} & & \downarrow{\scriptstyle\psi'} \\ R & \xrightarrow[c]{} & R \end{array}$$

commute.

Suppose now that $t = 1$, and $\varphi_n \ldots \varphi_2\varphi_1(1) \in (x_1,\ldots,x_n)$. Raising all the elements occuring in the construction of $*)$ to the p^m power, we see that we get a sequence of modifications

$$*_m) \qquad R \xrightarrow{\varphi_1^{(m)}} M_1^{(m)} \longrightarrow \cdots \xrightarrow{\varphi_n^{(m)}} M_n^{(m)} ,$$

with respect to the system of parameters $x_1^{p^m},\ldots,x_n^{p^m}$. These all have the property that

$$\varphi_n^{(m)} \cdot \ldots \cdot \varphi_1^{(m)} \in (x_1^{p^m},\ldots,x_n^{p^m}) .$$

However, by $**)$ we see that then

$$c^n \cdot 1 \in (x_1^{p^m},\ldots,x_n^{p^m}) .$$

Since this works for every m, we get a contradiction. Thus the modification procedure really does lead to a module M such that $x_1,\ldots,x_n$ is an M-sequence.

References

There are two excellent sources for a more detailed treatment of this material (and for a more complete bibliography!). They are

1) Hochster, M.: "Topics in the Homological Theory of Modules over Commutative Rings" (A.M.S. Regional Conference, June 1974, Lincoln, Nebraska).

2) Peskine, C., and Szpiro, L.: Dimension Projective Finie et Cohomologie Locale. Publ. Math. I.H.E.S., 42 (1973) Paris.

Too recent to be mentioned in the above is the following very interesting note on the intersection conjectures:

3) __________, ____________: Syzygies et Multiplicités. C. R. Acad. Sci., Paris, 27 May 1974.

III. "PERFECT IDEALS OF CODIMENSION 2 ARE DETERMINANTAL"- SOME APPLICATIONS

One of the active areas in commutative algebra is concerned with the structure of finite free resolutions and the consequences of that structure for ideal and module theory. Rather than go into the recent work on free resolutions here, I want to exhibit the archetypal result in this field (Theorem 1, below), and two of its more unusual applications (The reader who wishes a survey of work on free resolutions may find [5] useful.). The first of these applications, a theorem on factoriality, is not so new, but is a kind of mathematical "chestnut." The second represents our best current information on the Zariski-Lipman conjecture.

I): The structure of perfect ideals of codimension 2.

Hilbert seems to have arrived at the idea of classifying ideals in polynomial rings by their projective dimensions as soon as he had proved that their dimensions were finite [5]. He gives (as an example of the use of the Hilbert function) a special case of the following theorem, which gives a generic form for ideals of projective dimension 1 and their resolutions. For convenience we will work not with an ideal I in a ring R, but with the cyclic module R/I. We will always assume that R is a noetherian local ring, and that modules are finitely generated.

Theorem 1: Let

$$\underline{F} : 0 \longrightarrow F_2 \xrightarrow{\varphi_2} F_1 \xrightarrow{\varphi_1} R \longrightarrow R/I \longrightarrow 0$$

be an exact sequence of R-modules, with F_1 and F_2 free, and rank $F_1 = n$. Then rank $F_2 = n-1$, and φ_1 can be factorized as

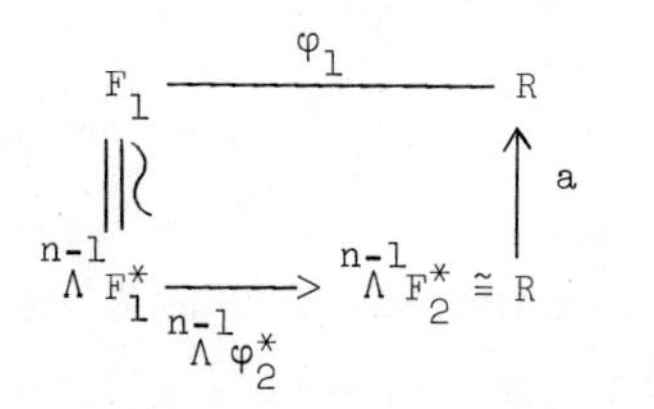

where $a \in R$ is a nonzerodivisor.

If we assume that I has height at least 2 (and thus exactly 2, since p.d. $R/I \geq$ ht I always holds), it follows by Krull's principal ideal theorem that a is a unit, which we may ignore. If we choose a basis for F_1, the map $\overset{n-1}{\Lambda}\varphi_2^*$ will have as its n coordinates the $n-1 \times n-1$ subdeterminants of φ_2. Remembering that an ideal I is said to be perfect if ht I = p.d. R/I, we can restate Theorem 1 more colloquially as "Perfect ideals of height 2 are determinantal."

Theorem 1 has a very useful converse:

Theorem 2: Let $\varphi: R^{n-1} \longrightarrow R^n$ be a map, and let

$$I = \text{image}\ (\overset{n-1}{\Lambda}\varphi^*: \overset{n-1}{\Lambda}R^{n*} \longrightarrow \overset{n-1}{\Lambda}R^{n-1*} \cong R)\ ,$$

the "ideal of $n-1 \times n-1$ subdeterminants of φ." Suppose I is contained in the maximal ideal of R. If $\text{depth}_I R \geq 2$, then I is perfect, and $\text{depth}_I R = 2$. Furthermore, if ψ is defined to be the composite

$$\underbrace{R^n \cong \overset{n-1}{\Lambda}R^{n*} \xrightarrow{\overset{n-1}{\Lambda}\varphi^*} \overset{n-1}{\Lambda}R^{n-1*} \cong R}_{\psi}\ ,$$

then the sequence

$$*)\qquad 0 \longrightarrow R^{n-1} \xrightarrow{\varphi} R^n \xrightarrow{\psi} R \longrightarrow R/I \longrightarrow 0$$

is exact. If the elements of φ are in the maximal ideal of R, then $*)$ is the minimal free resolution of R/I.

Here we are using the familiar isomorphisms $\overset{k}{\Lambda} R^n \cong \overset{n-k}{\Lambda} R^{n*}$, which are defined by the choice of a basis for R^n though, up to multiplication by a unit, it is independent of this choice. The fact that $*$) is a complex comes from the fact that the ith component of the map $\psi\varphi$ is the $n \times n$ determinant formed from the $n \times n-1$ matrix of φ by repeating the ith column. It should be remarked that our various hypotheses—the local and noetherian properties of R and the freeness of the F_i—could be eliminated; the theorem really concerns arbitrary finitely presented ideals of projective dimension 1.

Theorems 1 and 2 were proved in essentially the form given here by L. Burch [4]; the shortest proof is surely that of Kaplansky [8]. For various generalizations of the theorem to longer free resolutions, see [3]. Despite the existence of Kaplansky's bare-hands proof, I would like to sketch a slightly less elementary proof which, however, generalizes better. The point is that any good criterion for the exactness of a finite complex will yield a proof of Theorems 1 and 2. For, to prove Theorem 2, it suffices to show that the complex

$$0 \longrightarrow R^{n-1} \xrightarrow{\varphi} R^n \xrightarrow{\psi} R$$

considered there is exact. To prove Theorem 1, on the other hand, it would be enough to show that the dual complex

$$0 \longrightarrow R^* \xrightarrow{\psi^*} R^{n*} \xrightarrow{\varphi_2^*} R^{n-1*} \quad ,$$

(where ψ is constructed from $\varphi_2 = \varphi$ as in Theorem 2) is exact; for then the comparison theorem for resolutions would yield a map $a: R^* \longrightarrow R^*$ making the diagram

$$\begin{array}{ccccccc}
0 \longrightarrow & R^* & \xrightarrow{\varphi_1^*} & R^{n*} & \xrightarrow{\varphi_2^*} & R^{n-1*} \\
& a\downarrow & & \| & & \| \\
0 \longrightarrow & R^* & \xrightarrow[\psi^*]{} & R^{n*} & \xrightarrow[\varphi_2^*]{} & R^{n-1*}
\end{array}$$

commute. (It is then easy to see that a is induced by multiplication by a non-zero-divisor).

To state a suitable exactness criterion, we need two preliminaries: Let $\varphi: F \longrightarrow G$ be a map of free modules. Define $\operatorname{rank} \varphi = r$ to be the largest integer i such that $\overset{i}{\Lambda} \varphi \neq 0$, and let $I(\varphi)$ be the ideal generated by the $r \times r$ minors of a matrix representing φ. (Invariantly put, $I(\varphi)$ is the image of the natural map induced by $\overset{r}{\Lambda} G^* \otimes \overset{r}{\Lambda} F \longrightarrow R$ induced by φ.) We can now state an exactness criterion which is ample for the proofs of Theorems 1 and 2; it is a slightly special case of the main result of [2].

Theorem: Let $F: 0 \longrightarrow F_n \xrightarrow{\varphi_n} F_{n-1} \longrightarrow \cdots \longrightarrow F_1 \xrightarrow{\varphi_1} F_0$ be a complex of finitely generated free modules over a noetherian ring R. The complex F is exact if and only if

1) $\operatorname{rank} \varphi_r + \operatorname{rank} \varphi_{r+1} = \operatorname{rank} F_k \quad (k = 1,\ldots,n)$

and 2) for each k, $\operatorname{depth}_{I(\varphi_k)} R \geq k$.

(Note that by convention, $\operatorname{depth}_R R = \infty$.)

II): Two Applications.

Several people at this conference have already spoken about work on "subvarieties of low codimension," and the two theorems just described are important for some of this work. Another area of usefulness of the Hilbert-Burch theorems which has not yet been mentioned is deformation theory. But in this talk I want to speak of two other, somewhat more novel applications.

A) The local factoriality of surfaces in 3-space (after Andreotti-Salmon [1]).

We will say that an element f of a local ring R is a determinant if f is the determinant of some $n \times n$ matrix, $n \geq 2$, with entries in the maximal ideal of R. Equivalently, f could be the determinant of an $n \times n$ matrix whose $n-1 \times n-1$ minors do not generate the unit ideal of R.

Theorem: Let R be a 3-dimensional regular local ring, and let $f \in R$ be a prime element. Then $R/(f)$ is factorial if and only if f is not a determinant.

Proof: The factoriality of a 2-dimensional local domain S is equivalent to the statement that every unmixed ideal of dimension 1 of S is principal. Suppose that f is a determinant -- say $f = \det A$, where A is an $n \times n$ matrix whose entries are in the maximal ideal, $n \geq 2$. Let B be the matrix obtained from A by omitting the first column. By the Laplace expansion of $\det A$ along the first column, f is an element of the ideal I of $n-1 \times n-1$ minors of B. Moreover, since the elements of that first column are in the maximal ideal J of R,

$$f \in JI ;$$

in particular $I \neq (f)$, and since (f) is a prime, we see that $\operatorname{depth}_I R \geq 2$.

Theorem 2 now shows that the free resolution of R/I over R is

$$0 \longrightarrow R^{n-1} \xrightarrow{B} R^n \xrightarrow{\psi} R \longrightarrow R/I \longrightarrow 0 ,$$

where ψ is obtained from $B = \varphi$ as in Theorem 2. Since the entries of B are in the maximal ideal, I is minimally generated by $n \geq 2$ elements, and since $f \in JI$, $I/(f)$ is also minimally generated by n elements. Since I is perfect of codimension 2 in R, it is unmixed, and $I/(f)$ is of dimension 1 in $R/(f)$. By the remark at the beginning, the 2-dimensional domain $R/(f)$ is not factorial.

Conversely, suppose that f is not a determinant in R; we will show that if $I/(f)$ is an unmixed, one-dimensional ideal in $R/(f)$, then I can be generated by one element in addition to f. Let $f, g_1, \ldots, g_k$ be a set of generators for I. Since I is unmixed of dimension 1 in R, it is perfect of codimension 2; applying Theorem 1, and suppressing a (which must be a unit in this case) we see that I has a resolution

$$0 \longrightarrow R^k \xrightarrow{\varphi_2} R^{k+1} \xrightarrow{\varphi_1} R \longrightarrow R/I \longrightarrow 0 \quad ,$$

and that the determinant of some $k \times k$ submatrix A of φ_2 is f. Since f "is not a determinant," some $k-1 \times k-1$ minor of φ is a unit. Thus after a change of basis in R^k and R^{k+1}, the matrix of φ_2 has the form

$$\left(\begin{array}{ccccc|cc} 1 & & & & & 0 & 0 \\ & 1 & & 0 & & \vdots & \vdots \\ & & \ddots & & & \vdots & \vdots \\ & & & 1 & & & \\ & 0 & & & 1 & 0 & 0 \\ \hline 0 & \ldots & & & 0 & f & g \end{array}\right) ,$$

(the upper left block being $(k-1) \times (k-1)$), and $I = (f,g)$ is generated by one element, g, modulo f. //

B) The Zariski-Lipman conjecture for hypersurfaces (after Scheja-Storch [11,12]).

The Zariski-Lipman conjecture is the following:

Let R be a local domain essentially of finite type over a field k of characteristic 0. If the module $\mathrm{Hom}_R(\Omega_{R/k}, R)$ of k-derivations from R to R, is a free module, then R is regular. (Of course R is regular if $\Omega_{R/k}$ is free). Lipman proved in his Thesis [9] that under the hypothesis of the conjecture, R is at least normal. Since that time, Moen [10] gave a proof of the conjecture in case R is both the coordinate ring of a cone and a complete intersection, and his proof was subsequently simplified by Hochster [7]. (This setup is rather special, since a smooth cone is just a linear space!) Most recently, Scheja and Storch have proved the conjecture under the hypothesis that R is the coordinate ring of a hypersurface (and under slightly less restrictive hypotheses than those in the conjecture as given above). We will give a sketch of their proof.

We will assume for simplicity that the residue class field of R is k, and we will write

$$R = S/(f)\ ,\ S = k[x_1,\dots,x_n]_J\ ,$$

where J is the maximal ideal $(x_1,\dots,x_n)$. The map of k-algebras $S \longrightarrow R$ gives rise to an exact sequence

$$(f)/(f^2) \longrightarrow \Omega_{S/k} \otimes R \longrightarrow \Omega_{R/k} \longrightarrow 0\ .$$

Since $f/(f^2)$ is a free R-module of rank 1, and $\Omega_{S/k}$ is free of rank n on the elements dx_i, we may rewrite this as

$$*)\qquad R \xrightarrow{\begin{pmatrix} \frac{\partial f}{\partial x_1} \\ \cdot \\ \cdot \\ \cdot \\ \frac{\partial f}{\partial x_n}\end{pmatrix}} R^n \longrightarrow \Omega_{R/k} \longrightarrow 0\ .$$

By hypothesis, $\Omega^*_{R/k} = \mathrm{Hom}(\Omega_{R/k},R)$ is free, and since R is "generically smooth of dimension $n-1$", we must have $\Omega^*_{R/k} \cong R^{n-1*}$. Thus, dualizing *), we obtain an exact sequence

$$0 \longrightarrow R^{n-1*} \xrightarrow[\varphi]{} R^{n*} \xrightarrow[(\frac{\partial f}{\partial x})]{} R\ .$$

We may now apply Theorem 1 to this exact sequence. By Lipman's theorem, R is normal, and thus nonsingular in codimension 1 (it is easy to prove this directly), so the depth of the ideal generated by $(\frac{\partial f}{\partial x_i})$ in R is at least 2. Thus the element **a** of Theorem 1 must be a unit, and we ignore it. The conclusion of Theorem 1 is now that $\frac{\partial f}{\partial x_i}$ is (up to sign) the determinant obtained from φ be removing the ith row.

We now choose any matrix $\tilde{\varphi}$ with entries in S which reduces to φ modulo f, and let Δ_i be the determinant obtained from $\tilde{\varphi}$ by omitting the ith row. Since $\frac{\partial f}{\partial x_i} = \pm \Delta_i \pmod f$, we have

$$\#)\qquad df \in (\Delta_1,\dots,\Delta_n,\ f)\ \Omega_{S/k}\ .$$

We assert that #) implies that $f \in \sqrt{(\Delta_1,\dots,\Delta_n)}$. The condition # is stable under "base changes" $S \xrightarrow{\alpha} S'$, and if $f \notin \sqrt{(\Delta_1,\dots,\Delta_n)}$, we could choose S' to be a formal power series ring in one variable z, and α a map such that

$$\alpha((\Delta_1,\dots,\Delta_n)) = 0 \qquad \alpha(f) \neq 0\ .$$

However, we would then get

$$df \in f\,\Omega_{S'/k} \qquad (\Omega_{S'/k} \text{ is the module of } \underline{\text{continuous}} \text{ differentials!})$$

which is a contradiction because the order of the power series $\frac{df}{\partial z}$ is lower than the order of f. Thus $f \in \sqrt{(\Delta_1,\ldots,\Delta_n)}$.

If now $R = S/f$ is not regular, the ideal generated by the elements $\frac{\partial f}{\partial x_i}$ must be in the maximal ideal. Thus the $k-1 \times k-1$ subdeterminants of φ_1 and thus of $\tilde{\varphi}$, are in the maximal ideal. By Theorem 2, the height of the ideal $(\Delta_1,\ldots,\Delta_n)$ is ≤ 2. Since $f \in \sqrt{(\Delta_1,\ldots,\Delta_n)}$, the height in R of $\sqrt{(\Delta_1,\ldots,\Delta_n)}/(f) = \sqrt{\left(\frac{\partial f}{\partial x_1},\ldots,\frac{\partial f}{\partial x_n}\right)}/f$ is ≤ 1, contradicting the fact that, since R is normal, it is nonsingular in codimension 2. Thus R must be regular. //

REFERENCES

1) Andreotti, A., and Salmon, P.; Annelli con unica decomponibilita in fattori primi ed un problema di intersezioni complete. Monatsh. für Math. 61 (1957), 97-142.

2) Buchsbaum, D.A., and Eisenbud, D.; What makes a complex exact?, J. Alg. 25 (1973), 259-268.

3) ______________________; Some structure theorems for finite free resolutions, Adv. in Math. 12 (1974), 84-139.

4) Burch, L.; On ideals of finite homological dimension in local rings, Proc. Cam. Phil. Soc. 64 (1968), 949-952.

5) Eisenbud, D.; A survey of some results on finite free resolutions. Proceedings of the International Congress of Mathematicians, 1974.

6) Hilbert, D.; Über die Theorie der algebraischen formen, Math. Ann. 36 (1890), 473-534.

7) Hochster, M.;

8) Kaplansky, I.; "Commutative Rings," Allyn and Bacon, Boston, 1970.

9) Lipman, J.; Face derivation modules on algebraic varieties, Am. J. Math. 87 (1965), 874-898.

10) Moen,

11) Scheja, G., and Storch, U.; Über differentielle Abhängigkeit bei idealen analytischer algebren, Math. Z. 114 (1970), 101-112.

12) ____________________; Differentielle Eigenschaften der Lokaliserungen analytischer Algebren. Math. Ann. 197 (1972), 137-170.

Proceedings of Symposia in Pure Mathematics
Volume 29, 1975

EQUIVALENCE RELATIONS ON ALGEBRAIC CYCLES
and
SUBVARIETIES OF SMALL CODIMENSION

Robin Hartshorne[1]

CONTENTS:

PART I. EQUIVALENCE RELATIONS ON ALGEBRAIC CYCLES

§1. Introduction.

In studying an algebraic variety X, which we assume is non-singular and projective over an algebraically closed field k, it is natural to ask about its subvarieties. We make a group out of the set of subvarieties by taking formal linear combinations of them. Thus a <u>cycle</u> Z of codimension r on X is defined to be an element of the free abelian group generated by the irreducible subvarieties of codimension r of X. We denote by $C^r(X)$ the group of all cycles of codimension r on X.

This group is too big to be interesting, so we are led to introduce certain equivalence relations on the set of cycles, related to the geometry

---

AMS(MOS) subject classifications (1970). Primary 14-02, 14C10, 14C25, 14F25, 14K30, 14M05; secondary 14C15, 14C20, 14C30, 14M10, 14M15, 32J25.

[1]Partially supported by NSF Grant GP-37492X.

of X. We will only consider equivalence relations for which the set of cycles equivalent to 0 is a subgroup of $C^r(X)$. Then the quotient space will be a group, and we can ask for the structure of this group.

For example, consider the case $r = 1$. A cycle of codimension 1 is called a divisor, so $C^1(X)$ is the group of all divisors on X. Two divisors D_1 and D_2 are linearly equivalent if there is a rational function f on X such that $D_1 - D_2 = (f)_0 - (f)_\infty$, where $(f)_0$ denotes the divisor of zeros of f and $(f)_\infty$ denotes the divisor of poles of f. Let $C^1_{lin}(X)$ denote the group of divisors linearly equivalent to 0. Then the quotient group $C^1(X)/C^1_{lin}(X)$ is the group of linear equivalence classes of divisors on X, also called the Picard group of X, denoted Pic X.

Two divisors D_1 and D_2 are algebraically equivalent if there is an algebraic family of divisors $D(t)$, parametrized by an irreducible variety T, and two points $t_1, t_2 \in T$ such that $D_1 = D(t_1)$ and $D_2 = D(t_2)$. By an algebraic family parametrized by T we mean a divisor D on $X \times T$, which contains no fibres $X \times \{t\}$ for any $t \in T$. Then for any $t \in T$, we denote by $D(t)$ the divisor on X obtained by intersecting D with the fibre $X \times \{t\}$. In taking this intersection we must count multiplicities of the components properly (see remarks on intersection theory below).

We denote by $C^1_{alg}(X)$ the group of divisors algebraically equivalent to 0. One can show easily that linear equivalence implies algebraic equivalence, so $C^1_{alg} \supseteq C^1_{lin}$.

Now in the case of divisors, which we are considering here, we have good results about the structure of these quotient groups.

In the first place, $C^1(X)/C^1_{alg}(X)$ is a finitely generated abelian group, called the Néron-Severi group of X. In case $k = \mathbb{C}$, one can see this easily, by using the exponential sequence

$$0 \to \mathbf{Z} \to \mathcal{O} \to \mathcal{O}^* \to 0$$

for the complex topology on X, where $\mathcal{O}$ denotes the sheaf of germs of holomorphic functions, and $\mathcal{O}^*$ its group of units. From this sequence we obtain an exact sequence of cohomology

$$\cdots \to H^1(X,\mathbb{Z}) \to H^1(X,\mathcal{O}) \to H^1(X,\mathcal{O}^*) \to H^2(X,\mathbb{Z}) \to \cdots .$$

Now $H^1(X,\mathcal{O}^*)$ can be identified with Pic X, and the image of $H^1(X,\mathcal{O})$ becomes the subgroup C^1_{alg}/C^1_{lin} of Pic X. Thus C^1/C^1_{alg} is a subgroup of $H^2(X,\mathbf{Z})$, which is a finitely generated abelian group because X, as a topological space, is a compact polyhedron.

In the case of an arbitrary base field, one shows that C^1/C^1_{alg} is finitely generated by a study of the group of rational points on abelian varieties [35].

The second important fact about the structure of the divisor class group is that there is an abelian variety A, called the Picard variety of X, such that C^1_{alg}/C^1_{lin} is isomorphic to the group of k-rational points of A. Furthermore, this isomorphism is natural, in the sense that there is a "universal family" D of divisors on X, parametrized by A, such that for each rational point $a \in A$, $D(a)$ is the divisor corresponding to a under the above isomorphism.

In the complex case, one can construct the Picard variety of X as the complex torus $H^1(X,\mathcal{O})/H^1(X,\mathbb{Z})$, using the same exponential sequence above. In the abstract case there are several ways to construct the Picard variety, all fairly sophisticated.

In sum, one can say that the group of divisor classes modulo linear equivalence is well understood.

Coming back to the case of cycles of higher codimension, it is not hard to define some similar equivalence relations. Linear equivalence is replaced by rational equivalence, which turns out to be the same as linear equivalence in the case of codimension 1. Two cycles Z_1 and Z_2 of codimension r on X are rationally equivalent if there is a cycle Z on $X \times \mathbb{A}^1$, which intersects each fibre $X \times \{t\}$ in something of codimension r, and such that Z_1 and Z_2 are obtained respectively by intersecting Z with the fibres $X \times \{0\}$ and $X \times \{1\}$.

Note that this is in fact an equivalence relation: To show that $Z_1 \sim Z_1$, take $Z = Z_1 \times \mathbb{A}^1$. If $Z_1 \sim Z_2$, to show that $Z_2 \sim Z_1$, use an automorphism of $\mathbb{A}^1$ which interchanges 0 and 1. If $Z_1 \sim Z_2$, and $Z_2 \sim Z_3$, to show that $Z_1 \sim Z_3$, let $Z \subseteq X \times \mathbb{A}^1$ give the equivalence $Z_1 \sim Z_2$, and let Z' give the equivalence $Z_2 \sim Z_3$. Then take the cycle $Z + Z' - Z_2 \times \mathbb{A}^1$ on $X \times \mathbb{A}^1$, to get the equivalence $Z_1 \sim Z_3$. Note also that the cycles rationally equivalent to 0 form a subgroup.

To define algebraic equivalence of cycles, replace $\mathbb{A}^1$ and the points 0, 1 above by any irreducible curve and any two points. For divisors, this is the same as the previous definition, since any two points on an irreducible variety can be joined by an irreducible curve.

If $C^r(X)$ denotes the group of all cycles of codimension r on X, we let C^r_{alg} denote those algebraically equivalent to 0, and C^r_{rat} those rationally equivalent to 0.

In contrast to the case of divisors, we know very little about the groups of cycle classes modulo these equivalence relations for codimension > 1. For example, we do not know whether $C^r(X)/C^r_{alg}(X)$ is a finitely generated abelian group (although we do know that it is countable, because of the existence of the Chow variety parametrizing cycles on X). This problem is known as the problem of the "finiteness of the base".

One is tempted to hope that cycle classes of higher codimension will behave like divisors, and one can make numerous conjectures along these lines. However, I feel that the evidence so far available points rather in the other direction: namely that one should expect new and different phenomena in studying cycles of higher codimension. Therefore I would like to concentrate in these talks on what we do know in certain special cases, and let you draw your own conclusions about what to expect in general.

In §2 we will discuss the group of 0-dimensional cycles modulo rational equivalence, following work of Severi, Mumford, Mattuck, Roitman, Bloch, and Lieberman. In particular, we will see, as first shown by Mumford, that the group C_{alg}/C_{rat} can be much bigger than the group of rational points on any abelian variety.

In §3 we will discuss the connections between algebraic equivalence, homological equivalence, and numerical equivalence. In particular, we will see, following Griffiths, that the group C_{hom}/C_{alg} can have elements of infinite order. This represents another divergence from the case of divisors.

In §4 we will discuss the search for an "intermediate Jacobian variety", which should be an abelian variety whose rational points correspond to a suitable quotient group of the group C_{alg}/C_{rat}. In the case of a non-singular cubic 3-fold X in $\mathbb{P}^4$, there is an abelian variety which arises naturally in studying cycles of dimension 1 on X, and which plays the role of an intermediate Jacobian. The study of this abelian variety has led to the proof that X, which is unirational, is not a rational variety (Griffiths and Clemens in the case $k = \mathbb{C}$; Mumford and Murre in the abstract case).

We do not plan to discuss Grothendieck's "standard conjectures" and the theory of motives, which have been amply treated in the papers of Grothendieck [15], Kleiman [29, 33], and Manin [41].

Nor will we say anything about the connections with algebraic K-theory which have been treated in recent works of Bloch, Gersten, and Quillen.

In closing this introduction, let me say a few words about intersection theory. Intersection theory has two parts. One is to assign the correct multiplicity to each irreducible component of an intersection of cycles $Z \cap Z'$, in the case where Z and Z' intersect <u>properly</u>, which means that every irreducible component of the intersection has codimension equal to the sum of the codimensions of Z and Z'. This has been done e.g. by Weil [64] or Serre [59]. The other part of intersection theory is to show that given any two cycles Z and Z', one can move one of them to an equivalent cycle, for a suitable equivalence relations, so that they then intersect properly. This is "Chow's moving lemma", which depends heavily on a projective embedding (see Chevalley [10] or Roberts [54]).

Putting these together, one can show that the intersection of any two cycle classes modulo rational equivalence is well-defined, and this product makes the graded group $A(X) = \bigoplus_r C^r(X)/C^r_{rat}(X)$ into an associative, commutative ring, called the <u>Chow ring</u> of X.

For a general reference on equivalence relations on algebraic cycles, see Samuel [58] or Kleiman [31].

§2. <u>0-cycles modulo rational equivalence</u>.

Most of the results of this section are found in three papers of Roitman [55, 56, 57], which extend and generalize the results of Mumford [47].

Let X be a non-singular projective variety over $\mathbb{C}$. We denote by $B(X)$ the group of 0-dimensional cycles on X, modulo rational equivalence. If $Z = \Sigma n_i P_i$ is a 0-cycle, where the P_i are points of X, we define the <u>degree</u> of Z to be Σn_i, which is an integer. This defines a homomorphism $B(X) \to \mathbb{Z}$, which is surjective, and whose kernel $B_0(X)$, the cycles of degree 0, coincides with the group of 0-cycles algebraically equivalent to 0, modulo rational equivalence.

It is often convenient to regard an <u>effective</u> 0-cycle $Z = \Sigma n_i P_i$, i.e. one where all the $n_i > 0$, as a point on the nth symmetric product $S^n(X)$ of X, where $n = \deg Z$. Then by taking the associated rational equivalence class, we obtain a map $\gamma: S^n(X) \to B(X)$.

<u>Proposition 2.1</u>. If Z is an effective cycle of degree n, then the set of all effective cycles of degree n, rationally equivalent to Z (which is a fibre of the map γ) is a countable union of closed subsets of $S^n(X)$.

Note that even if Z and Z' are rationally equivalent effective cycles of degree n, there may not be a rational equivalence between them involving only effective cycles. In any case, there will be an integer r, and morphisms $f: \mathbb{A}^1 \to S^{n+r}(X)$ and $g: \mathbb{A}^1 \to S^r(X)$, such that $Z = f(0) - g(0)$ and $Z' = f(1) - g(1)$. The idea of the proof of the proposition is that the integer r ranges over a countable set. Furthermore, the image of $f \times g$ is a rational curve in $S^{n+r}(X) \times S^r(X)$, and the set of curves in any variety is parametrized by a countable union of irreducible algebraic varieties (the Chow variety). Thus the possible Z form a countable union of closed subsets of $S^n(X)$.

Roitman deals systematically with subset of an algebraic variety which are countable unions of closed sets, which he calls c-closed. A c-closed subset has a unique decomposition into a union of countably many irreducible closed subsets. We can define the dimension of a c-closed subset as the maximum dimension of its irreducible components. This is useful, because now we can define the dimension of the image $\gamma(S^n(X)) \subseteq B(X)$, even though it is not an algebraic variety, as

$$d_n = \dim S^n(X) - \min \{\dim \text{ of a fibre of } \gamma\}.$$

Because of the theorem on the dimension of the fibres of a morphism, this notion coincides with the usual dimension in case $\gamma(S^n(X))$ has a structure of algebraic variety. In particular, if $B_0(X)$ is the set of rational points on an abelian variety A, then we would have $d_n \leqslant \dim A$ for all n. This is the case if X is a curve. Then A is the Jacobian variety of X, which has dimension g, the genus of X, and we have $d_n = n$ for $n \leqslant g$, $d_n = g$ for all $n \geqslant g$.

This suggests that we introduce the following terminology in the general case: we say that $B(X)$ is finite-dimensional if the set of integers d_n is bounded, otherwise $B(X)$ is infinite-dimensional.

We can now state Mumford's result [47], which is that if X is a surface with geometric genus $p_g > 0$, then $B(X)$ is infinite-dimensional. In particular, $B_0(X)$ could not be isomorphic to the set of rational points on any algebraic variety.

Instead of proving Mumford's result directly, we give Roitman's generalizations.

Theorem 2.2. Let X be a non-singular projective variety over $\mathbb{C}$. Then there are integers $d(X), j(X) \geqslant 0$, and an integer n_0, such that for all $n \geqslant n_0$, we have

$$d_n = n \cdot d(X) + j(X).$$

Note that $d(X) \leqslant \dim X$, and that $d(X) = 0$ if and only if $B(X)$ is finite-dimensional, in the above terminology.

To prove this theorem, Roitman considers an effective cycle $Z \in S^n(X)$, and defines a subset of X,

$$V_Z^n = \{P \in X \mid \exists Z' \in S^{n-1}(X) \text{ with } P + Z' \sim Z\}.$$

Then V_Z^n is a c-closed subset of X, so it has a codimension δ_Z^n. Let δ_n be the maximum value of δ_Z^n as Z ranges over $S^n(X)$. Then one shows easily that $\delta_n \geqslant \delta_{n+1} \geqslant \ldots$, hence there is an integer n_0 such that $\delta_n = \delta_{n+1}$ for all $n \geqslant n_0$. Let $d(X)$ be this stable value of δ_n. Using the definition of δ_n one shows that

$$d_{n+1} = d_n + \delta_{n+1}$$

for all n. Since δ_{n+1} is eventually constant, the statement of the theorem follows. (We refer to Roitman's papers for more details of this and other proofs.)

Theorem 2.3. Suppose for some integer $q \geqslant 2$ that $H^0(X,\Omega^q) \neq 0$, i.e., that there is a non-zero global q-form on X. Then $d(X) \geqslant q$.

To prove this theorem, the basic idea, which goes back via Mumford to Severi, is that a q-form on X induces a q-form on $S^n(X)$, whose restriction to any orbit of γ must be zero. On the other hand, one shows by linear algebra that a q-form which is zero on a sufficiently large subvariety must be zero itself. We refer to Roitman's paper for details.

Now let us look at some examples. If X is a surface with $p_g = \dim H^0(X,\Omega^2) > 0$, then $d(X) = 2$, so $j(X) = 0$, so the minimum dimension of an orbit of $\gamma\colon S^n(X) \to B(X)$ is zero. Thus the typical orbit is a countable set of points. It would be interesting to describe these orbits more explicitly, for example in the case of an abelian variety of dimension 2.

We can show in a special case that an orbit of γ, while necessarily c-closed, need not be closed (this example is due to Samuel [58]). Let E be an elliptic curve, and let X be the Kummer variety, which is obtained as the non-singular model of the quotient by the automorphism $x \mapsto -x$ of the abelian variety $E \times E$. Then $p_g(X) = 1$, so the general orbit of $\gamma\colon X \to B(X)$ is a countable set of points. In particular, not all points of X are rationally equivalent. On the other hand, X contains infinitely many rational curves, namely the images of the sub-abelian varieties of dimension 1 of $E \times E$, and for each homomorphism $f\colon E \to E$, e.g. multiplication by n, the graph $\Gamma_f \subseteq E \times E$ is a sub-abelian variety. These rational curves in X all touch the blow-up of the image of $(0,0)$ in X, which is also a rational curve. Thus all the points of all these curves are rationally equivalent to each other. Thus we have in evidence at least one orbit of γ which contains infinitely many curves. Hence that orbit is not closed.

Looking at higher-dimensional varieties for a moment, we see that if X is an abelian variety of dimension $n \geq 2$ then $d(X) = n$. Taking a product with a projective space $\mathbb{P}^m$, we have $d(X \times \mathbb{P}^m) = n$. Thus for varieties of dimension n, $d(X)$ can take on all values between 0 and n except possibly for 1, and Roitman shows in fact $d(X) = 1$ is impossible.

If X is a ruled surface over a curve C of genus g, then any point of X can be moved rationally onto a section of X over C. Thus $B_0(X)$ is just the group of rational points on the Jacobian variety $J(C)$. Thus we have $d(X) = 0$, $j(X) = g$.

An open question, conjectured by Bloch [8], on the basis of some calculations in algebraic K-theory, is whether all surfaces with $p_g = 0$ have $d(X) = 0$. In this connection, Roitman shows

<u>Proposition 2.4</u>. Let X be a variety (of any dimension) with $d(X) = 0$. Then there is a morphism $f: X \to C$ of X to a non-singular curve C such that $B_0(C)$ is identified with the rational points on the Jacobian $J(C)$ (which is necessarily also the Albanese variety of X).

To prove this result, Roitman considers the natural map of X to its Albanese variety $\text{Alb } X$. The image of X must have dimension ≤ 1, because otherwise X would have a non-zero q-form for some $q \geq 2$, contradicting (2.3). He takes C to be the normalization of the image of X, and then proves that $B_0(X)$ is $J(C)$.

As a corollary, one finds that if X is a surface with $d(X) = 0$ (hence $p_g = 0$) and irregularity $q > 1$, then X is a ruled surface (this is a general theorem on the classification of ruled surfaces [61, IV §3, Thm 5]. So the only surfaces for which we have not yet determined $d(X)$ are the surfaces $p_g = 0$, $q = 1$ (which are quotients of certain products of curves by certain finite groups), and the surfaces with $p_g = q = 0$ (this latter class contains the Enriques surface, among others).

M. Artin has checked that for the Enriques surface, $d(X) = 0$. Since $q = 0$, we deduce from (2.4) that any two points of X are rationally equivalent to each other. This provides a counterexample to an old question of Severi, who asked if a surface, on which all points were rationally equivalent, was necessarily rational.

We close this section with an example, due to Bloch and Lieberman, of a surface with $p_g = 0$, $q = 1$, for which $d(X) = 0$.

Let E and F be elliptic curves. Let $\eta \in E$ be a point of order 2, and let $G = \mathbb{Z}/2\mathbb{Z}$ act on $E \times F$ by $(x,y) \mapsto (x+\eta,-y)$. Let $X = E \times F/G$. Then X is a surface with $p_g = 0$, $q = 1$. Let C be the elliptic curve $E/(\eta)$. Then there is a projection $f: X \to C$, whose fibre is isomorphic to F. Furthermore, f has a section $s: C \to X$ defined by $[x] \to [(x,0)]$. Here we use [ ] to denote passage to the quotient by the action of G on E (resp. $E \times F$).

We will show that $B_0(X)$ is isomorphic to the group of rational points on C. For this, it is sufficient to show, for any $x \in E$, that the homomorphism $\varphi: F \to B_0(X)$ defined by $y \mapsto [(x,y) - (x,0)]$ is zero. Since F is a divisible group, it will be sufficient to show that $4\varphi = 0$. Now $4[(x,y)-(x,0)]$ is the image in X of the cycle

$$2((x,y) + (x+\eta,-y) - (x,0) - (x+\eta,0))$$

on $E \times F$. But this is equal to

$$(2x,y) + (2x,-y) - (2x,0) - (2x,0)$$

$$= 2(2x,0) - 2(2x,0) = 0.$$

§3. Numerical, homological, and algebraic equivalence.

To further our study of the countable group $C^r(X)/C^r_{alg}(X)$, we introduce some more intermediate equivalence relations.

A cycle $Z \in C^r(X)$ is <u>numerically equivalent</u> to 0 if for every irreducible subvariety Y of dimension r in X, we have $\deg(Z.Y) = 0$, where $Z.Y$ is the intersection cycle of dimension 0, which is defined e.g. up to rational equivalence. The group of cycles numerically equivalent to zero is denoted by $C^r_{num}(X)$.

Clearly any cycle algebraically equivalent to zero is numerically equivalent to zero, so we have $C_{num} \supseteq C_{alg}$. Furthermore, since C/C_{num} has no torsion, we have $C_{num} \supseteq C_\tau \supseteq C_{alg}$, where C_τ denotes the group of cycles which are <u>torsion equivalent</u> to zero, meaning that some multiple is algebraically equivalent to zero.

Finally, if we have a Weil cohomology theory for X, we can define homological equivalence. For our purposes, a <u>Weil cohomology theory</u> means that we have cohomology groups $H^i(X)$, which are finite-dimensional vector spaces over a field K of characteristic zero, satisfying Poincaré duality, and we have a <u>cycle map</u> γ, which to each cycle $Z \in C^r(X)$ associates a cohomology class $\gamma(Z) \in H^{2r}(X)$. This cycle map should satisfy

$\gamma(Z.Z') = \gamma(Z)\cdot\gamma(Z')$ where one dot is the intersection of cycles, and the other dot is cup-product. Furthermore, if $P \in X$ is a point, then $\gamma(P) = 1 \in H^{2r}(X) \cong K$.

Given such a cohomology theory, we define $C^r_{hom}(X)$, the group of cycles homologically equivalent to zero, to be the kernel of $\gamma\colon C^r(X) \to H^{2r}(X)$. Then one can show, using some properties of the Weil cohomology which I may have forgotten to mention, that $C_{hom} \supseteq C_{alg}$. On the other hand, because of the compatibility of intersection theory with the cycle map and cup-product, we have $C_{num} \supseteq C_{hom}$. (See Kleiman [29] for more details on Weil cohomology and homological equivalence.)

Summing up, we have the following relations among these equivalence relations (where of course C_{hom} may depend on the particular choice of cohomology theory):

$$C^r \supseteq C^r_{num} \supseteq C^r_{hom} \supseteq C^r_{\tau} \supseteq C^r_{alg} .$$

Proposition 3.1. In the case of divisors $(r = 1)$ we have $C^r_{num} = C^r_{hom} = C^r_{\tau}$; furthermore C^r/C^r_{num} is a finitely generated free abelian group, and $C^r_{num}/C^r_{alg} = C^r_{\tau}/C^r_{alg}$ is a finite abelian group.

Proofs: In the case of varieties over $\mathbb{C}$, these results follow easily from the exponential sequence mentioned in §1, since C^1/C^1_{alg} is a subgroup of $H^2(X,\mathbb{Z})$, which is a finitely generated abelian group. In the abstract case, they depend on the finite generation of the Néron-Severi group, referred to above, and a theorem of Matsusaka [42] that $C^1_{num} = C^1_{\tau}$.

We naturally ask to what extent the results of the proposition remain valied in higher codimension. Here is what is known:

(1) It is still true for any r that $C^r(X)/C^r_{num}(X)$ is a finitely generated free abelian group. This follows easily from the existence of a Weil cohomology theory such as $H^*_{\acute{e}t}(X,\mathbb{Q}_\ell)$ (see Kleiman [29]).

(2) Lieberman [38] has shown in the classical case, using $H^*(\cdot,\mathbb{Q})$ as the cohomology theory, that $C^r_{num} = C^r_{hom}$ for $r = 0, 1, 2, n-1, n$, where $n = \dim X$, and for all r if X is an abelian variety. One expects that this should be true for all r, but that remains unknown.

(3) We will see below that one can have $C^r_{hom} \neq C^r_{\tau}$. This was shown by Griffiths, whose example we will discuss below. Analogous examples were constructed in the abstract case by Grothendieck and Katz [SGA7, exp XX].

(4) One does not know whether C^r_{hom}/C^r_{τ} is finitely generated, nor whether C^r_{τ}/C^r_{alg} is finite.

For the remainder of this section we will discuss one special case where one can calculate some of these groups, and that is the case of a non-singular complete intersection variety X in $\mathbb{P}^n$. We say that X is a complete intersection in $\mathbb{P}^n$ if it can be expressed as the intersection of exactly $n - \dim X$ hypersurfaces. These hypersurfaces need not be non-singular, even if X is, but in any case any complete intersection X is a deformation of a complete intersection of non-singular hypersurfaces. For simplicity, we will stick to the classical case of ground-field $\mathbb{C}$, and cohomology theory $H^*(\cdot,\mathbb{Q})$, even though many of these results are also valid in the abstract case using étale cohomology (see SGA5 and SGA7).

The principal tools used in deriving the following results are the theory of Lefschetz pencils, and the Hodge decomposition of complex cohomology.

Let X be a non-singular complete intersection variety in $\mathbb{P}^n$, and suppose that we can express it as a hypersurface section of a non-singular complete intersection variety W of one higher dimension. Then it is possible to consider X as a member of a Lefschetz pencil on W. A Lefschetz pencil is a 1-dimensional linear system of hypersurface sections of W, whose general member is non-singular, and whose special members have at most one conical double point each. Lefschetz was able to analyze completely the relationship between the cohomology of W, the cohomology of a general fibre X, and the action of the monodromy group on the cohomology of X, as the parameter t makes a loop around one of the singular fibres.

Here are some of the results one can obtain using these methods.

Theorem 3.2 (Lefschetz [37]). Let X be a non-singular complete intersection variety in $P = \mathbb{P}^n_{\mathbb{C}}$. Then

$$H^i(P,\mathbb{Z}) \to H^i(X,\mathbb{Z})$$

is an isomorphism for $i < \dim X$ (and injective for $i = \dim X$).

From this result, we conclude using Poincaré duality, that

$$H^{2i}(X,\mathbb{Q}) = \mathbb{Q},$$

for $0 \leqslant i \leqslant \dim X$, except for $i = \frac{1}{2} \dim X$, and that this vector space is generated by the image of a linear space section. Since the linear space section is an algebraic cycle, we conclude that

$$C^r(X)/C^r_{hom}(X) = \mathbb{Z}, \qquad r \neq \frac{1}{2} \dim X,$$

and that at least for $r < \frac{1}{2} \dim X$, it is generated by a linear space section.

Corollary 3.3 (Lefschetz). If X is a non-singular complete intersection variety of dimension $\geqslant 3$, then $\mathrm{Pic}\, X = \mathbb{Z}$.

Proof: Using the exponential sequence again, $H^1(X,\mathcal{O}) = 0$ because it is one piece of the Hodge decomposition of $H^1(X,\mathbb{C})$, which is 0 by (3.2). On the other hand, $H^2(X,\mathbb{Z}) = \mathbb{Z}$ by (3.2), and it is generated by a hyperplane section, which is in Pic X, so $\mathrm{Pic}\, X \cong \mathbb{Z}$.

The next case to consider is the case $r = \frac{1}{2} \dim X$. In this case things are not so simple, because $H^{2r}(X,\mathbb{Z})$ usually has rank bigger than 1, and not all of these cohomology classes need be algebraic. So we take into account the Hodge decomposition

$$H^{2r}(X,\mathbb{C}) = \bigoplus_{p+q=2r} H^{p,q}$$

of the complex cohomology. There is a natural map $H^{2r}(X,\mathbb{Z}) \to H^{2r}(X,\mathbb{C})$, and one knows that the algebraic cycles all land in the $H^{r,r}$ component.

Now one of the subtler consequences of the theory of Lefschetz pencils has to do with the vanishing cycles and the action of the monodromy transformation. Let $\{X_t\}_{t\in\mathbb{P}^1}$ be a Lefschetz pencil of hypersurface sections of a non-singular complete intersection variety W of dimension $2r+1$ in some projective space P. Let X_0 be a non-singular member of the pencil. Then $\dim X_0 = 2r$, and the middle-dimensional cohomology group $H^{2r}(X_0,\mathbb{C})$ splits into a direct sum of the image of $H^{2r}(P,\mathbb{C}) \cong \mathbb{C}$ and a subspace E generated by the so-called vanishing cycles. As we let $t \in \mathbb{P}^1$ move around, avoiding the critical points $t_1,\ldots,t_q$ of the pencil, we obtain the monodromy action of $\pi_1(\mathbb{P}^1 - \{t_i\})$ on $H^{2r}(X_0,\mathbb{C})$. One of Lefschetz' results says that the vanishing cycles are all conjugate to each other under the action of the monodromy. Because of the Picard-Lefschetz formula, this implies that the action of π_1 on E is irreducible. It follows that if we have any subspace $H^{2r}(X_0,\mathbb{C})$ stable under the action of the monodromy, it must be either 0 or the image of $H^{2r}(P,\mathbb{C})$, or E, or the whole space.

We apply this result to the subspace of $H^{2r}(X_0,\mathbb{C})$ generated by the classes of algebraic cycles, to obtain the following theorem (communicated by Griffiths):

Theorem 3.4. With W and X as above, assume furthermore that for some $p < r$, $H^{p,2r-p}(X_0,\mathbb{C}) \neq 0$ (a hypothesis which is satisfied whenever the degree of X_0 is sufficiently large). Then for all except countably many $t \in \mathbb{P}^1$, we have

$$C^r(X_t)/C^r_{hom}(X_t) = \mathbb{Z},$$

and it is generated by a linear space section.

Proof: First we must see how algebraic cycles on a fibre X_t behave when we vary t. Consider all closed subschemes of all the fibres X_t. These form algebraic families which are parametrized by a Hilbert scheme, which is a countable union of projective varieties. Each irreducible component of this Hilbert scheme will consist either of subschemes of a fixed X_t, or else it will be a family containing subschemes of X_t for all $t \in \mathbb{P}^1$. Thus, except for possibly countably many values of t, we can say that any subscheme of X_t belongs to a family of subschemes defined for all t.

Now let X_0 be one of these general fibres, and let $C \subseteq H^{2r}(X_0,\mathbb{C})$ be the subspace generated by the images of algebraic r-cycles on X_0. Then the subspace C is stable under the action of the monodromy, because any cycle on X_0 can be propagated along any loop in $\mathbb{P}^1$ which avoids the countably many values of t we have eliminated.

Now C contains the image of $H^{2r}(P,\mathbb{C})$. On the other hand, C is contained in the middle piece $H^{r,r}$ of the Hodge decomposition. Since by hypothesis $H^{p,2r-p} \neq 0$ for some $p < r$, we see that C is not all of $H^{2r}(X_0,\mathbb{C})$. Now by the result of Lefschetz referred to above, we conclude that C is just equal to the image of $H^{2r}(P,\mathbb{C})$, which is $\mathbb{C}$. It follows that $C^r(X_0)/C^r_{hom}(X_0) = \mathbb{Z}$.

As an example of this theorem, let $r = 1$, let $W = \mathbb{P}^3$, and let $\{X_t\}$ be a Lefschetz pencil of surfaces of degree $d \geqslant 4$. Then $H^{0,2}(X_t,\mathbb{C}) \neq 0$, so the theorem applies, and we find that for all but countably many $t \in \mathbb{P}^1$, we have $C^1(X_t)/C^1_{hom}(X_t) = \mathbb{Z}$. Combining with the exponential sequence above, we obtain

Corollary 3.5 (Noether, see also Moishezon [45]). If X is a sufficiently general surface in $\mathbb{P}^3$ of degree $d \geqslant 4$, then $\operatorname{Pic} X = \mathbb{Z}$.

Griffiths used a similar technique, combined with his theory of intermediate Jacobians, applied to the case W even-dimensional and X odd-dimensional, to find out something about cycles modulo algebraic equivalence on X. His result is this.

Theorem 3.6 (Griffiths [13]). Let W be a non-singular complete intersection of dimension $2r$ in $\mathbb{P}^n_{\mathbb{C}}$, let $\{X_t\}$ be a Lefschetz pencil of hypersurface sections, and assume that $H^{p,2r-p-1}(X,\mathbb{C}) \neq 0$ for some $p < r-1$. Then for all except countably many values of t, the following holds. Let $\varphi\colon C^r(W) \to C^r(X_t)$ be the restriction map. If $Z \in C^r(W)$ and if $\varphi(Z)$ is algebraically equivalent to 0, then Z is homologically equivalent to zero.

This theorem allows one to construct varieties X with $C^r_{hom} \neq C^r_{\tau}$. For example, let W be a quadric hypersurface of dimension 4 in $\mathbb{P}^5$, and let X_t be sections of W by hypersurfaces of degree $d \geqslant 5$. Then $r = 2$, $C^2(W)/C^2_{hom}(W) = \mathbb{Z} \oplus \mathbb{Z}$. On the other hand, $C^2(X)/C^2_{hom}(X) = \mathbb{Z}$ by (3.2). Hence one can find a cycle $Z \in C^2(W)$, $Z \notin C^2_{hom}(W)$, but such that $\varphi(Z) \in C^2_{hom}(X_t)$. On the other hand, by the theorem, no multiple of $\varphi(Z)$ could be algebraically equivalent to 0. Thus $C^2_{hom}(X_t) \neq C^2_{\tau}(X_t)$.

Note that in order for this example to work, X had to be a sufficiently general hypersurface section of a sufficiently special variety W. Because of (3.4), W had to be special to have $C^2/C^2_{hom} \neq \mathbb{Z}$. So we might expect that this phenomenon should not occur for X a sufficiently general complete intersection of hypersurfaces of large enough degrees.

Let me close this section with a question, which would be an analogue of the Lefschetz theorems for algebraic cycles. If X is a non-singular complete intersection variety, is $C^r_{hom}(X) = C^r_{alg}(X) = C^r_{rat}(X)$ for $r < \frac{1}{2}\dim X$? The first case to consider would be cycles of codimension 2 on a non-singular hypersurface X of dimension 5 in $\mathbb{P}^6$.

§4. The search for intermediate Jacobians.

Let X be a non-singular projective variety of dimension n over an algebraically closed field k. In the case of divisors, we have seen that the group $C^1_{alg}(X)/C^1_{rat}(X)$ is isomorphic to the group of k-rational points on an abelian variety $\mathrm{Pic}^{\circ}X$, called the Picard variety of X. This isomorphism is natural in the sense that there is a universal family of divisors on X, parametrized by $\mathrm{Pic}^{\circ}X$, which gives this isomorphism. Furthermore, $\mathrm{Pic}^{\circ}X$ is characterized by this universal property. In the language of schemes, we note that the association $X \mapsto \mathrm{Pic}\, X$ is a contravariant functor from schemes to the category of abelian groups. It is represented by a scheme $\mathrm{Pic}\, X$, of which $\mathrm{Pic}^{\circ}X$ is the connected component.

In the case of cycles of dimension 0 on X, we have seen that the group $C^n_{alg}(X)/C^n_{rat}(X)$ can be much too large to correspond to the points on any algebraic variety. However, there is an abelian variety Alb X, the Albanese variety of X, which is naturally associated to X, and there is a surjective homomorphism φ: $C^n_{alg}(X) \to$ Alb X, whose kernel contains $C^n_{rat}(X)$. We denote the kernel of φ by $C^n_{ab}(X)$, the 0-cycles abelian equivalent to zero. The map φ is regular in the sense that for any family $\{Z_t\}$ of 0-cycles, algebraically equivalent to zero, parametrized by a non-singular variety T, the corresponding map $T \to$ Alb X, defined by $t \mapsto \varphi(Z_t)$, is a morphism. In particular, if we fix a base point $P_0 \in X$, then the family $\{P-P_0\}$ parametrized by X itself gives a morphism ψ: $X \to$ Alb X. This morphism is universal for morphisms of X into abelian varieties sending P_0 to 0: given any morphism ψ' : $X \to A$ of X to an abelian variety A, which sends P_0 to 0, there is a unique homomorphism θ: Alb $X \to A$ such that $\psi' = \theta \circ \psi$. Clearly the abelian variety Alb X is uniquely characterized by the above properties. Furthermore, Alb X is a covariant functor in X.

The search for an abelian variety $J^r(X)$, together with a homomorphism $C^r_{alg}(X) \to J^r(X)$, for any r, having properties somehow analogous to the properties of $\mathrm{Pic}^o X$ and of Alb X described above, is what we call the search for intermediate Jacobians. No completely satisfactory answer to this problem has been found as yet, though several have been proposed.

In the classical case, Weil [65] constructed some abelian varieties $W^r(X)$, together with maps $C^r_{alg}(X) \to W^r(X)$, but they may not be generated by the image of the algebraic cycles, and they have the further defect that $W^r(X)$ does not vary holomorphically with X. Lieberman [39] continued the study of these abelian varieties.

Also in the complex case Griffiths [12], using a variation of Weil's method, defined some complex tori $G^r(X)$ associated with X, which have the advantage of varying holomorphically with X, but which may not be abelian varieties.

Lieberman [40] gave an axiomatic treatment of a theory of intermediate Jacobians, which should have both covariant and contravariant functorial properties, and which should coincide in the case of codimensions 1 and n with Pic^o and Alb, respectively. He shows in the classical case that such a theory exists, by taking the subgroup of Weil's or Griffiths' intermediate Jacobian generated by the image of the algebraic cycles, and showing that it is an abelian variety. However, we do not know how unique such a theory is.

Another approach, suggested by Mumford, is to look for a map of $C^r_{alg}(X)$ into an abelian variety, which is universal in the following sense. Let us say that a homomorphism φ: $C^r_{alg}(X) \to A$ to an abelian variety A is regular if for every family $\{Z_t\}$ of cycles in $C^r_{alg}(X)$ parametrized by a non-singular variety T, the induced map $T \to A$ defined by $t \mapsto \varphi(Z_t)$ is a morphism.

We will say that a regular homomorphism $\varphi\colon C^r_{alg}(X) \to A$ is <u>universal</u> if for any other regular homomorphism $\psi\colon C^r_{alg}(X) \to A'$ to an abelian variety A', there exists a unique homomorphism $\theta\colon A \to A'$ such that $\psi = \theta\circ\varphi$. If such a universal regular homomorphism exists, then clearly A is uniquely determined, and should be a good candidate for an intermediate Jacobian. However, we do not know if it exists in general.

For the remainder of this section, let us specialize to a particular case where we have some more satisfactory information, namely the case of non-singular projective 3-folds with $p_g = q = 0$. (Here we let $p_g = \dim H^0(X,\Omega^3)$, $q = \dim H^0(X,\Omega^1)$. This class of 3-folds includes the rational 3-folds, and it also includes the cubic hypersurface in $\mathbb{P}^4$. These 3-folds have been extensively studied by Fano, and more recently by Clemens, Griffiths, Manin, Tjurin, Mumford, and Murre.

We are interested in the group $C^2_{alg}(X)/C^2_{rat}(X)$ of curves on X modulo rational equivalence. (Because of the hypothesis $q = 0$, both $\mathrm{Pic}^\circ X$ and $\mathrm{Alb}\ X$ are zero.) In the classical case, because of the hypothesis $p_g = 0$, Weil's and Griffiths' intermediate Jacobians coincide, lending more weight to both as a candidate for the "right" intermediate Jacobian. Furthermore, in the case of the cubic 3-fold in $\mathbb{P}^4$, this abelian variety also is universal in the sense described above. In the abstract case, Murre has constructed an abelian variety with the universal property above. In both cases, a careful study of the resulting abelian variety has led to a proof that the cubic 3-fold, while unirational, is not a rational variety.

Let us describe these results in some more detail. We begin with the analytic approach of Clemens and Griffiths [11]. See also the lectures of Tjurin [63].

We let X be a non-singular 3-fold with $p_g = q = 0$ defined over $\mathbb{C}$. We define the <u>intermediate</u> <u>Jacobian</u> $J(X)$ to be the complex torus obtained by dividing $H^{1,2}(X,\mathbb{C})$ by the image of $H^3(X,\mathbf{Z})$ under the projection of $H^3(X,\mathbb{C})$ to its factor $H^{1,2}$. The homomorphism $\varphi\colon C^2_{alg}(X) \to J(X)$ is constructed by taking the periods of certain integrals. Then it is shown that $J(X)$ is an abelian variety, and that φ is a regular homomorphism.

Furthermore, $J(X)$ has a canonical polarization. In fact, in the discussion which follows, it is important to work always with polarized abelian varieties, and homomorphisms respecting the polarizations.

Next it is shown that if $X' \to X$ is a birational morphism, then $J(X)$ is isomorphic to a direct summand of $J(X')$. In particular, if X' is obtained from X by blowing up a non-singular curve C, then $J(X') \cong J(X) \oplus J(C)$, where $J(C)$ is the Jacobian of the curve C. From these results one can show, using the theorem of resolution of singularities, that if X is a rational 3-fold, i.e. if X is birational to $\mathbb{P}^3$, then $J(X)$ is isomorphic to a direct sum of Jacobians of curves.

Now let X be a cubic 3-fold in $\mathbb{P}^4$. In this case $J(X)$ is an abelian variety of dimension 5, and it turns out that $J(X)$ can be recovered abstractly from the geometry of X. The cubic 3-fold has a doubly infinite system of straight lines on it. They are parametrized by a subset S of the Grassmannian of lines in $\mathbb{P}^4$, and S is a non-singular surface with $q = 5$, called the <u>Fano surface</u>. A point on S determines a line on X, which we consider as a cycle in $C^2(X)$. Thus we obtain a homomorphism $\text{Alb } S \to J(X)$. Clemens and Griffiths show that this map is an isomorphism (which implies, by the way, that the map $C^2_{alg}(X) \to J(X)$ is universal). Then they show that $J(X)$ cannot be isomorphic to a sum of Jacobians of curves, and hence deduce that X is not rational.

The approach to Murre [49, 50, 51] in the abstract case is somewhat different. He constructs a special type of abelian variety, called a <u>Prym variety</u>, associated to the cubic 3-fold X as follows. Let ℓ_0 be a fixed (sufficiently general) line on X. Let C' be the curve on the Fano surface S consisting of the points corresponding to lines which meet ℓ_0. If ℓ is a line meeting ℓ_0, then the plane of ℓ_0 and ℓ meets X in a third line ℓ'. The map $\ell \mapsto \ell'$ defines an involution σ on C' without fixed points. Let $C = C'/\sigma$. Then the Prym variety of C'/C is defined as the connected component of the subgroup $\{x \in J(C') \mid \sigma(x) = -x\}$ of the Jacobian of C'. Murre takes this Prym variety to be the intermediate Jacobian $J(X)$. He constructes a regular map $C^2_{alg}(X) \to J(X)$, and shows that it is universal. He also shows that

$$C^2_{alg}(X)/C^2_{rat}(X) \cong J(X) \oplus T,$$

where T is a group consisting entirely of elements or order 2. In fact, T must be 0, because on any variety X, the group $C^r_{alg}(X)/C^r_{rat}(X)$ is <u>divisible</u>. Indeed, if Z is any cycle algebraically equivalent to zero, then there is a curve C, and a family $\{Z_t\}$ of cycles parametrized by $t \in C$, and there are two points $0, 1 \in C$ such that $Z_0 = 0$ and $Z_1 = Z$. The map $t \mapsto Z_t$ induces a homomorphism of the Jacobian variety $J(C) \to C^r_{alg}(X)/C^r_{rat}(X)$, whose image contains Z. Since any abelian variety is a divisible group, it follows that for any $n \neq 0$, there is a cycle $Z' \in C^r_{alg}(X)/C^r_{rat}(X)$ with $nZ' \sim Z$. Hence this group is divisible.

Murre's proof of non-rationality then depends

1) on showing that if X were rational, $J(X)$ would be isomorphic to a

sum of Jacobians of curves, and

2) some general results on Prym varieties showing that they are not sums of Jacobians of curves (see Mumford [48]).

PART II. SUBVARIETIES OF SMALL CODIMENSION

§5. Introduction

In these talks we will consider subvarieties of small codimension of a given non-singular projective variety X. The general philosophy here is that if Y is a subvariety of X whose codimension is small in comparison to the dimension of X, then there should be some strong connection between properties of X and properties of Y.

For example, take the case $X = \mathbb{P}^n_{\mathbb{C}}$, and let Y be a non-singular hypersurface in X. In this case we have the theorems of Lefschetz which tell us that various cohomological invariants of X are the same as those of Y:

a) $H^i(X,\mathbb{Z}) \to H^i(Y,\mathbb{Z})$ is an isomorphism for $i < \dim Y$,

b) if $n \geqslant 3$, then $\pi_1(Y) = 0$

c) if $n \geqslant 4$, then $\operatorname{Pic} X \to \operatorname{Pic} Y$ is an isomorphism.

Lefschetz' theorems also apply more generally to complete intersection varieties inside of X, but we will be interested also in subvarieties of X which are not necessarily complete intersections.

A remarkable discovery made by W. Barth is that even when Y is not a complete intersection, one can compare the cohomology of X to the cohomology of Y, and prove theorems similar to the Lefschetz theorems. In §6 we will discuss Barth's theorems and a number of variations of them which were inspired by his work.

In §7 we will examine subvarieties of Grassmann varieties, with applications to the problem of smoothing algebraic cycles. Here it is notable that smoothing theorems have been proved for cycles of small dimension (hence large codimension !) by Hironaka and by Kleiman, while for cycles of small codimension one can construct examples which are not smoothable.

§6. Barth's theorems.

The basic result, first proved by Barth [2] in 1970 is this:

Theorem 6.1. Let Y be a non-singular subvariety of dimension r of $X = \mathbb{P}^n_{\mathbb{C}}$. Then

$$H^i(X,\mathbb{C}) \to H^i(Y,\mathbb{C})$$

is an isomorphism for all $i \leqslant 2r - n$.

Let us look at a few examples to have an idea what the theorem means. First note that if $r < \frac{1}{2}n$, there is no conclusion. If $r = \frac{1}{2}n$, then the theorem tells us that $H^0(Y,\mathbb{C}) = \mathbb{C}$, in other words, Y is connected. This is obvious anyway, because any two r-dimensional subvarieties of $\mathbb{P}^{2r}$ always meet.

While the conclusion of the theorem is weak for small values of r, it becomes stronger and stronger as r gets larger with respect to n. In the extreme case $r = n - 1$, we find tht $2r - n = n - 2 = \dim Y - 1$, so we just have the statement of Lefschetz' theorem with $\mathbb{C}$ coefficients.

The first non-trivial case of the theorem between these two extremes is when $r = 3$ and $n = 5$. In this case it tells us that if Y is a non-singular 3-fold in $\mathbb{P}^5_{\mathbb{C}}$, then $H^1(Y,\mathbb{C}) = 0$. This is a non-trivial result. It implies, for example, that an abelian variety of dimension 3 cannot be embedded in $\mathbb{P}^5$.

In this particular example of the abelian variety, by the way, there is another way to see that it cannot be embedded in $\mathbb{P}^5$: let N be the normal bundle of Y in $\mathbb{P}^5$. Then N is a bundle of rank two, and so its third Chern class $c_3(N)$ must be zero. On the other hand, if Y is an abelian variety, its tangent bundle is trivial, so the Chern classes of N are equal to the Chern classes of $T_{\mathbb{P}^n}$ restricted to Y. Since $c(T_{\mathbb{P}^n}) = (1 + h)^{n+1}$, where h is the hyperplane class, we find that $c_3(N) = 20h^3.Y$ which is non-zero. This gives a contradiction.

This method shows more generally that there are necessary conditions on the Chern classes of a variety in order to embed it in a projective space of less than twice its dimension plus one. This is the approach taken by Holme [28], who finds necessary and sufficient conditions for a non-singular variety Y, given in a large projective space $\mathbb{P}^N$, to remain non-singular when it is projected into a smaller projective space $\mathbb{P}^n$.

Coming back to Barth's theorem, I would like first to give a very short proof which deduces it from the strong Lefschetz theorem (see Hartshorne [24]). After that we will examine some other proofs and variants of the theorem.

Proof of Theorem 6.1. Let $j: Y \to X$ be the inclusion map, and consider the following diagram of morphisms of cohomology

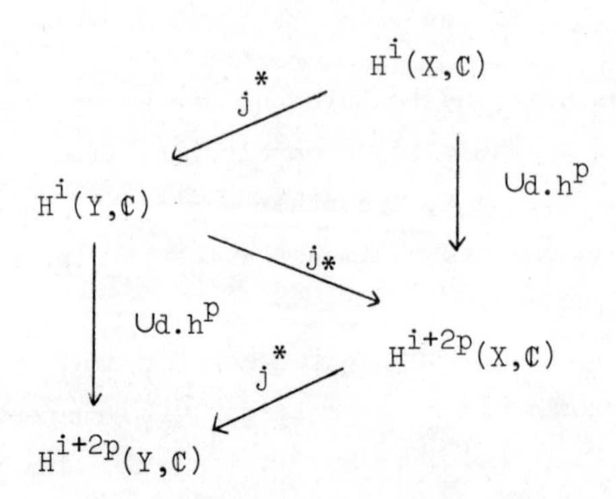

Here $p = n - r$ is the codimension of Y in X, h is the hyperplane class on X, and d is the degree of Y. We are using the fact that $j_* j^*$ is cup-product with the cohomology class of Y on X, which is $d.h^p$, and that $j^* j_*$ is cup-product with the same element in the cohomology of Y.

Now the strong Lefschetz theorem on Y tells us that $\cup d.h^p$ is injective, provided $i \leq r - p$. Since $p = n - r$, this says $i \leq 2r - n$. Now if $\cup d.h^p$ is injective on $H^i(X,\mathbb{C})$, then j_* is injective. But $\cup d.h^p$ is an isomorphism on $H^i(X,\mathbb{C})$, because $X = \mathbb{P}^n$, so we conclude that $j^*: H^i(X,\mathbb{C}) \to H^i(Y,\mathbb{C})$ is an isomorphism, as required.

<u>Remark</u>. Since Deligne has recently proved the strong Lefschetz theorem for the ℓ-adic étale cohomology of schemes in characteristic $p > 0$, this proof establishes Theorem 6.1 also for ℓ-adic cohomology if Y is a non-singular subvariety of $\mathbb{P}^n_k$, with k a field of characteristic $p \neq \ell$.

<u>Barth's original proof</u>. The method first used by Barth to prove this theorem is quite different. It is very interesting because of a delicate interplay between complex cohomology and coherent analytic sheaf cohomology. His basic idea was to "transplant" cohomology classes on Y by moving Y around using projective automorphisms. To make this more precise, let $G = SU(n+1)$ be the special unitary group, acting on $\mathbb{P}^n$. We define a map $\varphi: G \times Y \to \mathbb{P}^n$ by $\varphi(g,y) = g(y)$, and another map $p: G \times Y \to G$ by the first projection. Let F be a coherent analytic sheaf on $\mathbb{P}^n$, and consider the sheaf $F' = \varphi^{-1}(F)$ on $G \times Y$. Here we take φ^{-1} in the sense of pulling back an abelian sheaf by a continuous map between topological spaces. Thus F' is <u>not</u> a holomorphic sheaf on $G \times Y$. Rather, it is a sheaf which is "holomorphic vertically, and constant horizontally".

$$\begin{array}{ccc} G \times Y & \xrightarrow{\varphi} & \mathbb{P}^n \\ \downarrow & & \\ G & & \end{array}$$

Now Barth makes a careful study of the Leray spectral sequences for φ and for p, applied to the sheaf F'. This gives connections between the cohomology of F on $\mathbb{P}^n$, the cohomology of the (topological) restriction $F|_Y$, and the complex cohomology of the spaces involved. The final result is

Theorem 6.2: Let Y be a non-singular subvariety of dimension r of $X = \mathbb{P}^n_C$. Let $q \leq r$ be an integer. Then we have equivalent conditions

(i) $H^i(X,\mathbb{C}) \to H^i(Y,\mathbb{C})$ is an isomorphism for $i \leq q$

(ii) $H^i(X,\mathcal{O}_X) \to H^i(Y,\mathcal{O}_{X|Y})$ is an isomorphism for $i \leq q$

(iii) $H^i(X,F) \to H^i(Y, F|_Y)$ is an isomorphism for all coherent locally free analytic sheaves on X, and all $i \leq q$.

Furthermore, if $q = 2r - n$, (i), (ii), (iii) always hold.

The theorem is remarkable in that it establishes an equivalence between a problem of extending complex cohomology classes on Y to all of X, and a problem of extending sheaf cohomology classes. One implication, namely (iii) $\Rightarrow$ (i), is easily seen by using the holomorphic DeRham complex,

$$0 \to \mathbb{C} \to \mathcal{O}_X \to \Omega^1_X \to \dots \to \Omega^n_X \to 0.$$

This is a resolution of $\mathbb{C}$ by locally free coherent analytic sheaves. Thus there is a spectral sequence beginning $E_1^{pq} = H^q(X,\Omega^p_X)$ and ending with $E^m = H^m(X, \mathbb{C})$. On the other hand, restricting this complex to Y gives

$$0 \to \mathbb{C}|_Y \to \mathcal{O}_X|_Y \to \Omega^1_X|_Y \to \dots \to \Omega^n_X|_Y \to 0$$

which is a resolution of $\mathbb{C}|_Y$ by the restricted sheaves. So we have another spectral sequence beginning $E_1^{pq} = H^q(Y,\Omega^p_X|_Y)$ and converging to $E^m = H^m(Y,\mathbb{C})$. By (iii), the restriction map induces isomorphisms on the initial terms of these spectral sequences, so we conclude that the abutments are also isomorphic. This gives (i).

The other implications (i) $\Rightarrow$ (ii) $\Rightarrow$ (iii) require the full strength of the transplanting method.

Another interesting point about this theorem is that in the course of the proof it becomes necessary to know that the cohomology groups $H^i(Y, F|_Y)$ are finite-dimensional for $i < r$. This is proved by using duality, which relates them to $H^{n-i}(X-Y,F)$. Then one uses the Fubini metric on X to show that $X-Y$ has a pseudo-convexity property which allows one to use the finiteness theorems of Andreotti and Grauert [1] to prove that this cohomology is finite-dimensional. The non-singularity of Y is essential to prove pseudo-convexity here.

This same use of duality allows one to deduce some vanishing theorems for coherent analytic sheaf cohomology on $X-Y$. Namely, we have

Corollary 6.3. With the above hypotheses, (i), (ii) and (iii) of (6.2) are also equivalent to

(iv) $H^i(X-Y,F) = 0$ for all coherent analytic sheaves F on X, and all $i \geq n - q - 1$.

Homotopy groups and integral cohomology. Barth's original methods, suitably expanded, led to results about homotopy groups and integral cohomology. Barth and Larsen [3], using the transplanting method for coverings instead of cohomology classes, showed that Y was simply connected. Then Larsen [36], using Morse theory in addition to the result on π_1, proved the analogous theorem for integral cohomology.* The results are these

Theorem 6.4. Let Y be a non-singular variety of dimension r in $X = \mathbb{P}^n_{\mathbb{C}}$. Then

a) if $1 \leq 2r - n$, then $\pi_1(Y) = 1$.

b) $H^i(X,\mathbb{Z}) \to H^i(Y,\mathbb{Z})$ is an isomorphism for $i \leq 2r - n$.

Using the exact cohomology sequence associated to the exponential sequence

$$0 \to \mathbb{Z} \to \mathcal{O} \to \mathcal{O}^* \to 0,$$

we can deduce a result about the Picard group.

Corollary 6.5. With the same hypotheses, we have

c) if $2 \leq 2r - n$, then $\operatorname{Pic} Y = \mathbb{Z}$, generated by $\mathcal{O}_Y(1)$.

Algebraic proofs of Barth's theorems. So far the methods of proof we have discussed have been complex-analytic or topological. Naturally one wonders whether one can prove the theorem by purely algebraic methods, and express it in purely algebraic terms. This work was begun by Hartshorne [18], who followed Barth's transplanting method, but instead of using complex cohomology, used the algebraic DeRham cohomology. This cohomology is defined using algebraic differential forms on a non-singular variety in the Zariski

* See also Barth's article in this volume.

topology, and has been shown by Grothendieck [14] to coincide with the classical complex cohomology. By using the transplanting method with algebraic DeRham cohomology, it was possible to show that an algebraic version of (6.2) held, but the proof still made use of a comparison with the complex case.

The first purely algebraic proof of Barth's theorems (in characteristic 0) was given by Ogus [52, 53]. His method, which consists in first proving an analogous local theorem, and then deriving global results, has the added advantage that it gives a stronger result, since the hypotheses of non-singularity have been eliminated.

Let us first state his global results. To phrase the cohomology in purely algebraic terms, we use algebraic DeRham cohomology. Deligne (unpublished) and Herrera and Lieberman [26] were the first to notice that one could calculate the usual complex cohomology of a (singular) algebraic variety Y over $\mathbb{C}$ by embedding it in a smooth variety $X/\mathbb{C}$, taking the complex $\Omega^{\bullet}$ of algebraic differential forms on X, and taking the hypercohomology $H^{*}(Y,\hat{\Omega}^{\bullet})$ on Y of the formal completion of $\Omega^{\bullet}$ along Y. Hartshorne [19,22] wrote a general theory of this algebraic DeRham cohomology for singular varieties over a field k of characteristic zero, including a homology theory like Borel-Moore homology, and also a local theory for varieties of finite type over the spectrum of a complete local ring. We will denote the algebraic DeRham cohomology by H^{*}_{DR}.

Now we can state Ogus' global results (compare with (6.2) above).

Theorem 6.6. Let Y be a closed subscheme of dimension r, which is a locally complete intersection, of $X = \mathbb{P}^n_k$, where k is an algebraically closed field of characteristic zero. Then the following conditions are equivalent (where $q \leqslant r$ is a fixed integer):

(i) $H^i_{DR}(X) \to H^i_{DR}(Y)$ is an isomorphism for $i \leqslant q$.

(ii) $H^i(X,\mathcal{O}_X) \to H^i(\hat{X},\mathcal{O}_{\hat{X}})$ is an isomorphism for $i \leqslant q$.

(iii) $H^i(X,F) \to H^i(\hat{X},\hat{F})$ is an isomorphism for all $i \leqslant q$ and for all coherent locally free sheaves F on X.

Furthermore, if $q = 2r - n$, (i), (ii) and (iii) always hold.

He also has theorems of Barth type on the algebraic fundamental group $\pi_1^{alg}(Y)$ and on the Picard group $\mathrm{Pic}(Y)$ (see [53, 3.8 and 4.10]):

Theorem 6.7. With the same hypotheses as in (6.6), we have

a) $\pi_1^{alg}(Y) = 0$ if $1 \leqslant 2r - n$, and

b) In the case $k = \mathbb{C}$, and Y non-singular, we have $\operatorname{Pic} Y = \mathbb{Z}$, generated by $\mathcal{O}_Y(1)$, if $2 \leqslant 2r - n$.

Let us just mention some of the basic ingredients of Ogus' proofs. The first fundamental principle is the idea that global statements about a projective variety $Y \subseteq \mathbb{P}^n$ can be turned into statements of local algebra, by considering the cone $C(Y) \subseteq \mathbb{A}^{n+1}$ over Y in affine $n+1$ space. Let R be the local ring of the origin of $\mathbb{A}^{n+1}$, and let I be the ideal of $C(Y)$ in R. Then it is always profitable to translate properties of Y into properties of R and I. To preserve the geometric language, let $X' = \operatorname{Spec} R$, $Y' = \operatorname{Spec} R/I$, and let $P \in X'$ be the closed point. Then theorem (6.6) can be translated into a statement about the local DeRham cohomology of Y' with supports at P, and certain coherent sheaf cohomology of the formal completion of X' along Y'. We refer to Ogus [52] for the precise statement of the local theorem.

One other point which deserves mention is that Barth's transplanting method (described above) never appears explicitly in Ogus' work. In its place we have the notion of a module with integrable connection over a complete local ring R. Several of the local formal cohomology groups come equipped with natural integrable connections. Ogus' main technical lemma gives a criterion for a module with integrable connection to be free.

Applications. In closing this section, let us give three applications of the Barth type theorems.

The first is to vector bundles on projective space. Let E be a vector bundle of rank 2 on $X = \mathbb{P}^n_k$. Then, after twisting by $\mathcal{O}(1)$ a number of times, if necessary, we can find a global section $s \in H^0(X,E)$ such that the zero-set of s is a non-singular subvariety Y of codimension 2 in X. So it is natural to ask for a converse: given a subvariety Y of codimension 2 in X, can one find a vector bundle E and a section s, such that Y is the zero-set of s? An answer is given as follows.

Proposition 6.8. Let Y be a non-singular subvariety of codimension 2 of $\mathbb{P}^n_{\mathbb{C}}$, with $n \geqslant 6$. Then there exists a vector bundle E of rank 2, and a section $s \in H^0(E)$ whose zero-set is Y. Furthermore, Y is a complete intersection if and only if E is a direct sum of line bundles $E \cong \mathcal{O}(\ell_1) \oplus \mathcal{O}(\ell_2)$.

For the proof, one uses (6.7b) to deduce that $\operatorname{Pic} Y = \mathbb{Z}$. Therefore the canonical invertible sheaf ω_Y must be some multiple of $\mathcal{O}(1)$, say $\omega_Y = \mathcal{O}_Y(m)$. Then it is possible to construct E as the dual sheaf to an extension of $\mathcal{I}_Y$ by $\mathcal{O}_X(-m-n-1)$. (See for example Hartshorne [24, 6.1].)

In connection with this proposition, we should mention that it is not known whether there are any vector bundles of rank 2 on $\mathbb{P}^n$, $n \geqslant 6$ other than direct sums of line bundles. Nor is it known whether there are any non-singular sub-varieties Y of codimension 2 in $\mathbb{P}^n$, $n \geqslant 6$, which are not complete intersections. At least the two problems are equivalent to each other.

Our second is an application to local algebra due to Hartshorne and Ogus [20]. In this case the local version of the vanishing of Pic (which we haven't stated - see Ogus [52, 3.14]) allows one to conclude that certain rings are factorial, hence in certain cases Gorenstein, or even complete intersection. A typical result is the following

Proposition 6.9. Let A be a local ring of dimension r, having an isolated singularity, which is a quotient of a regular local ring R of dimension n, and of equicharacteristic zero. Assume that depth $A \geqslant 3$ and $r \geqslant \frac{1}{2}(n+3)$. Then A is factorial.

Our third application is to the problem of smoothing singularities by deformation. The problem is, given a closed subvariety V of $\mathbb{A}^n$, having an isolated singularity at a point P, say, can one find a flat family $\{V_t\}$ of subvarieties of $\mathbb{A}^n$, with $V_0 = V$, and V_t non-singular for $t \neq 0$. It turns out that the Barth theorems on cohomology give obstructions to the possibility of such smoothing, which can be phrased in terms of the cohomology of a punctured neighborhood of P in V. Rather than stating the general result, which can be found in Hartshorne [21], let us give a typical example.

Example 6.10. Let Y be a non-singular surface in $\mathbb{P}^4_{\mathbb{C}}$ with $H^1(Y,\mathbb{C}) \neq 0$ (such surfaces do exist!) . Let $V \subseteq \mathbb{A}^5$ be the cone over Y, and let P be its vertex. Then V cannot be deformed to a smooth subvariety of $\mathbb{A}^5$. The rough idea is that if it could be, then by taking the projective completion $W \subseteq \mathbb{P}^5$ of a nearby non-singular V_t, one would have a variety W with $H^1(W,\mathbb{C}) \neq 0$, which contradicts (6.1).

§7. Subvarieties of Grassmann varieties and the problem of smoothing algebraic cycles.

The smoothing problem. Let us begin with the problem of smoothing cycles.

Let X be a non-singular projective variety over k, and let $Y = \Sigma n_i Y_i$ be an algebraic cycle on X. Here the Y_i are irreducible subvarieties of X, which may have arbitrary singularities. But since non-singular subvarieties are somehow better than singular ones, it is natural to ask whether Y is equivalent, for any suitable equivalence relation, to a cycle $Y' = \Sigma n_i' Y_i'$, where the subvarieties Y_i' are all non-singular.

For example, let us consider the case of divisors, where the answer is "yes". Let D be any divisor on X, and let H be a hyperplane section of X. Bertini's theorem tells us that any sufficiently ample linear system contains an irreducible non-singular subvariety of codimension 1 in it. Thus for n sufficiently large, we can find Y non-singular, with $Y \sim D + nH$ (linear equivalence), and Z non-singular with $Z \sim nH$. Then $D \sim Y-Z$, so we have D linearly equivalent to a difference of two non-singular subvarieties.

At the other end of the scale, if we look at zero-dimensional cycles on X, they are already smoothed, by definition.

In between these two extremes, the problem is non-trivial. We will see below that for small dimensions one can smooth cycles in many cases, but for small codimension it may be impossible.

Grassmann varieties. Before discussing the smoothing theorems, we digress for a moment to describe Grassmann varieties. It turns out they are crucial for the proofs of smoothing where it is possible, and for making a counterexample where it is impossible.

Consider the set of all linear r-dimensional subspaces $\mathbb{P}^r$ inside of a fixed projective n-space $\mathbb{P}^n$. These are parametrized by a Grassmann variety which we will denote by $G(r,n)$. One sees from counting constants that it has dimension $(r+1)(n-r)$. One defines certain standard subvarieties of $G(r,n)$, called Schubert varieties, by imposing certain incidence conditions on the variable $\mathbb{P}^r$. For example, let L^{n-r-1} be a fixed linear subspace of dimension $n-r-1$. In general, a $\mathbb{P}^r$ will not meet L. So we define

$$\sigma_1 = \{\mathbb{P}^r \subseteq \mathbb{P}^n \mid \mathbb{P}^r \cap L^{n-r-1} \neq \emptyset\}.$$

This is a subvariety (in general singular) of codimension 1 in G. In codimension 2 there are two Schubert cycles, namely

$$\sigma_2 = \{\mathbb{P}^r \subseteq \mathbb{P}^n \mid \mathbb{P}^r \cap L^{n-r-2} \neq \emptyset\} \text{ and}$$

$$\sigma_2' = \{\mathbb{P}^r \subseteq \mathbb{P}^n \mid \dim \mathbb{P}^r \cap L^{n-r} \geqslant 1\},$$

where in each case L denotes a fixed linear space of the given dimension.

The group of algebraic cycles on the Grassmann variety can be completely described using the Schubert cycles. The Grassmann variety is a non-singular projective variety. Algebraic equivalence and rational equivalence coincide, and the group of cycles modulo either equivalence relation is freely generated by the Schubert cycles. Furthermore, if we work over $\mathbb{C}$ and use integral cohomology, then the Chow ring $A(G)$ of G is isomorphic to the cohomology ring $H^*(G,\mathbb{Z})$. In particular, algebraic and homological equivalence coincide, and every cohomology class is the class of some algebraic cycle. In low codimensions, we have

$$A^1(G) \cong H^2(G,\mathbb{Z}) \cong \mathbb{Z} \text{ is generated by } \sigma_1$$

$$A^2(G) \cong H^4(G,\mathbb{Z}) = \mathbb{Z} \oplus \mathbb{Z} \text{ is generated by } \sigma_2 \text{ and } \sigma_2'.$$

We will also need the universal bundles on the Grassmann variety $G(r,n)$. To give a $\mathbb{P}^r \subseteq \mathbb{P}^n$ is the same as to give a subvector space $\mathbb{A}^{r+1} \subseteq \mathbb{A}^{n+1}$. Thus, as the point $x \in G(r,n)$ varies, we get a variable $\mathbb{A}^{r+1} \subseteq \mathbb{A}^{n+1}$, hence we get the <u>universal subbundle</u> $E^{r+1} \subseteq \mathcal{O}^{n+1}$ on G. Here E is a vector bundle of rank $r+1$, which is a subbundle of a trivial bundle $\mathcal{O}^{n+1}$ of rank $n+1$. We consider the quotient $F^{n-r} = \mathcal{O}^{n+1}/E^{r+1}$ which is called the <u>universal quotient bundle</u>, and so we have a canonical exact sequence

$$0 \to E^{r+1} \to \mathcal{O}^{n+1} \to F^{n-r} \to 0.$$

It is well-known that the Chern classes of E and F can be expressed in terms of Schubert cycles. Thus for example

$$c_1(F) = \sigma_1 \qquad c_1(E) = -\sigma_1$$

$$c_2(F) = \sigma_2 \qquad c_2(E) = \sigma_2'.$$

The other important property of the universal quotient bundle F is that it has a universal property: Given any variety X, and a vector bundle F_0 on X of rank $n-r$, and given $n+1$ global sections of F_0, which generate the fibre at every point (i.e. given a surjection $\mathcal{O}_X^{n+1} \to F_0 \to 0$) then there exists a unique morphism $f: X \to G(r,n)$ such that $F_0 \cong f^*(F)$, and the $n+1$ global sections of F determined by the surjection $\mathcal{O}^{n+1} \to F$ pull back to the given sections of F_0.

This universal property gives a universal method for calculating the Chern classes of a vector bundle F_0 on X: use F_0 to map X to a Grassmann variety. There the Chern classes of the universal bundle are given by Schubert cycles. Then pull these Schubert cycles back to X.

Hironaka's smoothing method. Now we return to the problem of smoothing cycles, and say something about Hironaka's result [27] which is the following

Theorem 7.1. Let X be a non-singular projective variety of dimension n over a field k of characteristic 0. If Y is a cycle of dimension $\leqslant \min(3, \frac{1}{2}(n-1))$, then Y can be smoothed, for rational equivalence.

There are two main ideas for his proof. One is to use resolution of singularities to resolve the singularities of Y, then do something to the proper transform of Y upstairs, and gradually project down again. The restriction $\dim Y \leqslant \frac{1}{2}(n-1)$ is necessary to have enough freedom in projecting. This part of the proof we will not attempt to describe any further.

The other ingredient to the proof we might call the residual intersection method. If Y is a subvariety of a non-singular projective variety X, with $\operatorname{codim}(Y,X) = p$, we consider p hypersurface sections of X of some fixed large degree, $H_1,\dots,H_p$, passing through Y. If the H_i are sufficiently general, then their intersection $H_1\cap\dots\cap H_p$ will consist of Y and another component Z, in general irreducible. Now $H_1\cap\dots\cap H_p$ can be smoothed by Bertini's theorem, just as in the case of divisors, because it is a complete intersection. So if Z was non-singular, we would have $Y \sim H_1\cap\dots\cap H_p - Z$ which would be smoothable. But a curious thing happens: in general Z will have singularities in codimension 4! This accounts for the restriction $\dim Y \leqslant 3$ in Hironaka's theorem. To illustrate where these singularities come from, let us state a precise result which illustrates this phenomenon.

Lemma 7.2: Let $Y^r \subseteq \mathbf{P}^n$ be a non-singular subvariety with $r \geqslant 4$ and $p = n - r \geqslant 2$. Then for all sufficiently large integers d, one can find p hypersurfaces $H_1,\dots,H_p$ of degree d containing Y, such that $H_1\cap\dots\cap H_p = Y \cup Z$, where Z is irreducible of dimension r. Furthermore, Z will be singular in general, and its singular locus will have dimension $r - 4$.

The proof involves some Schubert cycles on a Grassmann variety. We will give an idea of the proof since it illustrates nicely how these Grassmann varieties arise.

Consider the linear system of all hypersurfaces of degree d (d fixed, large) containing Y. This linear system defines a birational map of $\mathbb{P}^n$ into a large projective space $\mathbb{P}^N$, whose image is the variety $\tilde{X}$ obtained by blowing up $\mathbb{P}^n$ along Y. Let $\tilde{Y}$ be the inverse image of Y in $\tilde{X}$. Then $\tilde{Y}$ is a $\mathbb{P}^{p-1}$-bundle over Y, where $p = n - r$, and the fibres are linearly embedded in $\mathbb{P}^N$. Thus we get a map $\varphi: Y \to G(p-1, N)$.

$$\begin{array}{ccccc} \tilde{Y} & \hookrightarrow & \tilde{X} & \hookrightarrow & \mathbb{P}^N \\ \downarrow & & \downarrow & & \\ Y & \hookrightarrow & \mathbb{P}^n & & \end{array}$$

Now the residual intersection Z is obtained by taking p hyperplanes in general position in $\mathbb{P}^N$, say $H_1',\ldots,H_p'$, intersecting them with $\tilde{X}$, and projecting back to $\mathbb{P}^n$. Let $Z' = \tilde{X} \cap H_1' \cap \ldots \cap H_p'$. Then Z is a non-singular subvariety of $\tilde{X}$, of dimension r. The singularities of Z will arise whenever Z' meets a fibre of $\tilde{Y}/Y$ in more than one point, or not transversally. If $L^{N-p} = H_1' \ldots H_p'$ in $\mathbb{P}^N$, we consider the Schubert cycle

$$\sigma_1 = \{\mathbb{P}^{p-1} \subseteq \mathbb{P}^N \mid \mathbb{P}^{p-1} \cap L^{N-p} \neq \emptyset\}$$

in $G(p-1,N)$. This has codimension 1, and the points of $\sigma_1 \cap \varphi(Y)$ are the images of those points of Y lying on Z. The singularities of Z occur when Z' meets one of the fibres of $\tilde{Y}/Y$ in more than one point. This corresponds to the Schubert cycle

$$\sigma_1^* = \{\mathbb{P}^{p-1} \subseteq \mathbb{P}^N \mid \dim \mathbb{P}^{p-1} \cap L^{N-p} \geqslant 1\}.$$

It is known that this cycle is the singular locus of σ_1, and it has codimension 4 in $G(p-1,N)$. Thus we find a subset of codimension 4 in Y (hence also codim 4 in Z) which consists of the singular points of Z. (For the last step, we need to use one of Kleiman's results, described below, which says that $\varphi: Y \to G(p-1,N)$ will be an embedding, and that its image will be transversal to a general Schubert cycle.)

<u>Kleiman's smoothing method</u>. Kleiman's smoothing method [30] is roughly this: express the given cycle Y as a Chern class of some vector bundle on X. Then use the vector bundle to map X into a Grassmann variety, where the Chern classes of the universal bundle are expressed in terms of certain Schubert cycles. Then recover Y (up to rational equivalence) by intersecting these Schubert cycles with X, while trying to avoid the singularities of the Schubert cycles. The result is this:

Theorem 7.3. Let X be a non-singular projective variety of dimension n, over a field k, and let Y be a cycle of codimension p on X. If $\dim Y < \frac{1}{2}(n+2)$, then $(p-1)!Y$ can be smoothed for rational equivalence.

Note that we don't smooth Y itself, only some multiple. This comes from the fact that one cannot in general express Y as the Chern class of a vector bundle. What one can do is to resolve the coherent sheaf $\mathcal{O}_Y$ on X by locally free sheaves. Let $\mathcal{O}(1)$ be a fixed ample line bundle on X. Then we can find a resolution

$$0 \to F_q \to \dots \to F_1 \to F_0 \to \mathcal{O}_Y \to 0$$

where all the sheaves except F_q are direct sums of sheaves of the form $\mathcal{O}(m_i)$. Then in the Grothendieck group $K(X)$, we have

$$\mathcal{J}(\mathcal{O}_Y) = \Sigma(-1)^i \mathcal{J}(F_i),$$

so the Chern classes of F_q are the same as those of $\mathcal{O}_Y$, up to adding hypersurface sections of X, which are harmless for the smoothing problem. But then we have a mysterious formula of Grothendieck [9]

$$c_p(\mathcal{J}(\mathcal{O}_Y)) = (p-1)!Y .$$

So to smooth $(p-1)!Y$ we are reduced to smoothing $c_p(F_q)$.

Next, Kleiman shows that after twisting $F = F_q$ if necessary, it gives rise to an embedding of X in some Grassmann variety $G(r',n')$ in such a way that it is transversal to a general Schubert cycle. Now $c_p(F)$ is obtained by intersecting the image of X with a Schubert cycle σ_p of codimension p, defined in a way similar to σ_1 with σ_2 above. Furthermore, Kleiman shows that the singular set of σ_p is another Schubert cycle, which has codimension exactly $p+2$ inside of σ_p. (See also Kleiman and Landolfi [32] for more information on the singularities of σ_p.) So if $\dim Y < p+2$ we can avoid the singular locus of σ_p, and we obtain a smoothing theorem. This gives the restriction $\dim Y < \frac{1}{2}(n+2)$ in the theorem.

Score card. Putting together what we know so far about the smoothing problem, let's see where we stand. X will denote a non-singular projective variety of dimension n over a field k of characteristic zero, and Y will denote a cycle. We ask for smoothing up to rational equivalence.

If $\dim X \leqslant 5$, then we can smooth everything.

If $\dim X = 6$, $\dim Y = 3$, then $2\cdot Y$ can be smoothed, but we don't know if Y itself can be smoothed.

If $\dim X = 6$, $\dim Y = 4$, we don't know even if any multiple of Y can be smoothed.

We will see next that if $\dim X = 9$, $\dim Y = 7$, it may not be possible to smooth Y, although it is conceivable that $4\cdot Y$ may be smoothable.

Example of non-smoothing. Let X be the Grassmann variety $G(2,5)$ of projective planes in $\mathbb{P}^5$. This is a projective variety of dimension 9. Let Y be the Schubert cycle σ_2 defined above, of codimension 2. This subvariety Y has dimension 7, and has a singular set of dimension 3. It was recommended by Kleiman and Landolfi [32] as a good candidate for a non-smoothable cycle, and it was shown by Hartshorne, Rees, and Thomas [23] that in fact it is not smoothable.

To state this result, we work over $\mathbb{C}$. Let $x \in H^2(X,\mathbb{Z}) = \mathbb{Z}$ be the generator given by the Schubert cycle σ_1, and let $y \in H^4(X,\mathbb{Z}) = \mathbb{Z} \oplus \mathbb{Z}$ be given by σ_2. Then x^2 corresponds to $\sigma_2 + \sigma_2'$, so that we can take x^2 and y as generators for $H^4(X,\mathbb{Z})$.

Theorem 7.4. Let M be a C^∞ submanifold of X of (real) codimension 4, and let the cohomology class of M in $H^4(X,\mathbb{Z})$ be $ax^2 + by$, $a,b \in \mathbb{Z}$. Then $b \equiv 0 \pmod 2$. If we require further that the normal bundle of M in X have the structure of a complex 2-plane bundle, then $b \equiv 0 \pmod 4$.

Since a non-singular algebraic subvariety $V \subseteq X$ of codimension 2 is in particular a C^∞ submanifold with complex normal bundle, its cohomology class must be $ax^2 + by$ with $b \equiv 0 \pmod 4$. The same is true of any linear combination of such V's. Hence the cycle Y given above, whose cohomology class is y, cannot be rationally equivalent to any cycle $\Sigma n_i V_i$ with the V_i non-singular.

The proof uses a method introduced by Thom [62]. The normal bundle of M is a real bundle of dimension 4 for the orthogonal group $O(4)$, hence determines a continuous map $g\colon M \to BO(4)$.

$$\begin{array}{ccc} X & \xrightarrow{f} & MO(4) \\ \uparrow & & \uparrow \\ M & \xrightarrow{g} & BO(4) \end{array}$$

This extends to a map $f\colon X \to MO(4)$ of X into the Thom space $MO(4)$, in such a way that $f^*U = ax^2 + by$, where $U \in H^4(MO(4),\mathbb{Z})$ is the Thom class. Then we use Steenrod squares, and the known cohomology of $MO(4)$ and the known cohomology of X, to show that such a map cannot exist unless $b \equiv 0 \pmod 2$.

The stronger result $b \equiv 0 \pmod 4$ in the case of complex normal bundle uses a similar argument with $MU(2)$, and also some higher cohomology operations.

If we ask whether some multiple of Y might be smoothable, this method will give no information. Indeed, Thom shows that for any cohomology class in $H^4(X,\mathbb{Z})$, some multiple can always be represented by a C^∞ sub-manifold, even with complex normal bundle if we like. So this leaves open the problem: is some multiple of Y smoothable? Since any positive multiple of x^2 can be represented by a non-singular subvariety, this problem is equivalent to the following: does there exist a non-singular subvariety $V \subseteq X$ of codimension 2 whose cohomology class is $ax^2 + by$ with $b \neq 0$?

Using the same techniques as above, with Steenrod reduced p^{th} powers, one can show that on larger Grassmann varieties there are non-smoothable cycles which require taking higher and higher multiples before they could conceivably be smoothed. To be precise, one can prove the following theorem. Here $X = G(2,n)$, and the cohomology classes x^2, y have the same significance as before.

Theorem 7.5. Given an integer $k > 0$, $\exists n_0$ such that if $X = G(2,n)$, with $n \geq n_0$, and if M is a C^∞ submanifold of real codimension 4 with cohomology class $ax^2 + by$, then $b \equiv 0 \pmod k$.

Subvarieties of Small Degree. Some closely related results about non-singular subvarieties of codimension 2 of Grassmann varieties have recently been obtained by Barth and van de Ven [5] using a combination of elementary geometric techniques, and a generalization of the Barth-type theorems of §1 to Grassmann varieties.

We work on a Grassmann variety $X = G(r,n)$ with $r \geq 2$ fixed. If Y is a subvariety of codimension 2 we denote its cohomology class by $\eta(Y) = ax^2 + by$. One sees easily that $a \geq 0$ and $a + b \geq 0$. The idea of their result is that if a and $a + b$ are small with respect to n, then the variety Y, supposed non-singular, must be a complete intersection.

Theorem 7.6. Let Y be a non-singular subvariety of codimension 2 with cohomology class $ax^2 + by$ of $X = G(r,n)$ over $\mathbb{C}$, with $r \geq 2$ fixed. Then there is a function $\varphi(n)$, which tends to ∞ with n, such that if

$n - r \geqslant 6$, and $a, a + b \leqslant \varphi(n)$, then in fact $b = 0$, and Y is a complete intersection of divisors on X.

There is an analogous theorem for subvarieties of $\mathbb{P}^n$ of small degree, proved independently by Barth and van de Ven [6] and by Hartshorne [25] (at least in case of codimension $\leqslant 4$):

Theorem 7.7. Given an integer d, $\exists n_0$ such that if Y is a non-singular subvariety of $\mathbb{P}^n_{\mathbb{C}}$ of degree d, which is not contained in any hyperplane $\mathbb{P}^{n-1}$, and if $n \geqslant n_0$, then Y is a complete intersection.

Actually we hope that something much stronger is true: if Y is a non-singular subvariety of dimension r of $\mathbb{P}^n_{\mathbb{C}}$, and if $r > \frac{2}{3}n$, is Y a complete intersection?

References

1. Andreotti, A. and Grauert, H. Théorèmes de finitude pour la cohomologie des espaces complexes. Bull. Soc. Math. France 90 (1962), 193-259.

2. Barth, W. Transplanting cohomology classes in complex-projective space. Amer. J. Math. 92 (1970), 951-967.

3. Barth, W. and Larsen, M.E. On the homotopy groups of complex projective algebraic manifolds. Math. Scand. 30 (1972), 88-94.

4. Barth, W. and Van de Ven, A. A decomposability criterion for algebraic 2-bundles on projective spaces (preprint).

5. Barth, W. and Van de Ven, A. On the geometry in codimension 2 of Grassmann varieties (preprint).

6. Barth W. Submanifolds of low codimension in projective space (preprint).

7. Bloch, S. K_2 and algebraic cycles. Annals of Math. 99 (1974), 349-379.

8. Bloch, S. K_2 of artinian $\mathbb{Q}$-algebras, with applications to algebraic cycles (preprint).

9. Borel, A. and Serre, J.-P. Le théorème de Riemann-Roch, Bull. Soc. Math. France 86 (1958), 97-136.

10. Chevalley, C. Anneaux de Chow et applications. Sém. Chevalley (1958), Paris.

11. Clemens, C. H. and Griffiths, P. A. The intermediate Jacobian of the cubic three-fold. Annals of Math. 95 (1972), 281-356.

12. Griffiths, P. A. Some results on algebraic cycles on algebraic manifolds. in Algebraic Geometry (Bombay colloquium) Oxford (1969), 93-191.

13. Griffiths, P. A. On the periods of certain rational integrals, I, II. Annals of Math. 90 (1969), 460-495, 498-541.

14. Grothendieck, A. On the DeRham cohomology of algebraic varieties. Publ. Math. I.H.E.S. 29 (1966), 95-103.

15. Grothendieck, A. Standard conjectures on algebraic cycles, in Algebraic Geometry (Bombay colloquium) Oxford (1969), 193-199.

16. Grothendieck, A. Séminaire de Géométrie Algebrique (SGA5) (to appear).

17. Grothendieck, A.; Raynaud, M.; Rim, D.S.; Deligne, P. and Katz, N. Groupes de monodromie en géométrie algébrique (SGA 7) Springer Lecture Notes 288 (1972), 340 (1973).

18. Hartshorne, R. Cohomology of non-complete algebraic varieties. Compos. Math. 23 (1971), 257-264 (also in Algebraic Geometry, Oslo 1970, ed. F. Oort).

19. Hartshorne, R. Algebraic DeRham cohomology, manus. math. 7 (1972), 125-140.

20. Hartshorne, R. and Ogus, A. On the factoriality of local rings of small embedding codimension, Comm. in Algebra 1 (1974), 415-437.

21. Hartshorne, R., Topological conditions for smoothing algebraic singularities. Topology 13 (1974), 241-253.

22. Hartshorne, R. On the De Rham cohomology of algebraic varities. Publ. Math. I.H.E.S. (to appear).

23. Hartshorne, R.; Rees, E., and Thomas, E. Non-smoothing of algebraic cycles on Grassmann varieties. Bull. A.M.S. (to appear).

24. Hartshorne, R. Varieties of small codimension in projective space. Bull. A.M.S. (to appear).

25. Hartshorne, R. Projective varieties of small degree (in preparation).

26. Herrera, M. and Lieberman, D. Duality and the DeRham cohomology of infinitesimal neighborhoods. Invent. math. 13 (1971), 97-124.

27. Hironaka, H. Smoothing of algebraic cycles of small dimensions. Amer. J. Math. 90 (1968), 1-54.

28. Holme, A. Embedding-obstruction for algebraic varieties, I. (preprint)

29. Kleiman, S.L. Algebraic cycles and the Weil conjectures. in Dix Exposés sur la cohomologie des schémas, North-Holland (1968), 359-386.

30. Kleiman, S.L. Geometry on Grassmannians and applications to splitting bundles and smoothing cycles. Publ. Math. I.H.E.S. 36 (1969), 281-298.

31. Kleiman, S.L. Finiteness theorems for algebraic cycles, Actes Cong. Int. Math (1970), vol. 1, 445-449.

32. Kleiman, S.L. and Landolfi, J. Geometry and deformation of special Schubert varieties. Compos. math. 23 (1971), 407-434.

33. Kleiman, S.L. Motives, in Algebraic Geometry, Oslo 1970, ed. F. Oort, Wolters-Noordhoff (1972), 53-82.

34. Kleiman, S.L. Finiteness theorems for algebraic cycles, in Algebraic Geometry, Oslo 1970, ed. F. Oort, Wolters-Noordhoff (1972), 83-88.

35. Lang, S. and Néron, A. Rational points of abelian varieties over function fields. Amer. J. Math. 81 (1959), 95-118.

36. Larsen, M.E. On the topology of complex projective manifolds, Invent. math. 19 (1973), 251-260.

37. Lefschetz, S. On certain numerical invariants of algebraic varieties with applications to Abelian varieties. Trans. A.M.S. 22 (1921), 327-482.

38. Lieberman, D. Numerical and homological equivalence of algebraic cycles on Hodge manifolds. Amer. J. Math. 90 (1968), 366-374.

39. Lieberman, D. Higher Picard varieties. Amer. J. Math. 90 (1968), 1165-1199.

40. Lieberman, D. Intermediate Jacobians, in Algebraic Geometry, Oslo 1970, ed. F. Oort, Wolters-Noordhoff (1972), 125-139.

41. Manin, Ju. I., Correspondences, motifs, and monoidal transformations. Mat. Sbornik 77 (1970), 475-507. (AMS translation)

42. Matsusaka, T. The criteria for algebraic equivalence and the torsion group. Amer. J. Math. 79 (1957), 53-66.

43. Mattuck, A. Ruled surfaces and the Albanese mapping. Bull. A.M.S. 75 (1969), 776-779.

44. Mattuck, A. On the symmetric product of a rational surface, Proc. A.M.S. 21 (1969), 683-688.

45. Moishezon, B. C. Algebraic homology classes on algebraic varieties. Math. U.S.S.R.-Izvestia 1 (1967), 209-251.

46. Mumford, D. Lectures on curves on an algebraic surface, Annals. of Math. Studies 59 (1966), Princeton.

47. Mumford, D. Rational equivalence of 0-cycles on surfaces. J. Math. Kyoto Univ. 9 (1969), 195-204.

48. Mumford, D. Prym varieties, I. in Contributions to analysis, ded. to Lipman Bers, Academic Press (1974).

49. Murre, J.P. Algebraic equivalence modulo rational equivalence on a cubic threefold. Compos. math. 25 (1972), 161-206.

50. Murre, J.P. Reduction of the proof of the non-rationality of a non-singular cubic threefold to a result of Mumford. Compos. math. 27 (1973), 63-82.

51. Murre, J.P. Some results on cubic threefolds. (preprint)

52. Ogus, A. Local cohomological dimension of algebraic varieties. Annals of math. 98 (1973), 327-365.

53. Ogus, A. On the formal neighborhood of a subvariety of projective space, Amer. J. Math. (to appear).

54. Roberts, J. Chow's moving lemma, in Algebraic Geometry Oslo 1970, ed. F. Oort, Wolters-Noordhoff (1972), 89-96.

55. Roitman, A.A. On Γ-equivalence of zero-dimensional cycles, Math. U.S.S.R.-Sbornik 15 (1971), 555-567.

56. Roitman, A.A. Rational equivalence of 0-dimensional cycles, Math. U.S.S.R.-Sbornik 18 (1972), 571-588.

57. Roitman, A.A. Algebraic surfaces with finite-dimensional non-zero group of zero-dimensional cycle classes of degree zero, modulo rational equivalence, are ruled. (in Russian). Funk. Anal. i evo prilog. 8 (1974), 88-89.

58. Samuel, P. Relations d'équivalence en géométrie algébrique. Proc. I.C.M. (1958), 470-487.

59. Serre, J.-P. Algèbre Locale• Multiplicités, Springer Lecture Notes 11 (1965).

60. Severi, F. Problèmes résolus et problèmes nouveaux dans la théorie des systèmes d'équivalence. Proc. I.C.M. (1954) vol. III, 529-541.

61. Shafarerich, I.R., et. al. Algebraic Surfaces. Proc. Steklov Inst. Math. 75, Trans. A.M.S. (1967).

62. Thom, R. Quelques propriétés globales des variétés différentiables, Comment. Helv. 28 (1954), 17-86.

63. Tyurin, A.N., Five lectures on three-dimensional varieties, Russian Math Surveys 27 (1972), 1-53.

64. Weil, A., Foundations of algebraic geometry, A.M.S. Colloquium Publ. 29 (1946), revised + enlarged (1962).

65. Weil, A. On Picard varieties, Amer. J. Math. 74 (1952), 865-893.

Mathematics Department
University of California at Berkeley

Proceedings of Symposia in Pure Mathematics
Volume 29, 1975

TRIANGULATIONS OF ALGEBRAIC SETS

Heisuke Hironaka[1]

INTRODUCTION

This note was mainly motivated by an impression (which may be wrong) that many of the modern algebraic geometers either do not at all know how to prove, or somehow have gotten a belief that it is very difficult to prove an old well-known theorem that an algebraic set (possibly with singularities) admits triangulations. Triangulations that we obtain are semi-algebraic in the sense made precise in the first section, so as to give a canonical PL-structure (up to isomorphisms) to each algebraic set. It should not be expected, of course, that an algebraic set has a unique PL-structure (up to isomorphisms) as a topological space.

A standard approach is to find triangulations by induction on dimensions, in that we take a suitable linear projection to a lower dimensional ambient space and apply induction assumption to the branch locus (or discriminant set) associated with such a projection. According to this inductive approach, it is more natural to consider a finite system of (rather than a single) semi-algebraic sets (rather than algebraic sets) in $\mathbb{R}^n$ and to search for a triangulation of $\mathbb{R}^n$ which is compatible with every member of the given system. This will become quite apparent from the proof given in the second section.

Needless to say, most of the important ideas (all more or less elementary) are classical. (See the papers listed for reference.) The present proof, in its basic structure, was extracted from the paper of Lojasiewicz [7], in which semi-analytic triangulations of a semi-analytic set was first formulated and proven. Semi-algebraicity, more special, has one big advantage over semi-analyticity, mainly due to the projection theorem of Seidenberg-Tarski which is stated as Proposition II in the first section. By this, the proof becomes substantially simpler in the semi-algebraic case than that of Lojasiewicz in the semi-analytic case. My effort in this note is above all to clarify the key ideas while describing a precise inductive procedure for triangulations. My hope is that this note is more easily understood by modern algebraic geometers at the least.

For those readers, interested in analytic sets and varieties as well, we shall indicate how essentially the same semi-algebraic method can be generalized to prove subanalytic triangulations of a subanalytic set, again extracting some idea from Lojasiewicz [7]. Subanalyticity is more general than semi-analyticity and, of course, closed real-analytic sets are semi-analytic. But, what is more interesting is that the proof of triangulation becomes simpler and much similar to the case of semi-algebraic sets, for the subanalytic case than the semi-analytic case, again due to the projection theorem which is valid in the former case but not in the latter. Though generalization does not mean strengthening, subanalytic triangulations are enough to determine a canonical PL-structure on each closed real-analytic set.

1. SEMI-ALGEBRAIC SETS

Let us recall some definitions and basic facts about semi-algebraicity. By a Boolean class of subsets, we shall mean a class of subsets which is closed by the following elementary set-theoretical operations.

a) finite intersection

b) finite union

c) complementary set (and difference of any two)

DEFINITION 1.1. A subset of $\mathbb{R}^n$ is said to be semi-algebraic if it belongs to the Boolean class of subsets of $\mathbb{R}^n$ which is generated by those of the form

$$\{x \in \mathbb{R}^n \mid f(x) \geq 0\}$$

where f is any polynomial function on $\mathbb{R}^n$.

REMARK 1.2. In the above definition, we may replace $\geq$ by $>$, because

$$\{x \in \mathbb{R}^n \mid f(x) > 0\} = \mathbb{R}^n - \{x \in \mathbb{R}^n \mid -f(x) \geq 0\}.$$

Moreover, the semi-algebraic Boolean class contains

$$\{x \in \mathbb{R}^n \mid f(x) = 0\}$$
$$= \{x \in \mathbb{R}^n \mid f(x) \geq 0\} \bigcap \{x \in \mathbb{R}^n \mid -f(x) \geq 0\} .$$

REMARK 1.3. A subset $X \subset \mathbb{R}^n$ is semi-algebraic if and only if there exist a finite number of polynomials f_{ij}, g_{ij} on $\mathbb{R}^n$ such that

$$X = \bigcup_i \{x \in \mathbb{R}^n \mid f_{ij}(x) > 0, \ \forall_j, \ g_{ij}(x) = 0, \ \forall_j\} .$$

The following propositions will be referred to as the basic three facts on semi-algebraic sets.

PROPOSITION I. If X is semi-algebraic in $\mathbb{R}^n$, then its interior (often denoted by int(X)) and its closure $\overline{X}$ are also semi-algebraic in $\mathbb{R}^n$.

PROPOSITION II. (Seidenberg-Tarski [8]) If $X \subset \mathbb{R}^n$ is semi-algebraic and if $h: \mathbb{R}^n \longrightarrow \mathbb{R}^m$ is a polynomial map, then $h(X)$ is semi-algebraic in $\mathbb{R}^m$.

We need this only in the case of a linear projection $\mathbb{R}^n \longrightarrow \mathbb{R}^{n-1}$. It should be noted that a linear projection of an algebraic set is not in general algebraic but semi-algebraic. (See the projection of a circle in $\mathbb{R}^2$ into $\mathbb{R}$.)

PROPOSITION III. Given a finite system of semi-algebraic sets $\{X_\alpha\}$ in $\mathbb{R}^n$, there exists a finite semi-algebraic stratification of $\bigcup_\alpha X_\alpha$ which is compatible with every X_α.

This statement requires some explanation. By a stratification of a subset A of $\mathbb{R}^n$, we mean a decomposition $A = \bigcup_\mu \Gamma_\mu$, which is disjoint, such that

a) $\{\Gamma_\mu\}$ is locally finite about every point of $\mathbb{R}^n$. (This condition is not only about points of A but also about every boundary point of A.)

b) Each Γ_μ is a smooth connected real-analytic submanifold, locally closed in $\mathbb{R}^n$.

c) (Frontier condition) If $\overline{\Gamma}_\mu \cap \Gamma_\nu \neq \emptyset$, then $\overline{\Gamma}_\mu \supset \Gamma_\nu$.

A stratification $A = \bigcup_\mu \Gamma_\mu$ is said to be finite if we have only finitely many strata Γ_μ. It is said to be semi-algebraic if every Γ_μ is so in $\mathbb{R}^n$.

Let Δ be a simplex of dimension r in $\mathbb{R}^m$. Namely, we have $r+1$ points $v_0, v_1, \cdots, v_r \in \mathbb{R}^m$, not all in any single linear subspace of dimension $r-1$, such that

$$\Delta = \left\{ \sum_{i=0}^{r} \alpha_i v_i \middle| \alpha_i \in \mathbb{R}, > 0, \sum_{i=0}^{r} \alpha_i = 1 \right\}.$$

Those v_i are called the vertices of Δ. The following example of semi-algebraic set will be used in our triangulation procedure.

EXAMPLE 1.4 (Semi-algebraic tent) Let $\pi: \mathbb{R}^n \longrightarrow \mathbb{R}^{n-1}$ be the projection: $(x_1, \cdots, x_n) \longrightarrow (x_1, \cdots, x_{n-1})$. Let Δ be a simplex of dimension r in $\mathbb{R}^{n-1}$, $0 \leq r \leq n-1$. Let Y be a closed semi-algebraic set in $\mathbb{R}^n$ such that π induces a homeomorphism $Y \xrightarrow{\sim} \overline{\Delta}$. Let B be an open ball, say centered at the origin 0 of $\mathbb{R}^{n-1}$, such that $B \supset \overline{\Delta}$. Let $S = \partial B$, an $(n-2)$-sphere in $\mathbb{R}^{n-1}$. Then there exists a closed

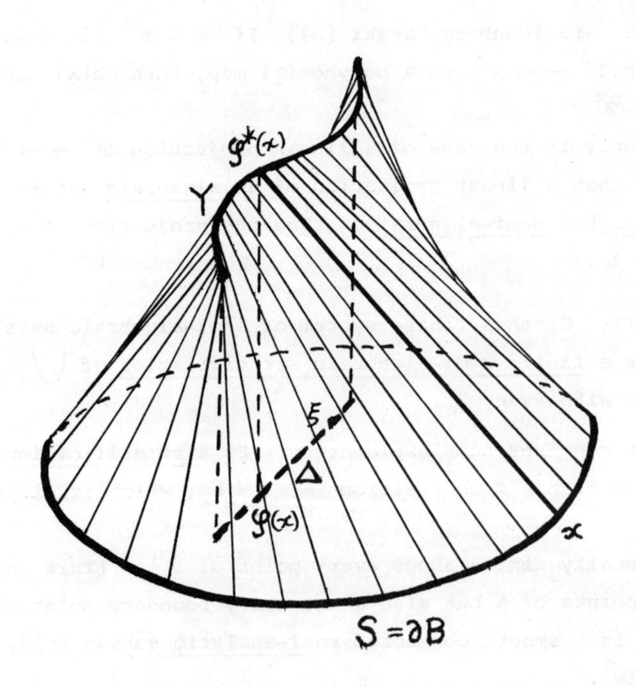

semi-algebraic set Z in $\mathbb{R}^n$ such that (with reference to $\mathbb{R}^{n-1} \subset \mathbb{R}^n$ by $y \longrightarrow (y,0)$)

a) $Z \supset Y \cup S$

b) π induces a homeomorphism $Z \xrightarrow{\sim} \overline{B}$.

We construct such Z explicitly as follows. Let L be the linear subspace of dimension r of $\mathbb{R}^{n-1}$, which contains Δ. Let s be the orthogonal projection from $\mathbb{R}^{n-1}$ to L. Pick a point $\xi \in \Delta$ such that if t denotes the translation in $\mathbb{R}^{n-1}$ which sends 0 to ξ, then $t(B) \supset \overline{\Delta}$. Define $\varphi: S \longrightarrow \overline{\Delta}$ as follows. For $x \in S$, let $H(x)$ be the (or any) half-line passing through $st(x)$ and ending at ξ. Let $h(x)$ be the length of $\overline{\Delta} \cap H(x)$. Then we define

$$\varphi(x) = \xi + \frac{h(x)}{N}(st(x)-\xi)$$

where N is the radius of B. Let $\varphi^*(x)$ be the unique point of $Y \cap \pi^{-1}(\varphi(x))$. Now let Z be the union of the closed segments connecting x and $\varphi^*(x)$ for all $x \in S$. It has the required properties a) and b). This Z will be called the <u>tent</u> spanned from Y to S.

EXAMPLE 1.5. (<u>Semi-algebraic cone</u>) Let L be any linear subspace of $\mathbb{R}^n$ and let Y be any semi-algebraic set in L. Let η be any point of $\mathbb{R}^n$ and let $C(\eta,Y)$ be the union of the closed segments connecting η and x, for all $x \in Y$. Then $C(\eta,Y)$ is semi-algebraic in $\mathbb{R}^n$. (This is an easy consequence of Prop. II.)

DEFINITION 1.6. Let A (resp. B) be a semi-algebraic subset of $\mathbb{R}^n$ (resp. $\mathbb{R}^m$). A map $\varkappa: A \longrightarrow B$ is said to be <u>semi-algebraic</u> if it is continuous and its graph is semi-algebraic in $\mathbb{R}^n \times \mathbb{R}^m$.

REMARK 1.7. A composition of two semi-algebraic maps, whenever defined, is semi-algebraic. The inverse of a semi-algebraic isomorphism is semi-algebraic. The image of a semi-algebraic set by a semi-algebraic map is semi-algebraic. (For this, we need Prop. II.)

EXAMPLE 1.8. Let $\varkappa: \mathbb{R}^2 \longrightarrow \mathbb{R}^2$ is the map defined by $(x,y) \longrightarrow (x, y-\sqrt[3]{x^2})$. This is semi-algebraic, as its graph is defined by $x'-x = 0$ and $(y-y')^3-x^2 = 0$, where (x',y') are the coordinates in the second $\mathbb{R}^2$. Note that the cusp $y^3 = x^2$ is mapped to the x'-axis (a straight line).

The following example gives the kind of semi-algebraic automorphisms which we use for the purpose of semi-algebraic triangulations.

EXAMPLE 1.9. (Semi-algebraic vertical shifting) Let $\pi: \mathbb{R}^n \longrightarrow \mathbb{R}^{n-1}$ be that of Ex. 1.4. Let Δ be a simplex in $\mathbb{R}^{n-1}$. Let M be a positive real number (or $+\infty$) and let $A = \{(y,t) \in \mathbb{R}^n \mid y \in \Delta, M \geq t \geq -M\}$. Let $\{\Gamma_\mu\}$, $1 \leq \mu \leq r$, and $\{E_\mu\}$, $1 \leq \mu \leq r$, be two systems of semi-algebraic sets in $\mathbb{R}^n$ such that

1) Γ_μ and E_μ are locally closed smooth real-analytic submanifolds of $\mathbb{R}^n$,

2) π induces real-analytic isomorphisms $\Gamma_\mu \xrightarrow{\sim} \Delta$ and $E_\mu \xrightarrow{\sim} \Delta$ for all μ,

3) if $\Gamma_\mu \cap \pi^{-1}(y) = \gamma_\mu(y)$ and $E_\mu \cap \pi^{-1}(y) = e_\mu(y)$ for $y \in \Delta$, then

$$-M < \gamma_1(y)_n < \cdots < \gamma_r(y)_n < M \quad \text{and}$$

$$-M < e_1(y)_n < \cdots < e_r(y)_n < M$$

for all $y \in \Delta$, where suffix n indicates the n-th coordinate. Define the map $h: A \longrightarrow A$ as follows. Write $\gamma_o(y) = e_o(y) = (y,-M)$ and $\gamma_{r+1}(y) = e_{r+1}(y) = (y,M)$ for each $y \in \Delta$. For $x \in A$, choose i such that

$$\gamma_i(y)_n \leq x_n \leq \gamma_{i+1}(y)_n \quad \text{with} \quad y = \pi(x),$$

and write $x_n = \alpha(x)\gamma_i(y)_n + \beta(x)\gamma_{i+1}(y)_n$ with $\alpha(x) \geq 0$, $\beta(x) \geq 0$ such that $\alpha(x) + \beta(x) = 1$. Define

$$h(x) = \alpha(x)e_i(\pi(x)) + \beta(x)e_{i+1}(\pi(x)).$$

Then h is a semi-algebraic automorphism of A (with reference to $A \subset \mathbb{R}^n$).

THEOREM. (Semi-algebraic triangulation) Given a finite system of bounded semi-algebraic sets $\{X_\alpha\}$ in $\mathbb{R}^n$, there exists a simplicial decomposition $\mathbb{R}^n = \bigcup_a \Delta_a$ and a semi-algebraic automorphism $\varkappa$ of $\mathbb{R}^n$ such that

1) each X_α is a finite union of some of the $\varkappa(\Delta_a)$,

2) $\varkappa(\Delta_a)$ is a locally closed smooth real-analytic submanifold of $\mathbb{R}^n$ and $\varkappa$ induces a real-analytic isomorphism $\Delta_a \xrightarrow{\sim} \varkappa(\Delta_a)$, for every a,

3) outside some compact subset of $\mathbb{R}^n$, $\varkappa$ is the identity.

REMARK 1.10. We lose little loss of generality by assuming the boundedness of the X_α. In fact, we have a real-algebraic embedding of $\mathbb{R}^n$ into some $\mathbb{R}^N$ (via $\mathbb{P}^n \subset \mathbb{R}^N$) which maps every semi-algebraic set in $\mathbb{R}^n$ to a bounded semi-algebraic set in $\mathbb{R}^N$.

REMARK 1.11. Since $\varkappa$ is semi-algebraic, all the $\varkappa(\Delta_a)$ are semi-algebraic in $\mathbb{R}^n$.

REMARK 1.12. Condition 1) means that the triangulation of $\mathbb{R}^n$ induces triangulation of X_α for every α. The conditions 2) and 3) are added mainly because they make our inductive proof easier.

2. SEMI-ALGEBRAIC TRIANGULATION

We now want to give a proof of the semi-algebraic triangulation theorem, stated in §1. The proof is by induction on n. It is trivial for $n \leq 1$. Assume $n \geq 2$.

STEP 1. The problem is reduced to the case in which every X_α is closed in $\mathbb{R}^n$. In fact, if X_α is not closed, then we replace X_α by a finite system of closed bounded semi-algebraic sets as follows. Let $X_{\alpha,0} = X_\alpha$, and $X_{\alpha,i+1} = \overline{X}_{\alpha,i} - X_{\alpha,i}$, $i = 0,1,2,\ldots$. Then $\dim X_{\alpha,i+1} < \dim X_{\alpha,i}$ so long as $X_{\alpha,i} \neq \emptyset$. Replace X_α by $\{\overline{X}_{\alpha,i}\}_{0 \leq i \leq s}$, where $X_{\alpha,s+1} = \emptyset$. (Cf. Basic fact I, §1.)

STEP 2. We may assume that each X_α has no interior points, i.e., $\mathrm{int}(X_\alpha) = \emptyset$, so that $\dim X_\alpha \leq n-1$. In fact, let $X'_\alpha = X_\alpha - \mathrm{int}(X_\alpha)$ (cf. Basic fact I, §1.) Since X_α is closed, $\mathrm{int}(X_\alpha)$ is a union of some connected component of $\mathbb{R}^n - X'_\alpha$. It is clear that a triangulation of $\mathbb{R}^n$ compatible with $\{X'_\alpha\}$ is automatically compatible with $\{X_\alpha\}$.

STEP 3. (Good projection) This is the most important step in the process of triangulations. It is the question of finding a "good" linear projection $\pi: \mathbb{R}^n \longrightarrow \mathbb{R}^{n-1}$ with respect to the given $\{X_\alpha\}$. We say that a linear projection π is "good" with respect to $\{X_\alpha\}$, if for every $y \in \mathbb{R}^{n-1}$, $\pi^{-1}(y) \cap X_\alpha$ is a discrete set of points for every α. In our case, the discreteness implies the finiteness because of the bounded semi-algebraicity of $\pi^{-1}(y) \cap X_\alpha$. (Cf. Basic fact III.)

The existence of good projections is proven as follows. By a "direction" in $\mathbb{R}^n$, we shall mean a point $v \in \mathbb{P}^{n-1}$, the real-projective space associated with $\mathbb{R}^n$. Each $v \in \mathbb{P}^{n-1}$ corresponds to a line through O in $\mathbb{R}^n$, denoted by $L(o,v)$, and we then choose a linear map $\pi: \mathbb{R}^n \longrightarrow \mathbb{R}^{n-1}$ such that the fibres $\pi^{-1}(y)$, $y \in \mathbb{R}^{n-1}$, are exactly those lines in $\mathbb{R}^n$ parallel to $L(o,v)$. For each $\xi \in \mathbb{R}^n$, we denote by $L(\xi,v)$ the line through ξ and parallel to $L(o,v)$. Given a subset Z in $\mathbb{R}^n$, we ask if the following condition is satisfied for each pair (ξ,v), $\xi \in Z$ and $v \in \mathbb{P}^{n-1}$:

(*) Z contains a neighborhood of ξ in the line $L(\xi,v)$. (This means that v is a "<u>bad</u>" direction for Z at ξ.)

Here is the key lemma due to Koopman and Brown, [3] and [7].

GOOD DIRECTION LEMMA. Let Z be a locally closed real-analytic subset of $\mathbb{R}^n$, which is nowhere dense in $\mathbb{R}^n$. Then the set

$$\mathcal{B}(Z) = \{v \in \mathbb{P}^{n-1} \mid \exists\ \xi \in Z \text{ such that } L(\xi,v) \text{ satisfies } (*)\}$$

does not contain any non-empty open subset of $\mathbb{P}^{n-1}$. In fact, it is a Baire category set (a countable union of nowhere dense closed subsets) in $\mathbb{P}^{n-1}$.

REMARK 2.1. The question is local. Namely it is enough to prove: <u>Given any</u> $x \in Z$, <u>there exists a neighborhood</u> N <u>of</u> x <u>in</u> Z <u>such that</u> B(N), <u>similarly defined, is nowhere dense in</u> $\mathbb{P}^{n-1}$. Note that if N is compact then B(N) is closed in $\mathbb{P}^{n-1}$. Z being countable at infinity, if the local question is affirmatively answered for every $x \in Z$, then B(Z) can be expressed as a countable union of nowhere dense closed subsets of $\mathbb{P}^{n-1}$. The proof of the local version is left to the readers. (Or refer to [3],[7].)

REMARK 2.2. The good direction lemma, in the form stated above, extends immediately to the case in which Z is a countable union of locally closed real-analytic subsets in $\mathbb{R}^n$.

Back to Step 3, it is now easy to find a "good" linear projection, $\pi: \mathbb{R}^n \longrightarrow \mathbb{R}^{n-1}$ with respect to the given $\{X_\alpha\}$. Let us pick one such π and, by change of coordinate, assume that π is given by $(x_1,\cdots,x_n) \longrightarrow (x_1,\cdots,x_{n-1})$.

Next we want to reduce the problem to the case in which, if

$X = \bigcup_\alpha X_\alpha$ and $D = \mathrm{int}(\pi(X))$, then

(**) $X \bigcap \pi^{-1}(D)$ is dense in X, π induces an <u>open</u> map $X \bigcap \pi^{-1}(D) \longrightarrow D$, and $X - \pi^{-1}(D)$ is contained in the hyperplane $x_n = 0$.

To achieve this reduction, the easiest way is to proceed along the next two steps without (**) and then come back to (**). So the next two steps will be repeated twice, once before (**) and again after (**).

STEP 4. There exists a "good" finite semi-algebraic stratification of $\{X_\alpha\}$ <u>with respect to</u> π. This means the following. Let $\{\Gamma_\mu\}$ be a finite semi-algebraic stratification of X, compatible with every X_α. It is said to be "<u>good</u>" with respect to π, if for every μ, π induces a real-analytic immersion $\Gamma_\mu \longrightarrow \mathbb{R}^n$, i.e., rank $d(\pi|\Gamma_\mu)_x = \dim \Gamma_\mu$ for all $x \in \Gamma_\mu$.

To find a "good" stratification as above, let us start with any finite semi-algebraic stratification $\{\Gamma_\mu\}$ of the $\{X_\alpha\}$. (This exists by Prop. III, §1.) If it is not "good", we make a modification as follows. By Step 3, π induces a finite-to-one map $\Gamma_\mu \longrightarrow \mathbb{R}^{n-1}$ for every μ. Let this be called $\pi_\mu : \Gamma_\mu \longrightarrow \mathbb{R}^{n-1}$. Let $S_\mu = \{x \in \Gamma_\mu | \mathrm{rank}(d\pi_\mu)_x < \dim \Gamma_\mu\}$ which is semi-algebraic in $\mathbb{R}^n$. If some of the S_μ are not empty, we make the following modification on $\{\Gamma_\mu\}$:

"Replace $\{\Gamma_\mu\}$ by a finite semi-algebraic stratification $\{\Gamma'_\nu\}$ of X which is compatible with every one of the Γ_μ and the S_μ."

Now examine $S'_\nu = \{x \in \Gamma'_\nu | \mathrm{rank}(d\pi'_\nu)_x < \dim \Gamma'_\nu\}$ for each ν, where $\pi'_\nu : \Gamma'_\nu \longrightarrow \mathbb{R}^{n-1}$ is induced by π. If $S'_\nu \neq \emptyset$, then there exists μ such that

$$S'_\nu \subset \Gamma'_\nu \subset S_\mu \subset \Gamma_\mu$$

so that $\dim S'_\nu < \dim \Gamma'_\nu \leq \dim S_\mu$. Hence if we repeat the above modification, at most n-1 times, we achieve the situation in which $S'_\nu = \emptyset$ for all ν. Namely we obtain a "good" stratification as required.

REMARK 2.3. A "good" stratification $\{\Gamma_\mu\}$ as in Step 4. does not imply that $\pi(\Gamma_\mu)$ is locally closed or smooth, in general.

STEP 5. (Induction assumption) Start with a "good" stratification $\{\Gamma_\mu\}$ as in Step 4. By Prop. II, §1, $\pi(\Gamma_\mu)$ is semi-algebraic in $\mathbb{R}^{n-1}$. Applying the induction assumption, we obtain a simplicial decomposition

$\mathbb{R}^{n-1} = \bigcup_a \Delta_a$ and a semi-algebraic automorphism $\varkappa$ of $\mathbb{R}^{n-1}$, which has the properties 1), 2) and 3) of the semi-algebraic triangulation theorem, §1, with respect to $\{\pi(\Gamma_\mu)\}$. We then replace $\{\Gamma_\mu\}$ by the family consisting of the connected components of $\Gamma_\mu \cap \pi^{-1}(\varkappa(\Delta_a))$ for all μ and a. This new family, again denoted by $\{\Gamma_\mu\}$, is a "good" finite semi-algebraic stratification as required in Step 4. Moreover, we can choose the above triangulation of $\mathbb{R}^{n-1}$ so as to have the following additional property:

(1) π induces a real-analytic isomorphism

$$\pi_\mu : \Gamma_\mu \xrightarrow{\sim} \varkappa(\Delta_{\gamma(\mu)})$$

for every μ, where $\gamma(\mu)$ is the unique index determined by this property.

Since X_α are bounded and closed, (1) implies

(1') π induces a homeomorphism

$$\overline{\pi}_\mu : \overline{\Gamma}_\mu \xrightarrow{\sim} \varkappa(\overline{\Delta}_{\gamma(\mu)}).$$

STEP 6. We now want to reduce the problem to the case in which (**) is satisfied. (See the paragraph preceding Step 4.). For this end, we use those $\varkappa$ and $\mathbb{R}^{n-1} = \bigcup_a \Delta_a$ obtained in Step 5. $\mathbb{R}^{n-1}$ will be identified with the hyperplane $x_n = 0$ as usual. Let J be the index set for $\{\Gamma_\mu\}$ and I that for $\{\Delta_a\}$. Let $\gamma : J \longrightarrow I$ be the map defined by (1) of Step 5. Pick a large enough ball B in $\mathbb{R}^{n-1}$ so that it contains all the $\overline{\Delta}_{\gamma(\mu)}$ for $\mu \in J$. Let S be the boundary of B. For each $\mu \in J$, let Z_μ be the <u>semi-algebraic tent</u> spanned from $\tilde{\varkappa}^{-1}(\Gamma_\mu)$ to S in the sense of Ex. 1.4, where $\tilde{\varkappa}$ is the semi-algebraic automorphism of $\mathbb{R}^n$ by $\tilde{\varkappa}(y,x_n) = (\varkappa(y),x_n)$ for $(y,x_n) \in \mathbb{R}^{n-1} \times \mathbb{R} = \mathbb{R}^n$. We then add all the $\tilde{\varkappa}(Z_\mu)$, $\mu \in J$, to our old family $\{X_\alpha\}$. After this, the new enlarged X satisfies condition (**).

At this point, of course, we must abandon the old $\{\Gamma_\mu\}$, $\varkappa$ and $\mathbb{R}^{n-1} = \bigcup_a \Delta_a$ which we had obtained by Steps 5 and 6. But then we can again apply these two steps to the new family $\{X_\alpha\}$ and obtain new $\{\Gamma_\mu\}_{\mu \in J}$, $\varkappa$ and $\mathbb{R}^{n-1} = \bigcup_{a \in I} \Delta_a$. This time, we not only regain (1) of Step 5, but also (thanks to (**)) we gain something more. Namely, we have:

(2) If $D = \mathrm{int}(\pi(X))$ and $X' = X - \pi^{-1}(D)$, then

$$X = \overline{X-X'} = \bigcup_{\mu\in J^o} \overline{\Gamma}_\mu$$

where $J^o = \{\mu \in J \mid \dim \Gamma_\mu = n-1\}$, and hence

$$\overline{D} = \pi(X) = \bigcup_{\mu\in J} \varkappa(\Delta_{\gamma(\mu)}) .$$

(3) If $J' = \{\mu \in J \mid \Gamma_\mu \subset X'\}$, then $X' = \bigcup_{\mu\in J'}\Gamma_\mu$ and $\Gamma_\mu = \varkappa(\Delta_{\gamma(\mu)})$ (via the identification $\mathbb{R}^{n-1} = \{x_n = 0\}$) for all $\mu \in J'$, so that $X' \subset \mathbb{R}^{n-1}$. Moreover, by (2), if $\Delta_{\gamma(\mu)} \subset \overline{\Delta}_a$ for $\mu \in J$ and $a \in \gamma(J)$, then there exists $\nu \in \gamma^{-1}(a)$ such that $\Gamma_\mu \subset \overline{\Gamma}_\nu$.

Applying a barycentric subdivision to $\mathbb{R}^{n-1} = \bigcup_a \Delta_a$ (and subdividing $\{\Gamma_\mu\}$ accordingly, but with the same $\varkappa$), we may further assume that

(4) for every $\mu,\nu \in J$, there exists $c \in I$ such that $\pi(\overline{\Gamma}_\mu \cap \overline{\Gamma}_\nu) = \varkappa(\overline{\Delta}_c)$, provided that the intersection is not empty.

Finally, viewing $\varkappa(\Delta_a)$'s as subsets of $\mathbb{R}^n$ (via $\mathbb{R}^{n-1} \subset \mathbb{R}^n$) and adding sufficiently many but only finite $\varkappa(\Delta_a)$, $a \in I-\gamma(J)$, to the above $\{\Gamma_\mu\}$ (and hence to X), we may assume that

(5) $\varkappa$ induces the identity in $\mathbb{R}^{n-1}-D$.

STEP 7 (Conclusion = vertical shifting) Under the assumptions (1)-(5) of Steps 5 and 6, we shall extend $\varkappa$ to a semi-algebraic automorphism $\hat{\varkappa}$ of $\mathbb{R}^n$ such that

(a) $\hat{\varkappa}^{-1}(\Gamma_\mu)$ is a simplex in $\mathbb{R}^n$ and $\hat{\varkappa}: \hat{\varkappa}^{-1}(\Gamma_\mu) \xrightarrow{\sim} \Gamma_\mu$ is a real-analytic isomorphism for all $\mu \in J$ and

(b) $\hat{\varkappa}$ induces the identity outside a compact subset G of $\mathbb{R}^n$.

Our extension $\hat{\varkappa}$ will be made explicitly so as to give a solution to the semi-algebraic triangulation for $\{X_\alpha\}$. First of all, G of (b) is chosen as follows. Pick a real number $M \gg 0$ so that all the Γ_μ are contained in

$$Q_o = \{x \in \pi^{-1}(\bar{D}) \mid -M < x_n < M\} .$$

Pick a pair of points in $\mathbb{R}^n$:

$$p_+ = (0, M+1) \text{ and}$$
$$p_- = (0, -M-1)$$

where 0 denotes the origin of $\mathbb{R}^{n-1}$. Let

$$Q_+ = \text{the cone from } p_+ \text{ to } \bar{Q}_o \cap \{x_n = M\}$$

$$Q_- = \text{the cone from } p_- \text{ to } \bar{Q}_o \cap \{x_n = -M\}.$$

Then our G will be $Q_o \cup Q_+ \cup Q_-$. We want to define $\hat{\kappa}$. For each $\mu \in J$, let E_μ be the image by $\tilde{\kappa}$ of the simplex in $\mathbb{R}^n$, which is spanned by

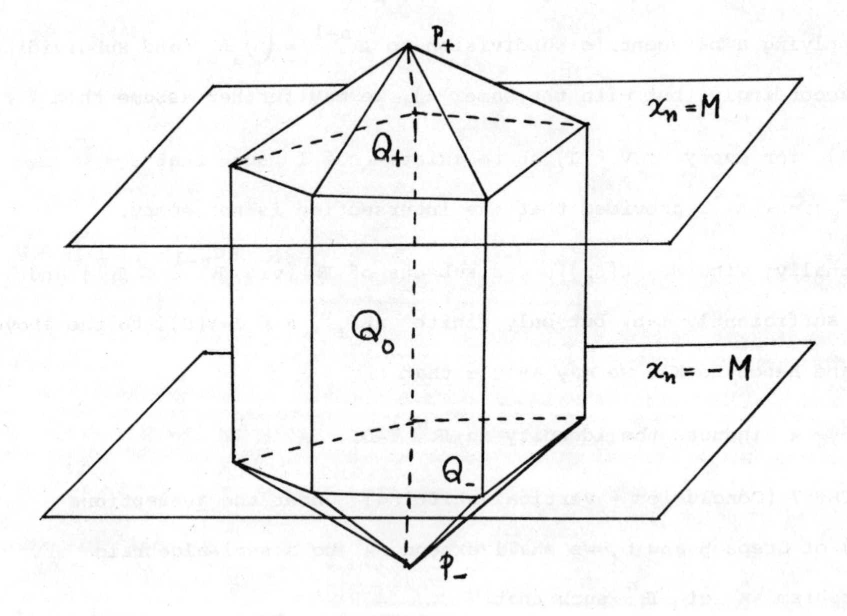

the "vertices" of $\tilde{\kappa}^{-1}(\Gamma_\mu)$. Here a point of $\tilde{\kappa}^{-1}(\bar{\Gamma}_\mu)$ is called a "vertex" if it is the unique point of $\tilde{\kappa}^{-1}(\bar{\Gamma}_\mu) \cap \pi^{-1}(v)$ for a vertex v of the simplex $\Delta_{\gamma(\mu)}$. Then, by (1) and (4), for every $a \in \gamma(J)$, the systems $\{\Gamma_\mu\}_{\mu \in \gamma^{-1}(a)}$ and $\{E_\mu\}_{\mu \in \gamma^{-1}(a)}$ satisfy the assumptions of Ex. 1.9. (Especially (4) is imperative to have all the E_μ distinct.) So, by the vertical shifting method of Ex. 1.9, we obtain a semi-algebraic automorphism h_a of $\overline{Q}_o \bigcap \pi^{-1}\kappa(\Delta_a)$. If $\Delta_b \subset \partial\Delta_a$ for $a,b \in \gamma(J)$, the condition (3) assures us that the combination of h_a and h_b defines a continuous automorphism of $\overline{Q}_o \bigcap \pi^{-1}\kappa(\Delta_a \cup \Delta_b)$. (Recall that for every $\mu \in \gamma^{-1}(a)$, $\overline{\Gamma}_\mu \bigcap \pi^{-1}\kappa(\Delta_b)$ is one of the Γ_ν with $\nu \in \gamma^{-1}(b)$, by (1) and by the closedness of X in $\mathbb{R}^n$.) Hence the $\{h_a\}_{a\in\gamma(J)}$ all together defines a semi-algebraic automorphism of $\overline{Q}_o$, which will be called h. Moreover, by (2) and (3), h induces the identity on $\partial\overline{Q}_o$. Hence, by (5), $h\tilde{\kappa}$ does the same on $Q_o \bigcap \partial\overline{Q}_o = Q_o - \pi^{-1}(D)$, where $\tilde{\kappa}$ is the extension of κ defined by $\tilde{\kappa}(y,x_n) = (\kappa(y),x_n)$. The restriction of $h\tilde{\kappa}$ to $\overline{Q}_o \bigcap \{x_n = M\}$ extends to an automorphism g_+ of Q_+ by linear suspension from p_+. Similarly, we obtain g_- of Q_-. Note that g_+ (resp. g_-) induces the identity on $\partial Q_+ \bigcap \{x_n > M\}$ (resp. $\partial Q_- \bigcap \{x_n < -M\}$). Define $\hat{\kappa}$ by letting

$$\hat{\kappa}|Q_o = h\tilde{\kappa}|Q_o$$

$$\hat{\kappa}|Q_\pm = g_\pm$$

$$\hat{\kappa}|\mathbb{R}^n-(Q_o \cup Q_+ \cup Q_-) = \text{the identity.}$$

Then $\hat{\kappa}$ is a semi-algebraic automorphism of $\mathbb{R}^n$ and has the properties (a) and (b). Moreover

(c) $\hat{\kappa}$ induces real-analytic isomorphisms in every one of the following ranges i) - iv):

i) from $\hat{\kappa}^{-1}(\Gamma_\mu)$ to Γ_μ for all $\mu \in J$,

ii) from $Q_o - \hat{\kappa}^{-1}(\Sigma)$ to $Q_o-\Sigma$, where Σ denotes the union of Γ_μ and $\pi^{-1}\kappa(\partial\Delta_{\gamma(\mu)})$ for all $\mu \in J$,

iii) from $[p^+,\Delta_a+M]$ to $[p^+,\kappa(\Delta_a)+M]$ for all $a \in \gamma(J)$, where Δ_a+M denotes the vertical translation of Δ_a by M (or $\Delta_a+M = \pi^{-1}(\Delta_a) \bigcap \{x_n = M\}$) and $[p^+,\Delta_a+M]$ denotes the cone from p^+ to Δ_a+M (p^+ itself excluded),

iv) from $[p^-,\Delta_a-M]$ to $[p^-,\kappa(\Delta_a)-M]$ for all $a \in \gamma(J)$.

Finally we take any simplicial decomposition $\mathbb{R}^n = \bigcup_d \Delta_d$ which is compatible with those simplices $\hat{\varkappa}^{-1}(\Gamma_\mu)$, $\mu \in J$, and with those subsets $\hat{\varkappa}^{-1}(\Sigma \cap \bar{Q}_o)$, $[p^+, \Delta_a + M]$ and $[p^-, \Delta_a - M]$, $a \in \gamma(J)$. These sets being all piecewise linear in $\mathbb{R}^n$, it is easy to find such a simplicial decomposition. It is clear that $\hat{\varkappa}$ and $\mathbb{R}^n = \bigcup_d \Delta_d$ have all the required properties for a semi-algebraic triangulation of the given $\{X_\alpha\}$ in $\mathbb{R}^n$, as was stated in the theorem of §1.

Q.E.D.

REMARK 2.4 (Canonicality of semi-algebraic triangulations) Let X be any bounded semi-algebraic set in $\mathbb{R}^n$. Pick any two semi-algebraic triangulations for the same X, say $(\varkappa_i, \mathbb{R}^n = \bigcup_a \Delta_{i,a})$, $i = 1,2$. Then let $\{X_\alpha\}$ be the collection of all those $\varkappa_i(\Delta_{ia})$ which are contained in X. It is a finite system of bounded semi-algebraic sets in $\mathbb{R}^n$. Hence there exists a semi-algebraic triangulation for $\{X_\alpha\}$, say $(\varkappa_3, \mathbb{R}^n = \bigcup_d \Delta_{3,d})$. Note that for each $i = 1,2$, and for each Δ_{ia}, $\varkappa_i(\Delta_{i,a})$ is a finite union of some of the $\varkappa_3(\Delta_{3,d})$. This shows that PL-structures induced on X by the first two triangulations are isomorphic to each other. In this sense, X carries a canonical PL-structure (up to isomorphisms).

3. SUBANALYTIC GENERALIZATION

DEFINITION 3.1. A subset A in $\mathbb{R}^n$ is said to be semi-analytic if for every point $x \in \mathbb{R}^n$ (not necessarily in A) the germ of A at x belongs to the Boolean class of germs of subsets at x which is generated by those of the form

$$\{\xi \in U \mid f(\xi) \geq 0\}$$

where U is an open neighborhood of x in $\mathbb{R}^n$ and f is a real-analytic function on U. (Cf. [13],[14])

REMARK 3.2. $A \subset \mathbb{R}^n$ is semi-analytic if and only if for every $x \in \bar{A}$ there exists an open neighborhood U of x in $\mathbb{R}^n$ and a finite system of real-analytic functions f_{ij}, g_{ij} on U, such that

$$A \cap U = \bigcup_i \{\xi \in U \mid f_{ij}(\xi) > 0,\ \forall j,\ g_{ij}(\xi) = 0, \forall j\}\ .$$

DEFINITION 3.3. $A \subset \mathbb{R}^n$ is said to be <u>subanalytic</u> if for every point $x \in \mathbb{R}^n$ (not necessarily in A) the germ of A at x belongs to the Boolean class of germs of subsets at x, which is generated by those of the form f(B) where B is a <u>bounded</u> semi-analytic set in some $\mathbb{R}^m$ and f is a real-analytic map $\mathbb{R}^m \longrightarrow \mathbb{R}^n$. (See [16],[17],[18] and [19].)

REMARK 3.4. It is easy to prove that $A \subset \mathbb{R}^n$ is subanalytic if and only if for every $x \in \bar{A}$, we can find an open neighborhood U of x in $\mathbb{R}^n$ and a finite number of <u>proper</u> real-analytic maps of real-analytic spaces $f_{ij}: Y_{ij} \longrightarrow U$, j = 1,2, such that

$$A \cap U = \bigcup_i (\mathrm{Im}(f_{i1}) - \mathrm{Im}(f_{i2}))\ .$$

REMARK 3.5. Using resolution of singularities, suitably formulated in the real-analytic case, we may assume that all the Y_{ij} of 3.4 are real-analytic <u>manifolds</u> (smooth and connected). (See [16].)

We need the following <u>three basic facts</u> about subanalytic sets, analogous to those about semi-algebraic sets.

PROPOSITION I. The interior and the closure of any subanalytic set in $\mathbb{R}^n$ are also subanalytic in $\mathbb{R}^n$.

PROPOSITION II. If $X \subset \mathbb{R}^m$ is subanalytic and bounded and if $h: \mathbb{R}^m \longrightarrow \mathbb{R}^n$ is a real-analytic map, then h(X) is subanalytic in $\mathbb{R}^n$.

PROPOSITION III. For a locally finite system of subanalytic sets $\{X_\alpha\}$ in $\mathbb{R}^n$, there exists a <u>subanalytic</u> stratification of $\bigcup_\alpha X_\alpha$ which is compatible with every X_α. (Cf. [16],[17] and [18].)

The definition of a stratification $A = \bigcup_\mu \Gamma_\mu$ is given in §1. It is said to be <u>subanalytic</u> if every Γ_μ is subanalytic in $\mathbb{R}^n$.

REMARK 3.6. Propositions I and III stay valid if we replace the adjective "subanalytic" by "semi-analytic". But II becomes false for the adjective "semi-analytic" (cf. [14]).

REMARK 3.7. Take a bounded subanalytic set Y in a linear subspace $\mathbb{R}^r \subset \mathbb{R}^n$, $r < n$. Let p be a point in $\mathbb{R}^n - \mathbb{R}^r$ and let $C(p,Y)$ be the cone with vertex p over base Y. Then $C(p,Y)$ is subanalytic in $\mathbb{R}^n$. (Immediate from Proposition II.) It can be proven that $C(p,Y)$ is semi-analytic if and only if it is semi-algebraic in $\mathbb{R}^n$ (or, equivalently Y is semi-algebraic in $\mathbb{R}^r$). It is, of course, the question of semi-analyticity of a cone at the vertex. At any rate, an important point here is that the subanalyticity (like semi-algebraicity) is preserved by taking cones or by taking barycentric subdivisions via a subanalytic automorphism of the ambient space, etc.

Ex. 1.4, Ex. 1.5 (for bounded Y in L), Def. 1.6, Rem. 1.7, Ex. 1.9, like many others, can be extended to subanalytic sets, simply by replacing the adjective "semi-algebraic" by "subanalytic" and also relaxing the condition "finite" by "locally finite (in $\mathbb{R}^n$)". The triangulation theorem becomes:

THEOREM. (Subanalytic triangulation) Given a locally finite (hence countable) system of subanalytic sets $\{X_\alpha\}$ in $\mathbb{R}^n$, there exists a simplicial decomposition $\mathbb{R}^n = \bigcup_a \Delta_a$ and a subanalytic automorphism κ of $\mathbb{R}^n$ such that

1) each X_α is a union (locally finite) of some of the $\kappa(\Delta_a)$, and

2) $\kappa(\Delta_a)$ is a locally closed smooth real-analytic submanifold of $\mathbb{R}^n$ and κ induces a real-analytic isomorphism $\Delta_a \xrightarrow{\sim} \kappa(\Delta_a)$ for every a.

Rem. 1.11 is valid for "subanalytic" instead of "semi-algebraic".

The proof of subanalytic triangulation will proceed, with some modifications, along the line of reasoning given in §2. The adjectives "semi-algebraic" and "finite" should be replaced by "subanalytic" and "locally finite" in every step. Let us indicate the points of necessary modifications step by step.

STEP 1. The reduction to "closed X_α" is the same. In addition, we may assume that each X_α is bounded. In fact, take the balls B_k of radii $k = 1,2,\cdots$, and replace X_α by

$$\{X_\alpha \cap (\bar{B}_k - B_{k-1})\}_{k=1,2,\cdots}$$

where $B_0 = \emptyset$.

Steps 2, 3, and 4 are valid for the subanalytic case as well.

Now, before we apply Step 5, we must make one essential modification. Here the new difficulty is that $\{\Gamma_\mu\}$ is only locally finite and not in general finite. This means that, in general, we cannot expect $\{\pi(\Gamma_\mu)\}$ to be locally finite in $\mathbb{R}^{n-1}$.

To overcome this difficulty, we apply the horizontal expansion lemma, stated below. First of all, we pick a point $\eta \in \mathbb{R}^{n-1}$ such that

(a) $\pi^{-1}(\eta) \cap \Gamma_\mu = \emptyset$ if $\dim \Gamma_\mu \leq n-2$

(b) if $\Gamma_\mu \cap \pi^{-1}(\eta) \neq \emptyset$ (and hence $\dim \Gamma_\mu = n-1$) then this intersection is transversal at every intersection point.

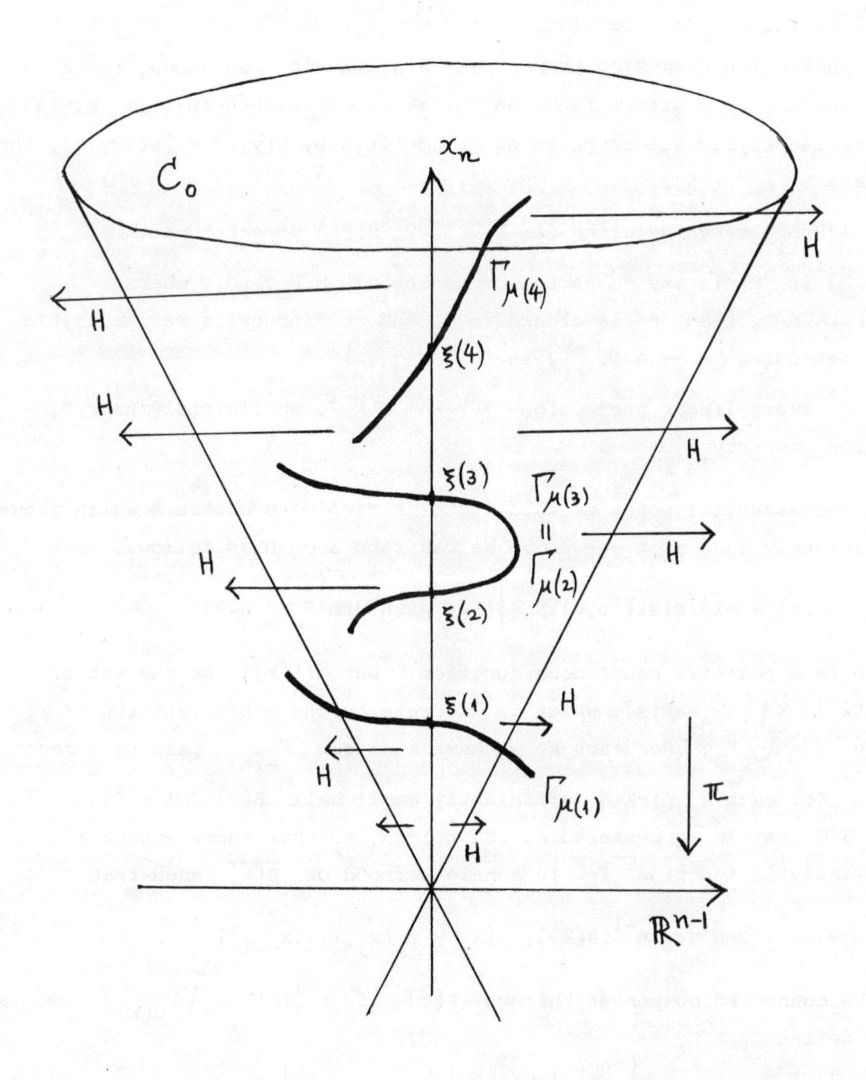

By a translation, we may assume $\eta = o$ (the origin). Let us denote by C_o the open cone

$$\left\{ x \in \mathbb{R}^n \;\middle|\; \sum_{i=1}^{n-1} x_i^2 < x_n^2 \right\}.$$

We shall identify

$$\mathbb{R}^n = \mathbb{R}^{n-1} \times \mathbb{R}$$

so that π is the projection, and write (y,t) for a point of $\mathbb{R}^{n-1} \times \mathbb{R}$.

HORIZONTAL EXPANSION LEMMA. For a given $\{\Gamma_\mu\}$ as above, there exists a real-analytic positive function $h\colon \mathbb{R} \longrightarrow \mathbb{R}_+$, such that if H is the real-analytic automorphism of $\mathbb{R}^n$ defined by $H(y,t) = (h(t)y,t)$, then we have

1) for every μ with $\dim \Gamma_\mu \leq n-2$, $H(\Gamma_\mu)$ does not meet C_o,

2) if F is any connected component of $H(\Gamma_\mu) \cap C_o$, where $\dim \Gamma_\mu = n-1$, then F is closed in C_o and π induces a real-analytic open imbedding $F \longrightarrow \mathbb{R}^{n-1}$, and

3) every linear projection $\mathbb{R}^n \longrightarrow \mathbb{R}^{n-1}$, sufficiently near π, has the property 2).

The essential point of this lemma is simply to choose h which grows sufficiently fast as $t \longrightarrow \pm\infty$. We can find such h as follows. Let

$$r(t) = \min\{\operatorname{dist}((o,t),\Gamma_\mu) \;|\; \mu \text{ with } \dim \Gamma_\mu \leq n-2\}$$

which is a positive continuous function. Let $\{\xi(k)\}$ be the set of points in $X \cap \{x_n\text{-axis}\}$, which is discrete by the subanalyticity of X, where $X = \bigcup_\alpha X_\alpha$. For each k, we have a unique $\Gamma_{\mu(k)}$ passing through $\xi(k)$. For each k, pick a sufficiently small ball $B(k)$ of radius $\varepsilon(k) > o$ in $\mathbb{R}^{n-1}$, centered at the origin, so that there exists a real-analytic function f_k in a neighborhood of $\overline{B(k)}$ such that

$$\pi^{-1}(B(k)) \cap \{x_n = f_k(x_1,\cdots,x_{n-1})\}$$

is the connected component through $\xi(k)$ of $\pi^{-1}(B(k)) \cap \Gamma_{\mu(k)}$. Let us then define

$$M_k = \max\left\{ \sum_{j=1}^{n-1} \left| \frac{\partial f_k}{\partial x_j}(x) \right|,\; x \in \overline{B(k)} \right\}.$$

Then pick a real-analytic function $\varphi(t)$ on $\mathbb{R}$ such that

a) $\varphi(t) > \max\{|t|, 1/r(t)\}, \ \forall t \in \mathbb{R}$

b) $\varphi(\xi(k)_n) > \max\left\{\frac{|\xi(k)_n|}{\varepsilon(k)}, 2M_k\right\}, \ \forall k$

c) $\varphi(t) = \varphi(-t)$ and $(d\varphi/dt)(t) > 0, \ \forall t \in \mathbb{R}_+$.

(It is easy to find such $\varphi(t)$.) Then we can find a constant $N \gg 0$ so that $h(t) = \exp(\varphi(t)) + N$ has the property required in H.E. Lemma.

Now, back to the triangulation problem, pick H of the H.E. Lemma. If π does not stay to be a "good" projection for $\{H(\Gamma_\mu)\}$, then by the G.D. Lemma of §2, we can choose a "good" linear projection π' for $\{H(\Gamma_\mu)\}$ which is as close to π as we want. Hence we can find π' such that if

$$S'_\mu = \{x \in H(\Gamma_\mu) \mid \operatorname{rank}(d\pi'_\mu)_x \leq n-2\}$$

(which is equal to $H(\Gamma_\mu)$ if $\dim \Gamma_\mu \leq n-2$), where $\pi'_\mu = \pi' | H(\Gamma_\mu)$, then the family $\{\pi'(S'_\mu)\}$ is locally finite in $\mathbb{R}^{n-1}$. If $\{\Gamma''_\nu\}$ is a "good" stratification of $\{S'_\mu\}$ with respect to π', then $\{\Gamma''_\nu$, connected components of $H(\Gamma_\mu) - S'_\mu\}$ form a "good" stratification of $\{H(X_\alpha)\}$ with respect to π'. This stratification with π' has a big advantage over the original stratification $\{\Gamma_\mu\}$ with π. Namely,

REMARK 3.8. For every compact subset K of $\mathbb{R}^{n-1}$, there exist only finitely many strata of dimensions $\leq n-2$ which meet with the inverse image of K by the projection $\mathbb{R}^n \longrightarrow \mathbb{R}^{n-1}$.

We shall assume, from now on, that $\{\Gamma_\mu\}$ and π of Step 4 have this Property 3.8.

STEP 5. Let J' be the set of those μ having $\dim \Gamma_\mu \leq n-2$. Then $\{\pi(\Gamma_\mu)\}_{\mu \in J'}$ is a locally finite system of subanalytic sets in $\mathbb{R}^{n-1}$, to which we can apply the induction assumption and obtain $\varkappa$ and $\mathbb{R}^{n-1} = \bigcup_a \Delta_a$ having the properties required for subanalytic triangulation.

STEP 6. In this case, instead of one ball B, we choose a <u>locally finite</u> system of balls $\{B_\mu\}_{\mu \in J'}$ such that for each $\mu \in J'$, B_μ contains $\overline{\Delta}_{\gamma(\mu)}$. We take the <u>tent</u> spanned from $\tilde{\varkappa}^{-1}(\Gamma_\mu)$ to $S_\mu = \partial B_\mu$, and call it Z_μ. Let $\{Y_\beta\}$ be the system $\{\tilde{\varkappa}^{-1}(Z_\mu)\}_{\mu \in J'}$ with $\mathbb{R}^{n-1}$ added to it, where $\mathbb{R}^{n-1}$ is identified as a subset of $\mathbb{R}^n$ as usual.

REMARK 3.9. $\{Y_\beta\}$ is π-locally finite. By this, we mean that for every compact $K \subset \mathbb{R}^{n-1}$, $\pi^{-1}(K)$ meet with only finitely many Y_β and $Y_\beta \cap \pi^{-1}(K)$ is bounded for all β. (In this case, this intersection is obviously compact.)

REMARK 3.10. If $\{Y_\beta\}$ is π-locally finite and if $\{T_b\}$ is locally finite in $\mathbb{R}^n$, then $\{Y_\beta \cap T_b\}$ is π-locally finite.

Let $J'' = \{\mu \mid \dim \Gamma_\mu = n-1\}$, and apply 3.10 to $\{T_b\} = \{\overline{\Gamma}_\mu\}_{\mu \in J''}$. If we denote $\{Y_\beta \cap \overline{\Gamma}_\mu\}_{\mu \in J''}$ by $\{Y'_{\beta'}\}$, then $\{Y_\beta, Y'_{\beta'}\}$ is also π-locally finite. If Y denotes the union of all the Y_β (which contains all the $Y'_{\beta'}$) then

(***) $\qquad \pi$ induces an <u>open</u> map $Y \longrightarrow \mathbb{R}^{n-1}$.

This is the analogue of (**) of §2, for the case of infinite support. We apply Steps 4 and 5, once again, to $\{Y_\beta, Y'_{\beta'}\}$, but this time without applying H.E. Lemma and without changing π. We obtain new $\varkappa$ and $\mathbb{R}^{n-1} = \bigcup_a \Delta_a$, and forfeit the old ones. About $\{\Gamma_\mu\}$, we make the following modification using the old $\{\Gamma_\mu\}_{\mu \in J''}$. We add $\{\Gamma_\mu \cap \pi^{-1}(\varkappa(\Delta_a))\}$ (with old Γ_μ, $\mu \in J''$, and with new $\varkappa(\Delta_a)$) to the new $\{\Gamma_\mu\}$ obtained for $\{Y_\beta, Y'_\beta\}$. After this addition is made, we forfeit the old $\{\Gamma_\mu\}$.

For the $\varkappa$, $\mathbb{R}^{n-1} = \bigcup_a \Delta_a$, $\{\Gamma_\mu\}$, finally obtained, we have (1) of Step 5 and (3) of Step 6. ((2) and the first statement of (3) are insignificant in our case.)

By the same reason as before, we may assume (4) of Step 6.

STEP 7. In our subanalytic case, this step is easier than the same in the semi-algebraic case. Namely, we take M to be infinity and $D = \mathbb{R}^{n-1}$. In other words, $Q_o = \mathbb{R}^n$, h_a is a subanalytic automorphism of $\pi^{-1}(\varkappa(\Delta_a))$ and h becomes a subanalytic automorphism of $\mathbb{R}^n$. We simply let $\hat{\varkappa} = h\tilde{\varkappa}$. $\mathbb{R}^n = \bigcup_d \Delta_d$ is required to be compatible only with $\hat{\varkappa}^{-1}(\Gamma_\mu)$ for all μ.

REMARK 3.11. The difference in Step 7 for the two cases is due to the special requirement in the semi-algebraic case that $\hat{\varkappa}$ be the identity outside some compact set. We do not (and in fact cannot) have such a requirement in the subanalytic case.

The details of the proof for subanalytic triangulation will be published elsewhere.

REFERENCES

Triangulation:

[1] B.L. van der Waerden, "Topologische Begründung des Kalküls der abzählenden Geometrie", Math. Ann. 102 (1929), 337-362.

[2] S. Lefschetz, "Topology", Amer. Math. Soc. Coll. Publications, New York (1930).

[3] B.C. Koopman and A.B. Brown, "On the covering of analytic loci by complexes", Trans. Amer. Math. Soc., 34 (1932), 231-251.

[4] S. Lefschetz and J.H.C. Whitehead, "On analytical complexes", Trans. Amer. Math. Soc., 35 (1933), 510-517.

[5] K. Sato, "Local triangulation of real analytic varieties", Osaka Math. J., 15 (1963), 109-125.

[6] B. Giesecke, "Simpliziale Zerlegung abzählbarer analytische Räume", Math. Zeit., 83(1964), 177-213.

[7] S. Lojasiewicz, "Triangulation of semi-analytic sets", Ann. Scu. Norm. di Pisa, 18(1964), 449-474.

Related materials:

[8] A. Seidenberg, "A new decision method for elementary algebra", Ann. Math., 60(1954), 365-374.

[9] H. Whitney, "Elementary structure of real algebraic varieties", Ann. Math., 66(1957), 545-556.

[10] F. Bruhat and H. Whitney, "Quelques propriétés fondamentales des ensembles analytiques-réels", Comment. Math. Helv., 33(1959), 132-160.

[11] A.H. Wallace, "Sheets of real-analytic varieties", Canad. J. Math., 12(1960), 51-67.

[12] R. Thom, "La stabilité topologique des applications polynomiales", Enseign. Math., 8(1962), 24-33.

[13] S. Lojasiewicz, "Une propriété topologique des sous-ensembles analytiques-réels", Coll. du CNRS (équations aux dérivées partielles), Paris (1962), 87-89.

[14] S. Lojasiewicz, "Ensembles semi-analytiques", I.H.E.S. Lecture note (1965); Reproduit No. A66.765, Ecole polytechnique, Paris.

[15] H. Hardt, "Slicing and intersection theory for chains associated with real analytic varieties", Acta Math., 129(1972), 57-136.

[16] H. Hironaka, "Subanalytic sets", Number theory, Algebraic geometry and Commutative algebra, in honour of Y. Akizuki, Kinokuniya, Tokyo (1973), 453-493.

[17] H. Hardt, "Stratification of real-analytic mappings and images", Preprint, School of Math., Univ. Minnesota.

[18] H. Hironaka, "Introduction to real-analytic sets and maps", Lecture note (Pisa, June, 1973), To appear in Ann. Scu. Norm. di Pisa.

[19] B. Teissier, "Théorèmes de finitude en géometrie analytique", Sem. Bourbaki, June 1974.

HARVARD UNIVERSITY

Proceedings of Symposia in Pure Mathematics
Volume 29, 1975

INTRODUCTION TO RESOLUTION OF SINGULARITIES

Joseph Lipman*

ABSTRACT

These lectures will be strictly introductory in nature. The object is to communicate some feeling for the resolution problem by focusing on a selected few of the many methods available for dealing with singularities of surfaces.

INTRODUCTION

The problem of resolution of singularities, simply put, is to prove (or disprove):

(RES) For any algebraic variety X over an algebraically closed field k there exists a proper map $f: Y \to X$, with Y non-singular (smooth), such that f is an isomorphism over some open dense subset U of X (i.e. f maps $f^{-1}(U)$ isomorphically onto U).

Of course we can generalize (RES) by allowing k to be, say, an excellent local ring[1], or by throwing k out altogether and letting X be an arbitrary reduced locally noetherian excellent scheme (in which case Y should be regular, i.e. the local rings of points on Y should be regular). We can also require that U be the set of all regular points of X, and that f be obtained by successive blowing up of nice subvarieties.[2]

AMS(MOS) subject classifications (1970). Primary 14B05, 14E15, 14H20, 14J15, 32C45.

*Supported by the National Science Foundation under grant GP29216.

[1] In Hironaka's work this is found to be necessary, even for proving (RES) when k is an algebraically closed field.

[2] For a definition and discussion of "blowing up" cf. [H1, chapter 0, §2].

But even as it stands (RES) poses a fundamental question, which is unanswered when k is of characteristic > 0 and X has dimension ≥ 4.[3]

In those cases where (RES) has been proved (cf. historical remarks below) the proof has usually been by induction on the dimension d of X. As part of the inductive step from $d - 1$ to d, it has been necessary to establish something akin to the following theorem ("Embedded Resolution" or "Simplification of the Boundary"), with $\dim. X' = d$:

(EMB) _Let_ X' _be a non-singular algebraic variety over_ k, _and let_ W _be a closed subvariety of_ X', _with dense complement_. _Then there exists a proper map_ $f: Y \to X'$, _with_ Y _non-singular_, _such that_ f _induces an isomorphism of_ $Y - f^{-1}(W)$ _onto_ $X' - W$ _and_ $f^{-1}(W)$ _is a divisor on_ Y _having only normal crossings_.

"Normal crossings" means that for each $y \in Y$, the defining ideal of $f^{-1}(W)$ is generated locally around y by an element of the form $\xi_1^{a_1}\xi_2^{a_2}\cdots\xi_d^{a_d}$ where $\{\xi_1\xi_2,\ldots,\xi_d\}$ is a regular system of parameters (= local coordinate system) at y, and each $a_i \geq 0$.

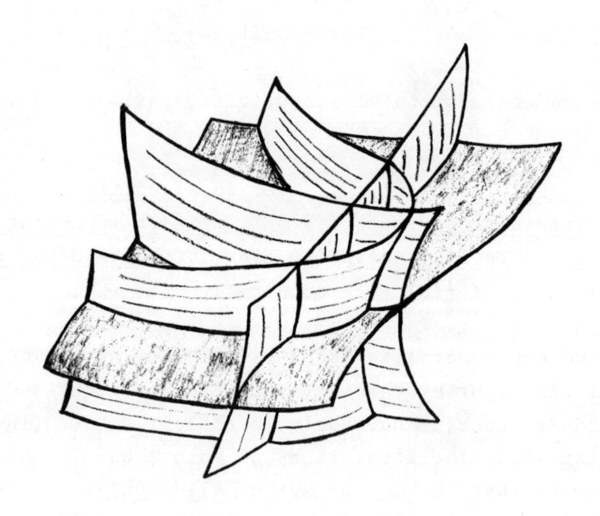

Divisor with normal crossings

[3] Nevertheless, at present - as in the past - not many people are working on the problem. At this Institute, the only report on recent progress in the area was Hironaka's lecture on Giraud's theory of "maximal contact".

* * *

Here are some brief historical indications.

(RES) was proved for curves over the complex numbers $\mathbb{C}$ in the last century by Kronecker, Max Noether, and others (cf. chapter VI in [N-B]). I will say a few things about the one-dimensional situation at the beginning of Lecture 1 below.

Several approaches to proving (RES) for X a surface over $\mathbb{C}$ were proposed by Italian geometers (cf. chapter 1 in [Z1]). One such approach, due to Albanese, has turned out to be quite practicable (Lecture 1, §5). The first completely rigorous proof was given by Walker in 1936 [W1]. What he did, basically, was to show how to patch together "local" resolutions constructed by Jung in 1908 [J]. His methods (and Jung's) were in part complex analytic; but the whole line of argument can be carried out algebraically, and this is what will be done in Lecture 2. Another proof based on Jung's work was given, in the context of analytic geometry, by Hirzebruch [Hz]. Jung's ideas have, in addition, influenced a number of other works on resolution (cf. end of Lecture 2).

Purely algebraic proofs for surfaces over fields of characteristic zero were first given by Zariski [Z2], [Z4][4] (Lecture 1, §§2, 3). Zariski's work on resolution in the late 1930's and early 1940's culminated in the memoir [Z6], in which he proved (RES) for surfaces, with f obtained by successively blowing up points and non-singular curves of maximal multiplicity ("theorem of Beppo Levi"), and then deduced (RES) for three-dimensional X (always with $\text{char.}k = 0$). More recently, Zariski found a simpler proof of the possibility of desingularizing surfaces by such a procedure [Z7]. A still simpler proof, due to Abhyankar [unpublished] will be presented in Lecture 3.

In his great 1964 paper [H1], Hironaka proved (RES) and (EMB) for any equicharacteristic zero excellent scheme. At the same time he solved the corresponding problems for real analytic spaces and locally for complex analytic spaces. In the last few years he has solved (affirmatively) the global resolution problem for complex spaces. The main ideas of this solution are outlined in the preprint [H6], and details have begun to appear [H7].

For k of characteristic > 0, most of the published results are due essentially to Abhyankar. (For an introduction to his work cf. [A4].) In his Harvard thesis (1954) Abhyankar proved (RES) for $\dim.X = 2$ and k of any characteristic; some years later [A1] he disposed of the more general case where k is an excellent Dedekind domain with perfect residue fields. (In fact he has announced a proof for $\dim.X = 2$ and k any excellent

[4] All of Zariski's papers on resolution are in volume 1 of his collected works [Z8].

domain [A5, §5].) In his book [A3], Abhyankar deduced (EMB) for $\dim.X' = 3$ from a previously developed algorithm of his on monic polynomials [A2]; and he also obtained (RES) for $\dim.X = 3$ and $\text{char}.k \neq 2,3,5$.

At the same time (January, 1967) that Abhyankar announced his most general results on surfaces [A5, §1 and §5], Hironaka announced a proof of (EMB) for X' an arbitrary excellent scheme and $\dim.W = 2$.[5] Hironaka's results include Abhyankar's; but full details of the proof have not yet been published (cf. however [H2], [H3], [H4], [H5]).

* * *

Apology. For lack of time, and because of its technical complexity, Hironaka's fundamental work will hardly be touched on. I can only express the hope that these lectures will help the reader to acquire the right frame of mind for exploring Hironaka's work on his own.

---

[5] Abhyankar and Hironaka were both at Purdue University during the Fall semester of 1966.

LECTURE 1: GENERALITIES ON CURVES AND SURFACES

§1. Curves.

Max Noether dealt with singularities of projective plane curves, to which he applied a succession of well-chosen quadratic transformations of $\mathbb{P}^2$. [If the curve C has a singular point at Q, choose coordinates so that (i) $Q = (1,0,0)$; (ii) C has no other singular point on any of the coordinate axes $X_0 = 0$, $X_1 = 0$, $X_2 = 0$; and (iii) these axes are not tangent to C at any point. Now apply the transformation $(x_0,x_1,x_2) \to (x_1x_2,x_2x_0,x_0x_1)$. For details cf. e.g. [W2, chapter III, §7].]

A careful analysis of the effect of such transformations on the singular points shows that eventually the curve is transformed into one having only ordinary multiple points (i.e. points at which C looks locally like several distinct lines through the origin).

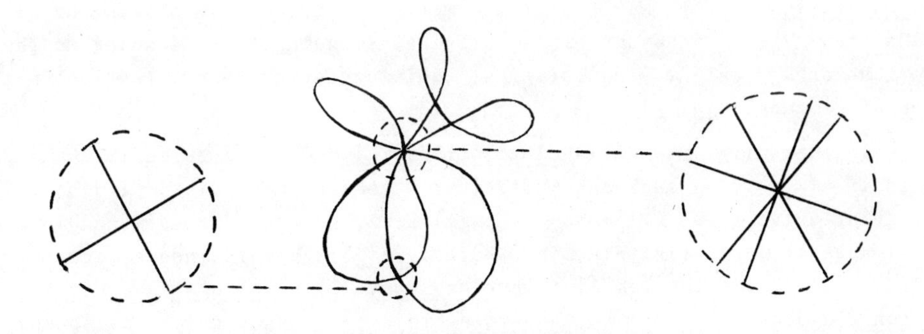

Ordinary multiple points

Such points are unavoidable because of the fact that global transformations are being used which blow the coordinate axes down to points; for each intersection of C with a coordinate axis, a branch of the transformed curve will pass through the corresponding point.

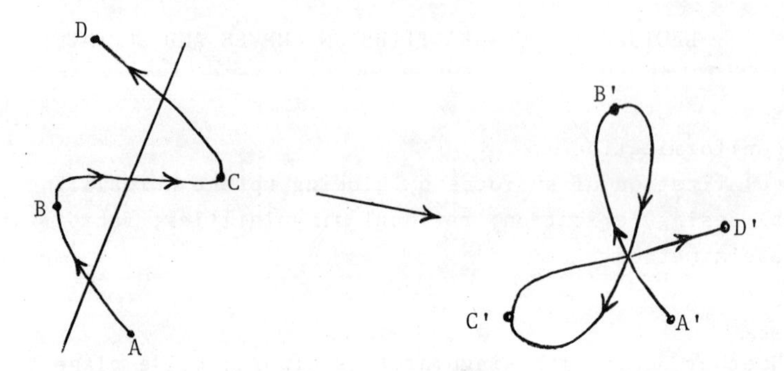

To avoid such irrelevancies, one can isolate the effect of the transformations on the points in which one is really interested (viz. the original singular points); this leads to the intrinsic (coordinate-free) notion of locally quadratic transformation (blowing-up points). Quite generally:

> successive blowing up of singular points transforms any projective curve X into a non-singular projective curve Y;

so we have (RES) in this case.

Conceptually, one can achieve the same thing by one operation, namely normalization (cf. [Z-M]).[6] But the method of successive blowing up gives us a step-by-step analysis of the original singularities, leading to the notions of equivalent singularities, equisingularity, etc., (notions which cannot be pursued here).

Incidentally, the desingularization Y may not be embeddable in $\mathbb{P}^2$, even if X is. However any sufficiently general projection of a non-singular curve from $\mathbb{P}^n$ into $\mathbb{P}^3$ will still be non-singular, and into $\mathbb{P}^2$ will have at worst finitely many ordinary double points (nodes), cf. [Z-M], [S, p. 46, Theorem 1]. Thus any curve is birationally equivalent to a non-singular curve in $\mathbb{P}^3$, and to a projective plane curve having only nodes.

* * *

---

[6] Both methods work equally well for any reduced excellent one-dimensional scheme X. The idea is that blowing up a singular point of X gets you a scheme lying "strictly closer" than X to the normalization $\bar{X}$; since $\bar{X}$ is finite over X, after finitely many blowing-ups you must get $\bar{X}$ itself, and $\bar{X}$ is regular.

We also have (EMB) <u>for a curve</u> W <u>in a non-singular surface</u> X'. (Here again the method will work for one-dimensional subschemes of arbitrary two-dimensional regular excellent schemes.). We proceed as follows. Let $f_1 : X_1 \to X'$ be obtained by blowing up a point $x \in X'$ at which W has a singularity. f_1 induces an isomorphism over $X' - \{x\}$. Let W_1 be the closure in X_1 of $f_1^{-1}(W - \{x\})$. Then the map $W_1 \to W$ induced by f_1 is identical with the one obtained by blowing up x, considered as a point of W. Hence by repeating the process sufficiently often (blow up a singular point x_1 of W_1 to get $f_2 : X_2 \to X_1$, let W_2 be the closure in X_2 of $f_2^{-1}(W_1 - \{x_1\})$, etc. etc.) we <u>resolve the singularities of W</u>, so that we have

$$\begin{array}{ccccccc} W_n & \to & \cdots & \to & W_1 & \to & W \\ \downarrow & & & & \downarrow & f_1 & \downarrow \\ X_n & \to & \cdots & \to & X_1 & \to & X' \end{array}$$

$$F_n = f_1 \circ f_2 \circ \cdots \circ f_n$$

with W_n <u>non-singular</u>. (In the following picture $n = 1$).

$f_1^{-1}(x)$, a non-singular rational curve

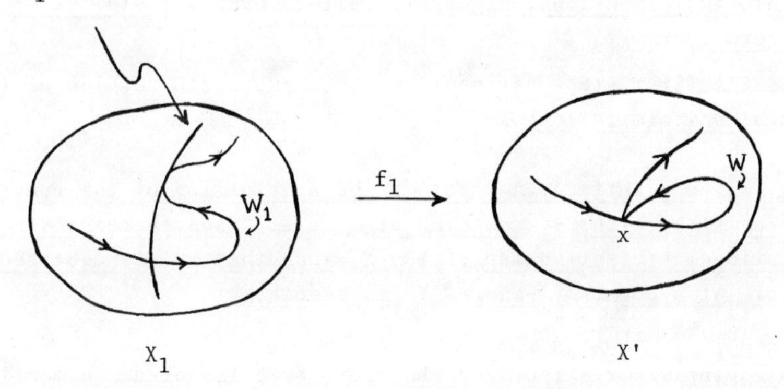

The irreducible curves in X_n whose image under F_n is a single point are all non-singular. So, after replacing X' by X_n and W by $F_n^{-1}(W)$, <u>we may assume that each irreducible component of</u> W <u>is non-singular</u>. But with this assumption on W it is easily seen that if $m \geq 1$ is such that W_m is non-singular (in other words, the irreducible components of W have become completely detached from each other in X_m), then $F_m^{-1}(W)$ has only normal crossings.

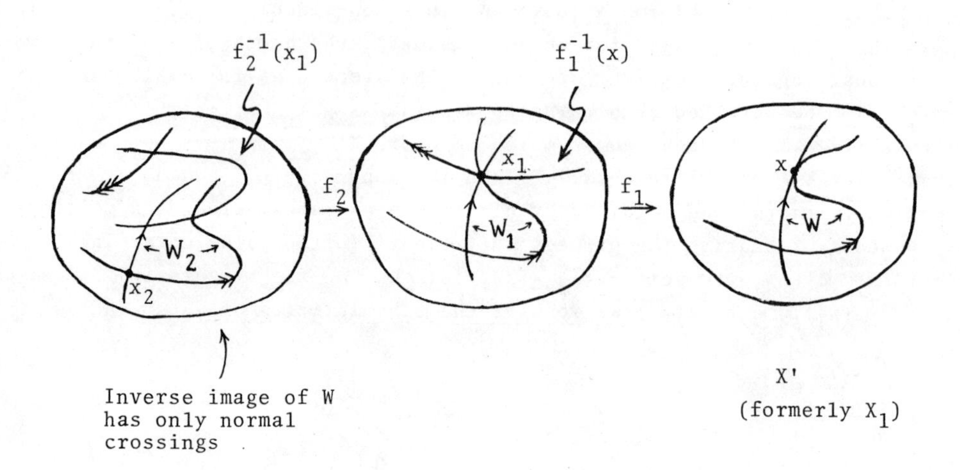

(Here, strictly speaking, we should blow up x_2 to detach the components of W_2 and obtain a non-singular W_3; but we needn't bother, since W_2 already has a normal crossing at x_2.)

§2. Local Uniformization.

We turn now to surfaces, and begin with Zariski's approach to the problem of desingularization. Until otherwise indicated, "surface" means "irreducible two-dimensional algebraic variety over a field k of characteristic zero".

Zariski's first step was to prove a much weakened version of (RES) called local uniformization:

(LU) Let X be a surface, and let v be a valuation of the function field $k(X)$, with v centered at $x \in X$. Then there exists a birational projective map $f: Y \to X$ such that v is centered on Y at a point y where Y is smooth.

(Subsequently Zariski proved (LU) for varieties of any dimension over a field of characteristic zero [Z3].)

For surfaces Zariski gave two ways of deducing (RES) from (LU) [Z2], [Z4]. The second is much shorter, but the first gives us a nice canonical desingularization process (blowing up and normalizing) which we discuss in §3 below.

* * *

To get some sort of picture of what (LU) means when $k = \mathbb{C}$, think of a valuation centered at x as being a little arc A on X emanating from x, and (after replacing Y by some complex neighborhood of y) think of Y

as being a bounded open disc around $y \in \mathbb{C}^2$, containing a little arc A' emanating from y, with f(A') = A. Here f is no longer birational, but rather a holomorphic map inducing an isomorphism from an open dense subset of Y onto a (complex) open set V of X, f^{-1} being given on V by rational functions.

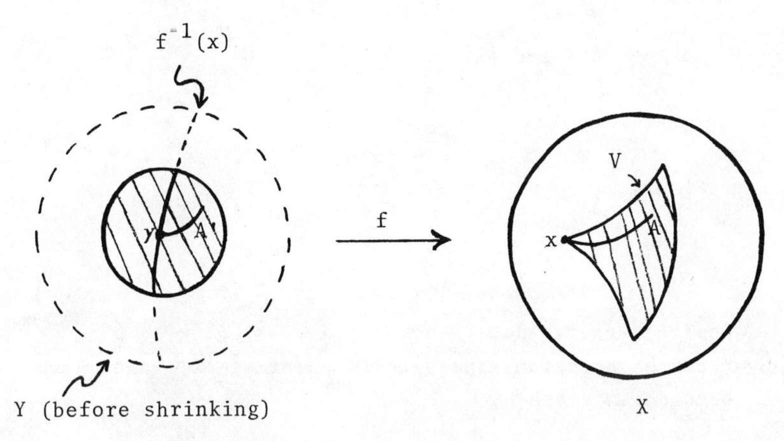

(LU) says that X can be covered - not only pointwise, but, so to speak, arcwise - by such "local parametrizations"[7] (in fact by finitely many of them, if X is compact, as one sees using the compactness of the "Riemann manifold" cf. [Z5] and also the introduction to [Z3]).

Example (Jung [J]). Let x be the origin on the affine surface

$$X = \{(\xi,\eta,\zeta) \in \mathbb{C}^3 \mid \zeta^5 = \xi^3\eta\}$$

Consider the following three parametrizations of X, (u,v) being coordinates in the unit disc $D \subset \mathbb{C}^2$.

I.	$\xi = u^5v^3$	$\eta = v$	$\zeta = u^3v^2$
II.	$\xi = u^3v$	$\eta = uv^2$	$\zeta = u^2v$
III.	$\xi = u$	$\eta = u^2v^5$	$\zeta = uv$

Together these three parametrizations cover a neighborhood of x. For any valuation v centered at x, one of them can serve as the desired f. (And they are all required: for example to lift the arc $\xi = t^5$, $\eta = t^5$, $\zeta = t^4$ to an arc centered at the origin of the unit disc take $u = t$, $v = t^2$ in II; no such lifting is possible via I or III.) Of course the unit disc is

[7] In [W1], these are called "wedges".

not an algebraic variety; but in I, for example, we can solve for u and v in terms of ξ, η, ζ:

$$u = \xi^2/\zeta^3, \qquad v = \eta \qquad (\zeta \neq 0)$$

so that I can be factored as

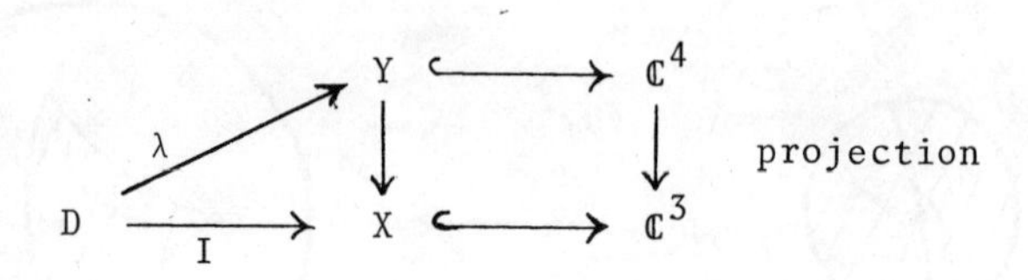

where Y is the variety whose general point is $(\xi,\eta,\zeta,\ \xi^2/\zeta^3)$ $[(\xi,\eta,\zeta)$ being, by abuse of notation, the general point of $X]$ and λ is the local analytic isomorphism given by

$$\lambda(u,v) = (u^5v^3,\ v,\ u^3v^2,\ u).$$

II and III can be treated similarly.

This example is too special in two respects. First of all, λ extends to an algebraic isomorphism from all of $\mathbb{C}^2$ onto Y; in general (for instance if we start with a non-rational surface X) this can't happen; our parametrizations will be given by convergent power series in u, v, rather than by polynomials.

Secondly, I, II and III happen to patch together into a single map $Z \to X$ which desingularizes an entire neighborhood of x. In general, if we cover a neighborhood of a point by local parametrizations, they will not patch together, nor will it be clear how to modify them so that they will patch. This is precisely the difficulty in deducing (RES) from (LU).

§3. Desingularization of Surfaces by Blowing up and Normalizing.

Let X be a surface, and let X_0 be the normalization of X. The set Σ_0 of singular points of X_0 is finite. Blow up Σ_0, and let X_1 be the normalization of the resulting surface. We have than a projective map $f_1: X_1 \to X_0$ inducing an isomorphism $X_1 - f_1^{-1}(\Sigma_0) \to X_0 - \Sigma_0$, and such that the inverse image of any point in Σ_0 is one-dimensional. Repeating the same process (blowing up the singular set and then normalizing) we get a sequence

$$X_n \xrightarrow{f_n} X_{n-1} \longrightarrow \cdots \longrightarrow X_1 \xrightarrow{f_1} X_0.$$

Using (LU), Zariski showed in [Z2] that <u>this process must terminate, i.e. X_n is non-singular for some n.</u>

<u>Example.</u> Let S be the normal surface in $\mathbb{C}^3$ given by the equation $Z^2 + X^3 + Y^7 = 0$. The origin is the only singular point (in other words the only point where the first partial derivatives of $Z^2 + X^3 + Y^7$ all vanish). Blowing up this point, we obtain a surface S_1 which is again normal; in fact, if R is the coordinate ring of S:

$$R = \mathbb{C}[X,Y,Z]/(Z^2 + X^3 + Y^7) \cong \mathbb{C}[x,y,z]$$

then S_1 has an open covering by two affine surfaces S_1', S_1'', whose coordinate rings are respectively

$$R_1' = R[y/x,\ z/x] = \mathbb{C}[x,\ y/x,\ z/x] \cong \mathbb{C}[X,Y,Z]/(Z^2 + X + Y^7X^5)$$

$$R_1'' = R[x/y,\ z/y] = \mathbb{C}[x/y,\ y,\ z/y] \cong \mathbb{C}[X,Y,Z]/(Z^2 + X^3Y + Y^5);$$

S_1' is smooth, while S_1'' has just one singular point, at the origin.

Blowing up the singular point of S_1'', we get a surface S_2 covered by two affine surfaces S_2', S_2'' whose coordinate rings are respectively

$$R_2' = R_1''[y/(x/y),\ (z/y)/(x/y)] = \mathbb{C}[x/y,\ y^2/x,\ z/x]$$

$$\cong \mathbb{C}[X,Y,Z]/(Z^2 + X^2Y + X^3Y^5)$$

$$R_2'' = R_1''[(x/y)/y,\ (z/y)/y] = \mathbb{C}[x/y^2,\ y,\ z/y^2]$$

$$\cong \mathbb{C}[X,Y,Z]/(Z^2 + X^3Y^2 + Y^3).$$

The element $(z/x)/(x/y)$ is integral over R_2', and

$$R_2'[(z/x)/(x/y)] = \mathbb{C}[x/y,\ y^2/x,\ zy/x^2] = \mathbb{C}[X,Y,Z]/(Z^2 + Y + XY^5).$$

Since the surface $Z^2 + Y + XY^5 = 0$ is smooth, S_2' has a smooth normalization. Similarly by adjoining $(z/y^2)/y$ to R_2'', we see that S_2'' has a smooth normalization. Thus the normalization $\bar{S}_2$ of S_2 is smooth, and so we have a desingularization $T \to S$ with $T = S_1' \cup \bar{S}_2$.

In this example, one finds that the inverse image on T of the singular point of S consists of a pair of rational curves E_1, E_2, with intersection numbers $(E_1 \cdot E_2) = 1$, $(E_1 . E_1) = -1$, $(E_2 \cdot E_2) = -2$. E_1 is non-singular, while E_2 has a simple cusp (with completed local ring isomorphic to $\mathbb{C}[[U,V]]/(U^2 - V^3)$).

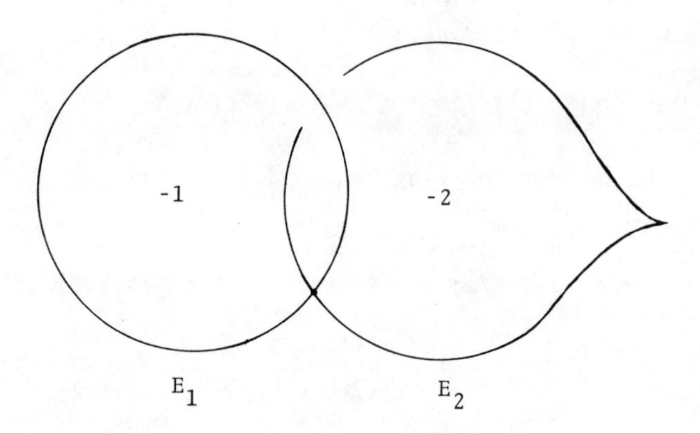

The curve E_1 can be blown down, so that $T \to S$ factors as $T \to T_0 \to S$, where T_0 is a smooth surface on which the inverse image of the singular point of S looks like this:

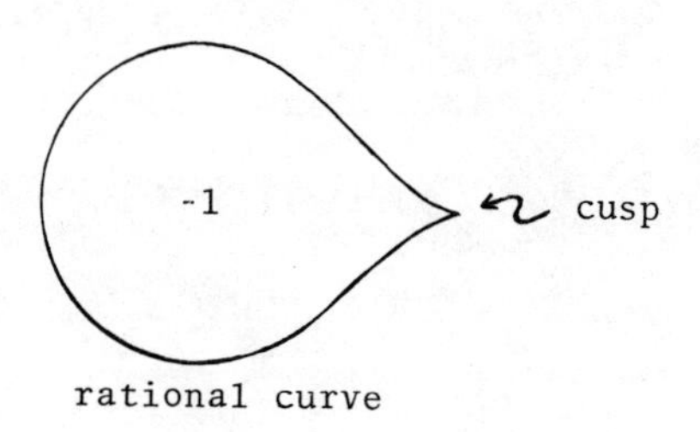

(T_0 can be obtained more directly from S by blowing up the ideal (x, y^2, z) and then normalizing.)

§4. Minimal Desingularization; Rational Singularities; Factorization Theorem.

It is a fact that any surface S over $\mathbb{C}$ has a unique minimal desingularization $S^* \to S$, i.e. a desingularization through which every other desingularization $f:S' \to S$ factors (uniquely):

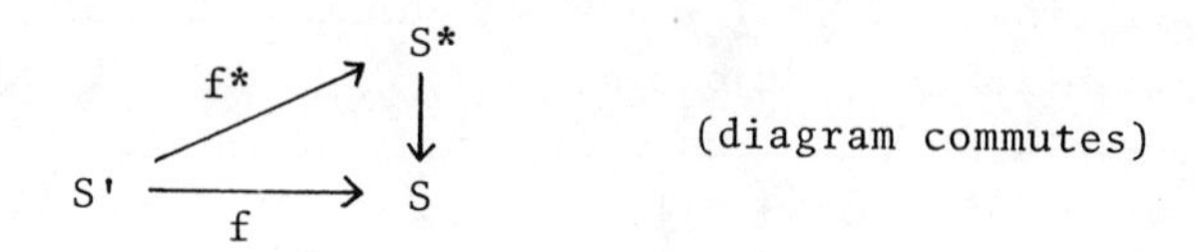

{This is mentioned in [H1], p. 151; a proof can be found in [B; Lemma 1.6]. For the same result in the case of arbitrary reduced two-dimensional excellent schemes cf. [L, p. 277, Cor. (27.3)]}.

The above example (§3) shows that successive blowing up and normalization does not always produce the minimal desingularization.

There is one class of normal singularities for which Zariski's process does give the minimal desingularization - the class of rational singularities. This fact, along with other applications of rational singularities in resolution questions, comes out of the following properties, [L, parts I and II].

Let X be a normal surface having only rational singularities[8] (X may be smooth) and let $f:Y \to X$ be a proper birational map. Then:

(A) If Y is normal, then Y has only rational singularities.

(B) If f is obtained by blowing up a point of X, then Y is normal.

(C) (Factorization Theorem) If all the local rings of points on Y are factorial, then f is obtainable by successively blowing up points.

[(A) and (B) enter into the proof of (C).]

(C), with Y the minimal desingularization of X, shows that X can be minimally desingularized by blowing up points (normalization being unnecessary).

Furthermore, using suitably formulated local versions of (A) and (C), one can make Zariski's proof that (LU) $\Rightarrow$ (desingularization by blowing up and normalization) work for any reduced excellent two-dimensional scheme [L, §2]. In this context (LU) can be taken to mean:

(LU)(bis) If R is an excellent two-dimensional local domain and v is a valuation of the quotient field of R such that the valuation ring R_v contains R, then there exist elements $x_1, x_2, \ldots, x_n$ in R_v such that the local ring

$$R[x_1, x_2, \ldots, x_n]_q \quad (q = \{x \in R[x_1, \ldots, x_n] \mid v(x) > 0\})$$

is regular.

[8] In case X is proper over a field, what this means is that $\chi(\mathcal{O}_X) \le \chi(\mathcal{O}_{X'})$ for any proper normal surface X' birationally equivalent to X (where $\chi(\mathcal{O}_X) = h^0(\mathcal{O}_X) - h^1(\mathcal{O}_X) + h^2(\mathcal{O}_X)$ = arithmetic genus of X.)

To prove (LU)(bis), even when R is complete (a special case to which the general one reduces without much trouble), is quite difficult. As indicated before, proofs were announced by Abhyankar and Hironaka in 1967, but the complete details have yet to appear in print.

§5. Albanese's Method.

A pretty method of Albanese [Alb] was revived in 1963 by M. Artin, who showed [unpublished] that it could be used to simplify greatly the resolution problem for surfaces over fields of <u>any characteristic</u>. The method extends in a straightforward way to irreducible projective varieties V of any dimension d over an algebraically closed field k. We sketch the basic idea[9].

For convenience we shall say that "$\pi: W \to W_1$ is a <u>permissible projection</u>" if W is a positive-dimensional irreducible closed subvariety of a projective space $\mathbb{P}^M$ over k, π is the projection of $\mathbb{P}^M$ into $\mathbb{P}^{M-1}$ from a point w on W of multiplicity $\mu < \deg.W$ ("deg" = "degree of"), and W_1 is the closure in $\mathbb{P}^{M-1}$ of $\pi(W-w)$. The condition $\mu < \deg.W$ guarantees that $\dim.W_1 = \dim.W$, so that we have a finite algebraic field extension $k(W)/k(W_1)$, whose degree we denote by $[W:W_1]$ (the generic covering degree of W over W_1). It is easily seen that

$$(5.1) \qquad [W:W_1](\deg.W_1) = \deg.W - \mu$$

(Cut W_1 by a generic linear space L of complementary dimension in $\mathbb{P}^{M-1}$; then consider $\pi^{-1}(L) \cap W \ldots$).

Assume that $V = V_0 \subseteq \mathbb{P}^N$ and that V is not contained in any hyperplane of $\mathbb{P}^N$. Let r be a fixed integer. We are going to try to find a succession of permissible projections

$$(5.2) \qquad V_0 \xrightarrow{\pi_0} V_1 \xrightarrow{\pi_1} V_2 \longrightarrow \ldots \xrightarrow{\pi_{n-1}} V_n$$

such that

$$(5.3) \qquad \text{every point on } V_n \text{ has multiplicity} \le r/[V_0:V_n].$$

By the "projection formula", it will then follow that on the normalization of V_n in $k(V_0)$ - a variety birationally equivalent to V - <u>every point has</u> multiplicity $\le r$.

[9] For the raw technical details - which are quite elementary - cf. [A3, §12]. Abhyankar uses the result in his proof of (LU) for 3-dimensional varieties over fields of characteristic > (3!).

Of course we want r to be as small as possible, and in fact, as we will now see, if we begin with a suitable projective embedding of V, then we can succeed with $r = d!$ $(d = \dim.V)$. In particular, for curves $(d = 1)$, this will give another proof of (RES).

We begin by projecting into $\mathbb{P}^{N-1}$ from a point v_0 on V_0 of multiplicity $> r$ and $<$ deg.V (if there is such a point); call this projection π_0, and let V_1 be the closure of $\pi_0(V_0 - v_0)$. Next, choose a point v_1 on V_1 of multiplicity $> (r/[V_0:V_1])$ and $< \deg.V_1$, and project into $\mathbb{P}^{N-2}$ from v_1; call this projection π_1, and let V_2 be the closure of $\pi_1(V_1 - v_1)$. Now choose a point v_2 on V_2 of multiplicity $> (r/[V_0;V_2])$ and $< \deg.V_2$, etc.etc.etc. Continue in this way as long as possible; obviously the process must stop after a finite number of steps: in other words we obtain a sequence (5.2) such that every point v on V_n of multiplicity $>(r/[V_0:V_n])$ must have multiplicity $= \deg.V_n$ (i.e. V_n is a cone with vertex v). So it will suffice to show that V_n cannot be a cone.

First of all, (5.1) gives

$$[V_0:V_1]\deg.V_1 = \deg.V_0 - (\text{mult. of } v_0) \leq \deg.V_0 - r - 1.$$

Similarly,

$$[V_1:V_2]\deg.V_2 \leq \deg.V_1 - (r/[V_0:V_1]) - 1$$

whence

$$[V_0:V_2]\deg.V_2 \leq [V_0:V_1]\deg.V_1 - r - [V_0:V_1] \leq \deg V_0 - 2r - 2.$$

Continuing in this way, we find that

$$[V_0:V_n]\deg.V_n \leq \deg.V_0 - n(r + 1).$$

So we have the crude inequality

$$n(r + 1) \leq \deg.V_0. \tag{5.4}$$

Next, if V_n is a cone with vertex v, then a generic linear space L through v of dimension $N - n - d + 1$ will intersect V_n in $\deg.V_n$ distinct lines. Hence the inverse image in $\mathbb{P}^N$ of L (under the composed projection $\pi_{n-1} \circ \pi_{n-2} \circ \cdots \circ \pi_0$) will cut V_0 in a curve having at least $\deg.V_n$ irreducible components. Setting

$$c = c(V_0) = \min\{\text{degree of irreducible curves on } V_0\}$$

we conclude that

$$(\deg.V_n)c \leq \deg.V_0 \tag{5.5}$$

But V_n is not contained in any hyperplane of $\mathbb{P}^{N-n}$ (otherwise, taking inverse images, we would find that V_0 is contained in a hyperplane of $\mathbb{P}^N$, contrary to assumption). An easy argument shows that consequently

$$\deg.V_n > N - n - d.$$

So from (5.5) we get

$$(N - n - d)c < \deg.V_0$$

i.e.

$$N - d - c^{-1}\deg.V_0 < n. \tag{5.6}$$

Combining (5.6) and (5.4), we have

$$N - d - c^{-1}\deg.V_0 < (r + 1)^{-1}\deg.V_0.$$

We conclude, therefore, that <u>if</u>

$$r + 1 \geq \deg.V_0/(N - d - c^{-1}\deg.V_0) \tag{5.7}$$

<u>then</u> V_n <u>is not a cone</u> (and hence every point on V_n has multiplicity $\leq r/[V_0:V_n]$).

Now finally consider the embedding

$$V \xrightarrow{\approx} V_0^\delta \subseteq \mathbb{P}^{N(\delta)}$$

of V via the linear system of sections of V by hypersurfaces in $\mathbb{P}^N$ of degree δ (or equivalently, for large δ, via the very ample sheaf $\mathcal{O}_V(\delta)$). V_0^δ is not contained in any hyperplane of $\mathbb{P}^{N(\delta)}$. Any curve on V_0^δ has degree $\geq \delta$, since cutting by hyperplanes in $\mathbb{P}^{N(\delta)}$ corresponds to cutting by hypersurfaces of degree δ in $\mathbb{P}^N$; thus $c_\delta = c(V_0^\delta) \geq \delta$. For large δ, $N(\delta) + 1$ is given by $\chi(\mathcal{O}_V(\delta))$, while $(1/d!)\deg.V_0^\delta$ is the coefficient of n^d in the polynomial $\chi(\mathcal{O}_V(n\delta))$. From these remarks, it follows that

$$\lim_{\delta\to\infty} \deg.V_0^\delta/(N(\delta) - d - c_\delta^{-1}\deg.V_0^\delta) = d!$$

So for large δ, if $V_0 = V_0^\delta$, then (5.7) holds with $r = (d!)$.

In summary, the argument has given us the <u>existence of a normal variety birationally equivalent to</u> V, <u>and on which each point has multiplicity</u> $\leq (d!)$ $(d = \dim.V)$.

* * *

If V is a surface $(d = 2)$, then we have a surface birational to V, and on which every point has multiplicity ≤ 2. To prove resolution for surfaces as in §3 above, all that is really needed is the weaker form of (LU) in which Y is required only to be birationally equivalent to X (the map f need not exist.) So we are reduced to "uniformizing" valuations centered at double points.

Taking advantage of the good behavior of blowing up and normalization vis-à-vis completion, we come down to proving (LU)(bis) for complete local rings R of multiplicity 2. In such an R there exist elements u, v, such that the field of fractions of R is the same as that of a ring

$$R' = k[[u,v]][w] \qquad w^2 = f(u,v)w + g(u,v)$$

($f,g \in k[[u,v]]$, $g \neq 0$; and $f = 0$ when char. $k \neq 2$). There is no harm in assuming that $R = R'$. An application of (EMB) to the curve $g = 0$ (resp. $fg = 0$ if $f \neq 0$) in the surface $\mathrm{Spec}(k[[u,v]])$ (cf. §1) gives us a further reduction to the situation where

$$g(u,v) = \mu(u,v)u^a v^b$$

and, in case $f \neq 0$,

$$f(u,v) = \nu(u,v)u^c v^d$$

where μ and ν are units in $k[[u,v]]$. At this point, if the characteristic of k is $\neq 2$, then after normalizing, at most one blowing up will be needed to resolve the singularity of R.

On the other hand, when k has characteristic 2, the analysis of the situation is much harder. It is recommended to the interested reader to play around a little with this case, as a gentle initiation into the intricacies of Abhyankar's algorithms.

LECTURE 2: JUNG'S METHOD FOR LOCAL DESINGULARIZATION

We noted in Lecture 1 that for surfaces resolution of singularities — for example by a succession of blowing-ups (of points) and normalizations — can be deduced from local uniformization. A fruitful method for achieving local uniformization on a surface X over an algebraically closed field k of characteristic zero was given by Jung in 1908 [J]. Jung's ideas have had a great influence on subsequent work on the resolution problem: last time the work of Walker and Hirzebruch was mentioned, and further instances will be indicated at the end of this lecture.

Actually, as observed by Walker [W1], a slight elaboration of Jung's method gives us a "local resolution" theorem which says considerably more than local uniformization:

THEOREM: *There exists a family of desingularizations $(f_x: Y_x \to U_x)_{x \in X}$,*[1] *where U_x is an open neighborhood of x, the restriction of f_x to $Y_x - f_x^{-1}(\{x\})$ is the normalization of $U_x - \{x\}$, and furthermore for all x outside some zero-dimensional subset S of X, f_x is just the normalization of U_x itself.*

Resolution for X follows at once from the Theorem: assuming, without loss of generality, that $U_x \cap S = \{x\}$ for each x in S, cover X by the open sets U_x $(x \in S)$ together with the open set $U_0 = X - S$; let $f_0: Y_0 \to U_0$ be the normalization of U_0; then because of the uniqueness properties of normalization it is evident that the local desingularizations f_x $(x \in S)$ and f_0 will patch together to give a global desingularization $f: Y \to X$, where the restriction of f to $Y - f^{-1}(S)$ is the normalization of $X - S$.[2]

---

[1] In other words Y_x is smooth, f_x is proper, and for some open dense subset V_x of U_x, f_x maps $f_x^{-1}(V_x)$ isomorphically onto V_x.

[2] cf. also [W1, p.343, Theorem 4].

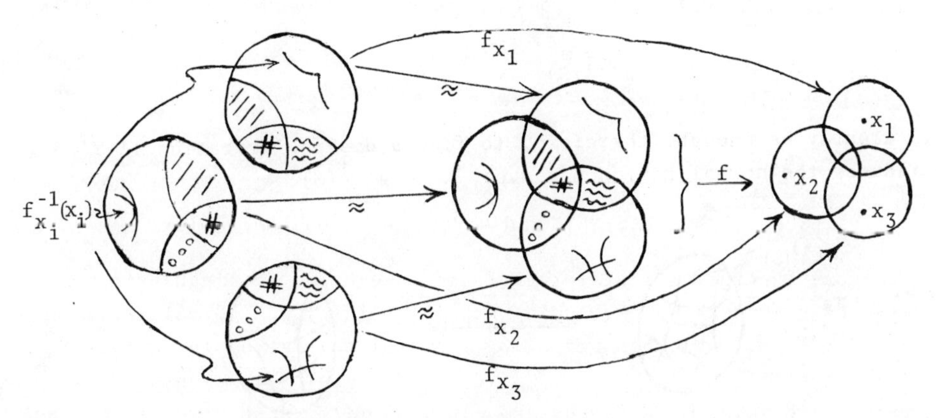

f_{x_1}, f_{x_2}, f_{x_3} induce same map (normalization) over overlaps; so they patch.

* * *

Now let us outline a <u>proof of the Theorem.</u> The Theorem is local on X, so we may assume that there exists a finite map (= branched covering) $\pi: X \to Z$ of X onto an open subset Z of affine 2-space over k. Let $D \subseteq Z$ be the <u>critical</u> (or <u>discriminant</u>) <u>variety</u> for π, i.e. the smallest closed subset of Z over whose complement π is an etale covering. [Equivalently: D consists of those z in Z such that the number of (geometric) points in $\pi^{-1}(z)$ is less than the covering degree of π.] D has dimension ≤ 1.

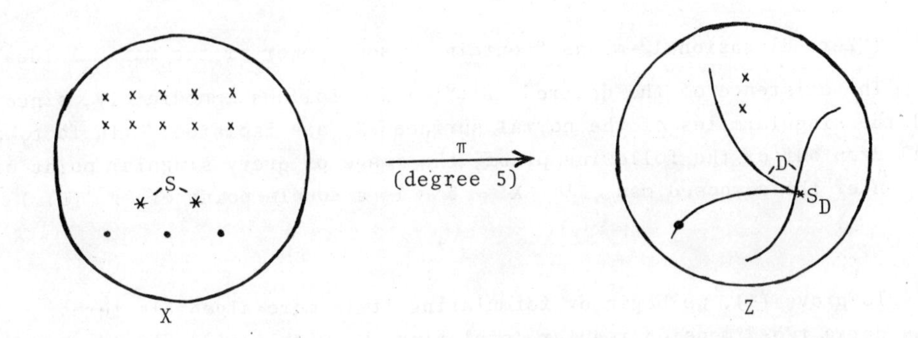

The first step is to reduce to the case where D has only normal crossings. This is done by applying embedded resolution to $D \subset Z$, as follows. Let S_D be the (zero-dimensional) set of singular points of D, and let $S = \pi^{-1}(S_D)$. As we saw in Lecture 1, there exists a map $g: Z' \to Z$, obtained as a succession of blowing-ups (of points), such that g induces an isomorphism $Z' - g^{-1}(S_D) \to Z - S_D$ and such that $g^{-1}(D)$ is a curve having only normal crossings, together with some isolated points (corresponding to the isolated points - i.e. the zero-dimensional components - of D). Let X' be the fibre product $X \times_Z Z'$ and let $g': X' \to X$ be the projection. Then g' induces an isomorphism

$$X' - g'^{-1}(S) \xrightarrow{\sim} X - S.$$

It will clearly be enough, therefore, to find a desingularization $f':Y' \to X'$ which induces the normalization map over $X' - g'^{-1}(S)$.

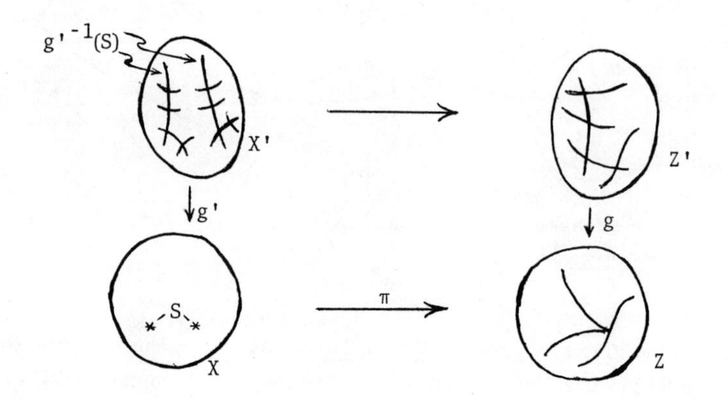

We will show that:

(*) Every singular point of $\bar{X}'$, the normalization of X', can be resolved by blowing up a zero-dimensional ideal in its local ring.

("Zero-dimensional" means "containing some power of the maximal ideal".)

The existence of the desired $f':Y' \to X'$ follows immediately, since all the singularities of the normal surface $\bar{X}'$ are isolated. (In fact, as will drop out of the following proof, the image of every singular point of $\bar{X}'$ under the composed map $\bar{X}' \to X' \to Z'$ is a double point of $g^{-1}(D)$.)

* * *

To prove (*), we begin by formulating it in more algebraic terms. Consider a two-dimension regular local ring Q with residue field k (algebraically closed, characteristic zero) and field of fractions K; let $\bar{Q}$ be the integral closure of Q in a finite field extension L of K, and let R be the localization of Q at one of its maximal ideals. Let u, v generate the maximal ideal of Q. The problem is then to show that if $\bar{Q}[1/uv]$ is étale over $Q[1/uv]$ (i.e. $\text{Spec}(\bar{Q})$ is étale over $\text{Spec}(Q)$ outside the two "lines" $u = 0$, $v = 0$), then $\text{Spec}(R)$ can be desingularized by blowing up a zero-dimensional ideal.

For this, we may pass to completions[3], i.e. we may assume that Q is a power series ring $k[[u, v]]$ and $R = \bar{Q}$.

[3]This would not be necessary if we were in the analytic category.

Suppose then that $R[1/uv]$ is étale over $Q[1/uv]$. A basic observation is that for some integer $\alpha > 0$

$$K = k((u,v)) \subseteq L \subseteq k((u^{1/\alpha}, v^{1/\alpha})).$$

(This was proved classically by topological methods, the point being that the complex plane minus two intersecting lines is topologically the product of two punctured discs, so its fundamental group is $\mathbb{Z} \times \mathbb{Z}$. A purely algebraic proof was given by Abhyankar [Amer. J. Math., 77 (1955) p. 585].) By Galois theory we find that L is generated over $k((u, v))$ by a collection of "monomials" $u^{\beta/\alpha}v^{\gamma/\alpha}$ where the pair (β,γ) runs through some subgroup of $\mathbb{Z} \times \mathbb{Z}$. Elementary considerations then show that

$$L = k((u^{1/c}, v^{1/d}))(u^{1/cn}v^{p/dn})$$

where c, d, n, p are non-negative integers (with nc and d both dividing α), $p < n$, and $(n, p) = 1$. Setting $\bar{u} = u^{1/c}$, $\bar{v} = v^{1/d}$, we have

$$L = k((\bar{u}, \bar{v}))(w) \qquad (w = \bar{u}^{1/n}\bar{v}^{p/n}).$$

From now on we write u for $\bar{u}$ and v for $\bar{v}$. Thus:
R <u>is the normalization of the ring</u>

$$R_0 = k[[u, v, w]] = k[[U, V, W]]/(W^n - UV^p);$$

<u>more explicitly</u>:

$$(**) \qquad R = k[[u^{1/n}, v^{1/n}]] \cap L = \text{free } k[[u, v]]\text{-module with basis } (u^{i/n}v^{p(i)/n})_{0 \le i < n}$$

where $0 \le p(i) < n$

and

$$p(i) \equiv pi \pmod{n}^{4}.$$

<u>Example</u> $(n = 5, p = 2)$.

$$R = k[[u,v,u^{1/5}v^{2/5},u^{2/5}v^{4/5},u^{3/5}v^{1/5},u^{4/5}v^{3/5}]]$$

$$= k[[u,v,u^{1/5}v^{2/5},u^{3/5}v^{1/5}]].$$

---

[4] R is the ring of invariants of the cyclic group of k-automorphisms of $k[[u^{1/n},v^{1/n}]]$ generated by ϕ, where $\phi(u^{1/n}) = \varepsilon^{n-p}u^{1/n}$, $\phi(v^{1/n}) = \varepsilon v^{1/n}$ (ε = primitive n-th root of unity). A detailed study of such "quotient singularities" has been carried out recently by Riemenschneider [Math. Ann. 209 (1974), pp. 211-248].

(Remark. We note in passing that if $R[1/v]$ is étale over $Q[1/v]$, then $n = 1$ and R is regular. Hence every point of $\bar{X}'$ which does not lie over a double point of $g^{-1}(D)$ is already smooth.)

It is convenient to work with R_0 instead of R. As we will soon see, it is quite simple to desingularize $\mathrm{Spec}(R_0)$; and of course every desingularization of $\mathrm{Spec}(R_0)$ factors through $\mathrm{Spec}(R)$, so we will have a desingularization of $\mathrm{Spec}(R)$ too. In fact the procedure to be used will indicate how to compute explicitly a zero-dimensional ideal in R whose blowing up is a desingularization, thereby proving (*).

One can avoid the explicit computation - at some cost, perhaps, in understanding - by using the fact that <u>any desingularization of a two-dimensional normal local ring is obtainable by blowing up a zero-dimensional ideal</u>. (Because of the negative-definiteness of the intersection matrix of the components of the closed fibre, there exists a relatively ample invertible sheaf supported on the closed fibre...)

Alternatively, since R is a quotient singularity we know - once some desingularization has been shown to exist - that R has a <u>rational singularity</u> (cf. E. Brieskorn, Inventiones math. <u>4</u> (1968), p. 340]). Hence (Lecture 1, §4) R <u>can be minimally desingularized by successively blowing up points</u>, and from this (*) follows at once.

* * *

<u>To desingularize</u> $\mathrm{Spec}(R_0)$, let

$$\cdots \to \Sigma_{j+1} \xrightarrow{\phi_j} \Sigma_j \to \cdots \xrightarrow{\phi_1} \Sigma_1 \xrightarrow{\phi_0} \Sigma_0 = \mathrm{Spec}(R_0)$$

be the sequence such that ϕ_j is obtained by blowing up the unique reduced irreducible subscheme L_j of Σ_j whose image in Σ_0 is the "line" $v = w = 0$. It is routine to verify (cf. following example) that there is, for each j, just one closed point σ_j on L_j, that all other closed points on Σ_j are regular, and that the maximal ideal in the local ring of σ_j on Σ_j is generated by three elements u_j, v_j, w_j satisfying a relation

$$w_j^{n_j} = u_j v_j^{p_j} \qquad (0 < n_j,\ 0 \le p_j).$$

Moreover, if $p_j \neq 0$, then

$$n_{j+1}p_{j+1} < n_j p_j .$$

Hence for some $J >> 0$, we must have $p_J = 0$; so σ_J is regular, and $\Sigma_J \to \Sigma_0$ is a desingularization.

The following example should make clear what's going on.

Example (n = 10, p = 7).

$R_0 = k[[u,v,w]]$ $\qquad\qquad$ $w^{10} = uv^7$

L_0 is given by $v = w = 0$; blowing this up, we get

$$\Sigma_1 = \mathrm{Spec}(R_1) \cup \mathrm{Spec}(R_1') \tag{1}$$

$$R_1 = R_0[v/w] \qquad\qquad R_1' = R_0[w/v].$$

Setting $u_1 = u$, $v_1 = v/w$, $w_1 = w$, we have

$$w_1^3 = u_1 v_1^7$$

$$u = u_1, \quad v = v_1 w_1, \quad w = w_1.$$

The inverse image of L_0 in $\mathrm{Spec}(R_1)$ has two components, given by $v_1 = w_1 = 0$, $u_1 = w_1 = 0$; the second maps onto $u = v = w = 0$, so $L_1 \cap \mathrm{Spec}(R_1)$ is $v_1 = w_1 = 0$. In a similar way, we see that $L_1 \cap \mathrm{Spec}(R_1')$ is empty.

Next, $(u,v,w)R_1' = vR_1'$, and the ideal $(u,v,w)R_1$ is invertible wherever $v_1 \neq 0$ (at such points u_1 is a multiple of w_1); thus (u,v,w) is <u>invertible</u> on $\Sigma_1 - L_1$. It follows that any point outside L_1 is regular (note that the quotient of the corresponding local ring by the principal ideal (u,v,w) is regular...). Finally, there is just one closed point σ_1 on L_1, defined by the maximal ideal $(u_1,v_1,w_1)R_1$.

Now blow up L_1 to get Σ_2. Since w_1/v_1 is integral over R_1, we find that

$$\Sigma_2 = \mathrm{Spec}(R_2) \cup \mathrm{Spec}(R_1') \tag{2}$$

$$R_2 = R_1[w_1/v_1] = R_0[v/w,\ w^2/v].$$

Setting $u_2 = u_1 = u$, $v_2 = v_1 = v/w$, $w_2 = w_1/v_1 = w^2/v$, we have

$$w_2^3 = u_2 v_2^4.$$

Here L_2 is the subscheme of $\mathrm{Spec}(R_2)$ defined by $v_2 = w_2 = 0$, and σ_2 is the point $u_2 = v_2 = w_2 = 0$.

Blowing up L_2, we get

$$\Sigma_3 = \mathrm{Spec}(R_3) \cup \mathrm{Spec}(R_1') \tag{3}$$

$$R_3 = R_2[w_2/v_2] = R_0[v/w,\ w^3/v^2]$$

and $$w_3^3 = u_3v_3 \qquad (u_3 = u,\ v_3 = v/w,\ w_3 = w^3/v^2).$$

$L_2 \subseteq \mathrm{Spec}(R_3)$ is defined by $v_3 = w_3 = 0$.

(4) $$\Sigma_4 = \mathrm{Spec}(R_4) \cup \mathrm{Spec}(R_4') \cup \mathrm{Spec}(R_1')$$

$$R_4 = R_3[v_3/w_3] = R_0[w^3/v^2,\ v^3/w^4]$$

$$R_4' = R_3[w_3/v_3] = R_0[v/w,\ w^4/v^3]$$

$$w_4^2 = u_4v_4 \qquad (u_4 = u,\ v_4 = v^3/w^4,\ w_4 = w^3/v^2)$$

$\mathrm{Spec}(R_4')$ is regular, and $L_4 \subseteq \mathrm{Spec}(R_4)$ is given by $v_4 = w_4 = 0$.

(5) $$\Sigma_5 = \mathrm{Spec}(R_5) \cup \mathrm{Spec}(R_5') \cup \mathrm{Spec}(R_4') \cup \mathrm{Spec}(R_1')$$

$$R_5 = R_4[v_4/w_4] = R_0[v^5/w^7]$$

$$R_5' = R_4[w_4/v_4] = R_0[v^3/w^4,\ w^7/v^5]$$

$$w_5 = u_5v_5 \qquad (u_5 = u,\ v_5 = v^5/w^7,\ w_5 = w^3/v^2).$$

The point $\sigma_5 \in \mathrm{Spec}(R_5)$ defined by the ideal $(u_5,v_5,w_5) = (u_5,v_5)$ is regular, and Σ_5 is a desingularization of $\mathrm{Spec}(R_0)$, as desired.

REMARKS 1. In the above desingularization process the map ϕ_j is <u>finite</u> precisely when $n_j < p_j$. When $n_j > p_j$, then $\phi_j^{-1}(\sigma_j) \cong \mathbb{P}^1$.

2. In the preceding example, the rings R_5, R_5', R_4', R_1' appearing in the expression for Σ_5 each give rise to a "wedge" of $\mathrm{Spec}(R_0)$ (cf. Lecture 1, §2). From R_5', for example, we get, with $v_* = v^3/w^4$, $w_* = w^7/v^5$,

$$u = v_*w_*^2 \qquad v = v_*^7w_*^4 \qquad w = v_*^5w_*^3.$$

This is precisely the kind of wedge which Jung constructed in his proof of local uniformization. (The example in Lecture 1, §2, corresponds to $n = 5$, $p = 3$.)

3. Walker observed that Jung's wedges (cf. remark 2) could always be pasted together to give a local desingularization. (This is illustrated by the preceding example.) A more explicit description of what happens (for any n, p) was given by Hirzebruch [Hz]: let

$$\frac{n}{n-p} = b_1 - \frac{1}{b_2 -}\frac{1}{b_3 -}\cdots\frac{1}{b_s} \qquad \text{(continued fraction; each } b_j \geq 2)$$

Then the above described desingularization $\sum$ of R_0 is covered by $\mathrm{Spec}(T_i)$, $0 \le i \le s$,

$$T_i = R_0[v^{\nu_i}/w^{\mu_i},\ w^{\mu_{i+1}}/v^{\nu_{i+1}}] = R[v^{\nu_i}/w^{\mu_i},\ w^{\mu_{i+1}}/v^{\nu_{i+1}}]$$

where the integers μ_j, ν_j $(0 \le j \le s + 1)$ are defined inductively by

$$\mu_0 = 0,\ \mu_1 = 1, \quad \mu_{j+1} = b_j\mu_j - \mu_{j-1}$$

$$\nu_0 = 1,\ \nu_1 = 1, \quad \nu_{j+1} = b_j\nu_j - \nu_{j-1}.$$

(It can be shown that $\mu_{s+1} = n$, $\nu_{s+1} = p$.)

Furthermore, the desingularization $\sum$ is minimal (cf. Lecture 1, §4). The reduced closed fibre, as given by Hirzebruch, is a chain of non-singular rational curves intersecting transversally, and with self-intersections $-b_1,-b_2,\ldots,-b_s$.

$$\underset{-b_1}{\circ}\!\!-\!\!-\!\!\underset{-b_2}{\circ}\!\!-\!\!-\!\!\underset{-b_3}{\circ}\!\!-\!\!-\cdots-\!\!-\!\!\underset{-b_s}{\circ}$$

Example (n = 10, p = 7, as in the above example.)

$$\frac{10}{3} = 4 - \frac{1}{2-}\,\frac{1}{2} \qquad\qquad (b_1 = 4,\ b_2 = 2,\ b_3 = 2)$$

μ: 0, 1, 4, 7, 10.

ν: 1, 1, 3, 5, 7 .

$$T_0 = R_1' \qquad T_1 = R_4' \qquad T_2 = R_5' \qquad T_3 = R_5.$$

4. (This was shown to me by E. Viehweg.) We can get $\sum$ from R by blowing up the ideal I generated by the elements $c_0,c_1,\ldots,c_s$ defined by

$$c_0 = v^{\nu_1}v^{\nu_2}\ldots v^{\nu_s}$$

$$c_i = c_{i-1}(w^{\mu_i}/v^{\nu_i}) \qquad\qquad (0 < i \le s).$$

Indeed,

$$T_i = R[Ic_i^{-1}] = R[\frac{c_0}{c_i}, \frac{c_1}{c_i},\ldots, \frac{c_s}{c_i}].$$

I is not zero-dimensional, but consider the following elements a_i, b_i $(0 \le i \le s)$:

let $\mu = \mu_1 + \mu_2 + \ldots + \mu_s$, choose an integer $r > \mu/n$, and set

$$a_i = c_i v^{rn-\mu-rp}$$
$$b_i = c_i(u^r/w^\mu) = a_i(w/v)^{rn-\mu} .$$

Using the fact that $n\nu_j > p\mu_j$ $(0 \le j \le s)$, one checks that all the a_i and b_i are in R; and the ideal $\bar{I}$ which they generate is <u>zero-dimensional</u> (since $b_s = u^r$ and $a_0 = v^\lambda$, $\lambda > 0$). Moreover, since

$$w/v \in T_0, \quad v/w \in T_i \qquad (0 < i \le s)$$

one sees that

$$\mathrm{Spec}(T_0) = \mathrm{Spec}(R[\bar{I}a_0^{-1}]) \supseteq \mathrm{Spec}(R[\bar{I}b_0^{-1}])$$
$$\mathrm{Spec}(T_i) = \mathrm{Spec}(R[\bar{I}b_i^{-1}]) \supseteq \mathrm{Spec}(R[\bar{I}a_i^{-1}]) \qquad (0 < i \le s).$$

Thus $\sum$ <u>is also gotten by blowing up</u> $\bar{I}$, and (*) is proved.

(This proof of (*) can be made somewhat shorter: just pull the ideal $\bar{I}$ out of a hat, and show directly that blowing it up gives a regular scheme!)

5. It is a curious fact, given without proof by Zariski in 1954 [Z8, p.521], that if $\sum' \to \mathrm{Spec}(R)$ is the blowing up of the closed point of $\mathrm{Spec}(R)$, then <u>every singular point on</u> $\sum'$ <u>is rational of type</u> A_m, i.e. the completion of its local ring is isomorphic to

$$k[[U,V,W]]/(W^{m+1} - UV) \qquad (m \ge 1).$$

More precisely, Zariski's result is that if we set (as we may, uniquely)

$$\frac{n}{n-p} = h_1 + \frac{1}{h_2+}\,\frac{1}{h_3+}\,\cdots\,\frac{1}{h_{2t+1}}$$

then the singularities on $\sum'$ are in one-one correspondence with the integers i such that $h_{2i} > 1$, the singularity corresponding to i being of type $A_{h_{2i}-1}$.

One way to see this is to begin by noting that if $E_1, E_2, \ldots, E_s$ are the components of the closed fibre on $\sum$ (remark 3), and if E is the cycle $E_1 + E_2 + \ldots + E_s$, then $E.E_i \le 0$ for all i $(1 \le i \le s)$, with equality if and only if $i \ne 1$, $i \ne s$, and $b_i = -2$. Thus E is the "fundamental cycle", and the theory of rational singularities tells us that

$$\textstyle\sum' = \mathrm{Proj}\big(\bigoplus_{n\ge 0} H^0(\mathcal{O}(-nE))\big) .$$

From this, and from the relation between the b's and h's, we can reach the desired conclusion.

Presumably however there is a more direct and elementary proof, starting from the explicit description (**) of R. Since - as is easily shown blowing up a singularity of type A_m leads to a scheme with just one singularity of type A_{m-2} (and no singularities at all if $m = 1$ or 2), this will give us another way to show that R can be desingularized by blowing up only points. This type of desingularization descends through completion, so we will have another proof of (*). [Cf. also Remark at end of this Lecture.]

6. There is an illuminating approach to the problem of desingularizing R, due to D. Lieberman:

Any unramified covering of $C' = \mathbb{C}^2 - \{xy = 0\}$ can be realized by a map $\theta: C' \to C'$ of the form

$$(v,w) \to (v^d, v^{-pc}w^{nc}) \qquad (0 \le p < n;\ (n,p) = 1).$$

From a complex analytic point of view, the problem is to find a manifold Y' containing C' as a dense open subset such that θ extends to a proper map of $Y' \to \mathbb{C}^2$, with $\theta(Y' - C') \subseteq \{xy = 0\}$. It is well-known that this can be accomplished by starting with $\mathbb{C}^2 \supseteq C'$ and successively blowing up points. Making things explicit, one comes up with essentially the same formal calculations as before, but with a different motivation. (Roughly, what we did before was, instead of successively blowing up points of indeterminacy of θ, to blow up their inverse image on the graph of θ.)

Algebraically speaking, let $\mathbb{C}\langle v,w\rangle$ be the ring of convergent power series, let $u = v^{-p}w^n$ and let $G = \mathbb{C}\langle v,w\rangle(u)$; then the completion of G at the maximal ideal generated by u, v, w is our old friend R_0 (with $k = \mathbb{C}$). Lieberman's approach produces a desingularization of G, and hence of R_0.

* * *

<u>Examples.</u> We illustrate Jung's method by describing local resolutions for some singularities of multiplicity two. Consider the origin (0,0,0) on the surface defined over k by $w^2 = f(u,v)$ $(f(0,0) = 0)$. If π is the projection to the (u,v)-plane, then the critical variety D is given by $f(u,v) = 0$.

A. $w^2 = (u - a_1v)(u - a_2v)\dots(u - a_nv)$

$(a_1, a_2, \dots, a_n,$ distinct elements of k).

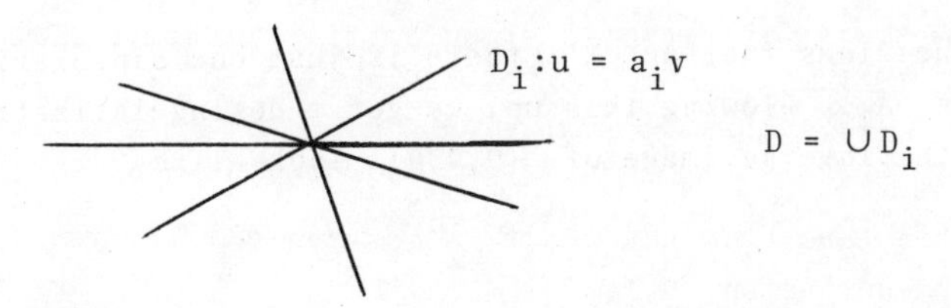

D is resolved by one blowing up:

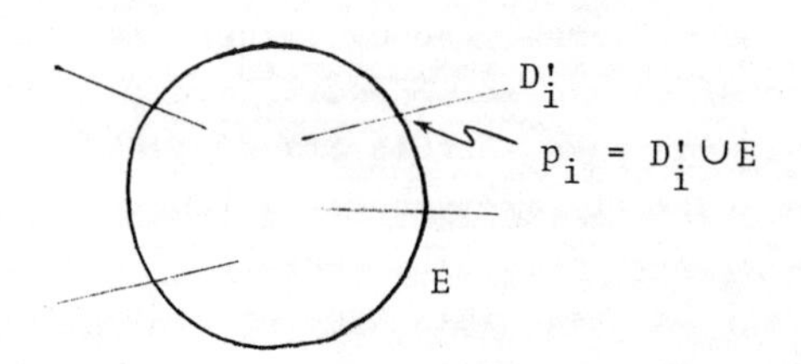

(Here E is the curve coming out of the point $u = v = 0$, and D_i' is the proper transform of D_i).

When n is even, say $n = 2g + 2$, $\bar{X}'$ is already non-singular (notation as at the beginning of the lecture.) The inverse image on $\bar{X}'$ of the original singularity $(0,0,0)$ is a non-singular double covering of E, ramified at the points $p_1, p_2, \ldots, p_{2g+2}$, hence hyperelliptic of genus $g = (n-2)/2$. The self intersection of this curve is -2.

When n is odd, the inverse image of E on $\bar{X}'$ is a non-singular rational curve mapping isomorphically onto E. The singularities of $\bar{X}'$ occur at the points lying over the p_i; and they are rational, of type A_1, so they are each resolved by one blowing-up. On the resulting desingularization, the inverse image of $(0,0,0)$ looks like

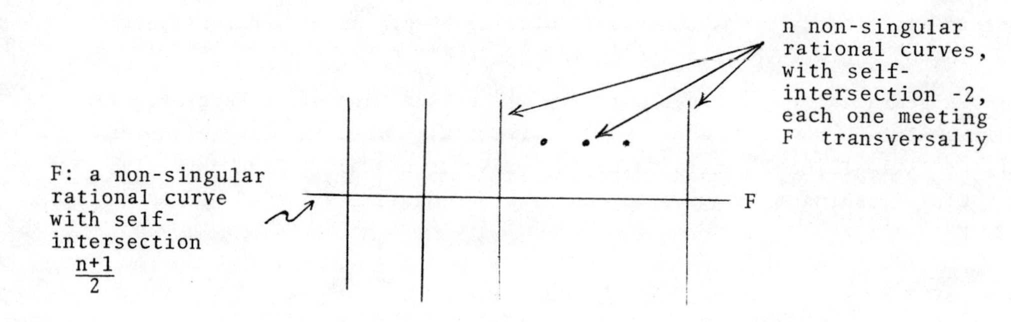

In either case, the resolution obtained is minimal.

B. $$w^2 = -(u^3 + v^7) \qquad \text{(cf. Lecture 1, §3).}$$

One finds that on $\bar{X}'$ there is just one singularity, which is rational of type A_1. Blowing this up, we get a desingularization of $(0,0,0)$, on which the inverse image of $(0,0,0)$ looks like

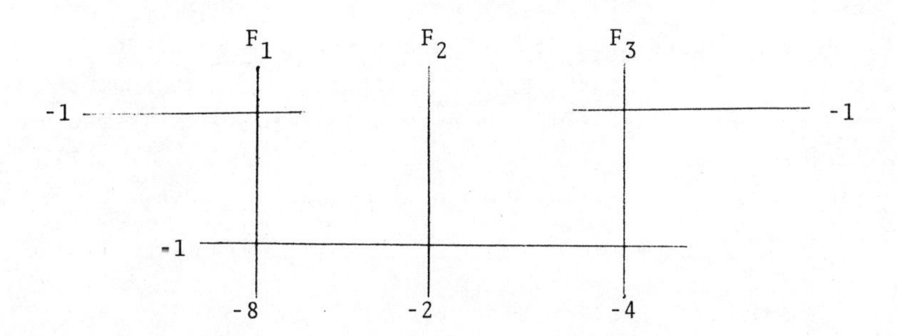

(All curves non-singular rational; all intersections transversal; self-intersections as shown.)

This is certainly not a minimal resolution. Blowing down the -1's, we get three non-singular rational curves meeting transversally at a point:

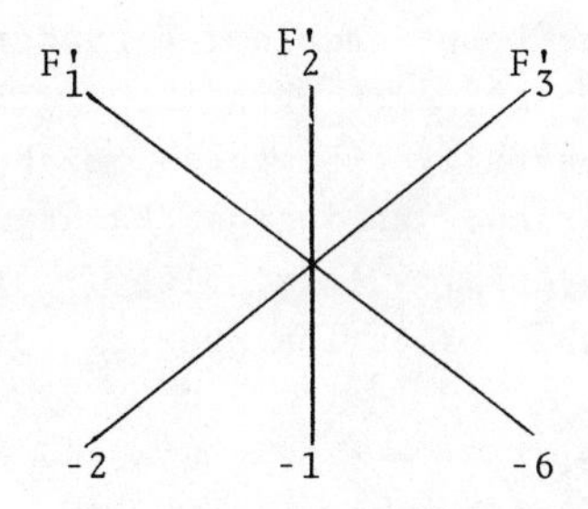

Blowing down F'_2, we get two non-singular rational curves meeting tangentially at one point (intersection multiplicity 2):

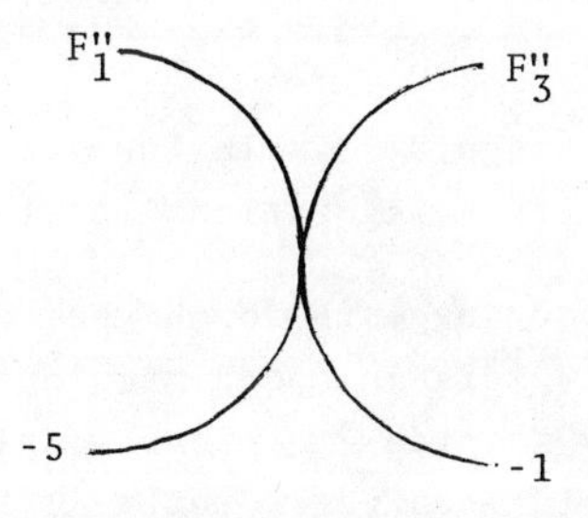

Finally, blowing down F''_3, we get the minimal desingularization on which (as in Lecture 1) the inverse image of the original singularity (0,0,0) is an irreducible rational curve with a cusp, having self-intersection -1.

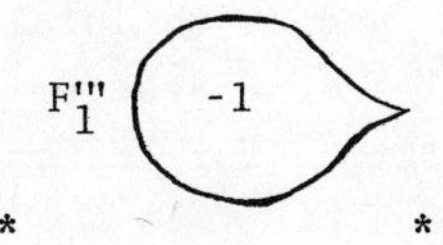

* * *

Here are three examples of the <u>influence of Jung's idea</u> of simplifying singularities by resolving critical varieties (for suitable projections).

I. As a graduate student, Abhyankar tried to adapt Jung's method to surfaces over fields of positive characteristic. He found (with notation as above) that even when $g^{-1}(D)$ has normal crossings, the structure of R may be quite horrible (cf. [Amer. J. Math. 77, (1955), 575-592]).

As an example of what may happen, consider the point $P = (0,0,0)$ on the surface

$$w^2 + wv^3 + u^5 = 0$$

over a field of characteristic 2. The critical variety for the projection into the (u,v)-plane is the line $v = 0$. P is an isolated singularity, hence normal, but one checks easily that on the blowing up of P, the inverse image of P is a singular curve; so P is not even a rational singularity.

But then about ten years ago, Abhyankar discovered that in certain cases (projections giving cyclic Galois coverings of degree = characteristic of the ground field) if one keeps on blowing up certain (possibly smooth) points of the critical variety, even when the critical variety has only normal crossings, then eventually a stage is reached where the structure of R becomes essentially as simple as in the characteristic zero case. (R is a "Jungian domain", in Abhyankar's terminology.) This discovery was a key point in his solution of the resolution problem for surfaces over excellent Dedekind domains.

(In the above example, for instance, blowing up the point $u = v = 0$ and normalizing in the function field of the given surface produces a surface covered by two affine surfaces whose equations are respectively

$$w^2 + wuv^3 + u = 0$$

$$w^2 + wv + vu^5 = 0.$$

The first of these is non-singular, while the second has just one singularity (at $(0,0,0)$), the singularity being rational of type A_9.)

II. What Jung's method suggests for higher dimensions is to apply embedded resolution for divisors in a d-dimensional non-singular variety to a certain critical variety in order to prove local uniformization on d-dimensional varieties. This idea works out well in characteristic zero, though the details are elaborate (cf. the main result (10.25) of chapter 10 in Abhyankar's book [A3]). In fact, using his generalization of Albanese's method (Lecture 1, §5), Abhyankar is able to make the idea work over fields of characteristic $> (d!)$. This is his approach to local uniformization on three dimensional varieties over fields of characteristic $> (3!)$. (The main difficulty is to prove embedded resolution for surfaces in non-singular threefolds.)

III. In 1954 Zariski proposed a <u>global</u> version of the local inductive procedure mentioned in II, as a method for resolution of singularities in characteristic zero [Z8, pp. 512-521]. In other words, apply embedded resolution to a certain $(d-1)$-dimensional critical variety in a non-singular d-dimensional variety, and then deduce embedded resolution for divisors in a non-singular variety of dimension $d + 1$. So far this idea has not met with much success, but it remains of interest as a possible source of a simpler alternative to Hironaka's proof.

* * *

What we have done in this lecture is to use embedded resolution of curves in non-singular surfaces to prove resolution for surfaces. But our resolution procedure is not an "embedded" one, in that it is not entirely made up of transformations of some non-singular ambient variety containing the surface X to be resolved. An embedded resolution procedure, in which Jung's method plays an important role, was given by Zariski in some Lincei notes in the 1960's (cf. [Z7]). But Zariski considered only the <u>proper</u> transform, and not the <u>total</u> transform of X, so that his result is still not as strong as (EMB) (Lecture 1) for surfaces in threefolds over fields of characteristic zero. However Zariski's work did stimulate Abhyankar to find a very simple proof of the stronger result. In Abhyankar's proof, embedded resolution is applied not to the <u>discriminant</u>, but rather to the <u>coefficients</u> of some Weierstrass polynomial. We will discuss these matters in detail in Lecture 3.

* * *

<u>Remark</u> ("Added in proof"). The quotient singularity R can be studied by the methods of Chapter I of Kempf, Knudsen, Mumford and Saint-Donat's <u>Toroidal Embeddings I</u> (Lecture Notes in Math., no. 339, Springer-Verlag), where higher-dimensional analogues of R are also treated. In <u>loc. cit.</u>, R is represented by the <u>plane sector</u> $\sigma = \langle \xi_0, \xi_{s+1} \rangle \subset \mathbb{R}^2$, where $\xi_0 = (1,0)$, $\xi_{s+1} = (p,n)$, and $\langle \xi_0, \xi_{s+1} \rangle$ consists of all points $c\xi_0 + d\xi_{s+1}$ with c, d real and ≥ 0. The desingularization $\sum = \bigcup_{i=0}^{s} \mathrm{Spec}(R[v^{\nu_i}/w^{\mu_i}, w^{\mu_{i+1}}/v^{\nu_{i+1}}])$ (cf. remark <u>3</u> above) corresponds to a "subdivision" $\sigma = \bigcup_{i=0}^{s} \langle \xi_i, \xi_{i+1} \rangle$, with $\xi_i = (\nu_i, \mu_i)$.

Blowing up the maximal ideal of R corresponds to a coarser subdivision $\sigma = \langle \xi_0, \eta_0 \rangle \cup \langle \eta_0, \eta_1 \rangle \cup \langle \eta_1, \eta_2 \rangle \cup \ldots \cup \langle \eta_{t-1}, \eta_t \rangle \cup \langle \eta_t, \xi_{s+1} \rangle$ where $\eta_0 = \xi_1$, $\eta_t = \xi_s$, and $\eta_1, \eta_2, \ldots, \eta_{t-1}$ are the "corners" (= vertices) lying strictly between ξ_1 and ξ_s on the boundary (B) of the convex hull of $(\sigma \cap \mathbb{Z}^2) - (0,0)$. [The "division points" η_i $(0 \leq i \leq t)$ can be characterized as being the points $(p,q) \in \mathbb{Z}^2$ for the which the line $pX + qY = 1$ has a segment (of positive length) in common with the boundary of the convex hull of $(\hat{\sigma} \cap \mathbb{Z}^2) - (0,0)$, where $\hat{\sigma} = \{(x,y) \mid ax + by \geq 0 \text{ for all } (a,b) \in \sigma\}$.]

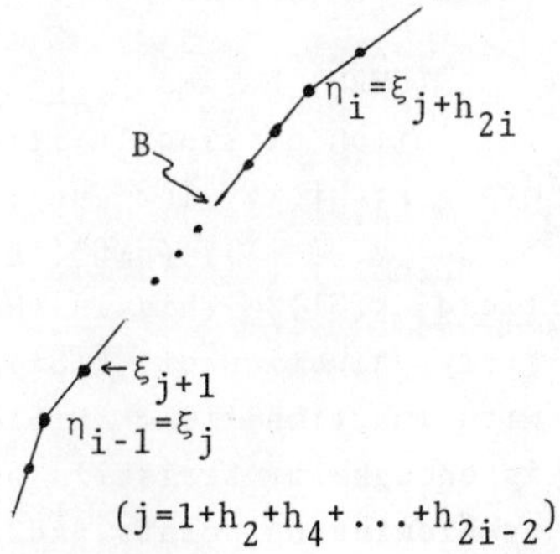

$\eta_i = \xi_{j+h_{2i}}$

B

$\leftarrow \xi_{j+1}$

$\eta_{i-1} = \xi_j$

$(j = 1 + h_2 + h_4 + \ldots + h_{2i-2})$

To establish the result of remark <u>5</u>, one shows that precisely h_{2i-1} of the ξ's are in the interior of the line segment joining η_{i-1} to η_i $(1 \leq i \leq t)$.

LECTURE 3: EMBEDDED RESOLUTION OF SURFACES (CHAR. 0)

The main topic of this lecture is the following weak form of (EMB) (cf. Lecture 1), for surfaces in non-singular threefolds:

(EMB*) Let X be a smooth irreducible three-dimensional variety over an algebraically closed field k of characteristic zero, and let S be a surface in X (i.e. S is a reduced pure two-dimensional closed subvariety of X). Then there exists a proper map $f:Y \to X$, with Y smooth, inducing an isomorphism

$$Y - f^{-1}(\mathrm{Sing}(S)) \xrightarrow{\approx} X - \mathrm{Sing}(S)$$

$$[\mathrm{Sing}(S) = \text{Singular locus of } S]$$

and such that S_Y, the closure in Y of $f^{-1}(S - \mathrm{Sing}(S))$, is smooth (whence the induced map $f:S_Y \to S$ is a desingularization).

We will see that f can in fact be obtained by successively blowing up points and "nice" curves.

As indicated earlier, (EMB*) was proved by Zariski, first in [Z6] and then again in [Z7]. (The latter proof is discussed briefly near the end of this lecture.) We shall present here a fairly detailed version of a previously unpublished proof due to Abhyankar (1966). This proof is in some vague ways influenced by Zariski's, but it is simpler, and accomplishes more (see end of lecture).

Of course Hironaka [H1] has proved far more general results than those to be mentioned in this lecture. But still Abhyankar's proof is worth being acquainted with, because it brings out very nicely - and quickly - the spirit and flavor of the subject.

* * *

(EMB*) has many applications. In the first place, it gives us resolution of singularities for surfaces (say, for simplicity, irreducible) over k; for, since any irreducible surface S' is birational to a surface S in $\mathbb{P}^3$ (= X), (EMB*) gives us a smooth surface (viz. S_Y) with function field $k(S')$. This is the weakest form of resolution; but it shows that every valuation of $k(S')/k$ is centered at a smooth point on some variety with function field $k(S')$; and even this weak form of local uniformization is enough for Zariski's proof that any surface over k can be desingularized by blowing up points and normalizing (cf. Lecture 1).

Secondly, (EMB*) is a central point in Zariski's proof of resolution of singularities of three-dimensional varieties [Z6], and even more so in Abhyankar's treatment [A3]. (Abhyankar works also over fields of positive characteristic, in which case the proof of (EMB*) is far more involved.) In order to deduce resolution from local uniformization (which he had already proved for varieties of any dimension [Z3]) Zariski needed the following theorem ("Dominance") on the elimination of the fundamental locus of a birational transformation:

(DOM) Let $h:X' \to X$ be a proper birational map (X as above). Then there exists a smooth variety Y and a commutative diagram of birational maps

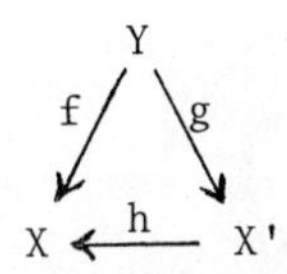

f being obtained as a succession of nice blowing-ups.

Zariski's approach to this theorem is first to reformulate it as a statement about the elimination of the base points of a linear system Λ of divisors on X. He then resolves the singularities of a generic member S of Λ, according to (EMB*). Because of a theorem of Bertini, this creates a situation where each base point of Λ is smooth on almost all members of Λ, and under this condition, the matter becomes relatively straightforward.

Abhyankar has extracted from Zariski's procedure its essential local algebraic content, which turns out to be quite elementary, so that he can give a much simpler deduction of (DOM) from (EMB*) (cf. [A3, p. 52 and p. 229]). Abhyankar's argument is valid also in characteristic $p > 0$, where Bertini's theorem may fail to hold.

Now (DOM) is (almost trivially) equivalent with the following statement ("Principalization"):

(PRIN) Let X be as before, and let $\mathcal{J} \neq 0$ be a coherent sheaf of $\mathcal{O}_X$-ideals. Then there exists a proper birational map $f:Y \to X$, with Y smooth, such that $\mathcal{J}\mathcal{O}_Y$ is locally principal; here f can be obtained as a succession of nice blowing-ups, and in such a way that f induces an isomorphism over the dense open subset of X consisting of those points x at which the stalk $\mathcal{J}_x$ is already principal.

(PRIN) reduces the strong embedded resolution theorem (EMB)(Lecture 1) for a subvariety W of X to the case where W is of pure codimension one, i.e. W is a surface (possibly reducible) in X. We will see at the end of this lecture that a simple modification of the following proof of (EMB*) allows us to add to the conclusion of (EMB*) the further condition that "$f^{-1}(S)$ has only normal crossings". In this way, we see that (EMB) holds for subvarieties of X.

Finally, recall that (EMB) was the basic point in Abhyankar's proof (via Jung and Albanese, cf. end of Lecture 2) of local uniformization on threefolds.

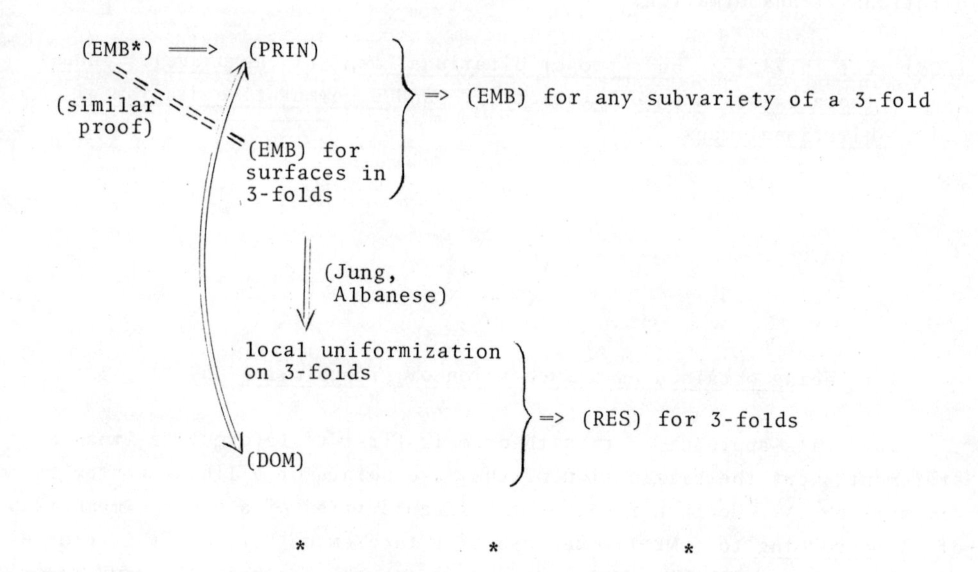

* * *

PROOF OF (EMB*). 1. First of all, recall that the multiplicity of a point s on S can be described as follows: we can represent the completed local ring of s on S in the form

$$\hat{\mathcal{O}}_{S,s} = k[[T,U,V]]/f(T,U,V)$$

for some power series f (T,U,V being local parameters at s on the smooth threefold X); then the multiplicity ν_s is the order of f, i.e. the unique integer ν such that

$$f = f_\nu + f_{\nu+1} + f_{\nu+2} + \dots$$

where $f_i = f_i(T,U,V)$ is a form of degree i, with $f_\nu \neq 0$. We say then that s is a ν-fold point of S. A point s is 1-fold if and only if S is smooth at s. For each ν, the set $S^{(\nu)}$ of points on S of multiplicity $\geqq \nu$ is closed in S [N, §40]. An irreducible curve C on S is said to be ν-fold if $C \subseteq S^{(\nu)}$, $C \not\subseteq S^{(\nu+1)}$ (equivalently: the generic point of C is ν-fold).

Since $S^{(\nu)}$ is empty for $\nu >> 0$ (S being compact), we can set

$$\nu_S = \max_{s \in S} (\nu_s).$$

To prove (EMB*), it is clearly enough to prove the weaker assertion obtained by replacing "S_Y is smooth" by the condition

$$\nu_{S_Y} < \nu_S \quad (\text{if } \nu_S > 1).$$

* * *

2. We are going to construct Y from X by a sequence of permissible transformations, a permissible transformation being one obtained by blowing up either a point, or a smooth curve. One good thing about permissible transformations is that they preserve smoothness, i.e. if $X' \to X$ is a permissible transformation, with X smooth (as above), then also X' is smooth (cf. [Z8; p. 241]).

Some other good properties of permissible transformations are contained in the following preliminary Lemma (cf. [Z6; §§3,6]):

LEMMA 1. Let P be a ν-fold point of S.

(a) Let $f:S' \to S$ be obtained by blowing up P. Then any point $P' \in f^{-1}(P)$ has multiplicity $\leq \nu$ on S'; and any irreducible ν-fold curve $C' \subseteq f^{-1}(P)$ is smooth.

(b) Let $g:S'' \to S$ be obtained by blowing up an irreducible ν-fold curve C on which P is a smooth point. Then $g^{-1}(P)$ is finite; any point $P'' \in g^{-1}(P)$ has multiplicity $\leq \nu$ on S''; and any irreducible ν-fold curve $C'' \subseteq g^{-1}(C)$ passing through P'' is smooth at P''.

Proof.[1] Let R (resp. R', resp. R'') be the completed local ring of P on S (resp. P' on S', resp. P'' on S'').

Using the Weierstrass preparation theorem we can write

$$R \cong k[[X,Y,Z]]/f(X,Y,Z) = k[[x,y,z]]$$

$$f(X,Y,Z) = Z^{\nu} + a_2(X,Y)Z^{\nu-2} + a_3(X,Y)Z^{\nu-3} + \ldots + a_{\nu}(X,Y)$$

where, for $2 \leq i \leq \nu$, $a_i(X,Y)$ is a power series of order $\geq i$. (To get rid of terms involving $Z^{\nu-1}$, we need that $\nu.1 \neq 0$ in k.) Along any ν-fold subscheme of $\operatorname{Spec}(R)$, we have that

[1] Zariski has somewhat stronger results in loc. cit.; his proof is valid also in char. $p > 0$. We give the following char. 0 argument because most of it is used again later on anyway. (By oversight "X" and "Y" are used in the sequel to denote indeterminates, and "f" for power series; no confusion should result.)

$$\nu! Z = \partial^{\nu-1} f/\partial Z^{\nu-1} = 0$$

so that

$$Z = 0 = a_\nu(X,Y).$$

It follows that if C is a ν-fold curve smooth at P, then we may assume that the "germ" of C at P is given by $Z = X = 0$, with $a_i(X,Y)$ divisible by X^i for $2 \leq i \leq \nu$. Hence z/x is integral over R, and we conclude that $g^{-1}(P)$ is finite, as asserted in (b).

Now it is easily seen that

$$R' \cong k[[X,Y,Z]]/f'(X,Y,Z)$$

where, for suitable elements $c, d \in k$, <u>either</u>

$$\text{(i):} \qquad f'(X,Y,Z) = \frac{1}{X^\nu} f(X, X(Y + c), X(Z + d))$$

<u>or</u>

$$\text{(ii):} \qquad f'(X,Y,Z) = \frac{1}{Y^\nu} f(YX, Y, Y(Z + d)).$$

Suppose, for example, that (i) holds. Then

$$f'(X,Y,Z) = Z^\nu + \nu d Z^{\nu-1} + \ldots$$

so that P' has multiplicity $\leq \nu$. Moreover, if some ν-fold curve $C' \subseteq f^{-1}(P)$ passes through P', then first of all $X = 0$ on C' (in fact $X = 0$ on $f^{-1}(P)$); secondly, $d = 0$ (otherwise $\nu d \neq 0$ and P' has multiplicity $< \nu$); and finally, as above, Z vanishes on C', so that C' is given at P' by $Z = X = 0$. This proves (a).

The rest of (b) is proved similarly, starting with the power series

$$f''(X,Y,Z) = \frac{1}{X^\nu} f(X, Y, X(Z + d)).$$

Q.E.D.

* * *

<u>3</u>. The underlying <u>idea of the proof</u> of (EMB*) is that by blowing up certain ν_S-fold points often enough, one achieves a situation where the maximal multiplicity ν_S can be lowered by blowing up smooth ν_S-fold curves.

To make this more precise, we introduce some convenient <u>terminology</u>. Let $\nu = \nu_S$. We say that a ν-fold point s of S is <u>good</u> if the following two conditions hold:

(i) Locally around s, $S^{(\nu)}$ is a <u>curve</u> having at s either a smooth point or an ordinary double point (node).

(ii) Let

(*) $$\cdots \to S_{i+1} \to S_i \to \cdots \to S_1 \to S_0 = S$$

be any sequence in which, for each $i \geq 0$, $S_{i+1} \to S_i$ is obtained by blowing up an irreducible ν-fold curve on S_i; and let $f_i : S_i \to S$ be the composed map $S_i \to S_{i-1} \to \cdots \to S$. Then, for any $i \geq 0$, every point of $f_i^{-1}(s)$ which is ν-fold for S_i lies on an irreducible ν-fold curve of S_i.

The usefulness of this notion is brought out by the following Proposition:

PROPOSITION 1. (a) The number of bad (= not good) ν_S-fold points on S is finite.

(b) If all ν_S-fold points on S are good, and if all the irreducible ν_S-fold curves on S are smooth, then for any sequence (*) as above there is an i such that $\nu_{S_i} < \nu_S$.

The Proposition results easily from the next Lemma. (Details are left to the reader; for (b) of Proposition 1 use also Lemma 1(b)).

LEMMA 2: For any S with $\nu_S > 1$, there exist only finitely many distinct sequences

$$S_n \to S_{n-1} \to \cdots \to S_0 = S$$

in which, for $0 \leq i < n$, S_{i+1} is obtained from S_i by blowing up an irreducible ν_S-fold curve whose image on S is a ν_S-fold curve of S.

(<u>Proof</u>: Consider the generic points of the irreducible ν_S-fold curves of S ...)

We can now reduce to proving the following:

THEOREM 1. <u>Let</u> $\nu = \nu_S$, <u>and let</u> $\cdots \to S'_{i+1} \xrightarrow{\psi_i} S'_i \to \cdots \to S'_0 = S$ <u>be the sequence such that</u> ψ_i <u>is obtained by blowing up all the</u> (finitely many) <u>bad ν-fold points on</u> S'_i $(i \geq 0)$. [Note that $\nu_{S'_i} \leq \nu$, (Lemma 1(a))]. <u>Then the sequence terminates, i.e. for some</u> n <u>all the ν-fold points on</u> S'_n <u>are good</u>.

Indeed, if S'_n is as in Theorem 1, then any irreducible ν-fold curve on S'_n has as its singularities at worst nodes; blowing up all such nodes, we get a surface S^* on which all irreducible ν-fold curves are smooth

(Lemma 1(a)); applying the Theorem to S^* instead of S, we obtain, again by Lemma 1(a), a surface on which all ν-fold points are good <u>and</u> on which all irreducible ν-fold curves are smooth; and then Proposition 1(b) completes the proof of (EMB*).

* * *

<u>4</u>. We sketch the <u>proof of Theorem 1</u>.

Theorem 1 is <u>essentially local</u>: since any surface $S_i^!$ has only finitely many bad ν-fold points, and since an inverse limit of non-empty finite sets is non-empty, we see that if Theorem 1 fails, then there exists an infinite sequence

$$(\sigma) \qquad R_0 \subseteq R_1 \subseteq R_2 \subseteq \dots$$

where R_i is the local ring of a bad ν-fold point on $S_i^!$; so we must show that such a sequence cannot exist.

Since multiplicities and blowing up behave well with respect to completion, we can consider instead of (σ) the corresponding sequence of complete local rings

$$(\hat{\sigma}) \qquad \hat{R}_0 \to \hat{R}_1 \to \hat{R}_2 \to \dots$$

Arguing as in the proof of Lemma 1, we find that

$$\hat{R}_n \cong k[[X,Y,Z]]/f_n(X,Y,Z)$$

$$f_n(X,Y,Z) = Z^\nu + a_{n,2}(X,Y)Z^{\nu-2} + a_{n,3}(X,Y)Z^{\nu-3} + \dots + a_{n,\nu}(X,Y)$$

where $a_{n,i}$ is a power series of order $\geq i$ $(2 \leq i \leq \nu)$; and furthermore, for each n, <u>either</u>

$$(\#) \qquad a_{n+1,i}(X,Y) = \frac{1}{X^i}\, a_{n,i}(X, X(Y + c_n)) \qquad (c_n \in k;\ 2 \leq i \leq \nu)$$

<u>or</u>

$$(\#\#) \qquad a_{n+1,i}(X,Y) = \frac{1}{Y^i}\, a_{n,i}(YX, Y) \qquad (2 \leq i \leq \nu).$$

Now let C_n be the algebroid curve $\mathrm{Spec}(k[[X,Y]]/\prod_j a_{n,j})$ where the product is taken over all j such that $a_{n,j} \neq 0$. Then the preceding remarks show that, modulo completion, C_{n+1} is a closed subscheme of the inverse image of C_n under the map obtained by blowing up the closed point of $\mathrm{Spec}(k[[X,Y]])$. By the <u>embedded resolution theorem for curves in surfaces</u> (Lecture 1), it follows that <u>for large</u> n, C_n <u>has a</u>

normal crossing at its closed point, so that

$$a_{n,i} = \mu_{n,i}(X,Y)X^{\alpha_{n,i}}Y^{\beta_{n,i}} \qquad (2 \le i \le \nu)$$

where each $\mu_{n,i}$ is either a unit in $k[[X,Y]]$, or identically zero, and

$$\alpha_{n,i} + \beta_{n,i} \ge i.$$

It is straightforward to check that since R_n is the local ring of a bad ν-fold point, we cannot have $\beta_{n,i} = 0$ for all i.

Similarly, since R_{n+1} is the local ring of a bad point, we cannot have $\beta_{n+1,i} = 0$ for all i; hence if the above relation (#) holds, we must have $c_n = 0$.

In summary, we have, for large n,

$$f_n(X,Y,Z) = Z^{\nu} + \sum_{i=2}^{\nu} \mu_{n,i}(X,Y)X^{\alpha_{n,i}}Y^{\beta_{n,i}}Z^{\nu-i}$$

$$f_{n+1}(X,Y,Z) = Z^{\nu} + \sum_{i=2}^{\nu} \mu_{n+1,i}(X,Y)X^{\alpha_{n+1,i}}Y^{\beta_{n+1,i}}Z^{\nu-i}$$

where

either $\quad (\alpha_{n+1,i}, \beta_{n+1,i}) = (\alpha_{n,i} + \beta_{n,i} - i, \beta_{n,i}) \qquad (2 \le i \le \nu)$

or $\quad (\alpha_{n+1,i}, \beta_{n+1,i}) = (\alpha_{n,i}, \alpha_{n,i} + \beta_{n,i} - i) \qquad (2 \le i \le \nu)$

* * *

5. The whole matter can now be reduced to a game played with the $(\nu-1)$-tuples of integer pairs $(\alpha_{n,i}, \beta_{n,i})$ $2 \le i \le \nu$.

For any real number ρ, $\{\rho\}$ denotes the fractional part of ρ, i.e. the number in $[0,1)$ which differs from ρ by an integer.

We write "$(\alpha,\beta) \le (\gamma,\delta)$" to signify "$\alpha \le \gamma$ and $\beta \le \delta$".

LEMMA 3. If for some i, we have $\mu_{n,i} \ne 0$ and

$$\left(\frac{\alpha_{n,i}}{i}, \frac{\beta_{n,i}}{i}\right) \le \left(\frac{\alpha_{n,j}}{j}, \frac{\beta_{n,j}}{j}\right) \quad \text{for all } j \text{ with } \mu_{n,j} \ne 0$$

and if furthermore

$$\left\{\frac{\alpha_{n,i}}{i}\right\} + \left\{\frac{\beta_{n,i}}{i}\right\} < 1$$

then R_n is the local ring of a good point.

Proof: Left to reader.

All that we need now is the following two numerical Lemmas, whose (simple) proofs are omitted.

LEMMA 4. Let

$$\delta_{n,i,j} = \left(\frac{\alpha_{n,i}}{i} - \frac{\alpha_{n,j}}{j}\right)\left(\frac{\beta_{n,i}}{i} - \frac{\beta_{n,j}}{j}\right)$$

Then $\delta_{n+1,i,j} \geq \delta_{n,i,j}$;

and if $\delta_{n,i,j} < 0$, then

$$\delta_{n+1,i,j} - \delta_{n,i,j} \geq 1/\nu^4.$$

COROLLARY. There exists n_0 such that for each $n \geq n_0$, and for each i, j, we have $\delta_{n,i,j} \geq 0$, so that the sequence of pairs

$$\left(\frac{\alpha_{n,j}}{j}, \frac{\beta_{n,j}}{j}\right)_{2 \leq j \leq \nu}$$

is totally ordered.

Remark. If n_0 is as in the Corollary, and i is such that $\mu_{n_0,i} \neq 0$

and $$\left(\frac{\alpha_{n_0,i}}{i}, \frac{\beta_{n_0,i}}{i}\right) \leq \left(\frac{\alpha_{n_0,j}}{j}, \frac{\beta_{n_0,j}}{j}\right) \quad \text{for all } j \text{ with } \mu_{n_0,j} \neq 0$$

then clearly for each $n \geq n_0$, we have $\mu_{n,i} \neq 0$

and $$\left(\frac{\alpha_{n,i}}{i}, \frac{\beta_{n,i}}{i}\right) \leq \left(\frac{\alpha_{n,j}}{j}, \frac{\beta_{n,j}}{j}\right) \quad \text{for all } j \text{ with } \mu_{n,j} \neq 0.$$

LEMMA 5. If n_0, i are as in the preceding Remark, then for some $n \geq n_0$, we have

$$\left\{\frac{\alpha_{n,i}}{i}\right\} + \left\{\frac{\beta_{n,i}}{i}\right\} < 1.$$

Hence (Lemma 3), R_n is the local ring of a good point.

This contradiction completes the proof.

* * *

Concluding Remarks.

A. We have previously mentioned Zariski's proof of (EMB*) [Z7]. This proof, though not as elementary as Abhyankar's, has some attractive features.

In Zariski's approach, the "good" singularities are those of "dimensionality type 1", i.e. those s at which S is <u>equisingular</u> along the singular locus (which is supposed to be a curve locally around s). If <u>all</u> the singularities are good (in this sense), and if

$$\cdots \to S_{i+1} \xrightarrow{\phi_i} S_i \to \cdots \to S_0 = S$$

is the sequence such that ϕ_i is obtained from S_i by blowing up the singular locus ($i \geq 0$), then for all i, the singular locus of S_i is a smooth curve, so that ϕ_i is induced by a <u>permissible transformation</u> of ambient threefolds; S_i <u>is finite over</u> S; and for some n, S_n <u>is smooth</u>. Thus the problem is to get rid of the (finitely many) "bad" singularities, those of dimensionality type > 1, which Zariski calls "exceptional singularities".

Let ν be the largest among the multiplicities of the exceptional singularities. Zariski uses a Jungian type argument to show that after the exceptional ν-fold singularities are blown up often enough, those whose multiplicity is not decreased become of a simple type, called "quasi-ordinary". (This means that some neighborhood can be projected into the affine plane in such a way that the branch locus has only normal crossings.) A direct analysis of quasi-ordinary singularities shows first of all that by blowing up the ν-fold locus of S sufficiently often, one reaches a situation where every ν-fold exceptional singularity is "isolated", in the sense that it does not lie on any ν-fold curve of S; and secondly, that the multiplicity of these isolated ν-fold exceptional singularities can be lowered by blowing them up often enough.

Repeating the process, one eventually eliminates the exceptional singularities.

For further insight into (EMB*), part III of [Z6] is recommended.

<u>B</u>. Finally, we indicate how Abhyankar modifies his proof of (EMB*) to obtain the stronger version in which $f^{-1}(S)$ has only normal crossings[2]. Actually it is convenient to show a little more; the precise - and somewhat lengthy - formulation of the result is as follows:

A <u>resolution datum</u> $\mathcal{R} = (E_0, E_1, S, X)$ consists of three surfaces E_0, E_1, S in a smooth threefold X, where each irreducible component of E_0 or E_1 is smooth, and $E = E_0 \cup E_1$ has only normal crossings in X. (Either or both of E_0, E_1 are allowed to be empty.) $\mathcal{R}$ is <u>resolved</u> at a point $s \in S$ if S is smooth at s, and if $E \cup S$ has only normal crossings at s.

Let $\nu = \nu(\mathcal{R})$ be the greatest among the multiplicities on S of points at which $\mathcal{R}$ is not resolved ($\nu = 0$ if $\mathcal{R}$ is resolved everywhere on S); and let Δ be the set of all such ν-fold points. Δ is a closed subset of S, of dimension ≤ 1. ($\Delta = S^{(\nu)}$ if $\nu > 1$).

---

[2] Zariski has recently found a proof of this stronger result, along the lines of [Z7]. (Oral communication).

An irreducible curve $C \subset \Delta$ is permissible at a point $s \in C$ if C has normal crossings with E at s [this means that there exist local coordinates x, y, z at s (in X) such that E is defined near s by $x^{\varepsilon_1}y^{\varepsilon_2}z^{\varepsilon_3} = 0$ ($\varepsilon_i = 0$ or 1) and C by $x = y = 0$], and if there is no irreducible curve $C' \subset \Delta$ which has normal crossings with E at s and which lies in more components of E_1 than does C. (Any curve lies on at most two components of E_1). C is permissible if it is so at all its points.

A permissible transformation of $\mathcal{R}$ is a map $g: X' \to X$ obtained by blowing up a subvariety B of Δ, where B is either a point or a permissible curve. The g-transform $\mathcal{R}' = (E_0', E_1', S', X')$ of $\mathcal{R}$ is defined by:

S' = proper transform of S [= closure in X' of $g^{-1}(S-B)$]

E_1' = proper transform of E_1

$E_0' = g^{-1}(E_0 \cup B)_{\text{reduced}}$

unless $\nu_{S'} < \nu_S$, in which case we set

$E_1' = g^{-1}(E_0 \cup E_1 \cup B)_{\text{reduced}}$

E_0' = empty set.

$\mathcal{R}'$ is a resolution datum, and $\nu(\mathcal{R}') \leq \nu(\mathcal{R})$. It is clear what is meant by the transform of $\mathcal{R}$ under a succession of permissible transformations[3].

The theorem to be proved is that a resolution datum (E_0, E_1, S, X) with E_0 empty can be resolved, i.e. there exists a succession of permissible transformations $f = g_1 \circ g_2 \circ \ldots$ under which $\mathcal{R}$ is transformed into a resolved datum (E_0^*, E_1^*, S^*, Y). (Then S^* is smooth, and $f^{-1}(S)$ has only normal crossings in Y.) It is clearly enough to show that if $\nu(\mathcal{R}) > 0$, then $\mathcal{R}$ can be transformed into a datum $\mathcal{R}^*$ with $\nu(\mathcal{R}^*) < \nu(\mathcal{R})$.

For the proof, one says that a point $s \in \Delta$ is pregood if locally around s, Δ is a curve having either a simple point or a node at s, each (formal) branch of Δ having a normal crossing with E at s. s is a good point if, roughly speaking, any ν-fold point t obtained from s by successively blowing up "locally" permissible curves is still pregood. (In other words there is a sequence $s = s_0, s_1, \ldots, s_n = t$ in which each s_{i+1} is a point lying over s_i on the surface obtained by blowing up a curve which is permissible at s_i ...). The number of bad points is finite. If all points of Δ are good, and if all components of Δ are smooth curves then we can lower $\nu(\mathcal{R})$, as desired, by successively blowing up components of Δ, subject to the following restriction: never blow up a component B of Δ if there is another component B' of Δ which lies in more components of E_1 than B does.

In this way (cf. §3 above) one reduces to proving the central local fact, viz. there cannot exist an infinite "quadratic sequence" of bad ν-fold points. The technique for showing this is similar to that of §§4,5 above except that one must work simultaneously with the local equations of E_0, E_1, S. Details are left to the interested reader.

[3] To understand fully the motivation behind all the foregoing definitions, one must work through the details of the proof indicated below.

REFERENCES

[A1] S. S. ABHYANKAR, Resolution of singularities of arithmetical surfaces, Arithmetical Algebraic Geometry, Harper and Row, New York, 1965 (edited by O. F. G. Schilling); 111-152.

[A2] __________, An algorithm on polynomials in one indeterminate with coefficients in a two dimensional regular local domain, Annali di Matematica Pura ed Applicata, Serie IV, 71 (1966), 25-60.

[A3] __________, Resolution of singularities of embedded algebraic surfaces, Academic Press, New York and London, 1966.

[A4] __________, On the problem of resolution of singularities, Proc. Internat. Congr. of Mathematicians (Moscow 1966), Izdat. "Mir", Moscow, 1968; 469-481.

[A5] __________, Resolution of singularities of algebraic surfaces, Algebraic Geometry, Oxford Univ. Press, London, 1969; 1-11.

[A1b] G. ALBANESE, Transformazione birazionale di una superficie algebrica qualunque in un'altra priva di punti multipli, Rend. Circolo Matematica Palermo, 48 (1924), 321-332.

[B] E. BRIESKORN, Über die Auflösung gewisser Singularitäten von holomorphen Abbildungen, Math. Annalen 166 (1966), 76-102.

[H1] H. HIRONAKA, Resolution of singularities of an algebraic variety over a field of characteristic zero, Annals of Math. 79 (1964), 109-326.

[H2] __________, Desingularization of excellent surfaces, Advanced Science Seminar in Algebraic Geometry, Bowdoin College, Summer 1967. (Notes by Bruce Bennett on Hironaka's lectures.)

[H3] __________, Characteristic polyhedra of singularities, J. Math. Kyoto Univ., 7 (1967), 251-293.

[H4] __________, Certain numerical characters of singularities, J. Math. Kyoto Univ. 10 (1970), 151-187.

[H5] __________, Additive groups associated with points of a projective space, Annals of Math. 92 (1970), 327-334.

[H6] __________, Bimeromorphic smoothing of complex analytic spaces, preprint, Warwick University, 1971.

[H7] __________, Introduction to the theory of infinitely near points, Publicaciones del Instituto "Jorge Juan" de matematicas, University of Madrid, 1974.

[Hz] F. HIRZEBRUCH, Über vierdimensionale Riemannsche Flächen mehrdeutiger analytischer Funktionen von zwei komplexen Veränderlichen, Math. Annalen 126 (1953), 1-22.

[J] H. W. E. JUNG, Darstellung der Funktionen eines algebraischen Körpers zweier unabhängigen Veränderlichen x, y in der Umgebung einer stelle x = a, y = b, J. Reine Angew. Math. 133 (1908), 289-314.

[L] J. LIPMAN, Rational singularities, with applications to algebraic surfaces and unique factorization, Inst. Hautes Etudes Sci., Publ. Math. no. 36(1969), 195-279.

[N] M. NAGATA, Local Rings, Interscience, New York, 1962.

[N-B] M. NOETHER, A. BRILL, Die Entwicklung der Theorie der algebraischen Funktionen in älterer und neuerer Zeit, Jahresbericht der Deutschen Math.-Verein. III (1892-93), 107-566.

[S] P. SAMUEL, Lectures on old and new results on algebraic curves, Tata Institute of Fundamental Research, Lectures on Mathematics, No. 36, Bombay, 1966.

[W1] R. J. WALKER, Reduction of the singularities of an algebraic surface, Annals of Math. 36 (1935), 336-365.

[W2] ________, Algebraic Curves, Princeton Univ. Press, Princeton, 1950. (reprinted: Dover, New York, 1962).

[Z1] O. ZARISKI, Algebraic Surfaces, Ergebnisse der Math. vol. 61, Springer-Verlag, Berlin-Heidelberg-New York, 1971.

[Z2] ________, The reduction of the singularities of an algebraic surface, Annals of Math. 40 (1939), 639-689.

[Z3] ________, Local uniformization on algebraic varieties, Annals of Math. 41 (1940), 852-896.

[Z4] ________, A simplified proof for the resolution of singularities of an algebraic surface, Annals of Math. 43 (1942), 583-593.

[Z5] ________, The compactness of the Riemann manifold of an abstract field of algebraic functions, Bull. Amer. Math. Soc. 45 (1944), 683-691.

[Z6] ________, Reduction of the singularities of algebraic three-dimensional varieties, Annals of Math. 45 (1944), 472-542.

[Z7] ________, Exceptional singularities of an algebroid surface and their reduction, Accad. Naz. Lincei Rend., Cl. Sci. Fis. Math. Natur. serie VIII, 43 (1967), 135-146.

[Z8] ________, Collected Papers, vol. 1 (edited by H. Hironaka and D. Mumford), MIT Press, Cambridge (Mass.) and London, 1972.

[Z-M] ________. H. J. MUHLY, The resolution of singularities of an algebraic curve, Amer. J. Math. 61 (1939), 107-114.

PURDUE UNIVERSITY
WEST LAFAYETTE, INDIANA
U.S.A.

Proceedings of Symposia in Pure Mathematics
Volume 29, 1975

EIGENVALUES OF FROBENIUS ACTING ON ALGEBRAIC VARIETIES OVER FINITE FIELDS

B. Mazur[1]

INTRODUCTION

What follows are some notes I wrote for myself, from which I prepared my two expository lectures. I have tried to make the beginning motivations, at least, accessible to anyone. No proofs are given. Some remarks about the technical vocabulary and prerequisites: The word variety is consistently used -- for which the reader can substitute the phrase system of equations. At times in the course of things it seemed appropriate to use the word scheme and I did so: in fact, it would have been obscure not to have done so. Some idea of cohomology, (coming from any elementary course in algebraic topology) is essential, simply because our project is to explain arithmetical questions by means of cohomology. I have purposefully left examples involving abelian varieties and modular forms for the end, expecting that these subjects speak to a smaller audience.

For those who know the general subject it may be useful to indicate what topics (which truly deserve a place) are not covered in my narrative:

§1. L functions (cf. [34] for an expository account which pays more attention to these companions of zeta functions) .

AMS (MOS) subject classicications (1970). Primary 14G10, 14G13; Secondary 14F20, 14F30.

[1]Work supported by the National Science Foundation under grant GP-43613X.

2. Questions over numberfields. Our subject has been strongly influenced by the analogy with numberfields.

(e.g., E. Artin was motivated to frame the conjecture which he called the Riemann hypothesis by analogy with that older conjecture bearing the same name).

Nevertheless I have consistently, and sometimes unnaturally, kept away from all numberfield considerations. The unnaturality is especially great when one attempts to treat almost any instructive particular example of a variety over a finite field; all the particular examples that come to my mind are in fact the reductions mod p of varieties over global fields and their instructive character usually derives from some global structure: a Grössencharakter or a modular form.

3. Exponential sums. Gauss sums.

4. Geometric applications of the Riemann hypothesis.

Omitting this aspect of the story gives our account of things a much too arithmetical bias. A reader of these pages might be surprised to find that the fruits of the most transcendental of methods over the complex numbers: the hard Lefschetz theorem, "a theory of Hodge structure", are obtainable in characteristic p only after one has a well-working theory of "eigenvalues of Frobenius" -- in particular, the Riemann hypothesis. The results of [24] in the direction of establishing algebraicity of certain cohomology classes also falls under this heading. I have left all this out because it is too difficult (for an expository lecture).

5. The Tate-Sato conjecture. I have no excuse for not mentioning this beautiful and pertinent matter.

6. A discussion of Deligne's proof of the Riemann hypothesis.

See, however, [23] and [35] for excellent brief accounts.

Eigenvalues of Frobenius for Varieties in Characteristic p .

1. Motivation

(1.1) The main question. Let k be a finite field of cardinality $q = p^f$. In a fixed algebraic closure $\bar{k}$, let $k_n \subset \bar{k}$ be the subfield of cardinality q^n. Let V be a variety over k. No matter how disguised by reformulation, our main arithmetical question concerning V will always be to understand the behavior of

$$N_n(V/k) = \text{cardinality } (V(k_n)) \qquad n > 0$$

where $V(k_n)$ is the set of points of V, rational over k_n.

(1.2) The link between arithmetic and topology, as illustrated in a trivial example. The deep aspects of our subject are wholly lacking in the example I am about to present. It is nevertheless a revealing case to experiment with.

If k is any field and V a complete variety defined over k, say that V is paved by affine spaces over k if there is a filtration

$$(*) \qquad V = V_d \supseteq V_{d-1} \supseteq \cdots \supseteq V_j \supseteq \cdots \supseteq V_0 ,$$

where $V_j - V_{j-1}$ is a disjoint union of a finite number (say B_{2j}) of copies of affine j-space over k. Examples of such varieties are given by projective space $\mathbb{P}^d/k$ or more generally Grassmannians over k.

We now compare the arithmetic and the cohomology of such varieties.

Arithmetic

If k is a finite field of cardinality q, and $\mathbb{A}^j$ is affine space over k, then $N_n(\mathbb{A}^j/k) = q^{jn}$. So, for a V paved - by - affine - spaces over k one has :

$$(I) \qquad N_n(V/k) = \Sigma B_{2j} \cdot q^{nj} .$$

Cohomology

If $k = \mathbb{C}$, the cohomology of (the classical topological space underlying the complex variety) V is readily computed from the filtration (*) (which gives that topological space the structure of a cell complex built with even-dimensional cells) and one gets:

$$\text{(II)} \quad \begin{aligned} H^{2j+1}(V, \mathbb{Z}) &= 0 \qquad \text{(for all } j) \\ H^{2j}(V, \mathbb{Z}) &= \text{free abelian group of rank } B_{2j} . \end{aligned}$$

In this example, all cohomology classes are algebraic, and, in particular, the odd Betti numbers vanish.

Now return to a V, as above, defined over any field k. In anticipation of a point of view due to Weil, we may define by analogy the <u>Betti numbers</u> of V by taking the odd Betti numbers to be zero, and letting the $2j$-th Betti number be the B_{2j} coming from the filtration (*). Formula (I) is then a relation between the "arithmetic" of V/k, and the "Betti numbers" of V.

The generalization of (I) for an arbitrary variety over a finite field, takes the form of a <u>Lefschetz fixed point formula</u>. That this might be the case was first surmised by Weil. To begin to explain this idea we must interpret $N_n(V/k)$ as the set of fixed points of some operator.

(1.3) <u>Frobenius</u>. Fixing k of cardinality q, the <u>Frobenius morphism</u> is the morphism $F : V \to V$ which "raises all coordinates to the q-th power". Somewhat more scheme-theoretically, if V is an algebraic variety over k, then $F : V \to V$ is the "identity map" on the underlying topological space of V, while $F : \mathcal{O}_V \to \mathcal{O}_V$ is the morphism (of sheaves of k-algebras) which raises sections to the q-th power. One also has the Frobenius automorphism in $\text{Gal}(\bar{k}/k)$, $\varphi : \bar{k} \to \bar{k}$, defined by the same rule $x \to x^q$. The action of φ on $\bar{k}$ induces an automorphism of $V(\bar{k})$; one sees easily that the action of F on V induces the same automorphism of $V(\bar{k})$. Since the fixed subfield of φ^n in $\bar{k}$ is k_n, the fixed subset of $V(\bar{k})$ under φ^n is $V(k_n)$.

Consequently:

$$V(\bar{k})^{F^n} = V(k_n) \qquad \text{(superscript meaning subset fixed by the indicated operator)}$$

of the Frobenius morphism.

(1.4) Zeta functions. It should already be apparent from what we have said, that we anticipate dealing systematically with all iterates of a certain operator, and more specifically, with their traces. Let $f : W \to W$ be any endomorphism of a finite dimensional vector space over a field of characteristic 0. Knowledge of the trace of f^n for all n is equivalent to knowledge of the characteristic polynomial of f, as is seen by the formal identity:

$$\det(1 - t \cdot f) = \exp\left[- \sum_{1}^{\infty} \mathrm{Trace}(f^n) \cdot t^n/n \right]$$

(which one verifies by first checking when W is of dimension 1, and then noting that both sides are multiplicative for short exact sequences of vector spaces endowed with endomorphisms).

In the light of our expressed anticipations, it is not unnatural to "package" the numbers $\{N_n(V/k); \forall\, n\}$ into the following expression, regarded first as a formal power series in $\mathbb{Q}[[t]]$:

$$Z(V/k, t) = \exp\left(\sum N_n(V/k) \cdot t^n/n \right).$$

This formal power series is called the zeta function of V/k and our main problem is simply: to understand it. As presented, it is a power series with coefficients in $\mathbb{Q}$. It admits, however, another expression as infinite product:

$$Z(V/k, t) = \prod_{v:\ \text{closed points of the scheme } V} \left(1 - t^{\text{degree } k(v)/k}\right)^{-1}$$

which shows that it lies in $\mathbb{Z}[[t]]$, (and also puts it in somewhat closer analogy to the classical ζ-function). One checks this formula by a calculation taking logs and noting that a point in $V(k_n)$ is a pair consisting in (1) a closed point v of V and (2) a k-isomorphism $k(v) \to k_n$. To work with our example, if V/k is paved-by-affine-spaces, then formula (I) gives:

$$Z(V/k, t) = \Pi(1 - q^j t)^{-B_{2j}}$$

and is therefore a rational function whose pole of smallest complex absolute value is at $t = q^{-d}$ $(d = \dim V)$. This is a special case of the following trivial general estimate.

If V/k is any variety of dimension d, then there is a constant C, independent of n such that $N_n(V/k, t) \leq C \cdot q^{nd}$ for all n. It follows that $Z(V/k, t)$ is holomorphic in the disc $t < q^{-d}$.

To see this just find a finite affine cover of V, each member of which admits a finite map to affine d-space, and use that $N_n(\mathbb{A}^d/k) = q^{nd}$.

But we are after much deeper information than this !

(1.5) Weil cohomology. From now on we assume V/k proper and smooth, and we shall indicate explicitly when an important result is valid without these hypotheses.

The detailed axiomatic description of a cohomology theory relevant to a "generalization" of formula (I) is worked out in [23]. The following hint of a description is adequate for our purposes.

Let $\bar{k}$ be any algebraically closed groundfield, and K a field of characteristic 0, called the coefficient field. A contravariant functor

$$X \longrightarrow H^*(X)$$

Category of:	complete connected smooth varieties over $\bar{k}$	augmented graded($\mathbb{Z}$ graded; vanishes for negative degrees); finite-dimensional; anti-commutative; K-algebras (multiplication called cup-product).

is called a Weil cohomology with coefficients in K if it comes along with three extra features:*

A. Poincaré duality: If $d = \dim X$, then one is given an (orientation) isomorphism $H^{2d} \cong K$, and cup-product induces a perfect pairing

$$H^j(X) \times H^{2d-j}(X) \longrightarrow K .$$

B. Künneth formula: For any X_1, X_2

$$H^*(X_1) \otimes H^*(X_2) \longrightarrow H^*(X_1 \times X_2)$$
$$a \otimes b \longrightarrow \mathrm{proj}_1 a \cdot \mathrm{proj}_2 b$$

is an isomorphism.

C. The existence of an adequate cycle theory: If $C^j(X)$ is the group of algebraic cycles of codimension j on X, there should be given a fundamental class homomorphism

$$\mathrm{FUND} : C^j(X) \longrightarrow H^{2j}(X) \qquad (\forall\ j)$$

functorial in X, compatible with products via the Künneth map, and which sends a 0-cycle in X to its degree regarded as an element of $K \cong H^{2d}(X)$.**

---

* An important fourth feature which we do not throw in, but is valid (in even greater strength) for the known examples of Weil cohomologies (and wished for by Weil in [46]) is a comparison theorem: If $X/\bar{k}$ comes by reduction from a proper smooth scheme $\tilde{X}/R$ where R is a ring endowed with an imbedding in $\mathbb{C}$, and if $X_{\mathbb{C}}$ denotes the complex variety induced via this imbedding, then one has an isomorphism $H^*(X) \cong H^*_{\text{classical}}(X_{\mathbb{C}}; \mathbb{Z}) \otimes K$, functorially dependent upon the above situation.

** For many purposes, a cohomology theory satisfying only A, B above suffices. Call such a cohomology theory a weak Weil cohomology. In a weak Weil cohomology, if $z \subseteq X$ is an irreducible algebraic cycle of codimension j which is the image of a smooth complete variety, then one can associate to z a natural class $\mathrm{Fund}(z) \in H^{2j}(X)$.

A formal consequence of the paraphernalia offered in A, B, is the following: Let $\Delta \subset X \times X$ be the diagonal, regarded as d-cycle. Let $u : X \to X$ be a morphism, and use the same letter to denote its graph $u \subset X \times X$ which is also a d-cycle. Then the intersection multiplicity (u, Δ) can be computed by means of any (weak) Weil cohomology,

$$(u, \Delta) = \sum_{j=0}^{2d} (-1)^j \operatorname{Trace}(u_* | H^j(X)) .$$

See [23] for a proof. Let V be a proper smooth variety over k and let $X = V \otimes_k \bar{k}$. Let F be the Frobenius endomorphism. We apply our formula $u = F^n$. First note that the support of the 0-cycle $F^n \cdot \Delta$ consists in those closed points v of V whose residue fields k(v) admit an imbedding into k_n. This follows from the discussion of (1.3). Next, the cycles F^n and Δ are transversal and consequently, <u>the multiplicities of these closed points</u> v <u>in the</u> 0-<u>cycle</u> $F^n \cdot \Delta$ <u>are all</u> 1. The reason for this is, briefly: the "derivative" of the map F^n (which sends the coordinate x to x^{q^n}) vanishes, since we are in characteristic p, whereas the identity map surely induces an isomorphism on tangent spaces.

Consequently, if we are given any (weak) Weil cohomology,

$$N_n(V/k) = \sum [k(v) : k] = (F^n, \Delta)$$

[the summation being over closed points v of V such that k(v) imbeds in k_n]

$$= \sum_j (-1)^j \operatorname{Trace}(F^n | H^j(\bar{V}))$$

and, from the discussion of (1.4) this gives us an expression

$$(*) \qquad Z(V/k, t) = \frac{P_1(t) \cdot P_3(t) \cdots P_{2d-1}(t)}{P_0(t) \cdot P_2(t) \cdots P_{2d}(t)} \in K(t)$$

where $P_j(t) = \det(1 - t \cdot (F | H^j(\bar{V}))$.

LEMMA: <u>(Granted the existence of any (weak) Weil cohomology)</u> $Z(V/k, t)$ <u>may be expressed as a rational function</u> $A(t)/B(t)$ <u>where</u> $A(t)$, $B(t) \in \mathbb{Z}[t]$ <u>are polynomials with constant term</u> 1.

<u>Proof</u>: Lot K be the coefficient field of the Weil cohomology. Then $Z(V/k, t)$ lies both in $\mathbb{Z}[[t]]$ and by (*) in $K(t)$. Now the necessary and sufficient criterion that a power series $\Sigma a_n t^n \in \mathbb{Z}[[t]]$ be the Taylor series development of a rational function in $K(t)$ or in $\mathbb{Q}(t)$ is the <u>same</u>: for N, M sufficiently large, all the Haenkel determinants

$$\det((a_{i+j+N}) \;;\; 0 \leq i, j, \leq M)$$

must vanish. Thus $Z(V/k, t) = A(t)/B(t)$ where $A(t), B(t) \in \mathbb{Q}(t)$ are relatively prime, with constant term 1. To prove the lemma, it suffices to show that $B(t) \in \mathbb{Z}[t]$. Suppose, on the contrary, that a prime r divides the denominator of some coefficient of $B(t)$. Writing $B(t) = \prod (1 - \alpha_j t)$ in some algebraic closure of $\mathbb{Q}_r$, the r-adic completion of $\mathbb{Q}$, we must have $\|\alpha_j^{-1}\| < 1$ for some j. Substituting $t = \alpha_j^{-1}$ in $A(t) = Z(t) \cdot B(t)$ we have that both A, and B admit α_j^{-1} as root, which is a contradiction.

COROLLARY: <u>If a (weak) Weil cohomology exists</u>, $Z(V/k, t)$ <u>is a rational function of</u> t; <u>its set of zeroes is a set of reciprocals of algebraic integers, closed under</u> $\mathbb{Q}$-<u>conjugation</u>.

<u>Reserve the letter</u> α to denote the <u>reciprocal of a zero or pole of</u> $Z(V/k, t)$. The set of zeroes of $Z(V/k, t)$ are <u>among</u> the set of eigenvalues of Frobenius acting on the odd-dimensional part of any Weil cohomology, while the set of poles are among the set of eigenvalues of Frobenius acation on the odd-dimensional part.

The reason we say <u>among</u> is that there is no a priori reason why cancellation may not occur in the fraction (*). That there <u>is no</u> cancellation occurring in (*) for the known (weak) Weil cohomologies is a consequence of the full story.

(1.6) The functional equation. Action of Frobenius on algebraic cycles.

Let us be given a (weak) Weil cohomology and let $f : V \to V$ be a finite morphism. One has the commutative diagram

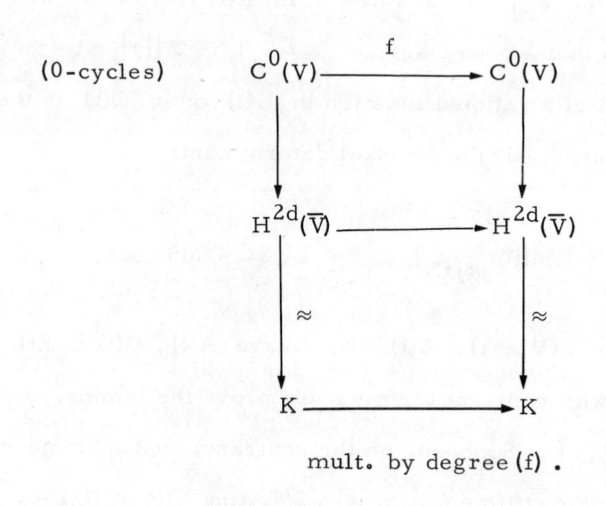

Since the (finite) morphism Frobenius is of degree q^d when acting on a variety of dimension d, its effect on $H^{2d}(\overline{V})$ is multiplication by q^d.* Comparing this with Poincaré duality, one obtains the functional equation: If α is an eigenvalue of Frobenius acting on $H^j(\overline{V})$, then q^d/α is an eigenvalue of Frobenius acting on $H^{2d-j}(\overline{V})$.

COROLLARY (Functional equation) : If a (weak) Weil cohomology exists, then $\alpha \longmapsto q^d/\alpha$ is a permutation of the "reciprocal zeroes" of the rational function $Z(V/k,t)$ and of its "reciprocal poles".

---

* Remark: For the ℓ-adic Weil cohomologies ($\ell \neq p$), one has a well-working theory of cohomology with compact supports and by studying the action of Frobenius on the compactly supported cohomology of the smooth locus of any irreducible cycle of codimension j defined over K, one may check that its fundamental class is an eigenvector for Frobenius with eigenvalue q^j. This accounts for the shape of the zeta function of a variety paved-by-affine-spaces.

(1.7) Statement of known results concerning the conjectures of Weil.

We let k be of cardinality $q = p^f$, $\overline{k}$ a fixed algebraic closure, and let X run through the category of smooth proper varieties over $\overline{k}$. We begin with assertions concerning Weil cohomologies, and go on to number-theoretic assertions.

THEOREM:

a. If ℓ is any prime different from p, the ℓ-adic cohomology

$$X \longmapsto [\lim H^*_{et}(X, \mathbb{Z}/\ell^{\nu})] \otimes \mathbb{Q}_\ell$$

is a Weil cohomology, with coefficients in $\mathbb{Q}_\ell$.

b. There does not exist a Weil cohomology with values in $\mathbb{Q}_p$, for varieties $X/\overline{k}$, but if K_p denotes the field of fractions of $W(\overline{k})$, the Witt vectors of $\overline{k}$, then crystalline cohomology

$$X \longmapsto H^*_{Cris}(X/W(\overline{k})) \otimes_{W(\overline{k})} K_p$$

is a weak Weil cohomology with coefficients in K_p. Refer to this cohomology theory as p-adic cohomology.**

c. (Independence of Betti number)

Let $X \to H^*(X)$ be any (weak) Weil cohomology. Then the integer $B_j = \dim_k H^j(X)$ (the j-th betti number of X) depends only on X and not on the (weak) Weil cohomology chosen. If V/k is proper and smooth, the polynomial

$$P_j(t) = \det(1 - (F|H^j(\overline{V}))$$

has integral coefficients, and is also independent of which (weak) Weil cohomology was chosen. In the expression (*)

** One expects that p-adic cohomology has an adequate cycle theory (so is a (strong) Weil cohomology), but there are technical difficulties yet to be overcome before one can establish this.

$$Z(V/k,t) = \prod_{j \text{ odd}} P_j / \prod_{j \text{ even}} P_j$$

the polynomial of the numerator is relatively prime to the polynomial of the denominator: thus the zeroes of $Z(V/k,t)$ are precisely the eigenvalues of Frobenius acting on odd-dimensional cohomology; its poles are precisely the eigenvalues of Frobenius acting on even-dimensional cohomology.

d. (Rationality). If V is any variety over k, $Z(V/k,t)$ is a rational function of t with coefficients in $\mathbb{Q}$ and whose reciprocal zeroes and reciprocal poles are algebraic integers.

e. (Functional equation). If V/k is connected, proper and smooth, of dimension d, then $\alpha \mapsto q^d/\alpha$ is a permutation of the reciprocal zeroes of $Z(V/k,t)$ and of its reciprocal poles.

f. (Riemann hypothesis). If α is an eigenvalue of Frobenius acting on $H^j(\overline{V})$, for H^j any of the listed Weil cohomologies, then $|\alpha| = q^{j/2}$ where $|\;|$ denotes any archimedian absolute value of α.

The last three assertions are known as the conjectures of Weil.

(1.8) A chart.*

Listed below are some of the principal works which have contributed to the the story. Before presenting the list, it is good to make some general remarks: remarks:

1. Proofs of rationality and the functional equation.

For curves, there is the method of proof (cf. Weil's Basic Number Theory) which treats curves over finite fields in strict analogy with global number fields, and deals with their zeta functions as special cases of L functions for

* Cf. also the chart of [23] .

quasi-characters on an idèle class group. This approach is, in spirit, related to Artin's thesis -- one of the oldest attacks on the problem -- in which quadratic extensions of the rational function field in one variable over a finite field (i.e., these are essentially hyperelliptic curves) are treated in strict analogy with quadratic number fields. In technique, however, this method is a product of a long historical development in global number theory involving Hecke's theory of Grössencharaktere and Tate's thesis.

Most other proofs (leaving aside [5], and special cases) proceed by developing some "cohomological device". Thus the ℓ-adic representations used by Weil in the course of his proof for curves and abelian varieties are isomorphic to the one-dimensional ℓ-adic cohomology of the abelian variety. Also, the particular machines developed by Dwork, Monsky-Washnitzer, and a variant by Lubkin, and the p-adic cohomology are all, in some way, related.*

2. Proofs of the Riemann hypothesis.

Some of the approaches to be found:

I. Explicit computation of the eigenvalues of Frobenius, in highly special cases, by means of Gauss and Jacobi sums.

This is mentioned first since it is, possibly, the oldest method and yet not well understood. For its complex history, see [49]. We have not included this line of development in our chart below, which accounts for the enormous

* These p-adic machines are by no means equivalent and there is much yet to be understood about the fine inter-relations between them. Some hints concerning their differences: at the moment, the p-adic cohomology is the most fluidly functorial, but for the purpose of exhibiting p-adic analytic variation of the eigenvalues of Frobenius (see [22], [30],) it yields results somewhat weaker than one might expect -- it provides analyticity in discs which are too small -- it is constrained to do so by virtue of the type of "divided power thickenings" it uses. Both Dwork's and Washnitzer-Monsky's machines, when they can be applied, are geared to provide sharper analytic information. One expects, in the future, a theory along the lines of Berthelot's which incorporates the type of "thickenings" used by Monsky and Washnitzer. Dwork's theory has the added complexity of being, in truth, not one theory, but an infinity of "nested" theories. See Chapter 8 of [30] for some comparison of these theories.

gap (1801-1923) in our table. This approach begins, as Weil mentions in [45], with Gauss: In art. 385 of the Disquisitiones, Gauss determines the number of solutions of certain cubic curves mod p by totally elementary means, and expresses this number in terms of cubic Gauss sums in order to understand those Gauss sums. These "elementary means" are, as to be expected, unfeasible in a more general setting. In later developments one inverts the procedure: for example, to treat diagonal hypersurfaces $(a_0 x^{n_0} + \cdots + a_r x^{n_r} = b)$ one first studies Jacobi sums and determines their complex absolute value, and then expresses the eigenvalues of Frobenius (in terms of these Jacobi sums) thereby obtaining the Riemann hypothesis in this case. What seems to be the key to the calculation for diagonal hypersurfaces is that the (global $= \mathrm{Gal}(\bar{\mathbb{Q}}/\mathbb{Q})$) ℓ-adic representation on the cohomology of a diagonal hypersurface "comes from" Grössencharaktere of a particularly simple type. Can one expect ultimately some analogue of this for a reasonably large class of Grössencharaktere - where the Riemann hypothesis, and possibly some analogue of Stickelberger's theorem, is readily visible?

In the spirit of Gauss's approach, one is also free to invert the procedure again and use the Riemann hypothesis to obtain estimates for sums which are generalizations of Gauss and Jacobi sums -- and one does this for Kloosterman sums, and those of (2.6) below.

II. The construction of a positive definite bilinear form.

In the notes [41], [42] Weil sketched two different proofs of the Riemann hypothesis for curves. In [41] (which Weil referred to later as a "transcendental proof" since it made use of the jacobian of the curve) the Riemann hypothesis was deduced from positivity of the "Rosatti involution". In [42], Weil obtained the Riemann hypothesis for curves by establishing the inequality of Castelnuovo-Severi for surfaces (over an arbitrary groundfield). In Weil's later treatment for curves and abelian varieties [43], [44] he relates this inequality to a study of traces of endomorphisms of Jacobians.

In effect for a (polarized) abelian variety A he obtains a positive definite bilinear form (,) on $\mathrm{End}_{\mathbb{Q}}(A)$ -- the ring of endomorphisms of A tensored with $\mathbb{Q}$ -- with respect to which Frobenius behaves as follows: If $F' \in \mathrm{End}_{\mathbb{Q}}(A)$ is the transpose of F, then $F \cdot F' = F' \cdot F = q$. From this -- especially from positive-definiteness of the form --Weil concludes the Riemann hypothesis for abelian varieties.

One may almost view Hasse's purely arithmetic proof of the Riemann hypothesis for elliptic curves (1934 [17]) as being a precursor of this method: the difference being that if A is an elliptic curve, $\mathrm{End}_{\mathbb{Q}} A$ can be made so extremely explicit that a general surface inequality need never be invoked. Hasse's earlier proof (1933 [16]) has a different approach: he considers (what will be called much later) the "canonical lifting" of the elliptic curve A (with its Frobenius) to characteristic 0, obtaining an elliptic curve to which the classical theory of complex multiplication might be applied. See [19].

Concerning the inequality of Castelnuovo-Severi, there was this line of development: Mattuck-Tate [28] showed by a beautiful geometric argument that it was a direct consequence of the Riemann-Roch theorem for surfaces. Grothendieck, studying the methods of Mattuck-Tate gave a still simpler proof of the inequality based on the Riemann-Roch theorem. In the Grothendieck-Mattuck-Tate analysis of the inequality, its relation to the Hodge index theorem is made clear: this returns to a suggestion of Weil made in [46] that the sought-for "positivity" be made to come from the Hodge index theorem. Indeed, in the same [46], Weil raised the question of whether this procedure has analogies in higher dimensions. In [33] Serre responded to this question by considering the analogues of the conjectures of Weil in the theory of Kählerian manifolds. He proved these analogous conjectures, making use of the positive bilinear forms provided by Hodge theory.

Before Deligne discovered his proof, it was commonly felt that any proof of the general Riemann hypothesis would necessarily be dependent on establishing a substantial part of the Hodge conjectures. The relations between the Hodge conjectures and the Riemann hypothesis are pursued in [25].

III. In dimension one: the elementary proof of Stepanov .

This beautiful method is unclassifiable, and would take longer to describe than to read (See [5]).

IV. Special cases:

a. intersections of two quadrics

b. K3 surfaces

c. complete intersections of Hodge level 1 .

Roughly speaking, the Riemann hypothesis in these cases is proved by reduction to the case of abelian varieties or curves. This reduction is quite ingenious, in the case of (b) .

V. Deligne's method in the general case.

This magnificent proof considers in detail the effect of Lefschetz pencils on ℓ-adic cohomology, and, influenced by a method of Rankin, obtains a crucial analytic lever on the problem by systematic study of cohomology of tensor powers of ℓ-adic representations.

TABLE

Author	Date	Reference	Description
C.F. Gauss	1801	[12]	computes the number of solutions mod p of (what is, in effect) certain elliptic curves with complex multiplication by $\sqrt[3]{1}$. This number is in terms of cubic Gauss sums.
E. Artin	1924	[1]	Introduces the zeta function for a quadratic extension of the rational function field in one variable over a finite field -- in analogy with a quadratic number field. Shows rationality and functional equation. Poses the question of the Riemann hypothesis. This line is directly pursued by F.K. Schmidt in a series of papers [32] .

H. Hasse	1933, 34	[16], [17]	First proofs of Riemann hypothesis for elliptic curves.
A. Weil	1940	[41], [42]	Sketch of first proofs of Riemann hypothesis for curves.
A. Weil	1948	[43], [44]	Proves Riemann hypothesis for curves; abelian varieties. For this purpose, develops the theory of abelian varieties.
A. Weil	1949	[45]	Determines the eigenvalues of "diagonal hypersurfaces" in terms of Jacobi sums; poses his celebrated conjectures, and in terms of them defines - by analogy "Betti numbers" for a smooth variety V/k.
A. Weil	1954	[46]	Suggests the possibility of Lefschetz fixed point formula, to prove rationality; and raises the question of the general Riemann hypothesis, and its relation to "positivity" statements. See [33] for Kählerian analogue.
A. Weil	1954	[47]	Proof of Riemann hypothesis for intersection of two quadrics.
A. Mattuck-J. Tate	1958	[28]	Proof of inequality of Castelnuovo-Severi based on Riemann-Roch.
A. Grothendieck	1958	[13]	Analysis of Mattuck-Tate proof; relations with Hodge index theorem.
		[25]	See the expository account for formulations of the conjectures of Hodge and Lefschetz type and their relations to the Weil conjectures. Also for a formal treatment of the notion of a Weil cohomology.
B. Dwork	1960	[11]	First general proof of rationality.
Artin-Grothendieck Verdier	1963-4	[3]	Construction of ℓ-adic cohomology. See [15] for an expository account of results concerning rationality that may be obtained by means of ℓ-adic cohomology -- treats L-functions along with zeta functions.

Monsky-Washnitzer	[31] pub. 1968	[31]	Develops a DeRham type cohomology for certain smooth affine varieties in in characteristic p. For an excellent expository account of some of the relations between the p-adic theories of Dwork, Washnitzer-Monsky, and Lubkin see Chapter 8 of [30].
Grothendieck	1966	[14]	Outlines a plan for "crystalline cohomology", influenced by Monsky-Washnitzer. This plan with certain important modifications has been carried through by Berthelot: [4].
Lubkin	1968	[27]	Gives a p-adic proof of rationality and functional equation for smooth projective liftable varieties by first showing that modulo torsion, the algebraic de Rham cohomology of any lifting of such a variety is independent of the lifting.
Verdier (d'aprѐs Deligne)	1972	[40]	Expository account of Deligne's proof of independence of the characteristic polynomial of Frobenius acting on r-dimensional ℓ-adic cohomology of V/k (under the hypothesis that V is smooth projective, and liftable).
Deligne	1972	[6]	Proof of Riemann hypothesis for K3 surfaces.
Deligne	1972	[7]	Proof of Riemann hypothesis for complete intersections of "Hodge-level 1".
Stepanov	1969	[36]	Elementary proof of the Riemann hypothesis for curves. (See [5]).
Deligne	1974	[8]	General proof of Weil conjectures cf. [35] for expository sketch.

(1.9) <u>The eigenvalues of Frobenius as algebraic numbers</u>.

Let α be an eigenvalue of Frobenius acting on $H^i(\overline{V})$ where $H^*(\overline{V})$ is any one of the known (weak)cohomologies. Then by Poincaré duality q^d/α is an eigenvalue of Frobenius acting on $H^{2d-i}(\overline{V})$. Thus, both α and q^d/α are algebraic integers. It follows that in the field $F = \mathbb{Q}(\alpha)$, α is a λ-*adic unit for any prime* λ *which does not lie over* $p = \text{char } k$.

Since, by the Riemann hypothesis, one has that *every complex absolute value of* α *is* $q^{i/2}$, one is left with the intriguing question: What are the P-*adic absolute values of* α, *for primes* P *lying over* p ?

This is a question, which can by hindsight be seen to have had a long tradition --- being related to Stickelberger's determination of the p-adic nature of Gauss sums and the Chevalley-Warning theorem which says that if N is the number of solutions of a polynomial equation mod p whose degree is less than the number of its variables, then $N \equiv 0 \mod p$.

It is reasonable to expect that the study of the above question should use a p-adic cohomology, and it does. To describe, briefly, the known results, we introduce the *Newton polygon* of a monotone increasing sequence $a_1, a_2, \cdots, a_m$ of nonnegative rational numbers: by definition, it is the *graph* of the real-valued continuous piece-wise linear function $\nu_{a_1, a_2, \cdots, a_m}$ on $[0, m]$ which takes the value 0 at 0 and whose derivative is the constant a_j on the interval $(j-1, j)$.

Now let $\tilde{V}/W(k)$ be a smooth projective scheme, where $W(k)$ is the ring of Witt vectors of k. Let V/k be its reduction to k. Make the hypothesis that the de Rham cohomology of V is extremely well-behaved: namely, $H^q(\tilde{V}, \Omega^p)$ is a free $W(k)$-module (whose rank we shall denote $h^{p,q}$ -- the (p,q)-th Hodge number of V) for all p, q. Fix an integer r, and define the r-th Hodge polygon of V to be the Newton polygon of the sequence of integers

$$\underbrace{0,0,\cdots,0}_{h^{0,r}},\quad \underbrace{1,1,\cdots,1}_{h^{1,r-1}},\quad \underbrace{2,2,\cdots,2}_{h^{2,r-2}},\quad \cdots\cdots\cdots \quad \underbrace{r,r,\cdots,r}_{h^{r,0}} \quad .$$

Define the r-th Newton polygon of V to be the Newton polygon of the sequence:

$$\mathrm{ord}_q\,\alpha_1,\ \mathrm{ord}_q\,\alpha_2,\ \cdots,\ \mathrm{ord}_q\,\alpha_m \qquad (m = r^{th} \text{ Betti number of } V)$$

where $\alpha_1, \cdots, \alpha_m \in \overline{K}$ are the complete set of eigenvalues (with multiplicities) of Frobenius acting on the r-dimensional p-adic cohomology of V ; ord_q denotes the p-adic ord function normalized so that $\mathrm{ord}_q\, q = 1$; and the α_j are indexed in such a manner that the above sequence is monotone increasing.

What is known about the r-th Newton polygon of V is the following:

the dotted line indicates a possible Newton polygon $(r = 2)$ for some chimerical surface with $h^{0,2} = h^{2,0} = 1$, $h^{11} = 3$.

The r-th Newton polygon and the r-th Hodge polygon both end at the point $(m, rm/2)$ in the euclidean plane. The break-points (non-differentiable points) of these polygon occur at <u>integral</u> lattice points. The r-th Newton polygon lies in the closed region bounded by the r-th Hodge polygon, and the straight line joining $(0,0)$ and $(m, rm/2)$ [10], [21], [29] .

<u>Note</u>: One has a further symmetry in the geometry of the Newton polygon implied by the strong Lefschetz theorem, and also the existence of algebraic cohomology classes in dimension r implies that a part of the

Newton polygon must have slope $r/2$.

<u>General remarks</u>: It was an important discovery* of Dwork, [10] that (in a certain, unobvious, sense) the eigenvalues of Frobenius vary p-adic analytically if one varies the variety V over a parameter space in characteristic p. A trace of this phenomenon may be seen in a theorem of Grothendieck: the Newton polygon <u>rises</u> under specialization of V.

If we fix r and are given a reasonable moduli space M in characteristic p, for varieties V, as considered above, one may construct a collection $(M_\nu)_\nu$ of closed subschemes indexed by possible Newton polygons -- the characteristic feature of M_ν being that any V_0 corresponding to a point v_0 in M_ν has its r-th Newton polygon above ν. Refer to these $(M_\nu)_\nu$ loosely, as the <u>stratification</u> of M defined by r-th Newton polygons.

It is an intriguing open question to analyze these stratifications in particular cases: What Newton polygons actually occur? What is the geometry of these stratifications? At the moment one lacks sufficient experience to be able to make a complete conjecture -- even in the case of curves --(but see : [26] for curves and hypersurfaces and [2] where a 20-fold stratification for the moduli space of K3 surfaces is described).

(1.10) <u>Open questions</u>:

a. <u>Semi-simplicity of Frobenius acting on ℓ-adic cohomology</u>?

The title of these lectures is "<u>Eigenvalues of Frobenius</u> ..." . One might imagine that subtler information of an arithmetic nature could be contained in the isomorphism class of the (ℓ-adic) representation of Frobenius on $H^r(\overline{V})$.

It is conjectured that no further information is to be had by these means: more precisely, that Frobenius acts <u>semi-simply</u> on $H^r(\overline{V})$. For this

* Answering a challenge of Tate.

conjecture one must take $\ell \neq p$, for the action of Frobenius on p-adic cohomology is, indeed, somewhat subtler. It is known that this conjecture follows from standard conjectures of Hodge type [25]. The conjecture is true for $r = 1$ [44]. We mention this explicitly, because this fact plays an especially important role in the full understanding of the eigenvalues of Frobenius in dimension one.

b. Determination of algebraic cohomology by means of eigenvalues of Frobenius.

Tate has conjectured that the part of ℓ-adic cohomology $H^{2j}(\overline{V})$ which is the eigenspace of Frobenius associated to the eigenvalue q^j is the $\mathbb{Q}_\ell$-subspace generated by the fundamental classes of algebraic cycles.* [37]

2. Examples

(2.2) Projective varieties, curves and hypersurfaces.

Let V be a smooth subvariety over k of projective space $\mathbb{P}^M$ Suppose V has dimension d.

Let us be given a (weak) Weil cohomology and consider the image of $H^*(\mathbb{P}^M)$ in $H^*(\overline{V})$. This image is isomorphic to $H^*(\overline{\mathbb{P}}^d)$ as a module under Frobenius. One obtains a short exact sequence of "Frobenius modules"

$$0 \longrightarrow H^*(\overline{\mathbb{P}}^d) \longrightarrow H^*(\overline{V}) \longrightarrow h^*(\overline{V}) \longrightarrow 0$$

where $h^*(V)$ is our notation for the cokernel. By multiplicativity of characteristic polynomials, one has:

$$\frac{Z(V/k,t)}{Z(\mathbb{P}^d/k,t)} = \prod_r \det(1 - F|h^r(\overline{V}))^{(-1)^{r+1}}$$

or, in terms of numbers of rational points,

* Tate even bets money on this, when $j = 1$.

$$N_n(V/k,t) - N_n(\mathbb{P}^d/k,t) = \sum_j (-1)^{j+1}(\Sigma\alpha^n)$$

α : eigenvalue of F acting on $h^j(\bar{V})$.

This way of viewing things will turn out to be a gain, since $N_n(\mathbb{P}^d/k)$ is given by an evident expression, an, after the Riemann hypothesis, provides the dominant contribution to $N_n(V/k,t)$.

Of particular simplicity in the case of curves or smooth complete intersections in projective space, where $h^j(\bar{V})$ vanishes except in the middle dimension $j = d$. If we set $b = \dim_K h^d(\bar{V})$, then if V is a curve of genus g , $b = 2g$, and if V is a smooth complete intersection, b is easily computed by formula in terms of the multi-degree of V .

In this case, the Riemann hypothesis gives us

$$\left|N_n(V/k,t) - N_n(\mathbb{P}^d/k,t)\right| \leq b \cdot q^{nd/2} \quad .^{*}$$

In the special case of an elliptic curve $b = 2$, the two eigenvalues of Frobenius are denoted $\alpha, \bar{\alpha}$, they are determined by the equations:

$$\alpha \cdot \bar{\alpha} = q \; ; \quad N_1(V/k) - (1+q) = \alpha + \bar{\alpha} \; .$$

The integer $a = \alpha + \bar{\alpha}$ (succinctly referred to as the trace of Frobenius) thus determines the entire zeta function, and its absolute value is constrained

---

* Knowledge of the p-adic ord of the eigenvalues of α would provide an estimate for $\|N_n(V/k,t) - N_n(\mathbb{P}^d/k)\|_p$. An example of such knowledge is when one is in the above case and one knows that the Hodge numbers $h^{0,d}, h^{1,d-1}, \ldots, h^{a-1,d+1}$ all vanish for some nonnegative integer a. In this case, the assertions of (1.9) concerning the r-th Newton polygon imply that $\mathrm{ord}_q \alpha \geq a$ if α is any eigenvalue of Frobenius acting on $h^d(V)$, giving

$$N_n(V/k,t) \equiv N_n(\mathbb{P}^d/k,t) \mod q^{na} \quad .$$

by the Riemann hypothesis to be $\leq 2 \cdot q^{1/2}$.

(2.3) Abelian varieties.

Two abelian varieties A, A' over k are said to be isogeneous if there is a homomorphism $\varphi : A \to A'$ defined over k which is surjective and has finite kernel. An elementary computation with Galois cohomology shows that if k is a finite field, and A, A' are isogeneous over k, then

$$N_n(A/k) = N_n(A'/k)$$

and therefore the zeta function of an abelian variety A/k depends only on the isogeny class of A. It is also true that (in a functorial way)

$$H^*(\bar{A}) = \text{Free exterior algebra generated by } H^1(\bar{A}) \text{ over } K$$

and therefore, the zeta function of A/k may be constructed, once one knows the eigenvalues of Frobenius acting on $H^1(\bar{A})$.

Thanks to an important theorem of Tate-Honda, [39] one has strikingly complete information concerning the relationship between the zeta-function of an abelian variety over k, and its isogeny class.

THEOREM: Let k be a finite field. 1. (Tate). The isogeny class of an abelian variety A/k is determined by the eigenvalues (counted with multiplicities) of Frobenius acting on $H^1(\bar{A})$. [38].

2. (Honda). Let α be an algebraic number such that the absolute value of every complex imbedding of α is $q^{1/2}$. Then there exists an abelian variety A/k possessing α as one of its eigenvalues of Frobenius acting on H^1. [20].

Remarks: The story is actually somewhat more complete than the assertions above: Tate determines the isomorphism class of the $\mathbb{Q}$-algebra $\mathrm{End}_{\mathbb{Q}} A$, and, for a number α as above, the multiplicity with which it must occur in a simple abelian variety over k. That there are some restrictions on the multiplicities is clear from the fact, noted in (1.9), that the break-points of the Newton polygon of A/k must occur at integral lattice-points.

The proof of Tate's theorem rests on two important facts:

(i) semi-simplicity of Frobenius acting on H^1,

(ii) existence of only a finite number of classes of abelian varieties over k of fixed dimension and degree of polarization.

(2.4) <u>Two examples from the theory of elliptic curves, and a conjecture</u>.

In some cases, one has amazing control of the zeta function. Consider the equations over $\mathbb{Z}$:

1. $y^2 = x^3 - x$ (reduces to an elliptic curve over $\mathbb{F}_p$, $p \neq 2$)

2. $y^2 + y = x^3 - x^2$ (reduces to an elliptic curve over $\mathbb{F}_p$, $p \neq 11$)

and let a_p be the trace (sum of the two eigenvalues) of Frobenius of either of these curves, reduced to $\mathbb{F}_p$.

One has:

<u>For Equation</u> 1.

If $p \equiv 3 \mod 4$, then $a_p = 0$.

If $p \equiv 1 \mod 4$, then p splits in $\mathbb{Q}(\sqrt{-1})$.

Find a splitting $p = \alpha \cdot \bar{\alpha}$ such that $\alpha \equiv \bar{\alpha} \equiv \mod 2 + 2i$.

Then $\alpha, \bar{\alpha}$ are the eigenvalues of Frobenius of Equation 1, over $\mathbb{F}_p$.

<u>For Equation</u> 2.

If $p \neq 11$, then a_p is the coefficient of $e^{2\pi(p-1)\tau}$ in the infinite product:

$$\prod_{n=1}^{\infty} \left(1 - e^{2\pi i n\tau}\right)^2 \left(1 - e^{2\pi i \cdot 11 n\tau}\right)^2 .$$

Remarks: The determination of a_p for Equation 1 is typical of the sort of information one has concerning the eigenvalues of Frobenius of an elliptic curve with complex multiplication. The determination of a_p for Equation 2 is hardly typical of anything, however both examples fall under the general heading of the relations between zeta functions and modular forms concerning which one has the following general (new) conjecture of Weil:

Given some equation which we will refer to as (*) with integral coefficients which determines an elliptic curve over $\mathbb{Q}$, let p run through the set S of primes for which (*) reduces to an elliptic curve over $\mathbb{F}_p$. Then S consists in almost all primes. Let $(a_p ; p \in S)$ denote the traces of Frobenius of (*) over $\mathbb{F}_p$.

Conjecture: There is a unique (primitive) parabolic modular form of weight two (for a congruence subgroup which can also be conjecturally made explicit) which is an eigenvector for the Hecke operators and whose Fourier expansion (q-expansion) has the form

$$e^{2\pi i\tau} + b_2 \cdot e^{2\pi i \cdot 2\tau} + \cdots + b_\nu e^{2\pi i \nu\tau} + \cdots$$

where $b_p = a_p$ for all $p \in S$.

(2.5) Applications to the study of Fourier coefficients of modular forms.

Although in certain circumstances one has quite a firm grasp of the Fourier coefficients of a modular form, one of the most important applications of the general Riemann hypothesis, at present, is to obtain information about these Fourier coefficients. Notably, let f be a modular form of integral weight $k \geq 1$ for $\Gamma_0(N)$ with character ϵ. This means that f is a holomorphic function on the upper half plane which satisfies

$$f\left(\frac{az+b}{cz+d}\right) = \epsilon(a)(cz+d)^k f(z)$$

for all $\begin{pmatrix} a & b \\ c & d \end{pmatrix} \in Sl_2(\mathbb{Z})$ such that $c \equiv 0 \mod N$. Here $\epsilon : \mathbb{Z}/N^* \to \mathbb{C}^*$ is a fixed character. One also supposes that f has a certain reasonable behavior at cusps.

THEOREM (conjecture of Ramanujan-Peterson):

If p is a prime not dividing N, and f is a parabolic modular form which is an eigenvector for the Hecke operators T_ℓ, $\ell \neq N$, such that $b_1 = 1$, then

$$|b_p| \leq 2 \cdot p^{\frac{k-1}{2}} .$$

Note: For weight $k \geq 2$, this is proved by "realizing" the b_p's as certain traces of Frobenius acting on a part of the (k-1)-dimensional cohomology of a certain algebraic variety over $\mathbb{F}_p$ (cf. [9] where Deligne does this,* following through a suggestion of Serre, and adapting construction of Kuga, Sato, Shimura.)

The original conjecture of Ramanujan had to do with the values of the Ramanujan τ-function. Namely, that the coefficient $\tau(p)$ of x^p in the infinite product expansion

$$\Delta = x \prod (1 - x^n)^{24}$$

satisfies $|\tau(p)| \leq 2 \cdot p^{11/2}$.

(2.6) Estimates for exponential sums.

Let $Q \in k[X_1, \cdots, X_n]$ be a polynomial of degree d, and suppose that Q_d, its homogeneous component of degree d defines a smooth hypersurface. Suppose further, that, p doesn't divide d.

Let ψ be a nontrivial homomorphism of k^+ to $\mathbb{C}^*$, where k^+ denotes the additive group of k. Then the following theorem suggested by Bombieri and proved by Deligne is in the classical tradition of obtaining upper bounds form exponential sums from bounds on the eigenvalues of

* For modular forms for the full modular group.

Frobenius. See [45] for a treatment of Kloosterman sums and discussion.

THEOREM:

$$\left| \sum_{x_1, \cdots, x_n \in k} \psi(Q(x_1, \cdots, x_n)) \right| \leq (d-1)^n q^{n/2}$$

Note: The typical additive character is

$$\psi(x) = \exp \frac{2\pi i \mathrm{Tr}_{k/\mathbb{F}_p}(x)}{p},$$

and after a scalar change of Q one may suppose that ψ is as above.

The key to studying the exponential sum in the theorem is to express it as a "Lefschetz number" coming from the action of Frobenius on the compact cohomology of a certain sheaf on affine $(X_1, \cdots, X_n)$ - space over k. This sheaf is "defined by" the Artin-Schreier covering

$$X^p - X = Q, \text{ and the character } \psi.$$

One then proceeds with proof of the theorem by determining the compact cohomology groups involved, and using an appropriate version of the (general) Riemann hypothesis [8].

BIBLIOGRAPHY

[1] E. Artin, Quadratische Körper im Gebeit der höheren Kongruenzen I, II, Collected Papers 1-94, Addison-Wesley.

[2] M. Artin, Supersingular K3 surfaces, to appear.

[3] Artin-Grothendieck-Verdier, Theorie des topos et cohomologie étale des schémas (SGA 4) L.N.S. 269, 270, 305. Springer-Verlag. 1972-1973.

[4] P. Berthelot. A series of notes in the Comptes Rendus, Acad. Sci. Paris, t, 269, pp. 297-300, 357-360, 397-400, t. 272, pp. 42-45, 141-144, 1314-1317, 1397-1400, 1571-1577.

[5] E. Bombieri. Counting points on curves over finite fields. Sém. Bourb. exp 430, 1972-1973.

[6] P. Deligne. La conjecture de Weil pour les surfaces K3. Inv. Math. 15 (1972) 206-226.

[7] ________, Les intersections complètes de niveau de Hodge un, Inv. Math. 15(1972) 237-250.

[8] ________, La conjecture de Weil I, Publ. Math. IHES, no. 43, P.U.F. 1974.

[9] ________, Formes modulaires et représentations ℓ-adiques. Sém. Bourb. exp. 355 L.N.S. 349, Springer-Verlag (1973)

[10] B. Dwork. A deformation theory for the zeta function of a hypersurface. Proc. Int. Cong. Math. Stockholm, 1962, pp. 247-259 259.

[11] ________, On the rationality of the zeta function of an algebraic variety. Amer. J. Math. 82 (1960) 631-648.

[12] C. F. Gauss. Disquisitiones Arithmeticae, English trans. Yale University Press, New Haven (1966) .

[13] A. Grothendieck. Sur une note de Mattuck-Tate. Journal für die reine und ang. Math. 200 (1958) 208-215.

[14] ________, Crystals and the deRham cohomology of schemes. Dix esposés sur la cohomologie des schémas, North-Holland Amsterdam (1968) 306-350.

[15] ________, Formule de Lefschetz et rationalité des fonctions L, Sém. Bourb. exp. 279, 1964-1965.

[16] H. Hasse, Beweis des Analogons der Riemannschen Vermutung für die Artinschen und F.K. Schmidtschen Kongruenz zeta Funktionen in gewissen elliptischen Fällen, Nachr. Ges. d. Wiss. Göttingen, Math.-Phys. Kl. Heft. 3, 1933, 253-262.

[17] ________, Abstrakte Begründung der komplexen Multiplikation und Riemannsche Vermutung in Funktionenkörpern, Abh. Math. Semin. Hamburg. Univ. 10 (1934) 325-348.

[18] H. Hasse, Zur Theorie des abstrakten elliptschen Funktionenkörper, I Journ. f.d.r.u. ang. Math. 175, 55-62, II 69-88, III 193-206.

[19] ________, Modular functions and elliptic curves over finite fields. Simposio Internazionale di Geometria Algebrica, Edizoni Cremonese, Roma 1967, pp. 248-266.

[20] T. Honda, Isogeny classes of abelian varieties over finite fields, J. Math. Soc. Japan 20 1968 (83-95).

[21] N. Katz, On a theorem of Ax, Amer. Journ. of Math. (1971) vol. 93 485-499.

[22] ________, Travaux de Dwork, Séminaire Bourbaki, exp. 409, 71/72 Lecture Notes in Math. no. 317 Springer-Verlag 1973.

[23] ________, An overview of Deligne's proof of the Riemann hypothesis for varieties over finite fields (to be published in the proceedings of the A.M.S. conference on Hilbert's Problems, DeKalb, Michigan 1974.

[24] ________ -W. Messing. Some consequences of the Riemann hypothesis for varieties over finite fields. Invent. Math, 23 (1974, 73-77.

[25] S. Kleiman, Algebraic cycles and the Weil conjectures. Dix exposés sur la cohomologie des schémas, p. 359-386, North-Holland Publ. Cy., Amsterdam, Paris, 1968.

[26] N. Koblitz, p-adic variation of the zeta-function over families of varieties defined over finite fields. Thesis, Princeton, 1974.

[27] S. Lubkin, A p-adic proof of the Weil conjectures, Am. of Math. 87 (1968) 105-194; ibid. 87 (1968) 195-225.

[28] A. Mattuck and J. Tate. On the inequality of Castelnuovo-Severi. Abh. Math. Sem. des Univ. Hamburg, B22 H. 3/4(1958) 295-299.

[29] B. Mazur. Frobenius and the Hodge Filtration, B.A.M.S. 78(1972), 653-667; (estimates) Annals of Math. 98 (1973), 58-95.

[30] P. Monsky, p-adic analysis and zeta functions, Lectures in Math. Kyoto Univ. published by Kinokuniya Book Store Co., Ltd. Tokyo, Japan (1970).

[31] ________ and G. Washnitzer. Formal Cohomology I. Annals of Math. 88(1968) 181-217.

[32] F.K. Schmidt,

a) Allgemeine Körper in Gebiet der höheren Kongruenzen, thesis Erlangen (1925).

b) Zur Zahlentheorie in Körpern von der Charakteristik p, Sitzungsberichte Erlangen 58/59 (1925).

c) Analytische Zahlentheorie in Körpern der Charakteristik p, Math. Zeitschrift 33(1931).

d) Die theorie der Klassenkörper uber einem Körper algebraischer Funktionen in einer Unbestimmten und mit endlichem Koeffizientenbereich, Sitzungsberichte Erlangen 62(1930).

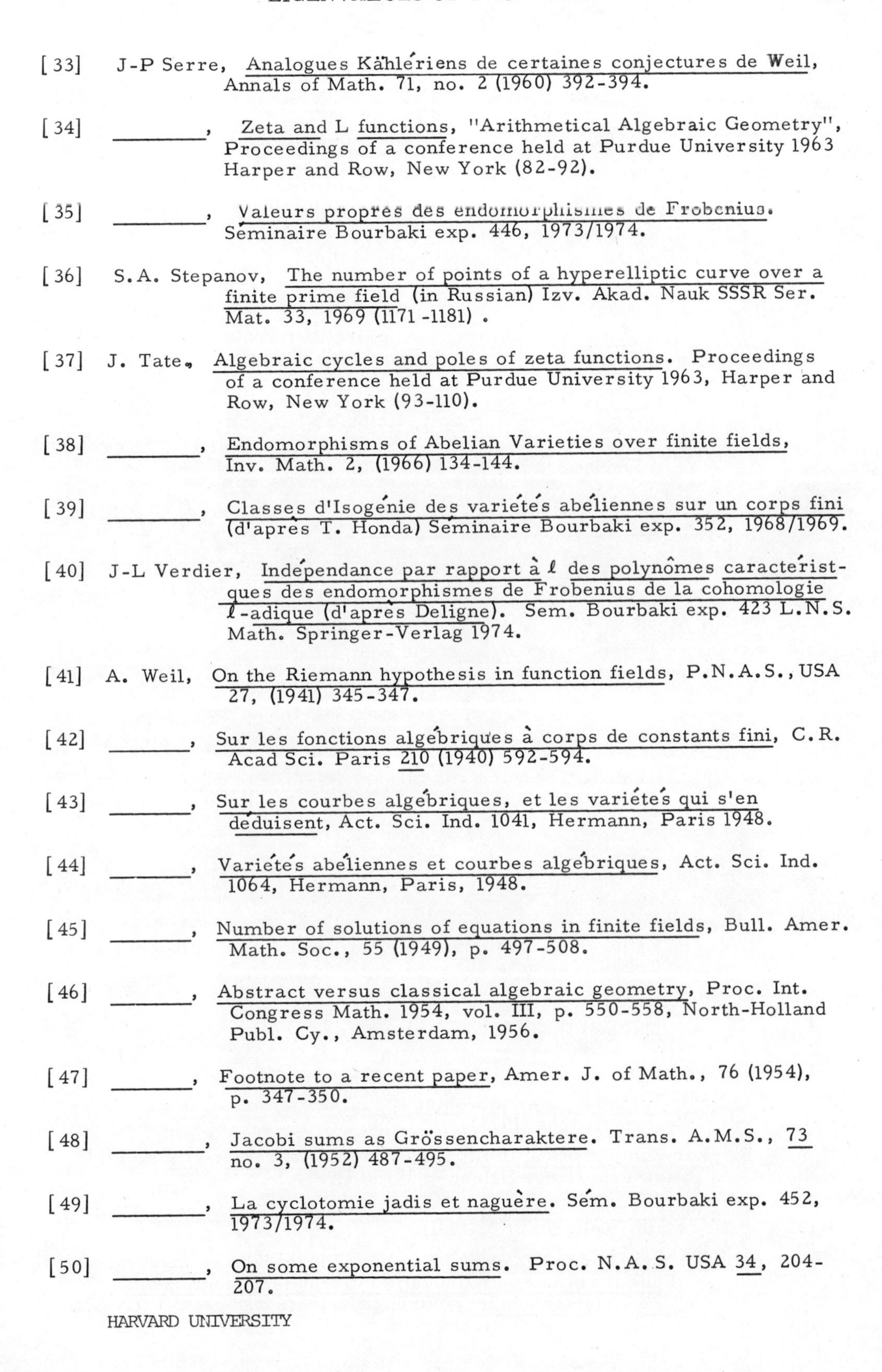

[33] J-P Serre, Analogues Kählériens de certaines conjectures de Weil, Annals of Math. 71, no. 2 (1960) 392-394.

[34] ________, Zeta and L functions, "Arithmetical Algebraic Geometry", Proceedings of a conference held at Purdue University 1963 Harper and Row, New York (82-92).

[35] ________, Valeurs propres des endomorphismes de Frobenius. Séminaire Bourbaki exp. 446, 1973/1974.

[36] S.A. Stepanov, The number of points of a hyperelliptic curve over a finite prime field (in Russian) Izv. Akad. Nauk SSSR Ser. Mat. 33, 1969 (1171-1181).

[37] J. Tate, Algebraic cycles and poles of zeta functions. Proceedings of a conference held at Purdue University 1963, Harper and Row, New York (93-110).

[38] ________, Endomorphisms of Abelian Varieties over finite fields, Inv. Math. 2, (1966) 134-144.

[39] ________, Classes d'Isogénie des variétés abéliennes sur un corps fini (d'après T. Honda) Séminaire Bourbaki exp. 352, 1968/1969.

[40] J-L Verdier, Indépendance par rapport à ℓ des polynômes caractéristiques des endomorphismes de Frobenius de la cohomologie ℓ-adique (d'après Deligne). Sem. Bourbaki exp. 423 L.N.S. Math. Springer-Verlag 1974.

[41] A. Weil, On the Riemann hypothesis in function fields, P.N.A.S., USA 27, (1941) 345-347.

[42] ________, Sur les fonctions algébriques à corps de constants fini, C.R. Acad Sci. Paris 210 (1940) 592-594.

[43] ________, Sur les courbes algébriques, et les variétés qui s'en déduisent, Act. Sci. Ind. 1041, Hermann, Paris 1948.

[44] ________, Variétés abéliennes et courbes algébriques, Act. Sci. Ind. 1064, Hermann, Paris, 1948.

[45] ________, Number of solutions of equations in finite fields, Bull. Amer. Math. Soc., 55 (1949), p. 497-508.

[46] ________, Abstract versus classical algebraic geometry, Proc. Int. Congress Math. 1954, vol. III, p. 550-558, North-Holland Publ. Cy., Amsterdam, 1956.

[47] ________, Footnote to a recent paper, Amer. J. of Math., 76 (1954), p. 347-350.

[48] ________, Jacobi sums as Grössencharaktere. Trans. A.M.S., 73 no. 3, (1952) 487-495.

[49] ________, La cyclotomie jadis et naguère. Sém. Bourbaki exp. 452, 1973/1974.

[50] ________, On some exponential sums. Proc. N.A.S. USA 34, 204-207.

HARVARD UNIVERSITY

Proceedings of Symposia in Pure Mathematics
Volume 29, 1975

THEORY OF MODULI*

C. S. Seshadri

1. INTRODUCTION

Let X be an algebraic scheme (resp. an analytic space) say over an algebraically closed field k (resp. $\mathbb{C}$).

DEFINITION 1.1. A deformation of X over (or parametrized by) an algebraic scheme S over k with base point s_0 (or more generally any scheme S over k) (resp. an analytic space over $\mathbb{C}$) consists of the following:

(i) A morphism $\underline{X} \xrightarrow{p} S$ which is flat and say of finite type. In case X is complete (resp. compact) p is proper over S.

(ii) A closed point $s_0 \in S$ (so that the residue field $k(s) = k$) such that the fibre $\underline{X}_s = \underline{X} \times_S k(s)$ of $\underline{X}$ over s is isomorphic to X.

Suppose S is connected. Take a closed point $s \in S$. Then we call the fibre of $\underline{X}$ over s a deformation of X and $\underline{X}$ the deformation space defining this deformation. The problem of moduli of X can be termed as the investigation of the set M of all isomorphism classes of deformations of X (say whether they could be parametrized by an algebraic scheme, if so its dimension, its properties, etc.).

Suppose X is smooth and complete (resp. compact). Then p is smooth over a neighbourhood of s_0 so that we get an open subset S' of S such that: $\underline{X}_s$ is smooth $\Longleftrightarrow$ $s \in S'$. Suppose that S' is also smooth; then p is of maximal rank over S', i.e., the differential map dp is surjective at every point. Conversely if dp is surjective over

---
*This written version has been revised and enlarged especially in §3.

a smooth open subset S' of S, then p is flat over S'. Thus if we are interested only in smooth deformations of X parametrized by smooth S, a deformation is merely a morphism $p: \underline{X} \longrightarrow S$ of maximal rank. This was the definition of Kodaira-Spencer (cf. [29]).

We see also that if $p: X \longrightarrow S$ is a smooth proper morphism of complex analytic spaces and S is connected any two fibres X_{s_1}, X_{s_2} are topologically isomorphic. Thus a deformation in this case is just the formalization of the notion of a complex analytic family of complex structures which are all topologically the same.

There are similar variants on deformations which have been studied, namely

(i) deformation over S of a line bundle L on X is a line bundle $\mathbb{L}$ on $\underline{X} = X \times S$ inducing L on $X = \underline{X} \times k(s)$. The problem of moduli in this case is known as the construction of Picard varieties or Picard schemes. Similarly one defines deformations of a vector bundle on X.
(ii) Let X be a closed subscheme of an algebraic scheme Y, e.g., $y = \mathbb{P}^n$. Then an embedded deformation (over S) of X (with respect to Y) is a deformation $p: \underline{X} \longrightarrow S$ and a closed immersion $X \longrightarrow Y \times_k S$. When $Y = \mathbb{P}^r$, this leads to the notion of Hilbert schemes in the sense of Grothendieck.
(iii) X affine scheme with an isolated singularity. This leads to the theory of deformations of isolated singularities.

The problem of moduli could be divided into two aspects:
(i) local theory, namely the investigation of M in the neighbourhood of the point corresponding to X and (ii) global theory, namely global properties of M once the local theory is done. In recent times, it is the papers of Kodaira-Spencer which could be considered as the starting point of this theory, especially the local one. The next influence came from Grothendieck's development of schemes, namely the introduction of nilpotent elements, flatness, functorial language and many important technical tools like prorepresentability. The local theory is in a fairly satisfactory state with the nice criterion of Schlessinger and the deep contributions of Kuranishi and M. Artin. As for global theory a beautiful general theorem has been proved recently by Matsusaka. For the deeper investigation of global moduli, as far as I see there are only two systematic methods: (i) Griffiths' period map and (ii) Mumford's geometric invariant theory which has recently been enriched by Haboush's proof of Mumford's conjecture. One should add that the global moduli

have been satisfactorily understood only in a few cases: (1) curves (2) polarized abelian varieties (3) vector bundles on algebraic curves and (4) recently K-3 surfaces.

In those talks, after giving the basic theorems on local theory, we will give a survey of those global moduli problems which have been solved using geometric invariant theory, namely the moduli of curves, polarized abelian varieties and vector bundles on curves. As for the work of Griffiths see for example [17], [18] and [9].

For simplicity we work mostly with algebraic schemes defined over an algebraically closed field k and by points we mean closed points.

2. LOCAL THEORY

Let X be a smooth algebraic variety over k (resp. analytic space over $\mathbb{C}$). Let $S = \text{Speck } k[\epsilon]$ (resp. spec $\mathbb{C}[\epsilon]$). Then we see easily

$$\left\{\begin{array}{c}\text{Isom. classes of deformations}\\ \text{of } X \text{ over } S\end{array}\right\} = H^1(X,\Theta) \qquad \left\{\begin{array}{l}\Theta \text{ tangent}\\ \text{bundle on } X.\end{array}\right.$$

Suppose now S is an algebraic scheme and $\underline{X} \xrightarrow{p} S$ is a deformation with base point x_0. Then we get a canonical linear map (called the infinitesimal deformation map)

$$\underline{t}_{S,s_0} \longrightarrow H^1(X,\Theta), \qquad \left\{\begin{array}{l}\underline{t}_{S,s_0} = \text{Zariski tangent space}\\ \qquad \text{to } S \text{ at } s_0\end{array}\right.$$

since an element ϕ of $\underline{t}_{S,s_0}$ is a morphism ϕ: Spec $k[\epsilon] \longrightarrow S$ (going to the point s_0) and $\phi \longrightarrow$ element of $H^1(X,\Theta)$ defined by base change of $\underline{X} \longrightarrow S$ by ϕ.

THEOREM 2.1 (Kuranishi, cf. [32],[11]). Let X be a compact analytic manifold. Then there exists a deformation $p: \underline{X} \longrightarrow S$ with base point s_0 such that

(i) $\underline{X}$ is complete at s_0, i.e., given a deformation $Y \longrightarrow T$ of X with base point t_0, there exists a morphism $f: T' \longrightarrow S$, $f(t_0) = s_0$, where T' is a neighbourhood of t_0, such that $\underline{Y}|_{T'}$ is base change of $\underline{X} \longrightarrow S$ by f and

(ii) the infinitesimal deformation map

$$\underline{t}_{S,s_0} \longrightarrow H^1(X,\Theta)$$

is an isomorphism.

(iii) If $H^0(X,\Theta) = 0$, i.e., X has no "infinitesimal automorphisms", then f is uniquely determined.

The existence of a "locally complete" family has been recently proved for any compact complex space (cf. [12],[16]).

REMARKS 2.1. (i) It can be seen easily that the germ of S at s_0 is determined up to a non-canonical isomorphism and if $H^0(X,\Theta) = 0$ up to a canonical isomorphism. We call the germ of S at s_0 the <u>local moduli space of</u> X.

(ii) If $H^2(X,\Theta) = 0$ then S is smooth at s_0 (this case was the starting point and was first proved by Kodaira, Spencer and Nirenberg, cf. [31]). Hence in this case dim S (at s_0) = dim $H^1(X,\Theta)$.

(iii) If $H^1(X,\Theta) = 0$ we see that S reduces to a point so that X is <u>locally rigid</u>, i.e., if $\underline{Y} \longrightarrow T$ is a deformation of X with base point t_0, then $\underline{Y}$ is <u>trivial</u> in a neighbourhood of t_0, i.e., it is a product over this neighbourhood; in particular all the fibres over this neighbourhood are <u>complex analytically isomorphic</u>.

Let $X = \mathbb{P}^n(C)$; then it is known that $H^1(X,\Theta) = 0$ (say because of Bott's vanishing theorems) so that X is locally rigid. We can ask the question: is <u>any</u> deformation Y of X isomorphic to X for $n \geq 2$ (this is classical for n = 1). Kodaira and Spencer have proved this (cf. [29]) for n = 2 (the deformation space could be even differentiable, cf. [29] for this more general notion of deformation). Kodaira and Hirzebruch (cf. [23]) have shown that if X is Kähler, has odd dimension and is differentiably isomorphic to $\mathbb{P}^n(C)$, then X is <u>analytically isomorphic</u> to $\mathbb{P}^n(C)$. <u>What about the case</u> n <u>even</u>, $n \geq 4$? Even for the case n = 2, I learned from Ramanujam that <u>if</u> X is a smooth complex analytic surface differentiably isomorphic to $\mathbb{P}^2$, it is not known whether X is complex analytically isomorphic to $\mathbb{P}^2$.

(iv) <u>Local moduli for curves</u>: Let X be a compact Riemann surface of genus g. Then $H^2(X,\Theta) = 0$ so that the local moduli space is smooth if dim = dim $H^1(X,\Theta)$ = dim $H^0(X,K^2)$ (by Serre duality, K being the canonical line bundle) = 3g-3 if $g \geq 2$ (if $g = 1$ it is =1 and if g = 0 it is 0).

(v) Take an S which represents the local moduli. Then it can be shown that there is a neighbourhood U of s_0 such that $\underline{X}|_U$ is complete at every point of U.

(vi) It is not true even if $H^0(X,\Theta) = 0$ that we have an injective parametrization in a neighbourhood of s_0 i.e. $\underline{X}_{s_1}$ is not isomorphic to $\underline{X}_{s_2}$ for s_1, s_2, $s_1 \neq s_2$, in a suitable neighbourhood of s_0. For example take the case of a compact Riemann surface X of genus $g \geq 2$ such that $\text{Aut } X \neq \text{Id}$. Then it can be seen that $\text{Aut } X$ acts (non-trivially) on the local moduli space and the fibres of X over an orbit are mutually isomorphic.

(vii) The method of Kuranishi seems to extend without much difficulty to the case of vector bundles (e.g., cf. [70]), say on a compact complex manifold X; in particular it applies to the case of line bundles. Let V be a vector bundle on X and take a deformation of V parametrized by S (with base point s_0 defining V). Then the infinitesimal deformation map in this case is a canonical linear map

$$\underline{t}_{S,s_0} \longrightarrow H^1(X, \text{Hom}(V,V)).$$

If S is now the local moduli space for V, then the infinitesimal deformation map $\underline{t}_{S,s_0} \longrightarrow H^1(X, \text{Hom}(V,V))$ is an isomorphism. Further if $H^2(X, \text{Hom}(V,V)) = 0$ then S is smooth at s_0. As an example let X be a compact Riemann surface of genus $g \geq 2$ and V a vector bundle such that $\text{End } V = H^0(X, \text{Hom}(V,V)) = \mathbf{C}$ (such vector bundles exist in arbitrary rank, e.g., take one associated to an irreducible unitary representation of $\pi_1(X)$ and irreducible unitary representations exist in arbitrary rank since $g \geq 2$). Since X is a curve we have $H^2(X, \text{Hom}(V,V)) = 0$ and we find by Riemann-Roch $\dim H^1(X, \text{Hom}(V,V)) = n^2(g-1) + 1$ where $n = \text{rank } V$. Thus the local moduli space S with base point s_0 is smooth at s_0 and its dimension $= n^2(g-1) + 1$. In this case it can be shown that for s_1, s_2 in a neighbourhood of s_0, the associated vector bundles are non-isomorphic.

(viii) Let X be a closed analytic subspace of a compact complex subspace Y (e.g., a projective variety). Then the existence of a local moduli space can be proved in this case. In fact, thanks to Douady (and Grothendieck in the algebraic case — Hilbert schemes) even the global existence theorem is true in this case and we have an injective parametrization. Let us take X,Y for simplicity to be smooth. Let S be the local moduli space of X with base point s_0. Let N be the normal bundle of X in Y. Then it can be seen that the infinitesimal deformation map in this case is a canonical isomorphism

$$\underline{t}_{S,s_0} \longrightarrow H^0(X,N).$$

Of course if $H^1(X,N) = 0$, S is smooth at s_0 but we may have even better conditions for this, as was done for the case when Codim $X = 1$ by Kodaira-Spencer (a condition called semi-regularity), cf. [30]. In this connection see also the paper of Spencer Bloch [4].

(ix) Examples when the local moduli space has singularities.

(a) Let X be a compact complex manifold. Then

$$H^*(X,\Theta) = \bigoplus_q H^q(X,\Theta)$$

has a graded Lie algebra structure. Kodaira-Spencer (cf. [29]) call an element $\alpha \in H^1(X,\Theta)$ obstructed if $[\alpha,\alpha] \neq 0$. It can be seen that if there exists an obstructed element then the local moduli space S cannot be smooth at s_0. They show for example that if $X = T_q \times \mathbb{P}^1$ (T_q torus of dim ≥ 2), there exist obstructed elements in $H^1(X,\Theta)$ so that the local moduli space for X is not smooth.

(b) Mumford (cf. 41): Let Γ be a smooth closed curve in $\mathbb{P}^3$. Let V = blowing up of $\mathbb{P}^3$ along Γ. Then one has

local moduli space of V = local moduli space of Hilb at the point corresponding to Γ.

Then it is shown that if A is the Hilbert scheme of curves of genus 24 and degree 14 in $\mathbb{P}^3$, there exists an irreducible component A_1 of A such that there is at least one smooth curve in the family parametrized by A_1 and A_1 is not reduced even at its generic point. Hence for a suitable choice of Γ and hence of V, the local moduli space S of V is not reduced and S_{red} is smooth. It follows easily that $H^1(V,\Theta)$ has obstructed elements.

(c) Kas (cf. [27]): Let X be an elliptic surface, i.e., X is a compact surface and we have a morphism $\psi: X \longrightarrow \Delta$ where Δ is a compact Riemann surface and the general fibre is an elliptic curve. Let j be the meromorphic function on Δ defined by $j(u) = J(C_u)$ - C_u fibre over u and $J(C_u)$ = modular invariant of C_u. Then Kas shows that if j = const $\neq$ 0,1 and X has not some properties (i.e., it is not a K-3 surface, has no multiple fibres, has no exceptional curves, not a fibre bundle of elliptic curves) then $H^1(X,\Theta)$ has obstructed elements.

(d) For more examples see the papers of Burns-Wahl and Kas (cf. [8],[28]) as well as Horikawa [24].

(x) Let X be a smooth projective variety over $\mathbb{C}$, S the local moduli space with base point s_o and $\underline{X} \xrightarrow{p} S$ the deformation space. It is not true in general that the fibres $\underline{X}_s$ are algebraic for s in some neighbourhood of s_o. An example (as given by Kodaira-Spencer (cf. [29])) is as follows: Let X be a g-dimensional torus defined by a lattice L in $\mathbb{C}^g$. A lattice L is defined by $2g$ vectors in $\mathbb{C}^g$ linearly independent over $\mathbb{R}$. Hence a lattice is defined by a $(g\times 2g)$ matrix Ω over $\mathbb{C}$ such that its columns are linearly independent over $\mathbb{R}$ and one sees that these conditions can be normalized to be

$$(*) \qquad i^g \det\begin{pmatrix}\Omega \\ \overline{\Omega}\end{pmatrix} > 0 \qquad (\overline{\Omega} \text{ complex conjugate of } \Omega).$$

One sees that $(*) \Longrightarrow \Omega$ is of rank g over $\mathbb{C}$. Let U be the set of $(g\times 2g)$ matrices satisfying $(*)$. We see easily that $\Omega_1, \Omega_2 \in U$ define the same torus $\Longleftrightarrow$

$$\Theta\Omega_1 L = \Omega_2, \qquad \Theta \in GL(g,\mathbb{C}), \qquad L \in SL(2g,\mathbb{Z}).$$

We see that $GL(g,\mathbb{C})\backslash U$ is an open subset of the Grassmannian $G(g,2g)$. Now $\dim G(g,2g) = g^2$ and hence $\dim(S = GL(g,\mathbb{C})\backslash U) = g^2$. We get a family of tori parametrized by S. It is intuitively clear that this family is complete at every point of S. Further we have $\dim H^1(X,\Theta) = g^2$ from which it is intuitively clear that the infinitesimal deformation map at every point of S is an isomorphism. Hence S is locally, at every one of its points, the local moduli space for the corresponding torus.

Let now X be an abelian variety and s_o the point of S representing X. It is clear that in every neighbourhood of s_o, there is a point which represents a non-algebraic torus. (We have only to find a point not satisfying the Riemann conditions.) This also gives an example when the local moduli is not injectively parametrized as one sees that the action of $SL(2g,\mathbb{Z})$ on S is not discontinuous. In fact every non-empty open subset D of S contains t such that $D \cap t\cdot SL(2g,\mathbb{Z})$ is infinite.

The above example shows that if X is a smooth projective variety over $\mathbb{C}$, S its local deformation space and $A = \hat{\mathcal{O}}_{S,s_o}$ (where $\mathcal{O}_S$ is the structure sheaf of S, $\mathcal{O}_{S,s_o}$ its stalk at s_o and $\hat{\mathcal{O}}_{S,s_o}$ its completion), to get A as the completion of the local ring of the parameter space of an algebraic deformation of X. This has led Grothendieck to the definition of formal deformations and pro-representability.

SCHLESSINGER'S THEOREM: Let X be an algebraic scheme $|k$, k algebraically closed. Let R be a complete local ring such that $R/m_R = k$, m_R-maximal ideal of R and $R_n = R/m_R^n$.

DEFINITION 2.1: A <u>formal deformation</u> X_R of X is a sequence $\{X_n\}$ such that

(i) $X_n = X_{R_n}$ where X_{R_n} is a <u>deformation</u> of X over R_n, i.e., X_{R_n} is a scheme which is flat and of finite type $|R_n$ and $X_{R_n} \otimes k = X$;

(ii) we are given a compatible sequence of isomorphisms $X_n \otimes_{R_n} R_{n-1} \xrightarrow{\sim} X_{n-1}$ for any n.

(<u>Note that</u> X_R <u>is not a scheme flat</u> R ... etc.).

Let A be a finite-dimensional local k-algebra. Then giving a k-algebra homomorphism $\phi: R \longrightarrow A$ ($\Longleftrightarrow$ Spec $A \longrightarrow$ Spec R morphism) $\Longleftrightarrow$ giving a compatible sequence of homomorphisms $\phi_n: R_n \longrightarrow A$ for $n >> 0$ ($\Longleftrightarrow$ Spec $A \longrightarrow$ Spec R_n, $n >> 0$). Hence given a formal deformation X_R of X and a homomorphism $\phi: R \longrightarrow A$, $X_n \otimes_{R_n} A$ (via $\phi_n: R_n \longrightarrow A$) is the same up to isomorphism for $n >> 0$. It is a deformation of X over A. We define this to be $X_R \otimes A$ — <u>called base change of</u> X_R <u>by</u> Spec $A \longrightarrow$ Spec R.

Let X_R be a formal deformation of X. Let F,G be functors:

$$\underline{\mathbb{C}} = (\text{Finite-dimensional local algebras}) \longrightarrow (\text{sets})$$

$$\| $$

$$(\text{Category of Spec } A)^{\circ} \qquad (A = \text{finite-dimensional local k-algebra})$$

defined by (1) $A \in \underline{\mathbb{C}}$, $F(A)$ = Isom classes/A of deformations X_A over A

(2) $G(A) = \mathrm{Hom}_{k\text{-alg}}(R,A)$.

We get a morphism of functors

$$j: G \longrightarrow F \quad \text{defined by} \quad j_A: G(A) \longrightarrow F(A)$$

$$\phi \in \mathrm{Hom}_{k\text{-alg}}(R,A) \longrightarrow X_R \otimes A.$$

DEFINITION 2.2: A formal deformation X_R is said to be versal if the functor j above is formally smooth (i.e., for any surjection $A' \longrightarrow A$ in $\mathbb{C}$, we have

$$G(A') \longrightarrow G(A) \times_{F(A)} F(A') \text{ is surjective).}$$

$$\begin{array}{ccccc} & & & & G \\ & & & & | \\ \operatorname{Spec} A & \xrightarrow[i]{} & \operatorname{Spec} A' & \longrightarrow & F \end{array}$$

This definition $\Longrightarrow$ that given a deformation X_A of X, $A \in \underline{\mathbb{C}}$ there exists a homomorphism $\phi\colon R \longrightarrow A$ (not necessarily unique) and an A-isomorphism $X_R \underset{R}{\otimes} A \xrightarrow{\sim} X_A$ and something more, namely given $A' \longrightarrow A \longrightarrow 0$ a surjective homomorphism, a deformation $X_{A'}$ over A' and a homomorphism $\phi\colon R \longrightarrow A$, there exists a homomorphism $\phi'\colon R \longrightarrow A'$ such that ϕ' lifts ϕ and $X_R \otimes_R A'$ is A'-isomorphic to $X_{A'}$.

A formal deformation X_R over R is said to be universal if the functor j is an isomorphism. We have $F(A) \xleftrightarrow{\sim} \operatorname{Hom}_{k\text{-alg}}(R,A)$ and F is said to be prorepresentable.

REMARK 2.2: Let F' be the functor

$$(\text{Schemes } |k)^{o} \longrightarrow (\text{Sets})$$

defined by $F'(T) = \{T\text{-isomorphism classes of schemes flat } /T\}$. Then we have $X \in F'(k)$. More generally given a functor F

$$F'\colon (\text{Schemes } |k)^{o} \longrightarrow (\text{Sets})$$

we can define the notion of a formal deformation of $\xi \in F(k)$ as well as the concept of a versal deformation of ξ. We denote by F the functor obtained by restricting F' to $\underline{\mathbb{C}}$ and such that $F(k) = X$.

THEOREM 2.2 (Schlessinger, cf. [63]): Let $F\colon \underline{\mathbb{C}} \longrightarrow (\text{Sets})$ ($\underline{\mathbb{C}}$ = category of finite-dimensional local k-algebras) be a covariant functor and $F(k)$ = a single element. Then there exists a "versal deformation" for F, i.e., there exists a complete noetherian local k-algebra R with residue field k such that $\operatorname{Hom}_{k\text{-alg}}(R,\cdot) \longrightarrow F(\cdot)$ is formally smooth if the following conditions are satisfied:

Consider the canonical map

$$(*) \qquad F(A' \times_A A'') \longrightarrow F(A') \times_{F(A)} F(A'')$$

where $A' \longrightarrow A$ and $A'' \longrightarrow A$ are morphisms in $\underline{\mathbb{C}}$. Then

(i) (*) is <u>surjective</u> if $A'' \xrightarrow{\phi} A \longrightarrow 0$ is exact and ker ϕ is of dim 1 over k. (In case $A' \longrightarrow A \longrightarrow 0$ is also exact, so that Spec A $\longrightarrow$ Spec A' and Spec A $\longrightarrow$ Spec A" are closed immersions (*) means that given deformations over Spec A' and Spec A" such that their restrictions over Spec A are isomorphic, then we get a deformation on the scheme obtained by patching up Spec A' and Spec A" along Spec A.)

(ii) (*) is <u>bijective</u> when $A = k$ and $A'' = k[\epsilon]$.

(iii) $\dim_k F(k[\epsilon]) < \infty$ (it can be seen that by the above axioms $F(k[\epsilon])$ has a natural structure of a k-vector space), $k[\epsilon]$ being the ring of dual numbers over k.

REMARK 2.3: Further it can be shown that there exists an R such that $m_R/m_{R^2} \times F(k[\epsilon])$, m_R = maximal ideal of R. Then it can be seen that R is determined up to a non-canonical isomorphism, and we call R the <u>hull</u> or (<u>local moduli space</u>) of F.

THEOREM 2.3 (cf. [63]): The conditions of Schlessinger's theorem are satisfied in the case of the following deformation functors:

(i) F = Pic or more generally deformation of vector bundles on a complete algebraic scheme X.

(ii) F = Deformation of a complete algebraic scheme X/k.

(iii) F = Deformation of an affine variety with an isolated singularity.

(iv) F = Hilb or more generally Quot in the sense of Grothendieck.

As for the proof of the theorem, the checking of the axioms (i) and (ii) in Schlessinger's theorem is valid more generally, e.g., this is true when F = deformation of an algebraic scheme X, X not being complete. As in the complex analytic case, we compute $F(k[\epsilon])$ for the different functors, e.g.,

(i) $F(k[\epsilon]) = H^1(X, Hom(V,V))$ (in particular for line bundles $= H^1(X,\mathcal{O})$) when F = deformation of a vector bundle V on S.

(ii) $F(k[\epsilon]) = H^1(X,\Theta)$, F = deformation of a smooth algebraic scheme X.

In addition, if X is complete it follows that $\dim F(k[\epsilon]) < \infty$. As for the case of isolated singularities and deformation of an arbitrary complete algebraic scheme, one has an exact sequence

$$0 \longrightarrow H^1(X,T^0) \longrightarrow F(k[\epsilon]) \longrightarrow H^0(X,T^1) \longrightarrow H^2(X,T^0)$$

where T^0 = sheaf of germs of derivations of X (i.e., dual of Kähler differentials) and T^1 = coherent sheaf of germs of deformations of X/k to $k[\epsilon]$. From this the required finite dimensionality follows easily.

REMARK 2.4: (1) We see that if F "comes from" a representable functor, i.e., if as above F comes from an F': (Schemes/k) $\longrightarrow$ (Sets) and F' is representable by an algebraic scheme S with base point s_0 corresponding to F(k), we see that $R = \mathcal{O}_{S,s_0}$.

(2) In order to check that the local moduli space R is smooth, it suffices to check that the functorial map $F \longrightarrow \operatorname{Spec} k$ is formally smooth since $G \longrightarrow F$ is formally smooth. This gives all the criteria of smoothness that we mentioned before, namely (a) if $H^2(X,\mathcal{O}) = 0$, Pic is smooth; (b) if $H^2(X,\Theta) = 0$ local moduli space S is smooth at s_0, etc.

Work of M. Artin (cf. [1],[2]):

Let F be a functor as in Schlessinger's theorem and suppose that it comes from a functor F': (Schemes/k) $\longrightarrow$ (Sets) like the moduli functors we have considered. We have seen that even if Schlessinger's conditions are satisfied, F need not be algebraisable as the example of moduli of tori (cf. (X), Remarks 2.1) shows. However, we may ask if, under some more conditions, F is algebraisable, e.g., if F' = Pic on a complete algebraic scheme this is true. This leads to M. Artin's work.

Let us call a versal deformation X_R of an algebraic scheme effective if in fact there exists a scheme X'_R which is of finite type and flat over R which induces $X_n = X_{R_n}$, i.e., $X_{R_n} = X_R \otimes_R R_n$. Similarly, we can more generally define the notion of effectivity of a versal deformation of a functor F": (Schemes/k) $\longrightarrow$ (Sets).

THEOREM 2.4 (Artin Cf. [2]): (1) Let F' be a functor which is locally of finite presentation. Then Schlessinger's axioms + effectivity $\Longrightarrow$ algebraisability, e.g., in the case of deformations of X, this means that there exists a scheme $\underline{Y} \longrightarrow S$ flat and of finite type over S with base point s_0 such that $\mathcal{O}_{S,s_0} = R$ and $\underline{Y} \otimes R_n \cong X_n$.

(2) We have also a uniqueness result for S, uniqueness in the étale topology at s_0 provided some conditions are satisfied (e.g., F prorepresentable).

The conditions in Artin's theorem are satisfied, e.g., when F' = Pic and proves the existence of a natural structure of an algebraic space (in the sense of M. Artin) on the set of isomorphism classes of line bundles algebraically equivalent to zero on a complete algebraic scheme. Since this structure is also a "group", it follows that this induces in fact a group structure. This gave for the first time a direct proof of the existence of a Picard variety by not appealing to projective methods (apart from the use of such methods for proving finiteness theorems for proper maps, etc.).

The effectivity theorem for the problem of isolated singularities has recently been proved by Elkik (cf. [14]) by generalizing Artin's approximation theorem (cf. [1]). In the analytic case this was proved by Grauert (cf. [15]).

3. GLOBAL MODULI

We have seen that a local deformation (in the analytic case) of a smooth projective variety need not be algebraic (cf. (X), Remark 2.1). Hence to stay completely in the category of projective varieties, we could consider for the moduli problem only deformations which are projective, i.e., a deformation Y of X together with an ample line bundle L' on Y which is a deformation of an ample line bundle L on X. This leads to the concept of a polarization (cf. [74],[36]).

DEFINITION 3.1: An (inhomogeneous) polarization on a projective scheme X is to give an ample line bundle L on X, determined up to numerical equivalence. Thus an isomorphism $f\colon (X_1,L_1) \longrightarrow (X_2,L_2)$ of polarized structures is just an isomorphism $f\colon X_1 \longrightarrow X_2$ such that $f^*(L_2) \equiv L_1$ ($\equiv$ means numerical equivalence; recall that $L_1 \equiv L_2$ if restriction of $L_1 \otimes L_2^{-1}$ has degree zero for every closed subcurve in X).

REMARK 3.1: Suppose that X is a smooth projective variety over $\mathbb{C}$. It is well known that $L_1 \equiv L_2 \iff$ the elements of $H^2(X,\mathbb{Z})$ determined by L_1,L_2 are the same when tensored by $\mathbb{Q}$. Recall that when we identify a line bundle with an element of $H^1(X,\mathcal{O}_X^*)$, the element of $H^2(X,\mathbb{Z})$ is obtained by writing the cohomology exact sequence associated to the well-known sequence $0 \longrightarrow \mathbb{Z} \longrightarrow \mathcal{O} \xrightarrow{\exp 2\pi i} \mathcal{O}^* \longrightarrow 1$. Thus a polarization on X is to give an element of $H^2(X,\mathbb{Q})$ coming from an ample line bundle on X.

DEFINITION 3.2: A polarized family $(X_s, L_s)_{s \in S}$ parametrized by a scheme S is a proper flat morphism $p: \underline{X} \longrightarrow S$ together with a relatively ample line bundle $\mathbb{L}$ on $\underline{X}$ such that if $\underline{X}_s$ is the fibre over s and $\mathbb{L}_s = \mathbb{L} | \underline{X}_s$, then $\underline{X}_s \approx X_s$ and $L_s \equiv \mathbb{L}_s$.

We define the Hilbert polynomial of a polarized structure (X,L) to be the polynomial $P_X(m) = \chi(X, L^m)$ (χ = Euler-Poincaré characteristic). If we have a polarized family parametrized by a connected S, we see that $P_{X_s}(m)$ is independent of $s \in S$.

Let us denote by M the set of isomorphism classes of polarized smooth projective varieties with a fixed Hilbert polynomial. Then we can ask the problem whether M has a natural algebraic structure (say algebraic space in the sense of Artin, algebraic scheme, etc.). In this direction we have the following:

THEOREM 3.1: (i) (Matsusaka, cf. [38],[34]). Let the characteristic of the ground field be zero. Then M is bounded, i.e., $\exists$ an integer m_0 independent of the choice of (X,L) in M such that $m_0 L$ is very ample for $m \geq m_0$ on X and in fact we can also suppose that $H^i(X, mL) = 0$ for $i > 0$ and $m \geq m_0$. We see then that the complete linear system $|m_0 L|$ gives a closed immersion $j: X \longrightarrow \mathbb{P}^N$ with $N = P(m_0) - 1$. In particular the Hilbert polynomial of $j(X)$ is $P_1(m) = P(mm_0)$ and hence independent of (X,L) in M.

(ii) (Matsusaka-Mumford, cf. [39]): When the ground field is of arbitrary characteristic, the same assertion is true for the case of surfaces, i.e., when dim X = 2.

REMARK 3.2: Regarding the determination of m_0 for surfaces of general type, cf. [5] .

Matsusaka has investigated in [37] the question of giving an "algebraic structure" to M. Roughly speaking, if one throws out the ruled varieties (we say that a variety X is ruled if the function field K(X) of X is a purely transcendental extension of one variable over a subfield), he seems to prove that one gets a "Q-variety" in his sense. One has to be careful about throwing out ruled varieties because it is not clear that the complement of this set is open. However in the case of surfaces, this is all right thanks to the rationality criterion $P_{12} = 0$ (Enriques-Kodaira in the classical case and Bombieri-Mumford in arbitrary characteristic, cf. [6]). Throwing out ruled varieties one expects to get a "separated algebraic space" since the "valuative criterion" for separation for non-ruled varieties is satisfied because of

THEOREM 3.2 (Matsusaka-Mumford, cf. [39]). Let A be a discrete valuation ring, say with an algebraically closed residue field k and quotient field K. Given a scheme X over A, we denote by X_K its generic fibre which is a K-scheme and by $\overline{X}$ the closed fibre which is a k-scheme. Let V, W be smooth projective schemes over A and L, M ample line bundles on V, W relative to A. Let $\varphi_K : V_K \longrightarrow W_K$ be an isomorphism of V_K onto W_K such that $L_K = \varphi^*(M_K)$. Suppose that either $\overline{V}$ or $\overline{W}$ is not ruled. Then φ_K extends to an isomorphism of V onto W such that $\varphi^*(M) = L$; in particular φ induces an isomorphism $\overline{\varphi}$ of $\overline{V}$ onto $\overline{W}$ such that $\overline{\varphi}^*(\overline{M}) = \overline{L}$.

The question whether M or the part obtained by throwing out the ruled ones (say the case of surfaces) is an algebraic space in the sense of Artin has not been investigated in general (under some hypotheses this question is treated in [59],[60]). The study of M can be reduced to a quotient problem. This is done in [37] but has to be redone if one is considering the problem of giving M the structure of an Artin algebraic space. We shall now sketch without much detail how the study of M can be reduced to a quotient problem.

Let Q be the Hilbert scheme (cf. [19]) which parametizes closed subschemes $\{Y_q\}_{q \in Q}$ in $\mathbb{P}^N$ with Hilbert polynomial $P_1(m)$ ($P_1(m)$ and N as in Theorem 3.1). Let R_1 be the subset of Q determined by points $q \in Q$ such that

(i) Y_q is smooth

(ii) $\exists$ a line bundle L_q on Y_q such that $m_o L_q = H|Y_q$ where H denotes the hyperplane bundle in $\mathbb{P}^N$

(iii) $H^i(Y_q, H|Y_q) = 0$, $i > 0$

(iv) the canonical map $H^0(\mathbb{P}^N, H) \longrightarrow H^0(Y_q, H|Y_q)$ is an isomorphism.

One has to show that R_1 is locally closed in Q; in fact it is clear that the conditions (i), (iii) and (iv) define open subsets of Q; as for the condition (ii), cf. [37]. However, for a proper investigation of M, one should show that R_1 is not merely locally closed in Q_1 but has a natural structure of a "subscheme" of Q. It is now clear that M is the quotient of R_1 by the equivalence relation: $q_1 \sim q_2 \Longleftrightarrow \exists$ an isomorphism $f: Y_{q_1} \longrightarrow Y_{q_2}$ such that $f^*(L_{q_2}) \equiv L_{q_1}$. We see that $f^*(L_{q_2}) \equiv L_{q_1} \Longleftrightarrow f^*(H|Y_{q_2}) \equiv H|Y_{q_1}$. If we had $f^*(H|Y_{q_2}) = H|Y_{q_1}$

(nor merely equivalent), because of (iv) above, we see that the isomorphism is induced by a global automorphism of $\mathbb{P}^N$, i.e., Y_{q_1} and Y_{q_2} are equivalent under PGL(N) or that q_1, q_2 are in the same orbit for the action of PGL(N) on R_1 induced from the action of PGL(N) on Q (we see easily that R_1 is stable for the action of PGL(N) on Q). Since the set of all line bundles on X, numerically equivalent to a fixed line bundle, is parametrized by a <u>complete</u> algebraic scheme, (the scheme $\mathrm{Pic}^{\tau}(X)$ in the sense of Grothendieck, cf. [19]), it follows that the equivalence relation $q_1 \sim q_2$ introduced above is the composite of the equivalence relation introduced by the action of PGL(N) on Q (i.e., of being in the same orbit under PGL(N)) and a "proper equivalence" relation.

Suppose that we have the following situation:

(*) Given $(X,L) \in M$, $\exists$ a well-determined line bundle L_X, $L_X \equiv L$ such that $f\colon (X,L_X) \longrightarrow (Y,L_Y)$ is a polarized isomorphism $\Longrightarrow f^*(L_Y) = L_X$.

Then we define a subset R of Q defined by conditions (i) to (iv) as above, but now (ii) being replaced by the following:

(ii)' $$m_0 L_{Y_q} = H|Y_q$$

Then we see that M is precisely the orbit space of R under the action of PGL(N) induced by the canonical action of PGL(N) on Q.

The condition (*) happens to be satisfied for the case of curves (genus ≥ 2) and for polarized abelian varieties but for the fact that in the latter case we have $L_X \equiv 2L$ instead of $L_X \equiv L$.

<u>Case of canonically polarized varieties</u>: Let X be a smooth projective variety. Let K_X be the canonical line bundle on X. Assume that K_X is <u>ample</u>. Take then the polarization defined by K_X. Then we say that X is <u>canonically polarized</u>. Now if $f\colon X_1 \longrightarrow X_2$ is an isomorphism of smooth varieties, we see that $f^*(K_{X_2}) = K_{X_1}$ so that (*) is satisfied in this case. Further in this case we see that R is not merely locally closed but has a natural structure of a subscheme of Q; this follows as an easy application of the semi-continuity theorems, observing that if we have a smooth morphism $\underline{X} \longrightarrow S$, the relative canonical bundle $K_{\underline{X}}$ induces the canonical line bundle on the fibres. It is known that the automorphism group of a canonically polarized X is finite (at least in char. 0; as a consequence of the Akizuki-Nakano vanishing theorem it follows that $H^0(X,\Theta) = 0$, cf. [53]).

Further action of PGL(N) is separated by Theorem 3.2 (cf. [53] for a proof of this fact over $\mathbb{C}$) so that in the complex analytic case it follows easily that M has a natural structure of a Hausdorff complex analytic space. In fact, over $\mathbb{C}$ it can be shown that M has a natural structure of an Artin algebraic space (cf. [59],[60]) and it is most likely that this holds in arbitrary characteristic as well.

The above consideration applies in particular to the case of curves of genus $g \geq 2$ since the canonical line bundle is ample in this case. We see that even in arbitrary characteristic the group of automorphisms of a curve X of genus ≥ 2 is finite. Further it can be seen as a consequence of the smoothness of the local moduli space for curves (cf. (iv), Remarks 2.1) that R is smooth of dimension = dim PGL(N) + 2g-3. Now PGL(N) operates on R with finite isotropies and it can be shown that the orbit space M = R mod PGL(N) certainly exists as a separated normal (Artin) algebraic space of dimension (3g-3).

<u>Case of abelian varieties</u>:

Let $\varphi: (X_1,L_1) \longrightarrow (X_2,L_2)$ be an isomorphism of <u>polarized abelian varieties</u>, i.e., X_i is an abelian variety with L_i an ample line bundle on X_i, i = 1,2 and φ is an isomorphism of the underlying varieties such that $\varphi^*(L_2) \equiv L_1$. One knows that numerical equivalence is the same as algebraic equivalence in this case. Now φ is a homomorphism up to a translation from standard results on abelian varieties and translation certainly preserves algebraic equivalence. Hence in the definition of a polarized isomorphism above, we can suppose that φ is a homomorphism, i.e., φ is an <u>isomorphism</u> for the structure of abelian varieties. Further, $\varphi^*(L_2)-L_1 = t_a^*(L_1)$ where t_a is translation by an element $a \in X_1$. Thus if X_1,X_2 are embedded in the same $\mathbb{P}^N$ by complete linear systems m_0L_1, m_0L_2 respectively, then a polarized automorphism induces a projective transformation taking X_1 to X_2. This perhaps explains why something like (*) could be true for abelian varieties.

Let us now see how (*) holds and explain the reduction to a problem of quotient spaces (cf. Chap. 7, [44]).

Let L be an ample line bundle on an abelian variety X. Let $\Lambda(L)$

$$\Lambda(L): X \longrightarrow \hat{X}, \quad \hat{X} \text{ dual abelian variety}$$

be the <u>isogeny</u> defined by the algebraic family of line bundles on $X \times X$ defined by $\mu^*(L)-p_1^*(L)-p_2^*(L)$ where $\mu: X\times X \longrightarrow X$ is the group law and

p_i: $X \times X \longrightarrow X$ are the canonical projections. Of course we can define $\Lambda(L)$ for any L in Pic X and we get a homomorphism

$$\Lambda: \text{Pic } X \longrightarrow \text{Hom}(X, \hat{X}) .$$

It is known that Ker $\Lambda = \text{Pic}^o X$ (i.e., points of $\hat{X}$) so that Λ induces an injection

$$\Lambda: NS(X) \longrightarrow \text{Hom}(X, \hat{X})$$

where NS(X) (Néron-Severi group of X) = Pic $X/\text{Pic}^o X$. Thus Λ is an <u>invariant</u> of the polarization, i.e., $L_1 \equiv L_2 \Longleftrightarrow \Lambda(L_1) = \Lambda(L_2)$.

Let (X,L) be a polarized abelian variety and

$$\Lambda(L) = \omega: \ X \longrightarrow \hat{X}$$

the isogeny defined by the polarization. Define the line bundle $L_X = L_X(\omega)$ on X as follows:

$$\underline{L_X = q^*(P), \ q: X \longrightarrow X \times \hat{X}, \quad q = Id_X \times \omega,}$$

$$\underline{P = \text{Poincaré line bundle on } X \times \hat{X}}$$

Then we have

PROPERTY 3.1. <u>$L_X \equiv 2L$ and $\Lambda(L_X) = 2\omega$.</u>

For proof, cf. Prop. 6.10, [44].

COROLLARY: $\varphi: (X_1, L_1) \longrightarrow (X_2, L_2)$ is a polarized isomorphism

$$\Longleftrightarrow \varphi^*(L_{X_2}) = L_{X_1} .$$

Thus the property (*) holds for polarized abelian varieties (X,L) but for the fact $L_X \equiv 2L$. From the foregoing one can equivalently define polarized abelian varieties as

DEFINITION 3.3: (i) A <u>polarization</u> on an abelian variety is an isogeny

$$\omega: \ X \longrightarrow \hat{X}$$

such that $\exists$ an ample line bundle L on X such that $\omega = \Lambda(L)$.

(ii) Let $\underline{X} \longrightarrow S$ be an abelian scheme. Then a polarization on $\underline{X}$ is an S-isogeny $\omega: \underline{X} \longrightarrow \hat{\underline{X}}$ ($\hat{\underline{X}}$ = dual Picard scheme) such that $\exists$ an S-ample line bundle $\underline{L}$ on $\underline{X}$ such that $\Lambda(\underline{L}) = \omega$ (we see that the above definitions of Λ etc. extend to arbitrary base. In the definition of a polarized family, weaker assumptions can be made which imply this, cf. Def. 6.3, [44]).

It is easily seen that the direct image $\omega_*(\mathcal{O}_X)$ of the structure sheaf of X is locally free (of finite rank) over $\mathcal{O}_X$.

DEFINITION 3.4: The degree of a polarization $\omega: X \longrightarrow \hat{X}$ is the rank of the locally free $\mathcal{O}_X$-module $\omega_*(\mathcal{O}_X)$.

We have then

PROP. 3.2: Let X be an abelian variety of dimension g, L an ample line bundle on X and $r = \dim H^0(X,L)$. Then

(i) L^3 is very ample and $H^i(X,L) = 0$, $i > 0$

(ii) $\deg \omega = r^2$, $\omega = \Lambda(L)$.

For proof cf. Prop. 6.13, [44].

COROLLARY: Let $\omega: X \longrightarrow \hat{X}$ be a polarization of degree d^2. Then if $P(m) = \dim H^0(X,(L_X(\omega))^{3m})$, we have

$$P(m) = 6^g d\, m^g .$$

Proof: Let $\Theta = \dim H^0((L_X(\omega))^{3m})$. Then by Prop. 3.2, it follows that $\deg(\Lambda((L_X(\omega))^{3m})) = \Theta^2$. But then by Prop. 3.1, we have $\Lambda((L_X(\omega))^{3m}) = 6m\omega$. The morphism $6m\omega: X \longrightarrow \hat{X}$ can be factored as

$$X \xrightarrow{6m} X \xrightarrow{\omega} \hat{X} .$$

It is well known that the degree of the covering $6m: X \longrightarrow X$ (multiplication by 6m) is $(6m)^{2g}$. Hence it follows that the degree of $(6m\omega)$ is $(6m)^{2g} = 6^{2g}m^{2g}.d^2$ so that $\Theta = 6^g m^g d$ which proves the corollary.

REMARK 3.3: If $d = 1$, then $\omega: X \longrightarrow \hat{X}$ is an isomorphism and in this case X is said to be principally polarized. If X is the Jacobian of a curve with the polarization induced by the Θ divisor, then it is principally polarized.

Let now $M = M_{g,d}$ denote the isomorphism classes of polarized abelian varieties $\omega: X \longrightarrow \hat{X}$ such that

(1) $\dim X = g$ (2) $\deg \omega = d^2$.

By Prop. 3.2 and its corollary, it follows that the line bundle $(L_X(\omega))^3$ on X is very ample and that the complete linear system $|L_X(\omega)^3|$ defines a closed immersion $j: X \longrightarrow \mathbb{P}^N$, with $N = 6^g d-1$. We have seen (Cor. Prop. 3.2) that if P(m) is the Hilbert polynomial of j(X), then

$$P(m) = 6^g d m^g$$

i.e., for all $X \in M$, j(X) has the same Hilbert polynomial. Let now Q be the Hilbert scheme which parametrizes closed subschemes of $\mathbb{P}^N$ with Hilbert polynomial $P(m)$. Let $Z \longrightarrow Q$ be the universal family over Q ($Z \subset Q \times \mathbb{P}^N$, the morphism $Z \longrightarrow Q$ is induced by the first projection and the fibres Z_q, $q \in Q$ give "all" closed subschemes of $\mathbb{P}^N$ with Hilbert polynomial $P(m)$). Let $W = Z \times_Q Z$, then W is the base change of $Z \longrightarrow Q$ by $Z \longrightarrow Q$

$$\begin{array}{ccc} Z \times_Q Z = W & \longrightarrow & Z \\ \Big| & & \Big| \\ Z & \longrightarrow & Q \end{array}$$

and the fibres of $W \longrightarrow Z$ are the fibres of $Z \longrightarrow Q$. But now note that $W \longrightarrow Z$ has a canonical section, namely the diagonal morphism $\Delta: Z \longrightarrow Z \times_Q Z$. It is not difficult to see that $W \longrightarrow Z$ is the universal family for all families $p: \underline{X} \longrightarrow S$, such that

(1) p is proper and flat

(2) $\underline{X}$ is a closed subscheme of $\mathbb{P}^N \times S$ and $p: X \longrightarrow S$ is induced from the canonical projection to S

(3) $p: \underline{X} \longrightarrow S$ has a section

(4) the Hilbert polynomial of the fibres of p is $P(m)$.

Hence given such a family $p: \underline{X} \longrightarrow S$, $\underline{X}$ is obtained by base change of $W \longrightarrow Z$ by a unique morphism $S \longrightarrow Z$ (the given section being base change of the canonical section of $W \longrightarrow Z$ given above).

Let now R be the subset of Z defined by points $q \in Z$ such that if W_q denotes the fibre of $W \longrightarrow Z$ over q, we have

(i) W_q is an <u>abelian variety with the identity element</u> given by the canonical section of $W_q \longrightarrow k(q)$ where $k(q)$ denotes the residue field of Z at q.

(ii) If H is the hyperplane line bundle on $\mathbb{P}^N$ then $H|X = L_X^3$ (X = abelian variety W_q) and $L_X = L_X(\omega)$ for a polarization ω on X of degree d^2.

(iii) The canonical map $H^0(\mathbb{P}^N, H) \longrightarrow H^0(X, L_X^3)$ is an isomorphism ($X = W_q$).

It can be shown that R has a natural scheme structure and that it is locally closed in Z. The conditions (i) and (iii) give open conditions; that (i) gives an open condition comes from the following theorem: Let $\underline{X} \longrightarrow S$ be a proper smooth morphism with S connected and such that it has a section. Then if for one point the fibre $\underline{X}_s \longrightarrow k(s)$ is an abelian variety (with identity induced by the given section) then $\underline{X} \longrightarrow S$ is an abelian scheme with the given section as the identity section.

To investigate condition (ii), note that we have the following: Let X be an abelian variety. Then an ample line bundle P on X is of the form $P \equiv L_X^3$ with $L_X = L_X(\omega)$ for a suitable polarization $\omega: X \longrightarrow \hat{X}$ $\Longleftrightarrow$ the isogeny $\nu = \Lambda(P): X \longrightarrow \hat{X}$ is divisible by 6. (This is an easy consequence of the above considerations, especially Prop. 3.1). It can be shown that the set of points $q \in Z$ such that W_q (fibre of $W \longrightarrow Z$ over q) is an abelian variety (i.e., satisfying (i)) and such that if $\nu_q = \Lambda(H|W_q)$ (H is the hyperplane bundle on $\mathbb{P}^N$) then $\nu_q = 6\omega_q$ for some isogeny $\omega_q: W_q \longrightarrow \hat{W}_q$ is a locally closed subset Z_1 in Z and has a natural scheme structure. Further one shows that the ω_q, $q \in Z_1$ patch up to define a Z_1-isogeny

$$\omega: W_1 \longrightarrow \hat{W}_1 \quad (W_1 \text{ inverse image of } Z_1 \text{ by } W \longrightarrow Z).$$

Then $L_{W_1}(\omega)$ defines a Z_1-ample line bundle on W_1 and the restriction of H to the fibres of $W_1 \longrightarrow Z_1$ also defines a Z_1-ample line bundle P on W_1. Now the set of points $q \in Z_1$ such that

$$P|W_q = (L_{W_1}(\omega))^3 \Big| W_q$$

is closed in Z_1 and has a natural structure of a subscheme. From these considerations it follows easily that the subset R of Z defined above is locally closed and has a natural structure of a subscheme (for more details, cf. Prop. 7.3 and Prop. 6.11, [44]).

We see easily that PGL(N) operates canonically on R and that the orbit space can be identified with $M = M_{g,d}$. Since the group of polarized automorphisms is finite, PGL(N) operates with finite isotropies on R. It is quite likely that one can prove with these considerations that $M_{g,d}$ has a natural structure of a separated Artin algebraic space (separated because an abelian variety is certainly not ruled and then use Theorem 3.2).

The question now arises whether $M_{g,d}$ has a natural structure of an algebraic scheme or moreover a quasi-projective algebraic scheme. In many moduli problems this seems to be a much subtler question than proving that the required moduli space is an Artin algebraic space. The difficulty comes from the fact that if G is a semi-simple algebraic group (say even PGL(N)) acting on a smooth quasi-projective variety X, say even freely and with separated action, the orbit space need not be an algebraic scheme though it can be shown that the orbit space has a natural structure of an (Artin) algebraic space (for counter-examples cf. [47] and §3, Chap. 4, [44]). This leads to Mumford's geometric invariant theory.

Geometric invariant theory

Let G be a reductive algebraic group defined over an arbitrary algebraically closed field k (G is reduced and the radical of G is a torus). The following result, conjectured by Mumford, has recently been proved by W. J. Haboush and marks a significant step. It was earlier proved in particular cases (cf. [57],[65],[66]).

THEOREM 3.3 (Haboush, cf. [20]): Let $\chi: G \longrightarrow \text{Aut } V$ be a rational representation on a finite-dimensional vector space V (over k) and $v_0 \neq 0$ be a G-invariant point. Then $\exists$ a G-invariant homogeneous polynomial F of degree ≥ 1 on V such that $F(v_0) \neq 0$.

Proof: It is easily seen that we can suppose without loss of generality that G is simply-connected and semi-simple. Now G is of the form $\mathbb{G} \times_{\mathbb{Z}} k$ where $\mathbb{G}$ is a split semi-simple group scheme over $\mathbb{Z}$ (cf. [7]). Further we have only to consider the case when $p = \text{char } k \neq 0$.

Let us fix a maximal torus T and a Borel subgroup B, $B \supset T$. Let us take roots in the usual way. Let ρ denote as usual half the sum of positive roots and $W_{m\rho}$ (m positive integer) the sections of the line bundle on G/B associated to $m\rho$; one knows that $W_{m\rho}$ can be identified with regular functions f on G such that

$$f(gb) = \chi(b)^m f(g)$$

where χ is the character of T (and hence of B) associated to ρ (cf. [7]). Then the proof consists of the following two steps (compare the proof for the case GL(2) together with (4) in Remarks on p. 355, [65]):

I. (a) The G-module $W_{m\rho} \otimes W_{m\rho}$ has a unique 1-dimensional subspace of G-invariant points

(b) Given V as in the statement of the theorem, $\exists$ a G-linear map φ (for all sufficiently large m)

$$\varphi: V \longrightarrow W_{m\rho} \otimes W_{m\rho}$$

such that $\varphi(v_0) \neq 0$.

II. Let $W'_{m\rho}$ be the irreducible representation with highest weight $m\rho$ (cf. [7]). Then

(a) $W'_{m\rho}$ is self dual (as a G-module)

(b) for $m = p^{\nu}-1$, $\dim W'_{m\rho} = p^{\nu N}$ (N = number of positive roots), in particular $\dim W'_{m\rho}$ is the same as the dimension of the irreducible representation of $\mathbb{G}\times_{\mathbb{Z}}\mathbb{C}$ with highest weight $m\rho$; and

(c) $W'_{m\rho} = W_{m\rho}$, $m = p^{\nu}-1$ and ν sufficiently large.

We see now how these two steps imply the theorem. If I and II hold, $\exists$ a G-linear map

$$\varphi: V \longrightarrow \mathrm{Hom}(W_{m\rho}, W_{m\rho}), \quad m = p^{\nu}-1, \quad \nu >> 0$$

$$\varphi(v_0) \neq 0.$$

Because of the assertion I(a) it follows that $\varphi(v_0) = \lambda \mathrm{Id}$, $\lambda \neq 0$ (Id = Identity map $W_{m\rho} \longrightarrow W_{m\rho}$). Take the determinant function F' on $\mathrm{Hom}(W_{m\rho}, W_{m\rho})$, then $F'(\mathrm{Id}) \neq 0$ and F' is G-invariant. Set $F = F'\circ\varphi$. Then $F(v_0) \neq 0$.

As for the proof of I and II let us take up II. The assertion II(a) is well known (cf. [7]). Further II(c) follows easily once II(b) is proved, for it is known that

$$W'_{m\rho} \subset W_{m\rho} \quad (\text{cf. [7]}).$$

Further it is easy to see that $W_{m\rho}$ is the reduction mod p of the irreducible representation in char. 0 with highest weight $m\rho$ whenever m is sufficiently large. This is so because if $L(\rho)$ is the line bundle on G/B associated to ρ, then $L(\rho)$ is ample and

$$H^1(G/B, mL(\rho)) = 0, \quad m >> 0$$

and then by application of the semi-continuity theorems the required assertion follows. The proof of II(b) is quite non-trivial; however it is known and is due to Steinberg (cf. [69],[7]).

Let us go to the proof of I. It is well known that $\exists$ a G-linear map $V \longrightarrow A(G)$ (A(G)-coordinate ring of G and say G acts on A(G) through the left regular representation) such that $v_0 \longrightarrow$ constant function 1. Make T operate on A(G) through the right regular representation, then A(G) is completely reducible as a T-module so that we have a canonical projection $A(G) \longrightarrow A(G/T)$ which is G-linear (for the action induced by left multiplication, i.e., the left regular representation). Thus we have a G-linear map

$$V \longrightarrow A(G/T)$$

such that $v_o \longrightarrow$ constant function 1. Thus we can assume that $V \subset A(G/T)$ and $v_o = 1$.

The proof of I is an immediate consequence of the following

LEMMA 3.1: We have a canonical increasing filtration of $A(G/T)$ by finite-dimensional G-submodules P_m

$$A(G/T) = \bigcup_m P_m, \quad P_m \subset P_{m+1}$$

and canonical G-isomorphisms

$$P_m \longrightarrow W_{m\rho} \otimes W_{m\rho}.$$

<u>Proof</u>: Consider the map

$$j: G/T \longrightarrow G/B \times G/B$$

induced by

$$j': G \longrightarrow G \times G, \quad g \longrightarrow (g, w_o g w_o^{-1})$$

where w_o is the unique element of the Weyl group of maximal length (<u>note that to define</u> j' <u>we have to choose a representative of</u> w_o <u>in</u> G, <u>we fix a representative in the sequel</u>). Since $T = B \cap B^-$, $B^- = w_o B w_o^{-1}$ (as well as Lie T = Lie B $\cap$ Lie B^-) we see that j (as well as dj) is injective. Further it is well known that dim G/T = 2 dim G/B. From these facts it follows easily that j is an <u>open</u> immersion. Define E by

$$E = (G/B \times G/B) - j(G/T).$$

Let Λ be the inverse image of E in $(G \times G)$ by the canonical map $G \times G \longrightarrow G/B \times G/B$. An easy computation shows that

$$\Lambda = \alpha^{-1}(D(\rho))$$

where $D(\rho) = G - B w_o B$ and α is the morphism

$$\alpha: G \times G \longrightarrow G, \quad (g_1, g_2) \longrightarrow g_2^{-1} w_o g_1.$$

(Now $D(\rho)$ is precisely the inverse image in G of the divisor in G/B associated to ρ). It is known that $\exists$ a regular function F on G vanishing precisely on $D(\rho)$ and with multiplicity 1 on all its irreducible components such that

$$F(b_1 g b_2) = \chi(b_1)^{-1} F(g)\chi(b_2),\ b_1, b_2 \in B$$

χ being the character of T associated to ρ. Let F_1 be the pull-back by α of the function F, i.e.,

$$F_1(g_1, g_2) = F(g_2^{-1} w_o g_1).$$

The function F_1 is easily checked to have the following properties

(i) $F_1(g_1 b_1, g_2 b_2) = \chi(b_1)\chi(b_2) F_1(g_1, g_2)$, $b_i \in B$

(ii) $L_s F_1 = F_1$ where for a function $f(g_1, g_2)$ on $G \times G$ we define $L_s f = f(s^{-1} g_1,\ w_o s^{-1} w_o^{-1} g_2)$.

Since $D(\rho)$ is a divisor in G it follows that E is a divisor (this also follows from the fact that G/T is affine). We now define the linear subspace P_m as the subset of regular functions on G/T with pole of order $\leq m$ on E. Hence P_m can be identified with functions f on $G \times G$ such that $f(g_1 b_1, g_2 b_2) = f(g_1, g_2)$ and such that f has pole of order $\leq m$ on Λ. Further the action of G on P_m is given by $s \cdot f(g_1, g_2) = L_s f$ where L_s is as in (ii) above. Set $Q_m = F_1^m \cdot P_m$. Then it follows that every element f of Q_m is a <u>regular</u> function on $G \times G$ and that it satisfies (because of (i) above)

(*) $\quad f(g_1 b_1, g_2 b_2) = \chi(b_1)^m \chi(b_2)^m f(g_1, g_2)$, f regular on G .

On the other hand it is immediate to see that functions f satisfying (*) are in Q_m so that $Q_m = Q_m'$ where $Q_m' = \{f | f \text{ satisfying } (*)\}$. Further one sees easily (by an application of Künneth formula for line bundles on $G/B \times G/B$) that

$$Q_m' = W_{m\rho} \otimes W_{m\rho}$$

so that we have a canonical identification of Q_m (and hence of P_m) with $W_{m\rho} \otimes W_{m\rho}$. Because of (ii) above, the action of G on $W_{m\rho} \otimes W_{m\rho}$ induced by this identification with P_m is given by

$$s \cdot (f_1(g_1) \otimes f_2(g_2)) = f_1(s^{-1} g_1) \otimes f_2(w_o s^{-1} w_o^{-1} g_2),\ s \in G$$

where $f_1 \otimes f_2$ is a decomposable element of $W_{m\rho} \otimes W_{m\rho}$. Thus if $W_{m\rho}^{w_o}$ is the G-module such that its underlying space is $W_{m\rho}$ and the action of G is given by $s \cdot f(g) = f(w_o s^{-1} w_o^{-1} g)$, $f \in W_{m\rho}$, $s \in G$, we have a canonical G-isomorphism

$$P_m \longrightarrow W_{m\rho} \otimes W_{m\rho}^{w_o} .$$

But $W_{m\rho}^{w_o}$ is G-isomorphic to $W_{m\rho}$. This concludes the proof of the lemma and hence of the theorem.

The above theorem says that $\exists$ a G-invariant polynomial F on V such that it separates the 2 distinct G-invariant points v_o and (0) of V. More generally, given two points v_1, v_2 in V such that $\overline{O(v_1)} \cap \overline{O(v_2)} = \phi$ ($\overline{O(v_1)}$-closure of the G-orbit $O(v_1)$), we see that $\exists$ a G-invariant polynomial F on V such that $F(v_1) = 1$ and $F(v_2) = 0$. To prove this, we see easily that $\exists$ a G-linear map $\varphi: V \longrightarrow W$ such that $\varphi(v_1) \neq (0)$ and $\varphi(v_2) = 0$. Then by Theorem 3.3, $\exists$ a G-invariant polynomial F on W such that $F(\varphi(v_1)) = 1$ and $F(\varphi(v_2)) = 0$ and pulling this function to V, we get the required assertion.

Let now $X = \text{Spec } A$ be an affine algebraic scheme on which a reductive algebraic group G acts. Then we see easily that $\exists$ a G-equivariant closed immersion

$$\varphi: X \longrightarrow \mathbb{A}^n$$

where G acts on $\mathbb{A}^n$ through a rational representation. Then from the foregoing it follows that given $x_1, x_2 \in X$ such that $\overline{O(x_1)} \cap \overline{O(x_2)} = \phi$ $\exists$ $f \in A^G$ separating x_1 and x_2. Nagata has shown (assuming Theorem 3.3, cf. [46]) that A^G is a finitely generated k-algebra. It can then be seen that the points of the algebraic scheme $Y = \text{Spec } A^G$ can be identified with the equivalence classes in X by the equivalence relation: $x_1 \sim x_2 \Longleftrightarrow \overline{O(x_1)} \cap \overline{O(x_2)} \neq \phi$. To put this in a more precise formulation, let us introduce the following concepts (cf. [44],[66]:

Let G be an algebraic group operating on an algebraic scheme X and $f: X \longrightarrow Y$ be a G-invariant morphism, i.e., for the trivial action of G on Y, f is a G-morphism.

DEFINITION 3.5: (1) We say that f is a categorical quotient if $\forall$ G-invariant morphism $g: X \longrightarrow Z$, $\exists$ a unique morphism $h: Y \longrightarrow Z$ such that $g = h \cdot f$.

We see that a categorical quotient is uniquely determined up to isomorphism.

(2) Suppose moreover that G is affine. Then we say that f is a good quotient if

(i) f is a surjective affine morphism and

$$f_*(\mathcal{O}_X)^G = \mathcal{O}_Y$$

(ii) if Z is a closed G-stable subset of X then f(Z) is closed in Y; further if Z_1, Z_2 are two closed G-stable subsets of X such that $Z_1 \cap Z_2 = \phi$ then $f(Z_1) \cap f(Z_2) = \phi$.

It is seen easily that a good quotient is a categorical quotient.

(3) Suppose that as in (2) above, G is affine. Then we say that f is a <u>geometric quotient</u> if

(i) f is a good quotient

(ii) $x_1, x_2 \in X$, $f(x_1) = f(x_2) \Longleftrightarrow O(x_1) = O(x_2)$

(equivalently because of (i), $\forall x \in X$ the G-orbit in X through x is closed in X).

THEOREM 3.4: Let G be a reductive algebraic group operating on an affine algebraic scheme X = Spec A. Let Y = Spec A^G. Then Y is an affine algebraic scheme, i.e., A^G is a k-algebra of finite type. Let further f: $X \longrightarrow Y$ be the canonical morphism induced by $A^G \subset A$. Then f is a good quotient. In particular if all the orbits are of the same dimension (this $\Longrightarrow$ all the orbits are closed), f is a geometric quotient.

For proof see §2, Chap. 1, [44] and [65],[66].

Let now X be a projective algebraic scheme with <u>a very ample</u> line bundle L on X. An action of a reductive algebraic group G on X is said to be <u>linear</u> (with respect to L) if it can be lifted to an action of the line bundle L or, to be more precise, comes from a rational representation on the affine space corresponding to the projective embedding of X. Let $R = \sum_{n \geq 0} \Gamma(X, L^n)$ so that X = Proj. R. Let Y = Proj. R^G and $\varphi: X \longrightarrow Y$ the <u>rational</u> morphism defined by $R^G \subset R$.

Let $\hat{X}$ be the cone over X defined by L. The action of G on X lifts to an action of G on $\hat{X}$ since it is linear with respect to L.

DEFINITION 3.6: Let X,G be as above and suppose that G is reductive.

(1) A point (i.e., a closed point) $x \in X$ is <u>semi-stable</u> if $\exists\ \hat{x} \in \hat{X}$ over x such that the closure (in $\hat{X}$) of the G-orbit through $\hat{x}$ does not pass through (0) ((0)-vertex of the cone).

(2) A point $x \in X$ is said to be <u>stable</u> (or <u>properly stable</u>) if $\exists\ \hat{x} \in \hat{X}$ ($\hat{x}$ over x) such that $O(\hat{x})$ is closed in $\hat{X}$ and dim G = dim $O(\hat{x})$.

(3) Let X^{ss} (resp. X^s) = set of semi-stable (resp. stable) points in S. Note that these are G-stable.

Then we have

THEOREM 3.5: (1) Given $x \in X^{ss}$, $\exists$ a homogeneous element $s \in R^G$ such that $s(x) \neq 0$ and conversely a semi-stable point is characterized in this manner. In particular it follows that X^{ss} is open.

(2) φ is regular in X^{ss}. Then the morphism

$$\varphi: X^{ss} \longrightarrow Y$$

is a good quotient.

(3) X^s is open, $\varphi(X^s) = Y^s$ is open in Y and

$$\varphi: X^s \longrightarrow Y^s$$

is a geometric quotient.

For proof cf. §4, Chap. 1, [44] and [65],[66].

Let X,G,L be as in Theorem 3.5. We call a 1-parameter subgroup of G a homomorphism $\lambda: \mathbb{G}_m \longrightarrow G$. Now λ induces a canonical linear action of $\mathbb{G}_m$ on X. Then we have

THEOREM 3.6: Let X,G,L be as in Theorem 3.5. Then a (geometric) point is stable (resp. semi-stable) if and only if it is so for the action of $\mathbb{G}_m$ induced by every non-trivial 1-parameter subgroup of G.

For proof cf. Theorem 2.1, Chap. 2, [44] and [66].

This is the result which Mumford says is inspired from Hilbert. It allows one to compute stable and semi-stable points for many actions which are explicitly given (cf. Chap. 4, [44]). These computations are key points in the construction of moduli (e.g., diagonal action of the special linear group on a product projective spaces or Grassmannians).

The idea of Mumford is to use Theorem 3.5 for proving the (existence and) quasi-projectivity of the moduli spaces for curves, polarized abelian varieties, line bundles on a projective variety and stable vector bundles (see later for the definition of stable vector bundles) on smooth projective curves (cf. [42] for an outline). In every one of these examples one is reduced to a problem of constructing a quotient of a locally closed subscheme R of a suitable Hilbert or Quot scheme (in the sense of Grothendieck cf. [19]) for the canonical action of a projective (or special) linear group on Hilb or Quot (as we have seen above for the case of curves and polarized abelian varieties). Note that Hilb (and Quot) are projective schemes and the canonical actions of the special linear group in question are linear (for suitable choice of ample line bundles). In every one of these cases it would suffice to show that the choice of Hilb (or Quot) and consequently R can be so made such that the points are stable. However it has not been possible to do this directly. Every one of these problems is reduced further to a quotient of a projective X under a linear action of PGL, where one knows X explicitly and for which the stable and semi-stable points can also be computed explicitly. This is in general rather subtle. One could however hope for more uniform and direct methods. For example, progress in the solution of the following problem would help in this direction.

PROBLEM 3.1: Let C be a smooth projective variety in $\mathbb{P}^n$. Consider the canonical action of PGL(n) (or SL(n+1)) on the corresponding Hilbert (or Chow) schemes. When is the point corresponding to C in the associated Hilbert (or Chow) scheme stable or semi-stable?

An answer to this problem is given in (Theorem 4.5, [44]) for the case of curves. We have

THEOREM 3.7: Let C be a smooth irreducible curve in $\mathbb{P}^n$ such that (i) if H is the hyperplane bundle on $\mathbb{P}^n$, then the canonical mapping $H^o(\mathbb{P}^n,H) \longrightarrow H^o(C, H|C)$ is an isomorphism and (ii) genus g of C is ≥ 1 and $\deg C \geq 3g$.

Then for the canonical action of PGL(n) (or SL(n+1)) on the Chow variety of curves in $\mathbb{P}^n$ of degree = deg C, the point corresponding to C is stable.

This theorem gives the existence of a natural structure of a quasi-projective algebraic scheme on the set $M_g = M$ of isomorphism classes of smooth projective curves of fixed genus $g \geq 2$. This is locally normal of dimension 3g-3 (normality follows since R is smooth). Note however that there is one technical complication; namely that when we reduced M to a quotient problem R/PGL(N), R was in a Hilbert scheme and not the corresponding Chow scheme. However it is not hard to overcome this fact. In fact one can show that there is a canonical morphism Hilb $\longrightarrow$ Chow which is equivariant for the canonical actions of the projective group in question and on the subset R this morphism is injective from which with the aid of Theorem 3.6 one concludes that the geometric quotient of R mod PGL exists and that it is quasi-projective (cf. Chap. 5, [44]).

Thus we get

THEOREM 3.8 (cf. Theorem 5.11, [44]). There is a natural structure of a quasi-projective (locally normal) scheme on the set M_g of isomorphism classes of smooth projective curves of genus $g \geq 2$.

Let us now see how one gets a natural structure of a projective scheme on $M = M_{g,d}$ - the set of isomorphism classes of polarized abelian varieties of dimension g and degree d^2. We have seen that M can be identified with the set of orbits in R for the action of PGL(N). We have also a canonical abelian scheme $A \longrightarrow R$ (A is the inverse image of R under the morphism $W \longrightarrow Z$, notations as above) and $A \subset R \times \mathbb{P}^N$. Choose n which is coprime to the characteristic of the ground field. Consider the map:

$q \in R$, $q \longrightarrow$ points of order n of the abelian variety A_q (fibre of A over q) which is a cycle in $\mathbb{P}^N$.

Since n is coprime to the characteristic the number of points in this cycle is n^{2g}. Thus we get a PGL(N) equivariant morphism

$$j: R \longrightarrow (\mathbb{P}^N)^{(n^{2g})} \text{— } n^{2g}\text{-fold symmetric product of } \mathbb{P}^N .$$

We have now a PGL(N) equivariant morphism

$$\theta: (\mathbb{P}^N)^{n^{2g}} \ (n^{2g}\text{-fold product}) \longrightarrow (\mathbb{P}^N)^{(n^{2g})}.$$

Thismorphism θ is finite and we see easily that z is stable (resp. semi-stable) $\Longleftrightarrow \theta(z)$ is stable (resp. semi-stable). One shows that j(q) is stable ($q \in R$) for $n \gg 0$ and for this it suffices to check that some point on $(\mathbb{P}^N)^{n^{2g}}$ representing j(q) is stable. The crucial step is

LEMMA 3.2: Let A be an abelian variety of dimension g, $A \subset \mathbb{P}^N$ and not contained in any hyperplane. Then for n such that $(n,p) = 1$ (p = characteristic of the ground field) and $n > \sqrt{(N+1)r}$, r = deg A, the set of points of order n on A arbitrarily ordered is a stable point in $(\mathbb{P}^N)^{n^{2g}}$.

For proof cf. Prop. 7.7, [44]. Here one uses the explicit computation for stable points in a product of projective spaces (cf. Prop. 4.3, [44]).

This lemma $\Longrightarrow j(R) \subset X^{ss}$ setting $X = (\mathbb{P}^N)^{(n^{2g})}$ (i.e., symmetric product of $\mathbb{P}^N$'s). Let $X_1 = (\mathbb{P}^N)^{n^{2g}}$. Then we have

$$\begin{array}{ccc} R_1 & \longrightarrow & X_1 \\ \downarrow & & \downarrow \\ R & \longrightarrow & X \end{array}$$

where $R_1 = R \times_X X_1$. Then on R_1 the symmetric group $S_{n^{2g}}$ and PGL(N) operate with commuting actions. We see that it suffices to prove that the geometric quotient $R_1/PGL(N) = M_1$ exists as a quasi-projective scheme for then $M = M_{g,d} = M_1/S_{n^2g}$. Now M_1 being quasi-projective, the geometric quotient of M_1 by the finite group S_{n^2g} certainly exists as a quasi-projective scheme. From Lemma 3.2 it follows that $j_1(R_1) \subset X_1^s$. From Theorem 3.5 it follows that the geometric quotient of $X_1^s/PGL(N)$ exists as a quasi-projective scheme and it can be seen that $X_1^s \longrightarrow X_1^s/PGL(N)$ is in fact a locally trivial principal PGL(N) fibration (cf. Theorem 3.8,[44]). The fact that now the geometric quotient $R_1/PGL(N)$ exists is an easy consequence of the following easily proved

LEMMA 3.3: Let an algebraic group G operate on algebraic schemes X, Y and $f: X \longrightarrow Y$ be a G-morphism. Assume that (1) the geometric quotient $Y \longrightarrow Q$ exists and that Y is a principal G-fibre space over Q, (2) Q is quasi-projective and (3) $\exists$ a line bundle L on X on which G operates (compatible with the action on X) and that L is relatively ample with respect to f.

Then the geometric quotient $X \longrightarrow P$ exists and P is quasi-projective.

For proof of Lemma, cf. Prop. 7.1,[44].

We now apply the lemma to the morphism $j_1: R_1 \longrightarrow X_1$. Except the condition (3) above, we have checked the other conditions. Since R_1 is itself quasi-projective, for checking (3) we have only to take a line bundle L on R_1 which is ample and on which G operates and such a line bundle certainly exists.

Thus we have shown

THEOREM 3.9: $\exists$ a natural structure of a quasi-projective algebraic scheme on the set $M_{g,d}$ of polarized abelian varieties of dimension g and degree d^2.

REMARK 3.4: One should note that the scheme $M_{g,d}$ (and similarly for the moduli space in Theorem 3.6) does not represent the functor

(*) $S \longrightarrow$ (isomorphism classes of polarized abelian schemes over S of relative dimension g and degree d^2)

i.e., if $A_{g,d}(S)$ denotes the R.H.S. in (*) in general we don't have

$$M_{g,d}(S) = A_{g,d}(S)$$

though for S = k this certainly holds. The difficulty essentially comes from the fact that a polarized abelian variety may have non-trivial automorphisms. To remedy this fact one should introduce additional structures such that the automorphism group is identity or a rigidification in the sense of Grothendieck (compare also the point of view from quotient spaces, cf. §6,[66]). In the case of polarized abelian varieties one introduces as additional structure "points of order n". Then one gets a moduli space called a <u>fine moduli</u> space $M_{g,d,n}$ (for suitable n) and in this case one gets a representable functor (cf. §2, Chap. 7 and Theorem 7.9,[44]). The moduli scheme $M_{g,d}$ is called a <u>coarse</u> moduli scheme.

REMARK 3.5: The existence and quasi-projectivity of the space of moduli for polarized abelian varieties implies the existence and quasi-projectivity for the moduli of smooth projective curves and we get another proof besides the one indicated as a consequence of Theorem 3.6. One associates to a smooth projective curve C of genus ≥ 1 its jacobian J with the principal polarization induced by the Θ divisor. One sees that this can also be done for a family. The theorem of Torelli says that two smooth projective curves (over an algebraically closed field) are isomorphic if and only if the corresponding principally polarized varieties are isomorphic. From this the required assertion for the moduli of curves follows easily and we get an injective morphism from the (coarse) moduli of curves to the (coarse) moduli of principally polarized abelian varieties of dimension g (cf. §4, Chap. 7,[44]).

REMARK 3.6: The moduli spaces considered above happen to be non-complete. One can ask whether there is a natural compactification by a suitable moduli problem. This is quite important and gives a better insight into the moduli problem. Historically the quasi-projectivity of the moduli of polarized abelian varieties (over $\mathbb{C}$) was first proved by Baily by getting a compactification and proving its projectivity and the question gets related to compactifications of quotients by discrete subgroups (cf. [3]). Further this is also related to the question of semi-stable reductions (cf. [72]). An illustration of the use of compactifications is the proof of Deligne-Mumford (cf. [10]) of the connectedness of the moduli space M_g of curves of genus g in arbitrary characteristic. In characteristic zero using transcendental methods it follows that M_g is connected, in fact irreducible. It can be shown that the characteristic zero variety specializes to characteristic p; hence if everything is complete, by Zariski's connectedness theorem, the required assertion follows. Unfortunately this is not the case. One then gets a canonical compactification $\overline{M}_g$ by means of <u>stable curves</u>. These moduli spaces specialize well and then by Zariski's connectedness theorem using connectedness in characteristic zero, it follows that $\overline{M}_g$ is connected in arbitrary characteristic. One shows that $\overline{M}_g$ is "nearly" smooth (i.e., quotient of a smooth scheme by a proper action of PGL) from which it follows easily that $\overline{M}_g$ is irreducible and M_g being open it is connected.

We shall now describe in some detail one moduli problem where geometric invariant theory gives the moduli at good points as well as the compactifications, namely the moduli of vector bundles on a curve.

Vector bundles on a curve (cf. [42],[52],[64]).

Let X be a smooth projective curve of genus $g \geq 2$ over an arbitrary algebraically closed field k. Fix an ample line bundle $\mathcal{O}_X(1)$ on X.

DEFINITION 3.7 (Mumford, cf. [42]): A vector bundle V on X is semi-stable (resp. stable) if $\forall$ sub-bundle (resp. proper sub-bundle) W of V we have

$$\mu(W) = \frac{\deg W}{\mathrm{rk}\, W} \leq \mu(V) = \frac{\deg V}{\mathrm{rk}\, V} \quad (\text{resp. } <)$$

where for example deg W is the degree of the vector bundle W (i.e., the degree of the line bundle $\overset{r}{\Lambda} W$, $r = \mathrm{rk}\, W$).

Let $\Sigma^{ss}_{n,d} = \Sigma^{ss}$ denote the category of semi-stable vector bundles of rank n and degree d and Σ^s the subcategory consisting of stable vector bundles. Then $\forall\ V \in \Sigma^{ss}$, $\mu(V) = \frac{d}{n}$. It can be shown easily that the category Σ^{ss} is abelian and in fact that every homomorphism $f: V \longrightarrow W$ in Σ^{ss} is of constant rank at every point and that moreover if one of V,W is stable, then either f is zero or an isomorphism. From these facts one deduces easily that given $V \in \Sigma^{ss}$, $\exists$ a Jordan-Holder series of sub-bundles V_i of V

$$V_0 \subset V_1 \subset V_2 \cdots \subset V_n = V$$

such that $V_0, V_1/V_0, \cdots, V/V_{n-1}$ are all stable and besides $\mu(V_0) = \mu(V_1/V_0) = \cdots = \mu(V/V_{n-1}) = \frac{n}{d}$. If we set

$$\mathrm{gr}\, V = V_0 \oplus V_1/V_0 \oplus \cdots \oplus V/V_{n-1}$$

then it can be seen that $\mathrm{gr}\,V$ is uniquely determined upto isomorphism. Let us introduce the equivalence relation $V_1 \sim V_2$, $V_i \in \Sigma^{ss}$ if $\mathrm{gr}\,V_1 = \mathrm{gr}\,V_2$ (i.e., the associated gradeds are isomorphic) and denote by $M = M_{n,d}$ the set of equivalence classes in Σ^{ss} under this equivalence relation. We see that if $V \in \Sigma^s$ then $\mathrm{gr}\, V = V$ and that if $V_1, V_2 \in \Sigma^s$ then $V_1 \sim V_2$ if and only if V_1 is isomorphic to V_2. Hence if M^s denotes the subset of M whose representatives come from stable bundles, M^s is in fact the set of isomorphism classes of stable vector bundles of rank n and degree d. Then we have

THEOREM 3.10 (cf. [51],[64]): (i) $\exists$ a natural structure of a normal projective variety on $M = M_{n,d}$ of dimension $n^2(g-1)+1$. Besides M^s is an open smooth subvariety of M. If $(n,d) = 1$, $M = M^s$.

(ii) If the base field is $\mathbb{C}$, the underlying topological space of $M_{n,d}$ can be identified with the isomorphism classes of unitary representations of rank n and of certain type of a certain Fuchsian group, for example if $d = 0$, the underlying topological space of $M_{n,o}$ can be identified with isomorphism classes of unitary representations of the curve X.

Thus in this moduli problem, M^s corresponds to the good part and M is the compactification of M^s.

Let us now sketch briefly a proof of this theorem. One shows first that the category Σ^{ss} is <u>bounded</u>, i.e., $\exists$ m_o such that for $m \geq m_o$ we have

(i) $V(m)$ is generated by the global sections $H^o(V(m))$

(ii) $H^i(V(m)) = 0$, $i > 0$.

Let us now denote by $\Sigma^{ss}(m_o)$ (resp. $\Sigma^s(m_o)$) the category of vector bundles $V(m_o)$ where $V \in \Sigma^{ss}$. We see that $\Sigma^{ss}(m_o) = \Sigma^{ss}_{n,d'}, \Sigma^s(m_o) = \Sigma^s_{n,d'}$ where $d' = d + m_o \deg. \mathcal{O}_X(1)$. We see that the study of Σ^{ss} is equivalent to that of $\Sigma^{ss}(m_o)$. Note that all the vector bundles in $\Sigma^{ss}(m_o)$ have the same Hilbert polynomial, denote this by P. The dimension of $H^o(V(m_o))$ is independent of $V \in \Sigma^{ss}$, denote this by p. Let E denote the trivial vector bundle (or the associated sheaf) on X of rank p. Then (i) above shows that every $V(m_o)$, $V \in \Sigma^{ss}$ has the following quotient representation

$$E \longrightarrow V(m_o) \longrightarrow 0.$$

Let $Q(E/P) = Q$ denote the Quot scheme (in the sense of Grothendieck) of all coherent quotients with Hilbert polynomial P. The projective group PGL(E) operates canonically on Q. Note that PGL(E) $\approx$ the projective group on a vector space of rank p.

Define the subset R of Q consisting of $q \in Q$ such that the associated quotient F_q of E has the following properties:

(i) F_q is locally free, i.e., corresponds to a vector bundle

(ii) the canonical map $H^o(E) \longrightarrow H^o(F_q)$ is an isomorphism and $H^i(F_q) = o$, $i > o$.

It is easy to see that R is a PGL(E) stable open subscheme of Q. Further by using the local smoothness criterion for the functor Q (similar to that of Hilb, cf. (viii), Remarks 2.1) it can be shown that R is smooth of dimension $= n^2(g-1)+1+(p^2-1)$ ($p^2-1 = \dim. PGL(E)$). For $q \in R$ we can write the associated vector bundle (which is a quotient of E) in the form $V_q(m_o)$ where V_q is a vector bundle of rank n and degree d. Let R^{ss} (resp. R^s) denote the subset of points $q \in R$ such that $V_q \in \Sigma^{ss}$ (resp. Σ^s), i.e., V_q is semi-stable (resp. stable). We shall see that these are PGL(E) stable open subsets of R. It can also be shown that R^{ss} is contained in a PGL(E) stable irreducible open subset of R. This comes from the fact that if V is a vector bundle generated by global sections, $\exists$ a trivial sub-bundle of rank = rk V-1 so that in particular for $q \in R^{ss}$ we have

$$(*) \qquad o \longrightarrow I_{n-1} \longrightarrow V_q(m_o) \longrightarrow L \longrightarrow o,$$

where I_{n-1} denotes the trivial vector bundle of rank(n-1) and L is a line bundle of degree d. Take all extensions of type (*) where L varies over J_d - the variety of line bundles of degree d $\approx$ J - Jacobian variety of X. Fixing L, the extensions of type (*) can be parametrized by the affine space. Hence the extensions of type (*) can be parametrized by an irreducible variety from which the required assertion follows easily.

Let us now take an ordered set of N points $P_1,\cdots,P_N$ on X. Let $H_{p,n}$ denote the Grassmannian of n-dimensional quotient linear spaces of $H^o(E)$ ($p = \dim H^o(E)$). For $q \in R$, $V_q(m_o)$ is a quotient of E and hence the fibre of $V_q(m_o)$ at P_i can be canonically identified with an element of $H_{p,n}$. Hence doing this at every P_i, we get a canonical map

$$\varphi\colon R \longrightarrow H^N_{p,n} \quad \text{(N-fold product of } H_{p,n}).$$

It is easily checked that for the diagonal action of PGL(E) on $H^N_{p,n}$, φ is a PGL(E) (or, to be correct, PGL($H^o(E)$)) morphism . Let $Z = H^N_{p,n}$. We take the ample line bundle on Z by tensoring the ample generators of the factors. We denote by Z^{ss} (resp. Z^s) the semi-stable (resp. stable) points of Z for the canonical diagonal linear action of SL(E).

The crucial step is now the following (cf. [64] and Lemma 2, [65]): The choice of m_o and points $P_1,\cdots,P_N$ can be so made such that

(i) for $q \in R$, $\varphi(q) \in Z^{ss}$ (resp. Z^s) $\Longleftrightarrow q \in R^{ss}$ (resp. R^s).

In particular since Z^{ss}, Z^s are open, it follows that R^{ss} and R^s are open in R.

(ii) the induced morphism

$$\varphi: \ R^{ss} \longrightarrow Z^{ss}$$

is a <u>closed</u> immersion.

In the checking of these facts the explicit determination of stable and semi-stable points on Z (cf. §4, Chap. 4, [44]) is used.

From Theorem 3.5, it follows that Z^{ss} has a good projective quotient mod PGL(E) and that its restriction to Z^s is a geometric quotient. From (ii) and the fact $Z^s \cap R^{ss} = R^s$ it follows that we have a <u>good</u> projective quotient mod PGL(E)

$$f: \ R^{ss} \longrightarrow M'$$

that $f(R^s) = (M')^s$ is open in M' and that $f: R^s \longrightarrow (M')^s$ is a geometric quotient. To identify M' with $M_{n,d} = M$ it suffices to show that for $q_1, q_2 \in R^{ss}$ the orbit closures intersect, i.e., $\overline{O(q_1)} \cap \overline{O(q_2)} \neq \emptyset \Longleftrightarrow \operatorname{gr} V_{q_1}(m_o) = \operatorname{gr} V_{q_2}(m_o)$. This can be shown without much difficulty. Besides one can show that $R^s \neq \emptyset$. Thus the proof of (i) in Theorem 3.7 follows.

To prove (ii) (cf. [52]), for example in the case d = 0 one shows first that a vector bundle associated to a unitary representation of the fundamental group is semi-stable of degree zero and that it is stable if the representation is irreducible. Let U be the subset of R^s coming from (i.e., isomorphic to vector bundles defined by) irreducible unitary representations. It is seen easily that it suffices to show that $R^s = U$. First of all one sees that $U \neq \emptyset$. From the fact that the set of all unitary representations of the fundamental group is compact it follows easily that U is <u>closed</u> in R^s. Since R^s is a variety (in particular <u>connected</u>), it suffices to show that U is <u>open</u> in R^s. This comes from the fact that if we have a family of vector bundles $\{V_t\}_{t \in T}$ on X and for one point t_o, V_{t_o} is (isomorphic to) a unitary vector bundle, then $\exists$ a neighbourhood (for the usual topology, not the Zariski one; note that we are over $\mathbb{C}$) D of t_o such that V_t is unitary. This fact is proved by showing that the topological dimension of the local moduli space for unitary representations of the fundamental group at an irreducible unitary representation is = 2·complex dimension of the local moduli space of the corresponding holomorphic vector bundle (cf. [51]).

This concludes an outline of the proof of Theorem 3.10.

REMARK 3.7: Let Γ be a discrete subgroup of $SL(2,\mathbb{R})$ so that Γ operates canonically on the upper half plane H. Suppose that $Y = H \bmod \Gamma$ has <u>finite</u> measure so that we have a compact Riemann surface X such that X-Y = a finite number of points. Then it can be shown that on the isomorphism classes of unitary representations of Γ of rank n and whose restrictions to the isotropy groups at parabolic and elliptic points lie in fixed conjugacy classes, there is a natural structure of a normal projective variety. The algebraic moduli problem posed by this question on X (and which can be defined abstractly on a smooth projective curve in any characteristic) is the following: Fix a finite number of points on the curve X (these points correspond to the elliptic and parabolic points), say one point P on X. Consider the category of vector bundles V on X together with an additional structure, namely a <u>flag</u> in the fibre V_P of V at P. It is easy to see what we understand by morphisms in this category. To define stability (and semi-stability) in this category we attach certain weights to this flag and the total structure we call a <u>parabolic structure</u> on V. We define degrees using this weight and the usual degree, which we call the parabolic degree. Using this we imitate the definitions of stability and semi-stability above and obtain the notion of stability and semi-stability in this category. One can prove the existence of these moduli varieties as a nice application of geometric invariant theory (cf. [68]).

REMARK 3.8: Geometric invariant theory can also be used to prove the existence of Picard varieties; these methods turn out particularly to be powerful when studying Picard schemes, i.e.,when we have a family (cf. [19]). The study of canonical compactifications of generalized Jacobians can also be done using geometric invariant theory. When we have a family of curves $f: X \longrightarrow S$ such that f is proper, flat, of relative dimension 1 and all the fibres are <u>irreducible</u>, say with ordinary double points, $\exists$ a natural compactification of the generalized Jacobian scheme (cf. [25],[40],[13]). The situation when the fibres are not necessarily irreducible, turns out to be quite interesting and in this case there could be many natural compactifications (cf. [58]).

REMARK 3.9: One can define semi-stable and stable vector bundles in the higher-dimensional case. In the case of a smooth projective curve X with an ample $O_X(1)$, the definition of semi-stability (resp. stability) of a vector bundle V is equivalent to

$$\frac{\chi(W(m))}{\text{rk } W} \leq \frac{\chi(V(m))}{\text{rk } V} \quad \forall \text{ sub-bundle (resp. proper sub-bundle) } W \text{ of } V.$$

Thus if X is a higher-dimensional projective variety with an ample $\mathcal{O}_X(1)$, one can define stability and semi-stability (with respect to $\mathcal{O}_X(1)$) as above. The study of these moduli has already begun (cf. [33],[35],[71]).

One can also ask the question of moduli on curves of principal fibre spaces with structure group more general than the linear group. This study has also started (cf. [62]).

REMARK 3.10: We have been working with an algebraically closed base field only for simplicity. All the above constructions (moduli of curves, polarized abelian varieties, vector bundles on curves etc.) can be carried out over a general base. First of all, one observes that geometric invariant theory can be done over a general base. For example, Theorem 3.3 generalizes to arbitrary base S as follows. Let G be a reductive group scheme over S (assume for simplicity $S = \text{Spec } B$), i.e., G is a smooth group scheme over S and the fibres are reductive algebraic groups of the same dimension. Let us be given a rational representation of G on an affine space $\mathbb{A}^n = \text{Spec } B[X_1,\cdots,X_n]$ over S. Then given a geometric point $x \in \mathbb{A}^n$, i.e., $x \in \mathbb{A}^n(k)$ (k an algebraically closed field) such that $\overline{O(x)}$ (closure of the $G\times_S k$ orbit in $\mathbb{A}^n\times_S k$) does not pass through (0), then $\exists$ a homogeneous G-invariant polynomial $F \in B[X_1,\cdots,X_n]$ such that $F(x) \neq 0$ (G-invariance could be defined via coalgebras or using invariance with respect to T-valued points $G(T)$ $\forall$ S-scheme T). One can then prove the analogues of Theorem 3.4 and Theorem 3.5. Using these theorems, going along the same lines as above one can prove the existence of the moduli spaces considered above over general base.

REMARK 3.11: We have outlined the methods of existence of several global moduli problems using geometric invariant theory. We have not touched upon those interesting aspects concerning the detailed study of these moduli varieties.

It is not necessary here to cite (in this respect) the extensive literature on Jacobian varieties (moduli of line bundles) and the moduli of curves and polarized abelian varieties. Concerning these moduli we content ourselves by mentioning the paper of Igusa (cf. [26]) on the explicit determination of moduli of curves for genus 2. As for the moduli of vector bundles on curves, one should mention the following papers ([21],[22],[45],[48],[49],[50],[54],[55],[56], and [61]).

REFERENCES

1. M. Artin - Algebraic approximation of structures over complete local rings - Volume dedicated to Professor Zariski - Publications Math. I.H.E.S., No. 38, p. 23-58

2. _______ - Algebraization of formal moduli I - A collection of mathematical papers published in honour of K. Kodaira - Tokyo University Press

3. W.L. Baily, Jr., and A. Borel - Compactification of arithmetic quotients of bounded symmetric domains - Ann. of Math. 84 (1966), p. 442-528

4. Spencer Bloch - Semi-regularity and deRham cohomology - Inventiones Math., 17 (1972), p. 51-66

5. E. Bombieri - Canonical models of surfaces of general type - Publications Math. I.H.E.S., No. 42, p. 171-219

6. E. Bombieri and D. Mumford - Enriques classification of surfaces (to appear)

7. A. Borel - Linear representations of semi-simple algebraic groups - Arcata lectures

8. D.M. Burns, Jr. and J.M. Wahl - Local contributions to global deformations of surfaces - Inventiones math., 26, (1974), p. 67-88

9. M. Cornalba and P. Griffiths - Arcata lectures

10. P. Deligne and D. Mumford - The irreducibility of the space of curves of given genus - Publications Math. I.H.E.S. No. 36

11. A. Douady - Exposé 277, Sem. Bourbaki (1964-65)

12. A. Douady and J. Hubbard - Le probleme des modules pour les espaces analytiques compactes - Comptes Rendus de l'academie des Sciences, Paris 1973 and to appear in Annales de l'Ecole normale Superieure

13. Cyril D'Souza - Compactification of generalized Jacobians (thesis to appear)

14. R. Elkik - Solutions d'equations a coefficients dans un anneau henselien - Ann. Scient. Ec. Norm. Sup., 4^{e} Serie t. 6 (1973), p. 553-604

15. H. Grauert - Über die Deformation isolierter Singularitäten analyticher Mengen - Inventiones math., 15 (1972), p. 171-198

16. H. Grauert - Der Satz von Kuranishi für compakte Komplexer Räume - Inventiones math., 25 (1974), p. 107-142.

17. P. Griffiths - Periods of integrals on algebraic manifolds: Summary of main results and discussion of problems - Bulletin of the AMS, vol. 76, No. 2, p. 228-296

18. P. Griffiths - Periods of integrals of algebraic manifolds - Publications Math. I.H.E.S., No. 38, p.

19. A. Grothendieck - Fondement de la gèometrie algèbrique - collected Bourbaki talks, Extraits du Sèminaire Bourbaki, 1957-1962, Paris 1962

20. W. J. Haboush - Proof of Mumford's conjecture (to appear)

21. G. Harder - Eine Bemerkung zu einer Arbeit von P. Newstead - Jour. für math. 242 (1970), p. 16-25

22. G. Harder and M.S. Narasimhan - On the cohomology groups of moduli spaces of vector bundles on curves (to appear)

23. F. Hirzebruch and K. Kodaira - On the complex projective spaces - Jour. de math., 36 (1957), p. 201-216

24. E. Horikawa - On deformations of quintic surfaces - Proc. Japan Acad., 49 (1973), p. 377-379

25. J. Igusa - Fibre systems of Jacobian varieties - Amer. J. Math., 78, (1956), p. 171-199

26. ________ - Arithmetic variety of moduli for genus two - Ann. of math., vol. 72 (1960), p. 612-649

27. A. Kas - On obstructions to deformations of complex analytic surfaces - Proc. Nat. Acad. Sci. USA 58 (1967), p. 402-404

28. ______ - Ordinary double points and obstructed surfaces (to appear).

29. K. Kodaira and D.C. Spencer - On deformations of complex analytic structures I, II - Ann. math., vol. 67 (1958), p. 328-466

30. ______________________ - A theorem of completeness of characteristic systems of complete continuous systems - Amer. J. Math. (1959)

31. K. Kodaira, L. Nirenberg and D.C. Spencer - On the existence of deformations of complex analytic structures - Ann. of Math., vol. 68 (1958), p. 450-459

32. Kuranishi - On the locally complete families of complex analytic structures - Ann. of Math., vol. 75 (1962), p. 536-577

33. Stacy Langdon - Valuative criteria for families of vector bundles (to appear)

34. D. Liebermann and D. Mumford - Proof of Matsusaka's big theorem - Arcata lectures

35. M. Maruyama - Stable vector bundles on an algebraic surface - to appear

36. T. Matsusaka - Polarized varieties, fields of moduli and generalized Kummer varieties of polarized abelian varieties - Amer. J. Math., vol. LXXX (1958), p. 45-82

37. ___________ - Algebraic deformations of polarized varieties - Nagoya math. J., vol. 31 (1968), p. 185-245

38. ___________ - Polarized varieties with a given Hilbert polynomial - Amer. J. Math., vol. 75 (1972), p. 1027-1077

39. T. Matsusaka and D. Mumford - Two fundamental theorems on deformations of polarized varieties - Amer. J., vol. LXXXVI (1964), p. 668-684

40. D. Mumford - Further comments on boundary points - mimeographed notes, Woods Hole Summer Institute, 1964

41. _________ - Further pathologies in algebraic geometry - Amer. J. 84 (1962), p. 642-647

42. _________ - Projective invariants of projective structures and applications - Proc. Intern. Cong. Math., Stockholm (1962), p. 526-530

43. _________ - Lectures on curves on an algebraic surface - Annals of Math. Studies, No. 59, Princeton Univ. Press (1966)

44. _________ - Geometric invariant theory - Springer Verlag (1965)

45. D. Mumford and P.E. Newstead - Periods of a moduli space of bundles on curves - Amer. J. Math. 90 (1968), p. 1201-1208

46. M. Nagata - Invariants of a group in an affine ring - Jour. math. Kyoto Univ. 3 (1963), p. 369-377.

47. ________ - Note on orbit spaces - Osaka math. J., 14, 21(1962)

48. M.S. Narasimhan and S. Ramanan - Moduli of vector bundles on a compact Riemann surfaces - Ann. of Math. 89 (1969), p. 19-51

49. ______________________________ - Vector bundles on curves - Bombay Colloq. alg. geometry, Oxford Univ. Press (1969), p. 335-346

50. ______________________________ - Some applications of Hecke correspondences (to appear)

51. M.S. Narasimhan and C.S. Seshadri - Holomorphic vector bundles on a compact Riemann surface - Math. Annalen, 155 (1964), p. 69-80

52. ______________________________ - Stable and unitary bundles on a compact Riemann surface - Ann. of math., 82 (1965), p. 540-567

53. M.S. Narasimhan and R.R. Simha - Manifolds with ample canonical class - Inventiones math. 5 (1968), p. 120-128

54. P.E. Newstead - Topological properties of some spaces of stable bundles, Topology 6(1967), p. 241-262

55. P.E. Newstead - Stable bundles of rank 2 and odd degree over a curve of genus 2 - Topology 7 (1968), p. 205-215

56. ____________ - Rationality of moduli spaces of stable bundles of rank 2 (to appear)

57. T. Oda - On Mumford conjecture concerning reducible rational representations of algebraic linear groups - Jour. math. Kyoto Univ. 3, 3(1964)

58. T. Oda and C.S. Seshadri - Compactifications of generalized Jacobians (to appear)

59. H. Popp - On moduli of algebraic varieties I - Inventiones math. 22, (1973), p. 1-40

60. ______ - On moduli of algebraic varieties II - Compositio math., vol. 28 (1974), p. 51-81

61. S. Ramanan - The moduli space of vector bundles over an algebraic curve, Math. Annalen 200 (1973), p. 69-84

62. A. Ramanathan - Stable principal bundles on a compact Riemann surface (to appear)

63. M. Schlessinger - Functors on Artin rings - Trans. Amer. Math. Soc., vol. 130 (1968), p. 208-222

64. C. S. Seshadri - Space of unitary vector bundles on a compact Riemann surface - Ann. of math. 82 (1965), p. 303-336

65. ______________ - Mumford's conjecture for GL(2) and applications, Proc. of the Bombay Colloq. on alg. geometry, Oxford Univ. Press (1969), p. 347-371

66. ______________ - Quotient spaces modulo reductive algebraic groups - Ann. of math., vol. 95 (1972), p. 511-556

67. ______________ - Moduli of π-vector bundles on an algebraic curve - Questions on algebraic varieties, C.I.M.E. (1969), III CICLO, p. 141-260

68. ______________ - Moduli of vector bundles with parabolic structures on curves (to appear)

69. R. Steinberg - Representations of algebraic groups - Nagoya math. J., vol. 22 (1963), p. 33-56

70. Sundararaman - Kuranishi spaces for vector bundles (to appear)

71. F. Takemoto - Stable vector bundles on algebraic surfaces - Nagoya math. J., 47(1972)

72. Toroidal Embeddings I (Kempf et al) - Springer lecture notes, 339

73. A. Tjurin - Analogs of Torelli's theorem for multi-dimensional vector bundles over an algebraic curve - Izv. Akad. Nauk SSSR, Ser. Mat. Tom 34(2)(1970), 338-365, Math. U.S.S.R. Izv. vol. 4(2), 343-370

74. A. Weil - On the theory of complex multiplication - Proceedings of the International symposium on algebraic number theory (1955), p. 9-22

Tata Institute of Fundamental Research

Harvard University

PART II: SEMINAR TALKS

Proceedings of Symposia in Pure Mathematics
Volume 29, 1975

LARSEN'S THEOREM ON THE HOMOTOPY GROUPS OF PROJECTIVE MANIFOLDS OF SMALL EMBEDDING CODIMENSION

Wolf Barth

ABSTRACT

A simplified version is presented of Larsen's proof for his theorem: If $A \subset \mathbb{P}_n(\mathbb{C})$ is algebraic, nonsingular, of pure dimension a, then $\pi_i(\mathbb{P}_n, A) = 0$ for $i \leqslant 2a-n+1$.

1. INTRODUCTION

Let $H \subset \mathbb{P}_n(\mathbb{C})$ be a complex-algebraic closed hypersurface (with or without singularities). Then H can be viewed as a hyperplane section on $\mathbb{P}_n$, so we have the

Lefschetz theorem [5, theorem 7.4]: The relative homotopy groups $\pi_i(\mathbb{P}_n, H)$ vanish for $i \leqslant n-1$.

Larsen's result is intermediate between this theorem and the well known fact that a subvariety $A \subset \mathbb{P}_n$ is connected, if its dimension is at least n/2 in all of its points:

THEOREM [4]: Let $A \subset \mathbb{P}_n$ be a closed, complex-algebraic submanifold of pure dimension a. Assume $s := 2a-n \geqslant 0$. Then

$$\pi_i(\mathbb{P}_n, A) = 0 \text{ for } i \leqslant s+1.$$

Contrary to the case of a hypersurface, the nonsingularity of A is essential here. (The union of three 3-planes in $\mathbb{P}_5$ in general position for example is not simply-connected.) I even do not know, whether the theorem remains valid for a locally complete intersection A.

By the relative Hurewicz theorem [6, Chap. 7.5, theorem 4], to prove Larsen's theorem, it suffices to show in the above situation

(a) $\pi_2(\mathbb{P}_n, A) = 0$ if $2a \geqslant n+1$

and

(b) $H_i(\mathbb{P}_n, A; \mathbb{Z}) = 0$ for $0 \leqslant i \leqslant s+1$.

By the universal coefficient theorem

AMS (MOS) subject classifications (1970). Primary 14F25, 14F35; Secondary 32F10, 57D70.

$$0 \longrightarrow \mathrm{Tor}(H_{i-1}(\mathbb{P}_n, A; \mathbb{Z})) \longrightarrow H^i_c(\mathbb{P}_n \setminus A; \mathbb{Z}) \longrightarrow \mathrm{Hom}(H_i(\mathbb{P}_n, A; \mathbb{Z}), \mathbb{Z}) \longrightarrow 0$$

the latter is equivalent to

(b') $\quad H^i_c(\mathbb{P}_n \setminus A, \mathbb{Z}_m) = 0$ for $0 \leq i \leq s+1$ and all primes m.

I shall give here the proof for (a) from [2] and (b') from [4] in a somewhat smoother form.

As to the notation, I shall not write the basepoint in $\pi_i(X,A)$. Also, if I use coefficients $\mathbb{Z}_m$ for cohomology, I shall suppress them. So $H^i(X)$ means cohomology with values in some $\mathbb{Z}_m$ and $R^i\varphi_*$ is a direct image sheaf of some constant sheaf $\mathbb{Z}_m$.

2. REDUCTION TO A LOCAL STATEMENT ABOUT A.

Put $G = SU(n+1, \mathbb{C})$ and write the operation of this group on $\mathbb{P}_n$ as $(\sigma, x) \longrightarrow \sigma x$, where $\sigma \in G$, $x \in \mathbb{P}_n$. Let $A \subset \mathbb{P}_n$ be as above and define two maps

$$\begin{array}{ccc} G \times A & \xrightarrow{\varphi} & \mathbb{P}_n \\ \downarrow \psi & & \\ G & & \end{array}$$

by $\varphi(\sigma, x) = \sigma x$ and $\psi(\sigma, x) = \sigma$, where $\sigma \in G$, $x \in A$. The map φ is G-equivariant, so it is locally trivial. Also $\varphi^{-1}A \subset G \times A$ is a submanifold, and ψ is proper.

The local statement to be proved in §4 is this

MAIN LEMMA: There exist arbitrarily small closed balls $B \subset G$ around the unit $1 \in G$ with the property: For all $\sigma \in G$

a) $\quad \pi_1(\psi^{-1}(B\sigma) \cap \varphi^{-1}A, \psi^{-1}(\sigma) \cap \varphi^{-1}A) = 0$,

b) $\quad H^i_c(\psi^{-1}(B\sigma) \cap \varphi^{-1}A \setminus \psi^{-1}(\sigma) \cap \varphi^{-1}A) = 0$ if $0 \leq i \leq s$.

The kind of neighborhood B used here will also be described later. $B\sigma$ is the ball B translated on the right by $\sigma \in G$. It is important that B does not shrink, if σ varies, because we shall need the following criterion for certain image sheaves to be constant.

Criterion: Let be S a sheaf on $G \times A$. Assume that there exist arbitrarily small balls B around $1 \in G$ such that

i) $\quad$ For each $\sigma \in G$ the restriction morphism $H^q(\psi^{-1}(B\sigma), S) \longrightarrow$ $\longrightarrow (R^q\psi_* S)_\sigma$ is bijective. Then $R^q\psi_* S$ is a constant sheaf.

ii) $\quad$ For each $\sigma \in G$ the restriction morphism $H^q(\psi^{-1}(B\sigma), S) \longrightarrow$ $\longrightarrow (R^q\psi_* S)_\sigma$ is injective. Then $\Gamma(G, R^q\psi_* S) \longrightarrow (R^q\psi_* S)_\sigma$ is injective.

Proof: i) Let $I(B, \sigma) \subset \Gamma(B\sigma, R^q\psi_* S)$ be the image of $H^q(\psi^{-1}(B\sigma), S)$. If $\sigma' \in G$ is sufficiently near to σ, then we can find balls B_1, B_2, B_1', and B_2' such that $B_1'\sigma' \subset B_1\sigma \subset B_2'\sigma' \subset B_2\sigma$. Then all the maps in the chain

$$I(B_2, \sigma) \longrightarrow I(B_2', \sigma') \longrightarrow I(B_1, \sigma) \longrightarrow I(B_1', \sigma')$$

have to be bijective. So $I(B_2, \sigma) \longrightarrow (R^q\psi_* S)_{\sigma'}$ is bijective too, which means that the sheaf $R^q\psi_* S$ is locally constant near σ. Since G is simply-connected, this sheaf is constant. (Note, that for the case $q = 0$, we might have started with a sheaf S of sets.)

The proof of ii) is very similar.

The proof of (a): In this case the main lemma reads

$$\pi_1(\psi^{-1}(B\sigma)\cap\varphi^{-1}A,\ \psi^{-1}(\sigma)\cap\varphi^{-1}A) = 0.$$

We claim that $\pi_1(\varphi^{-1}A, \{1\}\times A) = 0$. In fact, let $\gamma: C \longrightarrow \varphi^{-1}A$ be a (connected, non-ramified) covering of $\varphi^{-1}A$ corresponding to that subgroup of $\pi_1(\varphi^{-1}A)$, which is the image of $\pi_1(\{1\}\times A)$ under the inclusion $\{1\}\times A \longrightarrow$ $\longrightarrow \varphi^{-1}A$. View γ as a sheaf of sets, i.e., if $U \subset \varphi^{-1}A$ is open, $\Gamma(U, \gamma)$ is the space of continuous sections $U \longrightarrow C$ over U. We can form (as a sheaf of sets) the direct image sheaf $(\psi|\varphi^{-1}A)_*\gamma$. By the vanishing of the relative fundamental group above, $\Gamma(\psi^{-1}(B\sigma)\cap\varphi^{-1}A, \gamma)$ maps bijectively on the stalk $((\psi|\varphi^{-1}A)_*\gamma)_\sigma$. We apply our criterion to find that the sheaf $(\psi|\varphi^{-1}A)_*\gamma$ is constant. By construction of γ, the covering $\gamma|\{1\}\times A$ totally decomposes into sections, each of which can now be extended to a section over $\varphi^{-1}A$. This implies that the covering γ was trivial, i.e., $\pi_1(\{1\}\times A) \longrightarrow \pi_1(\varphi^{-1}A)$ was surjective.

Now use the exact homotopy sequence

$$\dots \longrightarrow \pi_2(G\times A,\ \{1\}\times A) \longrightarrow \pi_2(G\times A,\ \varphi^{-1}A) \longrightarrow \pi_1(\varphi^{-1}A,\ \{1\}\times A) \longrightarrow \dots$$

of the triple $\{1\}\times A \subset \varphi^{-1}A \subset G\times A$ and the vanishing of $\pi_2(G\times A, \{1\}\times A)$ $\simeq \pi_2(G)$ to find $\pi_2(G\times A, \varphi^{-1}A) \simeq \pi_2(\mathbb{P}_n, A) = 0$.

The proof of (b'): Let me put $B^* := B\setminus\{1\}$ and thus $B^*\sigma = \{\tau\sigma:$ $1 \neq \tau\in B\}$. The main lemma, part b) now reads $H^i_c(\psi^{-1}(B^*\sigma)\cap\varphi^{-1}A) = 0$ for $0 \leq i \leq s$. Since $H^i_c(\psi^{-1}(B^*\sigma)) = 0$ for all i, from the exact sequence

$$H^i_c(\psi^{-1}(B^*\sigma)) \longrightarrow H^i_c(\psi^{-1}(B^*\sigma)\cap\varphi^{-1}A) \longrightarrow H^{i+1}_c(\psi^{-1}(B^*\sigma)\setminus\varphi^{-1}A) \longrightarrow H^{i+1}_c(\psi^{-1}(B^*\sigma))$$

follows that $H^i_c(\psi^{-1}(B^*\sigma)\setminus\varphi^{-1}A) = 0$ for $0 \leq i \leq s+1$.

I claim that then $H^q_c(G\times A\setminus\varphi^{-1}A) = 0$ for $0 \leq q \leq s+1$. In fact, put $\tilde\psi := \psi|G\times A\setminus\varphi^{-1}A$ and let $R^q_c\tilde\psi_*$ be the direct image sheaves (of a constant sheaf $\mathbb{Z}_m$ over $G\times A\setminus\varphi^{-1}A$) with fibre-wise compact support, i.e., for each $\sigma\in G$, the stalk of this sheaf in σ is just the inductive limit over all $H^q_c(\tilde\psi^{-1}V)$, where V runs through the closed neighborhoods of σ in G. By [3, Chap. IV, prop. 4.2] the restriction maps $(R^q_c\tilde\psi_*)_\sigma \longrightarrow H^q_c(\tilde\psi^{-1}\sigma)$, $\tilde\psi^{-1}\sigma = \{\sigma\}\times A\setminus\varphi^{-1}A$, are isomorphisms. Furthermore, there is a Leray spectral sequence [3, Chap IV, theorem 6.1]

$$E_2^{p,q} = H^p(G,\ R^q_c\tilde\psi_*) \Longrightarrow H^{p+q}_c(G\times A\setminus\varphi^{-1}A).$$

So it suffices to show that $E_2^{p,q} = 0$ for $p+q \leq s+1$. But the vanishing of $H^i_c(\psi^{-1}(B^*\sigma)\setminus\varphi^{-1}A)$ means that the restriction morphism $H^q_c(\tilde\psi^{-1}(B\sigma)) \longrightarrow$ $\longrightarrow H^q_c(\tilde\psi^{-1}(\sigma)) = (R^q_c\tilde\psi_*)_\sigma$ is bijective if $0 \leq q \leq s$, and still injective if $q = s+1$. We apply our criterion to find that the sheaves $R^q_c\tilde\psi_*$ are constant in the range $q \leq s$ and that the morphism $E_2^{0,s+1} = H^0(G, R^{s+1}_c\tilde\psi_*)$ $\longrightarrow (R^{s+1}_c\tilde\psi_*)_1$ is injective. Now the stalks $(R^q_c\tilde\psi_*)_1$ at $1\in G$ vanish for all q, because $\{1\}\times A \subset \varphi^{-1}A$. This implies $E_2^{p,q} = 0$ for $p+q \leq s+1$ and the vanishing of $H_c^q(G \times A \setminus \varphi^{-1}A)$, $0 \leq i \leq s+1$, as claimed.

Put now $\tilde{\varphi} := \varphi | G \times A \setminus \varphi^{-1}A$. Let F be the typical fibre of the bundle φ over $\mathbb{P}_n$. The direct image sheaves $R^q\varphi_*$ are locally constant with stalk $H^q(F)$. Since $\mathbb{P}_n$ is simply connected, they are constant, and so are their restrictions $R^q\varphi_* | \mathbb{P}_n \setminus A \simeq R^q\tilde{\varphi}_*$. We use the Leray spectral sequence

$$E_2^{pq} = H_c^p(\mathbb{P}_n \setminus A,\ R^q\tilde{\varphi}_*) \Longrightarrow H_c^{p+q}(G \times A \setminus \varphi^{-1}A),$$

where the abutment terms vanish in dimensions $\leq s+1$. Since the coefficient group $\mathbb{Z}_m$ is a field, we have

$$E_2^{pq} = H_c^p(\mathbb{P}_n \setminus A,\ H^q(F)) = H_c^p(\mathbb{P}_n \setminus A) \otimes H^q(F).$$

Now it is easy to see that this special form of the E_2^{pq} - terms together with the vanishing of the abutment terms implies $E_2^{pq} = 0$ for $q \leq s+1$, i.e., $H_c^p(\mathbb{P}_n \setminus A) = 0$ for $0 \leq p \leq s+1$.

This proves the theorem.

3. MORSE THEORY AND q-PSEUDOCONVEXITY

Let me recall here the theory necessary to prove the main lemma.

Let X be a real $\mathcal{C}^\infty$-manifold and f some real $\mathcal{C}^\infty$-function on X. A point $x \in X$ is called critical point of f, if $df(x) = 0$. Then the Hessian of f

$$H_f(x) = \Sigma\, f_{x_i x_j}(x)\, dx_i\, dx_j$$

is well-defined as a symmetric form on the tangent space, independently of the real coordinates x_i chosen. In particular, $\text{Index}_{\mathbb{R}}(f,x) :=$ maximal dimension of a subspace, on which the Hessian is negative definite, is well defined. The critical point x is called nondegenerate, if $H_f(x)$ is nondegenerate. The $\mathcal{C}^\infty$-function f is called a Morse-function, if all its critical points are nondegenerate (and therefore isolated).

Theorem 1 [5, remark 3.3, p.19]: Let f be a Morse-function on X, mapping X properly and surjectively on the interval (a,b). For $\alpha \in (a,b)$ put $X_\alpha := f^{-1}(a,\alpha]$. Assume

$$\text{Index}_{\mathbb{R}}(f,x) \geq s+1$$

in all critical points x of f. Then for all $\alpha < \beta$, X_β has the homotopy type of X with finitely many cells of dimensions $\geq s+1$ attached. In particular

$$\pi_i(X_\beta, X_\alpha) = H_c^i(X_\beta \setminus X_\alpha) = 0 \quad \text{for } 0 \leq i \leq s.$$

Theorem 2 [5, Cor. 6.8, p.37]: Let f be a bounded $\mathcal{C}^\infty$-function on X, $K \subset X$ compact, and k a natural number. Then f can be uniformly, over K even in the $\mathcal{C}^k$-topology, approximated by a Morse-function g on X.

Let now X be a complex manifold and f some real $\mathcal{C}^\infty$-function on X. Then in each point $x \in X$, the Levi-form

$$L_f(x) = \sum f_{z_i \bar{z}_j} \, dz_i \, d\bar{z}_j$$

is well-defined as a hermitian form on the tangent space, independently of the complex coordinates z_i chosen. In particular, $\mathrm{Index}_{\mathbb{C}}(f,x) :=$ maximal complex dimension of a complex subspace, on which the Levi-form is negative definite, is well defined in each point x of X.

Proposition: If x is a critical point of f, then

$$\mathrm{Index}_{\mathbb{R}}(f, x) \geqslant \mathrm{Index}_{\mathbb{C}}(f, x).$$

Proof: Computing in local coordinates, one finds for each tangent vector u at x

$$L_f(x)(u) = 1/4 \ \{H_f(x)(u) + H_f(x)(iu)\} .$$

Counting real dimensions of maximal real subspaces on which the forms are positive semi-definite, one obtains on the left hand side $2n - 2\,\mathrm{Index}_{\mathbb{C}}(f,x)$ and on the right hand side at least $2(2n - \mathrm{Index}_{\mathbb{R}}(f,x)) - 2n$, which proves the proposition.

Corollary: Let $f: X \longrightarrow \mathbb{R}$ be as above and assume that f maps X properly onto the interval (a,b). Assume that $\mathrm{Index}_{\mathbb{C}}(f,x) \geqslant s+1$ in all points $x \in X$. Then for all $\alpha < \beta \in (a,b)$

a) $H_c^i(X_\beta \setminus X_\alpha) = 0$ for $0 \leqslant i \leqslant s$

b) $\pi_i(X_\beta, X_\alpha) = 0$ for $0 \leqslant i \leqslant s$, if X admits small neighborhoods V in X, which can be retracted on X_α within X (e.g., if (X, X_α) is triangulizable).

Proof: Approximate f by a Morse-function g as in theorem 2 such that for the sets $X' := g^{-1}(a,\alpha]$ holds: $X_\alpha \subset X'_\alpha$, $X_\beta \subset X'_\beta$, g maps the set $X'_{\beta+\varepsilon} \setminus X'_{\alpha-\varepsilon}$ properly onto $(\beta+\varepsilon, \alpha-\varepsilon]$, and $\mathrm{Index}_{\mathbb{C}}(g,x) \geqslant s+1$ whenever $x \in X'_{\beta+\varepsilon} \setminus X'_{\alpha-\varepsilon}$. Then by the proposition above $\mathrm{Index}_{\mathbb{R}}(g,x) \geqslant s+1$ in all critical points $x \in X'_{\beta+\varepsilon} \setminus X'_{\alpha-\varepsilon}$ of g. Then $H_c^i(X'_\beta \setminus X'_\alpha)$ vanishes for $0 \leqslant i \leqslant s$ by theorem 1 and in the limit $X'_\beta \longrightarrow X_\beta$, $X'_\alpha \longrightarrow X_\alpha$ we obtain a). To prove b), we may assume $X'_\alpha \subset V$. If B_r is the closed ball of real dimension $r \leqslant s$, then every continuous map $(B_r, \partial B_r) \longrightarrow (X_\beta, X_\alpha) \subset (X'_\beta, X_\alpha)$ can be deformed into (X'_α, X_α) and hence into X . This proves b).

4. PROOF OF THE MAIN LEMMA

A first step towards the main lemma is this

Lemma 1: Let $A \subset \mathbb{P}_n$ be as in 1., a = dim A, and $T \subset \mathbb{P}_n$ a sufficiently small tubular neighborhood of A with respect to Fubini's metric on $\mathbb{P}_n$. For each nonsingular complex $X \subset \mathbb{P}_n$ of dimension b then holds

$$\pi_i(T \cap X, A \cap X) = H_c^i(T \cap X \setminus A \cap X) = 0 \quad \text{if } 0 \leqslant i \leqslant s.$$

Proof: Let f(x), $x \in \mathbb{P}_n$, be the distance of x from A in Fubini's metric. Then $T \cap X = \{x \in X: f(x) \leqslant c\}$ for some c and $A \cap X = \{x \in X: f(x) = 0\}$. Whenever $V \subset T$ is a small neighborhood of A, then there exists a real $k > 0$ such that $\mathrm{Index}_{\mathbb{C}}(-\exp(-kf), x) \geqslant a+1$ for all $x \in T \setminus V$, see

[1, Satz 1]. Then for all $x \in X \setminus V$

$$\mathrm{Index}_{\mathbb{C}}(-\exp(-kf)|X, x) \geqslant \mathrm{Index}_{\mathbb{C}}(-\exp(-kf), x) - \mathrm{cod}_{\mathbb{P}}X \geqslant a+b-n+1.$$

The proof now follows from the corollary above.

Let me now specify the choice of $B \subset G$: Since $U(n+1)$ is compact, there exists a Riemann metric d_G on G which is invariant under inner automorphisms of $U(n+1) \supset G$. For real $t > 0$ define

$$B_t := \{\sigma \in G : d_G(\sigma, 1) \leqslant t\}.$$

Then $\sigma B_t \sigma^{-1} = B_t$ for all $\sigma \in U(n+1)$. As in [1, lemma 3], one can find a function $c(t)$ such that for all $x \in \mathbb{P}_n$:

$$B_t \cdot x = \{z \in \mathbb{P}_n : d_{\mathbb{P}}(x,z) \leqslant c(t)\},$$

d_P the Fubini distance. Choose t so small, that $B_t \cdot A$ is a tubular neighborhood of A. Since lemma 1 is invariant under the operation of G on $\mathbb{P}_n$, the main lemma follows, if we can prove (after suitably decreasing t)

Lemma 2: The map $\varphi : B_t \times A \longrightarrow B_t \cdot A$ admits a section s and the image $S = s(B_t \cdot A) \subset B_t \times A$ is a deformation retract of $B_t \times A$. The restriction of this retraction to each fibre $\varphi^{-1}(z) \cap B_t \times A$ is a deformation retract of this fibre onto the point $s(z)$.

Take some Riemann metric d_A on A. Then $B_t \times A$ is a tubular neighborhood of $\{1\} \times A$ in $G \times A$ with respect to the product metric $d_G \times d_A$. Since $\{1\} \times A$ is a section over A for φ, lemma 2 follows from the general

Lemma 2': Let $\varphi : X \longrightarrow Y$ be a $\mathcal{C}^\infty$-fibre bundle with nonsingular fibre, d some Riemann metric on X, $\tilde{A} \subset X$ a section over some compact submanifold $A \subset Y$, and $T \subset X$ a tubular neighborhood of $\tilde{A}$. If T is small enough, then there is a section $s : \varphi(T) \longrightarrow T$ for φ over T and $s(\varphi(T)) \subset T$ is a "fibrewise" deformation retract of T.

Proof: Let $f(x)$ be the distance of $x \in X$ from $\tilde{A}$ with respect to d. Then f^2 is $\mathcal{C}^\infty$ near $\tilde{A}$. If $x \in \tilde{A}$, then $f^2 | \varphi^{-1}\varphi(x)$ has a nondegenerate critical point in x. Hence also $f^2|F$ has just one critical point near $\tilde{A}$ for every fibre $F = \varphi^{-1}y$, $y \in Y$ near to A. Since A is compact, after possible shrinking of T, we may assume that the restriction $f^2 | T \cap \varphi^{-1}y$ to all fibres $T \cap \varphi^{-1}y$, $y \in \varphi(T)$, has just one critical point $s(y)$. This defines the section s. Now the gradients of $f^2 | T \cap \varphi^{-1}(y)$, $y \in \varphi(T)$, are vectorfields along the fibres, with zeroes in $s(y)$, which can be used to define a retraction of T on $s(\varphi(T))$ respecting the φ-fibres.

REFERENCES

1. Barth, W.: Der Abstand von einer algebraischen Mannigfaltigkeit im komplex-projektiven Raum. Math. Annalen 187, 150-162 (1970).

2. Barth, W. and M.E. Larsen: On the homotopy groups of complex projective manifolds. Math. Scand. 30, 88-94 (1972).

3. Bredon, G.E.: Sheaf Theory. New York: Mc Graw-Hill Book Company 1967.

4. Larsen, M.E.: On the topology of complex projective manifolds. Inv. math. 19, 251-260 (1973).

5. Milnor, J.: Morse Theory. Princeton: Princeton University Press 1963. Second printing 1965.

6. Spanier, E.H.: Algebraic Topology. New York: Mc Graw-Hill Book Company 1966.

MATH. INST. RIJKSUNIV.
LEIDEN
WASSENAARSEWEG 80
NETHERLANDS

Proceedings of Symposia in Pure Mathematics
Volume 29, 1975

SLOPES OF FROBENIUS IN CRYSTALLINE COHOMOLOGY

Pierre Berthelot[1]

If W is the ring of Witt vectors over a perfect field k of characteristic $p > 0$, σ its Frobenius endomorphism, $F : M \to M$ a σ-semilinear endomorphism of a free finitely generated W-module, the "eigenvalues" of F cannot be defined in general, but the valuation of such an "eigenvalue" is well defined : this is what is called a *slope* of F. Our purpose here is to discuss estimates for these valuations in the case of the representations of the Frobenius of an algebraic variety arising from crystalline cohomology, and to give interpretations of the decomposition of the cohomology according to given slopes, which tie together cohomology and K-theory, as well as various cohomology theories (crystalline, Witt vector, flat cohomologies), following B. Mazur ([15], [16], [17]) and S. Bloch ([3], [4]).

1. **F-crystals.** Let k be a perfect field of characteristic $p > 0$, $W = W(k)$ the ring of Witt vectors over k, $\sigma : W \to W$ the Frobenius endomorphism of W, K the fraction field of W.

(1.1) **Definition.** *An* F-crystal *(resp.* F-isocrystal) *on* k *is a free finitely generated* W-*module* M *(resp.* K-*vector space), together with a* σ-*semi linear injective map* $F : M \to M$ (cf. [27], VI §3).

A morphism of F-crystals (resp. isocrystals) is a W-linear (resp. K-linear) map commuting with F. Any F-crystal (M,F) defines an F-isocrystal (M_K, F_K) with $M_K = M \otimes_W K$, $F_K = F \otimes \sigma$. Two F-crystals such that the associated isocrystals are isomorphic are said to be *isogenous*.

(1.2) **Examples.**

a) Let G be a Barsotti-Tate group over k (also called p-divisible, cf. [23], [10], [11], [18]), M its Dieudonné module (cf. [10]), F, V the Frobenius and Verschiebung endomorphisms of M. Then (M,F) is an F-crystal.

b) Let X be a proper and smooth scheme over k, of dimension n, and let us consider for all m the free finitely generated W-module

$$H^m = H^m(X/W, \underline{O}_{X/W})/\text{torsion}$$

AMS(MOS) subject classifications (1970). Primary 14F30, 14G10, 14L05 ; Secondary 14F15, 14G13, 18F25.

[1] E.R.A. n° 451 du C.N.R.S., Université de Rennes, and Princeton University.

where $H^m(X/W, \underline{O}_{X/W})$ is the m - th crystalline cohomology module of X with respect to W (cf. [1], [12]). The absolute Frobenius endomorphism $\underline{f}_X$ of X defines by functoriality a σ-semilinear endomorphism $\underline{f}_X^*$ of H^m. Since k is perfect, we can use Poincaré duality to define the transpose $\underline{f}_{X*}$ of $\underline{f}_X^*$. These endomorphisms satisfy the relation

$$\underline{f}_{X*} \circ \underline{f}_X^* = p^n .$$

Therefore, $\underline{f}_X^*$ is injective, and $(H^m, \underline{f}_X^*)$ is an F-crystal.

The structure of F-isocrystals over an algebraically closed field is completely described by the following result of Dieudonné [7] and Manin [14].

(1.3) **Théorème.** i) *Assume* k *is algebraically closed, and let* $K_\sigma[T]$ *be the non-commutative polynomial ring in one indeterminate over* K, *with commutation rule* $T.a=\sigma(a).T$ *for all* $a \in K$. *The category of* F-*isocrystals is semi-simple, and its simple objects are*

$$H_{r,s} = K_\sigma[T]/(T^s - p^r) ,$$

where $r \in \mathbb{Z}$, $s \geq 1$, $(r,s) = 1$, *and* F *is given by left multiplication by* T.

The rationnal number $\alpha = r/s$ is called the *slope* of $H_{r,s}$. Any F-isocrystal M canonically breaks down in a direct sum

$$M = \bigoplus_{\alpha \in \mathbb{Q}} M_\alpha$$

of isotypic compenents M_α, where M_α is a direct sum of simple F-crystals of slope α. The α's for which $M_\alpha \neq 0$ are called the *slopes* of M; the dimension over K of M_α is called the *multiplicity* of α as a slope of M. If there exists an F-crystal N such that $M \simeq N_K$, the slopes of M are positive.

If M is an F-crystal, its slopes will be by definition the slopes of M_K. If (M,F) is an F-crystal over a perfect field k, with algebraic closure $\bar{k}$, we can define its slopes to be that of $(M \otimes_{W(k)} W(\bar{k}), F \otimes \sigma_{\bar{k}})$.

ii) [14, th. 2.2]. *Assume* k *is the finite field* $\mathbb{F}_q$ *with* $q = p^a$ *elements. Then the slopes of an* F-*isocrystal* (M,F) *are the* p-*adic valuations (normalized by* $v(q) = 1$*) of the eigenvalues of the linear endomorphism* F^a *of* M, *the multiplicity of a slope* α *being the number of eigenvalues of valuation* α.

(1.4) **Examples.**

a) If (M,F) is the Dieudonné module of a Barsotti-Tate group G over k, the existence of a V such that $F V = V F = p$ implies that all the slopes of M lie in the interval $[0,1]$. If it is the Dieudonné module of a p-divisible formal group, V is topologically nilpotent on M, which forces the slopes to lie in $[0,1[$.

b) If we consider the example of (1.2), b), the relation

$f_{X*} \circ f_X^* = p^n$ implies that the slopes lie in $[0,n]$. But, when X is projective , one can easily deduce from the "weak Lefschetz theorem" the following [2] :

- if $0 \leqslant m \leqslant n$, the slopes lie in $[0,m]$;
- if $n \leqslant m \leqslant 2n$, the slopes lie in $[m-n,n]$:

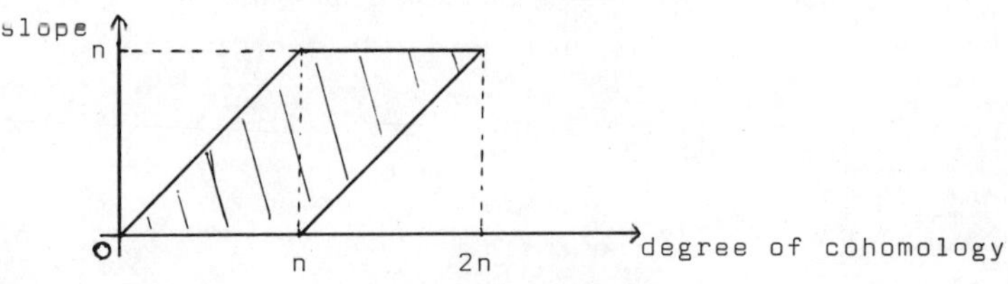

2. Newton polygon and Katz's conjecture.

(2.1) Let (M,F) be an F-crystal, $\alpha_1 < \alpha_2 < \dots < \alpha_k$ its slopes, λ_i the multiplicity of the slope α_i . The *Newton polygon* of (M,F) is the line joining the points P_i with coordinates

$$P_o = (0,0), \quad P_i = \left(\sum_{j=1}^{i} \lambda_j , \sum_{j=1}^{i} \lambda_j \alpha_j\right) , \quad i = 1, \dots, k .$$

If $\alpha_i = r_i/s_i$, with r_i , $s_i \in \mathbb{Z}$, $(r_i,s_i) = 1$, then λ_i is an integral multiple of s_i , so that the break-points of the Newton polygon have integral coordinates. If $k = \mathbb{F}_q$,it follows from (1.3), ii) that the Newton polygon of M is the ordinary Newton polygon of the polynomial $\det(t.\mathrm{Id}-F^a)$:

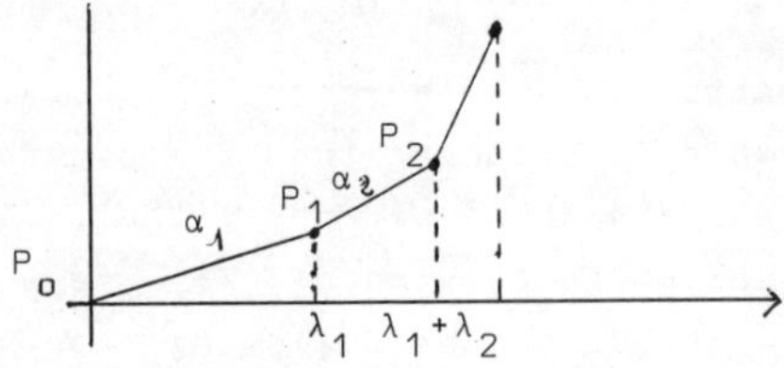

From now on, we assume we are in the geometrical situation of (1.2) b) ; let us fix an integer m , and let α be a slope of H^m . Then Poincaré duality implies that $n-\alpha$ is a slope of H^{2n-m} , with same multiplicity, and the "hard Lefschetz theorem" implies that $n-m+\alpha$ is a slope of H^{2n-m} , with same multiplicity. It follows that the extreme right point of the Newton polygon of H^m has coordinates $(\beta, \frac{m\beta}{2})$, where β is the m-th Betti number of X .

(2.2) **Example.** Let X be an elliptic curve over k (algebraically closed for simplicity) and $m = 1$. There are only two possibilities for the Newton polygon of H^1 :

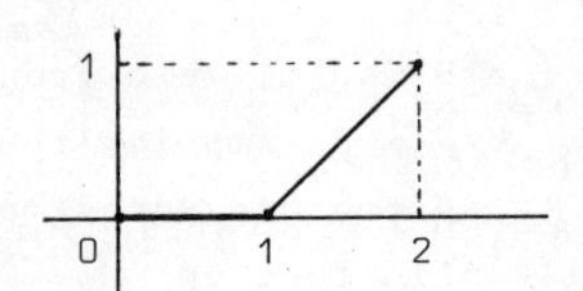

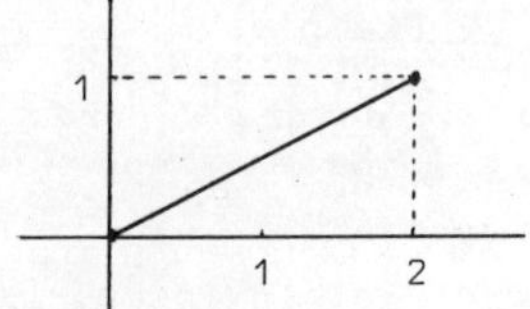

The first case occurs if and only if Frobenius has a fixed point in $H^1(X/W, \underline{O}_{X/W})$, or equivalently if and only if it has a fixed point in $H^1(X, \underline{O}_X)$. So the first polygon is the Newton polygon of ordinary elliptic curves, the second one the Newton polygon of supersingular elliptic curves.

(2.3) In the geometrical case, the *Hodge polygon* of X (in given degree m) is the line joining the points Q_i with coordinates

$$Q_i = \left(\sum_{j=0}^{i-1} h^{j,m-j} \ , \ \sum_{j=0}^{i-1} j \ h^{j,m-j} \right) \ , \ i = 0, \ldots, n+1 \ ,$$

where $h^{j,m-j} = \dim_k H^{m-j}(X, \Omega^j_{X/k})$ is the j-th Hodge number of X (in degree m) .

The Hodge and Newton polygons should be related by the following conjecture, due to Katz :

Conjecture. *The Newton polygon of* X *lies over its Hodge polygon.*

If we assume now that there exists a proper and smooth W-scheme X' such that $X \simeq X' \otimes_W k$, we can introduce in a similar way the Hodge polygon of $X'_K = X' \otimes_W K$, and conjecture :

Conjecture. *The Newton polygon of* X *lies over the Hodge polygon of* X'_K.

Since the Hodge Betti numbers increase by specialisation to characteristic p , this conjecture is stronger than the previous one. Note also that the Poincaré duality and the "hard Lefschetz theorem" imply that the Newton polygon of X and the Hodge polygon of X'_K have same extreme right point , so that the picture should be the following :

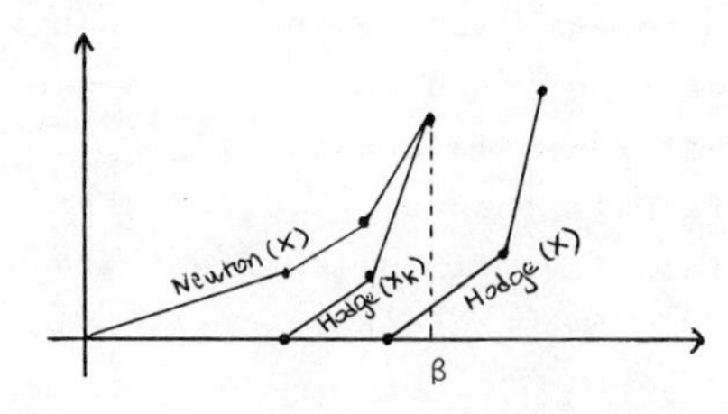

None of the above conjectures is proved in the general case, but they have been proved by Mazur ([15] , [16]) when they are identical.

(2.4) **Theorem.** *If there exists a projective and smooth* W*-scheme* X' *lifting* X *, and if for all* m, j, $H^{m-j}(X', \Omega^j_{X'/W})$ *is a free* W*-module, Katz's conjecture is true for all* m .

Mazur's approach is a subtle study of the arithmetical properties of the Frobenius endomorphism with respect to the Hodge filtration defined by the canonical isomorphism[1]

$$H^*(X/W,\underline{O}_{X/W}) \simeq \mathbb{H}^*(X',\Omega^{\cdot}_{X'/W})$$

and the Hodge-De Rham spectral sequence.

Precisely, let $M = H^m(X/W,\underline{O}_{X/W})$, $H = H^m(X/W,\underline{O}_{X/W}) \otimes_W W$ (where W is viewed as a W-module through σ), $\varphi : H \to M$ the W-linear factorisation of the σ-semilinear map f^*_X on M. Thanks to the classical invariant factor theorem we can find basis for H,M with respect to which φ has a diagonal matrix with entries p^i. The number h'^i of entries p^i is called the *i-th abstract Hodge number of* φ ; one can use these abstract Hodge numbers to construct an "abstract Hodge polygon", and one verifies easily that this last one lies under the Newton polygon. The previous theorem is thus reduced to the following one :

(2.5) **Theorem** (Mazur, [16] 7-6). *The abstract Hodge numbers of* φ *are the Hodge numbers of* X.

In some particular cases, this is easy to see. Assume the Frobenius endomorphism of X can be lifted to an endomorphism of X', or that $\dim X < p$. Then, using the degeneracy of the Hodge-De Rham spectral sequence and some results about crystalline cohomology, one can prove that

$$\varphi(H^i) \subset p^i.H^m(X/W,\underline{O}_{X/W}) ,$$

where H^i is the i-th submodule of the Hodge filtration, and deduce (2.5).

Although no counter-example has been written, computations of Katz make unlikely that the above divisibility property of φ might hold in general. So in the general case Mazur has to replace the $p^i.H^m(X/W,\underline{O}_{X/W})$ by a suitable (and more sophisticated) family of submodules of $H^m(X/W,\underline{O}_{X/W})$, and he still gets enough arithmetic information on the values of φ on the H^i to prove (2.5) (cf. [16]).

3. Crystalline cohomology and Witt vector cohomology.

(3.1) Let (M,F) be an F-isocrystal over k algebraically closed. For any subset $S \subset \mathbb{Q}$, M_S will denote the sub vector space

$$\bigoplus_{\alpha \in S} M_\alpha ,$$

where M_α is the isotypic component of slope α of M.

Problem. Let us consider the geometrical situation of ([1.2],b), with $M = H^m \otimes_W K$, where H^m is the m-th crystalline cohomology module of a proper and smooth k-scheme X. Crystalline cohomology thus canonically breaks into pieces on which Frobenius acts with a given slope. How can we

---

[1] Let us recall that $\mathbb{H}^*(X',\Omega^{\cdot}_{X'/W})$ is the De Rham cohomology of X' with respect to W', i.e. the hypercohomology (EGA 0_{III} 11.4.3) of X' with values in the complex $\Omega^{\cdot}_{X'/W}$. The Hodge-De Rham spectral sequence is the spectral sequence

$$E_1^{p,q} = H^q(X',\Omega^p_{X'/W}) \Rightarrow \mathbb{H}^n(X',\Omega^{\cdot}_{X'/W}) .$$

interpret them ? More precisely, what kind of relation is there between suitable pieces of crystalline cohomology and other theories, such as Witt vector cohomology [22] , flat cohomology [9] , Dieudonné theory, etc... ?

(3.2) The first results in this direction are due to Mazur [17] , who investigated the part of slopes lying in $[0,1[$. This is a very reasonnable question, since $M_{[0,1[}$ is a natural candidate to be the Dieudonné module of some formal group, at least up to isogeny.

Let $\mathcal{A}_k$ (resp. $\mathcal{A}_W$) be the category of artinian local k-algebras (resp. W-algebras) with residue field k . For fixed m , we define a functor Φ^m on $\mathcal{A}_k$ by

$$\Phi^m(A) = \operatorname{Ker}\left[H^m(X_A,\underline{O}^*_{X_A}) \to H^m(X,\underline{O}^*_X)\right] ,$$

with $A \in \operatorname{Ob}(\mathcal{A}_k)$, $X_A = X \times_k \operatorname{Spec}(A)$. For $m = 1$, Φ^1 is the formal completion of the Picard scheme of X along the origin. We can always define the Dieudonné module of Φ^m (for example using the construction in terms of "typical curves" which we recall below), and it is easy to see that this is $H^m(X,\underline{W}_X) = \varprojlim_n H^m(X,\underline{W}_n(\underline{O}_X))$, the cohomology of X with coefficients in the sheaf of Witt vectors over $\underline{O}_X$.

If $B \in \operatorname{Ob}(\mathcal{A}_W)$, let us denote $B_o = B/pB$. We define a functor Φ^m_{crys} on $\mathcal{A}_W$ by

$$\Phi^m_{crys}(B) = \operatorname{Ker}\left[H^m(X_{B_o}/B,\underline{O}^*_{X_{B_o}/B}) \longrightarrow H^m(X/k,\underline{O}^*_{X/k})\right] ,$$

where the first (resp. second) group is the crystalline cohomology of $X \times_k B_o$ with respect to B (resp. X with respect to k), with values in the units of its structural sheaf. One can then define the tangent space to Φ^m_{crys} , and show that it is isomorphic to $H^m(X/W,\underline{O}_{X/W})$, the crystalline cohomology of X with respect to W .

With some technical conditions on X (which include in particular the pro-representability and formal smoothness of Φ^m , Φ^m_{crys}), Mazur proved :

(3.3.) **Theorem.** *The part of the crystalline cohomology on which Frobenius acts with slopes in the interval* $[0,1[$ *is isomorphic, up to torsion, to the Witt vector cohomology, i.e., with the previous notations,*

$$(H^m(X/W,\underline{O}_{X/W}) \otimes K)_{[0,1[} \simeq H^m(X,\underline{W}_X) \otimes K .$$

Note that even for H^1 , the above conditions are restrictive, since we have to assume for example that the Picard scheme of X is reduced. One expects in fact that they are unnecessary, and this is really the case for H^1 .

The proof of (3.3) relies on (unpublished) results of Cartier, constructing an equivalence of categories between, on one hand, the category of formal Lie groups G over W , endowed with a σ^{-1}-linear endomorphism V whose reduction V_o mod p is the canonical Verschiebung of the reduction G_o of G , and, on the other hand, the category of free finitely generated W-modules M endowed with a σ-linear endomorphism F, the functor from formal Lie groups to W-modules being given by the "tangent space functor". From this one can derive the following facts, which are the key points in the proof of (3.3) when applied to $G = \Phi^m_{crys}$:

a) (G,V) being as above, and M being the tangent space of G, the Dieudonné module of G_o is isogenous to the part of $M \otimes K$ of slopes in $[0,1[$;

b) If Ψ^m is the restriction of Φ^m_{crys} to $\mathcal{a}_k$, Ψ^m and Φ^m have isogenous Dieudonné modules.

(3.4) **Remark.** Theorem (3.3) has a very interesting relation to Katz's conjecture discussed in §2. Namely, let us assume the conditions of (3.3) are satisfied. Then Φ^m is a formal Lie group, whose dimension is the rank over k of its tangent space ; this later being $H^m(X,\underline{O}_X)$, we get

$$\dim \Phi^m = \operatorname{rank}_k H^m(X,\underline{O}_X) = h^{o,m} .$$

On the other hand, the rank of the slope zero part of the Dieudonné module is the dimension of the multiplicative part of Φ^m, and therefore is less than $\dim \Phi^m$. Whence :

$$\dim_K(H^m(X/W,\underline{O}_{X/W}) \otimes K)_{[0]} \leqslant h^{o,m} ,$$

which proves Katz's conjecture for the horizontal parts of the Newton and Hodge polygons.

4. The slope spectral sequence.

We would like to have for parts of higher slope in crystalline cohomology a description analogous to that given in (3.3) for the part of slopes in $[0,1[$. For that purpose, Mazur suggested that one should study the "typical curves" associated to the sheaves of K-groups on X, in the sense of Quillen [21]. So let us first recall briefly Cartier's theory of typical curves ([5], [6]), which will be the underlying philosophy in the following constructions.

(4.1) **A brief survey of Cartier's theory of typical curves.** Let R be a commutative ring. The *ring* $\widetilde{W}(R)$ *of Witt vectors* with coefficients in R ("big Witt vectors") has, as underlying set, the set of families $(a_i)_{i\geqslant 1}$, $a_i \in R$. The map

$$(a_i)_{i\geqslant 1} \longmapsto \prod_{i\geqslant 1} (1-a_iT^i)^{-1}$$

gives a bijective correspondance between $\widetilde{W}(R)$ and the set $\Lambda(R)$ of formal series with coefficients in R and first coefficient equal to 1 ; the addition in $\widetilde{W}(R)$ corresponds then to the multiplication of formal series, and the multiplication to the composition law $\ast$ on $\Lambda(R)$ defined by

$$(1+\sum_{i\geqslant 1} a_iT^i) \ast (1+\sum_{i\geqslant 1} b_iT^i) = 1 + \sum_{i\geqslant 1} P_i(a_j,b_k)T^i ,$$

where $P_i(a_j,b_k)$ is an universal polynomial in a_j, b_k, $j,k\leqslant i$, with integer coefficients, determined by the symmetric functions theorem and the condition

$$(1-aT)^{-1} * (1-bT)^{-1} = (1-abT)^{-1} .$$

In particular,

$$(1-aT^m)^{-1} * (1-bT^n)^{-1} = (1-a^{n/d}\ b^{m/d}\ T^{mn/d})^{-1} ,$$

where $d=(m,n)$. The set of families $(a_i)_{i\geq 1}$ such that $a_i = 0$ for $i \leq n$ is an ideal I_n in $\widetilde{W}(R)$, and the quotient ring $\widetilde{W}(R)/I_n$ will be denoted $\widetilde{W}_n(R)$.

There exists a family of functorial ring endomorphisms $(F_n)_{n\geq 1}$ (Frobenius), and group endomorphisms $(V_n)_{n\geq 1}$ (Verschiebung) of $\widetilde{W}(R)$, with $F_n \circ V_m = V_m \circ F_n$ when $(m,n)=1$, and $F_n \circ V_n = n$. When R is a $\mathbb{Z}_{(p)}$-algebra, $\widetilde{W}(R)$ breaks into an infinite product of copies of the ordinary Witt vector ring $W(R)$. There exists a canonical projection of $\widetilde{W}(R)$ onto one of the factors $W(R)$, given by

$$\pi = \sum_{(n,p)=1} (\mu(n)/n)\ V_n \circ F_n ,$$

where $\mu(n)$ is the Möbius function, and the corresponding idempotent is the element in $\widetilde{W}(R)$ corresponding in $\Lambda(R)$ to the Artin-Hasse exponential

$$E(T) = \exp(\sum_{i\geq 0} T^{p^i}/p^i) = \prod_{(n,p)=1} (1-T^n)^{-\mu(u)/n} .$$

The formal line D over R is the functor on the category of R-algebras which associates to each R-algebra the set of its nilpotent elements. Let G be a formal commutative group over R. A *curve* on G is a morphism of functors $\gamma : D \to G$ such that $\gamma(0) = 0$; there is a canonical bijection

$$C(G) = \{\text{curves on } G\} \longleftrightarrow \varprojlim_n \operatorname{Ker}(G(R[T]/T^n) \xrightarrow{T \mapsto 0} G(R)) .$$

The set $C(G)$ has a natural group structure, and one can define on it a structure of $\widetilde{W(R)}$-module, and endomorphisms F_n, V_n. Suppose now R is a $\mathbb{Z}_{(p)}$-algebra. We can use the $\widetilde{W}(R)$-module structure of $C(G)$ to break it into pieces corresponding to the above decomposition of $\widetilde{W}(R)$. The factor $TC(G)=E(T)\cdot C(G)$ of $C(G)$ corresponding to the factor $W(R)$ of $\widetilde{W}(R)$ will be called the module of *typical curves* on G; it is a $W(R)$-module, and it is stable by F_p and V_p. If R is a perfect field of characteristic p, $TC(G)$ is the Dieudonné module of G [6].

(4.2) **The basic idea.** If we apply Cartier's construction to $\mathbb{G}_m$, viewed as a formal group, we get as typical curves the ring $W(R)$ itself. So we can consider the projective system of sheaves $\underline{W}_n(\underline{O}_X)$ as obtained by sheafifying the typical curves on $\mathbb{G}_m$ viewed as a formal group over the sections of $\underline{O}_X$.

But, for any local ring R, $\mathbb{G}_m(R) = R^* = K_1(R)$, whence the idea of considering the typical curves on the higher K_i, which should give parts of higher slopes in crystalline cohomology. Unfortunately, the K_i for $i\geq 2$ do not define formal groups in general, so that one cannot directly apply Cartier's construction, and one has to work out the whole process in terms of K-theory. This program has been carried out recently by S. Bloch ([3], [4]) who proved that the typical curves on K-theory even fit in a natural way into a *complex* , and got very interesting consequences.

(4.3) **Theorem** (Bloch). *Let* k *be an algebraically closed field of characteristic* $p > 0$, $p \neq 2$, *and* X *a proper and smooth* k-*scheme such that* $\dim X < p$. *Then there exists a projective system* $(\underline{C}_n^{\bullet})_{n \geq 1}$ *of complexes of sheaves on* X *(for the Zariski topology), with* $\underline{C}_n^q = 0$ *if* $q < 0$ *or* $q > \dim X$, *such that, if*

$$H^{*}(X,\underline{C}^q) = \varprojlim_n H^{*}(X,\underline{C}_n^q) \ , \ \mathbb{H}^{*}(X,\underline{C}^{\bullet}) = \varprojlim_n \mathbb{H}^{*}(X,\underline{C}_n^{\bullet}) \ ,$$

we get :

i) *There exists a canonical isomorphism*

$$\mathbb{H}^{*}(X,\underline{C}^{\bullet}) \xrightarrow{\sim} H^{*}(X/W,\underline{O}_{X/W})$$

between the hypercohomology of $\underline{C}^{\bullet}$ *and crystalline cohomology.*

ii) *The pro-complex* $(\underline{C}_n^{\bullet})_{n \geq 1}$ *has a canonical endomorphism* F , *and the induced endomorphism on the hypercohomology corresponds via* i) *to the action of Frobenius on crystalline cohomology.*

iii) *For all* q, $H^{*}(X,\underline{C}^q) \otimes_W K$ *is a finite dimensional* K-*vector space* .

iv) *There exists a spectral sequence*

$$E_1^{p,q} = H^q(X,\underline{C}^p) \Longrightarrow H^n(X/W,\underline{O}_{X/W})$$

(slope spectral sequence), *and it degenerates at* E_1 *modulo torsion.*

v) *The filtration induced on* $H^m(X/W,\underline{O}_{X/W})$ *by the previous spectral sequence has the following properties :*

a) *There exists a canonical isomorphism*

$$\mathrm{Fil}^i \, H^m(X/W,\underline{O}_{X/W}) \otimes_W K \simeq (H^m(X/W,\underline{O}_{X/W}) \otimes_W K)_{[i,n]}$$

where $n = \dim X$;

b) *On* $\mathrm{Fil}^i \, H^m(X/W,\underline{O}_{X/W})$, *Frobenius is divisible by* p^i .

In particular, one gets

$$H^q(X,\underline{C}^p) \otimes_W K \simeq (H^{p+q}(X/W,\underline{O}_{X/W}) \otimes_W K)_{[p,p+1[} \ .$$

Remark. One expects the hypothesis $p \neq 2$, $\dim X < p$ to be unnecessary.

The construction of $(\underline{C}_n^{\bullet})$ is very involved, and I will only give a vague outline of it.

(4.4) Let R be a commutative ring. We define

$$R_m = R\,[T]\,/(T^{m+1}) \ , \ R_\infty = R\,[[T]] \ .$$

For all $q \geqslant 1$, the curves on K_q are defined by

$$C_\infty K_q(R) = \operatorname{Ker}\left[K_q(R_\infty) \xrightarrow{T \mapsto 0} K_q(R)\right],$$

$$C'_m K_q(R) = \operatorname{Ker}\left[K_q(R_m) \xrightarrow{T \mapsto 0} K_q(R)\right],$$

$$C_m K_q(R) = \operatorname{Im}\left[C_\infty K_q(R) \longrightarrow C'_m K_q(R)\right].$$

Since we know very little about the structure of the higher K_i, we want to replace the $C_m K_q$ by smaller groups with which we can effectively compute. For this, we can use the multiplicative structure on $K_*(R)$ ([8], [13]) ; in particular, we have the *symbol maps*

$$\underbrace{R_m^* \times \ldots \times R_m^*}_{q \text{ times}} \subset K_1(R_m) \times \ldots \times K_1(R_m) \longrightarrow K_q(R_m),$$

$$(a_1, \ldots, a_q) \longmapsto \{a_1, \ldots, a_q\},$$

and, for $m \in \mathbb{N}$ or $n = \infty$, we will denote by $SC_m K_q(R)$ the subgroup generated by symbols

$$\{1+Ta, a_1, \ldots, a_{q-1}\}, \quad a \in R_m, \ a_i \in R_m^*.$$

When R is local and $q \leqslant 2$, $SC_m K_q(R) = C_m K_q(R) = C'_m K_q(R)$, but this is unknown in general.

a) There exists a natural structure of $\widetilde{W}(R)$-module on $C_m K_q(R)$. Let us describe multiplication by an element of $\widetilde{W}(R)$ corresponding to $P(T)^{-1}$ in $\Lambda(R)$, where $P(T)$ is a polynomial of degree d and constant coefficient 1 ; let $Q(T) = X^d P(X^{-1})$, $R' = R[X]/(Q(X))$, $j : R_m \subset R'_m$ given by $R \subset R'$, and $\varphi : R_m \to R'_m$ by $\varphi(T) = XT$. Then multiplication by $P(T)^{-1}$ is the composed map

$$K_q(R_m) \xrightarrow{\varphi_*} K_q(R'_m) \xrightarrow{j^*} K_q(R_m),$$

where φ_* is the functoriality map defined by φ, and j^* the tranfer map defined by j ([21], §4).

When R is a smooth local ring over a perfect field k of characteristic $p \neq 0,2$, with $\dim R < p$, Bloch gets a very precise description of the structure of $SC_m K_q(R)$, and deduces from it that $SC_m K_q(R)$ is a sub-$\widetilde{W}(R)$-module of $C_m K_q(R)$.

b) For all m, n let $\Psi_n : R_m \longrightarrow R_{mn+n-1}$ be the map defined by $\Psi_n(T) = T^n$. For all q, Ψ_n defines a functoriality map $V_n : K_q(R_m) \longrightarrow K_q(R_{mn+n-1})$ and a transfer map $F_n : K_q(R_{mn+n-1}) \to K_q(R_m)$. With the previous hypothesis, these maps define

$$V_n : SC_m K_q(R) \longrightarrow SC_{mn+n-1} K_q(R), \quad F_n : SC_{mn+n-1} K_q(R) \longrightarrow SC_m K_q(R).$$

c) Let $S = R[[T]]\,[T^{-1}]$. The product by $T \in K_1(S)$ defines maps $K_q(S) \longrightarrow K_{q+1}(S)$. When R is a regular local ring, these maps induce

$$\delta_q : SC_\infty K_q(R) \longrightarrow SC_\infty K_{q+1}(R).$$

d) With the hypothesis of a), we can use the $\widetilde{W}(R)$-structure on

$C_mK_q(R)$, to define the *typical curves on* K_q , by taking the direct factor corresponding to the direct factor $W(R)$ of $\widetilde{W}(R)$, namely

$$TC_mK_q(R) = E(T).SC_{p^{n-1}}K_q(R) \ .$$

For $q = 1$, $TC_nK_1(R) = W_n(R)$. We get the following structures on $(TC_nK_q(R))_{n\geqslant 1}$:

i) a $W(R)$-module structure ;

ii) endomorphisms F_p, V_p of the pro-objects $(TC_nK_q(R))_{n\geqslant 1}$, which we denote merely F, V, defined by b) ; $F \circ V = V \circ F = p$;

iii) differentials $\delta_q : TC_nK_q(R) \longrightarrow TC_nK_{q+1}(R)$ defined by c), such that $\delta_{q+1} \circ \delta_q = 0$. The key point, in view of the divisibility assertion of (4.3), is that F and V do not commute with δ_q , but instead we have the following commutative diagrams :

$$\begin{array}{ccc} (TC_nK_q) & \xrightarrow{\delta_q} & (TC_nK_{q+1}) \\ {\scriptstyle pV}\downarrow & & \downarrow{\scriptstyle V} \\ (TC_nK_q) & \xrightarrow{\delta_q} & (TC_nK_{q+1}) \end{array} \quad , \quad \begin{array}{ccc} (TC_nK_q) & \xrightarrow{\delta_q} & (TC_nK_{q+1}) \\ {\scriptstyle F}\downarrow & & \downarrow{\scriptstyle pF} \\ (TC_nK_q) & \xrightarrow{\delta_q} & (TC_nK_{q+1}) \end{array} \ .$$

d) Let X be proper and smooth over the perfect field k . Looking carefully, one can sheafify the previous constructions, and get the projective systems of sheaves $(\underline{TC}_n\underline{K}_q(\underline{O}_X))_{n\geqslant 1}$ on X (for the Zariski topology). Let us define the projective system of complexes $(\underline{C}^{\boldsymbol{\cdot}}_n)_{n\geqslant 1}$ of (4.3) by

$$\underline{C}^q_n = \underline{TC}_n\underline{K}_{q+1}(\underline{O}_X) \ ,$$

with differential δ^q , so that $(\underline{C}^o_n)_{n\geqslant 1} = (\underline{W}_n(\underline{O}_X))_{n\geqslant 1}$. The endomorphism F of $(\underline{C}^{\boldsymbol{\cdot}}_n)_{n\geqslant 1}$ is obtained by taking p^iF on $(\underline{C}^i_n)_{n\geqslant 1}$, thanks to the above diagram.

Finally, let us indicate that comparison with crystalline cohomology ((4.3),i) arises then from the following facts

i) $\underline{C}^{\boldsymbol{\cdot}}_1$ is isomorphic to the De Rham complex $\Omega^{\boldsymbol{\cdot}}_{X/k}$

ii) For all n , the map $\underline{C}^{\boldsymbol{\cdot}}_n/p\,\underline{C}^{\boldsymbol{\cdot}}_n \longrightarrow \underline{C}^{\boldsymbol{\cdot}}_1$ is a quasi-isomorphism,

and from the existence of a canonical homomorphism

$$H^*(X/W_n,\underline{O}_{X/W_n}) \longrightarrow \mathbb{H}^*(X,\underline{C}^{\boldsymbol{\cdot}}_n)$$

for all n .

(4.5) **Some applications of Bloch's theorem.** Let us assume $p \neq 2$ dim $X<p$, k algebraically closed.

a) *Relation between crystalline and Witt vector cohomology.* One gets another proof (under completely different hypothesis) of Mazur's theorem (3.3).

b) *Relation between crystalline cohomology and flat cohomology of* $\mathbb{Z}_p$. Using the previous result, and (1.3), one gets a canonical isomorphism between the invariants of Frobenius in $H^*(X/W, \underline{O}_{X/W}) \otimes \mathbb{Q}$ and $H^*(X_{fl}, \mathbb{Q}_p) = (\varprojlim_n H^*(X_{fl}, \mathbb{Z}/p^n\mathbb{Z})) \otimes \mathbb{Q}$.

c) *Relation between crystalline cohomology and flat cohomology of* $T_p(\mu)$. Bloch constructs an exact sequence of pro-sheaves for the etale topology

$$0 \to (\mathbb{G}_m/\mathbb{G}_m^{p^n}) \to (\underline{C}_n^1) \xrightarrow{1-F} (\underline{C}_n^1) \to 0 ,$$

which is the analog of the Artin-Schreier exact sequences

$$0 \to \mathbb{Z}/p^n\mathbb{Z} \to \underline{W}_n(\underline{O}_X) \xrightarrow{1-F} \underline{W}_n(\underline{O}_X) \to 0 .$$

One deduce from it a canonical isomorphism between $(\varprojlim_n H^*(X_{fl}, \mu_{p^n})) \otimes \mathbb{Q}$ and the subgroup of $H^*(X/W, \underline{O}_{X/W}) \otimes \mathbb{Q}$, on which Frobenius acts by multiplication by p.

Question. Is there a similar interpretation of the subgroup of $H^*(X/W, \underline{O}_{X/W}) \otimes \mathbb{Q}$ on which Frobenius acts by multiplication by p^i, for all i ? One could tentatively look at the flat cohomology of X with values in the tensor powers, or symmetric powers, of μ_{p^n}, but nothing is known about this cohomology.

d) *Application to the Artin-Tate conjecture.* Artin and Tate conjectured [25] that the Brauer group of a projective and smooth surface over a finite field is finite, and gave a formula which should compute its order. Using etale cohomology, they proved that, if Tate's conjecture on algebraic cycles [24] is true, then their conjecture holds for the part prime to p of the Brauer group. Using results on flat duality, the crystalline rationnality formula of the zeta function and the above result, Milne [19] proved that Tate's conjecture also implies the "p-part" of Artin-Tate's conjecture.

e) *Relation to Katz's conjecture.* On very interesting feature of Bloch's results is that they give rise to a lot of Dieudonné modules : for a fixed m, one can look at the filtration defined on $H^m(X/W, \underline{O}_{X/W})$ by the slope spectral sequence, divide by p^i the action of Frobenius on Fil^i, thanks to (4.3,v) and then take the part of slopes in $[0,1[$ of the F-crystal obtained in this way, which defines a Dieudonné module up to isogeny. To these Dieudonné modules correspond formal groups canonically attached to X ; for example, if $m = 1$, $i = 0$, one gets the formal group defined by the Picard variety of X. In general, it would be very interesting to know something about these formal groups, in particular to compute their dimension. In view of (3.4), one might hope to get in this way a new insight of Katz's conjecture, leading to a general proof of it.

BIBLIOGRAPHY

1. P. Berthelot, *Cohomologie cristalline des schémas de caractéristique* p > 0 , Lecture Notes in Math. 407, Springer Verlag.

2. P. Berthelot, *Sur le "théorème de Lefschetz faible" en cohomologie cristallino*, C.R. Acad. Sc. Paris 277 (12 novembre 1973), série A, 955-958.

3. S. Bloch, Letter to B. Mazur, May 1974.

4. S. Bloch, Secret papers.

5. P. Cartier, *Groupes formels associés aux anneaux de Witt généralisés*, C.R. Acad. Sc. Paris 265(10 juillet 1967), série A, 49-52.

6. P. Cartier, *Modules associés à un groupe formel commutatif. Courbes typiques*, C.R. Acad. Sc. Paris 265 (24 juillet 1967), série A, 129-132.

7. J. Dieudonné, *Groupes de Lie et hyperalgèbres de Lie sur un corps de caractéristique* p > 0 (VII), Math. Ann. 134(1957), 114-133.

8. S. Gersten, *Higher* K*-theory* I *: products*, preprint.

9. A. Grothendieck, *Crystals and the De Rham cohomology of schemes*, in Dix exposés sur la cohomologie des schémas, North-Holland.

10. A. Grothendieck, *Groupes de Barsotti-Tate et cristaux*, Act. Cong. Int. Math. (1970), t.1, 431-436.

11. A. Grothendieck, *Groupes de Barsotti-Tate et cristaux de Dieudonné*, Presses Univ. Montréal.

12. L. Illusie, *Report on crystalline cohomology*, in these proceedings.

13. J.L. Loday, *Structure multiplicative en* K*-théorie*, to appear.

14. Y. Manin, *The theory of formal groups over fields of finite characteristic*, Russ. Math. Surv. 18 n° 6 (1963), 1-84.

15. B. Mazur, *Frobenius and the Hodge filtration*, Bull. A.M.S. 78 n° 5 (1972), 653-667.

16. B. Mazur, *Frobenius and the Hodge filtration (estimates)*, Annals of Math. 98 (1973), 58-95.

17. B. Mazur, *Formal groups arising from algebraic varieties*, preprint, Harvard.

18. W. Messing, *The crystals associated to Barsotti-Tate groups : with applications to abelian schemes*, Lecture Notes in Math. 264 , Springer-Verlag.

19. J. Milne, *On a conjecture of Artin and Tate*, Lecture at M.I.T., May 1974.

20. T. Oda, *The first De Rham cohomology group and Dieudonné modules*, Ann. Scient. Ec. Norm. Sup. 4è série, t. 2 (1969), 63-135.

21. D. Quillen, *Higher algebraic* K*-theory* I , Algebraic K-theory I, Lect. Notes in Math. 341 , Springer-Verlag.

22. J.P. Serre, *Sur la topologie des variétés algébriques en caractéristique* p , Symp. Int. de Top. Alg. (1958), 24-53.

23. J. Tate, p*-divisible groups*, Proc. Conf. on Local Fields, Nuffic Summer School at Driebergen, Springer Verlag (1967).

24. J. Tate, *Algebraic cycles and poles of zeta functions*, in O. Schilling, Arithmetical algebraic geometry, Harper and Row.

25. J. Tate, *On a conjecture of Birch and Swinnerton-Dyer and a geometric analog*, Sém. Bourbaki, 1965/1966, exp. 306, Benjamin.

26. N. Katz, *Travaux de Dwork*, Séminaire Bourbaki, 1971/1972, exp. 409, Lecture Notes in Math. 317, Springer-Verlag.

27. N. Saavedra Rivano, *Catégories tannakiennes*, Lecture Notes in Math. 265, Springer Verlag.

E.R.A. 451 du C.N.R.S., DEPARTEMENT DE MATHEMATIQUES, UNIVERSITE DE RENNES, 35031 RENNES Cedex, FRANCE.

Proceedings of Symposia in Pure Mathematics
Volume 29, 1975

CLASSIFICATION AND EMBEDDINGS OF SURFACES

Enrico Bombieri and Dale Husemoller

The original classification of surfaces by F. Enriques, see [Enriques 1949], was for projective nonsingular algebraic surfaces over the complex numbers. Part of this was carried out in detail by Kodaira in two series of papers [An I - III] and [Am I - IV] and extended to the classification of compact two dimensional complex manifolds. Šafarevic [1965] in his Moscow seminar worked out the Enriques classification. In Mumford [1967] some of the results needed to extend the classification to characteristic p were developed and this has been completed in Bombieri and Mumford [1975].

The classification uses topological invariants together with intersection invariants associated with coherent sheaves on the surface. These numerical invariants and their relations are meaningful in both the algebraic case for arbitrary algebraically closed ground fields or in the complex analytic case. When there is both an algebraic and an analytic structure, then by GAGA [Serre 1956], the two kinds of coherent sheaves are naturally related along with their topological invariants. From the geometric point of view, surfaces are studied by examining their configurations of curves where curves on a surface are closely related to line bundles or locally free sheaves of rank one. The numerical invariants of these line bundles can be interpreted as certain invariants of families of curves.

The tangent bundle to a nonsingular algebraic variety or a complex manifold is intrinsically defined independent of any embeddings and is the infinitesimal form of the complex structure. The dual cotangent bundle can be viewed as a locally free sheaf of rank r, equal to the dimension of the variety, and its rth exterior power is a line bundle called the canonical line bundle $\mathcal{O}(K_X)$ of the variety X. As in the case of curves, the general classification theory of surfaces is in terms of the canonical intersection properties and homological invariants of the canonical line bundle. Again as in the case of curves we consider the embedding properties of the vector space $H^0(\mathcal{O}(nK_X))$ of cross sections of the nth tensor power $\mathcal{O}(K_X)^{n\otimes} = \mathcal{O}(nK_X)$ of the canonical line bundle.

This report has its origins in the Haverford Philips Lectures of E. Bombieri in December 1972 and April 1974. While this report was being prepared, the second author was a guest of the I.H.E.S. in Bures-sur-Yvette during the summers of 1973 and 1974 and of the S.F.B. at Bonn University during the year 1974-75. He also wishes to acknowledge the support given by Haverford College for the year 1974-75 and for the trip to Arcata.

TABLE OF CONTENTS

Part 1. Generalities on surfaces

In this section we collect the basic generalities on surfaces which we use in the classification theory. In this way we fix notations and cast the principal theorems into the form which they are used in the following parts.

References to this part include Mumford [1966, lectures 9-13] , Grothendieck [1958] , J.P. Serre [1955, GAGA] , and Shafarevich et al [1965] .

§ 1 Divisors, line bundles, and invertible sheaves

(1.1) Divisors. Let X/k denote a projective algebraic variety over an algebraically closed field k or a compact complex manifold.

Cartier divisor. A Cartier divisor is a section of the sheaf $\mathcal{D} = \mathcal{M}^*/\mathcal{O}^*$, i.e. a collection $\{f_i, U_j\}$ of rational (meromorphic) functions f_i defined on U_i such that $f_i/f_j \in H^0(U_i \cap U_j, \mathcal{O}^*)$ where f_i is determined mod $\mathcal{O}^*$, see also Mumford [1966, lectures 9,10] .

Weil divisor. A Weil divisor is a formal sum $\sum n_i D_i$ where n_i are integers and D_i are points (i.e. irreducible subvarieties) of codimension 1.

The exact sequence relating to Cartier divisors

(Cartier) $$0 \to \mathcal{O}^* \to \mathcal{M}^* \to \mathcal{D} = \mathcal{M}^*/\mathcal{O}^* \to 0$$

has the following form for X nonsingular

(Weil) $$0 \to \mathcal{O}^* \to \mathcal{M}^* \to \coprod_{x \text{ of codim } 1} \mathbb{Z}_x \to 0$$

where $f \in H^0(U, \mathcal{M}^*)$ is mapped to

$$\sum_{\substack{x \text{ of codim } 1 \\ x \in U}} \mathrm{ord}_x(f) \cdot \{\bar{x}\}$$

This includes an isomorphism, for X smooth, defined $\mathcal{D} \to \coprod_{x \text{ of codim } 1} \mathbb{Z}_x$ which yields an identification between Cartier and Weil divisor.

Let $Div(X)$ denote $H^0(\mathcal{D})$ the group of divisors with the additive notation for the group law. For $f \in H^0(\mathcal{M})$ and $f \neq 0$ i.e. $f \in H^0(\mathcal{M}^*)$ its image in $Div(X)$ is denoted by (f), so $(fg) = (f)+(g)$. Two divisors are linearly equivalent provided they differ by (f) for some $f \in H^0(\mathcal{M}^*)$ The subgroup $Div_\ell(X) = im(H^0(\mathcal{M}^*) \to H^0(\mathcal{D}))$ consists of divisors linearly equivalent to zero and the divisor class group $Div(X)/Div_\ell(X)$ is the group of linear equivalence classes of divisors.

The sequence of abelian monoids where $\mathcal{D}^+ \subset \mathcal{D}$

$$0 \to \mathcal{O}^* \to \mathcal{O}-\{0\} \to \mathcal{D}^+ \to 0$$

leads to notion of positive divisor which is an element of $Div^+(X) = im(H^0(\mathcal{D}^+) \to H^0(\mathcal{D})) \subset Div(X)$. With this notion of positive divisor $Div(X)$ is an ordered group.

(1.2) Line bundles and invertible sheaves. Roughly a line bundle is a map $L \to X$ with a one dimensional vector space structure on each fibre. Look at the sheaf of local cross sections (algebraic or analytic) also denoted L we see that it has the structure of an $\mathcal{O}_X$-module which is locally isomorphic to $\mathcal{O}_X$ itself. We call these invertible sheaves.

For an invertible sheaf L on X, it is clearly determined by isomorphisms $L|U_i \xrightarrow{\sim} \mathcal{O}|U_i$ and by the transition functions $\{f_{ij}, U_i \cap U_j\}$ with $f_{ij}f_{jk}f_{ki} = 1$. From $\{f_{ij}, U_i \cap U_j\}$, which is a 1-cocycle in $\mathcal{O}^*$, we can reconstruct a line bundle, unique up to isomorphism, which corresponds to the given invertible sheaf. The cohomology class $cl(L)$ of the 1-cocycle $\{f_{ij}, U_i \cap U_j\}$ in $H^1(\mathcal{O}^*)$ is independent of the local isomorphisms. If $Pic(X)$ denotes the group of isomorphism classes of invertible sheaves on X with group operations $L \otimes_{\mathcal{O}} M$ and $L^{-1} = Hom_{\mathcal{O}}(L, \mathcal{O})$, the we have an isomorphism $cl: Pic(X) \to H^1(X, \mathcal{O}^*)$.

(1.3) Divisors and invertible sheaves. From the cohomology exact sequence for $\mathcal{D} = \mathcal{M}^*/\mathcal{O}^*$

$$H^0(\mathcal{M}^*) \to H^0(\mathcal{D}) = \mathrm{Div}(X) \xrightarrow{\delta} H^1(\mathcal{O}^*) \to H^1(\mathcal{M}^*)$$

we have a monomorphism induced by δ and $\mathrm{cl}\colon \mathrm{Pic}(X) \to H^1(X,\mathcal{O}^*)$

$$\begin{array}{ccc} \mathrm{Div}(X)/\mathrm{Div}_\ell(X) & \longrightarrow & \mathrm{Pic}(X) \\ {\scriptstyle\delta}\searrow & & \nearrow{\scriptstyle \mathrm{cl}^{-1}} \\ & H^1(\mathcal{O}^*) & \end{array}$$

which assigns to a divisor D an invertible sheaf $\mathcal{O}(D)$ such that $\mathcal{O}(D_1)$ and $\mathcal{O}(D_2)$ are isomorphic if and only if D_1 and D_2 are linearly equivalent. A divisor D can be locally represented by $\{U_i,\phi_i\}$ where $\phi_i \in H^0(U_i,\mathcal{M}^*)$ and $\phi_i/\phi_j \in H^0(U_i \cap U_j,\mathcal{O}^*)$ and the invertible sheaf $\mathcal{O}(D)$ is defined by the 1-cocycle $\{\phi_i/\phi_j, U_i \cap U_j\}$. Then $H^0(U_i,\mathcal{O}(D))$ can be identified with the subspace of $f \in H^0(U_i,\mathcal{M}^*)$ such that $f\phi_i$ is regular. Since D is positive, we can choose $\phi_i \in H^0(U_i,\mathcal{O} - \{0\})$. This means that for $D \geq 0$, the sheaf $\mathcal{O}(-D)$ can be viewed as a subsheaf of $\mathcal{O}$ and we have the exact sequence

$$0 \to \mathcal{O}(-D) \to \mathcal{O} \to \mathcal{O}_D \to 0$$

where $\mathcal{O}_D$ is the structure sheaf on the scheme D and $\mathcal{I}_D = \mathcal{O}(-D)$ the sheaf of ideals defining D.

When $H^1(\mathcal{M}^*) = 0$, which is the case for X algebraic, then $D \mapsto \mathcal{O}(D)$ is an isomorphism $\mathrm{Div}(X)/\mathrm{Div}_\ell(X) \to \mathrm{Pic}(X)$. If $s \in H^0(L)$, then $\mathrm{ord}_x(s) \geq 0$ is well defined independent of the local chart for $s \neq 0$ where x is of codim 1. The divisor of zeros of s is $\mathrm{Div}(s) = \sum_{x \text{ codim } 1} \mathrm{ord}_x(s) \cdot \{\bar{x}\}$. Clearly L is isomorphic to $\mathcal{O}(\mathrm{Div}(s))$.

Divisors are very geometric in character and the geometry of an algebraic surface can studied using them instead of line bundles. On the other hand for compact complex manifolds line bundles must be used because there may be no curves, hence no divisors, on the surface.

(1.4) Linear system. The linear system $|D|$ determined by a divisor D is the set of all divisors $D' = D + (f) \geq 0$ for $f \in H^0(\mathcal{M}^*)$. We can also describe $|D|$ as the set of all $\mathrm{Div}(s)$ for $s \in H^0(\mathcal{O}(D))$. Since $\mathrm{Div}(s) = \mathrm{Div}(s')$ if and only if $s = \lambda s'$ for some $\lambda \in H^0(\mathcal{O}^*)$, we see that $|D| = \mathbb{P}H^0(\mathcal{O}(D))$, the projective space associated to the vector space $H^0(\mathcal{O}(D))$. Hence $\dim|D| = \dim H^0(\mathcal{O}(D)) - 1$. The rational map $\Phi_D : X \to |D| = \mathbb{P}H^0(\mathcal{O}(D))$ associated to the linear system $|D|$ assigns to $x \in X$ the point $(s_0(x),\dots,s_r(x))$ in $\mathbb{P}^r(k)$ where $s_0,\dots,s_r$ is a basis of $H^0(\mathcal{O}(D))$. The rational map Φ_D is defined at those points $x \in X$ where not all sections $s \in H^0(\mathcal{O}(D))$ vanish, and a change of basis shows that Φ_D is independent of the basis $s_0,\dots,s_r$.

(1.5) Remark of GAGA type. (Here GAGA means géometrie algebrique et géométrie analytique, see Serre [1955]). When X is a projective algebraic variety over $\mathbb{C}$, then the complex points $X(\mathbb{C})$ form a complex space and Serre proves that the natural morphism

$$H^1(X,\mathcal{O}_X^*) \to H^1(X(\mathbb{C}),\mathcal{O}_{X(\mathbb{C})}^*)$$

is an isomorphism.

§ 2 Intersection of divisors on a surface

For a curve X the degree of a divisor is given by a morphism $\deg : \mathrm{Div}(X) \to \mathbb{Z}$ of groups where $\deg(\sum_i n_i x_i) = \sum n_i$. Clearly $D \geq 0$ implies $\deg(D) \geq 0$ and it is a basic fact that $\deg(f) = 0$ so that $\deg : \mathrm{Div}(X)/\mathrm{Div}_\ell(X) \to \mathbb{Z}$ and $\deg : H^1(\mathcal{O}^*) \to \mathbb{Z}$ are defined.

(2.1) Intersection number for divisors on a surface. The intersection form $\mathrm{Div}(X) \times \mathrm{Div}(X) \to \mathbb{Z}$ for divisors on a surface X , denoted $D \cdot D'$, is symmetric and biadditive. For $D' = (f)$ we have $D\cdot(f) = 0$ for any D and thus the intersection pairing is defined $\mathrm{Pic}(X) \times \mathrm{Pic}(X) \to \mathbb{Z}$ or

$H^1(X,\mathcal{O}^*) \times H^1(X,\mathcal{O}^*) \to \mathbb{Z}$. The intersection number between two line bundles L and M can be defined by $L \cdot M = \chi(\mathcal{O}) - \chi(L^{-1}) - \chi(M^{-1}) + \chi(L^{-1} \otimes_{\mathcal{O}} M^{-1})$ for example.

Two divisors D and D' are called <u>disjoint</u> provided when written as linear combinations of irreducible curves they have no common summands. For D and D' disjoint and positive $D \cdot D' \geq 0$. If C is a curve with $C^2 \geq 0$, then $D \cdot C \geq 0$ for any $D \geq 0$. In effect, $D = nC + D'$ where D' is disjoint from C and then $D \cdot C = nC^2 + D' \cdot C \geq 0$. If two divisors D and D' are disjoint as subschemes of X , then $D \cdot D' = 0$.

For the hyperplane class H on an algebraic surface $H^2 > 0$ and $D \cdot H \geq 0$ for any $D \geq 0$.

(2.2) <u>Homological interpretation of intersection number</u>. In the classical topology over the complex numbers we have two short exact sequences of sheaves

$$\begin{array}{ccccccccc} 0 & \longrightarrow & \mathbb{Z} & \longrightarrow & \mathcal{O} & \xrightarrow{\exp(\ )} & \mathcal{O}^* & \longrightarrow & 0 \\ & & \downarrow & \exp(\frac{\ }{n}) & \downarrow & & \downarrow 1 & & \\ 0 & \longrightarrow & \mathbb{Z}/n & \longrightarrow & \mathcal{O}^* & \xrightarrow{n} & \mathcal{O}^* & \longrightarrow & 0 \end{array}$$

yielding the boundary morphisms.

$$H^1(X,\mathcal{O}^*) \begin{array}{l} \xrightarrow{\delta} H^2(X,\mathbb{Z}) \\ \qquad\qquad \downarrow \\ \xrightarrow{\delta_n} H^2(X,\mathbb{Z}/n) \end{array}$$

For L in $H^1(X,\mathcal{O}^*)$ its image $\delta(L)$ in $H^2(X,\mathbb{Z})$, denoted $c_1(L)$, is its first Chern class and its reduction mod n is denoted $c_1(L)(n)$. We denote $c_1(D)$ for $c_1(\mathcal{O}(D))$ and $c_1(D)(n)$ for $c_1(\mathcal{O}(D))(n)$.

For a <u>curve</u> X and a divisor D on X the following hold

$$\deg(D) = c_1(D)[X] \qquad \text{and} \qquad \deg(D) \bmod n = c_1(D)(n)[X]$$

and for a <u>surface</u> X and two divisors D , D' on X the following hold

$$D\cdot D' = (c_1(D).c_1(D'))[X] \qquad \text{and} \qquad D\cdot D' \bmod n = (c_1(D)(n).c_1(D')(n))[X]$$

where for m dimensional X Poincaré duality gives natural isomorphisms

$$[X] : H^{2m}(X,\mathbb{Z}) \to \mathbb{Z} \qquad \text{and} \qquad [X] : H^{2m}(X,\mathbb{Z}/n) \to \mathbb{Z}/n$$

The reason for including the $\mathbb{Z}/n$ sequence, which apparently yields less information, is that it works in characteristic p using the étale cohomology theory for n prime to p . Then the sequence is called the Kummer sequence and is denoted

$$1 \to \mu_n \to G_m \to G_m \to 1$$

(2.3) <u>Remark of GAGA type</u>. (Here GAGA means Grothendieck-Artin comparism theorem, see SGA 4). For X defined over the complex numbers the natural map $H^*(X(\mathbb{C}),\mu_n) \longrightarrow H^*_{et}(x,\mu_n)$ where $\mu_n \simeq \mathbb{Z}/n$ with the choice of an nth root of unity. Hence the mod n first Chern class is compatible with the étale class.

§ 3 <u>Riemann-Roch and Serre duality</u>.

The basic problem is to calculate $\dim_k H^o(X,L)$ for an invertible sheaf L or in fact any coherent sheaf. As a corollary we would have an expression for $\dim|D| = \dim H^o(\mathcal{O}(D)) - 1$, the dimension of a linear system. This is meaningful because $\dim_k H^i(X,L) < + \infty$ for all i and $H^i(X,L) = 0$ for all $i > \dim X$ on a complete or proper X . In the early 1950's Serre suggested that the Riemann-Roch theorem should be a formula for $\chi(L) = \sum_{o\leq i\leq n} (-1)^i \dim H^i(X,L)$, which has better formal properties than $\dim H^o(X,L)$, and indeed the first general Riemann-Roch theorem proved by Hirzebruch was a formula for $\chi(L)$ in terms of Chern classes of the tangent bundle for algebraic varieties of the complex numbers.

The solution to the problem of Riemann-Roch for our purposes comes in three parts (We consider only smooth varieties):

(1) A formula relating $\chi(L)$ to $\chi(\mathcal{O})$ in terms of invariants of L related to degree and intersection number. This part is relatively formal and a single argument works in characteristic p or on a complex manifold.

(2) A formula relating $\chi(\mathcal{O})$ to the Chern classes of X for its tangent bundle which in turn are related to topological invariants of the variety. For surfaces we will give both of the subsequent generalizations by Grothendieck to include characteristic p and by Atiyah-Singer to include nonalgebraic complex surfaces from their index theorem.

(3) A nondegenerate pairing $H^i(X,L) \times H^{n-i}(X,K \otimes L^{-1}) \to k$ where $n = \dim X$ which is Serre duality. This was formulated and proved by Serre for complex manifolds and then generalized to arbitrary smooth algebraic varieties by Grothendieck.

In the following discussion all curves and surfaces are assumed to be smooth.

(3.1) Definition. The canonical line bundle ω_X of an n dimensional smooth variety X or an n dimensional complex manifold X is the nth exterior power $\Lambda^n T^*_X$ of the cotangent bundle T^*_X of X. If $\omega_X = \mathcal{O}(K)$ for a divisor K, then K or K_X is the canonical divisor.

(3.2) Riemann-Roch I. For a line bundle L on a curve X

$$\chi(L) = \deg L + \chi(\mathcal{O}) = \deg L + 1 - p(X)$$

where $p(X) = \dim H^1(X,\mathcal{O})$ is the genus of the curve. So for a divisor D,

$$\chi(\mathcal{O}(D)) = \deg D + \chi(\mathcal{O})$$

For a line bundle L on a surface X

$$\chi(L) = \frac{1}{2} L \cdot (L \otimes \omega_X^{-1}) + \chi(\mathcal{O}_X)$$

or for a divisor D

$$\chi(\mathcal{O}(D)) = \frac{1}{2} D \cdot (D-K_X) + \chi(\mathcal{O}_X)$$

$$= \frac{1}{2} D \cdot (D-K_X) + 1 - q + p_g$$

where $q = \dim H^1(X,\mathcal{O})$ is the irregularity and $p_g = \dim H^2(X,\mathcal{O})$ is the geometric genus of the surface.

For a proof of this part see Mumford [1966, p. 87]

(3.3) Riemann-Roch II. In terms of the Chern classes of the tangent bundle we have:

$$\chi(\mathcal{O}_X) = \frac{1}{2} c_1 \qquad \text{for curves}$$

$$\chi(\mathcal{O}_X) = \frac{1}{12}(c_1^2 + c_2) \qquad \text{for surfaces (Noether's formula)}$$

In the general algebraic case this follows from Grothendieck's Riemann Roch theorem, see Borel and Serre [1958], and for general complex surfaces from the Atiyah, Singer index theorem.

In the next section we see that the top Chern class can be viewed as the Euler number $e(X)$ either in the sense of singular or ℓ-adic cohomology. Moreover $c_1^2 = \omega_X \cdot \omega_X = K_X \cdot K_X$ is the selfintersection number of the canonical line bundle. For curves $\chi(\mathcal{O}) = \frac{1}{2} c_1 = \frac{1}{2}(2-b_1(X)) = \frac{1}{2} e(X)$.

(3.4) Serre duality. For a line bundle L on an n dimensional compact variety we have a nondegenerate pairing

$$H^i(X,L) \times H^{n-i}(X,\omega_X \otimes L^{-1}) \to k$$

and in particular $\dim H^i(X,L) = \dim H^{n-i}(X,\omega_X \otimes L^{-1})$ or for divisors

$\dim H^i(X,\mathcal{O}(D)) = \dim H^{n-i}(X,\mathcal{O}(K-D))$.

For <u>curves</u> this means that

$$\begin{aligned}\chi(\mathcal{O}(D)) &= \dim H^0(\mathcal{O}(D)) - \dim H^1(\mathcal{O}(D))\\ &= \dim H^0(\mathcal{O}(D)) - \dim H^0(\mathcal{O}(K-D))\end{aligned}$$

Moreover $p(X) = \dim H^1(\mathcal{O}) = \dim H^0(\mathcal{O}(K))$, the dimension of the space of regular 1-forms. Hence $\chi(\mathcal{O}(K)) = \chi(\mathcal{O}) = 1 - p(X)$ and from Riemann Roch I, we deduce that

$$\deg(K) = 2p(X)-2 = -2\,\chi(\mathcal{O}) = -e(X)$$

For <u>surfaces</u> $p_g = \dim H^2(\mathcal{O}) = \dim H^0(\omega)$, the dimension of the space of regular 2-forms. Moreover

$$\begin{aligned}\chi(\mathcal{O}(D)) &= \dim H^0(\mathcal{O}(D)) - \dim H^1(\mathcal{O}(D)) + \dim H^0(\mathcal{O}(K-D))\\ &\leq \dim H^0(\mathcal{O}(D)) + \dim H^0(\mathcal{O}(K-D)) =\\ &= \dim |D| + \dim |K-D| + 2\end{aligned}$$

The term $\dim H^1(\mathcal{O}(D))$ is called the superabundance term in the classical formulation of Riemann-Roch. In the next section we will take the question of the duality theorem for a map.

(3.5) <u>Genus formula for a curve on a surface</u>. For $D > 0$ the exact sequence, where $\mathcal{O}_D$ is the structure sheaf of scheme D ,

$$0 \to \mathcal{I}_D = \mathcal{O}(-D) \to \mathcal{O} \to \mathcal{O}_D \to 0$$

leads to $\chi(\mathcal{O}_D) = \chi(\mathcal{O}) - \chi(\mathcal{O}(-D))$ and by Riemann Roch I the formula

$$\chi(\mathcal{O}_D) = -\tfrac{1}{2}D(D+K)$$

The relation $\omega_D = [\omega_X \otimes \mathcal{O}_X(D)] \otimes \mathcal{O}_D$ or $K_D = (K+D)$ restricted to D also holds. For $D = C$ an irreducible curve we have the genus formula

$$2p(C) - 2 = C \cdot K + C^2$$

In particular for any divisor D the integer $D\cdot(D+K)$ is even.

(3.6) Index theorem for complex surfaces. The pairing by cup product $H^2(X,\mathbb{R}) \times H^2(X,\mathbb{R}) \to H^4(X,\mathbb{R}) = \mathbb{R}$ has an index (or signature) $\sigma = b^+ - b^-$ where $b^\pm$ is the dimension of the largest subspace of $H^2(X,\mathbb{R})$ where $\pm\, v\cdot v > 0$ for $v \neq 0$. Then

$$\sigma = b^+ - b^- = \frac{1}{3}(c_1^2 - 2c_2)$$
$$= \frac{1}{3}(K^2 - 2e)$$

see Hirzebruch [1966] .

§ 4 Numerical invariants

The relation $c_2 = e(X)$ holds for any surface where

$$e(X) = b_o - b_1 + b_2 - b_3 + b_4 = 2 - 2b_1 + b_2$$

by Poincaré duality. This relation $c_2 = e(X)$ is proved by showing that $e(X) = \Delta\cdot\Delta$ in the 4-fold $X \times X$ and $c_2 = \Delta\cdot\Delta$ from the Lefschetz fromula. Recall (3.2)

$$12 - 12q + 12p_g = K^2 + 2 - 2b_1 + b_2$$

and rewrite in the form

$$\boxed{10 - 8q + 12p_g = K^2 + b_2 + 2\Delta}$$

where $\Delta = 2q - b_1$. For Kähler manifolds $\Delta = 0$ by Hodge theory more generally we have the following theorem.

(4.1) Theorem. For the relation $10 - 8q + 12p_g = K^2 + b_2 + 2\Delta$ where $\Delta = 2q - b_1$ we have:

(1) for complex surfaces $\Delta = 0$ if and only if b_1 is even

$\Delta = 1$ if and only if b_1 is odd ,

moreover $b^+ = 2p_g + 1 - \Delta$, $\dim H^0(\Omega^1) = q - \Delta$, and

(2) for algebraic surfaces $\Delta = 0$ in characteristic 0 and in characteristic p the term Δ is even and $0 \le \Delta \le 2p_g$.

For the proof of (1) we use the following series of elementary assertions.

(4.2) (a) Every holomorphic 1-form is d-closed. (This is false of three folds and surfaces in characteristic p).

(b) If $\omega_1, \ldots, \omega_r$ is a base for $H^0(\Omega^1)$, then $\omega_1, \ldots, \omega_r, \bar{\omega}_1, \ldots, \bar{\omega}_r$ are linearly independent in $H^1(\mathbb{C})$ so $2h^{1,0} \le b_1$ where $h^{1,0} = \dim H^0(\Omega^1)$.

(c) $2p_g \le b^+$, $2h^{1,0} \le b_1 \le h^{1,0} + h^{0,1}$, and $b_1 \le 2h^{0,1}$ where $h^{0,1} = \dim H^1(\mathcal{O}) = q$.

By eliminating K^2 from the Riemann-Roch formula (3.2) and the index theorem (3.5),we derive the relation

$$(b^+ - 2p_g) + (2q - b_1) = 1$$

From (4.2)(c) we have two possibilities $b^+ = 2p_g + 1$ and $2q = b_1$ i.e. $\Delta = 0$ or $b^+ = 2p_g$ and $2q = b_1 + 1$ i.e. $\Delta = 1$. Observe that $\Delta = 0$ when b_1 is even and $\Delta = 1$ when b_1 is odd.

Moreover for the exact sequence $0 \to \mathbb{Z} \to \mathcal{O} \xrightarrow{\exp} \mathcal{O}^* \to 0$ when $b_1 \equiv 0 \pmod 2$ we have a monomorphism

$$\mathbb{Z}^{2q} = H^1(X, \mathbb{Z}) \to H^1(\mathcal{O}_X) = \mathbb{C}^q$$

where the image is a lattice and when $b_1 = 1 \pmod 2$ a monomorphism

$$\mathbb{Z}^{2q-1} = H^1(X,\mathbb{Z}) \longrightarrow H^1(\mathcal{O}_X) = \mathbb{C}^q$$

where the image is discrete.

For the proof of (2) observe that $b_1 = 2s$ because

$$(\mathbb{Z}/\ell)^{b_1} = H^1_{et}(X,\mathbb{F}_\ell) = \{x \in \mathrm{Pic}(X) : \ell x = 0\} = \{x \in \mathrm{Pic}^o(X) : \ell x = 0\}$$
$$= (\mathbb{Z}/\ell)^{2s}$$

and $\Delta = 2q - b_1 = 2(q-s)$ is even. Since $H^1(\mathcal{O}_X) = \mathrm{Tang}_o(\mathrm{Pic}(X))$, the tangent space at 0, we have $q \geq s$ and so $\Delta \geq 0$. If β_i denotes the Bockstein operators $H^1(\mathcal{O}_X) \to H^2(\mathcal{O}_X)$, then

$$\mathrm{Tang}_o(\mathrm{Pic}^o_{red}(X)) \cong \bigcap_{1\leq i} \ker(\beta_i)$$

Hence $q-s = \dim(\mathrm{Tang}\ \mathrm{Pic}^o(X)) - \dim(\mathrm{Tang}\ \mathrm{Pic}^o_{red}(X))$

$$= \dim H^1(\mathcal{O}_X) - \dim(\bigcap_{1\leq i} \ker(\beta_i))$$
$$\leq \dim(\bigcup_{1\leq i} \mathrm{im}(\beta_i)) \leq \dim H^2(\mathcal{O}) = p_g$$

Thus $\Delta = 2(q-s) \leq 2p_g$. This completes our sketch of the theorem.

§ 5 Algebraic index theorem

Note that if $E_2 \in |D_2|$, then the function $E_1 \mapsto E_1 + E_2$ defined $|D_1| \to |D_1+D_2|$ is an injection and $\dim|D_1| \leq \dim|D_1+D_2|$.

(5.1) Proposition. If $D^2 > 0$ on X, then either

(i) $D\cdot H > 0$ and $\dim|nD| \to +\infty$ as $n \to +\infty$ (where H is the hyperplane class), or

(ii) $D\cdot H < 0$ and $\dim|nD| \to +\infty$ as $n \to -\infty$

Proof. By the Riemann-Roch inequality

$$\dim |nD| + \dim |K-nD| \geq \frac{1}{2} n^2D^2 - \frac{1}{2} n\cdot D\cdot K + \chi(\mathcal{O}) - 2 ,$$

we see that $\dim |nD| + \dim |K-nD| \to +\infty$ as $n \to +\infty$ or $-\infty$. If both $\dim |nD|$ and $\dim |K-nD| \to +\infty$ as $n \to +\infty$ or $-\infty$, then there exists an n with $E \in |nD|$ and $\dim |K| \geq \dim |K-nD|$ which contradicts the above remark since $K = nD + (K-nD)$. Moreover either $\dim |nD|$ or $\dim |K-nD|$ can not $\to +\infty$ for both $n \to +\infty$ and $n \to -\infty$ by the same argument.

If $E \in |nD|$ for $n >> o$ and $E > 0$, then $0 < E\cdot H = nD\cdot H$ so $D\cdot H > 0$ when $|nD| \to +\infty$ as $n \to +\infty$. If $E \in |nD|$ for $n << 0$ and $E > 0$, then $0 < E \cdot H = nD \cdot H$ so $D \cdot H < 0$. This proves the proposition.

(5.2) Corollary. If $H\cdot D = 0$, then $D^2 \leq 0$.

(5.3) Corollary. If $D^2 > o$ and $H\cdot D > 0$, then for some $n > o$, nD is linearly equivalent to a strictly positive divisor.

Corollary (5.2) contains the basic idea of the algebraic index theorem.

(5.4) Definition. Two divisors D_1 and D_2 are numerically equivalent provided $D_1\cdot E = D_2\cdot E$ for all $E \in \mathrm{Div}(X)$.

The set $\mathrm{Div}_n(X)$ of all divisors numerically equivalent to zero is a subgroup of $\mathrm{Div}(X)$ containing $\mathrm{Div}_\ell(X)$ the subgroup of all divisors linearly equivalent to zero.

(5.5) Definition. The Neron Severi group of X, denoted $NS(X)$, is $\mathrm{Div}(X)/\mathrm{Div}_n(X)$.

The intersection form $D\cdot D'$ is defined on $NS(X) \times NS(X) \to \mathbb{Z}$ and for $NS(X)$ it is a basic theorem that $NS(X)$ is a finitely generated group. The intersection form is also defined for pairs in $\mathbb{Q} \otimes NS(X)$ or in $\mathbb{R} \otimes NS(X)$ with values in $\mathbb{Q}$ or $\mathbb{R}$ respectively. Clearly on $K \otimes NS(X)$ the intersection form B nondegenerate for any field K.

(5.6) Theorem. (Algebraic Index Theorem). On $\mathbb{R} \otimes NS(X)$ the intersection form is of signature $(1,m)$.

Proof. Since $H^2 > 0$ we have one positive direction and on the orthogonal complement of $\mathbb{R} \cdot H$ the intersection form is negative by (5.2).

(5.7) Corollary. For $E \in Div(X)$ with $E^2 > 0$ it follows that $D^2 \leq 0$ for any divisor D with $E \cdot D = 0$.

(5.8) Definition. Let V be a real vector space with a nondegenerate bilinear form $x \cdot y$ on V. The dual X^* of a subset $X \subset V$ is the set of all $y \in V$ such that $x \cdot y \geq 0$ for all $x \in X$. A subset X is selfdual provided $X = X^*$.

The following is an exercise in linear algebra which is left to the reader.

(5.9) Proposition. Let V be a real vector space with a form $x \cdot y$ of type $(1,m)$ and let $h \cdot h > 0$. The cone Γ of all $x \in V$ with $x \cdot x \geq 0$ and $x \cdot h \geq 0$ is selfdual and is the closure of the set of all $x \in V$ with $x \cdot x > 0$, $x \cdot h > 0$.

(5.10) Remark. In $\mathbb{R} \otimes NS(X)$ let Γ denote the selfdual cone of all x with $x \cdot x \geq 0$ and $x \cdot h \geq 0$, the so called positive light cone. Let Γ_+ denote the cone of all D linearly equivalent to a positive divisor. Then by (5.3) the interior of Γ, and hence also Γ, are contained in Γ_+. Thus $\Gamma_+^* \subset \Gamma^* = \Gamma$ and Γ_+^* is the cone of ample divisors by the Nakai amplitude criterion. From $\Gamma_+^* \subset \Gamma$ we have the following theorem.

(5.11) Theorem. Let Δ be a divisor on X such that $\Delta \cdot D \geq 0$ for all positive divisors D. Then $\Delta^2 \geq 0$.

(5.12) Corollary. For the canonical divisor K, if $K \cdot D \geq 0$ for all positive D, then $K^2 \geq 0$.

See also 2(2.4) for another proof of this result.

§ 6 Albanese mapping and Picard group

(6.1) Definition. The Albanese variety $(\mathrm{Alb}(X),\theta)$ of a variety X is a pair where $\mathrm{Alb}(X)$ is an abelian variety and $\theta : X \to \mathrm{Alb}(X)$ is a morphism such that for any morphism $f:X \to A$ into an abelian variety there exist a unique morphism $\alpha : \mathrm{Alb}(X) \to A$ of abelian varieties and $a \in A$ such that $f = \alpha\theta + a$.

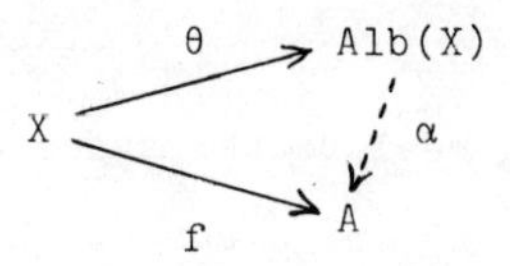

In Lang [1959] it is proved that the Albanese variety exists and is unique up to isomorphism and θ is unique up to translation. Moreover the image of θ generates $\mathrm{Alb}(X)$.

Over the complex numbers the Albanese variety and the mapping $\theta: X \to \mathrm{Alb}(X)$ has a rather concrete description. Let $\omega_1,\ldots,\omega_q$ be a basis for $H^0(\Omega^1)$ where $b_1 = 2q$. Then we define, choosing first a base point $x_0 \in X$, a map

$$\theta : X \longrightarrow \mathrm{Coker}\,(H_1(X,\mathbb{Z}) \longrightarrow \mathrm{Hom}(H^0(\Omega^1),\, \mathbb{C})) = T$$

by the relation $\theta(x)(\omega) = \int_{x_0}^{x} \omega$ modulo[image $H_1(X,\mathbb{Z})$] or in terms of coordinates

$$\theta(x) = (\int_{x_0}^{x} \omega_1,\ldots,\int_{x_0}^{x}\omega_q) \text{ modulo periods } (\int_\sigma \omega_1,\ldots,\int_\sigma \omega_q), \gamma \in H_1(X,\mathbb{Z})$$

in $\mathbb{C}^q/L$ where L is the lattice of integral periods.

If X is algebraic, then using the hyperplane class it is easy to show that Riemann's bilinear relations hold and the complex torus T is an abelian variety. Moreover $\theta : X \to T$ satisfies the universal mapping property in (6.1) hence this is the Albanese mapping.

Note there is a close relation between $\mathrm{Alb}(X)$ and part of $\mathrm{Pic}(X)$

for a Kähler manifold X. Let

$$\mathrm{Pic}^0(X) = \mathrm{coker}\,(H^1(\mathbb{Z}) \to H^1(\mathcal{O})) = \ker(H^1(\mathcal{O}^*) \to H^2(\mathbb{Z}))$$

the group of holomorphic (or algebraic) line bundles which are homologically trivial. In the Kähler case $H^1(X,\mathbb{C}) = H^0(\Omega^1) \oplus H^1(\mathcal{O})$ from the collapsing of the $\bar\partial$ - spectral sequence,and the operation of complex conjugation extended from complex conjugation on $\mathbb{C}$ interchanges the two factors. Then

$$H^1(\mathcal{O}) \times \mathrm{Hom}(H^0(\Omega),\mathbb{C}) \longrightarrow \mathbb{C}$$

given by $(x,\omega) \mapsto \omega(\bar{x})$ extends the usual pairing $H^1(\mathbb{Z}) \times H_1(\mathbb{Z}) \to \mathbb{Z}$ and hence defines a pairing

$$\mathrm{Alb}(X) \times \mathrm{Pic}^0(X) \longrightarrow \mathbb{C}/\mathbb{Z} \simeq \mathbb{C}^*$$

(6.2) <u>Remark</u>. This pairing between $\mathrm{Alb}(X)$ and $\mathrm{Pic}^0(X)$ is a duality between the two complex tori.

(6.3) <u>Remark</u>. The irregularity q is the dimension of the Albanese variety exactly when $\Delta = 0$. Since $\mathrm{Alb}(X \times Y) = \mathrm{Alb}(X) \times \mathrm{Alb}(Y)$, for $\Delta = 0$ we have

$$q(X \times Y) = q(X) + q(Y)$$

For example if $X = C_1 \times C_2$ the product of two curves, then $q(C_1 \times C_2) = p(C_1) + p(C_2)$ since the Albanese variety of a curve is just its Jacobian.

§ 7 Exceptional curves and minimal surfaces.

(7.1) Definition. A morphism $f : Y \to X$ is a blowing up of $x_0 \in X$ provided $f|Y - f^{-1}(x_0) : Y - f^{-1}(x_0) \to X - \{x_0\}$ is an isomorphism with $f^{-1}(x_0) = C$ a curve on Y. The map f is also called the blowing down of $C \subset Y$ to $x_0 \in X$.

The terminology σ-process applied to x_0 was used by Hopf to refer to $f : Y \to X$. A blowing up exists for any $x_0 \in X$ and in the case where x_0 is a smooth point of X we can view $C = f^{-1}(x_0)$ as the space of tangent lines in the tangent space of X to x_0. Hence C is a rational curve and it has the basic property that $C^2 = -1$. By the genus formula $C \cdot K_X = -1$.

(7.2) Definition. An exceptional curve C of first kind is an irreducible rational curve with $C^2 = C \cdot K = -1$ on a surface X.

While the existence of blowing up is relatively elementary, the existence of blowing downs of curves is a deeper result, due to Grauert in the complex analytic case and in the algebraic case to Zariski.

(7.3) Theorem. Let C be an exceptional curve of the first kind on X. Then there is a blowing down $f : X \to W$ of C.

(7.4) Remark. A blowing up or down is a birational map and preserves the plurigenera, irregularity, and $\chi(\mathcal{O}_X)$. If $f : Y \to X$ is a blowing up of x_0 to C, then $b_2(Y) = b_2(X) + 1$, $e(Y) = e(X) + 1$, and $K_Y^2 = K_X^2 - 1$.

(7.5) Definition. A surface X is minimal provided it has no exceptional curves of the first kind.

Now the following is a basic result of Zariski which was extended to 2 dimensional schemes by Lichtenbaum and Shafarevic.

(7.6) Theorem. Every surface X can be blown down to a minimal model. Moreover the minimal model is unique except in the case of ruled surfaces.

This theorem is a starting point for the classification theory since we classify minimal surfaces, and hence, all surfaces in terms of their minimal model.

General criteria for the contractibility or blowing down of a divisor D on a surface X to a (possibly singular) point on a surface $\bar{X}$ have been studied by Mumford [1961], Artin [1962] in the algebraic case, and by Grauert [1962] in the analytic case.

This question will come up again in part 5, (1.5) and (1.6), and we just mention the following necessary condition for contractibility contained in Mumford [1961].

(7.7) Proposition. Let $D = \sum n_i E_i$ be a connected divisor on a smooth surface X such that there exists a contraction $\pi : X \to \bar{X}$ of D to $\bar{x} \in \bar{X}$ where the restriction $X - D \to \bar{X} - \{\bar{x}\}$ is an isomorphism. Then the intersection matrix $(E_i \cdot E_j)$ is negative definite.

This result should be compared with 2(1.2) where a necessary condition is given for a divisor D to be the fibre of a map onto a curve.

Part 2. Elliptic and quasielliptic surfaces and fiberings

Elliptic and quasielliptic surfaces play a basic role in the theory of classification of surfaces because one has such a precise form for their canonical bundle. The canonical bundle formula was worked out by Kodaira for complex surfaces and extended to the general algebraic case by Bombieri-Mumford [1975]

In the final two sections we give a sketch of Kodaira's theory of elliptic fibrations in the complex analytic case, see [An II-III]. The general case is used by Artin in one of the two proofs of Castelnuovo's criterion which we outline in part 4. The general theory of elliptic families and fibrations is contained in Deligne-Rapoport [1974].

§ 1 Generalities on a surface fibred over a curve

The following algebraic result on symmetric bilinear forms leads to a necessary condition for a divisor to be a fibre of a map from a surface to a curve.

(1.1) Proposition. Let $x \cdot y$ by a symmetric bilinear form on a $\mathbb{Q}$-vector space V, and let $(e_i)_{i \in I}$ be elements generating V such that $e_i \cdot e_j \geq 0$ for $i \neq j$ and for some $z \in V$ where $z = \sum a_i e_i$ with $a_i > 0$ we have $z \cdot e_i = 0$ for all $i \in I$. Then $x \cdot x \leq 0$ for all $x \in V$, i.e. $x \cdot y$ is negative semidefinite.

Moreover, the set of $x \in V$ with $x \cdot x = 0$ is a subspace and its dimension equals the number of components in the 1-dimensional simplicial complex with vertices I and edges $\{i,j\}$ where $e_i \cdot e_j > 0$.

Proof. Recall that $bc \leq \frac{1}{2}(b^2+c^2)$ with equality if and only if $b = c$. Any $x \in V$ can be written $x = \sum c_i a_i e_i$ and from the hypotheses, it follows that

$$-\sum_i (c_i a_i e_i)\cdot(c_i a_i e_i) = \sum_{i,j}(c_i a_i e_i)\cdot(c_j a_j e_j) - \sum_i (c_i a_i e_i)(c_i a_i e_i)$$

$$= \sum_{i\neq j}(c_i a_i e_i)\cdot(c_i a_j e_j) = \tfrac{1}{2}\sum_{i\neq j} (c_i^2 + c_j^2)(a_i e_i)(a_j e_j)$$

$$\geq \sum_{i\neq j} c_i c_j (a_i e_i)\cdot(a_j e_j)$$

Hence $0 \geq \sum_{i,j} (c_i a_i e_i)(c_j a_j e_j) = x\cdot x$ with $0 = x\cdot x$ if and only if either $c_i = c_j$ or $e_i e_j = 0$ for all pairs $i \neq j$, thus in other words $0 = x\cdot x$ if and only if the funtion $i \mapsto c_i$ is constant on the component of I. This proves the proposition.

(1.2) Application I. Let $f : X \to B$ be a morphism of a surface onto a curve with fibre $D = \sum_i n_i E_i$ where the E_i are distinct irreducible curves and $n_i > 0$. Then $D\cdot E_i = 0$ because D is algebraically equivalent to a nearby fibre and $E_i \cdot E_j \geq 0$ for $i \neq j$. Hence by looking at the $\mathbb{Q}$-vector space generated by the E_i we deduce that $E_i^2 \leq 0$. This is a necessary condition for any divisor to be the fibre of such a morphism.

(1.3) Application II. Let X be a surface with $K^2 = 0$ and $K\cdot D \geq 0$ for all divisors $D \geq 0$. For $D \in |mK|$, $D \neq 0$, with $D = \sum_i n_i E_i$ and $n_i > 0$, we have $0 = (mK)\cdot K = D\cdot K = \sum n_i E_i \cdot K$. Since each $E_i \cdot K \geq 0$, we see that $E_i\cdot K = 0$ and $E_i \cdot D = 0$ for all i. Again (1.1) applies to the $\mathbb{Q}$-vector space generated by the E_i 's and we deduce that $E_i^2 \leq 0$ for $D = \sum n_i E_i \in |mK|$ on such a surface.

(1.4) Definition. A morphism $f : X \to B$ from a surface to a curve is called a fibering provided almost all fibres $f^{-1}(b) = X_b$ are irreducible curves. A fibering is called elliptic (resp. quasielliptic) when almost all fibres are nonsingular elliptic curves (resp. singular rational curves C with $p(C) = 1$).

The quasi-elliptic families exist only in characteristics 2 and 3 by a theorem of Tate [1952]. There are only a finite number of fibers $f^{-1}(b) = X_b$ of $f : X \to B$, called special fibres, such that X_b is not a nonsingular elliptic curve in the elliptic case or not a singular rational curve with $p(C) = 1$ in the quasielliptic case. These special fibres have been classified by Kodaira [An II, § 7] for surfaces over $\mathbb{C}$ and the classification has been extended to fiberings by Artin-Winters [1971].

(1.5) Remarks. For the general fibre F of a fibering $f : X \to B$ we can apply the genus formula 1 (3.5) since it is irreducible and we have

$$2p(F) - 2 = K \cdot F$$

because $F \cdot F = 0$ for a fibre. When $p(F) = 1$ as is the case for an elliptic or quasielliptic fibering, we deduce that $K \cdot F = 0$.

Going further, we assume that X has the property that $K \cdot D \geq 0$ for every $D \geq 0$. For an arbitrary fibre $D = \sum n_i E_i$ with $n_i > 0$ we have

$$K \cdot E_i = D \cdot E_i = 0 \qquad \text{and} \qquad E_i^2 \leq 0$$

as in the case (1.3).

This leads to the following class of divisors which the above considerations suggest are good candidates for fibres in an elliptic or quasi-elliptic fibration. We will see in part 3, § 4 that this is indeed the case.

(1.6) Definition. An effective divisor $D = \sum n_i E_i$ on X is of canonical type provided $K \cdot E_i = D \cdot E_i = 0$ for all i . In addition, D is indecomposable of canonical type provided D is connected and $\text{g.c.d.}(n_i) = 1$.

The following result is proved in Mumford [1969, pp. 332-333] and says that the scheme $D = \sum n_i E_i$ associated to an indecomposable divisor of canonical type has some of the properties of an irreducible curve with $p(C) = 1$.

(1.7) Proposition. Let $D = \sum n_i E_i$ be an indecomposable divisor of canonical type. If L is an invertible sheaf on the scheme D , and if $\deg(L \otimes \mathcal{O}_{E_i}) = 0$ for all i , then $H^0(D,L) \neq 0$ if and only if $L \cong \mathcal{O}_D$ and $H^0(D,\mathcal{O}_D) = k$. Moreover, we have an isomorphism

$$\mathcal{O}_X(K+D) \otimes \mathcal{O}_D = \mathcal{O}_D .$$

The last isomorphism should be compared with 1 (3.5).

Returning to elliptic and quasi-elliptic fibrations $f : X \to B$, we observe that among the special fibres there are those of the form nP where P is a divisor on X and $n > 1$. These are called <u>multiple fibres</u> and they play a basic role in the calculation of ω_X in terms of line bundles on B.

(1.8) <u>Definition</u>. We say $f : X \to B$ is reduced provided there are no multiple fibres.

(1.9) <u>Remark</u>. Kodaira shows in [An, III] how to remove multiple fibres by choosing a ramified covering $B^1 \to B$ in such a way that $X^1 = B^1 \times_B X \to B^1$ has no multiple fibres. More generally a simple proof of this stable reduction theorem can be found in Artin-Winters [1971].

It is the reduced families which lead themselves to classification as we shall see in § 3 and § 4.

(1.10) <u>Definition</u>. A surface X is elliptic (resp. quasi-elliptic) provided it has an elliptic (resp. quasi-elliptic) fibering $X \to B$.

§ 2. Canonical divisor of an elliptic fibration

Let $f : X \to B$ be an elliptic or quasi-elliptic fibration with $f_*\mathcal{O}_X = \mathcal{O}_B$. Then $R^1f_*\mathcal{O}_X = L \oplus T$ where L is a line bundle on B and T is a torsion sheaf. The support of T, $\mathrm{supp}(T)$, is called the set of <u>exceptional points</u> for f, and an <u>exceptional fibre</u> is one of the form $f^{-1}(b)$ for $b \in \mathrm{supp}(T)$. The results in this section will only be sketched and for the full details see the paper of Bombieri-Mumford.

(2.1) Proposition. A fibre $D = f^{-1}(b)$ is exceptional if and only if $\dim H^0(\mathcal{O}_D) \geq 2$.

Next we enumerate the multiple fibres $f^{-1}(b_i) = m_i P_i$ where $m_i \geq 2$. There are only finitely many such fibres and we choose m_i such that P_i is indecomposable. Each exceptional fibre is a multiple fibre. In general, all the fibres are divisors of canonical type, see (1.6), and the general fibre is smooth because $k(X)$ is separable over $k(B)$.

A fibration $f : X \to B$ is called relatively minimal provided there are no exceptional curves of the first kind contained in any fibre. Now we have the main theorem.

(2.2) Theorem. Let $f : X \to B$ be a relatively minimal elliptic or quasielliptic fibration with $f_*\mathcal{O}_X = \mathcal{O}_B$. Then

$$\omega_X = f^*(L^{-1} \otimes \omega_B) \otimes \mathcal{O}(\Sigma\, a_i P_i)$$

where (i) $0 \leq a_i \leq m_i - 1$, and

(ii) $a_i = m_i - 1$ if P_i is not exceptional, so that, length (T) $\geq$ number of $a_i < m_i - 1$. Moreover

$$\deg(L^{-1} \otimes \omega_B) = 2p(B)-2 + \chi(\mathcal{O}_X) + \mathrm{length}(T)$$

or

$$\deg(L) = -\chi(\mathcal{O}_X) - \mathrm{length}(T) .$$

Sketch of part of the proof

Step 1. The line bundle $\mathcal{O}(K_X) = f^*(M) \otimes \mathcal{O}(\Sigma a_i P_i)$ where $0 \leq a_i < m_i$ for some line bundle M on B , or in terms of divisors K_X is linearly equivalent to $f^{-1}(M) + \sum a_i P_i$ for some divisor M on B .

For this, let C_i be r nonsingular fibres of f . Then

$\mathcal{O}_{C_i} \otimes \mathcal{O}(K_X+C_i) \cong \mathcal{O}_{C_i}$ because C_i is an indecomposable divisor of canonical type (1.5); in fact, they are curves with $p(C_i) = 1$. Tensoring with $\mathcal{O}(K_X)$ the exact sequence

$$0 \to \mathcal{O} \to \mathcal{O}(\Sigma_i C_i) \longrightarrow \amalg_i \mathcal{O}_{C_i} \otimes \mathcal{O}(C_i) \longrightarrow 0 \quad ,$$

we obtain the exact sequence

$$0 \to \mathcal{O}(K_X) \to \mathcal{O}(K+\Sigma_i C_i) \longrightarrow \amalg_i \mathcal{O}_{C_i} \to 0 \quad .$$

From the exact cohomology sequence

$$0 \longrightarrow H^0(\mathcal{O}(K_X)) \longrightarrow H^0(\mathcal{O}(K+\Sigma_i C_i)) \longrightarrow \amalg_i H^0(C_i, \mathcal{O}_{C_i}) \to$$
$$\to H^1(\mathcal{O}(K_X)) \to \cdots \quad ,$$

we derive the inequality

$$\dim H^0(\mathcal{O}(K+\Sigma_i C_i)) \geq r + p_g - q = r - 1 + \chi(\mathcal{O})$$

or

$$\dim|K+\Sigma_i C_i| \geq r - 2 + \chi(\mathcal{O}) \ .$$

For r large enough we can find $D \in |K+\Sigma_i C_i|$.

From (1.5) we see that $D \cdot F = 0$ for each fibre F and hence D is supported in a union of fibres and we can write

$$D = \Delta + \Sigma_j m_j C_j^1 + \Sigma a_i P_i \quad \text{with} \quad 0 \leq a_i < m_i$$

where Δ does not contain any fibres or submultiples of P_i.

For each component Δ_ν of Δ and component E of a fibre, by (1.5), $E \cdot \Delta_\nu = 0$ and $\Delta_\nu^2 = K^2 = 0$; so Δ_ν is a rational multiple of a fibre containing it by (1.1), thus each $\Delta_\nu = 0$. Hence

$$K + \sum_i C_i = \sum_j m_j C_j^1 + \sum a_i P_i$$

which prove the assertion in the first step.

Step 2. The duality theorem for a map says that

$$M = f_*(\omega_X) = \mathrm{Hom}\,(R^1 f_* \mathcal{O}_X, \omega_B) = L^{-1} \otimes \omega_B$$

because the dual of the torsion part is zero. This can be found in Deligne-Rapoport pp. 19-20, formula (2.2.3). Hence

$$\omega_X = f^*(L^{-1} \otimes \omega_B) \otimes \mathcal{O}(\sum a_i P_i)$$

Step 3. From the spectral sequence of the map f we have

$$\begin{aligned} \chi(\mathcal{O}_X) &= \chi(\mathcal{O}_B) - \chi(\mathbb{R}^1 f_*(\mathcal{O}_X)) \\ &= 1 - p(B) - \chi(L) - \mathrm{length}(T) \\ &= (1-p(B)) - (\deg(L) + 1 - p(B)) - \mathrm{length}(T) \end{aligned}$$

Hence $-\deg(L) = \chi(\mathcal{O}_X) + \mathrm{length}(T)$ which is one form of the last formula.

Finally the assertions (i) and (ii) are proved by more intricate arguments which we do not include here.

§ 3 Outline of the classification of elliptic families

Our primary interest is in elliptic families over $\mathbb{C}$ where we present Kodaira's classification, but the general case will be used in a sketch of the proof of Castelnuovo's criterion, see part 4. For a discussion of the general classification theory see Deligne-Rapoport [1974].

(3.1) <u>Proposition</u>. Let $E \to B$ be a reduced elliptic fibration with a cross section $e : B \to E$, let $B^{\#}$ denote B minus all $b \in B$ where E_b is special, and let $E^{\#} = E|B^{\#}$. There is a unique group structure $E^{\#} \times_{B^{\#}} E^{\#} \to E^{\#}$ which induces the unique group structure on each fibre $(E_b, e(b))$ with base point for $b \in B^{\#}$.

In the complex analytic case see [An II, Th'm 9.1].

(3.2) <u>Theorem</u>. To each reduced elliptic (or quasielliptic) fibration $f : X \to B$ we can associate a unique elliptic (or quasielliptic) fibration $J_B(X) \to B$ of groups such that X/B is a principal homogeneous space over $J_B(X)/B$.

In the complex analytic case see [An II, § 8]. Given an elliptic (or quasielliptic) fibration $E \to B$ of groups we denote by $I(E/B)$ the set of isomorphism classes of fibrations X/B having $J_B(X) \cong E$. In order to give a cohomological description of the set $I(E/B)$ we introduce the sheaf $\mathcal{H}^0(E/B)$ of local cross sections of $f : E \to B$. The set $I(E/B)$ has a functoriality property relative to $u : B' \to B$, namely there is an induced function $u^* : I(E/B) \to I(u^*E/B')$ and $\mathcal{H}^0(u) : \mathcal{H}^0(E/B) \to \mathcal{H}^0(u^*E/B')$ is defined in an evident manner.

(3.3) <u>Theorem</u>. For any elliptic (or quasielliptic) fibration of groups E/B we have a function $\theta(E/B) : I(E/B) \to H^1(B, \mathcal{H}^0(E/B))$ such that for $u : B' \to B$ the following diagram is commutative

$$\begin{array}{ccc} I(E/B) & \xrightarrow{\theta(E/B)} & H^1(B, \mathcal{H}^0(E/B)) \\ {\scriptstyle u^*}\downarrow & & \downarrow{\scriptstyle u^*} \\ I(u^*E/B') & \xrightarrow{\theta(u^*E/B')} & H^1(B, \mathcal{H}^0(u^*E/B')) \end{array}$$

Moreover $\theta(E/B)$ is a bijection carrying the class of E/B into zero in $H^1(B, \mathcal{H}^0(E/B))$.

In the complex analytic case see [An II, § 10]. Thus the sheaf

$\mathcal{H}^o(E/B)$ plays the role of $\mathbb{G}_m$ where $H^1(\mathbb{G}_m)$ classifies the line bundles.

§ 4 Special features of the complex analytic case.

The following basic result distinguishes between algebraic and nonalgebraic compact complex surfaces fibered by elliptic curves.

(4.1) Theorem. Let $f : E \to B$ be a reduced family of elliptic curves where E is a group over B and E is algebraic. Then for the elliptic fibration $X \to B$ corresponding to $\eta \in H^1(B, \mathcal{H}^o(E))$ it follows that X is algebraic if and only if η is of finite order.

See Kodaira [An III, Th'm 11.5].

We saw in the previous section that $\mathcal{H}^o(E/B)$ played a role similar to that of $\mathbb{G}_m$ and now we consider the analogue of the sequence $0 \longrightarrow \mathbb{Z} \longrightarrow \mathcal{O} \xrightarrow{\exp} \mathcal{O}^* \longrightarrow 1$.

(4.2) Proposition. For a complex analytic, reduced family of elliptic curves $f : E \to B$ with group structure, we have an exact sequence of sheaves

$$0 \longrightarrow R^1 f_* \mathbb{Z} \longrightarrow \mathcal{H}^o(T(E)/B) \longrightarrow \mathcal{H}^o(E/B) \longrightarrow 0$$

where $\mathcal{H}^o(T(E)/B)$ is the sheaf of local cross sections to $T(E) \longrightarrow B$ and $T(E)_b$ is the tangent space to E_b at $e(b)$.

See Kodaira [An III, Th'm 11.2].

This sequence is useful for at least two things.

(4.3) Remark. The complex manifold $T(E)/B$ plays the role of a deformation space and the exact sequence

$$H^1(B,\mathcal{H}^\circ(T(E)/B)) \longrightarrow H^1(B,\mathcal{H}^\circ(E/B)) \xrightarrow{\delta} H^2(B,R^1f_*\mathbb{Z})$$

can be used to divide X/B and X'/B with respective classes η and η' in $H^1(B,\mathcal{H}^\circ(E/B))$ into equivalence classes where one is a deformation of the other if and only if $\delta(\eta) = \delta(\eta')$ in $H^2(B,R^1f_*\mathbb{Z})$, see Kodaira [An III, Th'm 11.4].

(4.4) Remark. Let S be a finite set in B. Given L a local system on $B - S$ locally isomorphic to $\mathbb{Z}^2$ and given a multivalued function on $B - S$ to the upper half plane such that on $\widehat{B - S}$ the universal covering space each element of $\pi_1(B-S)$ transforms $\mathbb{Z}^2$ and τ by the same element of $SL_2(\mathbb{Z})$, then Kodaira [An II, § 8] constructs a group elliptic fibration $f : E \to B$ such that $R^1f_*\mathbb{Z}|(B - S) \cong L$ and $j\tau(b)$ is the j-invariant for the elliptic curve E_b for $b \in B - S$.

Part 3. Classification by canonical dimension

In his book [1943] Enriques worked out a classification of surfaces where the canonical dimension was the first invariant dividing surfaces into four classes vaguely resembling the situation in botany. For certain canonical dimensions additional numerical invariants lead to a more refined classification.

In his two series of papers [An and Am] Kodaira extended many features of the Enriques classification to compact complex surfaces. Recently it has been shown in Mumford [1969] and Bombieri-Mumford [to appear] the classification described by Enriques in his book can be completely worked out in such a way as to include the case of characteristic p.

The key to our approach is the calculation of the canonical dimension of elliptic or quasi-elliptic surfaces and the proof that under certain conditions on numerical invariants the surfaces are elliptic or quasi-elliptic. Perhaps it should be noted that this approach does not depend on the Castelnuovo criterion.

§ 1 The canonical ring and dimension

The locally free sheaves Ω_X^r of germs of differential forms provide a rich source of birational invariants as we see in the following proposition which holds in both the algebraic and complex analytic case (modulo resolution of singularities).

(1.1) Proposition. (Kähler) Let $F_X = \otimes_{i=1}^m \Omega_X^{r(i)}$ for $r(i) \leq \dim X$. Then $\dim H^0(X,F_X)$ is a birational invariant for a given set $r(1),\ldots,r(n)$ of natural numbers.

Proof. If $X' \to X$ is separable dominating a rational map, then we have dominating morphisms and nonempty open sets $V \subset Y$ and $U' \subset X'$

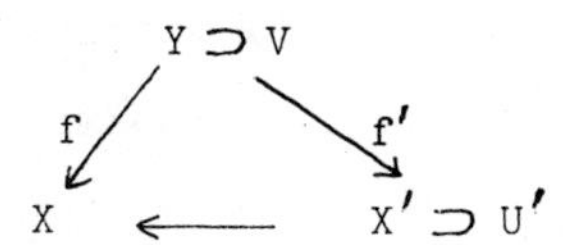

such that $X'-U'$ has codimension ≥ 2 in X' and $f'|V: V \rightarrow U'$ is an isomorphism. Then we have the following sequence of monomorphisms

$$H^0(X,F_X) \xrightarrow{f^*} H^0(Y,F_Y) \longrightarrow H^0(V,F_V) \underset{\sim}{\xrightarrow{g^*}} H^0(U',F) \underset{\sim}{\longleftarrow} H^0(X',F)$$

where $g = (f'|V)^{-1}$. The last isomorphism follows because codim $X'-U'$ in X' is ≥ 2 by Hartogs' theorem in the complex analytic case and by properties of normality in the algebraic case. Thus it follows that $\dim H^0(X,F_X) \leq \dim H^0(X',F_{X'})$ and the inverse morphism is defined in the case where $X' \to X$ is birational.

As an example, for surfaces, $H^2(T_X)^* = H^0(\Omega^1 \otimes \Omega^2)$ is a birational invariant while $\dim H^0(T_X)$ is not a birational invariant, i.e. the number of linearly independent vector fields can vary as one changes from one birational model to another.

For $s_1 \in H^0(X,L_1)$ and $s_2 \in H^0(X,L_2)$ the product $s_1 \otimes s_2 \in H^0(X,L_1 \otimes L_2)$ is given by $(s_1 \otimes s_2)(x) = s_1(x) \otimes s_2(x)$, and if $s_1 \otimes s_2 = 0$, then either $s_1 = 0$ or $s_2 = 0$. With this product $\coprod_{0\leq n} H^0(X,L^{n\otimes})$ is an integral domain containing the ground field k as a subfield. In particular we can speak of the transcendence degree over k .

(1.2) Definition. The canonical ring $R(X)$ of a variety X is $\coprod_{0\leq n} H^0(X,\omega_X^{n\otimes})$ and the canonical dimension $\kappa(X)$ of X is the transcendence degree of $R(X)$ over k minus 1 .

(1.3) Remarks. If $f : X \to Y$ is a separable dominating rational map, then $\kappa(X) \leq \kappa(Y)$. If moreover f is an étale covering of smooth complete varieties, then $\kappa(X) = \kappa(Y)$. Also the inequality

$$-1 \leq \kappa(X) \leq \dim X$$

holds for a smooth proper algebraic variety over k or a compact complex manifold X . The canonical ring $R(X)$ and the canonical dimension are birational invariants by (1.1). Observe that the dimension over k of the subspace of homogeneous polynomials of degree m in $k[Y_o,\dots,Y_r]$ is $\binom{m+r}{r}$. By comparing the canonical ring $R(X)$, which is an integral domain, with this polynomial ring we see that $\kappa(X) = r$ if and only if the plurigenus $P_m = \dim_k H^o(\omega_X^{m\otimes}) \sim m^r \cdot \frac{c}{r!}$ where c is a constant > 0 . Note $\kappa(X) = -1$ means all $P_m = 0$ and $\kappa(X) = 0$ implies $P_m \leq 1$.

Observe for integers $r_1,\dots r_n$, the multigenera $P(r_1,\dots,r_n) =$ $= \dim H^o(X,\Omega_X^{r_1} \otimes\dots\otimes\Omega_X^{r_n})$ are birational invariants by (1.1). For an m dimensional variety $P_n = P(m,\overset{(n)}{\dots},m)$.

The Kodaira dimension $\bar{\kappa}(X)$ of X is given by the relation $\bar{\kappa}(X) =$ $= \kappa(X)$ for $\kappa(X) \geq 0$ and $\bar{\kappa}(X) = -\infty$ for $\kappa(X) = -1$. Conjecturally $\bar{\kappa}(X)$ should be additive for fibrations. The study of the Kodaira dimension for complex manifolds has been the object of important work by Iitaka [1970], [1971], and Ueno [1973].

Let $\mathbb{C}(X)$ denote the field of meromorphic functions on a complex manifold X and view $\mathbb{C}$ as the subfield of constant functions.

(1.4) Remark. If $\dim \mathbb{C}(X)$ denotes the transcendence degree of $\mathbb{C}(X)$ over $\mathbb{C}$, then we have the relations

$$- 1 \leq \kappa(X) \leq \dim \mathbb{C}(X) \leq \dim X$$

The importance of $\dim \mathbb{C}(X)$ for the relation between complex analytic surfaces and complex algebraic surfaces is contained in the following basic result which we only quote.

(1.7.) Theorem. (Chow-Kodaira) For a complex surface X the following are equivalent:

(1) X is the underlying manifold of the complex points on a algebraic variety defined over $\mathbb{C}$,

(2) $\dim \mathbb{C}(X) = \dim X$, and

(3) there is a line bundle L on X with $L^2 > 0$.

We remark that (1) and (2) are not equivalent in dimensions ≥ 3 by Moišezon.

(1.8) Canonical dimension of curves. We have the following table.

κ	$p(X)$	$e(X)$	$\chi(\mathcal{O}_X)$	$\deg(K)$	structure
-1	0	$2>0$	$1>0$	$-2<0$	$X = \mathbb{P}^1$
0	1	0	0	0	$X =$ elliptic curve
1	$2 \leq p(X)$	$2-2p(X)<0$	$1-p(X)<0$	$2p(X)-2>0$	

(1.9) Examples of the canonical dimension of surfaces.

(1) For $\kappa = -1$ any $\mathbb{P}^1 \times C$ where C is a curve.

(2) For $\kappa = 0$ any $E \times E'$ product of two elliptic curves.

(3) For $\kappa = 1$ any $E \times C$ where E is an elliptic curve and C has genus ≥ 2.

(4) For $\kappa = 2$ and $C \times C'$ where C and C' are curves of genus ≥ 2.

These simple examples lead to our basic philosophy in the classification theory which is to look for pencils of elliptic curves on surfaces. From this we will be able to describe the cases of canonical dimension 0 and 1. After this the characterizations of canonical dimensions -1 and 2 will follow.

§ 2 The canonical dimension of an elliptic surface

Recall that for a line bundle M on a curve B we have

(i) $\dim H^0(B, M^{\otimes n}) \sim n \cdot \deg(M)$ for $n > 0$ and $\deg(M) > 0$

(ii)

$$\dim H^0(B, M^{\otimes n}) = \begin{cases} 1 & \text{for } M \cong \mathcal{O} \\ 0 & \text{for } \deg(M) < 0 \text{ or } \deg(M) = 0 \\ & \text{and } M \not\cong \mathcal{O} \text{ and for } n > 0 . \end{cases}$$

This remark yields immediately the statement about the canonical dimension of curves in (1.8) since $\deg(\omega_B) = 2p(B)-2$ and $\omega_B \cong \mathcal{O}_B$ for $p(B) = 1$. For $p(B) = 0$ we will make use of the fact that $M \cong \mathcal{O}(rP)$ for any point P and $r = \deg(M)$. The above remark is also the first step in our calculation of the canonical dimension of an elliptic or quasi-elliptic surface.

As in §2, part 2, we consider an elliptic or quasi-elliptic fibration $f : X \to B$ with $f_*\mathcal{O}_X = \mathcal{O}_B$ and multiple fibres $f^{-1}(y_i) = m_iP_i$ where $m_i \geq 2$. Then for $0 \leq a_i \leq m_i - 1$

$$\omega_X = f^*(L^{-1} \otimes \omega_B) \otimes \mathcal{O}(\Sigma a_iP_i)$$

and

$$\omega_X^{n\otimes} = f^*(L^{-n\otimes} \otimes \omega_B^{n\otimes}) \otimes \mathcal{O}(\Sigma a_inP_i)$$

$$= f^*(L^{-n\otimes} \otimes \omega_B^{n\otimes} \otimes \mathcal{O}(\Sigma \frac{a_i}{m_i} n\, y_i)) \quad \text{when all } m_i | n .$$

(2.1) Remarks. For natural numbers n such that all $m_i | n$, we have by a local calculation that

$$H^0(X, \omega_X^{n\otimes}) \xrightarrow{\sim} H^0(B, L^{-n\otimes} \otimes \omega_B^{n\otimes} \otimes \mathcal{O}(\sum_i \frac{a_i}{m_i} n\, y_i))$$

and

$$\deg(L^{-n\otimes} \otimes \omega_B^{n\otimes} \otimes \mathcal{O}(\sum_i \frac{a_i}{m_i} n\, y_i))$$

$$= n(2p(C)-2) + n\,\chi(\mathcal{O}_X) + n \text{ length } T + (\sum \frac{a_i}{m_i})n .$$

This formula contains most of the information that we will need to compute $\kappa(X)$ for elliptic or quasielliptic X. The formula for ω_X also implies that that $K_X^2 = c_1^2 = 0$ and thus the Riemann-Roch relation becomes

$$10 - 8q + 12p_g = b_2 + 2\Delta$$

where Δ is even and $0 \le \Delta \le 2p_g$ for algebraic surfaces ($\Delta = 0$ in characteristic zero) and $\Delta \equiv b_1 \pmod 2$ with $\Delta = 0,1$ for complex surfaces ($\Delta = 0$ for Kähler surfaces).

(2.2) Remarks and enumeration of cases. With the following preliminary remarks we see that $\kappa(X)$ can be calculated in many cases:

(i) The canonical dimension $\kappa(X) \le 1$ for $f:X \to B$ by (2.1).

(ii) If $p_g \ge 2$ or, more generally, if any $P_m \ge 2$, then $\kappa(X) = 1$.

(iii) If $p_g \le 1$, then $\chi(\mathcal{O}_X) \ge 0$ by $10-8q+12p_g = b_2+2\Delta$ also $q = 0,1$ for $p_g = 0$ and $q = 0,1,2$ for $p_g = 1$. Hence $\chi(X) = 0, 1, 2$. Now using (2.1), we will enumerate the various possibilities for $\kappa(X) \le 0$.

(a) $\chi(\mathcal{O}_X) = 2$ so $p(B) = 0$, length$(T) = 0$, and no multiple fibres. Then $p_g = 1$, $q = 0$, $\Delta = 0$, and $b_2 = 22$. These are K3 surfaces.

(b) $\chi(\mathcal{O}_X) = 1$ so $p(B) = 0$ and length$(T) \le 1$. These are three possibilities: $p_g = q = 1$, $\Delta = 0$; $p_g = q = 0$, $\Delta = 0$; and $p_g = q = 1$ and $\Delta = 2$.

(b1) $p_g = q = 1$, $\Delta = 0$ will not occur with canonical dimension $= 0$ Since $B = \mathbb{P}^1$ when $\chi(\mathcal{O}) = 1$, the canonical bundle formula shows that $\omega_X = \mathcal{O}_X$ and it can be shown, see Bombieri-Mumford, that this implies $\dim \mathrm{Alb}(X) = 0$ which contradicts $q = 1$ and $\Delta = 0$.

(b2) $p_g = q = 0$, $\Delta = 0$, so length$(T) = 0$ and $b_2 = 10$. Either there are no multiple fibres and $\kappa = -1$, or there are multiple fibres with $a_i = m_i - 1$ and

$$-1 + \sum_i \frac{m_i - 1}{m_i} = 0 \quad .$$

The only solution to this relation with $m_i \ge 2$ is $(m_1, m_2) = (2,2)$. In this case $\omega_X = \mathcal{O}(-F) \otimes \mathcal{O}(P_1+P_2)$ where F is a fibre since $L \otimes \omega_B = \mathcal{O}(-b)$ for $b \in B$. Observe that $\omega_X \not\cong \mathcal{O}_X$ while $\omega_X^{2\otimes} = \mathcal{O}(-2F) \otimes \mathcal{O}(2P_1+2P_2) \cong \mathcal{O}_X$ or in terms of the canonical divisor K,

$K \neq 0$ and $2K = 0$. These are the <u>classical Enriques surfaces</u>.

(b3) $p_g = q = 1$, $\Delta = 2$, so length(T) = 1 and $b_2 = 10$. There is one exceptional fiber $m_1 > 1$ supported over one point b_1. Since $\sum_i \frac{a_i}{m_i} = 0$, we see that $\omega_X = \mathcal{O}(0\, P_1) = \mathcal{O}_X$ and $K_X = 0$. This happens only in characteristic 2 and X is a <u>nonclassical Enriques surface</u>, see Bombieri-Mumford.

(c) $\chi(\mathcal{O}_X) = 0$ so either $p(B) = 0$ and length(T) ≤ 2 or $p(B) = 1$ and length(T) = 0.

(c1) $p_g = 1$, $q = 2$, $\Delta = 0$ so $b_2 = 6$. Then ω_X has a section and thus $L^{-1} \otimes \omega_B$ has a section which means it is of degree 0. For either $p(B) = 0$ or 1, we see that $L^{-1} \otimes \omega_B \cong \mathcal{O}_B$ and $\omega_X = \mathcal{O}(\sum a_i P_i)$. Since $\omega_X^{n\otimes} = \mathcal{O}(\sum n a_i P_i)$ and $P_m \leq 1$ for all n, it follows that $\omega_X \cong \mathcal{O}_X$. When length(T) = 0 there are no exceptional fibres with $a_i = 0$. From $\omega_X \cong \mathcal{O}_X$ or $K_X = 0$, it can be shown that $X \to \mathrm{Alb}(X)$ is an isomorphism, see Bombieri-Mumford.

(c2) $p_g = 1$, $q = 2$, $\Delta = 1$. Then X is a complex analytic non-algebraic surface with $b_1 = 3$ and $\omega_X = 0$. These surfaces form a one dimensional complex analytic family and are studied in detail in Kodaira [Am II]. We come back to them in § 6, and these are called <u>Kodaira surfaces</u>.

Any other algebraic case arising with $p(B) = 1$ reduces to $p(B) = 0$ by the following theorem in Bombieri-Mumford.

(2.3) <u>Theorem</u>. If $P_m \leq 1$, $\kappa(X) = 0$; if either $q = 1$, $p_g = 0$, $\Delta = 0$ or $q = 2$, $p_g = 1$, $\Delta = 2$; and if $X \to B$ is an elliptic fibration with $p(B) = 1$, then there is a second elliptic fibration $X \to \mathbb{P}^1$.

Note that for these two cases $b_2 = 2$ and $b_1 = 2$. We have three subcases (c3) - (c5) and $\omega_X = \mathcal{O}(rF) \otimes \mathcal{O}(\sum a_i P_i)$ where F is a general fiber of $X \to \mathbb{P}^1$. Moreover for all n with each $m_i | n$, we have again

$H^0(X,\omega_X^{n\otimes}) \longrightarrow H^0(B,\mathcal{O}(nry) \otimes \mathcal{O}(\sum n \frac{a_i}{m_i} y_i)$ where

$$r = -2 + \chi(\mathcal{O}_X) + \text{length}(T) = -2 + \text{length}(T)$$

in this case. If $\deg(\mathcal{O}(nry) \otimes \mathcal{O}(\sum_i n \frac{a_i}{m_i} y_i)) = n(r + \sum_i a_i/m_i) < 0$, then $\kappa(X) = -1$.

(c3) $p_g = 1$, $q = 2$, $\Delta = 2$ where length(T) = 1 or 2 . If length(T) = 1 , there is only on exceptional fibre, $1 = \sum \frac{a_i}{m_i}$, and this occurs only in characteristic 2 . If length(T) = 2 , there can be one point of multiplicity two in T or two points of multiplicity one, and this occurs only in characteristic 2 or 3 . These are the <u>nonclassical hyperelliptic surfaces</u> and are enumerated in the paper of Bombieri-Mumford.

(c4) $p_g = 0$, $q = 1$, $\Delta = 0$ and so length(T) = 0 .

(c4)' If $\sum_i (1 - \frac{1}{m_i}) < 2$, then $\kappa(X) = -1$, and

(c4)'' if $\sum_i (1 - \frac{1}{m_i}) = 2$, then $\kappa(X) = 0$, and there are four possible solutions:

(i) $(m_1,m_2,m_3,m_4) = (2,2,2,2)$ and $\omega_X^{2\otimes} \cong \mathcal{O}$ or $2K_X = 0$.

(ii) $(m_1,m_2,m_3) = (3,3,3)$ and $\omega_X^{3\otimes} \cong \mathcal{O}$ or $3K_X = 0$.

(iii) $(m_1,m_2,m_3) = (2,4,4)$ and $\omega_X^{4\otimes} \cong \mathcal{O}$ or $4K_X = 0$.

(iv) $(m_1,m_2,m_3) = (2,3,6)$ and $\omega_X^{6\otimes} \cong \mathcal{O}$ or $6K_X = 0$.

For example in case (iv) we have

$$\omega_X = \mathcal{O}(-2F) \otimes \mathcal{O}(P_1) \otimes \mathcal{O}(2P_2) \otimes \mathcal{O}(5P_3)$$

where $2P_1$, $3P_2$, and $6P_3$ are the three multiple fibres. Then

$$\begin{aligned}\omega_X^{6\otimes} &= \mathcal{O}(-12F) \otimes \mathcal{O}(6P_1) \otimes \mathcal{O}(12P_2) \otimes \mathcal{O}(30P_3) \\ &\cong \mathcal{O}(-12F) \otimes \mathcal{O}(3F) \otimes \mathcal{O}(4F) \otimes \mathcal{O}(5F) \cong \mathcal{O}\end{aligned}$$

These are the classical hyperelliptic surfaces.

(c5) $p_g = 0$, $q = 1$, $\Delta = 0$ with length(T) = 1 . There are at most two multiple fibres, one of which is exceptional. Using additional arguments, see Bombieri-Mumford, we can show that either $\kappa(X) = -1$; $\omega_X^{2\otimes} \cong \mathcal{O}_X$ i.e. $2K_X = 0$; or $\omega_X^{3\otimes} \cong \mathcal{O}_X$, i.e. $3K_X = 0$. These are further nonclassical hyperelliptic surfaces which occur only in characteristics 2 and 3.

(c6) $p_g = 0$, $q = 1$, $\Delta = 1$. Thus $b_1 = 1$ and $b_2 = 0$ and we have two subcases.

(c6)' If $\sum_i (1 - \frac{1}{m_i}) < 2$, then $\kappa(X) = -1$ and X is a Hopf surface, see Kodaira [Am II,III].

(c6)'' If $\sum_i (1 - \frac{1}{m_i}) = 2$, then $\kappa(X) = 0$ and there are four possible cases as in (c4) and again $\omega_X^{12\otimes} = \mathcal{O}_X$. These surfaces are quotients of Kodaira surfaces, see (c2), by fixed point free actions, see Kodaira [Am II].

(2.4) Summary. For a minimal elliptic or quasi-elliptic surface X we see that the above enumeration shows:

(1) $\kappa(X) = -1$ if and only if $H^0(X, \omega_X^{12\otimes}) = 0$ or $|12K| = \emptyset$.

(2) $\kappa(X) = 0$ if and only if $\omega_X^{12\otimes} = \mathcal{O}$ or $12 \cdot K = 0$. In particular, $D \cdot K_X \geq 0$ for all $D \geq 0$, and $K_X^2 = 0$.

(3) $\kappa(X) = 1$ if and only if $\omega_X^{12\otimes} \not\cong \mathcal{O}$ or $12K \not\equiv 0$ and $H^0(X, \omega_X^{12\otimes}) \neq 0$ or $|12K| \neq \emptyset$. The canonical bundle formula shows that $K_X \sim rC$ where $r > 0$ is a rational number and C is a fibre. Again $D \cdot K_X \geq 0$ for all $D \geq 0$ because $K_X^2 = r^2 C^2 = 0$.

With this information and with the existence theorems for elliptic and quasi-elliptic fibering, it will be easy to derive the main theorems of the classification theory.

§ 3 Characterization of elliptic nonalgebraic surfaces.

(3.1) Proposition. For a divisor D on a nonalgebraic surface X with $D^2 = 0$ it follows that $D \cdot K = 0$.

Proof. Consider $(K+mD)^2 = K^2+mD\cdot K$. If $D\cdot K \neq 0$, then for some m we have $(K+mD)^2 > 0$ and by 2(1.2), X is algebraic. Thus $K\cdot D = 0$.

(3.2) Corollary. If $f : X \to B$ is a morphism of a nonalgebraic surface onto a curve where almost all fibres are irreducible and nonsingular, then f is an elliptic fibration.

Proof. Let C be an irreducible fibre of f. Then $C^2 = 0$ and by (1.1) also $K\cdot C = 0$. Thus from $2p(C)-2 = K\cdot C+C^2 = 0$ we deduce that $p(C) = 1$.

(3.3) Theorem. For a nonalgebraic surface X,

(i) $\dim_{\mathbb{C}} \mathbb{C}(X) = 1$ if and only if X is elliptic, and

(ii) $\dim_{\mathbb{C}} \mathbb{C}(X) = 0$ if and only if X is nonelliptic.

Moreover if $f:X \to B$ is an elliptic fibration, then $f^*\mathbb{C}(B) = \mathbb{C}(X)$.

Proof. Assume $\dim_{\mathbb{C}} \mathbb{C}(X) = 1$, so we have a surjective morphism $\phi:X \to \mathbb{P}^1$. Taking the Stein factorization, we obtain $f:X \to B$ where almost all fibres are irreducible and nonsingular, and thus by (3.2) this $f:X \to B$ is an elliptic fibration.

To see $f^*\mathbb{C}(B) = \mathbb{C}(X)$, we consider a meromorphic function ψ on X which must be algebraic over $f^*\mathbb{C}(B)$ since $\dim \mathbb{C}(X) = 1$. If $\psi \in f^*\mathbb{C}(B)$, the Stein factorization $g:X \to B^1$ of ψ factors f by a finite covering $\pi : B^1 \to B$ of degree $= [f^*\mathbb{C}(B)(\psi), f^*C(B)]$ which contradicts the fact that $f : X \to B$ is a Stein factorization. So $\psi \in f^*\mathbb{C}(B)$ and $f^*\mathbb{C}(B) = \mathbb{C}(X)$. Thus if X is elliptic, then $\dim_{\mathbb{C}} \mathbb{C}(X) = 1$ and this proves the theorem.

(3.4) Remark. We put off a more complete discussion of the known nonalgebraic surfaces until part 6. For now, it suffices to say that there are K3 surfaces ($p_g = 1$, $q = 0$, $b_2 = 22$) and complex tori

($p_g = 1$, $q = 2$, $b_2 = 6$) with $\dim_{\mathbb{C}} \mathbb{C}(X)$ having any value 0,1, or 2 . For $\dim_{\mathbb{C}} \mathbb{C}(X) = 1$ and $\kappa(X) = 0$ there are Kodaira surfaces with $p_g = 0$, $q = 1$, $\Delta = 1$, $b_2 = 0$, $b_1 = 1$ and with $p_g = 1$, $q = 2$, $\Delta = 1$, $b_2 = 4$, $b_1 = 3$ which are like hyperelliptic surfaces, see (2.2) (c2), (c6). For $\kappa(X) = -1$ and either $\dim_{\mathbb{C}} \mathbb{C}(X) = 0$ or 1 there are Hopf surfaces with $p_g = 0$, $q = 1$, $\Delta = 1$, $b_2 = 0$, and $b_1 = 1$.

In view of (3.3) and (2.4) we have the following classification theorem similar to the Enriques classification theorem in (5.4).

(3.5) Theorem. For a minimal compact complex surface X with $\dim_{\mathbb{C}} \mathbb{C}(X) = 1$,

(a) $\kappa(X) = -1$ if and only if $H^0(\omega_X^{12\otimes}) = 0$.

(b) $\kappa(X) = 0$ if and only if $\omega_X^{12\otimes} = \mathcal{O}$.

(c) $\kappa(X) = 1$ if and only if $H^0(\omega_X^{12\otimes}) \neq 0$ and $\omega_X^{12\otimes} \not\cong \mathcal{O}$.

§ 4 Existence of algebraic elliptic and quasielliptic pencils

The following proposition is used both in this section and in the next part.

(4.1) Proposition. Let X be a surface with $P_2 = 0$, i.e. $|2K| = \emptyset$. Then $\dim|-K| \geq K^2 - q$.

Proof. The Riemann Roch inequality for $2K$ is

$$\dim|2K| + \dim|K - 2K| \geq \frac{1}{2} 2K\cdot(2K-K) + \chi(\mathcal{O}) - 2 .$$

The proposition follows from $\dim|2K| = -1$ and $\chi(\mathcal{O}) = 1-q$.

In view of the considerations in 2(1.5) and 2(1.6) we are led to the following existence statement for elliptic and quasielliptic fibrations.

(4.2) <u>Theorem</u>. (Mumford). Let X be a minimal surface with $K^2 = 0$ and $\kappa(X) \geq 0$. Then X is elliptic or quasielliptic if and only if it contains a curve D of canonical type. In fact, there is a natural number n such that the Stein factorization of $\phi_{nD} : X \to |nD|$ is a pencil of curves of canonical type.

For the proof, see Mumford [1969]; note that his argument applies also in the complex analytic case. The next result gives the existence of curves of canonical type on a surface.

(4.3) <u>Theorem</u>. (Enriques). Let X be an algebraic minimal surface with $K^2 = 0$ and $\kappa(X) \geq 0$. Then either $2K = 0$ or X contains at least one indecomposable divisor of canonical type.

<u>Sketch of proof</u>. We have $|-K| = \emptyset$ otherwise either $K = 0$ or $K \cdot E < 0$ for ample E so that $\kappa(X) = -1$. Hence by (4.1) we have $q \geq 1$. If $D \in |2K|$, then either $D = 0$ so $2K = 0$, or by 2(1.3), D is of canonical type, and by decomposing D, we obtain an indecomposable divisor of canonical type. If $|2K| = \emptyset$, then $p_g = 0$, $K^2 = 0$, and hence $\chi(\mathcal{O}_X) \geq 0$ as was already remarked. This leads to the subtle case of $p_g = 0$, $q = 1$ which is carried out in Mumford [1969, p. 331].

§ 5 The Enriques classification of algebraic surfaces

(5.1) <u>Definition</u>. A ruled surface X is one for which there exists a mao $f: X \to B$ onto a curve such that $f^{-1}(b)$ is a rational curve for each $b \in B$. The map f is called a ruling of X.

(5.2) <u>Definition</u>. Adjunction terminates on a surface X provided for every divisor D there exists an integer n_D such that $|D+nK| = \emptyset$ for all $n \geq n_D$.

(5.3) <u>Proposition</u>. For the following statements (1) implies (2), (2) implies (3) for minimal surfaces, and (3) implies (4).

(1) X is a ruled surface.

(2) There exists a curve C on X with $K\cdot C < 0$.

(3) Adjunction terminates on X.

(4) $K(X) = -1$, that is, all $|nK| = \emptyset$ for $n \geq 0$.

Proof. For (1) implies (2), let C be a fibre of a ruling. Then $C^2 = 0$ and by the genus formula $-2 = 2p(C)-2 = C^2+K\cdot C = K\cdot C$. For (2) implies (3) suppose that $K\cdot C < 0$ and choose n_D such that $nK\cdot C < -D\cdot C$ for all $n \geq n_D$. If $D_n \in |D+nK|$ for $n \geq n_D$, then $D_n\cdot C = D\cdot C+nK\cdot C < 0$ and so $C^2 < 0$ since $D_n \geq 0$.

For (3) implies (4) note that $|nK| = \emptyset$ for $n \geq n_K$. Since $E_1 \in |n_1K|$ and $E_2 \in |n_2K|$ implies that $E_1 + E_2 \in |(n_1+n_2)K|$ and $|nK| = \emptyset$ for all $n \geq 0$.

In the next part we will show that for a minimal surface (1), (2), (3), and (4) are equivalent and in this way we characterize surfaces of canonical dimension -1 as ruled surfaces.

Now we can state and prove the main theorem of the Enriques classification theory.

(5.4) Theorem. For a minimal algebraic surface X over an algebraically closed field

(a) $\kappa(X) = -1$ if and only if $K\cdot C < 0$ for some curve C.

(b) $\kappa(X) = 0$ if and only if $12K = 0$.

(c) $\kappa(X) = 1$ if and only if $K^2 = 0$, $K\cdot D \geq 0$ for all $D \geq 0$ and $12K \neq 0$.

(d) $\kappa(X) = 2$ if and only if $K^2 > 0$ and $K\cdot D \geq 0$ for all $D \geq 0$.

Proof. When $K^2 = 0$ and $K\cdot D \geq 0$ for $D \geq 0$, the classification of elliptic and quasielliptic fibrations (2.4) and the existence theorem (4.4) yields statements (b) and (c).

The converse of (a) follows from (5.3). The converse of (d) follows by applying Riemann Roch to $|mK|$

$$\dim |mK| + \dim|-(m-1)K| \geq \frac{m(m-1)}{2} K^2 + \chi(\mathcal{O})-2 \sim \frac{K^2}{2} m^2$$

as $m \to \pm \infty$. If $D \in |-(m-1)K|$, then $K \cdot D = -(m-1)K^2 < 0$ for $m > 1$ which is ruled out by the hypothesis $K \cdot D \geq 0$ for $D \geq 0$. Thus $\dim |mK| \sim m^2$ as $m \to +\infty$ and X is of canonical dimension 2 . By elimination of the various possibilities the theorem is now established.

(5.5) Remarks on the structure of algebraic surfaces. The structure of surfaces with $\kappa(X) = -1$ is considered in part 4 where they are characterized as ruled surfaces and the case $\kappa(X) = 2$ is dealt with in part 5. For $\kappa(X) = 1$ they are properly elliptic by (5.4)(c) and theorems (4.2) and (4.3).

For $\kappa(X) = 0$ we have the following table which should also be compared with the table 6(2.2).

	b_2	b_1	$e(X)$	q	p_g	$\chi(\mathcal{O})$	Δ
K 3 surfaces	22	0	24	0	1	2	0
Enrique surfaces classical	10	0	12	0	0	1	0
nonclassical (only in char. 2)	10	0	12	1	1	1	2
Abelian surfaces	6	4	0	2	1	0	0
Hyperelliptic surfaces classical	2	2	0	1	0	0	0
nonclassical (only in char. 2,3)	2	2	0	2	1	0	2

It should be mentioned that every classical Enriques surface is elliptic, see Enriques [1949], though this fact will not be proved here. We refer to Bombieri-Mumford for a closer analysis of algebraic surfaces with $\kappa(X) = 0$.

Part 4. Castelnuovo's criterion and the characterization of ruled surfaces

In 3(5.3) we saw that every minimal ruled surface X satisfies $\kappa(X) = -1$ or equivalently $|nK| = \emptyset$ for all $n \geq 1$. In this part we show that $\kappa(X) = -1$ implies that X is ruled or the projective plane $\mathbb{P}^2$. In the process we will establish the Castelnuovo criterion that X is rational ruled or $\mathbb{P}^2$ if and only if $P_2 = 0$ and $q = 0$. In this part all surfaces are algebraic over an arbitrary algebraically closed ground field k and usually minimal.

§ 1 Preliminaries on divisors D with $|K+D| = \emptyset$

The isomorphism $\mathcal{O}(K+D) \otimes \mathcal{O}_D = \omega_D$ suggests that divisors $D \geq 0$ with $|K+D| = \emptyset$ should be related to families of rational curves on a surface since an irreducible curve C is rational if and only if $|K_C| = \emptyset$. We consider two preliminary propositions about such divisors which used to construct fiberings of a surface by rational curves. Note these divisors D with $|D+K| = \emptyset$ exist on surfaces where adjunction terminates, definition 3(5.2).

(1.1) Proposition. Let D be an effective divisor such that $|K+D| = \emptyset$. Then $p_g = 0$, the natural map $\mathrm{Pic}^0(X) \to \mathrm{Pic}^0(D)$ is surjective, and $\mathrm{Pic}^0(D)$ is an abelian variety.

Proof. Since $|K+D| = \emptyset$, it follows that $|K| = \emptyset$, i.e. $p_g = 0$. Since $H^0(\mathcal{O}(K+D)) = 0$, we have $H^2(\mathcal{O}(-D)) = 0$ by Serre duality, and from the exact sequence

$$0 \longrightarrow \mathcal{O}(-D) \longrightarrow \mathcal{O} \longrightarrow \mathcal{O}_D \longrightarrow 0 \quad ,$$

we deduce that $H^1(\mathcal{O}) \to H^1(\mathcal{O}_D) = H^1(D,\mathcal{O}_D)$ is surjective. Since $p_g = 0$, the scheme $\mathrm{Pic}^0(X)$ is reduced, see Mumford [1966, lecture 22], and hence it is an abelian variety. Now $H^1(X,\mathcal{O})$ and $H^1(D,\mathcal{O}_D)$ are the tangent spaces to the identity on $\mathrm{Pic}^0(X)$ and $\mathrm{Pic}^0(D)$ respectively and thus $\mathrm{Pic}^0(X) \to \mathrm{Pic}^0(D)$ is surjective from the previous two statements. Finally, since $\mathrm{Pic}^0(X)$ is an abelian variety, $\mathrm{Pic}^0(D)$ is also.

(1.2) Proposition. Let D be an effective divisor such that $|K+D| = \emptyset$ and $D = \sum_i n_i E_i$ where $n_i > 0$ and E_i distinct. Then

(i) each E_i is nonsingular and $\sum_i p(E_i) \leq \dim \mathrm{Pic}^0(X) = q$,

(ii) the E_i are connected as a tree, and

(iii) if $n_j \geq 2$, then either

(a) E_j is rational,

(b) $E_j^2 < 0$, or

(c) E_j is elliptic with $E_j^2 = 0$ and $[E_j]_{E_j} \neq 0$, that is, E_j has a nontrivial normal bundle.

Proof. (i) Since $|K+D| = \emptyset$, it follows that $|K+E_j| = \emptyset$ and by (1.1) that $\mathrm{Pic}^0(D) \to \mathrm{Pic}^0(E_j)$ is surjective. Thus $\mathrm{Pic}^0(E_j)$ is an abelian variety, and E_j is nonsingular. The inequality follows from the relations $\sum_i p(E_i) = \sum \dim \mathrm{Pic}^0(E_i) \leq \dim \mathrm{Pic}^0(D) \leq \dim \mathrm{Pic}^0(X)$. Since $p_g = 0$, it follows that $q = \dim \mathrm{Pic}^0(X)$.

(ii) Any loop in the graph of the E_i's would contribute a subgroups subgroup G_m to $\mathrm{Pic}^0(D)$ which can not exist in an abelian variety.

(iii) For $n_j \geq 2$ we have $|2E_j+K| = \emptyset$, hence by (1.1) the morphism $\mathrm{Pic}^0(X) \to \mathrm{Pic}^0(2E_j)$ is surjective and $\mathrm{Pic}^0(2E_j)$ is an abelian variety. Thus $\mathrm{Pic}^0(2E_j) \to \mathrm{Pic}^0(E_j)$ is an isomorphism, and passing to the tangent space at 0 , we see that $H^1(E_j,\mathcal{O}_{2E_j}) \to H^1(E_j,\mathcal{O}_{E_j})$ is an isomorphism. This isomorphism and the exact sequence

$$0 \to \mathcal{O}_{E_j}(-[E_j]_{E_j}) \to \mathcal{O}_{2E_j} \to \mathcal{O}_{E_j} \to 0$$

yields $H^1(E_j,\mathcal{O}_{E_j}(-[E_j]_{E_j})) = 0$. Since $\deg \mathcal{O}_{E_j}(-[E_j]_{E_j}) = -E_j^2$, this $H^1 \neq 0$ by Serre duality and Riemann Roch except when E_j is rational with

$\deg \mathcal{O}_{E_j}(-[E_j]_{E_j}) = 0$ or -1 or when E_j is elliptic with $E_j^2 = 0$ and $\mathcal{O}_{E_j}(-[E_j]_{E_j}) = \mathcal{O}_{E_j}$. This proves the proposition.

We will make several uses of the following elementary proposition about linear systems having divisors passing through fixed points and having them as singular points.

(1.3) Proposition. Let $x_1,\ldots,x_s$ be s distinct points on a surface X , and let $\dim |D| \geq 3s$. Then there exists a divisor $D^* \in |D|$ with each x_i as a multiple point.

Proof. Consider the exact sequence
$0 \to \mathcal{O}(D) \otimes \mathcal{O}(-2x_1) \otimes \ldots \otimes \mathcal{O}(-2x_s) \to \mathcal{O}(D) \to \coprod_{1\leq i\leq s} k^3_{x_i} \to 0$ where $\mathcal{O}(-x)$ is the sheaf of ideals at x and k_x is the sheaf k concentrated just at x . Then $H^0(\mathcal{O}(D) \otimes \mathcal{O}(-2x_1)\otimes\ldots\otimes \mathcal{O}(-2x_s))$ is nonzero because $\dim H^0(\mathcal{O}(D)) > \dim H^0(\coprod k^3_{x_i}) = 3s$ and the divisor D^* of zeros of a nonzero element of $H^0(\mathcal{O}(D) \otimes \mathcal{O}(-2x_1)\otimes\ldots\otimes \mathcal{O}(-2x_s))$ is the element of $|D|$ which we are looking for, and this proves the proposition.

§ 2 Rulings arising from the existence of enough rational curves

In this section we show how to deduce that a surface is ruled by finding sufficiently many rational curves on it. The conditions depend on whether $q = 0$ or $q > 0$. This leads to the following classical terminology.

(2.2) Definition. A surface X is called regular when $q = 0$ and irregular when $q > 0$.

The following criterion is used for regular surfaces.

(2.2) Proposition. Let X be a minimal regular surface, and let C be a rational curve on X with $K\cdot C < 0$ and $|K+C| = \emptyset$. Then X is a ruled surface over $\mathbb{P}^1$ or isomorphic to $\mathbb{P}^2$. Moreover, the ruling or the isomorphism will be of the form $\phi_{|C_0|}$ where C_0 is a rational curve with $K\cdot C_0 < 0$ and $|K+C_0| = \emptyset$.

Proof. For a rational curve C on a minimal surface X note that $K\cdot C < 0$ if and only if $C^2 \geq 0$ by the genus formula. Let $c_o = \min C^2$ for C rational with $C^2 \geq 0$ and $|K+C| = \emptyset$, and let $\mathcal{C}$ denote the set of all rational curves C with $C^2 = c_o \geq 0$ and $|C+K| = \emptyset$. For the hyperplane class H, let $d_o = \min H\cdot C$ where $C \in \mathcal{C}$. We choose $C_o \in \mathcal{C}$ with $d_o = H\cdot C_o$ and hence also $C_o^2 \geq 0$ and $|C_o+K| = \emptyset$.

Next we show that every element of $|C_o|$ is a nonsingular rational curve. For this, let $D = \sum n_i E_i \in |C_o|$ where $n_i \geq 0$. Then $0 > K\cdot C_o = K\cdot D = \sum n_i K\cdot E_i$. Hence $E_j\cdot K < 0$ for some E_j and for these $E_j^2 \geq 0$. By (1.2) such a curve E_j is a nonsingular rational curve and we also have $|K+E_j| = \emptyset$. Thus $E_j^2 \geq c_o$, and we calculate

$$c_o = D\cdot D = \textstyle\sum_i n_i C_o\cdot E_i \geq n_j C_o\cdot E_j = n_j \sum_i n_i E_i\cdot E_j \geq n_j^2 E_j^2 \geq n_j^2 c_o$$

From this we see that $n_j = 1$ and $E_j \in \mathcal{C}$. Now calculate

$$H\cdot E_j \geq d_o = H\cdot C_o = H\cdot E_j + \textstyle\sum_{i\neq j} n_i H\cdot E_i$$

Since $H\cdot E_i > 0$ for all i, we see that $n_i = 0$ for $i \neq j$ and $D = E_j$. This proves that every member of $|C_o|$ is a nonsingular rational curve.

Now by (1.3) we see that $\dim|C_o| \leq 2$. From Riemann Roch we have $\dim|C_o| \geq 1+C_o^2$. For $C \in |C_o|$ and the restriction $\mathcal{O}_C(C_o)$ of $\mathcal{O}(C_o)$, we have the following exact sequence since $q = \dim H^1(\mathcal{O}) = 0$.

$$0 \to H^o(X,\mathcal{O}) \to H^o(X,\mathcal{O}(C_o)) \to H^o(C,\mathcal{O}_C(C_o)) \to 0$$

Thus the complete line system $|C_o|$ has no base points and $\dim|C_o| = C_o^2 + 1$, equals 1 or 2.

If $\dim|C_o| = 1$, then $\dim H^o(X,\mathcal{O}(C_o)) = 2$; $C_o^2 = 0$, and $\phi_{|C_o|} : X \to \mathbb{P}^1$ has as fibres the elements of $|C_o|$ which are nonsingular rational curves. So X is a rational ruled surface.

If $\dim|C_o| = 2$, then $\dim H^o(X,\mathcal{O}(C_o)) = 3$, $C_o^2 = 1$, and under

$\phi_{|C_0|} : X \to \mathbb{P}^2$ the inverse image of a line is an element of $|C_0|$. Thus $\phi_{|C_0|}$ is an isomorphism and X is isomorphic to $\mathbb{P}^2$. This proves the proposition.

(2.3) Remarks. For surfaces with $p_g = 0$ the basic Riemann Roch relations become

$$12(1-q) = K^2 + e(X) \qquad \text{or} \qquad 10 - 8q = K^2 + b_2$$

For regular surfaces $12 = K^2 + e(X)$ or $10 = K^2 + b_2$. Hence we deduce that $K^2 = 9$ for $\mathbb{P}^2$ and $K^2 = 8$ for a rational ruled surface. For irregular surfaces we have the following result which we bring in before the other ruling criterion.

(2.4) Proposition. Let X be a surface with $p_g = 0$ and $q > 0$. Then $K^2 \leqq 0$ and $K^2 < 0$ for $q \geqq 2$.

Proof. The Albanese map $X \to \mathrm{Alb}(X)$ is nontrivial and a fibre C has the property that $C^2 = 0$. This its image in $H^2(\mathbb{R})$ is linearly independent to the image of the hyperplane class. Thus $b_2 \geq 2$ and $K^2 \leq 0$. For $q \geq 2$ clearly $K^2 < 0$, and the proposition is proved.

The criterion for ruling an irregular surface is proven by using the Albanese mapping.

(2.5) Proposition. Let X be a minimal irregular surface such that each point $x \in X$ has a rational curve C through x with $|C+K| = \emptyset$. Then there is a factorization of $X \to \mathrm{Alb}(X)$

$$X \xrightarrow{\pi} B \to \mathrm{Alb}(X)$$

such that $\pi : X \to B$ is a ruling of X over a nonsingular curve B .

Proof. Under $X \xrightarrow{\alpha} \mathrm{Alb}(X)$ each rational curve maps to a point. We can factor α by $\pi : X \to B$ where B is nonsingular after normalization and the general fibre is irreducible with all fibres connected by Stein factorization.

A general fibre of $\pi : X \to B$ is a rational curve since it is irreducible and contains a rational curve by hypothesis. For an arbitrary fibre $D = \sum_{i=1}^{r} n_i E_i$, we have $D^2 = 0$ and $K \cdot D = -2$. By 2(1.2) and the connectivity of the fibres, $E_i^2 < 0$ unless $r = 1$. Then $K \cdot E_i \geq 0$ from the genus formula which contradicts $K \cdot D = -2$. Hence $D = nE$. Now $-2 = K \cdot D = nK \cdot E$ and $E^2 = 0$ from which wo deduce that $n = 1$. Each fibre D is a rational curve which proves the proposition.

§ 3 Preliminaries on surfaces with $K^2 \leq 0$

(3.1) Remark. By the classification theorem 3(5.4) the following are equivalent for a minimal surface X.

(a) There is an irreducible curve C with $K \cdot C < 0$.

(b) Adjunction terminates on X, see 3(5.2).

(c) $\kappa(X) = -1$, i.e. $|nK| = \emptyset$ for all $n \geq 1$.

(3.2) Lemma. If $K \cdot C < 0$ for an irreducible curve C on a minimal surface X, then for every $n > 0$ there exists an effective divisor D with $D^2 > 0$ and $K \cdot D \leq -n$.

Proof. Since X is minimal, $C^2 \geq 0$. For $m \geq 0$ and any divisor A with $A^2 > 0$, for example the hyperplane divisor,

$$(mC+A)^2 = m^2C^2 + 2m\, C \cdot A + A^2 \geq A^2 > 0$$

Since $K \cdot (mC+A) = mK \cdot C + K \cdot A \to -\infty$ as m gets large, some $D = mC+A$ has the desired properties.

(3.3) Proposition. Let X be a minimal surface with $\kappa(X) = -1$ and $K^2 \leq 0$. Then for all $n \geq 0$ there exists a divisor D with $|D+K| = \emptyset$, $D \cdot K < 0$, and $\dim |D| \geq n$.

Proof. By (3.2) there is an effective divisor Δ with $\Delta^2 > 0$ and $K\cdot\Delta \leq -2n-2q$ where $q = \dim H^1(\mathcal{O})$. By (3.1) there exists m with $D^* \in |\Delta+mK|$ and $|\Delta+(m+1)K| = \emptyset$. Hence $|D+K| = |\Delta+mK+K| = \emptyset$.

Now decompose $D^* = D+D'$ where each irreducible component C of D satisfies $K\cdot C < 0$ and of D' satisfies $K\cdot C \geq 0$. Since $D \geq 0$ and $|K| = \emptyset$, it follows that $|K-D| = \emptyset$. Now by Riemann-Roch

$$\dim|D| + \dim|K-D| \geq \frac{1}{2} D^2 - \frac{1}{2} D\cdot K + p_g - q - 1$$

and therefore

$$\begin{aligned}\dim|D| &\geq \frac{1}{2} D^2 - \frac{1}{2} D\cdot K - q \geq \frac{1}{2} D^2 - \frac{1}{2} D^*\cdot K - q = \\ &= \frac{1}{2} D^2 - \frac{1}{2} \Delta\cdot K - \frac{1}{2} mK^2 - q \\ &\geq \frac{1}{2} D^2 - \frac{1}{2} \Delta\cdot K - q \geq -\frac{1}{2} \Delta\cdot K - q \geq n\end{aligned}$$

Clearly $D\cdot K < 0$ since $D \neq 0$ and $|D+K| \subset |D^*+K| = \emptyset$ so that $|D+K| = \emptyset$. The divisor D has the desired properties.

(3.4) Remark. Independent of classification theory, one can prove that on a minimal surface where adjunction terminates and $K^2 \leq 0$ there exists $D > 0$ with $K\cdot D < 0$ and $|K+D| = \emptyset$. With this it is immediate to show that on a minimal surface X with $K^2 < 0$ there is an irreducible curve C with $K\cdot C<0$, and hence statements (a) (b) (c) of (3.1) hold by simple direct arguments. The last statement concerning surfaces with $K^2 < 0$ having $\kappa(X) = -1$ was deduced also from the algebraic index theorem in 1(5.12) and from the classification theorem in 3(5.4).

§ 4 Structure of irregular surfaces of canonical dimension -1

(4.1) Theorem. Let X be an irregular minimal surface. Then X has canonical dimension -1 if and only if X is ruled.

Proof. By 3(5.3) we know that if X is ruled, then $\kappa(X) = -1$. Conversely, by (2.5) we have only to show that through each point of X there is a nonsingular rational curve C with $|C+K| = \emptyset$. Suppose this does not hold, then there are only a finite number of curves E in each degree $E \cdot H$ such that either

(a) E is rational nonsingular,

(b) E is elliptic with $E^2 = 0$ and $\mathcal{O}_E(E) \not\cong \mathcal{O}_E$, or

(c) $E^2 < 0$.

These form a countable set $\mathcal{E}$ which can not cover X .

Let $x_1, \ldots, x_q$ be $q = \dim H^1(\mathcal{O})$ distinct points of X not on any of these curves. By (3.3) there is a divisor D with $D \cdot K < 0$, $|D+K| = \emptyset$ and $\dim |D| \geq 3q$. By (1.3) there is a divisor $\Delta \in |D|$ such that each x_i is a multiple point of Δ . For $\Delta = \sum n_i E_i$ each E_i with $n_i \geq 2$ is in $\mathcal{E}$ by (1.2), and consequently each $x_j \in E_{i(j)}$ where $n_{i(j)} = 1$. Again by (1.2) each such $E_{i(j)}$ is nonsingular and hence x_j will lie on two E_i with $n_i = 1$. Since the E_i connect together as a tree, we deduce that there are at least $q + 1$ of E_i containing some x_j . The genus $p(E_i) \geq 1$ and so we derive a contradiction

$$q = \dim(\mathrm{Pic}^0(X)) \geq \dim(\mathrm{Pic}^0(D)) \geq \textstyle\sum_i \dim(\mathrm{Pic}^0(E_i)) \geq q + 1 \quad .$$

This proves the theorem.

§ 5 Castelnuovo's criterion

The purpose of this section is to prove the following.

(5.1) Theorem. (Castelnuovo, Andreotti, Zariski, Nagata) A minimal surface X with $P_2 = 0$ and $q = 0$ is $\mathbb{P}_2$ or rational ruled.

The mapping proposition (2.2) gives the clue as to how we will proceed. First we consider the case $K^2 \leq 0$ which in fact never occurs.

(5.2) <u>Lemma</u>. On a minimal surface X with $P_2 = 0$ and $q = 0$ the canonical dimension equals -1 ..

<u>Proof</u>. For $K^2 < 0$ we use the results refered to in (3.4). For $K^2 \geq 0$, $q = 0$ we have $\dim |-K| \gtrsim 0$ by 3(4.1) since $|2K| = \emptyset$ and $K^2 - q \geq 0$. Thus there exists $D^* \in |-K|$, and so for any D there exists n_D with $|D+nK| = |D-nD^*| = \emptyset$ for $n \geq n_D$. Now the criterion in (3.1) applies.

(5.3) <u>Lemma</u>. On a minimal surface X with $P_2 = 0$, $q = 0$, and $K^2 \leq 0$, there exists a rational curve C with $K \cdot C < 0$ and $|K+C| = \emptyset$.

<u>Proof</u>. By (3.3) and (5.2) there exists an effective divisor D on X with $K \cdot D < 0$ and $|K+D| = \emptyset$. For some irreducible component C of D we still have $K \cdot C < 0$ and $|K+C| = \emptyset$ holds for any component of D. By (1.2) the curve is rational and the lemma is established.

In view of (5.3) and (2.2) theorem (5.1) is proven under the additional hypothesis that $K^2 \leq 0$.

Now only the case $K^2 > 0$ remains, and the next lemma reduces the proof of the theorem down to assertions about $\mathrm{Pic}(X)$, the linear system $|-K|$, and the size of K^2.

(5.4) <u>Lemma</u>. If $K^2 > 0$, $P_2 = 0$, and $q = 0$ on a minimal surface X, and if $K \cdot D \geq 0$ for all $D \geq 0$ with $|K+D| = \emptyset$, then $\mathrm{Pic}(X)$ is infinite cyclic generated by the class of K, every $D \in |-K|$ is irreducible with $p(D) = 1$, and $K^2 \leq 5$, $b_2 \geq 5$.

<u>Proof of the lemma</u>. First we show that all $D \in |-K|$ are irreducible. Such a divisor D satisfies $D \cdot K = -K^2 < 0$ and $D > 0$. If D were reducible, there would be a component C with $C \cdot K < 0$. If $E \in |K+C|$, then $0 < (D-C) + E \in |(K+C)+(D-C)| = |K+D| = \{0\}$ and so $|K+C| = \emptyset$ and $K \cdot C < 0$ contradicting our hypothesis concerning curves on X. Thus all $D \in |-K|$

are irreducible. For $D \in |-K|$ we can calculate $2p(D) - 2 = D\cdot D + K\cdot D = (-K)\cdot D + K\cdot D = 0$ and $p(D) = 1$.

Next, we show that $|K+D| = \emptyset$ for $D \geq 0$ implies $D = 0$. If $D > 0$, then there exists $C \in |-K|$ with $D \cap C$ nonempty, and therefore $D\cdot C > 0$ because $C^2 = K^2 > 0$. But $D\cdot C > 0$ implies $K\cdot D < 0$ contradicting our hypothesis on D with $|D+K| = \emptyset$.

Now for any divisor D^* on X there exists an integer n with $D \in |D^*+nK|$ and $|D^*+(n+1)K| = \emptyset$ by (5.2). Thus $|D+K| = \emptyset$ and $D \geq 0$ and by the argument in the previous paragraph $D = 0$. This means $D^*+nK = 0$ and $D^* = -nK$ in $\mathrm{Pic}(X)$.

Finally, by 3(4.1) we have $\dim |-K| \geq K^2$, and thus if $K^2 \geq 6$, then $\dim H^0(\mathcal{O}(-K)) \geq 7$. So (1.3) applies to show there is a divisor $D \in |-K|$ with $x \neq y$ as two multiple points. This is impossible for an irreducible D with $p(D) = 1$. Hence $K^2 \leq 5$ and $b_2 \geq 5$ from $10 = K^2 + b_2$. This proves the lemma.

(5.5) Reduction of the proof of (5.1). In view of (5.4) the last step of the proof of (5.1) is to show that the combination of the conditions $P_2 = 0$, $q = 0$, $1 \leq K^2$ together with any of the conclusions of (5.4) are impossible on a minimal surface. For this would mean that there is an irreducible curve C on X with $K\cdot C < 0$ and $|K+C| = \emptyset$ by the previous lemma. By (1.2) again C is rational, and now (2.2) applies to prove the theorem.

(5.6) Proof in characteristic zero, see Kodaira [Am IV]. From the exponential sequence $0 \to \mathbb{Z} \to \mathcal{O} \to \mathcal{O}^* \to 0$, we have part of the cohomology exact sequence

$$0 = H^1(\mathcal{O}) \to H^1(\mathcal{O}^*) \to H^2(\mathbb{Z}) \to H^2(\mathcal{O}) = 0$$

Hence $\mathrm{Pic}(X) = H^1(\mathcal{O}^*)$ and $H^2(\mathbb{Z})$ both have rank 1 assuming the hypothesis in (5.5). From $10 = K^2 + b_2$ we see $K^2 = 9$ which violates the conclusion $K^2 \leq 5$ in (5.4). This proves (5.1) in characteristic zero by this transcendental argument.

Alternatively, ω generates $H^1(\mathcal{O}^*)$ so $c_1(\omega)$ generates $H^2(\mathbb{Z})$ and $c_1(\omega)^2$ generates $H^4(\mathbb{Z})$. Thus $K^2 = c_1(\omega)^2[X] = 1$ which contradicts $K^2 = 9$ too.

For an algebraic proof which works in any characteristic, we use the Kummer sequence $1 \to \mu_n \to \mathbb{G}_m \xrightarrow{n} \mathbb{G}_m \to 1$ instead of the exponential sequence and ètale cohomology instead of ordinary cohomology. This brings us to a part of the cohomology exact sequence

$$0 \to \mathrm{Pic}(X)/n\ \mathrm{Pic}(X) \to H^2(\mu_n) \to {}_nH^2(\mathbb{G}_m) \to 0$$

where ${}_nH^2(\mathbb{G}_m) = \ker(H^2(\mathbb{G}_m) \xrightarrow{n} H^2(\mathbb{G}_m))$. Unlike in the transcendental case where $p_g = 0$ implies $H^2(\mathcal{O}) = 0$, we do not have such an easy vanishing result for $H^2(\mathbb{G}_m)$ or even its n torsion.

(5.7) Sketch of the proof for the general case. This proof was shown to us by M. Artin. The question was also taken up by G. Kurke [1972]. A complete treatment would lie beyond the scope of this article.

Recall by (5.4) the elements of $|-K|$ are irreducible curves C with $p(C) = 1$. We blow up β times the base points of $|-K|$ to get a new surface Y with a fibering $f: Y \to \mathbb{P}^1$ where $K_Y^2 = 0$. Then

$$0 = K_Y^2 = K_X^2 - \beta \quad , \quad \mathrm{rk}\ \mathrm{Pic}(Y) = \mathrm{rk}\ \mathrm{Pic}(X) + \beta \ , \quad \text{and} \quad b_2(Y) = b_2(X) + \beta \ .$$

We will argue that $H^2(Y,\mathbb{G}_m)$ is finite so that $b_2(Y) = \mathrm{rk}\ \mathrm{Pic}(Y) = 10$. Thus $10 = \mathrm{rk}\ \mathrm{Pic}(X) + \beta = 1 + \beta$ and $K_X^2 = \beta = 9$ again a contradiction which proves (5.1) by (5.5). There will be a possible exception in our proof: that $H^2(Y,\mathbb{G}_m)$ is finite which will be an assertion that Y is rational.

Let $\mathrm{Jac}(Y/B) \to B$ be the associated Jacobian fibering of $Y \to B = \mathbb{P}^1$. Then $\mathrm{Jac}(Y/B)$ contains a group scheme $A \to \mathbb{P}^1$ as an open subset which is the degree zero part of the relative Pic, namely $\mathrm{Pic}^0(Y/B)$. Using the Leray spectral sequence and the calculation of $R^1f(\mathbb{G}_m) = \mathrm{Pic}^0(Y/B)$ we obtain a morphism with finite kernel and cokernel

$$H^2(Y,\mathbb{G}_m) \longrightarrow H^1(\mathbb{P}^1, A)$$

Hence it suffices to show $H^1(\mathbb{P}^1, A) = 0$ in order to establish the finiteness of $H^2(Y,\mathbb{G}_m)$.

By a version of the theory outlined in part 2, § 3, the group $H^1(\mathbb{P}^1, A)$ classifies principal homogeneous spaces for $A \to \mathbb{P}^1$ or equivalently for $\mathrm{Jac}(Y/\mathbb{P}^1) \to \mathbb{P}^1$. The last step consists in showing that for a form Z of $A^* \to \mathbb{P}^1$, the principal homogeneous space Z either has a section and so is trivial or Z is rational. For this, let n be the minimum of all $D\cdot C > 0$ where $C \in |-K|$ and $D\cdot C = D(-K)$. Assume for D that $n = D\cdot C$. Adjusting D by $D \pm K$, we arrive at either $D\cdot C = 1$ where D is the image of a section or $D^2 > 0$ and $p(D) = 0$. In the last case we use (2.2) to show Z is rational and the original surface is also rational being a minimal model for Y .

Part 5. Surfaces of general type

In this part we survey the results in Kodaira [1968], in Bombieri [1970], [1973], and especially the 1973 article in the I.H.E.S. journal on Canonical models of surfaces of general type which we refer to as [CM]. There is a birational morphism $S \to X$ from a surface S of general type to its canonical model X and for each n a morphism $X \to X^{[n]}$ corresponding to the linear system $|nK|$ on S. The problem is to describe X in the birational isomorphism class of S and determine properties of the maps $S \to X$ and $X \to X^{[n]}$.

The other main topic concerns relations between the invariants K^2, p_g, q, $e(X)$, and the torsion subgroup of $\pi_1(X)$. For example $e(X) < 0$ implies that the canonical dimension is -1 and hence $q \le p_g$ on a surface of general type. As an application of the theory of canonical embeddings we will see that the number of homeomorphism classes of $X(\mathbb{C})$ where X is a surface with given numerical invariants is finite.

§ 1 Definition and geometric properties of the canonical model

Recall that for a graded ring $R = \coprod_{0 \le m} R_m$, there is a scheme, denoted $\mathrm{Proj}(R)$, associated to R and constructed from the homogeneous prime ideals of R. For a snappy introduction see Mumford [1966, lecture 5]. When R_0, and thus R, is a k algebra, $\mathrm{Proj}(R)$ is a scheme over k.

For any line bundle L on a scheme Y over k, we form the graded ring $R_L = \coprod_{0 \le m} H^0(Y, L^{m\otimes})$ where sections s_1 of $L^{m\otimes}$ and s_2 of $L^{n\otimes}$ multiply to give the section $s_1 \otimes s_2$ of $L^{(m+n)\otimes}$. In the case where Y is a variety R_L is an integral domain. Thus we can form $\mathrm{Proj}(R_L)$. We are interested here in the case where $L = \omega_S$ and $R = \coprod_{0 \le m} H^0(S, \mathcal{O}(mK))$.

(1.1) Remarks. Is S is a smooth surface of general type, then by Zariski [1962] and Mumford [1962] the following hold for $\pi : S \to X(S) = \operatorname{Proj}(R)$.

(a) The ring R is finitely generated and Noetherian.

(b) The morphism $\pi : S \to X(S)$ is birational. For each morphism $f : S \to S'$ there is an induced morphism $X(f) : X(S) \to X(S')$ such that the following diagram is commutative

$$\begin{array}{ccc} S & \xrightarrow{\pi = \pi(S)} & X(S) \\ f \downarrow & & \downarrow X(f) \\ S' & \xrightarrow{\pi(S')} & X(S') \end{array}$$

If f is birational morphism, then $X(f)$ is an isomorphism.

(c) The variety $X = X(S)$ is normal with a finite number of rational double points, see (1.6) also.

(d) There is a minimal desingularization $\bar{X}(S) \to X(S)$ and π is factored by $\bar{\pi} : S \to \bar{X}(S)$ where $\bar{\pi} : S \to \bar{X}(S)$ is an absolute minimal model for S.

(e) The inclusion $\coprod_{o \le m} R_{nm} \to \coprod_{o \le m} R_m = R$ where $R_m = H^o(S, \mathcal{O}(mK))$ induces an isomorphism

$$\operatorname{Proj}(R) = X(S) \longrightarrow \operatorname{Proj}(\coprod_{o \le m} R_{nm}) \quad .$$

(1.2) n-canonical image. Let $R^{[n]}$ denote the graded subring of R generated by R_n and $\phi_n : X(S) = \operatorname{Proj}(R) \to \operatorname{Proj}(R^{[n]}) = X^{[n]}$ the corresponding induced morphism. This morphism ϕ_n is the main object of study in this part. The variety $X^{[n]}$ is the n canonical image of S.

We will be interested in when $X^{[n]}$ is normal and when ϕ_n is a birational morphism or better an isomorphism.

(1.3) Remark. For a curve C general type means $p(C) \ge 2$. In this case $C \to X(C)$ is an isomorphism and $C \to C^{[n]}$ is an isomorphism for

$n \geq 3$ or for $n \geq 2$ and $p(C) \geq 3$.

Now we consider which curves in S map to a point in $X(S)$ under $\pi : S \to X(S)$. For a proof of the next proposition, see [CM, p. 175-6].

(1.4) Proposition. If C is an irreducible curve on S , then $K\cdot C \geq 0$,and if $K\cdot C = 0$, then $C^2 = -2$ and C is a nonsingular rational curve. Moreover, the irreducible curves E with $K\cdot E = 0$ form a finite set which are linearly independent in $NS(X) \otimes \mathbb{Q}$. The set of curves C in S with $C^2 < 0$ and $K\cdot C \leq n_0$ is also finite.

Let $\mathcal{E}$ denote the set of irreducible rational curves E on S with $K\cdot E = 0$. Roughly it is exactly the elements of $\mathcal{E}$ which map to points under $\pi : S \to X(S)$. More precisely, we have the following proposition due to Artin [1962], [1966].

(1.5) Definition. Let $\mathcal{C}$ be a maximal connected component of $\mathcal{E}$. The fundamental divisor Z associated with $\mathcal{C}$ is of the form $Z = \sum_{E \in \mathcal{C}} m_E E$, $m_E \geq 1$, which is minimal with respect to the property $Z\cdot E \leq 0$ for all $E \in \mathcal{C}$. A divisor (or cycle) Z is fundamental provided it is associated with some connected component $\mathcal{C}$ of $\mathcal{E}$.

(1.6) Proposition. For a divisor $Z > 0$ on S a smooth surface of general type the following are equivalent:

(a) Z is a fundamental divisor,

(b) Z is maximal with respect to $K\cdot Z = 0$ and $Z^2 = -2$, and

(c) Z is the fibre of $\pi : S \to X(S)$ over a singular point of $X(S)$.

Moreover, $\pi^* K_X = K_S$,and if Z is fundamental with $\pi(Z) = x$ and M_x is the maximal ideal of $\mathcal{O}_{X,x}$, then the canonical morphism $H^0(S,\mathcal{I}_Z/\mathcal{I}_Z^2) \to M_x/M_x^2$ is an isomorphism.

§ 2 Numerical connectivity of divisors

(2.1) Definition. Let $D > 0$ be a divisor on S. We say that D is (numerically) m-connected provided for each decomposition $D = D_1 + D_2$ with $D_1 > 0$, $D_2 > 0$ we have $D_1 \cdot D_2 \geq m$. A divisor is connected provided it is 1-connected.

We give briefly the argument for the following proposition which is also similar to lemma 11 [CM].

(2.2) Proposition. (Ramanujam - Lemma 3) If D is a 1-connected, then $\dim_k H^0(D, \mathcal{O}_D) = 1$.

Proof. The argument takes place in the Chow ring of S where the grading is by codimension. Now since $D_{red} = \Delta$ is connected, $H^0(D, \mathcal{O}_\Delta) = k$ and consider $\sigma \in H^0(D, \mathcal{O}_D)$ in the $\ker(H^0(D, \mathcal{O}_D) \to H^0(\Delta, \mathcal{O}_D))$. Let D_1 be a maximal divisor satisfying $0 < D_1 < D$ and $\sigma \in H^0(D, \mathcal{I}_{D_1} \mathcal{O}_D)$. Let $D = D_1 + D_2$. This yields two exact sequences

$$0 \to \mathcal{O}_{D_2} \xrightarrow{\sigma} \mathcal{O}_D \to \mathcal{O}_D / \sigma \mathcal{O}_D \to 0$$

$$0 \to \mathcal{F} \to \mathcal{O}_D / \sigma \mathcal{O}_D \to \mathcal{O}_{D_1} \to 0$$

where $\mathrm{supp}(\mathcal{F}) = \{P_i\}$ and $n_i = \mathrm{length}\ (\mathcal{F}_{P_i})$. Hence in the Chow ring the following hold

$$c(\mathcal{O}_D) = c(\mathcal{O}_{D_1}) \cdot c(\mathcal{O}_D / \sigma \mathcal{O}_D) = c(\mathcal{O}_{D_1}) \cdot c\ (\mathcal{O}_{D_2}) \cdot c\ (\mathcal{F})$$

or

$$\frac{1}{1-D} = \frac{1}{1-D_1} \cdot \frac{1}{1-D_2} \cdot c(\mathcal{F})$$

where $c(\mathcal{F}) = 1 - \ n_i P_i$. Thus it follows that

$$(1-D_1)(1-D_2) = (1-D)(1-\textstyle\sum_i n_i P_i)$$

which means that $D_1 \cdot D_2 = -\sum n_i \leq 0$. This is a contradiction to connectedness. Hence $\sigma = 0$ and $H^0(D,\mathcal{O}_D) = k$.

Lemmas 1-9, section 4 of [CM] list examples of divisors closely related to the canonical divisor which are connected, and on p. 184 these examples are summarized in a table. We consider one example. The notation $D \sim D'$ means D and D have the same image in $\mathbb{Q} \otimes NS(S)$.

(2.3) <u>Lemma</u>. If $D > 0$ and $D \sim mK$ for $m \geq 1$, then D is numerically 2-connected except for $K^2 = 1$, $m = 2$, $D = D_1 + D_2$ where $D_1 \sim D_2 \sim K$.

<u>Proof</u>. For $D = D_1 + D_2$ write $D_1 \sim rK + \xi$ and $D_2 \sim (m-r)K - \xi$ where $K \cdot \xi = 0$ and $r = K \cdot D_1/K^2 \geq 0$ in $\mathbb{Q} \otimes NS(S)$. Thus

$$D_1D_2 = r(m-r)K^2 - \xi^2$$

We can not have both $r(m-r) = 0$ and $\xi \neq 0$ since $D_1, D_2 > 0$. By the algebraic index theorem 1(5.6) either $\xi^2 < 0$ or $\xi = 0$, and hence $D_1D_2 \geq 1$.

From $D_1D_2 = (m+1)K \cdot D_1 - (D_1^2 + KD_1)$ where $D_2 \sim mK - D_1$, it follows that D_1D_2 is even unless both m is even and $K \cdot D_1$ is odd. Then $1/K^2 \leq r \leq m - (1/K^2)$ and

$$D_1 \cdot D_2 = r(m-r)K^2 - \xi^2 \geq m - \frac{1}{K^2} - \xi^2$$

which gives for $m \geq 2$ the inequality $D_1D_2 > 1$ unless $m = 2$, $r = 1$, $K^2 = 1$, and $\xi = 0$.

The above proof illustrates some of the techniques used to establish the connectivity assertions. We remark that in the table [CM, p. 184] refered to above $\pi : \widetilde{S} \to S$ is the blowing up of one point x to L or two points x,y to L,M, and Z, Z_μ, Z_ν denote fundamental cycles.

§ 3 Vanishing theorems

The following assertion contains two basic criterions for the vanishing of a first cohomology group.

(3.1) Theorem. Let S be a smooth complete algebraic surface S over an algebraically closed field of characteristic zero.

(a) If $D > 0$ is a connected divisor with $D^2 > 0$ on S, then $H^1(S,\mathcal{O}(-D)) = 0$.

(b) If L is a line bundle where $L^{n\otimes}$ is spanned by its sections and has three algebraically independent sections for some $n \geq 1$, then $H^1(S,L^{-1}) = 0$.

The second result is contained in Mumford [1967], and the first result is a corollary of the more general result of Ramanujam which we state in (3.2).

Recall that an algebraic family $\{C\}$ is composed of a pencil when there is a morphism $f : S \to B$ onto a smooth curve such that for every C in the family there is a divisor D on B with $C \leq f^{-1}(D)$. If $p(B) \geq 1$, then $\{C\}$ is said to be composed of an irrational pencil of genus $p(B)$.

(3.2) Theorem. Let $D > 0$ be a divisor on a smooth complete algebraic surface S over an algebraically closed field of characteristic zero. Assume for some $n > 0$ that $\dim |nD| \geq 1$ and nD is not composed of an irrational pencil. Then we have $\dim_K H^1(S,\mathcal{O}(-D)) = \dim_K H^0(D,\mathcal{O}_D) - 1$.

The proof of this theorem is contained in [CM, pp. 178-180].

To see that (3.2) implies (3.1), we first observe that for $D^2 > 0$ the Riemann Roch theorem gives

$$\dim |nD| \geq \frac{1}{2} n^2D^2 + O(n)$$

which grows like n^2 with n. If the linear system $|nD|$ was composed of a pencil, then the growth would be like n with n. Thus $\dim_k H^1(S,\mathcal{O}(-D)) = \dim_k H^0(D,\mathcal{O}_D) - 1 = 0$ by (2.2).

§ 4 Canonical mapping theorems

The following two theorems summarize the conditions under which $\phi_n : X \to X^{[n]}$ is an isomorphism and under which it is a birational map.

(4.1) Theorem. Let S be a surface of general type over k algebraically closed of characteristic zero. Then $\phi_n : X \to X^{[n]}$ is an isomorphism if

(i) $n \geq 5$,

(ii) $n \geq 4$ and $K^2 \geq 2$, or

(iii) $n \geq 3$, $K^2 \geq 3$, and $p_g \geq 2$.

This theorem is proved in [CM, theorems 3,4].

(4.2) Theorem. Let S be a surface of general type over k algebraically closed of characteristic zero. Then $\phi_n : X \to X^{[n]}$ is a birational morphism if:

(i) $n \geq 3$ except in the following cases for $n = 3$ or 4:

(a) $K^2 = 1$, $p_g = 2$ for $n = 3$ and 4 where $X^{[3]}$ is a rational ruled surface of degree 4 embedded in $\mathbb{P}^5(k)$ and $X^{[4]}$ is a quadric cone embedded in $\mathbb{P}^8(k)$,

(b) $K^2 = 2$, $p_g = 3$ for $n = 3$ where $X^{[3]}$ is the projective plane embedded in $\mathbb{P}^9(k)$ by means of the linear system of plane cubics, or

(c) some surfaces for $n = 3$ and 4 with $K^2 = 1$, $p_g = 0$ and for $n = 3$ with $K^2 = 2$, $p_g = 0$.

(ii) $n = 2$ and $K^2 \geq 10$, $p_g \geq 6$ except when S has the structure of a fiber space $f : S \to B$ with generic fiber a nonsingular curve of genus 2. Surfaces with such a fiber space structure and with $K^2 \geq 10$, $p_g \geq 6$ have the property that $X \to X^{[2]}$ is a ramified double covering and $X^{[2]}$ is birational equivalent to a rational or ruled surface.

This theorem occupies most of [CM], see in particular theorem 5, 6, 7, 8, and 16. A closely related theorem to the isomorphism theorem (4.1) is the following statement which is theorem 3A in [CM].

(4.3) Theorem. If $K^2 \geq 5$, $p_g \geq 3$, then $X^{[n]}$ is a projectively normal model for X when $n \geq 6$, and $R^{[n]} = \coprod_{0 \leq m} H^0(S, \mathcal{O}(mn))$.

An important step in the proof of (4.1) and (4.2) is theorem 2 [CM] which we state and give a proof of the first part of it as an illustration of the techniques used in the proof of (4.1) and (4.2).

(4.4) Proposition. For a minimal surface S in characteristic zero the line bundle $\mathcal{O}(nK)$ is spanned by its global sections for $n \geq 4$; for $n \geq 3$ with $K^2 \geq 3$ or $K^2 \geq 2$, $p_g \geq 1$; and for $n \geq 2$ with either $K^2 \geq 5$, $p_g \geq 3$ or $p_g \geq 3$, $q = 0$.

In order to see that $\mathcal{O}(nK)$ is spanned by its global sections for $n \geq 4$ we have to prove that for all $x \in S$ there is an element of $H^0(S, \mathcal{O}(nK))$ which is not zero at x. We blow up S at x by $\pi : \tilde{S} \to S$ and $\pi^{-1}(x) = L$ where $L^2 = -1$. Hence we have now to show that there is a section of $\mathcal{O}(n\pi^* K_S)$ not zero on L, i.e. not locally in $\mathcal{O}(-L) \subset \mathcal{O}_{\tilde{S}}$. Consider the exact sequence $0 \to \mathcal{O}(-L) \otimes \mathcal{O}(m\pi^* K_S) \to \mathcal{O}(m\pi^* K_S) \to \mathcal{O}_L(m\pi^* K_S) \to 0$ of sheaves where $\mathcal{O}(-L) \otimes \mathcal{O}(m\pi^* K_S) = \mathcal{O}(m\pi^* K_S - L)$ and $\mathcal{O}_L(m\pi^* K_S) = \mathcal{O}_L$ since $(\pi^* K)|_L = 0$ where L is the inverse image of a point. The cohomology sequence is of the form

$$0 \to H^0(\mathcal{O}(m\pi^* K - L)) \to H^0(\mathcal{O}(m\pi^* K)) \to k \to H^1(\mathcal{O}(m\pi^* K - L))$$

and we study the H^1 term with the Serre duality theorem.

$$H^1(\widetilde{S},\mathcal{O}(m\pi^* K_S - L)) \cong H^1(\widetilde{S},\mathcal{O}(K_{\widetilde{S}} + L - m\pi^* K_S))$$
$$H^1(\widetilde{S},\mathcal{O}(-\{(m-1)\}\pi^* K_s - 2L\}))$$

We used the relation $K_{\widetilde{S}} = \pi^* K_S + L$. There exists $D \in |(m-1)\pi^* K - 2L|$ because $P_{m-1} \geq 4$ for $m \geq 4$ using Riemann Roch. Thus from $L \cdot \pi^* K = 0$ we deduce that

$$D^2 = (m-1)^2 \pi^* K \cdot \pi^* K - 4 = (m-1)^2 K^2 - 4 > 0$$

for $m \geq 4$, and $D > 0$.

To apply (3.1) (a) we have to show that D is numerically 1-connected. By (2.3) any $\Delta \sim mK$ is 2-connected for $m \geq 4$. Suppose $D = D_1 + D_2$ where $D_1, D_2 > 0$. We can decompose

$$\Delta_i = D_i + (L \cdot D_i)L \geq 0 \quad \text{and} \quad \Delta_i \cdot L = 0$$

Hence $\Delta_i = \pi^*(E_i)$ where $E_1 + E_2 \in |(m-1)K|$. Since $L \cdot D_1 + L \cdot D_2 = 2$, we have

$$D_1 \cdot D_2 = \Delta_1 \cdot \Delta_2 - (L \cdot D_1)(L \cdot D_2) \geq C_1 \cdot C_2 - (L \cdot D_1)(2 - L \cdot D_1) \geq 2 - 1 \geq 1$$

since $(D_1 \cdot L)(2 - D_1 \cdot L) \leq 1$, and $C_1 + C_2 \in |mK|$ is a 2-connected by (2.3).

Now by (3.1) (a) the H^1 term is zero and the following sequence is exact

$$0 \to H^0(\mathcal{O}(m\pi^* K - L)) \to H^0(\mathcal{O}(m\pi^* K)) \to \mathbf{k} \to 0 \quad .$$

Thus $\mathcal{O}(m\pi^* K)$ has a section which is not zero on L.

§ 5 Applications of the mapping theorems.

(5.1) Proposition. Let S be a minimal surface of general type. Then $\chi(\mathcal{O}) \geq 1$, $e(S) \geq 0$, and the m - th plurigenus P_m of S is given by

$$P_m = \frac{1}{2} m(m-1)K^2 + \chi(\mathcal{O})$$

Proof. In theorem 13 [CM], a direct proof is given for the fact that $e(S) < 0$ implies S is birational equivalent to a ruled surface. Then $12\chi(\mathcal{O}) = K^2 + e(S) > 0$ for surfaces of general type means that $\chi(\mathcal{O}) \geq 1$.

Next $H^1(S,\mathcal{O}(-mK)) = 0$ for $m \geq 1$ by (3.1)(b) because $\mathcal{O}(mK)$ for $m \geq$ some m_o has always three algebraically independent sections. Now by duality and the vanishing of H^1 we have

$$\dim |mK| = \frac{1}{2} m (m-1)K^2 + \chi(\mathcal{O}) - 1$$

which is just Riemann-Roch.

(5.2) Proposition. The set of homeomorphism classes determined by $S(\mathbb{C})$ where S ranges over the isomorphism classes of surfaces with given $K^2 > 0$ and $e(S)$ is finite.

Proof. If S is not of general type, then $S(\mathbb{C})$ is diffeomorphic to either $\mathbb{P}^2(\mathbb{C})$ or $S^2 \times S^2$ since a $\mathbb{P}^1$ - bundle over $\mathbb{P}^1$ is diffeomorphically trivial.

For S of general type, consider $X^{[5]} \subset \mathbb{P}^N$ where $N = P_5 - 1$ is given by (5.1) and the degree of $X^{[5]}$ in $\mathbb{P}^N$ equals $25K^2$. The singularities are finitely many rational double points whose number is bounded by $b_2(S) \leq e(S) - 2$. There exists a minimal resolution of the singularities with a new degree d_o and embedding dimension n_o depending only on K^2 and $e(S)$. So each such surface has a nonsingular model of degree $\leq d_o$ in $\mathbb{P}^{n_o}$. Recall that K^2 decreases by one and $e(X)$ increases by one under blowing. These fall into a finite number of Chow families where all the smooth surfaces in a given family are homeomorphic.

§ 6 Relations between numerical invariants

We have had the relation $\chi(\mathcal{O}) > 0$ or $q \le p_g$ for all surfaces of general type, see (5.1). In the next theorem we have a basic inequality between p_g and K^2 for minimal surfaces of general type, see [CM, theorem theorem 9].

(6.1) Theorem. Let S be a minimal surface of general type. Then we have

$$p_g \le \frac{1}{2} K^2 + 2 \quad \text{for} \quad K^2 \text{ even}$$

$$p_g \le \frac{1}{2} K^2 + \frac{3}{2} \quad \text{for} \quad K^2 \text{ odd}$$

There are two parts of the proof depending on whether or not the variable part $|C|$ of $|K| = |C| + X$ is composed of a pencil or not. In the first case $C \sim aE$ where E is irreducible and $a \ge 1$ is an integer. Then $p_g \le a + 1$. Since $K \cdot X \ge 0$, it follows that $K^2 \ge aK\cdot E$ and since $E^2 > 0$, we have $K\cdot E > 0$ because $K^2 \ge 2$. This means $2a \le K^2$ and $p_g \le \frac{1}{2} K^2 + 1$. The other case uses the vanishing theorem (3.1).

Collecting theorems 10, 11, and 12 of [CM] together, we have the following criterion for a surface of general type to be regular, i.e. $q = 0$.

(6.2) Theorem. A minimal surface S of general type is regular if any of the following conditions hold:

(i) $K^2 = 2$ and $p_g = 2$,

(ii) $K^2 = 1$,

(iii) $p_g = \frac{1}{2} K^2 + 2$ or $p_g = \frac{1}{2} K^2 + \frac{3}{2}$.

For example if $K^2 = 1$, then $p_g = 0, 1, 2$ and $q = 0$ in all cases, and if $K^2 = 2$, then $p_g = 0, 1, 2, 3$ and $q = 0$ for $p_g = 0,2,3$. Surfaces with $K^2 = 1$ are considered in Bombieri [1970]. Using the order of the torsion subgroup of $\pi_1(S)$, we can sharpen the inequalities in (6.1).

(6.3) Theorem. Let S be a regular minimal surface of general type with torsion group $\mathrm{Tors}\ \pi_1(S)$ of order m . Then we have

$$p_g \leq \frac{1}{2} K^2 + \frac{3}{m} - 1$$

This theorem is [CM, theorem 14] and the corollary to it and theorem 15 in [CM] can be summarized as follows.

(6.4) Remark. If $K^2 = 1$, $p_g = 1$,then S has no torsion in $\pi_1(S)$. If $K^2 = 2$, $p_g = 1$, and $q = 0$, then $\mathrm{Tors}\ \pi_1(S)$ is 0 or $\mathbb{Z}/2$. If $K^2 = 2$, $p_g = 2$, then S has no torsion in $\pi_1(S)$.

For the invariants K^2 and $e(X)$ we have the relation

$$K^2 + e(X) = 0 \pmod{12}$$

and Van de Ven [1966] has consider for which $(x,y) \in \mathbb{N}^2$ with $x + y = 0 \pmod{12}$ does there exist a surface S with $e(S) = x$ and $K^2 = y$. Van de Ven proved that S always exists in the larger class of almost complex surfaces.

(6.5) Remarks on the relation between e and K^2 for algebraic surfaces of general type. In (6.1) we stated that $\chi(\mathcal{O}) \leq \frac{1}{2} K^2 + 2$ or $e \leq 5K^2 + 36$. For each value of K^2 there is an algebraic surface over $\mathbb{C}$ with $e = 5K^2 + 36$. Van de Ven [1966] proves that $K^2 \leq 8e$. It is conjectured that $K^2 \leq 9\chi(\mathcal{O})$ or equivalently $K^2 \leq 3e$. Hirzebruch [1956] and Borel [1963] combined to give a series of examples of surfaces of general type where $K^2 = 3e$ and $K^2 = 2e$. In the first case the surface is a quotient of D^2 by a discrete group action and the second case a quotient of $D^1 \times D^1$ by a discrete group action where D^n is the bounded domain of all $(z_j) \in \mathbb{C}^n$ with $|z_1|^2 + \ldots + |z_n|^2 < 1$.

The question of the maximum of K^2 for given $e(S)$ is still open; it would be of interest to determine all surfaces with minimum $e(S)$

for a given K^2. In [CM] it is proved that these spheres all have $q = 0$; for a surface of general type with $q > 0$ we have the better inequality $e(S) \leq 5K^2$.

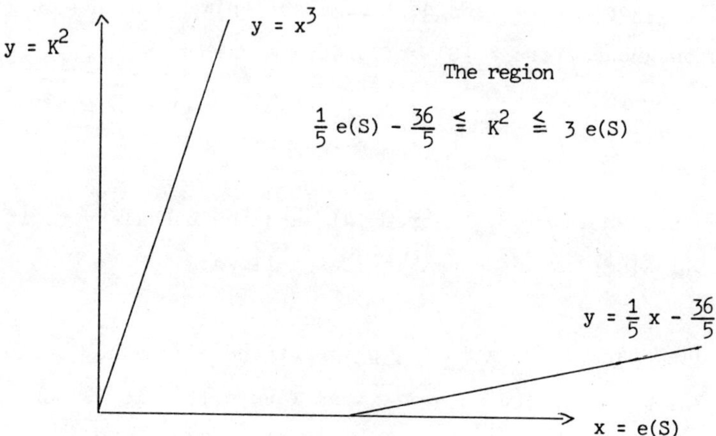

Part 6 Classification and examples of nonalgebraic surfaces

The nonalgebraic complex surfaces have the property that $\dim_{\mathbb{C}} \mathbb{C}(X) = 0$ or 1 which we saw 3(3.3) was equivalent to being nonelliptic or elliptic respectively. Since for complex curves X it is always the case that $\dim_{\mathbb{C}} \mathbb{C}(X) = 1 = \dim X$, the existence of nonalgebraic surfaces is a new feature of surface theory. Most of the early work on this subject was done by Kodaira.

The possible nonalgebraic elliptic surfaces were worked out from the canonical bundle formula and the discussion in § 2, part 3. After some prelimenaries we describe the nonalgebraic surfaces of canonical dimension 0. They are characterized by the relation $\omega_X^{12\otimes} = \mathcal{O}_X$ and they are either K3 surfaces, abelian surfaces, or quotients of a Kodaira surface by a finite group of automorphisms.

Then we turn to what is known about nonalgebraic surfaces of canonical dimension -1. The elliptic ones are all Hopf surfaces. Certain nonelliptic ones are Hopf surfaces and many of the surfaces without curves have been constructed and studied by Inoue, see [1974] and Bombieri. Recently Inoue and Hirzebruch have produced examples of nonelliptic surfaces of canonical dimension -1 where $b_2 > 0$.

§ 1 Preliminaries on nonalgebraic surfaces with $\kappa(X) = -1$ or 0

Note that if $\dim_{\mathbb{C}} \mathbb{C}(X) = 0$, then $\mathbb{C}(X) = \mathbb{C}$, i.e. the only meromorphic functions on X are constants since $\mathbb{C}(X)$ is finitely generated over $\mathbb{C}$ which is algebraically closed.

(1.1) Proposition. If $\dim_{\mathbb{C}} \mathbb{C}(X) = 0$ for a complex surface X and if L is a line bundle on X, then $\dim H^0(X,L) \leq 1$.

Proof. If s_1 and s_2 are two $\mathbb{C}$-linearly independent sections, then s_1/s_2 is a nonconstant meromorphic function on X contradicting the hypothesis $\dim_{\mathbb{C}} \mathbb{C}(X) = 0$.

(1.2) Corollary. If $\dim_{\mathbb{C}} \mathbb{C}(X) = 0$ for a complex surface X, then $p_g = 0$ or $p_g = 1$, all $P_m \le 1$, and $\kappa(X) \le 0$.

Since $\kappa(X) = 1$ means X is properly elliptic and comes under the classification of elliptic fibrations, we will consider the cases $\kappa(X) = -1$ and 0 in the remainder of this part.

(1.3) Proposition. If $p_g = 0$ on a complex surface X, then every $c \in H^2(X, \mathbb{Z})$ is of the form $c = c_1(L)$ for a holomorphic line bundle L.

Proof. From $0 \to \mathbb{Z} \to \mathcal{O} \xrightarrow{\exp} \mathcal{O}^* \to 0$ we have the exact sequence defining the first Chern class morphism

$$\mathrm{Pic}(X) = H^1(X, \mathcal{O}^*) \xrightarrow{c_1} H^2(X, \mathbb{Z}) \to H^2(X, \mathbb{Z}) = 0$$

from which the proposition follows.

(1.4) Theorem. Let X be a compact complex surface with $p_g = 0$. Then X is algebraic if and only if b_1 is even and so nonalgebraic if and only if b_1 is odd.

Proof. Recall from 1(4.1) that $b^+ = 2p_g + 1 - \Delta = 1 - \Delta$ in this case. By (1.3) there exists a holomorphic L with $L^2 > 0$ if and only if $b^+ = 1 - \Delta > 0$. Hence X is algebraic by 3(1.7) if and only if $\Delta = 0$, or equivalently by 1(4.1), b_1 is even. This proves the theorem.

(1.5) Proposition. If X is a nonalgebraic surface with $\kappa(X) \le 0$, then $\chi(\mathcal{O}_X) = 0$ or 2.

Proof. If $\dim_{\mathbb{C}} \mathbb{C}(X) = 1$, then X is elliptic and from the canonical bundle formula $K^2 = c_1(\omega_X)^2 = 0$. Hence from the basic formula $10 - 8q + 12p_g = K^2 + b_2 + 2\Delta$ and $p_g \le 1$ we see that $\chi(\mathcal{O}_X) = 0, 1, 2$.

If it were the case that $\chi(\mathcal{O}_X) < 0$, then $b_1 > 0$ since $q > 0$, and we have m fold coverings for all $m \geq 1$ defined $X_m \to X$, and $\chi(\mathcal{O}_{X_m}) = m\,\chi(\mathcal{O}_X) \leq -m < 0$. Since $\dim H^0(\Omega^1) = q - \Delta \geqq q-1$, it follows that $\dim H^0(\Omega^1_{X_m}) \geq m$ because $\kappa(X) = \kappa(X_m)$. Therefore we have linearly independent forms η_1 and η_2 on X_m with $\eta_1 \wedge \eta_2 = 0$ so that $\eta_1 = f\eta_2$ where f is a nonconstant meromorphic function. Thus $1 = \dim_{\mathbb{C}} \mathbb{C}(X_m) - \dim \mathbb{C}(X)$ which contradicts the relation $\chi(\mathcal{O}) \geq 0$ for elliptic X.

If $\chi(\mathcal{O}_X) = 1$, then either $p_g = 0$, $q = 0$, and $b_1 = 0$ or $p_g = q = 1$ and $b_1 = 1 > 0$. The first case is eliminated because X would have to be algebraic by (1.4). In the second case again there exists m fold coverings $X_m \to X$ with $\chi(\mathcal{O}_{X_m}) = m\,\chi(\mathcal{O}_X) = m$ which contradicts the inequality $\chi(\mathcal{O}_X) \leq 2$ for $\kappa(X) = \kappa(X_m) \leq 0$. This proves the proposition.

(1.6) Corollary. If X is a nonalgebraic surface with $p_g = 0$ and $\kappa(X) \leq 0$, then $q = 1$ and $b_1 = 1$.

(1.7) Remark. Using explicit constructions for elliptic surfaces, Kodaira [Am II, Theorem 28, p. 689] shows that the universal covering $\hat{X}$ of a nonalgebraic elliptic surface with $b_1(X) = 1$ depends only on $\kappa(X)$ and is given by the following table.

$\kappa(X)$	-1	0	1
$\hat{X}$	$W = \mathbb{C}^2 - \{0\}$	$\mathbb{C}^2$	$\mathbb{H} \times \mathbb{C}$

where $\mathbb{H}$ is the upper half plane. The condition $b_1 = 1$ implies $q = 1$ since $\Delta = 2q - b_1 = 1$ for b_1 odd by 1(1.4). Moreover the relation $\chi(\mathcal{O}) = 0$ or 2 implies that $p_g = 0$.

§ 2 Structure of nonalgebraic surfaces with $\kappa = 0$

Recall that we know for an elliptic nonalgebraic surface that $\kappa(X) = 0$ if and only if $\omega_X^{12\otimes} \cong \mathcal{O}_X$, by 3(3.5). This characterization holds more generally and we have the following structure theorem.

(2.1) Theorem. Minimal nonalgebraic surfaces X with $\kappa = 0$ are characterized by $\omega_X^{12\otimes} \cong \mathcal{O}_X$ and are classified as follows:

(i) K 3 surfaces. We have $\omega_X = \mathcal{O}_X$, $b_1 = 0$, $b_2 = 22$. The universal covering is a K 3 surface.

(ii) Complex tori. We have $\omega_X = \mathcal{O}_X$, $b_1 = 4$, $b_2 = 6$. The universal covering is $\mathbb{C}^2$.

(iii) Kodaira surfaces. We have $\omega_X = \mathcal{O}_X$, $b_1 = 3$, $b_2 = 4$. The universal covering is $\mathbb{C}^2$. These surfaces are fibre bundles of elliptic curves over an elliptic curve and differentiably a circle bundle over a 3-torus.

(iv) $b_1 = 1$, $b_2 = 0$. We have $\omega_X^{m\otimes} \cong \mathcal{O}_X$ with $m = 2,3,4,6$. These are elliptic surfaces which have a Kodaira surface as a finite unramified covering. The universal covering is $\mathbb{C}^2$.

(2.2) Remark. By (1.5) we know $\chi(\mathcal{O}_X) = 0$ or 2, and we have the following table of invariants which can be compared with 3(5.5). We shall show that the four categories in the table correspond to cases (i)...(iv) in our theorem.

	b_2	b_1	$e(X)$	q	p_g	$\chi(\mathcal{O})$	Δ
K 3 surfaces	22	0	24	0	1	2	0
Complex tori	6	4	0	2	1	0	0
Kodaira surfaces	4	3	0	2	1	0	1
Finite quotients of Kodaira surfaces	0	1	0	1	0	0	1

(2.3) <u>Proposition</u>. We have $\omega_X = \mathcal{O}_X$ in the first three cases. Also $\omega_X^{12\otimes} = \mathcal{O}_X$ in the last case.

<u>Proof</u>. Let $D \in |mK|$, which is possible for some $m > 0$ since $\kappa(X) = 0$. If $D = 0$, then $\omega^{m\otimes} \cong \mathcal{O}_X$. If instead $D > 0$, then D is of canonical type by 2(1.3). Now the argument in Mumford [1969, pp. 333-336], used in 3(4.3) and applies also in the nonalgebraic case, to show that X is elliptic. Our assertion follows from the classification of elliptic surfaces with $\kappa = 0$ given in 3(3.5).

It remains to show that the four cases have the structure specified in the theorem to complete the proof of (2.1).

(2.4) <u>Remark</u>. A surface with $q = 0$, $p_g = 1$, $b_1 = 0$, $b_2 = 22$, $\omega_X \cong \mathcal{O}_X$ is a K3 surface by definition.

(2.5) <u>Proposition</u>. A surface with $q = 2$, $p_g = 1$, $b_1 = 4$, $b_2 = 6$, $\omega_X = \mathcal{O}_X$ is a complex torus.

<u>Proof</u>. Since $b_1 = 4$ and $q = 2$, there are two independent holomorphic 1 - forms η_1, η_2 on X. Consider the Albanese map $X \xrightarrow{\phi} \mathrm{Alb}(X)$ given by

$$P \longmapsto \left(\int_{\underline{o}}^{P} \eta_1 , \int_{\underline{o}}^{P} \eta_2\right) / \text{periods}$$

where $\underline{o}$ is a base point on X. If $\phi(X)$ is a curve B, then B is of genus 2 since it generates $\mathrm{Alb}(X)$, and it is nonsingular. Any unramified covering $\widetilde{B}$ of B lifts to an unramified covering $\widetilde{X}$ of X, and clearly $q(\widetilde{X}) \geq p(\widetilde{B}) \geq 3$. This contradicts $\kappa(\widetilde{X}) = \kappa(X) = 0$. Hence ϕ is onto and thus $\eta_1 \wedge \eta_2$ is not identically 0 and is a holomorphic 2-form on X. Since $\omega_X \cong \mathcal{O}_X$, we see that $\eta_1 \wedge \eta_2$ never vanishes, hence so is the Jacobian of map ϕ. This shows that X is an unramified covering surface of the complex torus $\mathrm{Alb}(X)$, hence X is a torus.

The next result is due to Kodaira; our proof is based on his arguments (with some variations).

(2.6) Proposition. A surface X with $q = 2$, $p_g = 1$, $b_1 = 3$, $b_2 = 4$, $\omega_X \cong \mathcal{O}_X$ is elliptic.

We give a sketch of the proof to give the flavor of the finer arguments used in the study of complex surfaces.

Lemma a. Every curve on X is of canonical type. Hence X is elliptic if and only if it contains a curve.

Proof of lemma a: If C is a curve on X, then $K\cdot C = 0$ since $\omega_X \cong \mathcal{O}_X$. Hence either $C^2 = 0$ and C is of canonical type by 2(1.3), or $C^2 = -2$ and C is nonsingular rational. In the last case, let $\widetilde{X}$ be an unramified covering surface of X of degree m over X. Then $\widetilde{X}$ contains m disjoint curves $\widetilde{C}_1,\ldots,\widetilde{C}_m$ isomorphic to C and with $\widetilde{C}_i^2 = -2$. These curves are homologically independent, hence $b_2(X) \geq m$ and for m large enough, we get a contradiction to $\kappa(\widetilde{X}) = \kappa(X) = 0$.

The last statement of the lemma follows again from Mumford [1969, pp. 333-336].

Let η be a holomorphic 1-form on X.

Lemma b. If η vanishes at some point of X, then X is elliptic.

Proof of lemma b: The Euler number $e(X)$ of a complex manifold is equal to the number of zeros of a holomorphic 1-form provided its zeros are isolated. In our case $e(X) = 0$, we see that either η vanishes on some curve or does not vanish at all. When it vanishes on a curve, we use lemma a.

From now on, we shall assume that η never vanishes. Hence there are local coordinates (w_i, z_i) on a covering $\{U_i\}$ of X such that $\eta = dw_i$ on U_i.

The real $(1,1)$ - form $\psi = \sqrt{-1}\,\eta\wedge\bar{\eta}$ is exact. In fact, let ω be a holomorphic 2-form on X; ω never vanishes since $\omega_X \cong \mathcal{O}_X$. Let also $\omega_1 = \omega + \bar{\omega}$, $\omega_2 = \sqrt{-1}(\omega - \bar{\omega})$. If (u,v) denotes the quadratic pairing in $H^2_{DR}(X,\mathbb{R})$ (by integration on X of the wedge product $u \wedge v$), we get

$$(\omega_1,\omega_2) = (\omega_1,\psi) = (\omega_2,\psi) = (\psi,\psi) = 0$$

$$(\omega_1,\omega_1)\ ,\ (\omega_2,\omega_2) > 0$$

Now this quadratic form has only two positive eigenvalues since $p_g = 1$ and b_1 is odd. It follows that $\psi = 0$ in $H^2_{DR}(X,\mathbb{R})$ as asserted. Using this fact, one can easily show that there is a real closed 1-form σ such that

$$\sigma, \eta + \bar{\eta}, \sqrt{-1}(\eta - \bar{\eta}) \quad \text{is a basis of} \quad H^1_{DR}(X,\mathbb{R})$$

$$d\sigma_o = \eta \wedge \bar{\eta}$$

where σ_o is the (1,0) -component of σ . For this, see Kodaira [Am I, p. 761]. In terms of the local coordinates, let

$$\sigma_o = a_i dz_i + b_i dw_i \quad \text{on} \quad U_i$$

We have $d(\sigma_o + \bar{w}_i\, dw_i) = 0$, and it follows from this that a_i and $b_i + \bar{w}_i$ are holomorphic on U_i . Hence $\sigma_o \wedge \eta = a_i dz_i \wedge dw_i$ is a holomorphic 2-form on X which is not identically 0 since σ, η, $\bar{\eta}$ generate $H^1_{DR}(X,\mathbb{C})$. Hence a_i never vanishes, and changing the z_i - coordinate, we may assume $a_i = 1$ on U_i , thus

$$\begin{cases} \sigma_o = dz_i + b_i dw_i = dz_j + b_j dw_j & \text{on} \quad U_i \cap U_j \\ \eta = dw_i = dw_j & \text{on} \quad U_i \cap U_j \end{cases}$$

Also $f_i = b_i + \bar{w}_i$ is holomorphic on U_i . Integrating the last equation, we get

$$w_i = w_j + \alpha_{ij} \quad \text{on} \quad U_i \cap U_j$$

where $\{\alpha_{ij}, U_i \cap U_j\}$ is a nontrivial 1-cocycle with coefficients in $\mathbb{C}$. The first equation now becomes

$$dz_i + f_i dw_i = dz_j + f_j dw_j + \bar{\alpha}_{ij} dw_j$$

and

$$\sigma_o = dz_i + (f_i - \bar{w}_i)\, dw_i \quad .$$

Using $d\sigma_o = \eta \wedge \bar{\eta}$, we obtain that $df_i \wedge dw_i = 0$, hence $f_i = f_i(w_i)$ is a function of w_i alone. Replacing z_i by $z_i + \int^{w_i} f_i dw_i$, we now obtain $dz_i = dz_j + \bar{\alpha}_{ij} dw_j$. Hence we have local coordinates (w_i, z_i) on the covering $\{U_i\}$ such that on $U_i \cap U_j$

$$\begin{cases} w_i = w_j + \alpha_{ij} \\ z_i = z_j + \bar{\alpha}_{ij} w_j + \beta_{ij} \end{cases}$$

and $\eta = dw_i$, $\sigma = dz_i + d\bar{z}_i - \bar{w}_i dw_i - w_i d\bar{w}_i$ on U_i .

Now to study the topological structure of a Kodaira surface, we consider the map $\phi : X \to T^3$ of X onto the real 3-torus $\mathbb{C} \times \mathbb{R} / \Lambda$ where Λ is the lattice

$$\Lambda = (\textstyle\int_\gamma \eta, \int_\gamma \sigma) : \gamma \in H_1(X, \mathbb{Z})/\text{Tors } H_1(X, \mathbb{Z})$$

and ϕ is given by $P \mapsto (\int_{\underline{o}}^P \eta, \int_{\underline{o}}^P \sigma)$/periods where $\underline{o}$ is a base point on X .

Since $\eta \wedge \bar{\eta} \wedge \sigma$ never vanishes, we see that X is a circle bundle over a real 3-torus. We have a group extension

$$0 \to \mathbb{Z} \to \pi_1(X) \to \Lambda \to 0$$

where $\mathbb{Z}$ is the subgroup of $\pi_1(X)$ generated by the class g_o of the fibre $\phi^{-1}(\phi(0))$. If g_1 , g_2 , g_3 are generators of $\pi_1(X)$ mapped on generators of Λ , then we must have

$$g_i g_j g_i^{-1} g_j^{-1} = g_o^{n(i,j)}$$

for $i,j = 1,2,3$ and some integers $n(i,j)$ not all 0 because the group extension is nontrivial. Let γ_i be 1-cycles with class g_i in $\pi_1(X)$, $i = 1,2,3$ and let

$$\alpha_i = \int_{\gamma_i} \eta$$

Going from $\underline{o}$ to $\underline{o}$ on the cycle $\gamma_i \gamma_j \gamma_i^{-1} \gamma_j^{-1}$, we deduce from the coordinate transformations that

$$\overline{\alpha}_i \alpha_j - \alpha_i \overline{\alpha}_j = n_{ij} \beta_o$$

where $\beta_o = \int_{\gamma_o} \sigma o = 0$, for $i,j = 1,2,3$. It follows that

$$n_{23}\alpha_1 + n_{31}\alpha_2 + n_{12}\alpha_3 = 0$$

therefore

$$\Lambda_o = (\int_\gamma \eta) : \gamma \in H_1(X,\mathbb{Z})/\mathrm{Tors}\ H_1(X,\mathbb{Z})$$

is a lattice in $\mathbb{C}$. We conclude that the Albanese map $X \to \mathbb{C}/\Lambda_o$ maps X onto an elliptic curve. Hence X is elliptic.

In order to complete the analysis of case (iii), we show that the 1-form η never vanishes. Let $f : X \to B$ be the elliptic fibration. Since the canonical bundle is trivial, the canonical bundle formula shows that B is an elliptic curve and there are no multiple fibers. The Euler class of X is

$$e(X) = \sum_{b \in B} e(f^{-1}(b))$$

thus $e(f^{-1}(b)) = 0$ for all b and all fibres are nonsingular elliptic. Since the holomorphic 1-form η on X is obtained by lifting the corresponding 1-form η_B on B , f must be the Albanese map, and η never vanishes since η_B never vanishes and all fibres are nonsingular. Hence the analysis given before always holds for surfaces with $q = 2$, $b_1 = 3$, $b_2 = 4$, and $\omega_X \cong \mathcal{O}_X$.

The proof of the theorem is completed by noting that a surface X with $b_1 = 1$, $b_2 = 0$ and $\omega_X^{m\otimes} \cong \mathcal{O}_X$ has an unramified covering $\widetilde{X} \to X$ of degree m with $\mathcal{O}_{\widetilde{X}} = \omega_{\widetilde{X}}$. Since $b_1(\widetilde{X}) \geq 1$ and X is non Kähler, $\widetilde{X}$ must be a Kodaira surface which is elliptic. Hence X is elliptic too and the theorem is proved.

(2.7) Corollary. For nonelliptic nonalgebraic surface X , i.e. $\mathbb{C}(X) = \mathbb{C}$, $p_g = \kappa(X) + 1$ and $\kappa(X) = b_1(X) \bmod 2$.

Proof. For the theorem we see that only K3 surfaces and tori can be nonelliptic when $\kappa(X) = 0$ and hence $p_g = 1$. For $\kappa(X) = -1$ it is always the case that $p_g = 0$. For the other relation we know $b_1(X)$ is even for K3 surfaces and tori and must be one when $p_g = 0$ by (1.6).

§ 3 Hopf surfaces

In this and the next three sections we outline the present situation concerning the structure of minimal complex surfaces X with $\kappa(X) = -1$. Among these surfaces the Hopf surfaces form a large class including all elliptic surfaces (where $\dim_{\mathbb{C}} \mathbb{C}(X) = 1$) and a well defined class of surfaces without nonconstant meromorphic functions. We use the notation $W = \mathbb{C}^2 - \{0\}$ and the following definition is natural in view of (1.7). All surfaces are minimal in this and the remaining sections.

(3.1) Definition. A compact complex surface X is a Hopf surface provided W is a universal covering. A primary Hopf surface is a Hopf surface where $\pi_1(X) = \mathbb{Z}$.

By Hartogs theorem an analytic $T : W \to W$ prolongs to $\mathbb{C}^2 \to \mathbb{C}^2$ with $T(0) = 0$. Recall that T is a contraction means that $T^n(B) \to 0$ as $n \to +\infty$ for any bounded set B. Let $\{T\}$ denote the group generated by $T : W \to W$ which is fixed point free and acts properly discontinously on W.

(3.2) Remarks. If T is a contraction, then $W/\{T\}$ is a compact complex surface. Conversely if W/G is compact where G is a group acting fixed point freely and properly discontinuously, then G contains a contraction T and there is an integer n such that T^n is in the center of G. Thus $W/\{T\} \to W/G$ is a finite covering of a Hopf surface by a primary Hopf surface. In particular, the fundamental group of a Hopf surface contains an infinite cyclic subgroup of finite index. See Kodaira [Am II, pp. 694-5] for further details.

(3.3) Uniformization of Hopf surfaces I . Let $T : W \to W$ be a contraction as in (3.2). By a classical result, see Lattès [1911] and Sternberg [1957], there are coordinates of $\mathbb{C}^2$ such that

$$T(z_1, z_2) = (\alpha_1 z_1 + \lambda z_2^m, \alpha_2 z_2)$$

where m is a natural number and $\alpha_1, \alpha_2, \lambda$ are complex numbers with $(\alpha_1 - \alpha_2^m)\lambda = 0$ and $0 < |\alpha_1| \leq |\alpha_2| < 1$. We call this the normal form of T.

The curve $z_2 = 0$ is invariant under T in W and its image in the primary Hopf surface $W/\{T\}$ is an elliptic curve of the form $\mathbb{C}^*/\{\alpha_1\}$. When $\lambda = 0$, the curve $z_1 = 0$ is also invariant having as image an elliptic curve of the form $\mathbb{C}^*/\{\alpha_2\}$.

(3.4) Remark. For $T : W \to W$ in normal form, the primary Hopf surface $W/\{T\} = X$ is elliptic if and only if $\lambda = 0$ and $\alpha_1^a = \alpha_2^b$ for some natural numbers a and b. In this case z_1^a/z_2^b is invariant under T and induces a nonconstant meromorphic function f on X, see Kodaira [Am II, Theorem 31].

(3.5) Uniformization of Hopf surfaces II . For $X = W/G$ a nonelliptic Hopf surface with covering transformation group G, the group G is a direct product of an infinite cyclic group generated by a contraction T and a cyclic group of order ℓ generated by L where in suitable global coordinates of W

$$T(z_1, z_2) = (\alpha_1 z_1 + \lambda z_2^m, \alpha_2 z_2)$$

with $(\alpha_1 - \alpha_2^m)\lambda = 0$ and $0 < |\alpha_1| \leq |\alpha_2| < 1$, and

$$L(z_1, z_2) = (\varepsilon_1 z_1, \varepsilon_2 z_2)$$

where ε_1 and ε_2 are primitive ℓ-th roots of unity with $(\varepsilon_1 - \varepsilon_2^m)\lambda = 0$.

Again the curve $z_2 = 0$ in W is invariant by G and the image in W/G is a nonsingular elliptic curve. In particular every Hopf surface contains a nonsingular elliptic curve. For details see Kodaira [Am II, Theorem 32].

(3.6) Uniformization of Hopf surfaces III. For $X = W/G$ where G is nonabelian, there exists global coordinates for $\mathbb{C}^2$ such that the action of G is linear, i.e. $G \subset GL_2(\mathbb{C})$, for this see Kodaira [Am II, Theorem 47, p. 77].

Now we state two theorems characterizing Hopf surfaces. The first is in terms of analytic invariants like $\dim_{\mathbb{C}} \mathbb{C}(X)$, see Kodaira [Am II, Theorems 34, 35; pp. 699-708] and the second in terms of topological data like $\pi_1(X)$, see Kodaira [Am III, Theorems 41, 42, 43; pp. 56-63].

(3.7) Theorem. For a minimal compact complex surface X with $1 = \dim_{\mathbb{C}} \mathbb{C}(X)$, X is a Hopf surface if and only if $\kappa(X) = -1$, or equivalently $P_{12} = 0$. For $0 = \dim_{\mathbb{C}} \mathbb{C}(X)$, X is a Hopf surface if and only if $b_1(X) = 1$, $b_2(X) = 0$, and X contains a curve. In this case again $\kappa(X) = -1$.

(3.8) Theorem. For a minimal compact complex surface X the following assertions are equivalent.

(a) X is a Hopf surface.

(b) X has a finite covering which is a primary Hopf surface.

(c) $b_2(X) = 0$ and $\pi_1(X)$ has an ∞ cyclic subgroup of finite index.

The following assertions are also equivalent.

(a) X is a primary Hopf surface.

(b) X and $S^1 \times S^3$ are homeomorphic.

(c) $b_2(X) = 0$ and $\pi_1(X) \cong \mathbb{Z}$.

§ 4 Nonelliptic surfaces with $b_1(X) = 1$ and $b_2(X) = 0$

(4.1) Remarks. If $\mathbb{C}(X) = \mathbb{C}$, then the conditions $\kappa(X) = -1$. $b_1(X) = 1$, and $p_g = 0$ are all equivalent by (2.7). Also $\chi(\mathcal{O}) = 0$ and

$0 = c_1^2 + e(X) = c_1^2 + b_2(X)$. We study the case of $b_2(X) = 0$ in this section and mention some new results on the case $b_2(X) > 0$ in the next section. By (3.7) nonelliptic surfaces with $b_1(X) = 1$ and $b_2(X) = 0$ are Hopf surfaces if and only if they contain a curve. If a surface contains no curve, then clearly $\mathbb{C}(X) = \mathbb{C}$ so in this section we consider surfaces satisfying: $b_2(X) = 0$ and X has no curves.

This discussion is a survey of Inoue [1974] and Bombieri [unpublished].

(4.2) <u>The surface</u> X_m . Let $M \in SL_3(\mathbb{Z})$ with two eigenvalues $c, \bar{c} \in \mathbb{C} - \mathbb{R}$ and $(c\bar{c})^{-1}$ a third real eigenvalue with $(\alpha_1, \alpha_2, \alpha_3)$ a real eigenvector for the eigenvalue $(c\bar{c})^{-1}$ and $(\beta_1, \beta_2, \beta_2)$ a complex eigenvector for $\bar{c}$. Define the transformations

$g_o(z,w) = ((c\bar{c})^{-1}z, \bar{c}w)$ and $g_j(z,w) = (z+\alpha_j, w+\beta_j)$ and let $G = G(M)$ denote the group of automorphisms of $\mathbb{H} \times \mathbb{C}$ generated by g_o, g_1, g_2, g_3. The surface

$$X_M = \mathbb{H} \times \mathbb{C}/G(M)$$

where $\mathbb{H}$ is the upper half plane. One can check that the action is free and totally discontinuous.

(4.3) <u>Topological structure of</u> X_M . Let $\Lambda = \Lambda(M)$ be the subgroup of $G(M)$ generated by g_1, g_2, g_3 . Then $\phi : \mathbb{H} \times \mathbb{C}/\Lambda(M) \to \mathbb{R}_+ =$ the set of $y \in \mathbb{R}$, $y > o$ given by $\phi(z,w) = \mathrm{Im}(z)$ is well defined and of maximal rank. This induces $\phi : X_M = \mathbb{H} \times \mathbb{C}/G(M) \to \mathbb{R}_+/\langle g_o \rangle = S^1$ and the fiber is easily seen to be the real 3 dimensional tori $\phi^{-1}(t)$ for $t \in \mathbb{R}_+$. Hence X_M is a 3-torus bundle over the circle, and we have a group extension

$$1 \to \mathbb{Z}^3 = \Lambda(m) \to G(M) \to \mathbb{Z} \to 1$$

for the fundamental groups.

(4.4) <u>Line bundles on surfaces with</u> $b_2(X) = 0$. From the exponential exact sequence we have the usual cohomology exact sequence

$$H_1(\mathbb{Z}) \to H_1(\mathcal{O}) \to H_1(\mathcal{O}^*) = \mathrm{Pic}(X) \to H_2(\mathbb{Z}) \to 0$$

where $H_2(\mathbb{Z})$ is a torsion group. For $b_1(X) = 1$ it follows that $q = 1$ and the sequence becomes

$$1 \to \mathbb{C}^* \to \mathrm{Pic}(X) \to H_2(\mathbb{Z}) \to 0$$

Hence for each $a \in \mathbb{C}^*$ we have a line bundle F_a which is flat, and F_a can be described explicitly in terms of transition functions.

(4.5) <u>Theorem</u>. Let X be a surface without curves and $b_2(X) = 0$. If $\omega_X = F_a$ for some $a \in \mathbb{C} - \mathbb{R}$, then X is isomorphic to some X_M.

The proof of the theorem is in Inoue [1974, p. 280].

More generally for surfaces with $b_2(X) = 0$ it follows that $\mathrm{Tors}\, H^2(\mathbb{Z}) = H^2(\mathbb{Z})$ and thus every line bundle is flat. This means we can speak of $\Omega^1(F)$ and $\Omega^2(F)$ sheaves of differential forms with coefficients in F.

(4.6) <u>Examples</u>. For $N \in GL_2(\mathbb{Z})$ Inoue [1974] introduces a family of surfaces $S^{(+)}_{N,p,q,r;t}$ when $\det N = +1$ and another family $S_{N,p,q,r}$ when $\det N = -1$ by actions again on $\mathbb{H} \times \mathbb{C}$. Here $p, q, r \in \mathbb{Z}$, $r \neq 0$, and $t \in \mathbb{C}$, and he proves the following theorem.

(4.7) <u>Theorem</u>. The surfaces $S^{(+)}_{N,p,q,r;t}$ and $S^{(-)}_{N,p,q,r}$ have no curves and $b_2 = 0$. Moreover, if X is a surface without curves and $b_2 = 0$, and if for some line bundle F on X with $H^0(\Omega^1(F)) = 0$, then X is isomorphic to one of the following X_M, $S^{(+)}_{N,p,q,r;t}$, or $S^{(-)}_{N,p,q,r}$.

It is conjectured that the condition $H^0(\Omega^1(F)) = 0$ for some line bundle can be removed. If another type of surface X exists not in the above list, then $[\pi_1(X), \pi_1(X)]^{ab} \cong \mathrm{tors}\, \mathbb{Z}[c, c^{-1}]$ where $c \in Q$, $c \neq 0 \pm 1$. This property of the fundamental group should be contrasted with the assertion that $[\pi_1(X), \pi_1(X)]$ is finitely generated for X equal to one of the three families X_M, $S^{(+)}_N$, or $S^{(-)}_N$.

§ 5 Nonelliptic surfaces with $b_1(X) = 1$ and $b_2(X) > 0$

In this section we describe examples of nonalgebraic surfaces with $b_1(X) = 1$, $b_2(X) > 0$, and $\kappa(X) = -1$. They are related to the Hilbert modular group, see Hirzebruch [1973]. They were first discovered by Inoue in the case where K_X is one cyclic configuration of curves and extended by Hirzebruch to where K_X is the sum of two cyclic configurations of curves. These cyclic configurations of curves come from Hirzebruch's resolution of the cusp singularities of the Hilbert modular surface. We sketch here the Inoue-Hirzbruch construction briefly.

(5.1) Construction of $X(M,V)$. Let $K = \mathbb{Q}(\sqrt{D})$ be a real quadratic field, let $M \subset K$ be a module for the ring of integers in K , and let V be an infinite cyclic subgroup of U_M^+ , the totally positive units in K . Form the semidirect product $G(M,V) = V \propto M$ which can be viewed as the automorphism subgroup of $\mathbb{C}$ consisting of all $z \mapsto \varepsilon z + c$ for $\varepsilon \in V$ and $c \in M$. Now $G(M,V)$ acts on $\mathbb{H} \times \mathbb{C}$ by $(\varepsilon,c)(z_1,z_2) = (\varepsilon z_1+c, \varepsilon' z_2+c')$ where x' denotes the conjugate of $x \in K$ over $\mathbb{Q}$.

Let $Y(M,V) = \mathbb{H} \times \mathbb{C}/G(M,V)$ and observe that $(z_1,z_2) \mapsto y_1 y_2$ is $G(M,V)$ invariant and defines a function $\phi : Y(M,V) \to \mathbb{R}$ everywhere of maximal rank. The following assertion is easy to establish.

(5.2) Lemma. The fibers $\phi^{-1}(t) \subset Y(M,V)$ are 2 torus bundles over S^1 up to diffeomorphism.

Hence we compactify $Y(M,V) \cup \{s_\infty\} \cup \{s_{-\infty}\} = \bar{X}(M,V)$ where neighborhoods of s_∞ are of the form $\phi^{-1}(t,+\infty)$ and of $s_{-\infty}$ of the form $\phi^{-1}(-\infty,t)$. Hence topologically $\bar{X}(M,V)$ is the suspension of any fibre $\phi^{-1}(t)$. The complex surface $X(M,V)$ is the result of resolving the two cusp singularities s_∞ and $s_{-\infty}$.

(5.3) Theorem. The surface $X(M,V)$ has the property that $b_1 = 1$ and is a nonelliptic, nonalgebraic surface.

In order to calculate b_2 we look at the cyclic configurations of curves in the resolution . These have been described by Hirzebruch in terms of the arithmetic properties of M and V , see Hirzebruch [1973]. Near s_∞ we have $\mathbb{H} \times \mathbb{H}/G(M,V)$ and associated to the continued fraction expansion of a generator of V we have a set of numbers $(b_o,\ldots,b_{r-1})$. Near $s_{-\infty}$ we have $\mathbb{H} \times \mathbb{H}^-/G(M,V) \xrightarrow{\sim} \mathbb{H}\times\mathbb{H}/G(\sqrt{D}M,V)$ under the map induced by $(z_1,z_2) \longmapsto (\sqrt{D}z_1,-\sqrt{D}z_2)$ and again a set of numbers $(c_o,\ldots,c_{s-1})$. In $X(M,V)$ the inverse image of s_∞ is cycle of r curves with selfintersections $b_o,\ldots,b_{r-1}$ and of $s_{-\infty}$ is a cycle of r curves with selfintersections $c_o,\ldots,c_{s-1}$. Then

$$K_X = -(\text{sum of these two sets of curves}) = K_+ + K_-$$

Now $b_2(X) = r+s$ and in fact these curves from the resolution are all curves on the surface.

If there exists a unit ε such that $\varepsilon M = M$, $\varepsilon > 0$, $\varepsilon' < 0$, and $\varepsilon^2 \in V$, then $(z_1,z_2) \longmapsto (\varepsilon z_1, \varepsilon' z_2)$ induces an involution τ on $X(M,V)$ and $X(M,V)/\tau$ is also nonelliptic with $b_1 = 1$.

The question of what surfaces there are with $b_1(X) = 1$, $b_2(X) > 0$, and $\kappa(X) = -1$ is still completely open, but until 1974 there were not even any examples known.

TABLE FOR THE CLASSIFICATION OF SURFACES

$\kappa(X)$	-1	0	1	2
	Algebraic case - $\dim_k k(X) = 2$			
Classification	$K \cdot C < 0$ for some curve C	$K \cdot D \geqq 0$ for all divisors $D \geqq 0$ $K^2 = 0$		$K^2 > 0$
Enriques Classification	all $P_m = 0$ $\chi(\mathcal{O}) = 1 - q$	$12K = 0$ $P_{12} = 1$ and $P_m \leqq 1$ $\chi(\mathcal{O}) = 0, 1,$ or 2	$12K \neq 0$ $P_m \sim m$ $\chi(\mathcal{O}) \geq 0$	P_2 2 $P_m \sim m^2$ $\chi(\mathcal{O}) > 1$
Structure	$q = 0$ $\mathbb{P}^2$ $(K^2 = 9)$ ruled over $\mathbb{P}^1$ $(K^2 = 8)$ $q \geqq 1$ ruled over a curve of genus q with $K^2 = 8(1 - q) \leqq 0$	a) $\chi(\mathcal{O}) = 2$, $\begin{cases} b_2 = 22 \\ b_1 = 0 \end{cases}$ K3 surfaces b) $\chi(\mathcal{O}) = 1$, $\begin{cases} b_2 = 10 \\ b_1 = 0 \end{cases}$ Enriques surfaces (classical and nonclassical) c) $\chi(\mathcal{O}) = 0$, c1) $b_1 = 4$, $b_2 = 6$ Abelian surfaces c2) $b_1 = 2$, $b_2 = 2$ Hyperelliptic surfaces (classical and nonclassical)	properly elliptic	general type (Part 5)
	Nonalgebraic case over $\mathbb{C}$ - $\dim_{\mathbb{C}} \mathbb{C}(X) = 1 \iff$ X is elliptic			
Classification and Structure	$\chi(\mathcal{O}) = 0$, $\begin{cases} b_1 = 1 \\ b_2 = 0 \end{cases}$ Elliptic Hopf surfaces	a) $\chi(\mathcal{O}) = 2$, $\begin{cases} b_2 = 22 \\ b_1 = 0 \end{cases}$ K3 surfaces b) $\chi(\mathcal{O}) = 1$, does not occur c) $\chi(\mathcal{O}) = 0$, c1) $b_1 = 4$, $b_2 = 6$ Complex tori c2) $b_1 = 3$, $b_2 = 4$ Kodaira surfaces c3) $b_1 = 1$, $b_2 = 0$ (Kodaira)/finite G	properly elliptic	does not occur
	Nonalgebraic case over $\mathbb{C}$ - $\dim_{\mathbb{C}} \mathbb{C}(X) = 0 \iff$ X is nonelliptic			
Classification	$p_g = 0$	$p_g = 1$		
Structure	a) $\chi(\mathcal{O}) = 0$, $e = b_2 = 0$ a1) Nonelliptic Hopf surfaces a2) Inoue's surfaces S_M, $S_{N,p,q,t}$ a3) ? b) $\chi(\mathcal{O}) = 0$, $e = b_2 > 0$ b1) Inoue - Hirzebruch surfaces b2) ?	a) $\chi(\mathcal{O}) = 2$, $\begin{cases} b_2 = 22 \\ b_1 = 0 \end{cases}$ K3 surfaces b) $\chi(\mathcal{O}) = 1$, does not occur c) $\chi(\mathcal{O}) = 0$, $\begin{cases} b_2 = 6 \\ b_1 = 4 \end{cases}$ Complex tori	does not occur	

References

A. Andreotti, On the complex structures on a class of simply-connected manifolds, in Algebraic Geometry and Topology, Princeton University Press 1957.

M. Artin, On isolated rational singularities of surfaces, Am. J. Math. 88 (1966), 129-136.

- , Some numerical criteria for contractibility of curves on algebraic surfaces, Am. J. Math., 84 (1962), 485-496.

M.F. Atiyah and I.M. Singer, The Index of elliptic operators III, Ann. of Math. 87, (1968), 546-604.

E. Bombieri, The pluricanonical map of a complex surface, Springer Lecture Notes, 155 (1970), 35-87.

- , Canonical models of surfaces of general type, Publications Mathématiques n° 42, (1973), 441-445. This is refered to a [CM] in the text.

- , Superfici irregolari canoniche e pluricanoniche, Symposia Mathematica, Bologna - 1973.

- and D. Mumford, Classification of surfaces (to appear)

A. Borel et J.-P. Serre, Le Théorème de Riemann-Roch, Bull. Soc. math. France, 86 (1958) 97-136.

G. Castelnuovo and F. Enriques, Sopra alcune questioni fondamentali nella teoria della superficie algebriche, Annali di Matematica pura et applicata, III. s, 6 (1901), 165-225.

G. Castelnuovo, Sulle superficie di genere zero, Memorie della Società italiana delle Scienze detta dei XL, III. s, 10 (1896), 103-123.

W.L. Chow and K. Kodaira, On analytic surfaces with two independent meromorphic functions, Proc. of the Nat. Acad. of Sci., 38 (1952), 319-325.

F. Enriques, Le Superfici Algebriche, Zanichelli, Bologna, 1949.

P. Deligne et M. Rapoport, Les schémas de modules de courbes ellipiques, Antwerp Summer School, Springer Lecture Notes 349, 1973, 143 - 316.

A. Grothendieck, Théoreme de dualité pour faisceaux algébrique cohérents, Seminaire Bourbaki, 9, n° 149, 1956-57.

\- , La Théorie des Classes de Chern, Bull. Soc. math. France, 86 (1958), 137-154.

\- , Sur une note de Mattuck - Tate, Crelle, 1958 (20), 208-215.

F. Hirzebruch, New topological methods in algebraic geometry, Third Edition, Springer, Berlin, 1966.

\- , The Hilbert modular group, resolution of the singularities at the cusps and related problems, Séminaire Bourbaki 1970/71, exposé 396, pp. 275-288.

\- , Hilbert Modular Surfaces, L'Enseignement mathématique, T. XIX, fasc 3-4, 1973, 183-281.

J.I. Igusa, Betti and Picard numbers of abstract algebraic surfaces, Proc. Nat. Acad. Sci. U.S.A., 46 (1960), 724.

\- , On some problems in abstract algebraic geometry, Proc. Nat. Acad. Sci. U.S.A., 41 (1955), p. 964.

M. Inoue, On surfaces of class VII_0, Inventiones math., 24, (1974), 269-310.

K. Kodaira, On compact analytic surfaces in Analytic Functions, Princeton Univ. Press 1960.

K. Kodaira, On compact complex analytic surfaces I, II, III, Ann. of Math., 71; 77, 78 (1960; 1963) 111-152; 563-626, 1-40. This is denoted An I, II, III in the text.

K. Kodaira, On the structure of compact complex analytic surfaces, I, II, III, IV. Amer. J. of Math. 86; 88 (1964; 1966) 751-798, 682-721; 55-83, 1048-1066. This is denoted Am I, II, III, IV in the text.

K. Kodaira, Pluricanonical systems on algebraic surfaces of general type, J. Math. Soc. Japan, 20 (1968), 170-192.

G. Kurke, The Castelnuovo criterion of rationality, Matematicheskie Zametki, Vol 11, No. 1, 27-32, January, 1972.

S. Lang, Abelian varieties, Interscience, New York, 1958.

S. Lichtenbaum, Harvard thesis.

A. Mattuck and J. Tate, On the inequality of Castelnuovo-Severi, Ham. Abh., 1958(22), 295.

B. Moisezon, The criterion of projectivity of complete algebraic abstract varieties, Doklady Akad. Nauk., Math. Series 28, p. 179.

D. Mumford, The canonical ring of an algebraic surface, Annals of Math., 76 (1962), 612-615.

- , Lectures on curves on an algebraic surface, Annals of Math. Studies 59, Princeton, 1966.

- , Pathologies III, Am. J. Math., 89 (1967), 94-104.

- , Enrique's classification of surfaces in char p : I , Global Analysis, Papers in Honor of K. Kodaira, 1969.

M. Nagata, On rational surfaces I, Memoirs of the College of Science, Univ. of Kyota, series A, 32, (1960), 351-370.

A.P. Ogg, On pencils of curves of genus two, Topology 5 (1966), 355-362.

C.P. Ramanujam, Remarks on the Kodaira vanishing theorem, to appear.

I. Safarevic and others, Algebraic surfaces, Proc. of Steklov Inst. of Math. Moscow, 1965 or Am. Math. Soc., 1967.

J.-P. Serre, Géométrie algébrique et géométrie analytique Annales Inst. Fourier, 6, 1955, 1-42. This is refered to as [GAGA] in the text.

- , Faisceaux algébriques coherents, Annals of Math., 61, 1955, 197-278.

S. Sternberg, Local contractions and a theorem of Poincaré, American Journal of Mathematics, 79, (1957), pp. 809-824.

J.J. Tate, Genus change in inseparable extensions of function fields, Proc. Amer. Math. Soc. 3 (1952), 400.

T. van de Ven, On the Chern numbers of certain complex and almost complex manifolds, Proc. Nat. Acad. Sci. U.S.A., 55 (1966), 1624-1627.

J.-L. Verdier, Base change for twisted inverse image of coherent sheaves, Proc. Colloq. Alg.-Geom., Bombay, 1968, Oxford.

O. Zariski, Proof of a theorem of Bertini, Trans. Am. Math. Soc., 50 (1941), 81.

- , On Castelnuovo's criterion of rationality in the theory of algebraic surfaces, III J. Math. 2 (1958).

- , The problem of minimal models in the theory of algebraic surfaces, Am. J. of Math., 80 (1958), 146-184.

- , Introduction to the Problem of Minimal Models in the Theory of algebraic Surfaces, Tokyo 1958.

- , On Castelnuovo's criterion of rationality $p_a = p_2 = 0$ of an algebraic surface, Illinois J. of Math., 2 (1958), 303-315.

- , The theorem of Riemann-Roch for high multiples of an effective divisor on an algebraic surface, Annals of Math., 76 (1962), 550-612.

A. Borel, Compact Clifford - Klein forms of symmetric spaces, Topology, 2, (1963), 111 - 122.

H. Grauert, Über Modifikationen und exzeptionelle analytische Mengen, Math. Ann., 146, (1962), 331 - 368.

F. Hirzebruch, Automorphe Formen und der Satz von Riemann - Roch, Symp. Intern. Top. Alg., 1956, 129 - 144, Universidad de Mexico (1958).

S. Iitaka, Deformations of compact surfaces II, J. Math. Soc. Japan, 22, (1970), 247 -261.

S. Iitaka, On D - dimensions of algebraic varieties, J. Math. Soc. Japan, 23, (1971), 356 - 373.

D. Mumford, The topology of normal singularities of an algebraic surface and a criterion for simplicity, Publications Mathematiques de l'Institut des Hautes Études Scientifiques, No. 9, (1961).

K. Ueno, Classification of algebraic varieties, I., Compositiv Math., 27, (1973), 277 - 342.

Pisa
Bonn, Haverford

Proceedings of Symposia in Pure Mathematics
Volume 29, 1975

LINEAR REPRESENTATIONS OF SEMI-SIMPLE ALGEBRAIC GROUPS

Armand Borel

This paper is an outgrowth of two talks, whose aim was to give a survey of known results and of some open problems on rational representations of semi-simple groups over an algebraically closed ground field. I am grateful to A. Fauntleroy and W. Haboush, who made Notes from the talks which were very helpful to me in writing this up.

Throughout, k denotes an algebraically closed field, p the characteristic of k, and G a connected semi-simple group over k. All algebraic groups are affine and, unless otherwise stated, defined over k.

§1. CLASSIFICATION

1.1. We refer to [2] for the basic notions and facts on algebraic groups used here without further comment.

We recall that an (algebraic) torus T is a connected group isomorphic to a product of groups $\mathbb{GL}_1$. The character group $X(T) = \mathrm{Morph}(T, \mathbb{GL}_1)$ is a free abelian group of rank equal to $\dim T$. Given $t \in T$, $a \in X(T)$, we shall often write t^a for the value of a on t. It is then understood that the composition law in $X(T)$, which is given by taking products of values, is written additively: $t^{a+b} = t^a . t^b$.

1.2. ROOTS. The maximal tori of G are conjugate by inner automorphisms. Their common dimension is the rank $\mathrm{rk}\, G$ of G. Let T be one of them. It acts on the Lie algebra $L(G)$ of G (identified to the

AMS (MOS) subject classifications (1970). 20G05, 20G35, 20G40.

tangent space at the identity) via the adjoint representation. This representation is diagonalisable (as is any rational representation of a torus). For $a \in X(T)$, let

$$L(G)_a = \{X \in L(G) \mid \mathrm{Ad}\, t\, X = t^a . X \quad (t \in T)\} .$$

An element $a \in X(T)$ is a <u>root</u> of G with respect to T if $a \neq 0$ and $L(G)_a \neq \{0\}$. We let Φ or $\Phi(T, G)$ be the set of roots of G with respect to T. The space $L(G)$ is the direct sum of $L(T) = L(G)_o$ and of the $L(G)_a$ $(a \in \Phi)$. Moreover, $L(G)_a$ is one-dimensional for $a \in \Phi$.

Let $W = W(G)$ be the quotient of the normalizer $N_G(T)$ of T in G by T. Since T is its own centralizer in G, W operates faithfully, on T and $X(T)$ via inner automorphisms, on $L(T)$ by the adjoint representation.

1.3. PROPOSITION. The set Φ is a reduced root system in $X(T)_{\mathbb{R}} = X(T) \otimes \mathbb{R}$, with Weyl group W. Moreover, $X(T)$ contains the lattice $R(\Phi)$ spanned by Φ and is contained in the lattice $P(\Phi)$ of weights of Φ.

Briefly, this means the following (for more details, see [6; 7; 22; 28]): the set Φ is symmetric with respect to the origin, contains a basis of $X(T)_{\mathbb{R}}$ and is invariant under W; let $(\ , \ )$ be a positive non-degenerate scalar product on $X(T)$ which is invariant under W. Then W is generated by the reflections through the hyperplanes orthogonal to the roots; we have $2(a, b)/(b, b) \in \mathbb{Z}$ for all $a, b \in \Phi$. Moreover, $a, ra \in \Phi$ $(a \in \Phi, r \in \mathbb{R})$ imply $r = \pm 1$. (This last condition accounts for the word "reduced.") Finally the weights are the elements $r \in X(T)_{\mathbb{R}}$ such that $2(r, a)/(a, a) \in \mathbb{Z}$ for all $a \in \Phi$. They form the lattice $P(\Phi)$.

For brevity, let us call <u>diagram</u> a couple (Φ, Γ) formed of a reduced root system in some finite dimensional vector space and a lattice between $R(\Phi)$ and $P(\Phi)$.

1.4. THEOREM. The map $\delta : G \longmapsto (\Phi(T, G), X(T))$ induces a bijection between isomorphism classes of semi-simple groups over k and isomorphism classes of diagrams.

If $X(T) = R(\Phi)$, then G is of "adjoint type," i.e. the natural morphism $G \longrightarrow \mathrm{Ad}\, G \subset GL(L(G))$ is an isomorphism. If $X(T) = P(\Phi)$, then

G is "simply connected." The latter notion may be defined in several ways; we give two: (a) any rational projective representation $G \longrightarrow PGL(E)$ lifts to a rational linear representation $G \longrightarrow GL(E)$; (b) if G' is a semi-simple group, $f : G' \longrightarrow G$ an isogeny such that $\ker df : L(G') \longrightarrow L(G)$ is pointwise fixed under G' (acting by the adjoint representation), then f is an isomorphism (The requirement on $\ker df$ characterizes "central isogenies"; see [4: §2] for more details). Over $\mathbb{C}$, this is equivalent to simple-connectedness in the topological sense.

The theorem is in fact more functorial than stated. In particular, if G, G' correspond to (Φ, Γ) and (Φ, Γ') and if $\Gamma' \subset \Gamma$, then there is an (essentially unique) central isogeny $G \longrightarrow G'$ whose degree is equal to the index of Γ' in Γ, and in fact whose kernel and infinitesimal kernel can be read off Γ/Γ' [3: 3.3(2)]. Over $\mathbb{C}$, $P(\Phi)/\Gamma \xrightarrow{\sim} \pi_1(G)$, $\Gamma/R(\Phi) \xrightarrow{\sim} \text{Center}(G)$. Thus the reduced root systems parametrize the central isogeny classes of semi-simple groups over k, or the simply connected semi-simple groups over k. In fact they also parametrize the isogeny classes up to one exception: in characteristic two, the types $\mathbb{B}_n$ and $\mathbb{C}_n$ $(n \geq 3)$ are isogeneous, but not centrally isogeneous.

1.5. REFERENCES. Over $\mathbb{C}$, 1.4 goes back to results of Killing, Weyl, Cartan, proved however in a different context. Briefly, it may be viewed as the conjunction of the following:

(a) Classification of complex semi-simple Lie algebras by reduced root systems.

(b) Classification of connected complex semi-simple Lie groups with a given Lie algebra $\underline{g}$ with roots system Φ by means of lattices between $R(\Phi)$ and $P(\Phi)$.

(c) A complex connected semi-simple Lie group has one and only one structure of affine algebraic group compatible with its complex analytic structure.

(a) is in essence due to Killing and Cartan, although the connection with root systems emerged gradually only later. It is now standard (cf. e.g. [22; 28]).

It is more difficult to give a direct reference for (b). Results of H. Weyl and E. Cartan, as reformulated later by E. Stiefel (C.M.H. 14 (1942), 350-380; see also J. F. Adams, Lectures on Lie groups, Benjamin) show that diagrams also classify compact semi-simple Lie groups. One then uses the fact that the assignment: connected Lie group $\longmapsto$ maximal compact subgroup induces a bijection between isomorphism classes of connected complex semi-simple Lie groups and of connected compact semi-simple Lie groups (see e.g. [19]). In the course of proving this, one also sees that a complex connected complex semi-simple Lie group always has a faithful finite dimensional representation [19: p. 200].

Finally, in view of this last fact, (c) amounts to showing that the $\mathbb{C}$-algebra of holomorphic functions on G whose translates span a finite dimensional space (the "representative functions") is finitely generated. It is then the coordinate ring for the desired structure of algebraic group [20].

In positive characteristics, 1.4 is due to C. Chevalley. There are two parts to the proof:

(i) surjectivity of the map δ. More precisely, Chevalley associates to each diagram (Φ, Γ) a smooth group scheme G_o over $\mathbb{Z}$ such that $G_o \otimes_{\mathbb{Z}} k$ is the k-group with diagram (Φ, Γ), for every k. This construction is given first in [11]; see also [3] and, for a more general existence theorem over schemes [15; 16]. It will be sketched in §5.

(ii) injectivity of δ. This is proved in [10; Vol. 2]; see also [15; 16].

1.6. REMARKS. (1) In characteristic zero, the Lie algebra is an invariant of the isogeny class, but it is not necessarily so in positive characteristic. For example, in characteristic two, the Lie algebra of $\mathbb{SL}_2$ is nilpotent, with a one-dimensional center, namely $L(T)$, whereas the Lie algebra of $\mathbb{PSL}_2$ is solvable, not nilpotent, with a two-dimensional commutative ideal.

(2) With a slight extension of the notion of reduced root system, 1.4 can also be stated and proved for connected algebraic groups which are <u>reductive</u>, i.e. groups which are isogeneous to products of semi-simple groups and tori, or, equivalently, groups whose radical if a torus [15; 16].

§2. IRREDUCIBLE REPRESENTATIONS

2.1. DOMINANT AND FUNDAMENTAL WEIGHTS. Let Δ be a basis of Φ. Thus, Δ is a basis of $X(T)_{\mathbb{R}}$ contained in Φ, such that every element of Φ is linear combination of elements in Δ with coefficients in $\mathbb{Z}$, all of the same sign. The set of such positive linear combinations will be denoted Φ^+; it is the set of positive elements in Φ for some ordering.

An element $r \in P(\Phi)$ is <u>dominant</u> if $(r, a) \geqq 0$ for all $a \in \Delta$. For $a \in \Delta$, let d_a be the dominant weight such that $2(d_a, b)/(b, b) = \delta_{ab}$ (Kronecker symbol) for all $b \in \Delta$. The d_a's are the <u>fundamental dominant weights</u>. The fundamental dominant weights form a basis of $P(\Phi)$, and generate the semi-group of dominant weights. Any weight d can be written uniquely

$$d = \Sigma_{a \in \Delta} m_a(d) d_a , \quad (m_a \in \mathbb{Z}) ,$$

and is dominant if and only if $m_a(d) \geqq 0$ for all $a \in \Delta$.

2.2. To Δ there is associated a maximal connected solvable subgroup B of G containing T. It is the semi-direct product of T by a normal unipotent subgroup U, whose Lie algebra is the direct sum of the $L(G)_\alpha$ $(\alpha \in \Phi^+)$. This yields in fact a bijection between bases of Δ and maximal solvable subgroups containing T.

Every rational character of B is trivial on U, whence a natural bijection $X(B) \cong X(T)$.

2.3. Let E be a finite dimensional vector space over k and $\pi : G \longrightarrow GL(E)$ a rational representation of G. Its restriction to T can be put in diagonal form. For $r \in X(T)$, let $E_r = \{e \in E, \pi(t).e = t^r e\}$ and let $P(\pi) = \{r \mid E_r \neq \{0\}\}$. The elements of $P(\pi)$ are the <u>weights of</u> π. They form a finite set, invariant under W, since the latter obviously permutes the E_r's.

By the Lie-Kolchin theorem, $\pi(B)$ can be put in triangular form, in suitable coordinates, which implies in particular that there is at least one line $D \subset E$ stable under B. We then get a one-dimensional representation of B, which is described by a rational character of B (or T).

2.4. THEOREM. Let $\pi : G \longrightarrow GL(E)$ be a rational irreducible representation of G.

(i) There is a unique line D stable under B. The weight d_π of T on D is dominant, with multiplicity one. Every other weight $r \in P(\pi)$ is of the form

$$(1) \qquad r = d_\pi - \Sigma_{\alpha \in \Delta} m_\alpha(r)\alpha \text{ with } m_\alpha(r) \geqq 0 \text{ integral.}$$

(ii) The map $\pi \mapsto d_\pi$ induces a bijection between isomorphism classes of rational irreducible representations of G and dominant weights belonging to $X(T)$.

The weight d_π will be called the <u>highest weight</u> of π. In general, $P(\pi)$ may contain several dominant weights, but only one corresponding to a line stable under B.

Thus the dominant weights contained in $X(T)$ parametrize the irreducible linear representations of G. If G is simply connected, then $X(T) = P(\Phi)$ and all dominant weights in $P(\Phi)$ occur. In the general case, we can also say that the dominant weights in $P(\Phi)$ parametrize the rational irreducible <u>projective</u> representations of G.

2.5. REFERENCES. Over $\mathbb{C}$, this again goes back to E. Cartan and H. Weyl. In fact they proved these results in the complex analytic framework. See e.g. [22; 28]. The transition to algebraic groups is then automatic, if one takes into account the fact that a holomorphic representation of G is always rational for the associated algebraic group structure, which is clear from its description (see 1.5).

2.6. REMARKS. (1) Like the classification, the parametrization of irreducible representations is independent of k. However some features of the representation with a given highest weight may depend on the characteristic of k; this is a source of problems, which are only partially solved and will be mentioned in §5.

(2) In characteristic zero, all representations are fully reducible, but this is not so in positive characteristic (except for tori). In this connection we recall Mumford's conjecture*: "Let $\pi : G \longrightarrow GL(E)$ be a rational representation of G and D a line stable under G. Then there exists a homogeneous polynomial P on E which is invariant under G but not zero on D." If we had full reducibility, we could take for P a linear form defining an invariant hyperplane supplementary to D. In general, Mumford's conjecture calls for an invariant cone not containing D. In fact, this conjecture is usually stated more generally for representations of algebraic groups whose identity component is reductive (1.6(2)), the requirement on P being that the line $k.P$ be stable under G. It would have many interesting

* (Added October 30, 1974) This conjecture has recently been proved in the general case by W. Haboush - see Seshadri's article in this volume.

consequences [27; 30]; so far, it has been proved only for groups whose identity component is isogeneous to a product of $\mathbb{SL}_2$ and tori [30].

2.7. THE HIGHEST WEIGHT OF THE CONTRAGREDIENT REPRESENTATION. There exists a unique element $w_o \in W$ such that $w_o(\Phi^+) = \Phi^- = -\Phi^+$. We let i be the automorphism $r \longmapsto -w_o(r)$ of $X(T)$. It leaves Φ^+, Δ and the set of fundamental dominant weights stable. Both w_o and i have order $\leqq 2$. The latter is called the <u>opposition involution</u>. If $-Id. \in W$, then it is equal to w_o and i is the identity. This happens for the types $\mathbb{B}_n$, $\mathbb{C}_n$, $\mathbb{D}_n$ (n even), $\mathbb{G}_2$, $\mathbb{F}_4$, $\mathbb{E}_7$, $\mathbb{E}_8$.

Given a rational representation $\pi : G \longrightarrow GL(E)$ we denote by $\check{\pi}$ the contragredient representation to π. Thus π operates on the dual space E^* to E and $\check{\pi}(g) = {}^t\pi(g^{-1})$ for all $g \in G$. Diagonalizing $\pi(T)$, we see immediately that $P(\check{\pi}) = -P(\pi)$. If π is irreducible, then so is $\check{\pi}$, and it follows from the above that

$$d_{\check{\pi}} = i(d_\pi) . \tag{1}$$

In particular, if $w_o = -Id.$, then every irreducible representation is self-contragredient.

§3. INFINITESIMALLY IRREDUCIBLE REPRESENTATIONS. THE TENSOR PRODUCT THEOREM.

3.1. Let $\pi : G \longrightarrow GL(E)$ be a rational representation. Its differential $d\pi$ at e defines a homomorphism of Lie algebras $L(G) \longrightarrow g\ell(E)$, which is also a homomorphism of restricted Lie algebras if $p > 0$. We say that π is <u>infinitesimally irreducible</u> if $d\pi$ is irreducible. If π is infinitesimally irreducible, then it is irreducible. The converse is true in characteristic zero (and follows from the connection between Lie algebras and Lie groups given by the exponential mapping), but not in positive characteristic.

In this section, we assume $p > 0$, and let $M(G)$ be the set of rational infinitesimally irreducible representations of G. The results will be stated for simply connected groups and linear representations. Alternatively, we could drop the assumption G simply connected, but then consider projective representations.

3.2. THEOREM. Let G be simply connected. An irreducible

representation π of G is infinitesimally irreducible if and only if its highest weight d_π is of the form $d_\pi = \Sigma_{a\in\Delta} m_a(\pi) d_a$ with $0 \leq m_a(\pi) < p$.

This shows in particular that $M(G)$ has p^r elements $(r = \mathrm{rk}\, G)$. This theorem is due to C. Curtis and R. Steinberg. More precisely, C. Curtis determined (with some restrictions on the characteristic) all restricted irreducible representations of $L(G)$ [12], and Steinberg showed all these representations to be differentials of representations of G [34], see also [3].

Let $Fr : x \longmapsto x^p$ be the Frobenius automorphism of k. Given a rational representation π of G on E, we get another rational representation by identifying $GL(E)$ to $\mathbb{GL}_n$ $(n = \dim E)$ via a basis of E and applying Fr to the coefficients of $\pi(g)$ $(g \in G)$. This representation will be denoted π^{Fr}; its equivalence class depends only on that of π. Clearly, $P(\pi^{Fr}) = p.P(\pi)$. If π is irreducible, so is π^{Fr} and its highest weight is $p.d_\pi$. Thus, the representation with highest weight $p^i.d_\pi$ is obtained by applying Fr^i to π and has the same degree as π (whereas this degree tends to infinity with i, in characteristic zero, unless $\pi = 1$).

3.3. THEOREM. *Let G be simply connected. Let m be a positive integer, and $\pi_i \in M(G)$ $(0 \leq i \leq m)$. Then the tensor product $\pi_o \otimes \pi_1^{Fr} \otimes \ldots \otimes \pi_m^{Fr^m}$ is an irreducible representation, with highest weight $d = \Sigma_{0 \leq i \leq m}\, p^i.d_{\pi_i}$. Any non-trivial irreducible representation of G can be written uniquely in this way, for a suitable m, with $\pi_m \neq 1$.*

This theorem is due to R. Steinberg [34]. (See also [3; 35].)

Let k_o be a finite subfield of k. Assume that G is obtained by extension of the ground field from a group over k_o, also to be denoted G, which is split over k_o (i.e. contains a T defined over k_o and isomorphic over k_o to a product of $\mathbb{GL}_1$'s). The groups $G(k_o)$ obtained in this way are the finite Chevalley groups.

3.4. THEOREM. *Let G be simply connected. Let*

$p^m = q = \mathrm{Card}\, k_o$. The restrictions to $G(k_o)$ of the q^r representations $\pi_o \otimes \pi_1^{Fr} \otimes \ldots \otimes \pi_{m-1}^{Fr^{m-1}}$ $(\pi_i \in M(G))$ form a complete set of irreducible representations of $G(k_o)$ over k. In particular, the elements of $M(G)$ define irreducible representations of $G(k_o)$ and all irreducible representations of $G(k_o)$ over k are restrictions of rational irreducible representations of G.

This is due to C. Curtis [13] and Steinberg [34]. For other accounts, see [3; 35]. In [34; 35], there are analogues for the so-called "twisted Chevalley groups": groups $G(k_o)$, where G is defined over k_o, but not necessarily split over k_o, or groups of Ree or Suzuki type. For a systematic discussion of those groups, see e.g. [7; 35]. A further extension to finite groups with "split BN pairs" was given by F. Richen and C. Curtis (see e.g. [14]).

The representations occurring in 3.4 are those with highest weight of the form $d = \Sigma_{a \in \Delta} m_a . d_a$ with $0 \leqq m_a < q$. The one for which the m_a are all equal to $q-1$ has degree p^{Nm}, where $N = \mathrm{Card}\, \Phi^+$, and any irreducible representation of $G(k_o)$ non-equivalent to it has a strictly smaller degree [34]. The number p^{Nm} is also the order of the group $U(k_o)$, which is a p-Sylow subgroup of $G(k_o)$. A similar representation exists over any field F, and is irreducible if the characteristic of F is zero, or prime to the index of $B(k_o)$ in $G(k_o)$ [33]. It is now called the Steinberg representation of $G(k_o)$. Again, it also exists for twisted Chevalley groups; it has been given a cohomological interpretation in terms of the Tits building of $G(k_o)$ by Solomon and Tits [31].

§4. CONSTRUCTION OF IRREDUCIBLE REPRESENTATIONS.

Our main purpose in this section is to outline the realization of irreducible representations via cross sections of line bundles on G/B (4.3; 4.4), which leads to the existence assertion in 2.4(ii). However, since this is quite short, we shall also sketch the proof of the remaining assertions of 2.4. For more details, see [3; 10; 35].

4.1. Let $B^- = w_o . B . w_o$, where w_o is as in 2.7. The group B^- is the semi-direct product of T and of $U^- = w_o . U . w_o$. The Lie algebra of U^-

is spanned by the $L(G)_\alpha$ $(\alpha < 0)$. The set $U^-.B = B^-.B$ is Zariski-open in G. As an example, if $G = \mathbf{SL}_n$, we may take for B the group of upper triangular matrices. Then B^- is the lower triangular group, and $U^-.B$ the set of $g = (g_{ij}) \in \mathbf{SL}_n$ for which the principal minors $\det(g_{ij})_{1 \leqq i, j \leqq m}$ are $\neq 0$ for $1 \leqq m \leqq n$.

For each $a \in \Phi$, there is an injective morphism x_a of the additive group of k into G, whose image is normalized by T, and which satisfies

$$(1) \qquad t.x_a(z).t^{-1} = x_a(t^a.z), \qquad (t \in T;\ z \in k) .$$

The group U (resp. U^-) is generated by the images of the x_a for $a \in \Phi^+$ (resp. $a \in \Phi^-$).

4.2. Let now $\pi : G \longrightarrow GL(E)$ be a rational representation of G and $r \in P(\pi)$. We claim first that

$$(1) \qquad \pi(x_a(z)).E_r \subset \Sigma_{i \geqq 0} E_{r+i.a}, \qquad (z \in k;\ a \in \Phi) .$$

Let $y \in E_r$. Then we may write $x_a(z).y = \Sigma_s e_s(z)$, where $s \in P(\pi)$ and $z \longmapsto e_s(z)$ is a regular function from k to E_s, hence an E_s-valued polynomial in z. We have then

$$(2) \qquad \pi(x_a(z)).y = \Sigma_{i \geqq 0, s}\, z^i.e_{s,i}, \qquad (e_{s,i} \in E_s) .$$

Since $\pi(t) = t^s.\mathrm{Id}$ on E_s for $t \in T$, we have

$$\pi(t.x_a(z).t^{-1}).y = t^{-r}\Sigma_{s,i}\, z^i.t^s.e_{s,i},$$

and on the other hand

$$\pi(x_a(t^a.z)).y = \Sigma_{s,i}\, z^i.t^{ia}.e_{s,i} .$$

Comparison of coefficients, and 4.1(1), then show that $e_{s,i} \neq 0$ implies $s = r+i.a$.

Let now D be a line stable under B and d the corresponding weight. Iterated application of (1) yields

$$(3) \qquad \pi(U^-.B)(D) = \pi(U^-).D \subset D + \Sigma_s E_s ,$$

where s runs through weights of the form

(4) $\quad s = d - \Sigma_{a\in\Delta} m_a(s).a,$ with the $m_a(s) \geqq 0,$ integral, not all zero.

Since $U^-.B$ is Zariski dense in G, it follows that the Zariski closure of $G(D)$ is contained in the right-hand side of (3), hence so is the smallest G-invariant subspace containing D. In particular, if E is irreducible, then E is equal to the right-hand side, which proves that the weight d of T on D has multiplicity one and all other weights are of the form (4). On the other hand, if s_a is the reflection of the Weyl group associated to $a \in \Delta$, then $s_a(d) = d - 2(d, a).(a, a)^{-1}a$, whence $(d, a) \geqq 0$; this shows that d is dominant and concludes the proof of 2.4(i).

4.3. We shall denote by A the coordinate ring $k[G]$ of G. The group G acts on A by left and right translations. For $g \in G$, $f \in A$, let $\ell_g f$ (resp. $r_g f$) be defined by

(1) $$\ell_g f(x) = f(g^{-1}.x), \qquad (\text{resp. } r_g f(x) = f(x.g)) .$$

We have then $\ell_{g.g'} = \ell_g.\ell_{g'}$ (resp. $r_{g.g'} = r_g.r_{g'}$). It is elementary that every $f \in A$ is contained in a finite dimensional subspace of A which is left and right invariant under G.

4.4. DEFINITION. For $r \in X(T)$, let $A_r = \{f \in A \mid r_b f = b^r f, \quad b \in B)\}$. This is a G-module with respect to left-translations. It can be canonically identified to the space of regular cross-sections of the line bundle ξ_{-r} on G/B associated to $-r$, on which G acts by left translations. (We recall that the total space of ξ_s $(s \in X(B))$ is the quotient of $G \times k$ by the equivalence relation $(g, x) \sim (g.b, b^{-s}.x)$ $(g \in G; x \in k)$.) Since G/B is complete, this space is finite-dimensional. However, this fact is not really needed in the sequel.

4.5. PROPOSITION. *Assume $A_{i(r)} \neq \{0\}$. Then the space $A_{i(r)}$ contains a unique line D_r stable under B. The weight of B on D_r is r and is dominant. The smallest non-zero G-invariant subspace E_r of $A_{i(r)}$ contains D_r and is an irreducible module with highest weight r. Every irreducible G-module with highest weight r is* isomorphic to E_r. If $p = 0$, then $A_{i(r)} = E_r$. If r is not dominant, then $A_r = \{0\}$.

Let $D \subset A_{i(r)}$ be a line stable under B and s the corresponding weight. Let $f \in D$. We have then

$$f(b.x.b') = b^{-s}.f(x).b'^{i(r)} \quad .$$

Since $B.w_o.B = w_o B^- B$ is open in G (4.1), this shows that f is completely determined by its value at w_o. In particular $f \neq 0$ if and only if $f(w_o) \neq 0$. There exists therefore at most one B-stable line. Moreover

$$f(t.w_o) = t^{-s}.f(w_o) = f(w_o.w_o^{-1}.t.w_o) = f(w_o).(w_o^{-1}.t.w_o)^{i(r)} = f(w_o).t^{-r} \quad ,$$

whence $s = r$.

Since every non-zero G-stable subspace of $A_{i(r)}$ contains a line stable under B, it contains D, which proves the third assertion. The second and the last one then follow from 4.2. Assume $p = 0$. By full reducibility, every proper G-invariant submodule has a G-invariant complement. They cannot contain both D, hence

$$(1) \qquad E_r = A_{i(r)} = H^o(G/B; \xi_{-i(r)}) \ \text{ if } \ p = 0 \quad .$$

Let now (π, E) be an irreducible representation with highest weight r. Given $e \in E$, $f \in E^*$, let c_{e,e^*} be the "coefficient" of π, given by $c_{e,e^*}(g) = \langle e^*, g^{-1}(e)\rangle$ where $\langle\, ,\rangle$ is the canonical bilinear form. We have $c_{g.e,h.e^*} = \ell_g . r_h . c_{e,e^*}$. In particular, for fixed e^*, the map $q_{e^*} : e \longmapsto c_{e,e^*}$ sends G-equivariantly E into A (acted upon by left translations). Take now for e (resp. e^*) a highest weight vector of π (resp. $\check{\pi}$). The corresponding weight is then r (resp. $i(r)$), from which it follows that q_{e^*} maps E injectively into $A_{i(r)}$; the uniqueness assertion of 2.4(ii) follows.

In order to complete the proof of 2.4, there remains only to show the existence of an irreducible representation whose highest weight is a given dominant weight $r \in X(T)$. This amounts to proving that $A_r \neq \{0\}$ for every such r. This can be done directly (see e.g. [3: §5]) or follows by "reduction mod p" from a construction over $\mathbb{Z}$, which is sketched in §5 (see 5.4).

§5. GROUP SCHEMES OVER $\mathbb{Z}$. REDUCTION mod p OF IRREDUCIBLE REPRESENTATIONS.

5.1. We shall first outline Chevalley's construction of a group scheme over $\mathbb{Z}$ already referred to in 1.5.

The starting point is, for $k = \mathbb{C}$, a suitable $\mathbb{Z}$-structure on $L(G)$, defined by means of what is now called a Chevalley basis [3; 7; 9; 22; 35]. It consists of elements $e_a \in L(G)_a$ $(a \in \Phi)$ and of a suitable basis of $L(T)$. (In fact, there may be several possibilities for the $\mathbb{Z}$-structure of $L(T)$: the space $L(T)$ may be identified to the dual of $X(T) \otimes \mathbb{C}$, and the dual lattice to any lattice between $R(\Phi)$ and $P(\Phi)$ is suitable.)

Let now (π, E) be a faithful finite dimensional complex representation of $L(G)$. It may also be viewed as a rational representation of G, assumed to be simply connected. A $\mathbb{Z}$-form $E_{\mathbb{Z}}$ of E is said to be <u>admissible</u> if

$$\pi(e_a^j/j!)(E_{\mathbb{Z}}) \subset E_{\mathbb{Z}}, \quad \text{for all } a \in \Phi \text{ and } j \geqq 0 . \tag{1}$$

This implies that $E_{\mathbb{Z}}$ is the direct sum of its intersections with the weight spaces E_s $(s \in P(\pi))$ [3: 2.3]. Chevalley has shown that any E has an admissible $\mathbb{Z}$-form (This is stated in [11], proved in [3; 22; 35]). This really amounts to showing the existence of a nice $\mathbb{Z}$-form $U_{\mathbb{Z}}$ of the universal enveloping algebra U of $L(G)$ ([26]; see also [22; 35]).

Identify E to $\mathbb{C}^n$ $(n = \dim E)$ via a basis of $E_{\mathbb{Z}}$. We have then

$$\pi(x_a(t)) = \Sigma_{j \geqq 0}\, t^j\, \pi(e_a^j/j!) \in \mathbb{SL}_n(\mathbb{Z}[t]) ,$$

[$\pi(e_a)$ is nilpotent, hence the exponential series is in fact a finite sum.] and this makes sense in fact for any indeterminate t. Let now F be a commutative field, and let

$$G_{\pi, F} = \langle \pi(x_a(z)) \mid a \in \Phi,\ z \in F \rangle \tag{2}$$

be the subgroup of $\mathbb{SL}_n(F)$ generated by the elements $\pi(x_a(z))$ $(a \in \Phi,\ z \in F)$. The groups so constructed are the "Chevalley groups." If G is simple, then $G_{\pi, F}$ is simple modulo its center, except in a few cases where F has two or three elements [6; 9; 35]. This construction was first made by Chevalley when π is the adjoint representation [9] and yielded the first new finite simple groups since Dickson.

5.2. THEOREM. Let Γ_π be the lattice of $X(T)_{\mathbb{R}}$ spanned by the weights of π. Then $G_{\pi, k}$ is a connected semi-simple k-group with diagram (Φ, Γ_π).

More precisely, if $A_{\pi, \mathbb{Z}}$ is the ring of regular functions on $G_{\pi, \mathbb{C}}$ generated by the matrix coefficients, the statement is in fact that $A_{\pi, \mathbb{Z}}$ represents a smooth group scheme over $\mathbb{Z}$, and that $G_{\pi, k} = \mathrm{Hom}(A_{\pi, \mathbb{Z}}, k)$ [11; see also [3; 15; 16]]. There is also a functorial complement to it: if π' is another faithful representation of $L(G)$, and $\Gamma_{\pi'} \subset \Gamma_\pi$, then there is a canonical morphism $A_{\pi', \mathbb{Z}} \longrightarrow A_{\pi, \mathbb{Z}}$ (an isomorphism if $\Gamma_{\pi'} = \Gamma_\pi$) such that the associated morphism

$$G_{\pi, k} = \mathrm{Hom}(A_{\pi, \mathbb{Z}}, k) \longrightarrow \mathrm{Hom}(A_{\pi', \mathbb{Z}}, k) = G_{\pi', k}$$

is the central isogeny mentioned in 1.4.

5.3. The existence part of 1.4 (see 1.5(i)) is then a consequence of the fact that any Γ between $R(\Phi)$ and $P(\Phi)$ is a Γ_π. In particular, if $\Gamma_\pi = P(\Phi)$, then $G_{\pi, k}$ is the simply connected group with root system Φ. In the sequel we denote it by G_k, and let $G/\mathbb{Z}$ be the corresponding group scheme over $\mathbb{Z}$. Then, for any π, we have a rational representation $\pi \otimes k : G_k \longrightarrow G_{\pi, k}$ obtained by reduction mod k from a $\mathbb{Z}$-form of $\pi : G_{\mathbb{C}} \longrightarrow GL(E)$. In particular, it has the same degree as π. We may also say that it has the "same character." In fact, Chevalley's construction of $G/\mathbb{Z}$ also yields schemes $T/\mathbb{Z}$ and $B/\mathbb{Z}$ which give our earlier $T, B \subset G_k$ by specialization. In particular, there are natural identifications

$$X(T/\mathbb{Z}) = X(T_{\mathbb{C}}) = X(T_k) = P(\Phi) ,$$

and the character of π or $\pi \otimes k$ may be written $\Sigma s . \dim E_s$, $(s \in P(\pi) \subset P(\Phi))$. In this connection, we note that if $R_k(G_k)$ (resp. $R_{\mathbb{Z}}(G/\mathbb{Z})$) is the Grothendieck group of rational representations of G_k (resp. of $G/\mathbb{Z}$-rational modules, of finite type over $\mathbb{Z}$), then $R_k(G_k)$ and $R_{\mathbb{Z}}(G/\mathbb{Z})$ are canonically isomorphic [29: 3.6, 3.7]. In particular $R_k(G_k)$ and $R_{\mathbb{C}}(G_{\mathbb{C}})$ are isomorphic by an isomorphism which is compatible with the identification of characters.

5.4. Assume from now on $G_{\mathbb{C}}$ to be simple. Let d be a dominant weight, and (π_d, E) the corresponding irreducible representation of $G_{\mathbb{C}}$. If x_o is a non-zero highest weight vector, then $E_{\mathbb{Z}} = U_{\mathbb{Z}}(x_o)$ is an admissible $\mathbb{Z}$-form of E and $(\pi \otimes k, E_k)$, where $E_k = E_{\mathbb{Z}} \otimes k$, is a rational representation of G_k. It is not necessarily irreducible (as is already clear from the end remark of 3.2, and 3.3); however, it has a biggest proper invariant subspace F and the representation $\pi_{d,k}$ of G_k in E_k/F is irreducible, with highest weight d. This then yields the existence part of 2.4(ii), and also shows that $A_d \neq \{0\}$ if d is dominant (4.5). Furthermore we see that $P(\pi_{d,k}) \subset P(\pi_d)$.

In some cases, $\pi \otimes k$ is irreducible for all k. This happens if all weights of π are extremal (i.e. form only one orbit of W) [3: 5.12]. Examples are the spinor or half-spinor representations of Spin(n) or the exterior power representations of $\mathbf{SL}_n$. However, if $G_{\mathbb{C}}$ is also adjoint, no non-trivial representation of $G_{\mathbb{C}}$ has only extremal weights since every representation has the weight zero. For instance, the seven-dimensional representation of $\mathbf{G}_2$ has the zero weight with multiplicity one, besides extremal weights, and is in fact reducible in characteristic two.

There arises therefore the problem of knowing when $\pi \otimes k$ is irreducible, or more generally describing a composition series for $\pi \otimes k$ and finding a formula for the character and the degree of $\pi_{d,k}$, which would be the analogues of H. Weyl's formulae in characteristic zero [22; 28]. We note that the equivalence class of $\pi \otimes k$ may depend on the choice of an admissible $\mathbb{Z}$-form of π; however, the composition factors do not (and depend only on the characteristic of k); moreover, there is, up to dilation, a natural "minimal" one [21].

Given d, the representation $\pi \otimes k$ is irreducible if p is big enough [32], but p may depend on d, and not just on the isomorphism class of $G_{\mathbb{C}}$. There are a number of conjectures and results on this general problem, due notably to B. Braden, C. W. Curtis, R. Carter and G. Lusztig, J. E. Humphreys, J. C. Jantzen, D. Verma, W. J. Wong, for which we refer the reader to [8; 21; 23; 36].

§6. COHOMOLOGY OF LINE BUNDLES ON G/B.

6.1. To a rational character r of B there is associated a line bundle ξ_r on G/B (see 4.4), and hence coherent sheaf cohomology spaces

$H^i(G/B; \xi_r)$; these are finite-dimensional (since G/B is complete) vector spaces over k, on which G acts rationally, the action stemming from left translations, and we are interested in identifying the representations thus obtained. We already saw (4.5) that $A_r \cong H^o(G/B; \xi_{-r})$ is zero if r is not dominant and contains the irreducible representation with highest weight $i(r)$ if r is dominant.

6.2. Take $k = \mathbb{C}$. If r is dominant, then (4.5(1)), $H^o(G/B; \xi_{-r})$ is the irreducible representation of G with highest weight $i(r)$. (Theorem of Borel-Weil); moreover, it follows readily from Kodaira's vanishing theorem that $H^i(G/B; \xi_{-r}) = 0$ for $i \geqq 1$.

Let ρ be half the sum of the positive roots. Let r be a weight and assume $r + \rho$ to be regular, i.e. not orthogonal to any root. There are then a unique $w \in W$ and a unique dominant weight s such that $r + \rho = w(s + \rho)$. Let $n(w) = \mathrm{Card}\,(w(\Phi^-) \cap \Phi^+)$. Then $H^i(G/B; \xi_{-r}) = 0$ except for $i = n(w)$, in which case it is an irreducible G-module with highest weight $i(s)$. If $r + \rho$ is singular (i.e. not regular), then $H^i(G/B); \xi_{-r}) = 0$ for all i's. These results are due to R. Bott [5]; for other proofs, see [17; 25].

6.3. For a general k we want, as in 5.4, to compare the situations over k and over $\mathbb{C}$. We revert to the notation of 5.3, hence write G_k, B_k instead of G, B, and view these groups as obtained by change of basis from group schemes over $\mathbb{Z}$. Since r is defined over $\mathbb{Z}$, the line bundle ξ_r is also a specialization of a line bundle over $\mathbb{Z}$. Let

$$\chi_F(\xi_r) = \chi(G_F/B_F; \xi_r) = \Sigma_i\, (-1)^i \dim H^i(G_F/B_F; \xi_r)\ , \qquad (F = k,\ \mathbb{C})\ .$$

It follows from general principles that $\chi_{\mathbb{C}}(\xi_r) = \chi_k(\xi_r)$. Now these two alternating sums can be viewed as elements of $X(T_{\mathbb{C}})$ and $X(T_k)$ respectively, or also as elements of the Grothendieck groups of rational representations of G_k and $G_{\mathbb{C}}$. It seems rather likely that, as such, they correspond to each other under the isomorphisms mentioned in 5.3.

A natural question is whether $H^i(G_k/B_k; \xi_r) = 0$ for $i \geqq 1$ if $-r$ is dominant. If so, the invariance of the Euler-characteristic would show that $\dim H^o(G_k/B_k; \xi_r) = \dim H^o(G_{\mathbb{C}}/B_{\mathbb{C}}; \xi_r)$. So far, this has been proved for $G_k = \mathbb{SL}_n(k)$ in [23], and for G_k classical or isomorphic to $\mathbb{G}_2$ in [1]; moreover [18] shows that it would follow in general from a conjecture on admissible $\mathbb{Z}$-forms.

It seems therefore rather likely that one has in characteristic p rather straightforward analogues of the results in characteristic zero for r dominant. However, this is certainly not so otherwise: for $G = \mathbb{SL}_3$, $p = 2$, Mumford has given an example of an $r \in X(T)$ such that $r + \rho$ is singular and $H^i(G_k/B_k; \xi_{-r})$ has dimension one for $i = 1, 2$; I am told that recently, Larry Griffiths has found many examples in arbitrary characteristic, for $G = \mathbb{SL}_3(k)$, of $r \in X(T)$, with $r + \rho$ either singular or regular, for which $H^i(G_k/B_k; \xi_r) \neq 0$ for two dimensions. I do not know whether the representations of G_k occurring in these groups have been determined. This might be helpful in trying to see whether there is a general pattern.

6.4. REMARK. Let P be a parabolic subgroup of G, i.e. a closed subgroup containing a conjugate of B. Let r be an irreducible rational representation of P and ξ_r the associated line bundle on G/P. The cohomology spaces $H^i(G/P; \xi_r)$ are again finite dimensional G-modules. The results of [5; 25] give more generally the structure of these G-modules for any P if $k = \mathbb{C}$. The speaker does not know of results pertaining to the analogous problem when $p > 0$, $P \neq B$ and r is not one-dimensional.

REFERENCES

[1] L. Bai, C. Musili, and C. S. Seshadri, "Cohomology of line bundles on G/B," Annales E. N. S. (to appear).

[2] A. Borel, "Linear algebraic groups," Benjamin, New York 1969.

[3] ———, "Properties and linear representations of Chevalley groups," Seminar in algebraic groups and related finite groups, Springer Lecture Notes 131 (1970), 1-55.

[4] ——— et J. Tits, "Compléments à l'article: Groupes réductifs," Publ. Math. IHES 41 (1972), 253-276.

[5] R. Bott, "Homogeneous vector bundles," Annals of Math. (2) 66 (1957), 203-248.

[6] N. Bourbaki, "Groupes et algèbres de Lie," Chap. IV, V, VI, Paris Hermann 1968.

[7] R. Carter, "Simple groups of Lie type," Pure and Applied Math. XXVIII, J. Wiley and Sons, New York (1972).

[8] ——— and G. Lusztig, "On the modular representations of the general linear and symmetric groups," Math. Z. 136 (1974).

[9] C. Chevalley, "Sur certains groupes simples," Tôhoku Math. J. (2) 7 (1955), 14-66.

[10] ———, "Classification des groupes de Lie algébriques, "Notes polycopiées, Inst. H. Poincaré, Paris (1956-58).

[11] ———, "Certain schémas de groupes semi-simples," Sém. Bourbaki 13è année, (1960-61), Exp. 219.

[12] C. W. Curtis, "Representations of Lie algebras of classical type with applications to linear groups," J. Math. Mech. 9 (1960), 307-326.

[13] ———, "On projective representations of certain finite groups," Proc. A.M.S. 11 (1960), 852-860.

[14] ———, "Modular representations of finite groups with split (B, N) pairs," Seminar on algebraic groups and related finite groups, Springer Lecture Notes 131 (1970), 58-95.

[15] M. Demazure, "Schémas en groupes réductifs," Bull. Soc. Math. France 93 (1965), 369-413.

[16] ———, "Schémas en groupes III," Springer Lecture Notes 153 (1970).

[17] ———, "Une démonstration algébrique d'un théorème de Bott, Invent. Math. 5 (1968), 349-356.

[18] ————, "Désingularisation des variétés de Schubert généralisées, Annales E.N.S. (to appear).

[19] G. Hochschild, "The structure of Lie groups," Holden-Day Inc. 1965.

[20] ———— and G. D. Mostow, "On the algebra of representative functions on an analytic group," Amer. J. M. LXXXIII (1961), 111-136.

[21] J. E. Humphreys, "Modular representations of classical Lie algebras and semisimple groups," J. of Algebra 19 (1971), 51-79.

[22] ————, "Introduction to Lie algebras and representation theory," Graduate Texts in Math. 9, Springer, 1972.

[23] J. C. Jantzen, "Darstellungen halbeinfacher algebraischer Gruppen und zugeordnete Kontravariante Formen," Bonner math. Schriften Nr. 67, Universität Bonn, 1973.

[24] G. Kempf, "Schubert methods with an application to algebraic curves," Stichting mathematisch centrum, Amsterdam (1971).

[25] B. Kostant, "Lie algebra cohomology and the generalized Borel-Weil theorem," Ann. of Math. 74 (1961), 329-387.

[26] ————, "Groups over $\mathbb{Z}$," Algebraic groups and discontinuous subgroups, Proc. Symp. pure math. 9, A.M.S. Providence, R. I. (1966).

[27] M. Nagata, "Invariants of a group in an affine ring," J. Math. Kyoto Univ. 3 (1964), 369-77

[28] J-P. Serre, "Algèbres de Lie semi-simples complexes," Benjamin 1966.

[29] ————, "Groupes de Grothendieck des schémas en groupes réductifs déployés," Publ. Math. I.H.E.S. 34 (1968), 37-52.

[30] C. S. Seshadri, "Mumford's conjecture for GL(2) and applications," Proc. Bombay Colloquium on algebraic geometry, 1968, 347-371.

[31] L. Solomon, "The Steinberg character of a finite group with BN-pair," Theory of finite groups. A Symposium, ed. by R. Brauer and C. H. Sah, Benjamin 1969, 213-221.

[32] T. A. Springer, "Weyl's character formula for algebraic groups," Invent. Math. 5 (1968), 85-105.

[33] R. Steinberg, "Prime power representations of finite linear groups II," Canadian J. M. 9 (1957), 347-351.

[34] ————, "Representations of algebraic groups," Nagoya M. J. 22 (1963), 33-56.

[35] ——————, "Lectures on Chevalley groups," Notes by J. Faulkner and R. Wilson, Yale University (1967)

[36] D. Verma, "Role of affine Weyl groups in the representation theory of algebraic Chevalley groups and their Lie algebras," Proc. Budapest Summer School on group representations 1971 (to appear).

THE INSTITUTE FOR ADVANCED STUDY
Princeton, New Jersey 08540

Proceedings of Symposia in Pure Mathematics
Volume 29, 1975

KNOT INVARIANTS OF SINGULARITIES

Alan H. Durfee[1]

In this lecture we'll survey what knot theory has to say about the topology of an isolated singularity of a complex analytic hypersurface. Suppose that f is an analytic function on some neighborhood U of the origin 0 in $\mathbb{C}^{n+1}$, that $f(0) = 0$, and that the derivative of f is nonzero in $U - \{0\}$. Then $f^{-1}(0)$ transversally intersects a suitably small sphere S^{2n+1} about 0 to give a smooth compact $(2n-1)$-manifold K, the link of the singularity. (As a general reference for this and the following see [16].) We'd like to know what manifolds K can occur, and how they embed in S^{2n+1}. K has a natural orientation as the boundary of $f^{-1}(0)$ intersected with the corresponding small ball D^{2n+2}. This is useful since it is easier to classify manifolds with a fixed orientation. K is also $(n-2)$-connected. Furthermore, for complex numbers δ close to 0, the intersection of $f^{-1}(\delta)$ and S^{2n+1} is still transverse and hence diffeomorphic to K. Thus K is the boundary of an $(n-1)$-connected parallalizable manifold $F = f^{-1}(\delta) \cap D^{2n+2}$.

Now what manifolds K satisfying the above conditions actually occur as links of singularities? When $n = 1$, K is a collection of one-spheres corresponding to the branches of f at 0. Thus all one-manifolds occur.

When $n = 2$ this is no longer true; the manifold $K = S^1 \times S^1 \times S^1$ satisfies the above conditions but is not the link of a surface singularity in codimension 1, or any codimension. Suppose it were. We resolve the singularity, and let T be a "tubular neighborhood" [17] of the exceptional locus in

AMS(MOS) Subject Classifications (1970): Primary 14-02, 14B05, 55-02, 55A25, 57-02, 57D40.

[1]Research partially supported by National Science Foundation grant MPS 72-05055 A02.

the resolution; ∂T is diffeomorphic to K. In the long exact sequence

$$\to H^1(T) \to H^1(\partial T) \to H^2(T, \partial T) \xrightarrow{\iota} H^2(T) \to$$

the map ι is injective since it may be identified with the intersection pairing, which is negative definite. Let α, β, γ be generators of $H^1(\partial T)$; then $\alpha \cup \beta \cup \gamma$ is nonzero. However, there are α', β', γ' in $H^1(T)$ mapping to α, β, γ, and since T retracts to the two (real) dimensional exceptional locus, we have $\alpha' \cup \beta' \cup \gamma' = 0$, a contradiction. (This example is due to D. Sullivan.)

Furthermore, there are manifolds, such as the lens space $L(3,1)$, that occur as links of singularities in codimension greater than one, but not one: A singularity whose link has finite abelian fundamental group $\mathbb{Z}/3\mathbb{Z}$ has a resolution whose dual graph is $\underset{-3}{\bullet}$ or $\underset{-2}{\bullet}\!\!-\!\!\underset{-2}{\bullet}$, where the vertices represent nonsingular rational curves [6]. Any such singularity is a quotient singularity. The link of the first is $L(3,1)$, but its embedding dimension, computed via the fundamental cycle, is 4. The second embeds in dimension 3, but has link $L(3,2)$ which is not orientation-preserving diffeomorphic to $L(3,1)$.

For $n \geq 3$, $(n-2)$-connected $(2n-1)$-manifolds K that bound parallelizable manifolds are well understood [8, 23].

<u>Problem</u> 1. Is there such a manifold that is not the link of a hypersurface singularity?

Some subclasses of these manifolds do occur as links, for instance, all exotic spheres that bound parallelizable manifolds [5; see also 8].

It is too restrictive to look at the link K above, so let's examine its embedding in the sphere S^{2n+1} We'll call this an <u>algebraic</u> <u>knot</u>. Already when $n = 1$ there is much more information. For example, let

$f(z_0, z_1) = z_0^k + z_1^2$, where $k \geq 3$ is odd. The variety $f^{-1}(0)$ is a rational curve with a singularity at the origin, and is parameterized by $\mathbb{C}$ under the map $t \mapsto (-t^2, t^k)$. As t travels once about the unit circle S^1, its image travels on the torus $S^1 \times S^1$ in $\mathbb{C}^2$, twice one way and k times the other. Mapping the unit three-sphere in $\mathbb{C}^2$ to $\mathbb{R}^3 \cup \{\infty\}$ such that $S^1 \times S^1$ goes onto a standard torus, we see that the link of our singularity is a _torus knot of type_ $(2, k)$ (Fig. i).

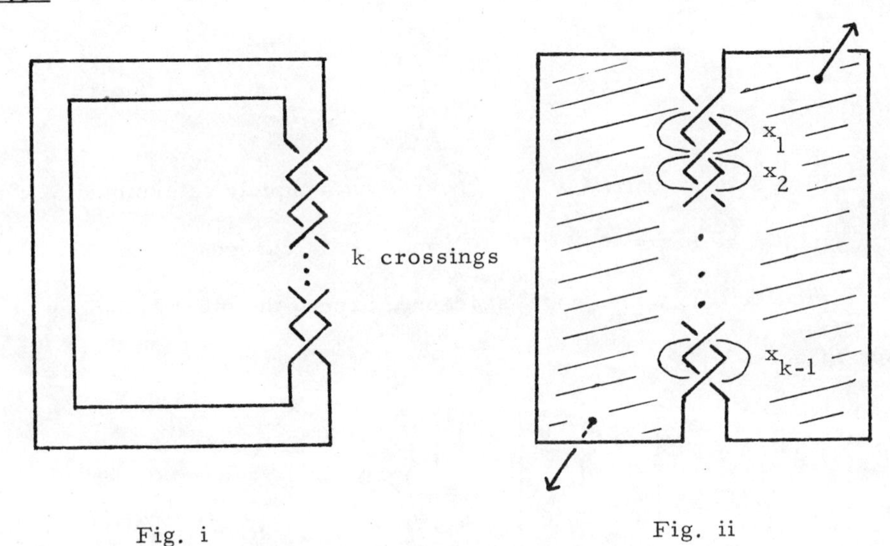

Fig. i Fig. ii

Another way of writing this equation is $z_1 = z_0^{k/2}$. The addition of a higher fractional power of z_0 perturbs the value of z_1, hence perturbs the above torus knot into a torus knot built upon the core of the old torus knot. Iterating this process shows that the Puiseux expansion of the function f determines explicitly the embedding of K in S^3 as a _compound torus knot._ (See [1, 14, 16]. Good references for classical knot theory are [10] and the appropriate sections of [22].)

Returning to arbitrary $n \geq 1$, any oriented $(2n-1)$-manifold K in S^{2n+1} gives rise to a _Seifert matrix_ V defined as follows: Choose an embedded oriented $2n$-dimensional surface F (_the Seifert surface_) in S^{2n+1}

whose boundary is K, and find a basis $x_1, \ldots, x_\mu$ of the torsion-free part of $H_n(F)$. The ij entry of the $\mu \times \mu$ matrix V is defined to be the linking number in S^{2n+1} of the translate of x_i in the positive normal direction with x_j. For example, if we rearrange the torus knot of type $(k, 2)$ as in Figure ii, take F to be the shaded portion, take the positive normal direction as indicated by the arrows, and take a basis $x_1, \ldots, x_{k-1}$ of $H_1(F)$ as shown, we then compute V to be

$$\begin{bmatrix} -1 & 1 & & 0 \\ & \ddots & \ddots & \\ & & \ddots & 1 \\ 0 & & & -1 \end{bmatrix}$$

The Seifert matrix V is of course not uniquely determined by the knot. Changing the basis of $H_n(F)$ by a matrix T changes V to T^tVT. Adding a small handle whose core is wrapped around the other homology classes changes V to

$$\begin{bmatrix} V & 0 & 0 \\ \alpha & 0 & 0 \\ 0 & 1 & 0 \end{bmatrix} \quad \text{or} \quad \begin{bmatrix} V & \beta & 0 \\ 0 & 0 & 1 \\ 0 & 0 & 0 \end{bmatrix}.$$

These are called <u>elementary enlargements</u> of V; V is an <u>elementary reduction</u>. Congruence, elementary enlargements and reductions generate an equivalence relation on the class of Seifert matrices called S-<u>equivalence</u>. It was shown by Murasugi for $n = 1$, and by Levine [15] (a good general reference for higher dimensional knots) for all n, that isotopic knots have S-equivalent Seifert matrices. Levine also obtains the converse for a certain subclass of knots with $n \geq 2$; this is the key classification result.

It was essentially known to Seifert that this converse is not true for $n = 1$. The knot in [21, p. 591] has Seifert matrix $\begin{bmatrix} 1 & -1 \\ -2 & 2 \end{bmatrix}$, which is congruent to $\begin{bmatrix} 0 & 1 \\ 0 & 0 \end{bmatrix}$, an elementary enlargement of the trivial matrix. However, its fundamental group admits a surjective map to a group of rigid motions of the hyperbolic plane, and hence is not free cyclic. Thus the knot is nontrivial. As another example, the torus knot of type $(2, 3)$ and one of its doubles (Figure iii)

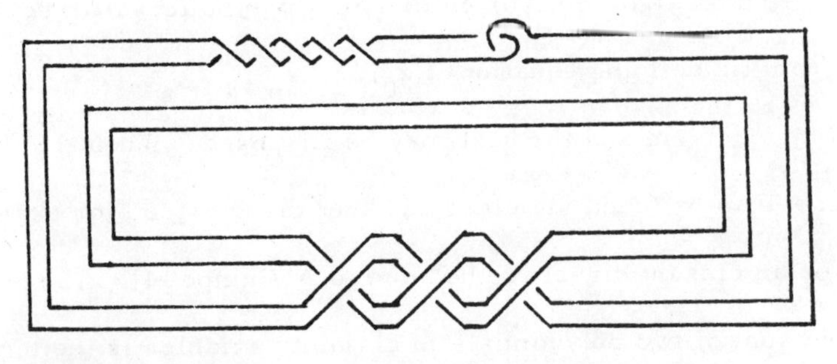

Fig. iii

have the same Seifert matrix but are not isotopic, since the torus knot has no nontrivial companions but the doubled knot does [20, p. 196].

There is added information about our knot K in S^{2n+1} that we haven't used yet, namely, that the complement of K in S^{2n+1} is a smooth fiber bundle $\phi : S^{2n+1} - K \to S^1$ over the circle [16]. We call this a <u>fibered knot</u>; a general reference for these is [9]. Thus $F = \phi^{-1}(1)$ is a canonical Seifert surface, and the Seifert matrix changes only by congruence, hence yielding an integral bilinear form on $H_n(F)$. This form is unimodular, since it is the composition

$$H_n(F) \otimes H_n(F) \to H_n(F) \otimes H_n(\phi^{-1}(-1)) \to H_n(F) \otimes H_n(S^{2n+1} - F) \to \mathbb{Z}$$

where the first map is $1 \otimes$ (map induced by the action of π_1), the second is $1 \otimes$ (map induced by inclusion), and the third is linking in S^{2n+1}, which expresses Alexander duality. It is not hard to adapt Levine's arguments to show for $n \geq 3$ that isotopy classes of fibered knots in S^{2n+1} are in one-one correspondence with isomorphism classes of integral unimodular bilinear forms [9, 12]. Thus we would expect the Seifert matrix to determine all the other knot invariants, such as monodromy, intersection form, and Alexander polynomials, and this is indeed true. (See for example [9, p. 51] and [16,

Lemma 10.1].)

In conclusion, here are some more unsolved problems.

Problem 2. Find an algorithm for determining the Seifert matrix of an algebraic knot from its defining equation $f(z_0, \ldots, z_n)$.

This problem and the next may be ill-posed. When $n = 1$, the Seifert matrix may be found via classical knot theory (see, for example, [9, p. 50]) or an elegant direct method due to A'Campo [4]. The Seifert matrix of the sum of two polynomials in disjoint variables is the tensor product of the Seifert matrices [11, 13, 19]. Furthermore, algorithms are known for the monodromy over the complex numbers [7] and its characteristic polynomial, which is the same as the Alexander polynomial of the knot [3]. An easier problem might be to compute the homology groups of the link K, or the signature of the knot.

Problem 3. Find a necessary and sufficient condition for a unimodular integral bilinear form to occur as the Seifert matrix of an algebraic knot.

It is known that the matrix V must be upper triangular with respect to some basis [9], and that the monodromy $(-1)^{n+1}V^{-1}V^t$ must have trace $(-1)^{n+1}$ [2] and be quasi-unipotent (see the bibliography in [7]).

Problem 4. Are algebraic knots prime?

A knot is prime if it is not the connected sum of two nontrivial knots. Compound torus knots are prime [20, p. 250]; furthermore, an algebraic knot is not the connected sum of two algebraic knots [2]. On the other hand, the integral monodromy may be reducible [1, Theorem 3].

Problem 5. Are cobordant algebraic knots (with K homeomorphic to a sphere) isotopic?

This is true when $n = 1$ [14, Cor. 2.5.7].

Finally, Milnor's problem on the Überschneidungszahl κ [16, p. 92] has apparently not yet been solved, although the inequality $\kappa \leq \delta$ has been checked [18].

Bibliography

[1] N. A'Campo, "Sur la monodromie des singularités isolées d'hypersurfaces complexes", Inventiones Math. 20 (1973), 147-169.

[2] ___________, "Le nombre de Lefschetz d'une monodromie", Indag. Math. 35 N° 2 (1973), 113-118.

[3] ___________, "La fonction zêta d'une monodromie", preprint.

[4] ___________, "Le groupe de monodromie du déploiement des singularités isolées de courbes planes", preprint.

[5] E. Brieskorn, "Beispiele zur Differentialtopologie von Singularitäten", Inventiones Math. 2 (1966), 1-14.

[6] ___________, "Rationale Singularitäten komplexer Flächen", Inventiones Math. 4 (1968), 336-358.

[7] ___________, "Die Monodromie der isolierten Singularitäten von Hyperflächen", Manuscripta Math. 2 (1970), 103-161.

[8] A. Durfee, "Diffeomorphism classification of isolated hypersurface singularities", Ph. D. thesis, Cornell University (1971) (revised version to appear).

[9] ___________, "Fibered knots and algebraic singularities", Topology 13 (1974), 47-59.

[10] R. Fox, "A quick trip through knot theory", in Topology of 3-Manifolds, M. K. Fort, Jr., ed., 120-167, Prentice-Hall, 1962.

[11] A. Gabrielov, "Intersection matrices of certain singularities", Funk. anal. i jewo pril 7 (1973), 18-32 (in Russian).

[12] M. Kato, "A classification of simple spinnable structures on a 1-connected Alexander manifold", preprint.

[13] L. Kauffman, "Products of knots", preprint.

[14] Lê Dũng Tráng, "Sur les noeuds algébriques", Compositio Math. 25 (1972), 281-321.

[15] J. Levine, "An algebraic classification of some knots of codimension two", Comm. Math. Helv. 45 (1970), 185-198.

[16] J. Milnor, Singular points of complex hypersurfaces, Ann. of Math. Studies No. 61, Princeton University Press, 1968.

[17] D. Mumford, The Topology of Normal Singularities of an Algebraic Surface and a Criterion for Simplicity", Publ. Math. N° 9, l'Inst. des Hautes Études Sci., Paris, 1961.

[18] H. Pinkham, "On the Überschneidungszahl of algebraic knots", preprint.

[19] K. Sakamoto, "Milnor fiberings and their characteristic maps", in Proc. International Conference on Manifolds, Tokyo, 1973.

[20] H. Schubert, "Knoten und Vollringe", Acta Math., 90 (1953), 131-286.

[21] H. Seifert, "Über das Geschlecht von Knoten", Math. Annalen, 110 (1935), 571-592.

[22] N. Steenrod, Ed., Reviews of Papers in Algebraic and Differential Topology, Topological Groups, and Homological Algebra, Amer. Math. Soc., 1968.

[23] C. T. C. Wall, "Classification problems in differential topology VI: Classification of (s-1)-connected (2s+1)-manifolds", Topology, 6 (1967), 273-296.

University of Washington, Seattle

Proceedings of Symposia in Pure Mathematics
Volume 29, 1975

RIEMANN-ROCH FOR SINGULAR VARIETIES

William Fulton

1. INTRODUCTION

Let D be a divisor on a compact Riemann surface of genus g, and let $L(D)$ be the space of meromorphic functions f with poles in D, i.e. $\operatorname{div}(f) + D \geq 0$. Riemann's inequality

$$\dim L(D) \geq \deg(D) + 1 - g$$

has grown during the last century to the problem of finding the dimension of the space of sections $\Gamma(X,E) = H^0(X,E)$ of an algebraic vector-bundle E on a variety X.

Roch found that the error term in Riemann's inequality was $\dim L(K - D)$, where K is a divisor of a meromorphic form on the Riemann surface. By Serre duality [FAC], this can be interpreted as a dimension of a cohomology group. Serre conjectured that there was a formula for

$$\chi(X,E) = \sum(-1)^i \dim H^i(X,E)$$

in terms of the Chern classes of E and the tangent bundle T_X of X. Such a formula was found by Hirzebruch, following clues in the work of Todd. His proof, valid for non-singular complex projective varieties, used methods from differential topology [H].

The Chern character $\operatorname{ch}(E)$ of E is a cohomology class (with rational coefficients) on X, determined by the properties:

(1) If $f:Y \to X$, then $\operatorname{ch}(f^*E) = f^*\operatorname{ch}(E)$.

(2) If E is the line bundle of a divisor D, i.e. $E \cong \mathcal{O}(D)$, so D represents the cohomology class $c_1(E)$, then $\operatorname{ch}(E) = \exp(D) = 1 + D + \frac{1}{2}D^2 + \ldots$.

AMS(MOS) subject classifications (1970). 14F05, 14C15, 18F25.

(3) $ch(E \oplus E') = ch(E) + ch(E')$.

Similarly the Todd class $td(E)$ is determined by: (1) as above, (2) $td(\mathcal{O}(D)) = D/(1 - \exp(-D))$, and (3) $td(E \oplus E') = td(E) \smile td(E')$. Hirzebruch's formula is then

$$\chi(X,E) = ch(E) \smile td(T_X)[X]$$

where the right side means that the top-dimensional cohomology class is evaluated on the fundamental cycle $[X]$.

To answer Riemann's original problem one needs, besides an answer to the Riemann-Roch problem of computing $\chi(X,E)$, more knowledge about the higher cohomology groups. The fact that Riemann's inequality becomes an equality for $\deg(D) \geq 2g - 1$ has been generalized by vanishing theorems of Kodaira [K] and Serre [FAC]. For discussion on these matters and applications of Riemann-Roch, see [H]; for purely analytic approaches to the problem see [C-G].

Grothendieck [B-S] extended Hirzebruch's result to mappings $f:X \to Y$ of non-singular projective varieties, in arbitrary characteristic, by saying that the mapping from the Grothendieck group $K^0(X)$ of algebraic vector-bundles on X to the Chow ring $A(X)_{\mathbb{Q}}$ given by $E \to ch(E) \smile td(T_X)$ is a natural transformation of covariant functors. To make the normally contravariant functor K^0 into a covariant functor, Grothendieck shows that, when X is non-singular, $K^0(X)$ coincides with the Grothendieck group $K_0(X)$ of coherent algebraic sheaves on X. The functor K_0 is covariant, for all varieties and proper morhpisms, by the rule: if $f:X \to Y$, and F is a sheaf on X, then $f_!(F) = \sum(-1)^i R^i f_*(F)$, where $R^i f_*$ are the higher direct images.

Besides giving Hirzebruch's formula (by letting Y be a point), this also gives a formula for $ch(f_! E)$. This latter point of view was extended by Berthelot, Grothendieck, Illusie, et al. [SGA 6] to singular varieties when f is a complete intersection morphism, giving a formula

$$ch(f_! E) = f_*(td(T_f) \smile ch(E)).$$

Here T_f is the virtual tangent bundle of f: if we factor f into an inclusion $i:X \hookrightarrow V$ followed by a smooth morphism $g:V \to Y$, then $T_f = i^* T_g - N$, where N is the normal bundle of imbedding i, and T_g is the relative tangent bundle of V over Y. The values are in

$\mathrm{Gr}^{\bullet}Y_{\mathbb{Q}}$, a cohomology theory constructed from the λ-filtration on K^0Y. The restrictive hypotheses are needed so that the naturally contravariant functors K^0 and $\mathrm{Gr}^{\bullet}_{\mathbb{Q}}$ can be made covariant for these maps.

To get a Riemann-Roch theorem and Hirzebruch formula for arbitrary projective varieties it is necessary to use the naturally covariant functors K_0 and a suitable covariant "homology" theory $H_{\bullet}$ (with rational coefficients). Suitable homology theories include singular homology (for complex varieties), or ℓ-adic homology [SGA 5], or the graded theory $\mathrm{Gr}_{\bullet}$ associated to the filtration of K_0 by dimension of support [SGA 6]. It is also necessary to have a contravariant "cohomology" theory $H^{\bullet}$ to accompany the homology theory, with cap products $H^{\bullet} \otimes H_{\bullet} \xrightarrow{\frown} H_{\bullet}$, the projection formula, a cycle map from the group of algebraic cycles to the homology, and Chern classes for vector-bundles in the cohomology. All the above theories have such cohomology theories.

The finest such pair of theories is the Chow theory $A_{\bullet}$, $A^{\bullet}$, which is discussed briefly in the next section. It maps to any of the above theories, so it gives the strongest Riemann-Roch theorem and the finest invariants; it is not needed if one wants Riemann-Roch only for one of the other theories, however.

2. CHOW HOMOLOGY AND COHOMOLOGY

For simplicity we work in the category of quasi-projective schemes over an algebraically closed field. The Chow homology group $A_{\bullet}X$ of cycles modulo rational equivalence is the free abelian group on the reduced, irreducible subvarieties of X, modulo the subgroup generated by all $[\mathrm{div}(r)]$ for r a rational function on a subvariety V of X; here $\mathrm{div}(r)$ is the Cartier divisor of r on V, and $[\mathrm{div}(r)]$ is the associated Weil divisor, regarded as a cycle on X. That such a definition gives a covariant theory for proper maps has been known for some time ([C],[S]).

An auxiliary cohomology ring $A^{\bullet}X$ can be constructed from the known Chow ring on non-singular varieties as follows [F]: Fix X and consider the category $\mathcal{C}(X)$ of pairs (Y,f), where Y is non-singular and f is a morphism from X to Y. The known Chow theory $A^{\bullet}$ gives a contravariant functor from $\mathcal{C}(X)$ to rings, and we set

$$A^{\bullet}X = \varinjlim A^{\bullet}Y$$

Thus $A^{\bullet}X$ is the disjoint union of the $A^{\bullet}Y$, for all (Y,f), modulo the identification generated by setting $c' = g^*c$ whenever $g:Y' \to Y$ is a morphism from (Y',f') to (Y,f), $c \in A^{\bullet}Y$, $c' \in A^{\bullet}Y'$. One makes $A^{\bullet}X$ into a ring by using the cartesian product $Y_1 \times Y_2$ to add or multiply classes in $A^{\bullet}Y_1$ and $A^{\bullet}Y_2$.

The $A^{\bullet}Y$ act compatibly on $A_{\bullet}X$ by using Serre's intersection theory [S] and the moving lemma [C], so we have a "cap product" $A^{\bullet}X \otimes A_{\bullet}X \xrightarrow{\frown} A_{\bullet}X$. Note that if X is non-singular, $A^{\bullet}X \xrightarrow{\frown [X]} A_{\bullet}X$ is an isomorphism.

The construction of Chern classes in $A^{\bullet}X$ follows from two basic facts: (1) Any vector-bundle on X is the pull-back of a vector-bundle on some non-singular Y. (2) (Landman) Whenever two bundles on Y pull back to isomorphic bundles on X, then there is a factorization $X \to Y' \to Y$, with Y' non-singular, so that the bundles pull back to isomorphic bundles on Y'. There results a natural transformation

$$ch : K^{0} \to A^{\bullet}{}_{\mathbb{Q}}$$

of contravariant functors from quasi-projective schemes to rings.

3. RIEMANN-ROCH FOR SINGULAR VARIETIES

THEOREM [B-F-M]. <u>There is a unique natural transformation of covariant functors from the category of quasi-projective schemes and proper morphisms to the category of graded abelian groups</u>

$$\tau : K_0 \to A_{\bullet\mathbb{Q}}$$

<u>such that if</u> X <u>is a point,</u> $\tau(\mathcal{O}_X) = 1 \in \mathbb{Q} = A_{\bullet}X_{\mathbb{Q}}$, <u>and satisfying</u>

(1) $\tau(E \otimes F) = ch(E) \frown \tau(F)$ <u>for</u> $E \in K^0X$, $F \in K_0X$.

Define $\tau(X) \in A_{\bullet}X_{\mathbb{Q}}$ by $\tau(X) = \tau(\mathcal{O}_X)$, the "homology" <u>Todd class</u> of X. If X is complete, we deduce from this part of the theorem that $\chi(X,E)$ is the degree of the zero-dimensional part of $ch(E) \frown \tau(X)$. The rest of the theorem gives other properties of this Todd class.

(2) <u>If</u> X <u>is a local complete intersection</u>, <u>then</u>

$$\tau(X) = td(T_X) \frown [X]$$

where T_X *is the virtual tangent bundle of* X. This includes Grothendieck's Riemann-Roch theorem for non-singular varieties.

(3) *The Todd class of a variety restricts to the Todd class of an open subvariety.*

(4) *The Todd class of a cartesian product* $X \times Y$ *is the product of the Todd classes of the factors.*

(5) *If* $f : X \to Y$ *is a complete intersection morphism, with virtual tangent bundle* T_f, then $f_* \tau(X) = f_* td(T_f)) \cap \tau(Y)$.

(6) *If* $X \to C$ *is a flat morphism from* X *to a non-singular curve* C, *then the Todd class of the general fibre specializes to the Todd class of the special fibre.*

If we take the degree of the zero-dimensional part of the Todd class, we find familiar properties of the arithmetic genus.

COROLLARY. $A_\bullet X_{\mathbb{Q}} \cong Gr_\bullet X_{\mathbb{Q}}$ *for any quasi-projective scheme* X.

If fact, the Riemann-Roch map τ becomes an isomorphism modulo torsion. The uniqueness assertion in the Riemann-Roch theorem then reduces to a result of Landman that there are no non-trivial natural transformations on $A_{\bullet \mathbb{Q}}$ (cf. [F]).

The proof begins by imbedding X in a non-singular variety M. Given a coherent sheaf F on X, chose a resolution $E_\bullet \to F$ of F by vector-bundles on M. There is an exact sequence

$$A_\bullet X_{\mathbb{Q}} \to A_\bullet M_{\mathbb{Q}} \to A_\bullet (M-X)_{\mathbb{Q}} \to 0$$

and $\sum (-1)^i ch(E_i) \cap [M]$ goes to zero in $A_\bullet (M-X)_{\mathbb{Q}}$ An essential step in the proof is to construct, in a canonical way, a class $ch_X^M E_\bullet$ in $A_\bullet X_{\mathbb{Q}}$ which maps to $\sum (-1)^i ch(E_i) \cap [M]$. Then

$$\tau(F) = td(T_M) \cap ch_X^M E_\bullet$$

will give the Riemann-Roch map from $K_0 X$ to $A_\bullet X_Q$.

4. MACPHERSON'S GRAPH CONSTRUCTION

This localization of the Chern character of $E_\bullet$ to X is achieved by a construction of MacPherson [M] which is useful in studying maps of varying ranks between vector-bundles. Let

$$0 \to E_r \xrightarrow{d_r} E_{r-1} \xrightarrow{d_{r-1}} \cdots \xrightarrow{d_1} E_0 \xrightarrow{d_0} E_{-1} = 0$$

be a complex of vector-bundles on a variety M, exact off a subscheme X, $e_i = \operatorname{rank} E_i$, and let $G_i = \operatorname{Grass}_{e_i}(E_i \oplus E_{i-1})$ be the grassmann bundle (over M) of e_i-dimensional planes in $E_i \oplus E_{i-1}$. Each boundary map d_i gives a section $s(d_i)$ of G_i over M which takes a point P in M to the point in G_i represented by the graph of d_i in the fibre of $E_i \oplus E_{i-1}$ over P. Let

$$G = G_r \times_M G_{r-1} \times_M \cdots \times_M G_0$$

with projection map $\pi: G \to M$. For each λ in the ground-field, let $s_\lambda: M \to G$ be the section obtained by taking the section $s(\lambda d_i)$ in the factor G_i. Let $Z_\lambda = s_\lambda(M) \subset G$. We complete this family of cycles $\{Z_\lambda\}$ at infinity by imbedding

$$M \times \mathbb{A}^1 \hookrightarrow G \times \mathbb{P}^1$$

by $(P,\lambda) \to (s_\lambda(P), (1:\lambda))$. Let W be the closure of $M \times \mathbb{A}^1$ in $G \times \mathbb{P}^1$, and let $\phi: W \to \mathbb{P}^1$ be the projection. Then let $Z_\infty = \phi^*[\infty]$ as a cycle on $G = G \times [\infty]$; if M is non-singular, then $Z_\infty \times [\infty]$ is the intersection-cycle of W and $G \times [\infty]$.

Over $M-X$ the above imbedding extends to an imbedding $(M-X) \times \mathbb{P}^1 \hookrightarrow G \times \mathbb{P}^1$. It follows that the cycle Z_∞ decomposes into a sum $Z_\infty = [M_*] + Z$, where M_* is an irreducible subvariety of G which maps birationally to M, and Z is a cycle which maps by π into X.

Now let ξ_i be the tautological e_i-plane bundle on G_i, and let $\xi = \sum(-1)^i \xi_i \in K^0 G$. Then $ch\xi \cap Z$ is in $A_\bullet(\pi^{-1}X)_{\mathbb{Q}}$, and we define

$$ch_X^M E_\bullet = \pi_*(ch\xi \cap Z) \quad \text{in } A_\bullet X_{\mathbb{Q}} .$$

EXAMPLE. Let X be a local complete intersection in M, with normal bundle N, and let $E_\bullet$ be a resolution of a locally free $\mathcal{O}_X$-module F on X. Then one can show that

$$\begin{array}{ccc} & Z \cong P(N \oplus 1) & \\ \pi \searrow & & \swarrow p \\ & X & \end{array}$$

M_* is the blow-up of M along X (and Z and M_* intersect along

$\mathbb{P}(N))$. If H is the universal hyperplane bundle on $\mathbb{P}(N \oplus 1)$, then ξ restricts to $\Lambda^{\bullet} H \otimes p^{*}F$ on $Z = \mathbb{P}(N \oplus 1)$. In this case, $ch_X^M E_{\bullet} = p_{*}(ch(\Lambda^{\bullet} H \otimes p^{*}F)) \cap [X] = (td(N)^{-1} \smile ch(F)) \cap [X]$ has a natural lifting to cohomology. (For this one needs Gysin maps in cohomology - cf. [SGA 6], [F], for $Gr^{\bullet}_{\mathbb{Q}} \cong A^{\bullet}_{\mathbb{Q}}$, and [B-F-M] for singular cohomology on complex varieties.) The Riemann-Roch theorem of [SGA 6], as well as Riemann-Roch without denominators, can be deduced from this description, at least for quasi-projective schemes over a field.

For general singular varieties the construction produces homology classes. In fact, examples show that the Todd class of X may not be of the form $\eta \cap [X]$ for any cohomology class η.

To show $\tau(F) = td(T_M) \cap ch_X^M E_{\bullet}$ is independent of the imbedding, an essential step is to compare the imbedding $M \subset P$ (say of non-singular varieties) with the imbedding $M \subset N$ as the zero-section of the normal bundle N. In [B-S] and [SGA 6] this is done by blowing P up along M to reduce to the case of hypersurface. In [A-H] a local diffeomorphism (with a suitable complex analytic property) was used. Using the graph construction, one can construct an algebraic family of imbeddings

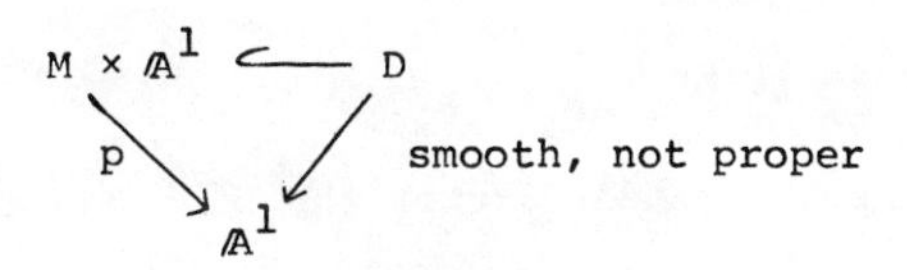

which for $t \neq 0$ is isomorphic to the given imbedding of M in P, and for $t = 0$ is the imbedding of M in N. One must then show that, although the cycles constructed in the graph construction may not form a family, the resulting classes in the Chow group don't vary.

5. RELATIONS WITH K-THEORY

For complex varieties there is a commutative diagram of contravariant ring-valued functors

$$\begin{array}{ccc} K^0 & \longrightarrow & K^0_{top} \\ {\scriptstyle ch}\downarrow & & \downarrow{\scriptstyle ch} \\ A^{\bullet}_{\mathbb{Q}} & \longrightarrow & H^{ev}(\ ;\mathbb{Q}) \end{array}$$

where $K^0_{top}X$ is the Grothendieck group of topological vector-bundles on X. Dual to this diagram is a diagram

$$\begin{array}{ccc} K_0 & \longrightarrow & K_0^{top} \\ \tau\downarrow & & \downarrow ch \\ A_{\bullet\mathbb{Q}} & \longrightarrow & H_{ev}(\ ;\mathbb{Q}) \end{array}$$

The factoring of the Riemann-Roch map through topological homology K-theory K_0^{top} was suggested by Atiyah. Note that all horizontal maps translate algebraic geometry to topology.

On the contravariant side there is a natural transformation from Qullen's higher K-groups K^i of the category of vector-bundles [Q] (K_i in his notation) to K^i_{top}, so one expects a higher-K-Riemann-Roch map from Quillen's higher K-groups K_i of the category of coherent sheaves (K_i' in his notation) to K_i^{top}. (This has been constructed for non-singular varieties.) What are the images of the higher K_i's in K_0^{top} and K_1^{top} (or singular homology) when Bott periodicity is used?

Finally, what "higher Chow groups" should fill in the diagrams

$$\begin{array}{ccc} K^i & \longrightarrow & K^i_{top} \\ \vdots & & \downarrow ch \\ ? & \dashrightarrow & H^{\bullet}(\ ;\mathbb{Q}) \end{array} \qquad\qquad \begin{array}{ccc} K_i & \longrightarrow & K_i^{top} \\ \vdots & & \downarrow ch \\ ? & \dashrightarrow & H_{\bullet}(\ ;\mathbb{Q}) \end{array}$$

(Here take the even or odd part of the homology and cohomology depending on the parity of i.)? The fact that all vertical maps of our diagrams are isomorphisms when we tensor with $\mathbb{Q}$ (and the lower ones become isomorphic to associated graded groups of the upper ones) should provide some clues. The isomorphism $A^pX = H^p(X,\underline{K}_p)$ of Bloch-Gersten-Quillen for non-singular varieties is certainly suggestive of what ? might be like.

REFERENCES

[A-H] M. F. Atiyah and F. Hirzebruch, The Riemann-Roch theorem for analytic embeddings, Topology Vol. 1 (1961) 151-166.

[B-F-M] P. Baum, W. Fulton, and R. MacPherson, Riemann-Roch for singular varieties, to appear.

[B-S] A. Borel and J.-P. Serre, Le théorème de Riemann-Roch, Bull. Soc. Math. France 86 (1958), 97-136.

[C] C. Chevalley, A. Grothendieck, and J.-P. Serre, Anneaux de Chow et applications, Séminaire C. Chevalley 2e année (1958), Secr. Math., Paris.

[C-G] M. Cornalba and P. Griffiths, Some transcendental aspects of algebraic geometry, lectures at this conference.

[F] W. Fulton, Rational equivalence on singular varieties, appendix to [B-F-M], to appear.

[FAC] J.-P. Serre, Faisceaux algébriques cohérents, Ann. Math. 61 (1955) 197-278.

[H] F. Hirzebruch, Topological methods in algebraic geometry, Springer, Berlin-Heidelberg-New York (1956, 1966).

[K] K. Kodaira, On a differential-geometric method in the theory of analytic stacks, Proc. Nat. Acad. Sci. U.S.A. 39 (1953) 1268-1273.

[M] R. MacPherson, Analytic vector-bundle maps, to appear.

[Q] D. Quillen, Higher algebraic K-theory, Springer Lecture Notes 341 (1973), 85-147.

[S] J.-P. Serre Algebre locale• multiplicités, Springer Lecture Notes 11 (1965).

[SGA 6] P. Berthelot, A. Grothendieck, L. Illusie, et. al., Théorie des intersections et théorème de Riemann-Roch, Springer Lecture Notes 225 (1971).

BROWN UNIVERSITY

Proceedings of Symposia in Pure Mathematics
Volume 29, 1975

REPORT ON CRYSTALLINE COHOMOLOGY

Luc ILLUSIE

The purpose of this talk is to give a survey of the definitions and main results in the crystalline cohomology of varieties, with the emphasis on the theory in characteristic $p > 0$ as developed in Berthelot's thesis [1]. For historical considerations and comparison with other approaches to a "p-adic cohomology" we refer the reader to the introduction of [1].

I wish to thank P. Berthelot and P. Deligne for their help in the preparation of this report.

1. CRYSTALLINE COHOMOLOGY OVER $\mathbb{C}$. Fix a variety X over $\mathbb{C}$ (*). If $i : X \to Y$ is a closed immersion of X into a smooth variety, let $\Omega^\bullet_Y$ be the (algebraic) De Rham complex of Y, denote by $(\Omega^\bullet_Y)^\wedge_{/X}$ its formal completion along X, which can be viewed as a complex of (Zariski) sheaves on X. We then have the following basic fact, noted first, I think, by Deligne :

PROPOSITION 1.1. *Up to canonical isomorphism, the hypercohomology* $H^*(X,(\Omega^\bullet_Y)^\wedge_{/X})$ *does not depend on* i.

For a proof, see [18] or [17]. The key fact it uses - which is also a particular case of 1.1 when X is taken to be a point - is the *formal Poincaré lemma*, which says that the continuous De Rham complex of $\mathbb{C}[[t_1,\dots,t_n]]$ is a resolution of $\mathbb{C}$.

In view of 1.1 it is natural to ask if it is possible to attach to X an algebraically defined site whose cohomology with value in some natural sheaf of rings would give the $H^*(X,(\Omega^\bullet_Y)^\wedge_{/X})$ above. It is indeed the case. For a precise statement we first have to recall Grothendieck's definition of the crys-

(*) The word "variety" will be used throughout to mean "scheme of finite type".

AMS classifications : 14 F 10, 14 F 25, 14 F 30, 14 G 10.

talline site of X.

A closed, nilpotent immersion $U \hookrightarrow \bar{U}$, where U is an open Zariski subset of X, will be called a _thickening_ of X. The thickenings of X form a category in an obvious way, a map from $(U \overset{i}{\hookrightarrow} \bar{U})$ to $(V \overset{j}{\hookrightarrow} \bar{V})$ being a pair of maps $(f : U \to V , g : \bar{U} \to \bar{V})$ such that $gi = fj$. One defines a Grothendieck topology on this category by taking for covering families the families $((U_\alpha \hookrightarrow \bar{U}_\alpha) \to (U \hookrightarrow \bar{U}))$ such that $(\bar{U}_\alpha \to \bar{U})$ is an open cover of $\bar{U}$. The resulting site is called the _crystalline site_ of X and will be denoted by X_{crys}. A sheaf F on X_{crys} is the same as the following data : for each thickening $(U \hookrightarrow \bar{U})$ a sheaf $F_{\bar{U}}$ on $\bar{U}$, for each map $(f,g) : (U \hookrightarrow \bar{U}) \to (V \hookrightarrow \bar{V})$ a map $g^*F_{\bar{V}} \to F_{\bar{U}}$ which is an isomorphism when g is an open immersion, these maps satisfying an obvious compatibility. In particular, the data $(U \hookrightarrow \bar{U}) \mapsto \underline{O}_{\bar{U}}$ defines a sheaf of rings $\underline{O}_{X_{crys}}$ on X_{crys}. We can now state :

THEOREM 1.2. _There is a canonical isomorphism_

$$H^*(X_{crys}, \underline{O}_{X_{crys}}) \simeq H^*(X, (\Omega^\bullet_Y)^\wedge_{/X}) .$$

This result to some extent explains 1.1. The proof (see [1]) consists in a Čech calculation plus an application of the formal Poincaré lemma. As a particular case 1.2 contains the following result, previously established by Grothendieck [15] and which was the starting point of the whole theory :

COROLLARY 1.2.1. _If_ X _is a smooth variety over_ $\mathbb{C}$, _there is a canonical isomorphism_

$$H^*(X_{crys}, \underline{O}_{X_{crys}}) \simeq H^*(X, \Omega^\bullet_X) .$$

The principal interest of crystalline cohomology over $\mathbb{C}$ is to yield in some cases an algebraic interpretation for transcendental cohomology. Let again X be an arbitrary variety over $\mathbb{C}$, and let X^{an} be the corresponding analytic space. Then one has the striking result, conjectured by Grothendieck in [15] :

THEOREM 1.3 (Deligne [6]). <u>There is a canonical isomorphism</u>

$$H^*(X^{an},\mathbb{C}) \simeq H^*(X_{crys},\underline{O}_{X_{crys}}) .$$

In view of 1.2.1, it implies the former result of Grothendieck [14] :

COROLLARY 1.3.1. <u>If</u> X <u>is a smooth variety over</u> $\mathbb{C}$, <u>there is a canonical isomorphism</u>

$$H^*(X^{an},\mathbb{C}) \simeq H^*(X,\Omega^{\bullet}_X) .$$

In [6] Deligne gives various generalizations of 1.2 and 1.3, inspired by the comparison theorems of [7]. For instance, $\underline{O}_{X_{crys}}$ can be replaced by a <u>coherent crystal</u> F on X , i.e. a sheaf of $\underline{O}_{X_{crys}}$-modules F such that each $F_{\overline{U}}$ as above is locally free of finite type and each map $g^*F_{\overline{V}} \to F_{\overline{U}}$ is an isomorphism (it can be shown [6] that a coherent crystal is the same as a coherent Module on X equipped with a stratification in the sense of Grothendieck [15] - if X is smooth, "stratification" = "integrable connection"). With the notations of 1.2, F defines a coherent Module $\underline{F}$ on the formal completion $Y^{\wedge}{}_{/X}$ endowed with an integrable connection (with respect to $(\Omega^{\bullet}_Y)^{\wedge}{}_{/X}$) , and there is a canonical isomorphism

$$H^*(X_{crys},F) \simeq H^*(X,\underline{F}\otimes(\Omega^{\bullet}_Y)^{\wedge}{}_{/X}) , \tag{1.4}$$

generalizing 1.2. On the other hand, if the restriction of F to each smooth stratum of a stratification of X_{red} is regular in the sense of ([7] 4.5) - in which case we shall say that F is <u>regular</u> - then F defines a local system F^o of $\mathbb{C}$-vector spaces of finite rank on X^{an} , and there is a canonical isomorphism

$$H^*(X^{an},F^o) \simeq H^*(X_{crys},F) \tag{1.5}$$

generalizing 1.3. For X smooth and $X = Y$ a proof of 1.4 (resp. 1.5) is given in [15] (resp. [7]).

Let $f : X \to S$ be a proper and smooth map of varieties over $\mathbb{C}$. Then f defines a map of sites $f_{crys} : X_{crys} \to S_{crys}$ and one has a Leray spectral sequence

1.6 $$E_2^{pq} = H^p(S_{crys}, R^q f_{crys*} \underline{O}_{X_{crys}}) \Longrightarrow H^{p+q}(X_{crys}, \underline{O}_{X_{crys}}) .$$

On the other hand f defines a map of analytic spaces $f^{an} : X^{an} \to S^{an}$ and one has a Leray spectral sequence

1.7 $$E_2^{pq} = H^p(S^{an}, R^q f_*^{an} \mathbb{C}) \Longrightarrow H^{p+q}(X^{an}, \mathbb{C}) .$$

The abutments of 1.6 and 1.7 are canonically isomorphic by 1.2. It can be shown, moreover, that $R^q f_{crys*} \underline{O}_{X_{crys}}$ is a regular coherent crystal on S_{crys} whose associated local system is $R^q f_*^{an} \mathbb{C}$ (when S is smooth this crystal is just $R^q f_* \Omega^\bullet_{X/S}$ equipped with the Gauss-Manin connection [23], which is known to be regular, see [7] or [20]). Therefore, by 1.5 the E_2 terms of 1.6 and 1.7 are canonically isomorphic, and it seems reasonable to expect that 1.6 is indeed an algebraic portrait of the entire transcendental spectral sequence 1.7. It should be added, however, that if one drops the hypothesis of smoothness on f then no algebraic interpretation of 1.7 is known (even if S is the projective plane and f is the blowing-up of one point !).

2. CRYSTALLINE COHOMOLOGY IN CHARACTERISTIC $p > 0$. Let k be a perfect field of characteristic $p > 0$, and W the ring of Witt vectors on k. We want to attach functorially to any variety X/k reasonable "p-adic" cohomology groups. We would like these to be finitely generated modules over W and coincide with the De Rham cohomology groups of X'/W when X is the reduction of a proper, smooth variety X'/W. By analogy with 1.1 we might think of embedding X in a smooth variety Y/W and taking the cohomology groups of X with value in the formal completion of $\Omega^\bullet_{Y/W}$ along X. Unfortunately, these groups do depend on the embedding (as one can see already for $X = k$), essentially because the formal Poincaré lemma is no longer true over W. However, a

slight modification of this construction, involving divided powers, turns out to yield good invariants, at least for X proper and smooth over k.

2.1. Let A be a commutative ring with unit, and $I \subset A$ be an ideal. Recall that a structure of <u>divided powers</u> (in short, DP) on I is a family of maps $(\gamma_n : I \to A)_{n \in \mathbb{N}}$ such that $n!\gamma_n(x) = x^n$ for all $x \in I$, and satisfying the same formal identities as $x^n/n!$, e.g. $\gamma_n(x+y) = \sum_{p+q=n} \gamma_p(x)\gamma_q(y)$, etc., see [32] or [1]. For example, on pW, the maximal ideal of W, there is a unique structure of DP : indeed, for $n \geqslant 1$, one has $\mathrm{ord}_p(p^n/n!) > 0$. An ideal equipped with DP will be called a DP <u>ideal</u>.

There is a useful construction, due to Berthelot ([1] I 2), which functorially associates to a pair (A,I) as above a pair $(D_A(I),\bar{I})$, where $\bar{I}$ is a DP ideal, and a map $(A,I) \to (D_A(I),\bar{I})$ (i.e. a map of rings $A \to D_A(I)$ sending I into $\bar{I}$), this map being universal among all maps $(A,I) \to (B,J)$ where J is a DP ideal. One has $D_A(I)/\bar{I} = A/I$, and $D_A(I)$ is generated as an A-algebra by the $\gamma_n(x)$ for $x \in I$. $D_A(I)$ is called the DP <u>envelope</u> of I in A.

Let M be an A-module, and $\Gamma_A(M)$ denote the divided power algebra on M [23]. $\Gamma_A(M)$ is a graded A-algebra, whose piece of degree n is generated by the monomials $\gamma_{r_1}(x_1)\ldots\gamma_{r_k}(x_k)$, $x_i \in M$, $r_1+\ldots+r_k = n$. On the augmentation ideal $\Gamma_A^+(M) = \bigoplus_{n>0} \Gamma_A^n(M)$ there is a unique DP structure extending the canonical maps $\gamma_n : M \to \Gamma_A^n(M)$. If $(S_A(M),S_A^+(M))$ denotes the pair consisting of the symmetric algebra on M and its augmentation ideal, there is a canonical map $(S_A(M),S_A^+(M)) \to (\Gamma_A(M),\Gamma_A^+(M))$, which is easily seen to make $\Gamma_A(M)$ the DP envelope of $S_A^+(M)$. If M is free with basis $(x_i)_{i\in I}$ we shall write $\Gamma_A(M) = A\langle x_i \rangle_{i \in I}$.

Let R denote either W or some truncation $W_n = W/p^{n+1}W$. There is a variant of the DP envelope construction for pairs (A,I) where A is an R-algebra. Namely, one can functorially associate to any such pair (A,I) a pair $(D_{A/R}(I),\bar{I})$ where $D_{A/R}(I)$ is an R-algebra and $\bar{I}$ an ideal in it equipped with DP compatible with the canonical DP on pR (i.e. such that $\gamma_n(px) = (p^n/n!)x^n$ for $x \in D_{A/R}(I)$, $px \in \bar{I}$), and a map of R-algebras $A \to D_{A/R}(I)$ sending I into $\bar{I}$ and universal among all maps of R-algebras $A \to B$ sending I into a DP ideal J whose DP are compatible with those of pR . $D_{A/R}(I)$ is called the DP <u>envelope of</u> I <u>in</u> A <u>with respect to</u> R . It is again true that $D_{A/R}(I)/\bar{I} = A/I$ and that $D_{A/R}(I)$ is generated as an A-algebra by the $\gamma_n(x)$ for $x \in I$.

The constructions of DP envelopes sheafify in an obvious way.

2.2. Let X be a variety over k . Fix an integer n , let $W_n = W/p^{n+1}W$, assume there is a closed W_n-immersion i of X into a smooth variety Y/W_n (this is the case for instance if X is affine, or projective). Let $(\underline{D}_X(Y/W_n),\bar{I})$ be the DP envelope with respect to W_n of the Ideal I of i : $\underline{D}_X(Y/W_n)$ is a sheaf of quasi-coherent $\underline{O}_Y$-algebras. Observe that :

a) $\underline{D}_X(Y/W_n)$ <u>is concentrated on</u> X . Indeed, $\underline{D}_X(Y/W_n)/\bar{I} = \underline{O}_X$, and $\bar{I}$ is a nil-ideal, since any DP ideal in a torsion ring is a nil-ideal (if A is a ring such that $N.A = 0$ for some integer N , and J A is a DP ideal, then, for any $x \in J$, $x^N = N!\gamma_N(x) = 0$).

b) $\underline{D}_X(Y/W_n)$ <u>has a canonical integrable connection with respect to</u> W_n . Indeed, this is the connection D on $\underline{D}_X(Y/W_n)$ such that $D(\gamma_k x) = \gamma_{k-1}x \otimes dx$ for all $x \in I$, $k \in N$.

Thanks to b), it makes sense to form the DR complex of Y/W_n with coefficients in $\underline{D}_X(Y/W_n)$, which is a W_n-linear complex of $\underline{O}_Y$-modules, concentrated on X (*). Abstracted from the "yoga" of crystalline cohomology, a fundamental result of [1] is the following analog to 1.1 :

(*) DR means De Rham ; recall that if Y is a scheme over Z , and E a Module on Y endowed with an integrable connection $D : E \to E \otimes_{\Omega_Y} \Omega^1_{Y/Z}$ then D extends to a complex $\Omega^{\cdot}_{Y/Z} \otimes E$ called the DR complex of Y/Z with coefficients in E .

PROPOSITION 2.3. *For X/k and n fixed, the W_n-modules of hypercohomology $H^r(X, \Omega^{\bullet}_{Y/W_n} \otimes \underline{D}_X(Y/W_n))$ do not depend, up to canonical isomorphism, on the immersion $i : X \to Y$.*

No direct proof is given in [1], but it should not be hard to find one. The main ingredient is the following easy lemma :

LEMMA 2.3.1 (DP Poincaré lemma). *For any ring A and integer n, the DR complex of $A[t_1,\dots,t_n]/A$ with coefficients in the DP algebra $A\langle t_1,\dots,t_n\rangle$ (2.1), viewed as an $A[t_1,\dots,t_n]$-module, and endowed with the integrable connection D defined by $D(\gamma_k t_i) = \gamma_{k-1} t_i \otimes dt_i$, is a resolution of A, in other words the sequence*

$$0 \to A \to A\langle t_1,\dots,t_n\rangle \xrightarrow{D} \bigoplus A\langle t_1,\dots,t_n\rangle dt_i \to \dots \to \bigoplus A\langle t_1,\dots,t_n\rangle dt_{i_1}\dots dt_{i_r} \to \dots,$$

where the differential is the natural extension of D and the first map is the natural inclusion, is exact.

Like 1.1, 2.3 suggests that $H^*(X, \Omega^{\bullet}_{Y/W_n} \otimes \underline{D}_X(Y/W_n))$ can be expressed as the cohomology of a suitable ringed site intrinsically attached to X. This is indeed possible.

Let the variety X/k and the integer n be fixed. A closed immersion $U \hookrightarrow \bar{U}$ of W_n-schemes where the Ideal defining U is given a structure of DP compatible with the canonical DP of W_n and U is an open Zariski subset of X will be called a DP W_n-*thickening* of X. As we have seen above, the Ideal of a DP W_n-thickening is automatically a nil-Ideal. The DP W_n-thickenings of X form a category in an obvious way, a map from $(U \overset{i}{\hookrightarrow} \bar{U})$ to $(V \overset{j}{\hookrightarrow} \bar{V})$ being a pair of W_n-maps $(f : U \to V,\ g : \bar{U} \to \bar{V})$, compatible with the DP, such that $gi = fj$. This category, equipped with the Grothendieck topology for which the covering families are the families $((U_\alpha \hookrightarrow \bar{U}_\alpha) \to (U \hookrightarrow \bar{U}))$ such that $(\bar{U}_\alpha \to \bar{U})$ is an open cover of $\bar{U}$, is called the *crystalline site* of X with respect to W_n, and will be denoted by $(X/W_n)_{crys}$ or simply X/W_n if no confusion ensues. Sheaves on this site are described, as in §1, by families of Zariski sheaves $(U \hookrightarrow \overline{U}) \mapsto F_{\overline{U}}$, in particular $(U \hookrightarrow \overline{U}) \mapsto \underline{O}_{\overline{U}}$ is a sheaf of rings on X/W_n which we shall denote by $\underline{O}_{X/W_n}$. Now 2.3 is a corollary of the following

THEOREM 2.4 ([1] V 2.3). *With the notations of 2.2, there is a canonical isomorphism*

$$R\Gamma(X/W_n, \underline{O}_{X/W_n}) \simeq R\Gamma(X, \Omega^{\bullet}_{Y/W_n} \otimes \underline{D}_X(Y/W_n))$$

in the derived category $D(W_n)$, *hence a canonical isomorphism*

$$H^*(X/W_n, \underline{O}_{X/W_n}) \simeq H^*(X, \Omega^{\bullet}_{Y/W_n} \otimes \underline{D}_X(Y/W_n))$$

of graded W_n*-modules (which is moreover compatible with the natural structures of algebras on the two sides).*

The proof is similar to that of 1.2 and is based on the DP Poincaré lemma.

COROLLARY 2.4.1. *Assume* $X = X' \otimes k$, *where* X' *is a smooth scheme over* W_n . *Then there is a canonical isomorphism*

$$R\Gamma(X/W_n, \underline{O}_{X/W_n}) \simeq R\Gamma_{DR}(X'/W_n) \quad (*) .$$

Indeed, since the Ideal of X in X' is just $p\underline{O}_{X'}$, one has $\underline{D}_X(X'/W_n) = \underline{O}_{X'}$ (because of the compatibility between the DP).

As in §1, define a *coherent crystal* on X/W_n as an $\underline{O}_{X/W_n}$-Module F such that each $F_{\overline{U}}$ as above is a coherent sheaf on $\overline{U}$ and each map $g^*F_{\overline{V}} \to F_{\overline{U}}$ is an isomorphism. Under the assumptions of 2.4.1 it can be shown ([1] IV 1.6) that a coherent crystal on X/W_n is the same as a coherent sheaf on X' equipped with an integrable connection with respect to W_n whose reduction mod p is nilpotent. Moreover ([1] V 2.3), if F is such a crystal, there is a canonical isomorphism (where $E = F_{X'}$)

(*) If $Y \to Z$ is a morphism of schemes we shall sometimes write $R\Gamma_{DR}(Y/Z)$, $H^*_{DR}(Y/Z)$ instead of $R\Gamma(Y, \Omega^{\bullet}_{Y/Z})$, $H^*(Y, \Omega^{\bullet}_{Y/Z})$.

2.4.2 $$R\Gamma(X/W_n, F) \simeq R\Gamma(X', E \otimes \Omega^{\bullet}_{X'/W_n}) ,$$

which generalizes 2.4.1.

2.5. Let X be a variety over k, abbreviate $H^*(X/W_n, \underline{O}_{X/W})$ into $H^*(X/W_n)$. For $n \leqslant n'$ there is a natural map $H^*(X/W_{n'}) \to H^*(X/W_n)$, and these maps satisfy an obvious compatibility for $n \leqslant n' \leqslant n''$. By definition, for $i \in \mathbb{Z}$, the ith crystalline cohomology group of X/W_n is :

2.5.1 $$H^i(X/W) = \varprojlim_{n \in \mathbb{N}} H^i(X/W_n) .$$

The $H^i(X/W)$ are W-modules, which are zero for i large (e.g. $i > \dim(X) + N$, if X can locally be embedded as a closed W-subscheme of the affine space $\mathbb{A}^N_W$). $H^*(X/W) = \oplus H^i(X/W)$ is an anti-commutative graded W-algebra. In the absence of hypotheses on X/k, the only thing one can say about it is that it depends functorially on X/k, i.e. if

2.5.2 $$\begin{array}{ccc} X & \longleftarrow & X' \\ \downarrow & & \downarrow \\ \mathrm{Spec}(k) & \longleftarrow & \mathrm{Spec}(k') \end{array}$$

is a commutative square (where k'/k is an extension of perfect fields) then there is a canonical map of W-algebras

2.5.3 $$H^*(X/W) \to H^*(X'/W') ,$$

where $W' = W(k')$ (and these maps satisfy an obvious compatibility for a composition of squares). If $X \to X'$ is "embedded" in a $(Y \to Y')/W$, with Y/W, Y'/W smooth, then 2.5.3 can be realized as a functoriality map on DR cohomology by means of 2.4.

In particular the pth-power map on X induces a W-linear map $H^*(X/W)^{(p)} \to H^*(X/W)$ (where the exponent (p) means extension of scalars to W according to the pth-power map), the so-called canonical lifting of Frobenius (see 3.4).

3. PROPERTIES OF $H^*(X/W)$ FOR X PROPER AND SMOOTH OVER k.

3.0. We keep the notations of §2 : k is a fixed perfect field of characteristic $p > 0$, $W = W(k)$ is the ring of Witt vectors on k. We denote by K the field of fractions of W. If X is a variety over k, we shall set $H^*(X/W)_K = H^*(X/W) \otimes_W K$.

If X/k is not proper ans smooth, $H^*(X/W)$ turns out to be "pathological". Indeed, if $\mathbb{A}^1_k$ is the affine line, $H^1(\mathbb{A}^1_k/W)$, which by 2.4.1 is $\varprojlim H^1_{DR}(\mathbb{A}^1_{W_n}/W_n)$, is non zero, even $H^1(\mathbb{A}^1_k/W)_K$ is not of finite type over K. Also $H^1(X/W)_K$ for X a union of two projective lines crossing at one point is not of finite type over K. However, for X/k proper and smooth, $H^*(X/W)$ is indeed a good "p-adic cohomology". We shall sum up what is known of it so far.

3.1. FINITENESS. Let X/k be smooth, proper, of dimension d. Then the $H^i(X/W)$ are of finite type over W, and $H^i(X/W) = 0$ for $i > 2d$. If X is projective or lifts to W, the rank $b_i(X)$ of $H^i(X/W)$ is equal to the ℓ-adic i^{th}-Betti number of X for any prime $\ell \neq p$ (which is by definition $\dim H^i(\bar{X}, \mathbb{Q}_\ell)$ where $\bar{X}$ is deduced by extension of scalars to an algebraically closed field). For a liftable X this is indeed, as we shall see (3.4b)), an easy consequence of the comparison of crystalline cohomology with DR cohomology. However, in the projective case it is a deep fact, whose only known proof uses the Weil conjectures (see 3.7).

There is a more precise result than the finitesess of the $H^i(X/W)$, namely one can define [1] a perfect complex of W-modules $R\Gamma(X/W)$, having $H^*(X/W)$ for cohomology, and depending functorially on X/k (i.e. in the situation of 2.5.2 there is a canonical map $R\Gamma(X/W) \otimes_W W' \to R\Gamma(X/W)$ in $D(W')$, plus a compatibility for a composition of squares). (Briefly, the construction is as follows : one replaces (up to quasi-isomorphism) the inverse system of the $R\Gamma(X/W_n)$ by an inverse system of bounded complexes of projective modules of finite type, and then one takes its limit).

3.2. BASE CHANGE. Let

$$\begin{array}{ccc} X & \longleftarrow & X' \\ \downarrow & & \downarrow \\ \mathrm{Spec}(k) & \longleftarrow & \mathrm{Spec}(k') \end{array}$$

be a cartesian square, where $k \to k'$ is an extension of perfect fields and X/k is proper and smooth. Then the canonical map

$$H^*(X/W) \otimes_W W' \to H^*(X'/W')$$

is an isomorphism.

3.3. KÜNNETH FORMULA. Let X, Y be proper and smooth varieties over k. Then there is a canonical isomorphism in $D(W)$:

$$R\Gamma(X/W) \overset{L}{\underset{W}{\otimes}} R\Gamma(Y/W) \simeq R\Gamma(X \times_k Y/W),$$

yielding exact sequences

$$0 \to \bigoplus_{p+q=n} (H^p(X/W) \otimes H^q(Y/W)) \to H^n(X \times Y/W) \to \bigoplus_{p+q=n+1} \mathrm{Tor}_1^W(H^p(X/W), H^q(Y/W) \to 0 .$$

3.4. COMPARISON WITH DE RHAM COHOMOLOGY.

a) Let X/k be proper and smooth. Then there is a canonical isomorphism in $D(k)$

3.4.1 $$R\Gamma(X/W) \overset{L}{\otimes}_W k = R\Gamma(X, \Omega^{\bullet}_{X/k}),$$

yielding exact sequences

$$0 \to H^i(X/W) \otimes_W k \to H^i_{DR}(X/k) \to \mathrm{Tor}_1^W(H^{i+1}(X/W), k) \to 0 .$$

In particular, we have $\mathrm{rk}\, H^i(X/W)$ (the ith-Betti number) $\leqslant \dim H^i_{DR}(X/k)$, with equality if and only if $H^i(X/W)$ and $H^{i+1}(X/W)$ have no torsion. More precisely, if we write

$$H^i(X/W) \simeq W^{b_i} \oplus \bigoplus_j (W/p^{s_{ij}})^{t_{ij}} , \quad s_{ij} \geqslant 1 ,$$

and set $t_i = \sum_j t_{ij}$, we get

$$\dim H^i_{DR}(X/k) = b_i(X) + t_i + t_{i+1} .$$

In short, the torsion in the crystalline cohomology of X/W explains why the DR cohomology of X/k may be "too big" - thus confirming a guess of Grothendieck in [14] (footnote 12).

For example, the surface constructed by Serre in [33] has $b_1 = 0$, $\dim H^1_{DR} = 1$, $t_1 = 0$ (this is true for any X as a corollary of the exact sequence above for $i=0$), hence $t_2 = 1$; Berthelot has indeed verified that the torsion of H^2 is W/p.

b) Suppose X is lifted to a proper and smooth X'/W. Then there is a canonical isomorphism

3.4.2 $$H^*(X/W) \simeq H^*_{DR}(X'/W) .$$

In other words, for X'/W proper and smooth $H^*_{DR}(X'/W)$ depends only on the reduction of X' mod p. This is an analog of the invariance result of Washnitzer and Monsky in the affine case [30].

Thanks to 3.4.2, the map $H^*(X/W)^{(p)} \to H^*(X/W)$ can be viewed as a map $H^*_{DR}(X'/W)^{(p)} \to H^*_{DR}(X'/W)$, which justifies the terminology "canonical lifting of Frobenius". As is well known, the Frobenius endomorphism of X itself seldom lifts to an endomorphism of X', yet through crystalline cohomology it defines a canonical endomorphism of the DR cohomology of X'/W. This operation is beautifully analyzed by Mazur ([17],[25]) in answer to a conjecture of Katz, see Berthelot's exposé in this volume.

3.5. **POINCARE DUALITY.** Let X/k be proper and smooth, of pure dimension d. Then a canonical map

3.5.1 $$\mathrm{Tr} : H^{2d}(X/W) \to W$$

is defined, called the <u>trace map</u>. Its reduction mod p coincides via 3.4.1 with the trace map for the DR cohomology of X/k, and when X is lifted to an X' as in 3.4 b), 3.5.1 is the trace map on the DR cohomology of X'/W. If X is geometrically connected, the trace map is an isomorphism.

The $H^i(X/W)_K$ (notation of 3.0) satisfy the Poincaré duality, namely the pairing

3.5.2 $$H^i(X/W)_K \otimes H^{2d-i}(X/W)_K \to K$$

given by the cup-product followed by the trace map (tensored with K) is a perfect duality.

There is a finer statement, which does not neglect the torsion. Observe that, since $H^i(X/W) = 0$ for $i > 2d$, Tr can be viewed as a map $Tr : R\Gamma(X/W)[2d] \to W$. Then the map

3.5.3 $$R\Gamma(X/W) \overset{L}{\otimes}_W R\Gamma(X/W)[2d] \to W$$

defined by the cup-product followed by Tr is a perfect pairing between perfect complexes. One deduces from that universal coefficient exact sequences

3.5.4 $$0 \to Ext^1_W(H^{2d-i+1}(X/W),W) \to H^i(X/W) \to Hom(H^{2d-i}(X/W),W) \to 0 .$$

3.6. GYSIN MORPHISM, CYCLE CLASS. Let $f : X \to Y$ be a map between proper, smooth varieties/k, of pure dimensions m, n. Put $d = m-n$. Then, by Poincaré duality, $f^* : R\Gamma(Y/W) \to R\Gamma(X/W)$ defines a transposed map

3.6.1 $$f_* : R\Gamma(X/W) \to R\Gamma(Y/W)[-2d] ,$$

called the <u>Gysin map</u>. It induces

$$f_* : H^i(X/W) \to H^{i-2d}(Y/W) .$$

For $Y = Spec(k)$, $i = 2d$, it is just the trace map.

Suppose f is a closed immersion, of codimension n. Then

3.6.2 $$c(X/Y) \overset{dfn}{=} f_*1 \in H^{2n}(Y/W)$$

is called the <u>class of</u> X. An important result of [1] is that, for fixed Y, these classes satisfy an <u>intersection formula</u>, namely : if X', X'' are transversal, smooth, closed subvarieties of Y, with intersection X, then

3.6.3 $$c(X'/Y)c(X''/Y) = c(X/Y) .$$

Whereas the results of 3.1 to 3.5 are comparatively easy consequences of 2.4, 3.6.3 is much harder to prove. Berthelot derives it from a fine analysis of the DR cycle classes, based on the formalism of Ext of complexes of differential operators of order 1 developed by Lieberman and Herrera [18].

Assuming Y to be projective, and denoting by $A(Y)$ the Chow ring of Y, it is not known if $X \mapsto c(X/Y)$, defined on smooth, proper subvarieties of Y, extends to a graded ring homomorphism

$$c : A(Y) \to H^*(Y/W) .$$

Using the theory of Chern classes with value in crystalline cohomology [5] (which is constructed in the usual way by calculating the cohomology of a projective bundle), one can at least define a map $A(Y) \to H^*(Y/W)_K$ (of graded rings), but it has not yet been verified that it coincides with the map deduced from the c above by extension of scalars to K.

3.7. ZETA FUNCTION, CHARACTERISTIC POLYNOMIAL OF AN ENDOMORPHISM. It is easy to derive from the formalism above a Lefschetz fixed point formula for an endomorphism of a proper, smooth X/k with transversal, isolated fixed points. By the well-known argument one deduces from it the rationality of the zeta function for proper, smooth varieties over finite fields. More precisely, let X be a proper, smooth variety over the finite field $\mathbb{F}_q$ with q elements, suppose X of dimension d. Then the Frobenius endomorphism F of X given by $x \mapsto x^q$ on $\underline{O}_X$ induces an endomorphism (actually an automorphism because of Poincaré duality) of $H^*(X/W)_K$ (where $W = W(\mathbb{F}_q)$, K = field of fractions of W), call it again F. Put

3.7.1 $$P_i(t) = \det(\mathrm{Id}-Ft\, , H^i(X/W)_K) .$$

Then we have the following formula for the zeta function of X:

3.7.2 $$Z(X,t) = \frac{P_1(t)\dots P_{2d-1}(t)}{P_0(t)\dots P_{2d}(t)} .$$

For X projective, it was proven by Katz and Messing [22] as a consequence of the Weil conjectures [8] that $P_i(t)$ has in fact integral coefficients and coincides with $\det(\mathrm{Id}-Ft, H^i(\bar{X},\mathbb{Q}_\ell))$ for $\ell \neq p$, where $\bar{X}$ denotes the variety deduced by extension of scalars to an algebraically closed field.

Using an argument of specialisation (see 3.10), one can deduce from that the following generalization. Let k be any perfect field, $W = W(k)$, and let u be an endomorphism of a smooth, projective variety over k. Then $\det(\mathrm{Id}-ut, H^i(X/W)_K)$ (where K = field of fractions of W) has integral coefficients, more precisely

3.7.3 $$\det(\mathrm{Id}-ut, H^i(X/W)_K) = \det(\mathrm{Id}-ut, H^i(\bar{X},\mathbb{Q}_\ell))$$

(the right-hand side being the ℓ-adic characteristic polynomial for $\ell \neq p$, which is known, by the Weil conjectures [8] to have integral coefficients and be independent of ℓ.

3.8. LEFSCHETZ TYPE THEOREMS. Let X be a fixed smooth, absolutely irreducible, proper subvariety of some projective space P over the perfect field k. Then we have :

a) Weak Lefschetz theorem [3] : There exists an integer d_0 depending only on X such that for any hypersurface section Y of X of degree $d \geqslant d_0$ the restriction map $H^i(X/W) \to H^i(Y/W)$ is an isomorphism for $i < \dim Y$, and injective for $i = \dim Y$. In [22] it is shown that if k is a finite field then the corresponding assertion for $H^i(\;)_K$ is true with $d_0 = 1$, and thanks to the specialization argument of 3.10 one can drop the hypothesis "finite". However, for $H^i(\;/W)$ itself it is not known if one can always take $d_0 = 1$. It is doubtful, since the question seems closely related to Kodaira's vanishing theorem, which is not true in char. $p > 0$ (I've heard Paršin has constructed a smooth, projective surface on which it does not hold).

b) Hard Lefschetz theorem : Let $L \in H^2(X/W)$ be the class of a smooth hyperplane section. Then, for each i, the cup-product by L^i defines an isomorphism $H^{d-i}(X/W)_K \xrightarrow{\sim} H^{d+i}(X/W)_K$, where $d = \dim X$. In the case k is finite, this is deduced in [22] from the Weil conjectures and the Hard Lefschetz theorem in ℓ-adic cohomology [9]. The general case can be reduced to the finite case by means of the specialization argument of 3.10.

3.9. A PROBLEM ON TORSION. Let X/W be a proper and smooth scheme, put $X_0 = X \otimes k$, $X_K = X \otimes K$, and let $X_{\mathbb{C}}$ denote a (proper and smooth) variety over $\mathbb{C}$ deduced from X_K by descent to a field K' of finite type over $\mathbb{Q}$ followed by extension of scalars to $\mathbb{C}$ via the choice of an embedding $K' \hookrightarrow \mathbb{C}$. Fix $i \in \mathbb{Z}$. We know that $H^i_{DR}(X/W)$ depends only on X_0/k , namely $H^i_{DR}(X/W) = H^i(X_0/W)$. By specialisation of DR cohomology, the rank of $H^i_{DR}(X/W)$ is equal to the ith-Betti number of $X_{\mathbb{C}}$, $b_i(X_{\mathbb{C}}) = \mathrm{rk}\, H^i(X_{\mathbb{C}},\mathbb{Z})$, so, with the notations of 3.1, we have

$$b_i(X_0) - b_i(X_{\mathbb{C}}) .$$

For any prime $\ell \neq p$, we have, by the comparison theorem between etale and classical cohomology, plus the specialization theorem,

$$H^i(\overline{X}_0,\mathbb{Z}_\ell) \simeq H^i(X_{\mathbb{C}},\mathbb{Z}) \otimes \mathbb{Z}_\ell ,$$

hence the ℓ-torsion of $H^i(X_{\mathbb{C}},\mathbb{Z})$ is isomorphic to the torsion of $H^i(\overline{X}_0,\mathbb{Z}_\ell)$. It is then natural to ask : how does the p-torsion of $H^i(X_{\mathbb{C}},\mathbb{Z})$ relate to the torsion of $H^i(X_0/W)$ (i.e. the torsion of $H^i_{DR}(X/W)$). What makes the question difficult is that we know no way of comparing DR and integral cohomologies. It should be interesting to investigate first a few examples, e.g. the Serre surface (3.4) or the Igusa surface (quotient of a product of two elliptic curves in char. 2 by the group $\mathbb{Z}/2$ acting by translation by a point of order 2 on the first factor and by the symmetry on the second), which both lift to W and have torsion in H^2 and H^3 .

3.10. CRYSTALLINE COHOMOLOGY OF PROPER AND SMOOTH FAMILIES. Let S be a smooth, affine variety over k and let S' be a p-adically complete, smooth, formal scheme over W lifting S. For $n \in \mathbb{N}$, set $S'_n = S \otimes W_n$. Let $f : X \to S$ be a proper and smooth morphism. Then, for each n, f induces a morphism f_n from the crystalline site X/W_n to the Zariski site of S'_n, and the $R^i f_{n*} \mathcal{O}_{X/W_n}$ are coherent Modules on S'_n. For i fixed and n variable these Modules form an inverse system, whose limit is a coherent Module on S', which we shall denote by $H^i(X/S')$. Taking into account the fact that f_n is underlying a map of crystalline sites $X/W_n \to S'_n/W_n$, one sees that the $H^i(X/S')$ come equipped with an integrable connection with respect to W, whose reduction mod p is nilpotent, an analog of the Gauss-Manin connection. Furthermore, if one lifts the Frobenius endomorphism F of S to an endomorphism F' of S' (which is always possible), then the Frobenius of X defines a canonical map $F'^*H^i(X/S') \to H^i(X/S')$, which is an isomorphism "modulo torsion" (i.e. after extension of scalars to the field of fractions K of W), and the maps corresponding to various F' satisfy a certain compatibility with the above connection, precisely each $H^i(X/S')$ is an F-crystal on S' in the sense of ([19] 1.3)(*). One can even define an "F-crystal in perfect complexes" $R\Gamma(X/S')$ having $H^*(X/S')$ as cohomology. If $t : \mathrm{Spec}(k_1) \to S$ is a k-point of S with values in a perfect field k_1, and $t' : \mathrm{Spec}(W_1) \to S'$ (where W_1 = Witt vectors on k_1) is the F'-Teichmüller lifting of t ([19] 1.1.2), then one has a canonical isomorphism $R\Gamma(X_t/W_1) \simeq t'^*R\Gamma(X/S')$ (hence $H^i(X_t/W_1) \simeq t'^*H^i(X/S')$ if $H^i(X/S')$ is locally free). The "specialization argument" invoked passim above is based on this fact, the standard technique of passage to the limit of EGA IV, and the following remark, due to Deligne. Let E be a coherent module on S'. Then there exists an open dense Zariski subset U of S such that on the completion of U in S' (i.e.

(*) except that the value of this crystal is not necessarily locally free (as was assumed in loc. cit.).

$\mathrm{Specf}(\varprojlim A_n)$ where A_n is the ring $\Gamma(S'_n, \underline{O}_{S'_n})$ localized with respect to the functions whose reduction mod p vanish on $S-U$) E is isomorphic to $M \otimes_W \underline{O}_{S'}$ where M is a finitely generated W-module. This comes from the fact that the local rings of S' at the generic points of S are discrete valuation rings whose maximal ideals are generated by p , plus the structure theorem for modules of finite type over a principal ring.

For a discussion of F-crystals (Newton polygons, slopes, zeta functions) we refer the reader to [19] and the exposes of Mazur [26] and Berthelot [4] in this volume.

It would be nice if one could work over larger liftings of S than S', e.g. Monsky-Washnitzer liftings [30], and carry over to general proper and smooth families the finer aspects of Dwork's theory, developed so far for families of hypersurfaces, see Katz's expository article [22] and Dwork [10], [11], [12], [13].

3.11. CRYSTALLINE H^1 . In the wake of Grothendieck [16], Mazur and Messing ([27], [28], [29]) have thoroughly investigated crystalline cohomology in dimension one and its relations with the Dieudonné theory of p-divisible groups. For lack of place I'll mention only one of their results [27] : let A be an abelian variety over k and G its associated p-divisible group, then there is a canonical isomorphism between $H^1(A/W)$ and the Dieudonné module of G . Over a "general" base S of char. $p > 0$, they define via crystalline techniques a functor which generalizes in a natural way the Dieudonné module functor. Except for the case S is the spectrum of a perfect field it is not known, however, how their constructions relate to Cartier's theory of formal groups.

REFERENCES

[1] BERTHELOT P.- Cohomologie cristalline, Lecture Notes n° 407 (1974).

[2] BERTHELOT P.- A series of Comptes Rendus Notes : t. 269, pp. 297-300, 357-360, 397-400, t. 272pp. 42-45, 141-144, 254-257, 1314-1317, 1397-1400, 1574-1577.

[3] BERTHELOT P.- Sur le "théorème de Lefschetz faible " en cohomologie cristalline, Comptes Rendus t. 277 pp. 955-958 (1973).

[4] BERTHELOT P.- The slope filtration on crystalline cohomology, this volume.

[5] BERTHELOT P. and ILLUSIE L.- Classes de Chern en cohomologie cristalline, Comptes Rendus t. 270, pp. 1695-1697, 1750-1752 (1970).

[6] DELIGNE P.- Lectures at the IHES, Spring 1970 (unpublished).

[7] DELIGNE P.- Equations différentielles à points singuliers réguliers, Lecture Notes n° 163.

[8] DELIGNE P.- La conjecture de Weil I, Pub. IHES n° 43 (1974), pp. 273-307.

[9] DELIGNE P.- La conjecture de Weil II, to appear.

[10] DWORK B.- p-adic cycles, Pub. IHES n° 37 (1969), pp. 27-115.

[11] DWORK B.- On Hecke Polynomials, Inv. Math. 12 (1971), pp. 249-256.

[12] DWORK B.- Normalized Period Matrices I, Ann. of Math. 94 (1971), pp. 337-388.

[13] DWORK B.- Normalized Period Matrices II, Ann. of Math. 98 (1973), pp.1-57.

[14] GROTHENDIECK A.- On the De Rham cohomology of algebraic varieties, IHES n° 29.

[15] GROTHENDIECK A.- Crystals and the De Rham cohomology of schemes, Notes by Coates, I. and Jussila, O., in Dix exposés, Advanced studies in pure Mathematics, North Holland 1968.

[16] GROTHENDIECK A.- Groupes de Barsotti-Tate et cristaux, Actes Congrès intern. Math. 1970, I, pp. 431-436, Gauthier-Villars.

[17] HARTSHORNE R.- Algebraic De Rham cohomology, to appear in Pub. IHES.

[18] HERRERA M. and LIEBERMAN D .- Duality and the De Rham cohomology of infinitesimal neighborhoods, Inv. Math. 13 (1971), pp. 97-124.

[19] KATZ N.- Travaux de Dwork, Séminaire Bourbaki n° 409, février 1972, Lecture Notes n° 383.

[20] KATZ N.- The regularity theorem in algebraic geometry, Actes Congrès Intern. Math. 1970, I, pp. 437-443, Gauthier-Villars.

[21] KATZ N.- Introduction aux travaux récents de Dwork, 1970.

[22] KATZ N. and MESSING W.- Some Consequences of the Riemann Hypothesis for Varieties over Finite Fields, Inventiones Math. 23, 73-77 (1974).

[23] KATZ N. and ODA T.- On the differentiation of De Rham cohomology classes with respect to parameters, Kyoto Math. J., Vol. 8, nº 2, 1968.

[24] MAZUR B.- Frobenius and the Hodge filtration, Bull. Amer. Math. Soc., Vol. 78, nº 5, 1972.

[25] MAZUR B.- Frobenius and the Hodge filtration (estimates), Ann. of Math., 98 (1973), pp. 58-95.

[26] MAZUR B.- Eigenvalues of Frobenius acting on algebraic varieties over finite fields, this volume.

[27] MAZUR B. and MESSING W.- Universal Extensions and One Dimensional Crystalline Cohomology, Lecture Notes nº 370.

[28] MESSING W.- The Crystals associated to Barsotti-Tate Groups, Lecture Notes nº 264.

[29] MESSING W.- The Universal Extension of an Abelian Variety by a Vector Group, Istituto Nazionale di Alta Matematica, Symposia Mathematica, Vol. XI (1973).

[30] MONSKY P. and WASHNITZER G.- Formal Cohomology I, Annals of Math. 88 (1968), pp. 181-217.

[31] ROBY N.- Lois polynômes et lois formelles en théorie des modules, Ann. de l'ENS, 80, 1963, pp. 213-348.

[32] ROBY N.- Les algèbres à puissances divisées, Bull. Soc. Math. France, 89, pp. 75-91, 1965.

[33] SERRE J.P.- Sur la topologie des variétés algébriques en caractéristique p, Symposium Internacional de Topologia algebraica, Mexico 1958.

CNRS PARIS

Proceedings of Symposia in Pure Mathematics
Volume 29, 1975

P-ADIC L-FUNCTIONS VIA MODULI OF ELLIPTIC CURVES

Nicholas M. Katz

The first half of the paper is quite elementary. We explain the relation between p-adic congruences for zeta-values and p-adic measures, and then we give an Eulerian construction of the p-adic measure required for zeta and L-functions of $\mathbb{Q}$. Unfortunately, such elementary constructions are unknown for other number fields.

The second half is less elementary. We explain the basic facts about the moduli of p-adic elliptic curves. We then show how to use the resulting theory of "generalized modular functions" to construct p-adic measures and explain how these measures lead to the p-adic interpolation of certain "L-series with grossencharacter" of quadratic imaginary fields.

I would like to thank Neal Koblitz for preparing a preliminary version of this paper based on my lecture at the conference.

I. p-adic congruences for zeta (Euler, Kummer, Kubota-Leopoldt)

The theory of p-adic zeta functions may be said to have begun when Euler discovered that the Riemann zeta function $\zeta(s)$ assumes rational values (essentially Bernoulli numbers) at negative integers.

Some two centuries later, Kummer was led to look at Bernoulli numbers when he discovered that the question of whether a given prime p was "regular" depended upon the p-adic shape of certain Bernoulli numbers. He discovered the "Kummer congruences" between Bernoulli numbers, which we will record here as congruences between ζ-values.

<u>Kummer Congruence</u> If $k \geq 1$ is an integer not divisible by $p - 1$, then $\zeta(1-k)$ is p-integral, and the value of $\zeta(1-k) \bmod p$ depends only on the value of $k \bmod p - 1$. (If k is divisible by $p - 1$, $\zeta(1-k)$ is <u>not</u> p-integral).

In the late 1950's, Kubota and Leopoldt, in trying to understand this, discovered that much more was true. To state their results, it is convenient to introduce the auxiliary functions

$$\zeta^*(s) = (1 - p^{-s})\zeta(s)$$

and, for each integer $a \geq 2$ prime to p,

$$\zeta^{*,a}(s) = (1 - a^{1-s})\,\zeta^*(s).$$

<u>Kubota-Leopoldt congruences</u>:

I. If $k \geq 1$ is an integer, then $\zeta^{*,a}(1-k)$ is p-integral, and the value of $\zeta^{*,a}(1-k) \bmod p^{n+1}$ depends only on the value of $k \bmod (p-1)p^n$.

II. If $k \geq 1$ is an integer not divisible by $p - 1$, then $\zeta^*(1-k)$ is p-integral, and its value mod p^{n+1} depends only on the value of $k \bmod (p-1)p^n$.

In fact, II may be directly deduced from I by choosing a to be a primitive root mod p. Notice that we recover the Kummer congruences as the special case $n = o$ of II.

The viewpoint of Kubota-Leopoldt is this. For each integer $b \bmod p - 1$, let $S(b)$ denote the set of strictly positive integers $\equiv b \bmod p - 1$. Then $S(b)$ is dense in $\mathbb{Z}_p$. By congruence I, the $\mathbb{Z}_p$-valued function on $S(b)$ defined by $k \mapsto \zeta^{*,a}(1-k)$ extends to a <u>continous</u> $\mathbb{Z}_p$-valued function on all of $\mathbb{Z}_p$. Such functions on $\mathbb{Z}_p$ are called p-adic zeta and L functions. In what follows, we will supress this point of view, and emphasize the actual congruences which underlie it.

II. p-adic congruences from p-adic measures (Mazur)

Let us begin by recalling the notion of a p-adic measure. Let X be a compact and totally disconnected space. In the applications, it will always be either $\mathbb{Z}_p$, $\mathbb{Z}_p^{\times}$, a finite space, a galois group, or a finite product of the preceding. Let R be any ring which is complete and separated in its p-adic topology. For the time being, R will be $\mathbb{Z}_p$ itself, but later R will be very large.

Consider the ring $\mathrm{Contin}(X, \mathbb{Z}_p)$ of all continuous $\mathbb{Z}_p$-valued functions on the space X. A $\mathbb{Z}_p$-linear map (not necessarily a ring homomorphism)

$$\mu : \mathrm{Contin}(X, \mathbb{Z}_p) \to R$$

is called an R-valued p-adic measure on X. The identity $\mathrm{Contin}(X,R) = \mathrm{Contin}(X, \mathbb{Z}_p) \hat{\otimes} R$ shows that we can also view μ as an R-linear map $\mathrm{Contin}(X,R) \to R$. For $f : X \to \mathbb{Z}_p$ an element of $\mathrm{Contin}(X, \mathbb{Z}_p)$, we denote its image $\mu(f) \in R$ symbolically as

$$\int_X f d\mu \ , \text{ or } \int_X f(x) d\mu(x)$$

Obvious lemma Suppose that $f,g : X \to \mathbb{Z}_p$ are continuous functions, and that $f(x) \equiv g(x) \mod p^n$ for all $x \in X$. Then

$$\int_X f d\mu \equiv \int_X g d\mu \quad \mod p^n R$$

proof We have $f - g = p^n h$, with $h \in \mathrm{Contin}(X, \mathbb{Z}_p)$. Thus

$$\int_X f d\mu - \int_X g d\mu = p^n \int_X h d\mu \qquad \text{QED}$$

Let us specialize to the case $X = \mathbb{Z}_p^\times$, and the power functions x^k. A congruence $k \equiv k' \mod (p-1)p^n$ between exponents implies a congruence $x^k \equiv x^{k'} \mod p^{n+1}$ for all $x \in \mathbb{Z}_p^\times$ (since $(\mathbb{Z}/p^{n+1}\mathbb{Z})^\times$ has order $(p-1)p^n$). Then the obvious lemma gives

$$\int_{\mathbb{Z}_p^\times} x^k d\mu \equiv \int_{\mathbb{Z}^\times} x^{k'} d\mu \mod p^{n+1} \quad \text{if } k \equiv k' \mod (p-1)p^n$$

for any p-adic measure on $\mathbb{Z}_p^\times$.

These are exactly the congruences that the function $k \mapsto \zeta^{*,a}(-k)$ satisfies. So it is natural to expect that we can explain the Kubota-Leopoldt congruences by constructing a $\mathbb{Z}_p$-valued measure $\mu^{(a)}$ on $\mathbb{Z}_p^\times$, one for each integer $a \geq 2$ prime to p, such that

$$\int_{\mathbb{Z}_p^\times} x^k d\mu^{(a)} = \zeta^{*,a}(-k) = (1-a^{k+1})(1-p^k)\zeta(-k) \; ; \quad k = 0,1,2,\ldots$$

In fact, we will first construct a measure $\mu^{(a)}$ on all of $\mathbb{Z}_p$ which satisfies

$$\int_{\mathbb{Z}_o} x^k d\mu^{(a)} = (1-a^{k+1})\zeta(-k) \quad \text{for } k = 0,1,2,\ldots$$

[Notice that we cannot expect the moments $\int_{\mathbb{Z}_p} x^k d\mu$ of a measure on $\mathbb{Z}_p$ to satisfy the same sharp congruences. The point is that if k and k' are congruent modulo $(p-1)p^n$, then the functions x^k and $x^{k'}$ are congruent mod p^{n+1} on $\mathbb{Z}_p^\times$, but on $p\mathbb{Z}_p$ they are only congruent modulo $p^{\min(k,k')}$.]

The measure $\mu^{(a)}$ on $\mathbb{Z}_p^\times$ is then obtained by restriction:

$$\int_{\mathbb{Z}_p^\times} f(x)d\mu^{(a)} \underset{=}{\underline{dfn}} \int_{\mathbb{Z}_p} \left(\text{the contin. fct. } x \to \begin{cases} f(x) & \text{if } x \in \mathbb{Z}_p^\times \\ 0 & \text{if not} \end{cases} \right) d\mu^{(a)}$$

(It is a general phenomenon in interpolating L functions that the first step is to construct a measure on $\mathbb{Z}_p$ whose "moments" give the entire L-function. When the measure is restricted to $\mathbb{Z}_p^\times$, the effect upon the moments is to remove the p-Euler factor of the corresponding L-value.)

This brings us to another point. Once we have constructed a measure $\mu^{(a)}$ on $\mathbb{Z}_p^\times$, we can integrate any continuous function. Thus let $\chi(x)$ be a Dirichlet character mod p^n, which we view as a locally constant function on $\mathbb{Z}_p^\times$. It turns out (though it is by no means obvious), that

$$(*) \qquad \int_{\mathbb{Z}_p^\times} \chi(x) \cdot x^k d\mu^{(a)} = (1 - a^{k+1}\chi(a))\, L(-k,\chi)$$

where $L(s,\chi)$ is the Dirichlet series

$$L(s,\chi) = \sum_{\substack{n \geq 1 \\ (n,p) = 1}} \chi(n) \cdot n^{-s}$$

[The value $L(-k,\chi)$ lies in the field $\mathbb{Q}$(values of χ). In order to view χ as a continuous p-adic function on $\mathbb{Z}_p^\times$, we must choose a prime $\mathcal{Y}$ of this field lying over p, and view χ as having values in its $\mathcal{Y}$-adic completion. The integral then lies in this completion, and it is in that completion that the equality (*) holds.]

III. p-adic measures on $\mathbb{Z}_p$

To understand what a p-adic measure on $\mathbb{Z}_p$ is, we must understand what the ring Contin($\mathbb{Z}_p$, $\mathbb{Z}_p$) looks like. For example, the binomial coefficient functions

$$\binom{x}{n} = \frac{x(x-1)\ \dots\ (x-n+1)}{n!} = \sum_{i=o}^{n} C_{i,n} x^i$$

assume $\mathbb{Z}_p$-values on $\mathbb{Z}_p$ (by continuity : they take $\mathbb{Z}$-values on $\mathbb{Z}$).

Theorem (Mahler) Let R be p-adically complete and separated. Then any $f \in \text{Contin}(\mathbb{Z}_p, R)$ can be uniquely written

$$f(x) = \sum_{n \geq o} a_n \binom{x}{n}, \quad a_n \in R, \quad a_n \to o$$

The a_n may be recovered as the higher differences of f :

$$a_n = (\Delta^n f)(o) = \sum_{i=o}^{n} (-1)^{n-i} \binom{n}{i} f(i).$$

Conversely, any series

$$\sum_{n \geq o} a_n \binom{x}{n}, \quad a_n \in R, \ a_n \to o$$

converges to an element of $\text{Contin}(\mathbb{Z}_p, R)$.

Corollary An R-valued measure μ on $\mathbb{Z}_p$ is uniquely determined by the sequence $b_n(\mu) = \int_{\mathbb{Z}_p} \binom{x}{n} d\mu$ of elements of R, and any sequence b_n determines an R-valued measure μ by the formula

$$\int_{\mathbb{Z}_p} f(x) d\mu = \sum_{n \geq o} a_n b_n = \sum_{n \geq o} b_n (\Delta^n f)(o)$$

Corollary Suppose that p is not a zero divisor in R. Then an R-valued measure μ on $\mathbb{Z}_p$ is uniquely determined by the sequence $m_n(\mu) \in R$ of its moments

$$m_n(\mu) = \int_{\mathbb{Z}_p} x^n d\mu.$$

A sequence m_n of elements of R arises as the moments of an R-valued measure μ if and only if the quantities

$$b_n \stackrel{\text{dfn}}{=\!=} \sum_{i=0}^{n} C_{i,n} m_i \quad , \text{ a priori in } R[1/p]$$

all lie in R, in which case we have

$$\int_{\mathbb{Z}_p} \binom{x}{n} d\mu = b_n$$

IV. Elementary construction of the zeta measure $\mu^{(a)}$ (Euler, Leibniz)

We begin by recalling Euler's computation of $(1-2^{k+1})\zeta(-k)$.

Euler writes

$$(1-2^{k+1})\zeta(-k) = (1-2^{k+1}) \sum_{n \geq 1} n^k = \sum_{n \geq 1} n^k - 2 \sum_{n \geq 1} (2n)^k$$

$$= \sum_{\substack{n \geq 1 \\ \text{odd}}} n^k - \sum_{\substack{n \geq 1 \\ n \text{ even}}} n^k = - \sum_{n \geq 1} (-1)^n \cdot n^k$$

and applies Abel summation to this last expression:

$$- \sum_{n \geq 1} (-1)^n n^k = - \sum_{n \geq 1} (-1)^n n^k T^n \Big|_{T=1} = \left(T\frac{d}{dT}\right)^k \left(\frac{T}{1+T}\right) \Big|_{T=1}.$$

If we make the substitution $T \to 1/T$ in this last formula, we see that $\zeta(-k) = o$ if $k \geq 2$ is even. If we make the substitution $T = e^x$ then a short calculation gives the usual expression in terms of Bernoulli numbers. More generally, let $a \geq 2$ be any integer. Then

$$(1 - a^{k+1})\zeta(-k) = \sum_{n \geq 1} n^k - a \sum_{n \geq 1} (an)^k = \sum_{n \geq 1} f(n) \cdot n^k$$

where

$$f(n) = \begin{cases} 1 & \text{if } n \not\equiv o \mod a \\ 1 - a & \text{if } n \equiv o \mod a \end{cases}$$

The important points are that f is periodic mod a, and $\sum_{n \bmod a} f(n) = 0$.

Thus

$$(1 - a^{k+1})\zeta(-k) = (T\frac{d}{dT})^k \left(\sum_{n \geq 1} f(n)T^n \right)_{|T = 1}$$

$$= (T\frac{d}{T})^k \left(\frac{\sum_{n=1}^{a} f(n)T^n}{1 - T^a} \right)_{|T = 1}$$

$$= (T\frac{d}{dT})^k \left(\frac{\sum_{n=1}^{a} f(n)(T^n - 1)}{1 - T^a} \right)_{|T = 1}$$

$$= (T\frac{d}{dT})^k \left(\frac{-\sum_{n=1}^{a} f(n)(1+T+\ldots+T^{n-1})}{1+T+\ldots+T^{a-1}} \right)_{|T = 1}$$

$$= \frac{\text{element of } \mathbb{Z}[T]}{(1+T+\ldots+T^{a-1})^{k+1}} \; {|T = 1}$$

This last expression shows that $a^{k+1}(1 - a^{k+1})\zeta(-k) \in \mathbb{Z}$, a fact equivalent to the Lipschitz-Sylvester theorem on Bernoulli numbers (cf [15]).

We are now ready to prove

<u>Theorem</u> For each integer $a \geq 2$ prime to p, there exists a $\mathbb{Z}_p$-valued measure $\mu^{(a)}$ whose moments are given by

$$\int_{\mathbb{Z}_p} x^k d\mu^{(a)} = (1 - a^{k+1})\zeta(-k) \quad , \quad k = 0,1,\ldots$$

<u>Proof</u> Our Eulerian calculation showed that

$$(1 - a^{k+1})\zeta(-k) = (T\frac{d}{dT})^k \left(\frac{\text{elt. of } \mathbb{Z}[T]}{1+T+\ldots+T^{a-1}}\right)\Big|_{T = 1}$$

What is important here is that the denominator is an integral polynomial which assumes a p-adic unit value (namely a) at $T = 1$.

Let us denote by A the subring of $\mathbb{Q}_p(T)$ consisting of all ratios $P(T)/Q(T)$ with $P,Q \in \mathbb{Z}_p[T]$, and $Q(1) \in \mathbb{Z}_p^{\times}$. Since this is an algebraic geometry conference, let me point out that A is the local ring of $\mathbb{Z}_p[T]$ at the maximal ideal $(p,T-1)$.

<u>Theorem</u> (bis) Given any element $F(T) \in A$, there is a $\mathbb{Z}_p$-valued measure μ_F on $\mathbb{Z}_p$ whose moments are given by the formula

$$\int_{\mathbb{Z}_p} x^k d\mu_F = (T\frac{d}{dT})^k (F)\Big|_{T = 1}$$

<u>Proof</u> What must be shown is that

$$\sum_{i=0}^{n} c_{i,n} (T\frac{d}{dT})^i (F)\Big|_{T = 1} \quad \text{lies in } \mathbb{Z}_p \quad \text{for } n = 0,1,\ldots$$

or equivalently that

$$\binom{T\frac{d}{dT}}{n}(F)\Big|_{T = 1} \in \mathbb{Z}_p \qquad \text{for } n = 0,1,\ldots$$

In fact, we have

<u>Lemma</u> The operators $\binom{T\frac{d}{dT}}{n}$ act stably on the ring A, i.e.,

$$\binom{T\frac{d}{dT}}{n}(F) \in A$$

<u>Proof</u> If we notice that

$$\binom{T\frac{d}{dT}}{n}(T^a) = \binom{a}{n}T^a$$

then we see that they act stably on $\mathbb{Z}_p[T]$.

If we notice that

$$\binom{T\frac{d}{dT}}{n} = \frac{T^n(\frac{d}{dT})^n}{n!}$$

then we see that Leibniz's formula is satisfied.

$$\binom{T\frac{d}{dT}}{n}(F\cdot G) = \sum_{i+j=n} \binom{T\frac{d}{dT}}{i}(F)\cdot \binom{T\frac{d}{dT}}{j}(G)\cdot$$

Let $Q \in \mathbb{Z}_p[T]$ have $Q(1) \in \mathbb{Z}_p^{x}$. Applying Leibniz's formula to the product $Q \cdot \frac{1}{Q} = 1$, we find inductively that $\binom{T\frac{d}{dT}}{n}(\frac{1}{Q}) \in A$. Applying the same formula to the product $P\cdot\frac{1}{Q}$ then shows that A is stable by the $\binom{T\frac{d}{dT}}{n}$.

QED

V. Examples of μ_F, and relations to Iwasawa's approach

$$F = T^n \quad : \quad \int f d\mu_F = f(n)$$

$$F = (T-1)^n \quad : \quad \int f d\mu_F = (\Delta^n f)(o)$$

This second example permits us to identify $\mathbb{Z}_p$-valued measures on $\mathbb{Z}_p$ with the elements of $\mathbb{Z}_p[[T-1]]$, the measure μ corresponding to the series $\Sigma\, b_n(T-1)^n$, where $b_n = \int \binom{x}{n} d\mu$. The measures μ_F we considered above correspond exactly to the rational functions in $\mathbb{Z}_p[[T-1]]$ (the "rationality of the zeta function"!). The multiplication of power series corresponds to convolution of measures on the additive group $\mathbb{Z}_p$:

$$\int f(x)d(\mu * \nu) \stackrel{\text{dfn}}{=\!=} \iint f(x + t)d\mu(x)d\nu(t).$$

This identification of measures is the link to Iwasawa's approach. Let f_s, for $s \in \mathbb{Z}_p$, denote the function $f_s(x) = (1+p)^{sx}$; then

$$(\Delta^n f_s)(o) = ((1+p)^s - 1)^n,$$

so that

$$\int f_s(x)d\mu = \sum_{n \geq o} b_n ((1+p)^s - 1)^n$$

when the measure μ corresponds to the series $\Sigma\, b_n (T-1)^n$.

VI. Relation to L-functions

Let Ψ be a locally constant function on $\mathbb{Z}_p$, say constant on cosets modulo p^n. For $F \in A$, we define the function $[\Psi](F)$ by

$$[\Psi](F)(T) \stackrel{dfn}{=\!=} \frac{1}{p^n} \sum_{b \bmod p^n} \Psi(b) \sum_{\zeta^{p^n} = 1} \zeta^{-b} F(\zeta T)$$

Let us check that $[\Psi]F$ lies again in A. We have

$$[\Psi](T^n) = \Psi(n) T^n$$

and

$$[\Psi](GF) = G\cdot [\Psi](F) \text{ if } G(T) = G(\zeta T) \text{ for all } \zeta^{p^n} = 1.$$

The first formula shows that $[\Psi]$ preserves $\mathbb{Z}_p[T]$. If we notice that every element of A may be written with a denominator which is invariant by $T \to \zeta T$ for $\zeta^{p^n} = 1$ (simply multiply numerator and denominator of $P(T)/Q(T)$ by $\prod Q(\zeta T)$, the product extended to the non-trivial p^n'th roots of unity), then the second formula shows that $[\Psi]$ is stable on A.

This operation $[\Psi]$ gives A the structure of module over the ring of locally constant functions on $\mathbb{Z}_p$. On the other hand, the set of <u>all</u> measures on $\mathbb{Z}_p$ is a module over the ring of all continuous functions, the module structure defined by $\int g(x)d(f\mu) = \int f(x)g(x)d\mu$. These structures are compatible, for we have the

<u>Integration Formula</u> $\int f(x)d\mu_{[\Psi]F} = \int \Psi(x)f(x)\cdot d\mu_F$

<u>Proof</u> Suppose first that $F = T^n$. Then $[\Psi](T^n) = \Psi(n)T^n$, and both integrals

are $\psi(n)f(n)$. By linearity, the two integrals then coincide for all $F \in \mathbb{Z}_p[T]$.

Now consider the general case $F \in A$. Suppose that ψ is <u>constant</u> on cosets mod p^n. By the already used trick of writing elements of A with denominators which are polynomials in T^{p^n}, we may suppose F is so written. Separating the exponents occurring in its numerator into congruence classes mod p^n, we may suppose (by additivity in F) that F is of the form $T^a G(T^{p^n})$, where $G(T) \in A$.

Since the $(p,T-1)$-adic completion of A is $\mathbb{Z}_p[[T-1]]$, the subring $\mathbb{Z}_p[T]$ is dense in A. Thus we may write $G(T) = \lim G_k(T)$, with $G_k \in \mathbb{Z}_p[T]$. Putting $F_k = T^a G_k(T^{p^n})$, we obtain

$$\begin{cases} F_k \in \mathbb{Z}_p[T],\ F = \lim F_k \\ [\psi](F_k) = \psi(a)F_k \\ [\psi](F) = \psi(a)F \end{cases}$$

Because the $(p,T-1)$-adic topology on $\mathbb{Z}_p[[T-1]]$ is the strong topology on measures, we may compute

$$\begin{aligned} \int \psi(x)f(x)d\mu_F = \lim \int \psi(x)f(x)d\mu_{F_k} &= \lim \int f(x)d\mu_{[\psi]F_k} \\ &= \lim \psi(a) \int f(x)d\mu_{F_k} \\ &= \psi(a) \int f(x)d\mu_F \\ &= \int f(x)d\mu_{[\psi]F} \end{aligned}$$

QED

<u>Corollary</u> For any locally constant function ψ on $\mathbb{Z}_p$, we have

$$\int_{\mathbb{Z}_p} \psi(x)x^k d\mu^{(a)} = L(-k, \psi - a^{k+1}\psi_a)$$

where ψ_a is the function $\psi_a(x) = \psi(ax)$.

In particular, if ψ is a <u>multiplicative</u> function on $\mathbb{Z}_p$,

$$\int_{\mathbb{Z}_p} \psi(x)x^k d\mu^{(a)} = (1 - a^{k+1}\psi(a))\cdot L(-k,\psi).$$

(Notice that for ψ = the characteristic function of $\mathbb{Z}_p^x$, $L(s,\psi)$ is $\zeta^*(s)$).

<u>Proof</u> Recall that $\mu^{(a)}$ is of the form μ_F for $F \in A$ the rational function

$$F = \frac{\sum_{n=1}^{a} f(n)\, T^n}{1 - T^a} \qquad f(n) = \begin{cases} 1 & n \not\equiv 0 \bmod a \\ 1-a & n \equiv 0 \bmod a \end{cases}$$

written here in "non-A" form.

Suppose that ψ is constant on cosets modulo p^M. Then we can rewrite F in another "non-A" form,

$$F = \frac{\sum_{n=1}^{ap^M} f(n)\, T^n}{1 - T^{ap^M}} .$$

The denominator is now invariant under $T \to \zeta T$ for $\zeta^{p^M} = 1$, so we get a "non-A" representation of $[\psi]F$:

$$[\psi]F = \frac{\sum_{n=1}^{ap^M} \psi(n)f(n)T^n}{1 - T^{ap^M}} = \sum_{n \geq 1} \psi(n)f(n)T^n$$

$$= \sum_{n \geq 1} \psi(n)T^n - a \sum_{n \geq 1} \psi(an)\cdot T^{an}$$

$$= \frac{\sum_{n=1}^{p^M} \psi(n)T^n}{1 - T^{p^M}} - a\, \frac{\sum_{n=1}^{p^M} \psi(an)T^{an}}{1 - T^{ap^M}}$$

By the integration formula, we have

$$\int \psi(x)x^k d\mu^{(a)} = \int \psi(x)x^k d\mu_F = \int x^k d\mu_{[\psi]F} = (T\frac{d}{dT})^k([\psi]F)|_{T=1}.$$

If we now substitute into this last formula the expression we found above for the rational function $[\psi]F$, we obtain the desired value à la Euler, namely $L(-k,\psi - a^{k+1}\psi_a)$.

VII. The relation with modular forms (Siegel, Serre)

We will now begin an extremely indirect approach to the values of ζ at negative integers. We will first obtain these values as the constant terms of the q-expansions of certain modular forms. We will then interpret these modular forms as functions on a certain moduli problem for p-adic elliptic curves. Finally, we will explain how geometric information about this moduli problem gives information about zeta values, culminating in a new construction of the measure $\mu^{(a)}$.

The point of this method, as opposed to the elementary one, is that it works equally well for studying the values at negative integers of the Dedekind zeta function of any totally real number field (if the number field K is not totally real, its zeta function _vanishes_ at all negative integers). See Ribet's talk in this volume for more details. In fact, it is at present the _only_ method known for general totally real fields (but cf [1] for an explicit approach to real quadratic fields). As a bonus, it also gives information on the values of certain L-series with "grossencharacter of type A_0" attached to quadratic imaginary fields.

VIII. Modular forms and their q-expansions

Over any ring R, we have the notion of an elliptic curve E/R (an abelian scheme of dimension one), together with the notion of a nowhere vanishing invariant differential $\omega \in H^0(E,\Omega^1_{E/R})$. Given E/R, such an ω always exists at least _locally_ on R, and if ω is one such, any other is of the form $\lambda\omega$ for some $\lambda \in R^{\times}$.

When $R = \mathbb{C}$, such a pair (E,ω) is equivalent to the giving of a <u>lattice</u> $L \subset \mathbb{C}$: to (E,ω) we associate the lattice of all periods of ω over the elements of $H_1(E^{an}, \mathbb{Z})$, and to the lattice $L \subset \mathbb{C}$ we associate the complex torus $\mathbb{C}/L$ with the invariant differential dz (z the parameter on $\mathbb{C}$). This pair $(\mathbb{C}/L, dz)$ is given algebraically by $(y^2 = 4x^3 - g_2x - g_3, dx/y)$, where x is the Weierstrass $\wp$-function $\wp(z;L)$ associated to the lattice L, $y = \wp'(z;L)$, and the constants g_2 and g_3 are given by the formulas

$$g_2 = 60 \sum_{\substack{\ell \in L \\ \ell \neq 0}} 1/\ell^4 \quad , \quad g_3 = 140 \sum_{\substack{\ell \in L \\ \ell \neq 0}} 1/\ell^6 .$$

Under this correspondence, the replacement of (E,ω) by $(E,\lambda\omega)$ corresponds to the replacement of L by λL.

An (unrestricted) modular form of weight k over $\mathbb{C}$ is classically defined to be a holomorphic function of lattices $F(L)$ satisfying $F(\lambda L) = \lambda^{-k}F(L)$. Since any lattice is multiple of one of the form $\mathbb{Z} + \mathbb{Z}\tau$ with $\mathrm{Im}(\tau) > 0$, F is uniquely determined by the holomorphic function on the upper half-plane $f(\tau) = F(\mathbb{Z} + \mathbb{Z}\tau)$. The function f arises from an F of weight k if and only if it satisfies the functional equation

$$f\left(\frac{a\tau + b}{c\tau + d}\right) = (c\tau + d)^k f(\tau) \qquad \forall \begin{pmatrix} a & b \\ c & d \end{pmatrix} \in SL_2(\mathbb{Z})$$

An example of such a modular form is

$$A_k(L) = \sum_{\substack{\ell \in L \\ \ell \neq 0}} 1/\ell^k \qquad \text{for } k \geq 3\text{; it is zero for } k \text{ odd.}$$

The <u>q-expansion</u> of a classical modular form is the possibly infinite-tailed Laurent series expansion in $q = e^{2\pi i\tau}$ of the (periodic with period one) function $\tau \mapsto F(2\pi i\,\mathbb{Z} + 2\pi i\,\mathbb{Z}\tau) = (2\pi i)^{-k}f(\tau)$. For example, the q-expansion of

$$G_k = \frac{(-1)^k (k-1)!}{2} A_k \qquad , \quad k \geq 4$$

is given by

$$G_k(q) = \begin{cases} \frac{1}{2}\zeta(1-k) + \sum_{n \geq 1} q^n \sum_{d|n} d^{k-1} & \text{for } k \geq 4 \text{ even} \\ 0 \text{ for } k \text{ odd} \end{cases}$$

This formula is the basis of the connection between modular forms and ζ-values.

We may algebraically define an (unrestricted) modular form of weight k over any ring R to be a rule which assigns to __any__ pair (E,ω) over __any__ R-algebra R' a value $F(E,\omega) \in R'$, in such a way that

$$\begin{cases} F(E,\lambda\omega) = \lambda^{-k} F(E,\omega) \qquad \forall\ \lambda \in (R')^{\times} \\ F \text{ commutes with all extensions of scalars } R' \to R'' \\ F(E,\omega) \text{ depends only on the } R'\text{-isomorphism class of } (E,\omega) \end{cases}$$

The q-expansion of a modular form F over R is a finite-tailed Laurent series $F(q) \in R((q))$ obtained as follows. Over $\mathbb{C}$, the lattice $2\pi i\,\mathbb{Z} + 2\pi i\,\mathbb{Z}\tau$ defines the elliptic curve with differential

$$(\mathbb{C}/2\pi i\,\mathbb{Z} + 2\pi i\,\mathbb{Z}\tau \ , \ dz)$$

which is isomorphic, by the exponental map $z \to t = \exp(z)$, to

$$\left(\mathbb{C}^{\times}/q^{\mathbb{Z}} \ , \ \frac{dt}{t}\right)$$

The Weierstrass equation defining this curve has coefficients in the ring $\mathbb{Z}[1/6][[q]]$, and with a bit of rearranging can be turned into an equation over $\mathbb{Z}[[q]]$ which defines an elliptic curve with differential over $\mathbb{Z}((q))$. This curve we call the Tate-Jacobi curve $\mathrm{Tate}(q)$ with its canonical differential ω_{can}. By extension of scalars, we obtain $(\mathrm{Tate}(q), \omega_{can})$ over $R((q))$. The q-expansion of a modular forms over R is defined to be its value on this curve:

$$f(q) \overset{\text{dfn}}{=\!=} F(\mathrm{Tate}(q), \omega_{can}).$$

Let us state without proof the GAGA-type relations between these different concepts of modular forms (cf [4], [7]).

1. A classical modular form $F^{c\ell}$ over $\mathbb{C}$ of weight k comes from an algebraic one F over $\mathbb{C}$ of weight k if and only if $F^{c\ell}$ has a finite-tailed q-expansion. In this case $F(q) = F^{c\ell}(q)$, and F is uniquely determined.
2. (q-expansion principle) If the q-expansion of a modular form F over R of weight k has all of its coefficients in a subring $R_o \subset R$, then there exists a unique modular form F_o of weight k over R_o which gives rise to F by extension of scalars. In particular, a modular form of given weight k is uniquely determined by its q-expansion.

The modular forms with holomorphic q-expansions are really rather down to earth objects. For example, over a ring R in which both 2 and 3 are invertible, the graded ring of all (holomorphic) modular forms is just $R[g_2, g_3]$, with g_2 of weight four and g_3 of weight six.

IX. Heuristic

As a corollary of the GAGA principle, the modular forms $2G_k$ are defined over $\mathbb{Q}$, in fact over $\mathbb{Z}[\zeta(1-k)]$. By looking at their coefficients, we obtain many measures on $\mathbb{Z}_p$, as follows. For $n \geq 1$, the coefficient of q^n in G_k, $k \geq 3$, is given by

$$\text{coef of } q^n \text{ in } 2G_k = \sum_{d|n} (d^{k-1} - (-d)^{k-1})$$

This suggests that we define a measure μ_n on $\mathbb{Z}_p$ by the formula

$$\int_{\mathbb{Z}_p} f(x)d\mu_n = \sum_{d|n} (f(d) - f(-d))$$

Thus we have

$$\int_{\mathbb{Z}_p} x^{k-1}d\mu_n = \text{coef. of } q^n \text{ in } 2G_k$$

So we would like to "let n go to zero" and define a measure μ_0 on $\mathbb{Z}_p$ such that

$$\int_{\mathbb{Z}_p} x^{k-1}d\mu_0 = \text{constant term in } 2G_k = \zeta(1-k).$$

Of course this is not possible as stated, simply because $\zeta(1-k)$ need not be p-integral. We could at best hope to obtain $\mu^{(a)}$ as the "limit" in some sense of the measures $\mu_n^{(a)}$ defined by

$$\int_{\mathbb{Z}_p} f(x)d\mu_n^{(a)} = \int_{\mathbb{Z}_p} (f(x) - af(ax))d\mu_n.$$

Let us explain how to realize this hope, by studying a moduli problem on which the series $G_k - \frac{1}{2}\zeta(1-k)$ themselves have an intrinsic meaning.

X. Generalized p-adic modular functions and trivialized elliptic curves

Henceforth, p is a fixed prime number, and we consider only rings R which are p-adically complete and separated. Rather than consider pairs (E,ω) over R, we will now consider "trivialized elliptic curves" (E,φ) over R:

$$\begin{cases} E \text{ an elliptic curve over } R \\ \varphi : \hat{E} \xrightarrow{\sim} \hat{\mathbb{G}}_m \text{ an isomorphism between the formal} \\ \quad \text{group of } E \text{ and the formal multiplicative group.} \end{cases}$$

Given an elliptic curve E/R, it cannot admit a trivialization φ unless all the fibres of E/R are ordinary elliptic curves. If the fibres are all ordinary, then in order to construct a trivialization we will in general have to extend scalars from R to a vast pro-ind-etale over-ring R_∞ (i.e., R_∞ will be p-adically complete and separated, and for each n, $R_\infty/p^n R_\infty$ will be an increasing union of finite etale over-rings of $R/p^n R$).

Once we have one trivialization φ, then any other is of the form $\alpha\varphi$ for some $\alpha \in \mathbb{Z}_p^\times = \mathrm{Aut}(\hat{\mathbb{G}}_m)$, at least over a ring R such that $\mathrm{Spec}(R/pR)$ is connected.

Notice that a trivialized elliptic curve (E,φ) over R gives rise to an elliptic curve with differential $(E, \varphi^*(\frac{dt}{t}))$, simply by pulling back the standard invariant differential dt/t on $\hat{\mathbb{G}}_m$. Notice also that the Tate curve Tate (q), considered over $\widehat{\mathbb{Z}_p((q))}$, the p-adic completion of $\mathbb{Z}_p((q))$, admits a canonical trivialization φ_{can}, simply because Tate(q) was obtained as the quotient of $\mathbb{G}_m$ by the discrete subgroup $q^{\mathbb{Z}}$. Under this trivialization, we have $\varphi_{can}^*(dt/t) = \omega_{can}$.

We now define a generalized p-adic modular function over a p-adically complete and separated R to be rule F which assigns to any trivialized elliptic curve (E,φ) over any p-adically complete and separated R-algebra R' a value $F(E,\varphi) \in R'$ in such a way that

$$\begin{cases} F \text{ commutes with extension of scalars } R' \to R'' \\ \quad \text{of } p\text{-adically complete and separated } R\text{-algebras} \\ F(E,\varphi) \text{ depends only upon the } R'\text{-isomorphism class} \\ \quad \text{of } (E,\varphi) \end{cases}$$

It is of the utmost importance that we do not require that F have a weight, i.e., a transformation property under the action of $\mathbb{Z}_p^\times$, defined by

$$[\alpha]F(E,\varphi) = F(E,\alpha^{-1}\varphi)$$

The ring of all generalized p-adic modular functions defined over R, noted $\mathbb{V}_R$, is itself a p-adically complete and separated R-algebra. It represents the functor

$$M^{\mathrm{triv}}(R') = \text{isomorphism classes of trivialized elliptic curves } (E,\varphi) \text{ over } R'$$

on the category of p-adically complete and separated R-algebras.

Geometrically, the functor M^{triv} is formed from the stack of trivialized elliptic curves $\mathcal{M}^{\mathrm{triv}}$ by passing to isomorphism classes. The stack $\mathcal{M}^{\mathrm{triv}}$ sits over the stack $\mathcal{M}^{\mathrm{ord}}$ of ordinary elliptic curves as an ind-etale covering, with structural group $\mathbb{Z}_p^\times$ (we need stacks because the functor M^{ord} is not representable):

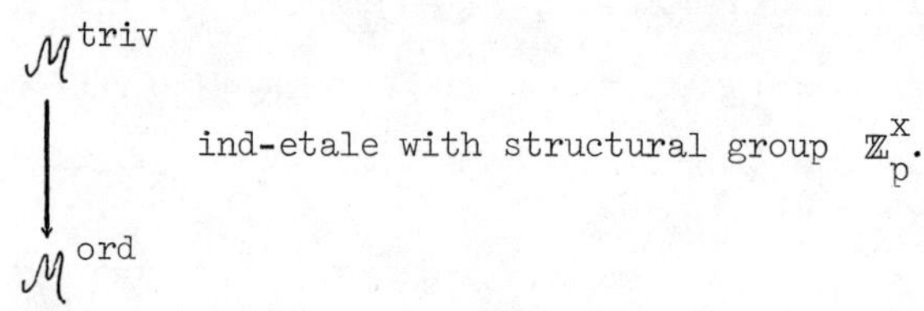

Indeed, if $E \xrightarrow{\pi} \mathcal{M}^{\mathrm{ord}}$ is the tautological "universal elliptic curve" over the stack $\mathcal{M}^{\mathrm{ord}}$, then $\mathcal{M}^{\mathrm{triv}}$ is obtained as the bundle of frames of the locally free rank one $\mathbb{Z}_p$-etale sheaf $R^1\pi_*\mathbb{Z}_p$ on $\mathcal{M}^{\mathrm{ord}}$.

This sheaf corresponds to a character χ of the fundamental group of $\mathcal{M}^{\mathrm{ord}}$ with values in $\mathbb{Z}_p^\times$. The surjectivity of this character is equivalent to the irreducibility of $\mathcal{M}^{\mathrm{triv}}$ (since $\mathcal{M}^{\mathrm{ord}}$ is itself irreducible). There are two proofs of this surjectivity. The first, due to Igusa, studies what happens in the neighborhood of a "missing point", corresponding to a supersingular elliptic curve. Igusa proves that even after restricting χ to the local monodromy group (inertia group) at such a point, χ remains surjective. The second proof, found recently by Ribet, is <u>arithmetic</u> in nature, based on the interpretation of the value of X on the Frobenius element attached to a closed point of $\mathcal{M}^{ord}$ (i.e., to an ordinary elliptic curve over a finite field) as the unit root of the zeta function of that elliptic curve. Ribet's proof works also for Hilbert modular varieties, and is discussed, along with its applications, in his article in these Proceedings.

<u>Remark</u> To avoid stacks, we could instead choose an integer $N \geq 3$ prime to p and rigidify our moduli problems with level N structures. Then $M^{ord}(N)$ is itself representable, though no longer geometrically irreducible, and the covering $M^{triv}(N) \to M^{ord}(N)$ is irreducible over each irreducible component of $M^{ord}(N)$. The coordinate ring $\mathbb{V}(N)$ of $M^{triv}(N)$ has a canonical action of $GL(2, \mathbb{Z}/N\mathbb{Z})$ on it, under which the subring of invariants is $\mathbb{V}$.

XI. <u>Consequences of the irreducibility of M^{triv}</u>

Evaluation at $(\text{Tate}(q), \varphi_{can})$ defines the q-expansion homomorphism $\mathbb{V}_R \to \widehat{R((q))}$ for any R. Suppose first that R is a field k of characteristic p. Then $\mathbb{V}_k$ is the coordinate ring of $M^{triv} \otimes k$, which is (essentially) a smooth irreducible affine curve over k. Then $(\text{Tate}(q), \varphi_{can})$ defines a $k((q))$-valued point of this curve which obviously does not lie over any closed point, so therefore must lie over the generic point. Thus we have proven

<u>Lemma</u> When k is a field of characteristic p, the q-expansion homomorphism $\mathbb{V}_k \to k((q))$ is injective.

As an immediate corollary, we get

<u>Lemma</u> Let R be a complete discrete valuation ring with uniformizing parameter π, and residue field k of characteristic p. Then the q-expansion homomorphism

$$\mathbb{V}_R \to \widehat{R((q))}$$

is injective, and its cokernel $\widehat{R((q))}/\mathbb{V}_R$ is R-flat, i.e., has no p-torsion.

Proof Since $\mathbb{V}_R$ is flat over R (being formally ind-smooth over R), and $\mathbb{V}_R \otimes k = \mathbb{V}_k$ (because $M_k^{triv} = M_R^{triv} \otimes k$), the injectivity of $\mathbb{V}_R \to \widehat{R((q))}$ follows from its injectivity modulo π, and the fact that $\mathbb{V}_R$ is p-adically separated. Given the injectivity of $\mathbb{V}_R \to \widehat{R((q))}$, the flatness of its cokernel results from (and is equivalent to) its injectivity modulo π.

QED

Technical as it looks, this last lemma is the key to everything, for it allows us to construct a large number of elements of $\mathbb{V}$ ($= \mathbb{V}_{\mathbb{Z}_p}$):

1. Let f be an (unrestricted) modular form of weight k, defined over $\mathbb{Z}_p$. Then f may be viewed as an element of $\mathbb{V}$, as being the rule

$$(E,\varphi) \to f(E,\varphi^*(dt/t)).$$

The q-expansion of f remains the same before and after we view f as lying in V. Under the action of $\mathbb{Z}_p^\times$ on $\mathbb{V}$ defined by $[\alpha]F(E,\varphi) = F(E,\alpha^{-1}\varphi)$, f transforms by

$$[\alpha]f = \alpha^k f$$

2. Let $\{f_i\}$ be a finite set of modular forms, f_i of weight i, defined over $\mathbb{Z}_p$, and suppose that $\sum f_i(q) \in p^N \mathbb{Z}_p[[q]]$, say $\sum f_i(q) = p^N h(q)$. Then there is a unique element $h \in \mathbb{V}$ whose q-expansion is $h(q)$. (For by hypothesis, the series $h(q)$ is a p^M-torsion element in $\widehat{\mathbb{Z}_p((q))}/\mathbb{V}$, hence itself is the q-expansion of a necessarily unique element $h \in V$).

   For example, $G_k - \frac{1}{2}\zeta(1-k)$ lies in $\mathbb{V}$. Elements of this sort are called divided congruences, and form a subring of $\mathbb{V}$ which can be proven to be dense in all of $\mathbb{V}$ (cf [8]).

3. Let h_i be a sequence of elements of $\mathbb{V}$ such that $\lim h_i(q)$ exists in $\widehat{\mathbb{Z}_p((q))}$. Then there is an element $h_\infty \in \mathbb{V}$ such that $h_\infty(q) = \lim h_i(q)$. (Because $p^n\mathbb{V} = \mathbb{V} \cap p^n \widehat{\mathbb{Z}_p((q))}$, the h^i are a Cauchy sequence in $\mathbb{V}$).

Thus $\mathbb{V}$ contains all "p-adic modular forms" in the sense of Serre [17] and all divided congruences between them. In fact, Serre's "p-adic modular forms of weight χ" are exactly the elements $F \in V$ satisfying $[a]F = \chi(a)F$ for all $a \in \mathbb{Z}_p^{\times}$ (cf [8]).

4. Let θ denote the differential operator $q\frac{d}{dq}$. If $h \in \mathbb{V}$, then $\theta h \in \mathbb{V}$ in the sense that $\theta(h(q))$ is the q-expansion of an element of $\mathbb{V}$. (Use the fact that θ operates stably on Serre's p-adic modular forms, stably on all divided congruences between them, and then invoke 3. above together with the density of divided congruences in $\mathbb{V}$).

XII. Construction of the $\mathbb{V}$-valued measure $\mu^{(a)}$ on $\mathbb{Z}_p$

The modular forms G_k, $k \geq 3$, have already been defined over $\mathbb{Q}_p$ albeit by a transcendental summation followed by an appeal to the q-expansion principle. Serre [17] has shown that the series

$$G_2(q) = \frac{1}{2}\zeta(-1) + \sum_{n \geq 1} q^n \sum_{d|n} d \qquad ; \zeta(-1) = \frac{-1}{24} .$$

is the q-expansion of a p-adic modular form of weight two over $\mathbb{Q}_p$. We define G_1 to be zero. We view all the G_k, $k = 1,2,\ldots$, as elements of $\mathbb{V}[\frac{1}{p}]$. The fundamental theorem is

<u>Theorem</u> For $a \in \mathbb{Z}_p^{\times}$, there exists a $\mathbb{V}$-valued measure on $\mathbb{Z}_p$ whose moments are given by

$$\int_{\mathbb{Z}_p} x^k d\mu^{(a)} = 2(1 - a^{k+1})G_{k+1} = 2G_{k+1} - 2[a]G_{k+1} \qquad \text{for } k = 0,1,2,..$$

<u>Proof</u> Let us begin by noting that $G_{k+1} - [a]G_{k+1}$ does indeed lie in $\mathbb{V}$: Certainly we have $G_{k+1} \in \mathbb{V}[\frac{1}{p}]$, and G_{k+1} has integral q-expansion except possibly for its constant term. Thus we may apply the

Key Lemma Suppose $h \in \mathbb{V}[\frac{1}{p}]$, and h has p-integral q-expansion except possibly for its constant term. Then for any $a \in \mathbb{Z}_p^{\times}$, $h - [a]h \in \mathbb{V}$.

Proof of Key Lemma Let us write $h(q) = A + f(q)$, with $A \in \mathbb{Q}_p$ and $f(q) \in \widehat{\mathbb{Z}_p((q))}$. Suppose that $p^M h \in \mathbb{V}$. Then $p^M A \in \mathbb{Z}_p$, and $p^M f(q) = (p^M h)(q) + p^M A$, which shows that $p^M f(q)$, and hence $f(q)$ itself, is the q-expansion of an element of $\mathbb{V}$. Thus

$$h = A + \text{an elt. of } \mathbb{V} \qquad \text{in } \mathbb{V}[\tfrac{1}{p}]$$

Applying $[a]$ yields

$$[a]h = A + [a](\text{an elt. of } \mathbb{V}) = A + \text{an elt. of } \mathbb{V}.$$

Subtracting gives $\quad h - [a]h \in \mathbb{V}$. QED

Let us conclude the construction. We must show that

$$\sum_{i=0}^{m} C_{i,m} \int_{\mathbb{Z}_p} x^i d\mu^{(a)} \in \mathbb{V} \qquad \text{for } m = 0,1,\ldots$$

where the coefficients $C_{i,m}$ are defined by

$$\binom{x}{m} = \sum_i C_{i,m}\, x^i.$$

Thus we must show that

$$2\sum_i C_{i,m}\, G_{i+1} \; - \; [a]\left(2\sum_i C_{i,m}\, G_{i+1}\right) \in V.$$

By the Key Lemma, it suffices to check that the element

$$2\sum_{i=0}^{m} C_{i,m}\, G_{i+1} \in \mathbb{V}\,[\tfrac{1}{p}]$$

has integral q-expansion except for its constant term. But we have already computed the coefficient of q^n in $2G_{i+1}$: it was

$$\int_{\mathbb{Z}_p} x^i d\mu_n \text{ , where } \int_{\mathbb{Z}_p} f(x) d\mu_n = \sum_{d|n} (f(d) - f(-d)).$$

Thus

$$2 \sum_{l=o}^{m} C_{i,m} G_{i+1}(q) = \text{constant} + \sum_{n \geq 1} q^n \int_{\mathbb{Z}_p} \binom{x}{m} d\mu_n. \qquad \text{QED}$$

XIII. Applications to quadratic imaginary fields: Hurwitz numbers

The constant term of the q-expansion of the $\mathbb{V}$-valued measure $\mu^{(a)}$ is a $\mathbb{Z}_p$-valued measure on $\mathbb{Z}_p$, which for $a \in \mathbb{Z}$, $a > o$, differs from the measure we constructed by elementary means, only by a single point mass at the origin (because the G_k's ignore the "accident" that $\zeta(o) = -\frac{1}{2}$ is $\neq o$, while $G_1 = o$).

However, as we have a $\mathbb{V}$-valued measure, we may evaluate it at any trivialized elliptic curve (E, φ) over any p-adically complete and separated ring R, and obtain an R-valued measure on $\mathbb{Z}_p$.

To fix ideas, consider the gaussian curve $y^2 = 4x^3 - 4x$, which has complex multiplication by the gaussian integers; i acts as $(x,y) \to (-x, iy)$. We view this curve as defined over $\mathbb{Z}_p$, but because we want it to be ordinary mod p, we must suppose that $p \equiv 1 \mod 4$.

What about a trivialization? None exists over $\mathbb{Z}_p$. If we extend scalars to W, the Witt vectors of the algebraic closure of $\mathbb{F}_p$, then there will be trivializations. In fact, if φ is a trivialization, then $\varphi^*(dt/t) = c \cdot dx/y$ with $c \in W^\times$, and c satisfies the equation

$$c^\sigma / c = u$$

where σ is the Frobenius automorphism of W, and $u \in \mathbb{Z}_p^\times$ is the unit root of Frobenius. Conversely, if $c \in W^{x}$ satisfies $c^{\sigma} = uc$, there exists a unique trivialization φ with $\varphi^{*}(dt/t) = cdx/y$. In fact, the discovery by Tate of the relationship between trivializations and unit roots of Frobenius was the starting point of the application of p-adic analysis to any sort of zeta function (cf [3], p. 257).

Let us fix a choice of c, and denote by φ_c the corresponding trivialization. Then by evaluating $\mu^{(a)}$ at $(y^2 = 4x^3 - 4x, \varphi_c)$ we obtain:

<u>Theorem</u> There exists a W-valued measure $\mu^{(a)}$ on $\mathbb{Z}_p$ whose moments are given by

$$\int_{\mathbb{Z}_p} x^{k-1} d\mu^{(a)} = (1 - a^k)\, G_k(y^2 = 4x^3 - 4x,\ cdx/y) \quad \text{for } k = 1,2\ldots$$

Let us make explicit the interpretation of these moments as values of L-series with grossencharacter. For $k \geq 4$, we can write

$$G_k(y^2 = 4x^3 - 4x,\ cdx/y) = c^{-k} G_k(y^2 = 4x^3 - 4x,\ dx/y)$$

$$= \frac{(-1)^k (k-1)!}{2}\, c^{-k}\, A_k(y^2 = 4x^3 - 4x,\ dx/y).$$

Over $\mathbb{C}$, the lattice corresponding to $(y^2 = 4x^3 - 4x, dx/y)$ is easily seen to be $\Omega\,\mathbb{Z}[i]$, where

$$\Omega = 2\int_0^1 \frac{dt}{\sqrt{1-t^4}} = 2.622057\ldots$$

Thus we obtain an equality of complex numbers

$$A_k(y^2 = 4x^3 - 4x,\ dx/y) = A_k(\Omega\,\mathbb{Z}[i]) = \Omega^{-k} A_k(\mathbb{Z}[i]) = \Omega^{-k} \sum_{(a,b) \neq (0,0)} \frac{1}{(a+bi)^k}$$

which shows that the A_k are essentially the Hurwitz numbers [5].

The series $\sum_{(a,b)\neq(o,o)} \frac{1}{(a+bi)^k}$ converges absolutely for $k \geq 3$, but sums to zero unless $k = 4\ell$, in which case it is

$$4 \sum_{\text{ideals } \mathfrak{A}} \chi^{-k}(\mathfrak{A}) = L(0,\chi^{-k})$$

where χ^{-k} is the grossencharacter of $\mathbb{Z}[i]$ defined by

$$\chi^{-k}(\mathfrak{A}) = 1/\alpha^k \quad \text{if } \mathfrak{A} = (\alpha)$$

Thus we obtain the p-adic interpolation the function $k \to L(0,\chi^{-k})$.

XIV. Manin's function of two variables

By quite different techniques, Manin and Vishik [12] have shown that series of the form

$$\sum_{(a,b)\neq(o,o)} \frac{(a-bi)^r}{(a+bi)^{k+r}} = \begin{cases} L(0,\overline{\chi}^r\chi^{-k-r}) & \text{if } k+2r \equiv 0(4) \\ 0 \text{ if not} \end{cases}$$

once their transcendental factors are removed, can be p-adically interpolated to continuous p-adic functions of two variables (k,r), provided that $p \equiv 1\ (4)$.

By using the techniques explained here, we are able to construct a $\mathbb{V}$-valued measure $\mu^{(a)}$ on $\mathbb{Z}_p \times \mathbb{Z}_p$ whose moments

$$\int\int_{\mathbb{Z}_p \times \mathbb{Z}_p} x^{k-1}\, y^r\, d\mu^{(a)}(x,y) \in \mathbb{V}$$

when evaluated on $(y^2 = 4x^3 - 4x, \varphi_c)$, give essentially these same L-values. The proofs are long, and will appear elsewhere.

A Short Bibliography

The following list is very far from being complete. The interested reader should also consult the bibliographies of the works cited here.

1. J. Coates and W. Sinnott, On p-adic L-functions over real quadratic fields. Inv. Math. 25 (1974), 253-279.

2. Z. I. Borevich and I. R. Shafarevich, Number Theory, esp. pp. 355-389. Academic Press, New York and London, 1966.

3. B. Dwork, A deformation theory for the zeta function of a hypersurface. Proc. Intl. Cong. Math. (1962), Stockholm, 247-259.

4. P. Deligne and M. Rapoport, Les schemas de modules de courbes élliptiques. Proceedings of the 1972 Antwerp Summer School, Springer Lecture Notes in Mathematics 349 (1973), 143-316.

5. A. Hurwitz, Uber die Entwicklungskoeffizienten der lemniscatischen Functionen. Math. Ann. 51 (1899), 196-226.

6. K. Iwasawa, Lectures on P-adic L-Functions, Annals of Math. Studies 74, Princeton Univ. Press, 1972.

7. N. Katz, P-adic properties of modular schemes and modular forms, Proc-ceedings of the 1972 Antwerp Summer School, Springer Lecture Notes in Mathematics 350 (1973), 70-189.

8. N. Katz, Higher congruences between modular forms, to appear in Annals of Math.

9. N. Katz, The Eisenstein measure and p-adic interpolation, to appear in Amer. J. Math.

10. T. Kubota and H. W. Leopoldt, Eine p-adische Theorie der Zetawerte I, J. Reine Ang. Math. 214/215 (1964), 328-339.

11. K. Mahler, An interpolation series for a continuous function of a p-adic variable. J. Reine Ang. Math. 199 (1958), 23-34.

12. J. Manin and S. Vishik, P-adic Hecke series for quadratic imaginary fields (in Russian), to appear in Math. Sbornik.

13. B. Mazur, Analyse p-adique. secret Bourbaki redaction, 1973.

14. B. Mazur and H. P. F. Swinnerton-Dyer, Arithmetic of Weil curves. Inv. Math. 25 (1974), 1-61.

15. J. Milnor and J. Stasheff, Characteristic Classes, esp. Appendix B, Annals of Math. Studies 76, Princeton Univ. Press, 1974.

16. K. Ribet, P-adic interpolation via Hilbert modular forms, this volume.

17. J.-P. Serre, Formes modulaires et fonctions zeta p-adiques. Proceedings of the 1972 Antwerp Summer School, Springer Lecture Notes in Mathematics 350 (1973), 191-268.

PRINCETON UNIVERSITY

Proceedings of Symposia in Pure Mathematics
Volume 29, 1975

TOPOLOGICAL USE OF POLAR CURVES

Lê Dũng Tráng

In his lectures B. Teissier has mainly spoken about the use of polar curves in the case of isolated singularities of complex hypersurfaces. Then in the case of complex plane curves, M.Merle has shown the relations between the topology of generic polar curves and that of the related plane curve. In this lecture we shall deal with polar curves in the case of hypersurfaces with not necessarily isolated singularities and we shall give the different theorems one can prove involving polar curves.

1. DEFINITION

Let $f: U \subset \mathbb{C}^{n+1} \longrightarrow \mathbb{C}$ be an analytic function defined on the open neighbourhood U of $0 \in \mathbb{C}^{n+1}$. Let $\phi: U \longrightarrow \mathbb{C}^2$ be the mapping defined by $\phi(z) = (x_0, f(z))$, where x_0 is a linear form of $\mathbb{C}^{n+1}$. We call $c(\phi)$ the critical locus of ϕ. We may suppose x_0 is a coordinate of $\mathbb{C}^{n+1}$ and that $x_1, \cdots, x_n$ are the others, then:

$$C(\phi): = \{z \in U \,|\, \partial f/\partial x_1 = \cdots = \partial f/\partial x_n = 0\}.$$

(1.1) Notice that $C(\phi)$ always contains the critical locus $\Sigma(f)$ of f, when U is sufficiently small:

$$\Sigma(f): = \{z \in C(\phi) \,|\, \partial f/\partial x_0 = 0\}.$$

In [4] we have proved:

LEMMA (1.2) There is a Zariski open dense set Ω of the projective space of linear hyperplanes of $\mathbb{C}^{n+1}$ passing through 0, such that for any $L \in \Omega$, say defined by some $x_0 = 0$, choosing U sufficiently small, one has

$$C(\phi) = \Sigma(f) \cup \Gamma$$

where Γ is either void or a curve not contained in $H: = \{f = 0\}$.

Moreover if $\Gamma \neq \emptyset$, at any $x \in \Gamma - \{0\}$ the ideal $(\partial f/\partial x_1, \cdots, \partial f/\partial x_n)$ defines a <u>reduced</u> curve.

DEF. (1.3) When $L \in \Omega$, if $\Gamma \neq \emptyset$, we say that Γ is the polar curve of L.

REMARKS (1.4) $\Gamma = \emptyset$ when, for instance, $f = 0$ is an analytic product, but $\Gamma \neq \emptyset$, when f has an isolated singularity at O.

(1.5) In our definition f need not be reduced.

(1.6) The second part of our lemma means that the tangent hyperplane $T(x, f^{-1}(f(x)))$, parallel to L, of $f^{-1}(f(x))$ at x has a quadratic contact with $f^{-1}(f(x))$ or equivalently the restriction of x_o to $f^{-1}(f(x))$ has an isolated ordinary double point at x.

We have the alternative definition of the polar curve:

Let $\pi: Z \longrightarrow U$ be the blowing-up of $J(f) = (\partial f/\partial x_o, \cdots, \partial f/\partial x_n)$. Namely let $\psi: U - |J(f)| \longrightarrow \mathbb{P}^n$, where $|J(f)|$ is the set where $J(f)$ vanishes, be defined by $\psi(z) = (\partial f/\partial x_o : \cdots : \partial f/\partial x_n)$. Let Z be the closure of the graph of ψ in $U \times \mathbb{P}^n$. Then $\pi: Z \longrightarrow U$ is defined by the first projection $U \times \mathbb{P}^n \longrightarrow U$. Consider the Gauss map $Z \longrightarrow \mathbb{P}^n$ defined by the second projection $U \times \mathbb{P}^n \longrightarrow \mathbb{P}^n$. Then for almost every point ℓ of $\mathbb{P}^n$, if U is sufficiently small, either $G^{-1}(\ell)$ is void or $G^{-1}(\ell)$ is a reduced curve. The image $\pi(G^{-1}(\ell))$ by π is the polar curve of L, the hyperplane with direction orthogonal to the conjugate of ℓ.

Notice that from this definition it is clear that $\Gamma = \emptyset$ iff $\pi^{-1}(0)$ is different from $\{0\} \times \mathbb{P}^n$.

We shall show that using polar curves we obtain:

1. Equisingularity theorems;
2. Lefschetz type theorems;
3. A way to study the geometric local monodromy of complex hypersurfaces.

We are not going to give any proofs.

2. EQUISINGULARITY THEOREMS

Using results of [6] or [6 bis] and [10] one can prove:

THEOREM (2.1) If the singular locus Σ of $H := \{f = 0\}$ has dimension one, and if for a Zariski open dense set of linear hyperplanes L the polar curve is void, then Σ is smooth and H satisfies "μ constant" along Σ.

DEF. (2.2) When a hypersurface H of $\mathbb{C}^{n+1}$ has a smooth singular locus Σ of dimension d we say that H satisfies "μ constant" along Σ if any (n+1-d) affine subspace L transverse to Σ at a point x defines a hypersurface $L \cap H$ having at x an isolated singular point with its Milnor number independent of x and L.

In [9] relations are obtained between the condition "H satisfies "μ constant" along Σ, when dim $\Sigma = 1$", and equisingularity.

3. LEFSCHETZ TYPE THEOREMS

We can use the polar curve to prove local Lefschetz theorems on open varieties. One finds in [7] the following theorem:

THEOREM (3.1) Let Ω be as in Lemma (1.2). Let $L \in \Omega$. Let $r > 0$ be small enough and $0 < \epsilon \ll r$. Then:

$$B_r \cap \{f=\epsilon\} \text{ has the homotopy type of } (B_r \cap \{f=\epsilon\} \cap L) \cup (e_1^n \cup \cdots \cup e_k^n),$$

where e_i^n are n-dimensional cells, B_r is the open ball of radius r centered at 0 and $k = (H.\Gamma)_0 = \dim \mathcal{O}_{\Gamma,0}/(\bar{f})$, where $\bar{f}$ is the image of f in $\mathcal{O}_{\Gamma,0}$, the local ring of the reduced curve Γ at 0.

REMARKS (3.2) The homotopy type of $B_r \cap \{f = \epsilon\}$ is a topological invariant of the hypersurface H at 0 when f is reduced, $r > 0$ sufficiently small and $0 < \epsilon \ll r$.

(3.3) In [7] the preceding theorem is stated with the assumption f is reduced at 0 but H. Hironaka pointed out to me that this isn't needed.

(3.4) The proof of this theorem is based on (1.6).

COROLLARY (3.5) The inclusion $(B_r - H) \cap L \longrightarrow B_r - H$ gives isomorphisms

$$\pi_i((B_r-H) \cap L, x) \xrightarrow{\sim} \pi_i(B_r-H, x) \quad \text{for } i \leq n-2$$

and an epimorphism:

$$\pi_{n-1}((B_r-H) \cap L, x) \longrightarrow \pi_{n-1}(B_r-H, x)$$

with $x \in (B_r-H) \cap L$.

The proof of this corollary was Theorem (3.1) and Milnor's fibration theorem (cf. [12]).

As a corollary we got a theorem stated by O. Zariski in [15] and proved by other methods in [14] and [3]:

COROLLARY (3.6) If H is a projective hypersurface in $\mathbb{P}^n$, there is a Zariski open dense set of hyperplanes L of $\mathbb{P}^n$ such that the inclusion $(\mathbb{P}^n - H) \cap L \longrightarrow \mathbb{P}^n - H$ gives an isomorphism:

$$\pi_1((\mathbb{P}^n - H) \cap L_x) \xrightarrow{\sim} \pi_1(\mathbb{P}^n - H, x)$$

with $x \in (\mathbb{P}^n - H) \cap L$, provided $n \geq 3$.

This corollary reduces the computation of fundamental groups of complements of projective hypersurfaces to that of complements of plane curves.

Using a finer analysis of the situation, H. Harmon and I have proved in [4] the following theorem:

THEOREM (3.7) Let $L \in \Omega$ as in (3.1) and L_t parallel to L, $0 \notin L_t$ and L_t sufficiently near to L. Then:

> $B_r - H$ has the homotopy type of $((B_r - H) \cap L_t) \cup (e_1^{n+1} \cup \cdots \cup e_s^{n+1})$ where e_j^{n+1} are (n+1)-dimensional cells and $s = m_0(\Gamma) =$ multiplicity of Γ at 0.

4. GEOMETRIC LOCAL MONODROMY

In [8] we recently proved:

THEOREM (4.1) Let $L \in \Omega$, r and ϵ as in (3.1). If $f(0) = 0$ and $df(0) = 0$, there is a diffeomorphism φ of $\{f = \epsilon\} \cap \bar{B}_r$ on itself such that $\varphi(\{f=\epsilon\} \cap \bar{B}_r \cap L) = \{f=\epsilon\} \cap \bar{B}_r \cap L$, φ has no fixed points and it induces the local monodromy on the homology and the cohomology of Milnor's fiber.

We may say that the geometric local monodromy of Milnor's fibration has no fixed points.

This theorem has the following corollary from N.A' Campo (cf. [1]):

COROLLARY (4.2) If $f(0) = 0$, $df(0) = 0$, the local monodromy of f has its Lefschetz number equal to 0. (Recall that the Lefschetz number of the local monodromy is the alternating sum of the traces of the endomorphisms defined on each dimension of the cohomology of Milnor's fiber).

The essential notion used for the proof of the theorem (4.1) is the Cerf diagram (compare with [5]):

DEF. (4.3) If Γ is the polar curve of L, $\phi(\Gamma)$ is the Cerf diagram of f relative to L (compare with [18]). Then:

LEMMA (4.4) If $f(0) = 0$, $df(0) = 0$, then either $\Gamma = \phi$, or $\phi(\Gamma)$ is a curve tangent to the x_0-axis at 0.

Moreover I. Iomdine [16] and J. Mather [17] proved:

LEMMA (4.5) There is a Zariski open subset Ω' of Ω such that ϕ induces a 1-1 mapping of Γ onto $\phi(\Gamma)$.

REMARKS (4.6) A'Campo's proof of (4.2) uses the resolution of singularities of H. Hironaka. The one of (4.1) does not use it. As A'Campo gets results (cf. [2]) on the Lefschetz number of iterated of the local monodromy, we expect to get them here.

(4.7) The non-transverse first Puiseux exponents of the components of Cerf diagram $\phi(\Gamma)$ are the $\dfrac{m_0(\Gamma_i)}{(\Gamma_i \cdot H)_0}$.

In his exposé B. Teissier has introduced them (in the case of isolated singularities as $\dfrac{m_i}{e_i + m_i}$ (cf. [13]).

(4.8) Following B. Teissier in the case of isolated singularities one may conjecture that the non-transverse first Puiseux exponents of the Cerf diagram are topological invariants of H at 0 (cf. [13]).

In connection with (3.1) and (3.2), because B. Teissier conjectured that the homotopy type of $B_r \cap \{f = \epsilon\} \cap L$ should be a topological invariant, a weaker conjecture is that $(\Gamma . H)_0$ is a topological invariant.

(4.9) These results have been obtained by M. Merle in the case of analytic irreducible plane curves (cf. [11]).

(4.10) General results about the topology of the Cerf diagram haven't been obtained yet. Nevertheless it has been used to prove the quasi-unipotence of local monodromy (cf. [5]).

The proof of A. Landman of the quasi-unipotence of the local monodromy uses the resolution of singularity. The polar curves and the Cerf diagrams have been introduced by R. Thom in an unpublished paper about the relative local monodromy in the case of hypersurfaces with isolated singularities. One should find a topological proof of the quasi-unipotence of the local monodromy without using the resolution of singularity.

BIBLIOGRAPHY

[1] N. A'Campo, Le nombre de Lefschetz d'une monodromie, Ind. Mat., Proc. Kon. Ned. Akad. Wet., serie A, 76, 113-118 (1973).

[2] N. A'Campo, La fonction Zeta de la monodromie, Preprint de l'IHES, 1974.

[3] D. Cheniot, Sur le groupe fondamental du complementaire d'une hypersurface complexe, Compos. Mat., 1973.

[4] H. Hamm-Lê Dũng Tráng, un théorème de Zariski du type de Lefschetz, Ann. Sc. de l'Ecole Norm. Sup., t. 6, 1973, 317-366.

[5] A. Landman, On the Picard-Lefschetz transformation for algebraic manifold acquiring general singularities, Trans. Amer. Math. Soc., Vol. 181, 1973, 89-126.

[6] F. Lazzeri, On the local monodromy, Singularités a Cargèse, Asterisque, 1974.

[6]bis Lê Dũng Tráng, Une application d'un théorème d'A'Campo à l'équisingularité, Ind. Mat., Proc. Kon. Ned. Akad. Wet., Serie A, (1973).

[7] Lê Dũng Tráng, Calcul des cycles evanouissants des hypersurfaces complexes, Ann. Inst. Fourier, 1973, Vol. 4.

[8] Lê Dũng Tráng, La monodromie n'a pas de points fixes, preprint du Centre de Mathématique de l'Ecole Polytechnique, 1974.

[9] Lê Dũng Tráng-C.P. Ramanujam, The invariance of Milnor's number implies the invariance of the topological type, to be published in Amer. J. Math.

[10] Lê Dũng Tráng-K. Saito, La constance du nombre de Milnor donne de bonnes stratifications, C.R.Acad. Sc., Serie A, 1973.

[11] M. Merle, Ideal Jacobien, courbes polaires et équisingularité, Preprint, Ecole Polytechnique, Centre de Mathematiques, 1974.

[12] J. Milnor, Singular points of complex hypersurfaces, Ann. Math. Stud. 61, Princeton.

[13] B. Teissier, Variétés polaires des hypersurfaces et invariants topologiques, to be published.

[14] A.N. Varchenko, Theorems on the topological equisingularity of families of algebraic varieties and families of polynomial mappings, Isv. Akad. Naouk, t. 36, 1972.

[15] O. Zariski, On the Poincare group of a projective hypersurface, Ann. of Math., vol. 38, 1937, 131-141.

[16] I. Iomdine, unpublished.

[17] J. Milnor, Singular points of complex hypersurfaces, Ann. Math., 1973, Vol. 98, 226-245.

[18] J. Cerf, Topologie de certains espaces de plongements, Bull. Soc. Math. Fr., Vol. 89, 1961, 227-380.

CENTRE DE MATHEMATIQUES
DE L'ECOLE POLYTECHNIQUE
17, RUE DESCARTES

75230 PARIS CEDEX 05 - FRANCE

HARVARD UNIVERSITY
DPT. OF MATHS. - SCIENCE CENTER
1, OXFORD STREET

CAMBRIDGE (MASS.) 02138 - U.S.A.

Proceedings of Symposia in Pure Mathematics
Volume 29, 1975

MATSUSAKA'S BIG THEOREM

D. Lieberman and D. Mumford

§1. INTRODUCTION

The goal of this note is to present an outline of Matsusaka's proof [4],[5] of the following Theorem:

THEOREM 1. Let $P(k)$ be a rational polynomial with integral values for all integers k. Then there is a k_o such that for every non-singular complex projective variety V, and every ample line bundle L on V with

$$\chi(V, L^{\otimes k}) = P(k),$$

then $L^{\otimes k}$ is very ample if $k \geq k_o$.

The proof can be divided into 2 parts. The more difficult part consists in the proof of Theorem 2 below, and a somewhat easier but still subtle part consists in checking that Theorem 2 implies Theorem 1.

THEOREM 2. Given constants $\epsilon > 0$, $\gamma, k_o, n \in \mathbb{Z}$ and $t \in \mathbb{Q}$, there is a $k_1 = k_1(\epsilon, \gamma, k_o, t, n)$ for which the following holds: Let V be any normal projective variety of dimension n over any algebraically closed field k; let C be an ample divisor on V and let D be a codimension 1 cycle on V; assume $\gamma = (C^n)$, and

$$t = \frac{(D \cdot C^{n-1})}{(C^n)}.$$

Assume

$$\dim H^o(\mathcal{O}_V(kD)) \geq \frac{(\frac{1}{2}+\epsilon)}{n!} t^n((kC)^n), \text{ for all } k \geq k_o.$$

Then for every $k \geq k_1$, one can find a subspace

$$\Lambda \subset H^o(\mathcal{O}_V(kD))$$

such that the induced rational map

$$\phi_\Lambda: V \longrightarrow \mathbb{P}^N$$

AMS (MOS) subject classifications (1970). Primary 14D20.

is birational and does not blow down any codimension one subvarieties. Moreover:

$$\deg \phi_\Lambda(V) \leq \gamma k^n t^n.$$

Many special cases of Theorem 1 are well known. If V is a curve of genus g, then L is ample if and only if $\deg L \geq 1$, and it is well known that in this case $L^{\otimes k}$ is very ample if $k \geq (2g+1)/\deg L$. If V is an abelian variety, L ample, then $L^{\otimes k}$ is very ample if $k \geq 3$ (Lefschetz: cf. Mumford [8], §17). If V is a K3-surface, L ample, then again $L^{\otimes k}$ is very ample if $k \geq 3$ (Mayer [7], Saint-Donat [9]). If V is a normal surface of general type with its rational curves E with $(E^2) = -2$ blown down, and $L = \mathcal{O}_V(K)$, then $L^{\otimes k}$ is very ample if $k \geq 5$ (Enriques, Kodaira [3], Bombieri [2]). For arbitrary surfaces and ample L's in any characteristic, the Theorem was proven in Matsusaka-Mumford [6].

Once Theorem 1 is established, one may apply the theory of Hilbert schemes or of Chow varieties to conclude that the set of polarized varieties (V,L) with given Hilbert polynomial $P(k)$ may all be parametrized by a quasi-projective scheme. (In particular this family contains all deformations of the polarized variety (V,L) because the Hilbert polynomial is invariant under deformation.)

Matsusaka's proof of Theorem 1 is non-cohomological, unlike for instance Bombieri's approach to canonically polarized surfaces. Theorem 1 would follow immediately if, for instance, one could solve directly the following:

PROBLEM: Given $P(k)$, find k_o such that for every (V,L) with Hilbert polynomial $P(k)$ and every $x,y \in V$,

$$H^i(V, \mathfrak{m}_x \cdot \mathfrak{m}_y \cdot L^{\otimes k}) = (0), \quad \text{all } i \geq 1, k \geq k_o.$$

Conversely, Matsusaka's result implies that such a k_o exists because it implies that the quadruples (V,L,x,y) form a "bounded family".

We want to add a word about the completeness of our presentation of Matsusaka's proof. We believe the careful reader can reconstruct the whole proof from what we say. However in some places we have not written out fully various details. In particular, a more complete version would include a whole section working out the elementary properties of Matsusaka's operation $\Lambda^{[j]}$ (cf. §2 below): instead we simply introduce these without proof where they are needed.

§ 2. PROOF OF THEOREM 2

(I.): The first step is to find a k_2 depending only on γ, k_o, t such that for all $k \geq k_2$, the rational map

$$\phi_{kD}: V \longrightarrow \mathbb{P}(H^o(\mathcal{O}_V(kD)))$$

satisfies $\dim \phi_{kD}(V) = n$. We shall in fact prove:

LEMMA 2.1. If $\Lambda \subset H^o(\mathcal{O}_V(kD))$ and

$$\dim \Lambda > \max_{1 \leq i \leq n-1} (i + \gamma t^i k^i)$$

then $\dim \phi_\Lambda(V) = n$.

Because of our assumed lower bound on $h^o(kD)$ one gets immediately a $k_2(\gamma, k_o, t)$ such that $\Lambda = H^o(\mathcal{O}_V(kD))$ satisfies this for $k \geq k_2$.

To prove the lemma, let $W = \phi_\Lambda(V)$. We show in fact that if $\dim W = i$, then

$$\dim \Lambda \leq i + \gamma t^i k^i.$$

Firstly, recall the well known fact (cf. [6], Th. 3) that for any projective variety $X \subset \mathbb{P}^n$, X not in any hyperplane:

$$n+1 \leq \deg X + \dim X.$$

In particular, if $X = W$, $\mathbb{P}^n = \mathbb{P}(\Lambda)$, then

$$\dim \Lambda \leq \deg W + i,$$

so it suffices to prove

$$\deg W \leq \gamma t^i k^i.$$

To transform the inequality on deg W into an estimate on V itself, Matsusaka introduces an interesting new concept of the <u>variable</u> j-<u>fold intersection cycle</u> $\Lambda^{[j]}$ of a linear system $\Lambda \subset \Gamma(V,M)$. This is the codimension j cycle, defined only up to rational equivalence on V, obtained in either of the following ways: let $B \subset V$ be the base points of the linear system Λ so that Λ defines a morphism

$$\phi_\Lambda: \ V-B \longrightarrow \mathbb{P}(\Lambda).$$

Take the closure in V of $\phi_\Lambda^{-1}(L)$, $L \subset \mathbb{P}(\Lambda)$ a general codimension j linear space; or take the closure in V of the intersection cycle $V(s_1). \cdots .V(s_j)$ in V-B, where the s_i are j general element of

Λ. If $\Lambda = \Gamma(V, \mathcal{O}_V(E))$ for some divisor E, write $(E)^{[j]}$ for $\Lambda^{[j]}$. Note that if $\Lambda_1 \subset \Lambda_2$ are 2 linear systems, then

$$\Lambda_2^{[i]} \underset{\text{rat.eq.}}{\sim} \Lambda_1^{[i]} + \text{eff. cycle.}$$

With this concept, we find:

$$\begin{aligned} \deg W &\leq \# \text{ of components of } \Lambda^{[i]} \\ &\leq (C^{n-i}.\Lambda^{[i]}) \\ &\leq (C^{n-i}.(kD)^{[i]}) , \end{aligned}$$

hence Lemma 2.1 follows by taking E = kD in the following key result:

PROPOSITION 2.2: Let V be a normal projective variety of dimension n, C an ample divisor and E a codimension one cycle on V such that $\dim \phi_{kE}(V) \geq i$ for $k \gg 0$. Then

$$(E^{[i]}.C^{n-i}) \leq (C^n)\cdot\left(\frac{(C^{n-1}\cdot E)}{(C^n)}\right)^i$$

Proof: Replacing C by ℓC multiplies both sides by ℓ^{n-i} so we may assume C very ample. Let V' be a general intersection of n-i divisors $C_1,\ldots,C_{k-i} \in |C|$, let C' = V'.C and let E' = V'.E. Then one sees easily that

$$\begin{aligned} (E^{[i]}.C^{n-i})_V &= (E')^{[i]}_{V'} \\ (C^n)_V &= (C'^i)_{V'} \\ (C^{n-1}.E)_V &= (C'^{i-1}.E')_{V'} , \end{aligned}$$

hence replacing V by V', we may assume i = n. Now $\dim \phi_{kE}(V) = n$ for $k \gg 0$, hence in fact ϕ_{kE} is birational for $k \gg 0$ [i.e., if $W_k = \phi_{kE}(V) \subset \mathbb{P}_{N_k}$ and if $\pi: W_k' \longrightarrow W_k$ is the normalization of W_k in the field k(V), then $\pi^*(\mathcal{O}_{W_k}(1))$ is ample on W_k' and

$$\Gamma(W_k', \pi^*(\mathcal{O}_{W_k}(\ell)) \subset \Gamma(V, \mathcal{O}(k\ell E)).]$$

We may also replace E by kE to prove the Proposition because:

$$\begin{aligned} (kE)^{[n]} &\geq k^n(E)^{[n]} \\ (C^{n-1}.kE)^n &= k^n(C^{n-1}.E)^n \end{aligned}$$

(because the base locus of $|kE|$ is contained in the base locus of $|E|$, and the variable intersection of n general divisors in $|kE|$ specializes to k^n times the variable intersection of n general divisors in $|E|$ plus some components in the base locus of $|E|$.) So we may assume ϕ_E is birational.

Now let $W = \phi_E(V)$. Since ϕ_E is birational, $\deg W = (E^{[n]})$. Moreover, if $k >> 0$:

$$\begin{aligned} h^0(V,\mathcal{O}_V(kE)) &\geq h^0(W,\mathcal{O}_W(k)) \\ &= \deg W\cdot\frac{k^n}{n!} + \text{lower terms} \\ &= E^{[n]}\cdot\frac{k^n}{n!} + \text{lower terms.} \end{aligned}$$

The Proposition now follows from considering the upper bound on $h^0(\mathcal{O}_V(kE))$ as $k \longrightarrow \infty$, which is given by

PROPOSITION 2.3 ("Q-estimate"): Let V be a projective variety of dimension n, let C be a hyperplane section of V and let $\mathfrak{I}$ be a torsion-free rank 1 sheaf on V. Then

$$h^0(\mathfrak{I}) \leq \binom{[t]+n}{n}\gamma + \binom{[t]+n-1}{n-1}$$

where

$$t = \deg \mathfrak{I}/\deg V \qquad \gamma = \deg V .$$

(degree measured via C as in Kleiman, Annals, 1966)

<u>Proof</u>: For n = 1, the inequality reads

$$h^0(\mathfrak{I}) \leq ([t]+1)\gamma + 1$$

which follows from the Riemann-Roch estimate:

$$h^0(\mathfrak{I}) \leq \deg \mathfrak{I} + 1 = t\gamma + 1.$$

We proceed by induction, assuming the result true on a general hyperplane section C. First we need to find a hyperplane C such that C is again a variety and $\mathfrak{I}\otimes\mathcal{O}_C$ is still torsion-free. Indeed almost all C's are varieties (Seidenberg's Theorem) and for $\mathfrak{I}\otimes\mathcal{O}_C$ to be torsion-free, it suffices to make sure $\operatorname{depth}_{\mathcal{O}_x}(\mathfrak{I}_x) \geq 2$ for all $x \in C$ except the generic point of C. Since there are only finitely many $x \in V$ with $\operatorname{codim}_V\{\overline{x}\} \geq 2$ and $\operatorname{depth}_{\mathcal{O}_x}(\mathfrak{I}_x) = 1$ (cf. for instance EGA, Ch. 4, §10.8), this is possible. Then one has the exact sequences:

$$0 \longrightarrow \mathfrak{J}(-k-1) \longrightarrow \mathfrak{J}(-k) \longrightarrow \mathfrak{J}\otimes\mathcal{O}_C(-k) \longrightarrow 0,$$

hence

$$h^0(\mathfrak{J}(-k)) \leq h^0(\mathfrak{J}(-k-1)) + h^0(\mathfrak{J}\otimes\mathcal{O}_C(-k)).$$

But $h^0(\mathfrak{J}(-k)) \neq 0$ implies there is a homomorphism

$$0 \longrightarrow \mathcal{O}_V(kC) \longrightarrow \mathfrak{J}$$

hence $\deg \mathfrak{J} \geq k \deg V$, hence $[t] \geq k$. Thus

$$h^0(\mathfrak{J}) \leq \sum_{k=0}^{[t]} h^0(\mathfrak{J}\otimes\mathcal{O}_C(-k)).$$

Using the estimate on C, we get:

$$h^0(\mathfrak{J}) \leq \sum_{k=0}^{[t]} \binom{[t]-k+n-1}{n-1}\gamma + \binom{[t]-k+n-2}{n-2}$$

$$= \binom{[t]+n}{n}\gamma + \binom{[t]+n-1}{n-1} \qquad \text{QED}$$

(II.): This completes the first step: if $W_k = \phi_{kD}(V)$, we have a k_2 such that if $k \geq k_2$, $\dim W_k = n$. The second step is to find a k_3 also depending only on ϵ, γ, k_0, t such that if $k \geq k_3$, then ϕ_{kD} is <u>birational</u>. We will in fact produce an ℓ_0 such that $\phi_{\ell_0 k_2 D}$ is birational. Note that since $k \geq k_0$ implies kD is effective, then for $k \geq k_3 = \ell_0 k_2 + k_0$, $\Gamma(\mathcal{O}(kD)) \supseteq \Gamma(\mathcal{O}(\ell_0 k_2 D))$, hence ϕ_{kD} is also birational.

To produce ℓ_0, consider for each $k \geq k_2$, $\ell > 1$, the diagram of rational maps:

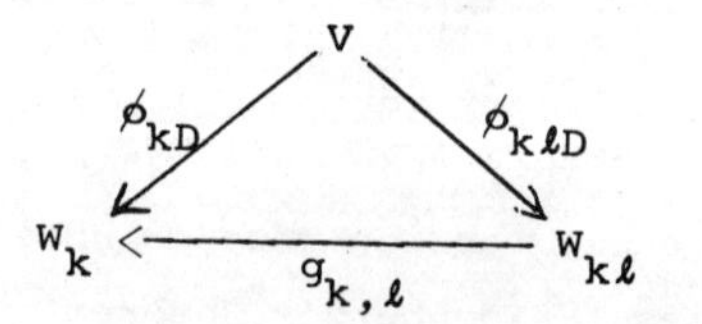

Note that $\deg(\phi_{k\,D}) = \deg(\phi_{k\ell D}) \cdot \deg(g_{k,\ell})$.

LEMMA 2.4: There is an integer ℓ (depending on ϵ, n and t) such that if $\deg(\phi_{kD}) > 1$ and $\deg(g_{k,\ell}) = 1$, then one must have

$$\frac{(k\ell D)^{[n]}}{k^n \ell^n} > (1+\epsilon)^{1/n} \cdot \frac{(kD)^{[n]}}{k^n} .$$

<u>Proof</u>: Choose ℓ such that

$$\binom{\ell(1+\epsilon)^{1/n}+n+1}{n} + \frac{2}{t^n}\binom{\ell(1+\epsilon)^{1/n}+n}{n-1} < \frac{1+2\epsilon}{n!}\ell^n.$$

This is possible because for $\ell >> 0$, the left hand side grows like $(1+\epsilon)\ell^n/n!$. Now W_k and $W_{k\ell}$ both have explicit projective embeddings:

$$W_k \subset \mathbb{P}(H^0(\mathcal{O}_V(kD)))$$

$$W_{k\ell} \subset \mathbb{P}(H^0(\mathcal{O}_V(k\ell D))).$$

Since $g_{k,\ell}$ is birational by assumption, let $U \subset W_k$ be the domain of definition of $g^{-1}_{k,\ell}$. Then the morphism

$$U \longrightarrow W_{k\ell} \subset \mathbb{P}(H^0(\mathcal{O}_V(k\ell\, D)))$$

is defined by an invertible sheaf $\mathfrak{F}'$ on U and $h^0(k\ell D)$ sections s_i' of $\mathfrak{F}'$ generating $\mathfrak{F}'$. There is a unique torsion free sheaf $\mathfrak{F}$ on W_k plus sections s_i generating $\mathfrak{F}$, which restrict on U to $\{\mathfrak{F}', s_i'\}$. Thus $h^0(\mathfrak{F}) \geq h^0(k\ell D)$. On the other hand, for the given projective embedding of W_k, we calculate:

$\deg \mathfrak{F}$ = # of intersections on V outside base locus of $|kD|$ of (n-1) general sections of $\mathcal{O}(kD)$, one section of $\mathcal{O}(k\ell D)$,

hence:

$$\ell^{n-1}.\deg \mathfrak{F} \leq (k\ell D)^{[n]} .$$

Now combine the assumed lower bound on $h^0(k\ell D)$, and the upper bound on $h^0(\mathfrak{F})$ given by the Q-estimate to get:

$$\frac{\frac{1}{2}+\epsilon}{n!} t^n (k\ell)^n \gamma \leq h^o(k\ell D)$$
$$\leq h^o(\mathfrak{F})$$
$$\leq \binom{[\frac{\deg \mathfrak{F}}{\deg W_k}]+n}{n} \deg W_k + \binom{[\frac{\deg \mathfrak{F}}{\deg W_k}]+n-1}{n-1} .$$

Moreover:

$$\deg W_k = (kF)^{[n]}/\deg(\phi_{kD})$$
$$\leq \frac{t^n k^n \gamma}{2} \quad \text{by Prop. 2.2}$$

and

$$[\frac{\deg \mathfrak{F}}{\deg W_k}] \leq \frac{(k\ell D)^{[n]}}{\ell^{n-1}(kD)^{[n]}} + 1$$
$$= \ell R + 1$$

if

$$R = \frac{(k\ell D)^{[n]}}{\ell^n (kD)^{[n]}} .$$

Hence

$$\frac{\frac{1}{2}+\epsilon}{n!} t^n k^n \ell^n \gamma \leq \binom{\ell R+n+1}{n} \frac{t^n k^n \gamma}{2} + \binom{\ell R+n}{n-1}$$

hence

$$\frac{1+2\epsilon}{n!} \ell^n \leq \binom{\ell R+n+1}{n} + \frac{2}{t^n} \binom{\ell R+n}{n-1} .$$

If $R \leq (1+\epsilon)^{1/n}$, this contradicts the inequality that ℓ was chosen to satisfy. Thus $R > (1+\epsilon)^{1/n}$. QED

However, for all k,

$$\frac{(kD)^{[n]}}{k^n} \leq \gamma t^n$$

by Prop. (2.2). Hence starting at any ϕ_{kD}, we see that:

$$\deg \phi_{(k\ell^e D)} < \deg \phi_{kD} \quad \text{if} \quad e > \frac{n \log(\gamma t^n k^n)}{\log(1+\epsilon)} .$$

Since we know that ϕ_{k_2D} is finite-to-one and

$$\deg \phi_{k_2D} \leq (k_2D)^{[n]} \leq \gamma t^n k_2^n,$$

it follows that $\phi_{(k_2 \ell e)D}$ is birational if

$$e > n\gamma t^n k_2^n \frac{\log(\gamma t^n k^n)}{\log(1+\epsilon)} .$$

(III): This completes the second step: we have a k_3 such that if $k \geq k_3$, ϕ_{kD} is birational. The third step is to find a k_4 also depending only on ϵ, γ, k_o, t such that if $k \geq k_4$ then there is a $\Lambda \subset |kD|$ such that ϕ_Λ is birational and does not blow down any codimension 1 subvarieties of V. We will in fact only produce a k_5 such that if $k \geq k_5$, then there is a $\Lambda' \subset |kD|$ such that $\dim \phi_{\Lambda'}(V) = n$ and $\phi_{\Lambda'}$ does not blow down any codimension 1 subvarieties. Setting $k_4 = k_3+k_5$ and Λ = minimal sum of $|k_3D|$ and Λ', we get the Λ with <u>all</u> the properties. The proof is very similar to the beautifully simple <u>Method of Albanese</u> by which for any projective n-dimensional variety X one constructs a $\Lambda \subset H^o(\mathcal{O}_X(k))$ such that $\phi_\Lambda(X)$ is birational to X and has no points of multiplicity $> n!$ (cf. [1],(12.4.4)) We show in fact:

LEMMA (2.5). Choose k such that

$$tk \geq 2n \cdot n!(1+\tfrac{1}{\gamma}) \quad \text{and} \quad k \geq k_o.$$

Then there is a $\Lambda \subset H^o(\mathcal{O}_V(kD))$ such that $\dim \phi_\Lambda(V) = n$ and ϕ_Λ does not blow down any codimension 1 subvarieties.

<u>Proof</u>: Note by the assumption on k,

$$n\gamma(tk)^{n-1}+n \leq \gamma(tk)^{n-1}(n+\tfrac{n}{\gamma}) \leq \frac{\gamma(tk)^n}{2n!} < h^o(kD).$$

Also $tk \geq 1$, so by Lemma (2.1) any $\Lambda \subset H^o(kD)$ for which

$$\dim \Lambda \geq n + \gamma(tk)^{n-1}$$

has the property $\dim \phi_\Lambda(V) = n$. So any Λ such that

$$\operatorname{codim} \Lambda \leq (n-1)\gamma(tk)^{n-1}$$

has this property too. What we will do is this: starting

---

* See also Lecture 1, §5 of Lipman's article, " Introduction to resolution of singularities", in this volume.

with $\Lambda_o = H^o(\mathcal{O}_V(kD))$, choose a sequence of subspaces

$$\Lambda_o \supset \Lambda_1 \supset \Lambda_2 \supset \cdots .$$

with $\dim \Lambda_i/\Lambda_{i+1} = 1$, until we reach the desired Λ. In fact, say Λ_r is chosen but there is still an $E \subset V$, $\dim E = n-1$, such that

$$\dim \phi_{\Lambda_r}(E) = i-1, \qquad n-1 \geq i \geq 1.$$

If s is the multiplicity to which E occurs as a fixed component of Λ_r, let $\Lambda_r' \subset H^o(\mathcal{O}_V(kD-sE))$ be the linear system such that $\Lambda_r = (e^{\otimes s})\otimes\Lambda_r'$, e = canonical section of $\mathcal{O}_V(E)$. Let x be a general point of E. Define

$$\Lambda_{r+1}' = \{s \in \Lambda_r' \mid s(x) = o\},$$

$$\Lambda_{r+1} = (e^{\otimes s}) \otimes \Lambda_{r+1}' .$$

Note that ϕ_{Λ_r} is defined at x, so define

$$Z = \text{closure of } \phi_{\Lambda_r}^{-1}(\phi_{\Lambda_r}(x)).$$

Then $\dim Z = n-i$ if x is sufficiently general, and if $s \in \Lambda_r'$ vanishes at x, it vanishes on all of Z. Thus

$$(\Lambda_r')^{[i]} \underset{\text{rat.eq.}}{\sim} (\Lambda_{r+1}')^{[i]} + Z + \text{eff. cycle}.$$

But $\Lambda_r^{[i]} = (\Lambda_r')^{[i]}$ and $\Lambda_{r+1}^{[i]} = (\Lambda_{r+1}')^{[i]}$, so it follows

$$(\Lambda_r^{[i]}.c^{n-i}) > (\Lambda_{r+1}^{[i]}.c^{n-i}).$$

Since for each j,

$$(\Lambda_r^{[j]}.c^{n-j}) \geq (\Lambda_{r+1}^{[j]}.c^{n-j}),$$

it follows that the invariant

$$\delta(\Lambda) = \sum_{j=1}^{n-1} (\Lambda^{[j]}.c^{n-j})$$

decreases when you pass from Λ_r to Λ_{r+1}. But by Prop. (2.2)

$$\delta(\Lambda) \leq \delta(kD) \leq \sum_{j=1}^{n-1} \gamma(kt)^j \leq (n-1)\gamma(kt)^{n-1}.$$

Since we have "this much room" in $H^o(\mathcal{O}(kD))$, we can find a Λ for which no E is blown down. QED

§3. TH. 2 ⟹ TH. 1

This is the part of the proof that involves char. 0 because we want to apply Kodaira's Vanishing Theorem. The idea is to apply Theorem 2 to V with C,D chosen so that

$$L = \mathcal{O}_V(C)$$

$$L^{m_o} \otimes \Omega_V^n = \mathcal{O}_V(D).$$

Here m_o will be chosen below depending only on P so as to make Theorem 2 apply. Note that for $m > 0$ by Kodaira Vanishing and Serre duality:

$$(*) \qquad \begin{aligned} \dim H^o(L^m \otimes \Omega_V^n) &= \chi(L^m \otimes \Omega_V^n) \\ &= (-1)^n \chi(L^{-m}) \\ &= (-1)^n P(-m): \text{ call this } P'(m). \end{aligned}$$

Moreover, by Riemann-Roch,

$$P(k) = (C^n)\frac{k^n}{n!} - \frac{(K_V.C^{n-1})}{2(n-1)!} k^{n-1} + \text{lower degree terms}$$

hence P determines the integer $\gamma = (C^n)$ and, once m_o is chosen, P determines $(D.C^{n-1})$ and hence $(D.C^{n-1})/(C^n)$ too. Finally, we need a lower bound for $\dim H^o(\mathcal{O}_V(kD))$ of the type used in Th. 2. This is obtained as follows -

a) Say $P'(m_1) > 0$, so that by (*) in divisor notation m_1C+K is an effective divisor. Then:

$$\begin{aligned} kD &= k(m_oC + K) \\ &= (k-1)(m_1C+K) + ((k(m_o-m_1)+m_1)C + K). \end{aligned}$$

The first term is an effective divisor, so

$$\begin{aligned}\dim H^{o}(\mathcal{O}_V(kD)) &\geq \dim H^{o}(\mathcal{O}_V((k(m_o-m_1)+m_1)C+K)) \\ &= P'(k(m_o-m_1)+m_1) \\ &= \frac{(C^n)}{n!}(k(m_o-m_1)+m_1)^n + \text{lower degree terms in } k \\ &= \frac{((kC)^n)}{n!}(m_o-m_1)^n + \text{lower degree terms in } k.\end{aligned}$$

b) But

$$\begin{aligned} t \underset{\text{def}}{=} \frac{(D.C^{n-1})}{(C^n)} &= \frac{((m_oC+K).C^{n-1})}{(C^n)} \\ &= m_o + \frac{(K.C^{n-1})}{(C^n)}, \end{aligned}$$

i.e.,

$$\dim H^{o}(\mathcal{O}_V(kD)) \geq \frac{((kC)^n)}{n!}\left[t-m_1-\frac{(K.C^{n-1})}{(C^n)}\right]^n + \text{lower degree terms in } k.$$

If m_o and hence t is large enough, the term $[\ ]^n$ is at least $\frac{3}{4}t^n$ and then for k_o large enough, we certainly obtain:

$$\dim \Gamma(\mathcal{O}_V(kD)) \geq \frac{((kC)^n)}{n!}\cdot\frac{5}{8}t^n, \quad \text{if } k \geq k_o.$$

Thus Theorem 2 applies for some m_o and k_o readily computed in terms of the polynomial P alone. Thus we can find k, so that for every (V,L)

$$\begin{aligned}\exists\ \Lambda \subset \Gamma(V,\mathcal{O}_V(k_1D)) &= \Gamma(V,\mathcal{O}_V(k_1(m_oC+K))) \\ &= \Gamma(V,L^{k_1m_o}\otimes(\Omega_V^n)^{k_1})\end{aligned}$$

for which ϕ_Λ is birational and does not blow down any divisors - we abbreviate this to "Λ is quasi-ample".

Now let's analyze the projective variety $U = \phi_\Lambda(V)$. By Prop. (2.2) we know:

$$(**) \qquad \deg U \leq \gamma k_1^n t^n.$$

Automatically then, the ambient space $\mathbb{P}(\Lambda)$ has its dimension bounded as follows:

$$\dim \mathbb{P}(\Lambda) \leq \deg U + n - 1 \underset{\text{def}}{=} N.$$

It follows that the set of varieties U lies in a bounded family when

(V,L) varies over all pairs with Hilbert polynomial P! This is the key point, from which we want to argue backwards, obtaining eventually the boundedness of the set of pairs (V,L). From this point on, we leave the area in which we can make <u>explicit</u> estimates, and rely on general results asserting that various numbers are bounded when calculated for some set of varieties and divisors in a bounded family. The first point is that if U_{nor} is the normalization of U, then there is a k_2 such that for all U with degree bounded by (**), the pullback of $\mathcal{O}_U(k_2)$ to U_{nor} is very ample. It follows that if we choose a suitable $\Lambda' \subset \Gamma(V,\mathcal{O}_V(k_1k_2D))$, then $\phi_{\Lambda'}(V) \cong U_{nor}$. Replacing k_1 by k_1k_2 and Λ by Λ', this means we may assume that U is always normal. Call these Λ "normally quasi-ample". In that case, working with "Weil"-divisors on U, i.e., cycles of codimension 1,we may define the total transform $\phi_\Lambda(E)$ for every divisor E on V; and because ϕ_Λ does not contract any divisors, this defines an injection of the groups of Weil-divisors

$$\mathrm{Div}(V) \xhookrightarrow[\phi_\Lambda]{} \mathrm{Div}(U)$$

such that

a) $\phi_\Lambda(E)$ eff. $\Longleftrightarrow$ E eff.

b) $\phi_\Lambda((f)_V) = (f)_U$.

Thus ϕ_Λ^* sets up an isomorphism between

$$\Gamma(V,\mathcal{O}_V(E)) \xrightarrow{\approx} \Gamma(U,\mathcal{O}_U(\phi_\Lambda E)).$$

Moreover, if $U_o \subset U$ is the maximal open set such that

$$\phi_\Lambda^{-1}: \ U_o \longrightarrow V$$

is a morphism, then codim $U-U_o \geq 2$. ϕ_Λ^{-1} then defines an injection:

$$(\phi_\Lambda^{-1})^*: \ \Omega_V^n \longrightarrow \Omega_U^n|_{U_o} .$$

This implies that

$$\phi_\Lambda(K_V) = K_U + (\text{eff. divisor}).$$

It follows that if $E \in |\ell C+K_V|$, then

$$\deg \phi_\Lambda(E) = \ell \deg \phi_\Lambda(C) + \deg \phi_\Lambda(K_V)$$
$$\leq \ell(C.\Lambda^{[n-1]}) + \deg K_U$$
$$\leq \gamma k_1^{n-1} t^{n-1} \ell + \deg K_U.$$

Of course, $\deg K_U$ is bounded when U varies over all U's with degree bounded by (**): call this bound κ.

We can now reveal the diagram on which the rest of the proof is based. We consider 3 sets, related by 2 maps, as follows:

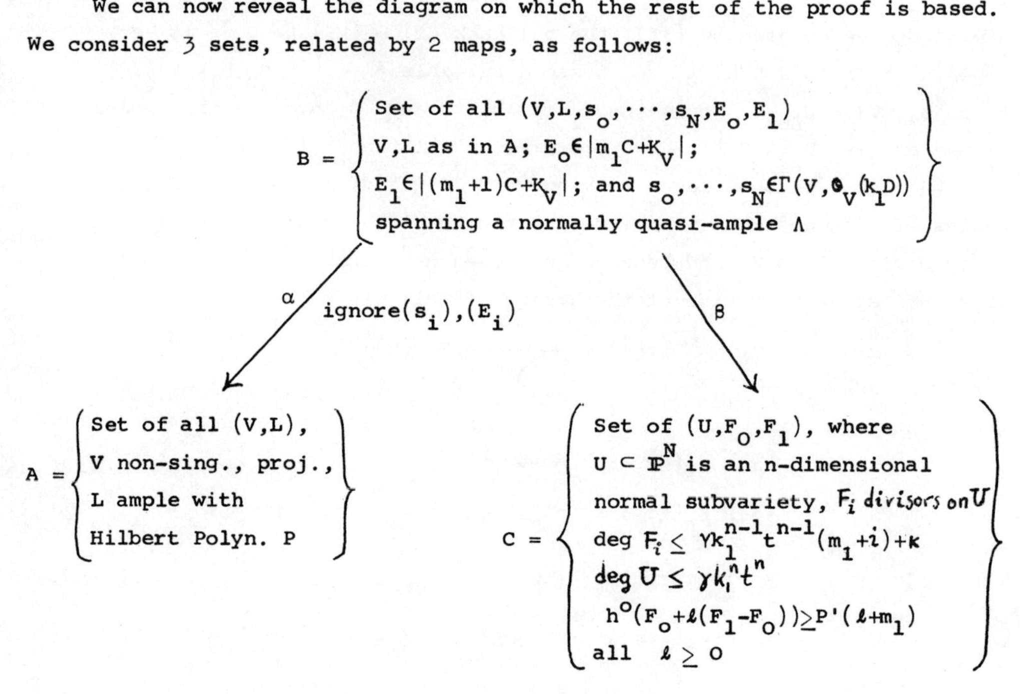

B =
Set of all (V,L,s_0,···,s_N,E_0,E_1)
V,L as in A; E_0∈|m_1C+K_V|;
E_1∈|(m_1+1)C+K_V|; and s_0,···,s_N∈Γ(V,O_V(k_1D))
spanning a normally quasi-ample Λ
α
ignore(s_i),(E_i)
β
A =
Set of all (V,L),
V non-sing., proj.,
L ample with
Hilbert Polyn. P
C =
Set of (U,F_0,F_1), where
U ⊂ P^N is an n-dimensional
normal subvariety, F_i divisors on U
deg F_i ≤ γk_1^{n-1}t^{n-1}(m_1+i)+κ
deg U ≤ γk_1^n t^n
h^0(F_0+ℓ(F_1-F_0))≥P'(ℓ+m_1)
all ℓ ≥ 0

Here m_1, k_1, N, γ, and t are chosen as above, in particular so that α is surjective. β is defined by

$$U = \phi_\Lambda(V)$$
$$F_i = \phi_\Lambda(E_i).$$

Note that this is OK because

$$H^0(U, \mathcal{O}_U(F_0+\ell(F_1-F_0)) \cong H^0(U, \mathcal{O}_U(\phi_\Lambda(E_0+\ell(E_1-E_0))))$$
$$\cong H^0(V, \mathcal{O}_V((m_1+\ell)C+K_V))$$

which has dimension exactly $P'(m_1+\ell)$. Also, set C is isomorphic to the set of points on a <u>locally closed subset of a union of 3-way products of Chow Varieties</u>, i.e., each U, F_1, F_2 has a Chow form, normal is an open condition and $h^0(\cdots) \geq c$ are all closed conditions. Thus C has a natural structure of a (reducible) variety.

LEMMA 3.1: β is injective.

Proof: In fact, to recover V from (U,F_1,F_2), let $\psi_\ell: U \longrightarrow \mathbb{P}^M$ be the rational map defined by $H^o(U,\mathcal{O}_U(\ell(F_1-F_o)))$. Then if $\ell >> o$, $V = \psi_\ell(U)$ (using the fact that

$$H^o(U,\mathcal{O}_U(\ell(F_1-F_o))) \cong H^o(V,L^{\otimes \ell})$$

and that L is ample on V). Moreover $\phi_\Lambda = \psi_\ell^{-1}$, L is the line bundle associated to $\psi_\ell(F_1-F_o)$, $E_i = \psi_\ell(F_i)$, and $(s_o,\cdots,s_N)$ are the sections of $\mathcal{O}_V(k_1D)$ corresponding to the canonical sections $(X_o,\cdots,X_N)$ of $\mathcal{O}_U(1)$ via

$$\phi_\Lambda^*: H^o(\mathcal{O}_U(1)) \longrightarrow H^o(\mathcal{O}_V(k_1D)).$$ QED

LEMMA 3.2: The image of β is Zariski-open.

Proof: It is elementary to see that the image of β is a countable union of locally closed subsets of the set C. Therefore it is enough to show that for any valuation ring R and morphism ϕ: Spec R $\longrightarrow$ C, if the closed point is in the image of β, then so is the generic point. Then over R, we get a flat family of normal varieties (by Hironaka's lemma) $\mathcal{U} \longrightarrow$ Spec R, plus divisors $\mathfrak{F}_1,\mathfrak{F}_2$ on $\mathcal{U}$. For every ℓ,m, let

$$M_{\ell,m} = H^o(\mathcal{U},\mathcal{O}(m\mathfrak{F}_o+\ell m(\mathfrak{F}_1-\mathfrak{F}_o))).$$

$M_{\ell,m}$ is a finitely generated torsion-free and hence free R-module, and $\sum_m M_{\ell,m} = R_\ell$ is an R-algebra. If $k = R/M$ is the residue field, $\overline{U},\overline{F}_o,\overline{F}_1$ is the induced triple over k, we get an injection:

$$\sigma_{\ell,m}: M_{\ell,m} \otimes_R k \hookrightarrow H^o(\overline{U},\mathcal{O}_{\overline{U}}(m\overline{F}_o+\ell m(\overline{F}_1-\overline{F}_o))).$$

Let $(\overline{U},\overline{F}_o,\overline{F}_1) = \beta(\overline{V},\overline{L},\overline{s}_o,\cdots,\overline{s}_N;\overline{E}_1,\overline{E}_2)$. If K is the fraction field of R, U^*,F_o^*,F_1^* is the induced triple over K, we get an isomorphism:

$$M_{\ell,m}\otimes_R K \xrightarrow{\approx} H^o(U^*,\mathcal{O}_{U^*}(mF_o^*+\ell m(F_1^*-F_o^*))).$$

But then it follows that

$$\begin{aligned}\dim_k M_{\ell,1} \otimes_R k &\leq \dim_k H^0(\overline{U}, \mathcal{O}_{\overline{U}}(\overline{F}_0 + \ell(\overline{F}_1 - \overline{F}_0))) \\ &= \dim_k (H^0(\overline{V}, \mathcal{O}_{\overline{V}}((m_1+\ell)\overline{C} + K_{\overline{V}})) \\ &= P'(m_1+\ell) \\ &\leq \dim_K H^0(U^*, \mathcal{O}_{U^*}(F_0^* + \ell(F_1^* - F_0^*))) \\ &= \dim_K M_{\ell,1} \otimes_R K.\end{aligned}$$

Since $M_{\ell,1}$ is free, the 2 extremes are equal, so equality holds everywhere. In particular, $\sigma_{\ell,1}$ is an isomorphism

$$\sigma_{\ell,1}: M_{\ell,1} \otimes_R k \xrightarrow{\approx} H^0(\overline{U}, \mathcal{O}_{\overline{U}}(\overline{F}_0 + \ell(\overline{F}_1 - \overline{F}_0))).$$

Now on $\overline{V}$, since $\overline{C}$ is ample, for $\ell >> 0$ it follows that the ring

$$\sum_{m=0}^{\infty} H^0(\overline{V}, \mathcal{O}_{\overline{V}}((m_1+\ell)\overline{C} + K_{\overline{V}})^{\otimes m})$$

is generated by its elements of degree 1 and that $\overline{V}$ is its Proj. This implies that the ring

$$\sum_{m=0}^{\infty} H^0(\overline{U}, \mathcal{O}_{\overline{U}}(m\overline{F}_0 + m\ell(\overline{F}_1 - \overline{F}_0)))$$

is generated by its elements of degree 1. But since $\sigma_{\ell,1}$ is surjective, this implies that $\sigma_{\ell,m}$ is surjective too: i.e., if $\ell >> 0$, there is an isomorphism of rings:

$$\sigma_\ell: R_\ell \otimes_R k \xrightarrow{\approx} \sum_{m=0}^{\infty} H^0(\overline{U}, \mathcal{O}_{\overline{U}}(m\overline{F}_0 + m\ell(\overline{F}_1 - \overline{F}_0))).$$

Therefore $\mathrm{Proj}(R_\ell \otimes_R k) \cong \overline{V}$. So $\mathcal{V} = \mathrm{Proj}(R_\ell)$ itself is a flat family of schemes of Spec R with special fibre $\overline{V}$. Moreover since $R_\ell \otimes_R k$ is generated by its elements of degree 1, R_ℓ is also generated by its elements of degree 1. Therefore $\mathrm{Proj}(R_\ell)$ comes equipped with a line bundle $\mathcal{O}_{\mathcal{V}}(1)$, which on the closed fibre $\overline{V}$ is just $\mathcal{O}_{\overline{V}}((m_1+\ell)\overline{C} + K_{\overline{V}})$, i.e., $\overline{L}^{m_1+\ell} \otimes \Omega^n_{\overline{V}}$. Since $\overline{V}$ is non-singular, $\mathcal{V}$ is smooth over R. Moreover by deformation theory $\overline{L}$ lifts to a unique invertible sheaf $\mathcal{L}$ on $\mathcal{V}$ such that

$$\mathcal{O}_{\mathcal{V}}(1) \cong \mathcal{L}^{m_1+\ell} \otimes \Omega^n_{\mathcal{V}/R}.$$

Let (V^*, L^*) be the generic fibre of $(\mathcal{V}, \mathcal{L})$. It is now easy to see that the rational map

$$\mathcal{V} \longrightarrow \mathcal{U}$$

defines $s_0, \cdots, s_N, \mathcal{E}_1, \mathcal{E}_2$ on $\mathcal{V}$, hence $s_0^*, \cdots, s_N^*, E_0^*, E_1^*$ on V^* such that $(U^*, F_0^*, F_1^*) = \beta(V^*, L^*, s_0^*, \cdots, s_N^*, E_0^*, E_1^*)$. QED

Heuristically, this shows that $\beta(B)$ is a "limited family", hence so is B, hence so is A. To be precise, note that all elements of B can be parametrized a suitable countably infinite set of families each defined over a base space B_α which is an algebraic variety. Then $\beta(B_\alpha)$ is at least a constructible subset of $C_0 = \text{Im}\ \beta$. But assuming the ground field k is uncountable*, then a (reducible) variety C_0 which is a countable union of constructible subsets $\beta(B_\alpha)$ is also a finite union of them: hence B is a finite union of B_α's.

---

*The other way of arguing is to look at 2 countable algebraically closed ground field $\overline{\mathbb{Q}} \subset k$, where $\overline{\mathbb{Q}}$ = field of algebraic numbers and k has infinite transcendence degree over $\mathbb{Q}$. Considering k-rational points, we get a bijection

$$\beta:\ B(k) \longrightarrow C_0(k)$$

but each B_α may be assumed to be defined over $\overline{\mathbb{Q}}$. Apply the elementary compactness assertion: if any set of $\overline{\mathbb{Q}}$-rational constructible sets covers $C_0(k)$, a finite subset already covers $C_0(k)$.

[1] S. Abhyankar, Resolution of singularities of embedded algebraic surfaces, Academic Press, 1966.

[2] E. Bombieri, Canonical models of surfaces of general type, Publ. IHES, 42 (1973), p. 171.

[3] K. Kodaira, Pluricanonical systems on algebraic surfaces of general type, J. Math. Soc. Japan, 20 (1968), p. 170.

[4] T. Matsusaka, On canonically polarized varieties II, Am. J. Math., 92 (1970), p. 283.

[5] T. Matsusaka, Polarized varieties with a given Hilbert polynomial, Am. J. Math., 94 (1972), p. 1027.

[6] T. Matsusaka and D. Mumford, 2 fundamental theorems on deformations of polarized varieties, Am. J. Math., 86 (1964), p. 668.

[7] A. Mayer, Families of K3 surfaces, Nagoya Math. J., 48 (1972), p. 1.

[8] D. Mumford, Abelian Varieties, Tata Studies in Math., Oxford Univ. Press, 1970.

[9] B. Saint-Donat, Projective models of K-3 surfaces, to appear.

BRANDEIS UNIVERSITY

HARVARD UNIVERSITY

Proceedings of Symposia in Pure Mathematics
Volume 29, 1975

UNIQUE FACTORIZATION IN COMPLETE LOCAL RINGS

Joseph Lipman[1]

§0. Introduction: U.F.D.'s and algebraic geometry.

In this lecture I will report on a number of themes which can be traced back to a large extent to the work of Samuel in the early 1960's on the topic of unique factorization domains. Through his work, and especially through some fertile conjectures, Samuel stimulated a great deal of research in an area which was more or less dormant. As will become evident , a remarkable feature of the subsequent research was the extent to which methods of algebraic geometry were employed to give deep insight into what were apparently purely algebraic questions. (Major credit for this profitable synthesis belongs to Grothendieck.)

Recall that a unique factorization domain (U.F.D.), or factorial ring, is a commutative integral domain in which every non-zero element can be factored into irreducible ones in an essentially unique way. Every U.F.D. is normal (integrally closed in its field of fractions). We will deal only with noetherian U.F.D.'s. A noetherian normal domain is a U.F.D. if and only if every height one prime ideal is principal.

We can reformulate this criterion for noetherian U.F.D.'s in the following useful way. To each noetherian normal domain R we associate the group of divisors $\mathrm{Div}(R)$, which is defined to be the free abelian group generated by the height one prime ideals in R. Among the divisors we have the principal divisors, which are those of the form

$$(f) = \sum_{p} v_p(f) \qquad (p \text{ - a height one prime ideal of } R)$$

AMS subject classifications (1970). Primary: 13F15, 13H99, 13J05, 14K30; Secondary: 14M99.

[1]Supported by the National Science Foundation under grant GP29216 at Purdue University.

where f is a non-zero element of the fraction field of R, and for each height one prime p, v_p is the associated discrete valuation (with valuation ring R_p). We have

$$(f/g) = (f) - (g),$$

so the principal divisors form a subgroup of $\mathrm{Div}(R)$. The quotient

$$C\ell(R) = \mathrm{Div}(R)/(\text{principal divisors})$$

is called the <u>divisor class group</u> of R. To say that every height one prime is principal is to say that <u>all</u> divisors are principal; thus:

$$R \text{ is a U.F.D.} \iff C\ell(R) = (0)$$

(For details, cf. [2, §3]).

* * *

The concept of U.F.D. arose in connection with number theory (Euler, Gauss, Kummer,...), and has played an important role in that subject up to the present day. We have to pass by this line of development, in favor of the geometric aspects of the study of U.F.D.'s. Here is one:

Let V be a <u>projectively normal</u> closed subvariety of some projective space over a field k (so that the corresponding homogeneous coordinate ring $k[V]$ is a normal domain). Let R be the local ring at the vertex of the projecting cone over V (R is obtained from $k[V]$ by localizing at the maximal ideal generated by all homogeneous elements of positive degree). Samuel showed [30, §2] that

$$C\ell(R) \cong C\ell(V)/(\text{hyperplane section}).$$

[$C\ell(V)$ is the group of linear equivalence classes of codimension one cycles on V; and we are factoring out the subgroup "generated" by a hyperplane section, i.e. the least subgroup containing the equivalence classes of those cycles which are (scheme-theoretically) the complete intersection of V with a hypersurface of the ambient projective space.] Consequently:

R is a U.F.D.

$\iff$ (#) every irreducible codimension one subvariety of V is cut out (scheme-theoretically) by a hypersurface of the ambient projective space.

In case V is non-singular (or even locally factorial) the condition (#) means that <u>every invertible $\mathcal{O}_V$-module is isomorphic to $\mathcal{O}_V(n)$ for some integer</u> n. Some examples of non-singular projectively normal V's having this property are:

-(i) Grassmann varieties.

(These were treated by Severi and, over fields of positive characteristic, by Igusa. Cf. also [20, p. 124].)

-(ii) Non-singular complete intersections of dimension ≥ 3.

(This is the theorem of Lefschetz, mentioned in §3 of Hartshorne's talk on equivalence of cycles (these Proceedings); cf. also §5 below.)

-(iii) "Most" non-singular two-dimensional complete intersections.

(Lefschetz' generalization of Noether's theorem; cf. Hartshorne's talk and also [11, exposé XIX].)

Assuming that V is projectively normal, non-singular, and satisfies (#), one can show that for the completion $\hat{R}$ to be a U.F.D. a sufficient condition (which is also necessary if $\text{char}.k = 0$, or if $\dim.V = 1$) is that

(##) $$H^1(V, \mathcal{O}_V(n)) = 0 \qquad \text{for all } n > 0$$

(cf. [7, §4])[2]. (##) is satisfied e.g. if V is a complete intersection and $\dim.V \geq 2$.

So we see that questions about local U.F.D.'s can have non-trivial global geometric significance.

§1. Unique factorization in formal power series rings.

It is elementary (going back to Gauss) that if R is a U.F.D. then so is any polynomial ring $R[X_1, X_2, \ldots, X_n]$.

In this section, we survey some highlights in the history of the corresponding question for formal power series rings:

For which noetherian U.F.D.'s R is it true that:

(*) any power series ring $R[[X_1, X_2, \ldots, X_n]]$ is a U.F.D.?

The story begins in 1905 with Lasker's proof that (*) is true if R is an infinite field. In 1938, Krull [21] showed the same thing with R any

---

[2] When k = complex numbers and $\dim.V \geq 2$, we have $H^1(V, \mathcal{O}_V(-n)) = 0$ for $n > 0$ (Kodaira's vanishing theorem), and (#) implies that $H^1(V, \mathcal{O}_V) = 0$ (since V has discrete Picard scheme). So (##) says that $H^1(V, \mathcal{O}_V(n)) = 0$ for all $n \in \mathbb{Z}$, which means precisely that $\text{depth}(R)(= \text{depth}(\hat{R})) \geq 3$.

discrete valuation ring having infinite residue field, but he could not settle the finite residue field case[3]. This was done by Cohen in 1946 [6, p. 94, Theorem 18]. Another proof of Cohen's result, making more explicit use of the Weierstrass preparation theorem - and thus quite close to the proofs of Lasker and Krull - is given by Bourbaki in [2, §3.9, Proposition 8]. (What Krull missed was a simple lemma on automorphisms of power series rings [2, §3.7, Lemma 3]). Krull also expressed doubt that (*) would hold for R an arbitrary principal ideal domain [21, p. 770]; but in this he was mistaken [2, §3, exercise 9]. In fact in 1961 Samuel [29, pp. 3-4] and Buchsbaum [5, p. 753] showed more generally that *(*) is true for any locally regular U.F.D.* R. (The main ingredient of their proofs is the Auslander-Buchsbaum theorem that every regular local ring is a U.F.D.)

At the same time, Samuel gave the following result [29, p. 5]:

(I) *If* R *is a U.F.D. whose localizations at maximal ideals are all Cohen-Macaulay, and if the power series ring in one variable* $R_p[[X]]$ *is a U.F.D. for all height two prime ideals* p *in* R, *then* $R[[X]]$ *is a U.F.D.*

(Samuel's proof of (I) is complicated, and is said by Danilov [9, p. 368] to be incomplete. However, Danilov [*loc. cit*] gives a more general result, which we will describe in §3.)

(I) brings the study of power series rings $R[[X]]$ over *two-dimensional local U.F.D.'s* R to the foreground. Samuel found examples of such R for which $R[[X]]$ is *not* a U.F.D [29], [30]. These examples were, however, suspect, in that their *completions* were not U.F.D.'s: to a geometer, properties of local rings which are not preserved under completion somehow lack authenticity. Anyway, Samuel conjectured [30, p. 171]:

(II) *If* R *is a* complete *local U.F.D., then so is* $R[[X]]$ (?)

The first progress on (II) was due to Scheja, who showed [32, p. 128, Satz 2] that *(II) holds whenever* R *has depth* ≥ 3. A nice consequence of this is that *if* $R[[X]]$ *is a U.F.D.*, (R as in (II)) *then so is* $R[[X_1,X_2,\ldots,X_n]]$ *for any* n. (For, if depth $R[[X]] < 3$, then R is either a field or a discrete valuation ring.)

Of course Scheja's result says nothing about (II) in case R is two-dimensional. One difficulty in attacking this case was that when (II) was formulated, there was only one example known of a *non-regular two-dimensional henselian local U.F.D.*, namely the analytic ring $C\{\{x,y,z\}\}/(x^2 + y^3 + z^5)$,

---

[3]Krull's idea, to deduce the finite case from the infinite one [21, p. 778], works well once one has available - which Krull didn't - some basic facts on faithfully flat ring extensions. (Cf. [12, p. 35, Cor. 6.11].)

which Mumford had investigated with transcendental methods [25, §IV].[4] Such was the state of knowledge (or rather ignorance) ten years ago.

Scheja tried to find a two-dimensional counterexample to (II). He proved that for any three-dimensional regular local ring S with maximal ideal generated by u, v, w, the two-dimensional ring $R = S/(u^2 + v^3 + w^5)$ is a U.F.D., but so is $R[[X]]$. He then discovered a number of previously unknown complete local two-dimensional U.F.D.'s; but each one of them satisfied Samuel's conjecture (cf. [32]). So after a while, he gave up. Apparently he did so too soon, for, as it turned out, he had actually found essentially all the two-dimensional R for which (II) holds - there are very few[5], whereas there are many others for which (II) fails.

The first counterexample to (II) was found by Salmon [28] - it is the ring

$$R = k(U)[[X,Y,Z]]/(X^2 + Y^3 + UZ^6)$$

where k is any field, and U is an indeterminate. Here $R[[X]]$ is not a U.F.D., whereas R is. But, if in the description of R we replace the field $k(U)$ by its algebraic closure, the resulting ring is no longer a U.F.D.. This again should arouse the scepticism of any geometer; it means that R is not a genuine U.F.D., in that $C\ell(R)$ has many non-zero elements which happen to be thinly concealed, i.e. defined over an algebraic extension of the residue field! Later on, a whole series of counterexamples was given by Danilov [8, §1] and Grothendieck [unpublished]; these all had the same deficiency: they lost their U.F.D. property when the residue field was extended to its algebraic closure.

And indeed, (II) is true if the residue field of R is algebraically closed.

In fact, a better result holds. To formulate it, we need the notion of a discrete divisor class group (DCG). If R is any normal noetherian domain, there is a canonical map $i: C\ell(R) \to C\ell(R[[X]])$ defined as follows: for any height one prime ideal p in R, $pR[[X]]$ is a height one prime ideal in $R[[X]]$; there is a unique map $\mathrm{Div}(R) \to \mathrm{Div}(R[[X]])$ sending each p to the corresponding divisor $pR[[X]]$;

---

[4]Subsequently Brieskorn showed that there are no other analytic examples! [4]. (In [22, §25], Brieskorn's result is extended to arbitrary two-dimensional henselian local rings with algebraically closed residue field.) A propos, the ring in question has a distinguished history, going back to Klein's lectures on the icosahedron.

[5]Precisely those which have rational singularities (provided the singularity is resolvable), cf. Theorem 3 in §3. A complete list of rational U.F.D.'s is given in [22, §25].

one checks that this map takes principal divisors of R to principal divisors of $R[[X]]$, and hence induces a map $i: C\ell(R) \to C\ell(R[[X]])$ (cf. [12, §6]). i is easily seen to be injective [12, p. 35, Cor. 6.13] (in fact i has a left inverse [ibid, p. 130, remarks]). R *is said to have DCG if* i *is bijective*.

Note that if R is a U.F.D. (i.e. $C\ell(R) = (0)$), and R has DCG, then $R[[X]]$ is a U.F.D. (and conversely).

So (II) may be looked at from a more general point of view (due to Danilov): *study rings with DCG*.[6]

Here is the result:

THEOREM 1. *Let* R *be a complete normal noetherian local ring with algebraically closed residue field, and suppose that* $C\ell(R)$ *is a finitely generated abelian group. Then* R *has DCG. (In particular, if* R *is a U.F.D., then so is* $R[[X]]$.)

The Theorem was proved by Danilov [10], with some additional restrictions on R, in the equicharacteristic case (i.e. when R contains a field);[7] his proof uses resolution of singularities and the theory of the Picard scheme of a scheme which is proper over a field. A much more elementary proof, using the theorem of Ramanujam-Samuel (cf. §5 below), was given by Storch [34], under the assumptions that R contains a field and that the residue field of R is uncountable and of cardinality strictly greater than that of $C\ell(R)$. Following some hints of Grothendieck, I worked out a theory of Picard schemes for schemes proper over any complete local ring; with this machinery, it was possible to make Danilov's and Storch's arguments apply to any R whose singularities can be resolved [23, Theorem 1']. Finally, Boutot showed how to get rid of the ungainly condition of resolvable singularities [23, Theorem 1].

§2. The local Picard scheme.

To get a proper feeling for the preceding results (and in particular for the terminology "DCG") one can adopt a philosophy due to Grothendieck. Let R be a normal complete local noetherian ring. Assume for simplicity that $\operatorname{Spec}(R)$ is regular outside the closed point, and that R contains a field of representatives k. Following some work of Mumford [25] on two-dimensional analytic local rings Grothendieck proposed a method for giving $C\ell(R)$ a natural structure of locally algebraic group - the "local Picard scheme" - over the residue field k of R. (For details, cf. [16, pp. 189-191]. Roughly speaking, there should be a locally algebraic k-group P, and

[6] A survey of Danilov's excellent work on rings with DCG is given in §19 of [12].

[7] Under these conditions he also proves the *converse*: DCG $\Rightarrow$ $C\ell(R)$ finitely generated.

for k-algebras K a natural (K-functorial) map

$$\theta_K : C\ell(R \hat{\otimes}_k K) \to P(K) \quad (= \text{K-valued points of } P)$$

which, as a first approximation, can be thought of as being bijective. (This is not true for all K, but never mind; at this moment we are just describing a philosophy, and don't want to get involved with technicalities.)

If P is discrete (i.e. zero-dimensional) then $P(k) = P(k[[X]])$, so $C\ell(R) = C\ell(R[[X]])$ and R has DCG. Conversely, if P has positive dimension, then there is a tremendous number of points in the kernel of $P(k[[X]]) \to P(k)$ (think of them as little analytic arcs on P emerging from the zero-point). So $C\ell(R[[X]])$ is bigger than $C\ell(R)$, i.e. R does not have DCG.

Now R is a U.F.D. if $P(k)$ $(= C\ell(R)) = (0)$, i.e if P has just one k-rational point. If k is algebraically closed, this means that P is zero-dimensional, so R has DCG. (This is the "explanation" of Theorem 1, §1).

If k is not algebraically closed, then $P(k)$ can be (0) even if P has positive dimension. In Salmon's example, for instance, P is the projective plane cubic curve defined over $k(U)$ by

$$X^2 Z + Y^3 + UZ^3 = 0,$$

whose only $k(U)$-rational point is (1,0,0). The above-mentioned counter-examples of Danilov and Grothendieck were constructed by completing the local rings at the vertices of projecting cones over certain curves, namely principal homogeneous spaces over elliptic curves having just one rational point over their field of definition; then P turns out to be the elliptic curve itself. As mentioned before, in each of these examples, the U.F.D. property is destroyed by extending the residue field to its algebraic closure; this appears now as a reflection of the fact that an elliptic curve over an algebraically closed field has many rational points.

Thus we can say that (II) (§1) is basically a geometric statement; the counterexamples of Salmon, Danilov and Grothendieck have arithmetic, but not geometric, significance.

It should be emphasized that the local Picard scheme, *per se*, plays no role, except for motivation, in the proof of Theorem 1. However the main lines of the proof are similar to - or suggested by - those in Grothendieck's proposed construction.

Actually, it is only recently that the theory of local Picard schemes has really been developed, by Boutot [3], who uses a different approach than the one outlined by Grothendieck. Boutot considers a local ring R with maximal ideal $\underline{\underline{m}}$, such that R contains a field of representatives k (i.e. k maps canonically onto $R/\underline{\underline{m}}$). Let $U = \mathrm{Spec}(R) - \{\underline{\underline{m}}\}$. (Note that for normal R, $C\ell(R) = \mathrm{Pic}(U)$ if R has an isolated singularity, or more generally if U is locally factorial.) The idea is to make $\mathrm{Pic}(U)$ into a locally algebraic k-group in a natural way. To this end, for each noetherian k-algebra A, let $\hat{R}_A$ be the $\underline{m}$-adic completion of $R \otimes_k A$, let $\hat{U}_A$ be the inverse image of U in $\mathrm{Spec}(\hat{R}_A)$, and consider the functor

$$P(A) = \mathrm{Pic}(\hat{U}_A).$$

Using M. Artin's representability criteria, Boutot shows ([3], and oral communication):

> *Assume that* depth $R \geq 2$ *and that the* k-*vector space* $H^1(U, \mathcal{O}_U)$ *is finite-dimensional* (these assumptions hold, for example, if R is complete, normal, and of dimension ≥ 3). *Then the étale sheaf associated to* P *is a locally algebraic k-group, whose Zariski tangent space at the origin is* $H^1(U, \mathcal{O}_U)$.

It seems reasonable to anticipate further interesting developments in the study of local Picard schemes. After all, they should be no less important for local rings than global Picard schemes are for varieties. One can hope, for example, that the local Pic. will enter in a significant way into the theory of classification and deformation of singularities.

§3. Depth and discrete divisor class groups.

This section is centered around Danilov's generalization of Samuel's theorem (I) (§1). Danilov's result [9, p. 374, Theorem 1] asserts that the DCG property lives in depth 2:

THEOREM 2. *If the normal noetherian ring* R *is such that the localizations* R_p *have* DCG *for all prime ideals* p *such that* $\mathrm{depth}(R_p) = 2$, *then* R *itself has* DCG.

(Danilov's proof of Theorem 2 uses some of the deep cohomological results of [16], and he needs some additional mild hypothesis on R; but there is a quite elementary proof which does not require this additional condition [24].)

Recall that a *normal* noetherian ring R satisfies the Serre condition (S_3) if for any prime ideal p with $\dim.(R_p) \geq 3$ we have also $\mathrm{depth}(R_p) \geq 3$.

COROLLARY 1. *If the noetherian normal ring* R *satisfies* (S_3), *and if all its two-dimensional localizations* R_p *have* DCG - *for example if they are regular - then* R *has* DCG.

As before, this focuses attention on the two-dimensional case, for which Danilov shows:

THEOREM 3 [9, §4]. For a two-dimensional normal local ring R with resolvable singularity and perfect residue field the following are equivalent:

(i) R has DCG.

(ii) R has a rational singularity.

(iii) The strict henselization of R has a finite divisor class group[8].

Danilov also shows [9, §5] that:

The converse of Corollary 1 holds if R is excellent and contains a field of characteristic zero.

(Here again he uses some of the heavy machinery from [16], plus the following fact: if R is an excellent $\mathbb{Q}$-algebra and $f: X \to \mathrm{Spec}(R)$ is a resolution of singularities, then R has DCG if and only if $H^1(X, \mathcal{O}_X) = (0)$.[9] (Hence the DCG property localizes for excellent $\mathbb{Q}$-algebras; whether this is true for more general R doesn't seem to be known.))

In view of Scheja's result (immediately following (II)(§1)), the preceding gives:

If R is a complete equicharacteristic zero U.F.D. of depth ≥ 3, then R satisfies (S_3). Furthermore, the two-dimensional localizations of R have at worst rational singularities (so they are explicitly known [22, §25]).

Here, if the residue field is algebraically closed (char. 0), the "depth ≥ 3" hypothesis is superfluous. For, the U.F.D. property implies that Boutot's local Picard scheme (§2) is zero-dimensional (it has just one rational point), so its tangent space $H^1(U, \mathcal{O}_U) = 0$, i.e. R has depth ≥ 3. Alternatively, as indicated just after Theorem 1 (§1), Danilov showed that a normal local ring R with algebraically closed char. 0 residue field, and with finitely generated divisor class group, has DCG provided it is complete (or, more generally, provided that a reduction to the complete case via Artin's approximation theorems is possible, for example if R is the

---

[8]Cf. also [22, §17]. The resolvability of the singularity of R is equivalent to analytic normality [ibid. §16.2]. Without this assumption, and without any assumption on the residue field, it is still true that R has DCG if and only if R has a pseudo-rational singularity [ibid, §9].

[9]"Explanation": $H^1(X, \mathcal{O}_X)$ is the tangent space at the origin of the local Picard scheme. (Danilov does not use this.)

local ring of an algebraic variety over $\mathbb{C}$, or if R is analytic). Hence, by the converse of Corollary 1, any such R satisfies (S_3).

A different approach to this last result is given by Hartshorne and Ogus, in [18, §2]. For excellent normal local rings R with residue field $\mathbb{C}$, whose completion is algebraizable, they show that the (S_3) property follows from the vanishing of $H^1(X, \mathcal{O}_X)$, where $f: X \to \mathrm{Spec}(R)$ is a resolution of singularities. Their methods are analytic, being based on a dualized version of the Grauert-Riemenschneider vanishing theorem.

In conclusion, we note that for the converse of Corollary 1, characteristic zero is essential. Serre has given examples (over fields of positive characteristic) of non-singular projective surfaces V whose Picard scheme is discrete (zero-dimensional) but not reduced. Danilov points out [9, pp. 376-377] that the projecting cone over a suitable projective embedding of such a V has a vertex whose completed local ring is normal, with DCG, and three-dimensional, but of depth two (so it doesn't satisfy (S_3)).

§4. U.F.D.'s which are not Cohen-Macaulay.

In Samuel's theorem (I)(§1) one assumption on the U.F.D. R is that the localizations of R are Cohen-Macaulay (C-M). At the end of the paper [29] where (I) is given, Samuel states that all the examples of U.F.D.'s known to him are locally C-M, and asks whether this is true in general.

In this connection, Murthy showed [26] (cf. also [12, §12] or [19, (7.18)]) that any C-M U.F.D. R which is a homomorphic image of a regular local ring is in fact a Gorenstein ring. The reason for this is that the "canonical" (or "dualizing") module of R is a reflexive fractionary ideal [19, (6.7), (7.29)], hence invertible, since R is a U.F.D.; and the invertibility of the canonical module characterizes Gorenstein rings [ibid, (5.9)].

Using a variant of local duality, Hartshorne and Ogus [18, §1] have improved Murthy's result by weakening the hypothesis that R be C-M to the two conditions:

(i) R satisfies (S_3) and

(ii) $\mathrm{depth}(R_p) \geq \frac{1}{2}\dim.(R_p) + 1$ for all primes p with $\dim.(R_p) \geq 5$.

For example, if R is a complete local U.F.D. with algebraically closed residue field of characteristic zero, then R satisfies (S_3) (cf. §3); hence if R has dimension ≤ 4, then both (i) and (ii) are satisfied, so R is a Gorenstein ring. (Cf. [18, p. 428]; this result is originally due to M. Raynaud (unpublished).)

The first example of a U.F.D. which is not locally C-M was given by Bertin [1]; it is the ring of invariants of a cyclic group G of order 4 acting linearly on a polynomial ring B in 4 indeterminates over a field k of characteristic 2. The U.F.D. property is established by a "Galois descent" technique of Samuel [12, §16], which gives an injective map of the divisor class group of the fixed ring B^G into $H^1(G,B^*)$ (B^* = units of B); but $B^* = k^*$, on which G acts trivially, so that

$$H^1(G,B^*) = H^1(G,k^*) = \mathrm{Hom}_{\mathbb{Z}}(G,k^*) = 0$$

(since k^* has no 2-torsion). (The same argument applies to a cyclic group of order p^r acting on any polynomial ring over a field of char.p.)

To show that her example was not locally C-M, Bertin constructed explicitly a system of homogeneous parameters which is not a regular sequence (cf. [12, §16.8]). Then Hochster and Roberts noticed [20, p. 127] that Bertin's example was very closely related to some surfaces previously studied by Serre (the same surfaces used in Danilov's example at the end of §3), and that Serre's computations led to another proof of the failure of the C-M property; since Serre's surfaces exist over fields of any characteristic ≥ 5, one gets examples similar to Bertin's over any such field.

In analogy with the situation discussed in §1, one could still reasonably ask whether any complete local U.F.D. is C-M.

This question was open until quite recently, when the answer was found to be "no". Freitag and Kiehl constructed a class of analytic local rings (over the complex numbers) which are U.F.D.'s of dimension 60 and depth 3, hence certainly not C-M. [15, p. 144, Thm. 5.8] (These examples arise in connection with the study of the cusps of Hilbert modular groups associated with totally real Galois extensions of $\mathbb{Q}$, whose Galois group is the alternating group A_5 (of order 60) on five elements; the methods used involve complex analysis, cohomology of groups, number theory,...). An argument of Danilov [10, p. 235, remarks 5 and 3], making use of Artin's analytic approximation theorem, implies that the completions of the Freitag-Kiehl examples are U.F.D.'s (which are not C-M).

Over fields of characteristic > 0, some complete local non C-M U.F.D.'s have been found even more recently by Fossum and Griffith. In fact they show that the completion of Bertin's example is a U.F.D. [13]. (Bertin's example is a graded subring of a polynomial ring, and completion is with respect to the powers of the irrelevant ideal.) More generally, in [14] they treat the following situation:

Let k be a field of characteristic $p > 0$, let $n > 0$, and let G be a cyclic group of order p^n operating on the polynomial ring $B = k[X_1, X_2, \ldots, X_{p^n}]$ by cyclically permuting the indeterminates. This action of G extends to the power series ring $\hat{B} = k[[X_1, X_2, \ldots, X_{p^n}]]$. Then:

(i) The fixed ring B^G is a graded U.F.D. which is not C-M.

(Proof: essentially the same as for the above-mentioned Hochster-Roberts examples.)

(ii) The fixed ring $\hat{B}^G$ is the completion of B^G at its irrelevant ideal (so that $\hat{B}^G$ is not C-M.)

(Proof: elementary.)

(iii) $C\ell(\hat{B}^G) \subseteq H^1(G,\hat{B}^*)$ ($\hat{B}^*$ = units of B)

(Proof: by Samuel's Galois descent.)

(iv) (Main Result). If $p^n = 4$, or $n = 1$ and $p \geq 5$, then $H^1(G,\hat{B}^*) = 0$ (whence, by (ii) and (iii), $\hat{B}^G$ is a complete local U.F.D. which is not C-M).

(Proof: elaborate computation.)

* * *

I hope the matter will not rest here, for there does not seem to be any real insight yet into the connection between the U.F.D. and C-M properties for complete local rings (not even for those obtained from cones over non-singular projective varieties (§0)). By real insight, I mean the sort of understanding which the local Picard scheme gives us for the subjects treated in §1.

§5. Parafactoriality

The notion of parafactoriality was introduced by Grothendieck in [16], and used effectively there to study factorial rings.

Recall that a noetherian local ring R with maximal ideal $\underline{\underline{m}}$ is parafactorial if depth $R \geq 2$ and $\mathrm{Pic}(\mathrm{Spec}(R) - \{\underline{\underline{m}}\}) = 0$.

The connection of this notion with factoriality is the following simple fact [16, p. 130, Cor. 3.10]: when $\dim.R \geq 2$, R is factorial if and only if: R is parafactorial and R_p is factorial for every prime ideal $p \neq \underline{\underline{m}}$.

An early application was to the following conjecture of Samuel [30, p. 172], which generalizes a classical global theorem of Lefschetz (cf. §0, example (ii)): a local complete intersection which is factorial in codimension ≤3 is factorial. In [16, p. 132] Grothendieck showed:

A local complete intersection of dimension ≥ 4 is parafactorial.

(From this, and from the above characterization of factorial rings, Samuel's conjecture follows immediately, by induction on the dimension.)

A result along similar lines has been proved by Ogus [27, p. 350, Cor. 3.14]. Let R, $\underline{\underline{m}}$ be as above, with $R/\underline{\underline{m}}$ of characteristic zero. Let $d = \dim.(R)$, and assume that R is a homomorphic image of a regular local ring A of dimension $d + r$. Let $U = \mathrm{Spec}(R) - \{\underline{\underline{m}}\}$, $V = \mathrm{Spec}(A)$ - {closed point}. Assume further that:

(i) U is locally a complete intersection in V.

(ii) depth(R) ≥ 3.

(iii) $r \leq d - 3$ ("small embedding codimension").

Then, under these conditions, R is parafactorial.

Hartshorne and Ogus derive a number of corollaries from this result [18, §3]. Here is one: if, furthermore, R is locally factorial in codimension ≤ 3 (so that by assumption (i) and Grothendieck's above theorem, U is locally factorial) then R itself is factorial.

* * *

In response to a conjecture of Grothendieck, Samuel proved the following theorem [31]:

Let R, $\underline{\underline{m}}$, be as above, and assume that the completion $\hat{R}$ is normal. Let A be the power series ring R[[X]], and let p be the prime ideal $\underline{\underline{m}}A$. Then the canonical map of divisor class groups $C\ell(A) \to C\ell(A_p)$ is bijective.

The same result for power series rings in any finite number of variables was proved independently by Ramanujam [33, appendix]. (He reduces the question to the one-variable case, then uses -as does Samuel- the Weierstrass preparation theorem.)

An easy consequence is that A is parafactorial. (In fact one can derive many other parafactorial rings, cf. [17, §(21.14)].)

This Ramanujam-Samuel theorem plays an important role in the proof of Theorem 1 (§1). It is also essential in Boutot's theory of local Pic. (§2), for showing that the zero-section of the "local Picard functor" P is represented by a closed immersion (this being one of Artin's representability conditions). In fact Boutot proves and uses the following strong form of the Ramanujam-Samuel theorem:

Let R and A be noetherian local rings, $\underline{\underline{m}}$ the maximal ideal of R, $\phi: R \to A$ a local homomorphism making A formally smooth over R (for the usual topologies), and such that the residue field of A is finite over that of R. Let $q \in \mathrm{Spec}(A)$ be such that $q \not\subset \underline{\underline{m}}A$ and depth $A_q \geq 3$. Then A_q is parafactorial.

Having used this as input for the existence of local Pic., Boutot then obtains further refinements as output, for example:

Let $\phi: R \to A$ be a local homomorphism of equicharacteristic local rings making A formally smooth over R. Then A is parafactorial under either of the following two sets of conditions:

(i) dim.(A) > dim.(R) and depth(A) ≥ 3.

(ii) dim.(A) = dim.(R), R is parafactorial and strictly henselian.

As a corollary of this last generalized form of Ramanujam-Samuel's theorem, Boutot gets the following result, which is manifestly related to the subject matter of §1:

> *Let* $f:X \to S$ *be a regular morphism of equi-characteristic locally noetherian schemes. Suppose that the strict henselization of the local ring of each point of* S *is factorial. Then the same holds for* X.

Details will appear in Boutot's thesis.

REFERENCES

[1] M.-J. BERTIN, *Anneaux d'invariants d'anneaux de polynomes, en caractéristique p*, C.R. Acad. Sci. Paris 264 (Série A)(1967), 653-656.

[2] N. BOURBAKI, *Algèbre Commutative, Chapitre 7,* Act. Sci. Ind. no. 1314, Hermann, Paris, 1965. (English translation, 1972).

[3] J.-F. BOUTOT, *Schéma de Picard local*, C.R. Acad. Sci. Paris, 277 (Série A)(1973), 691-694.

[4] E. BRIESKORN, *Rationale Singularitäten komplexer Flächen*, Inventiones Math. 4 (1968), 336-358.

[5] D. A. BUCHSBAUM, *Some remarks on factorization in power series rings,* J. Math. Mech. 10 (1961), 749-753.

[6] I. S. COHEN, *On the structure and ideal theory of complete local rings*, Trans. Amer. Math. Soc. 59 (1946), 54-106.

[7] V. I. DANILOV, *The group of ideal classes of a completed ring*, Math. USSR Sbornik 6 (1968), 493-500 (Mat. Sb. 77(119) (1968), 533-541.)

[8] __________, *On a conjecture of Samuel*, Math. USSR Sbornik 10 (1970), 127-137. (Mat. Sb. 81(123) (1970), 132-144.)

[9] __________, *Rings with a discrete group of divisor classes*, Math. USSR Sbornik 12 (1970), 368-386. (Mat. Sb. 83(125) (1970), 372-389.)

[10] __________, *On rings with a discrete divisor class group*, Math. USSR Sbornik 17 (1972), 228-236. (Mat. Sb. 88(130) (1972), 229-237.)

[11] P. DELIGNE, N. KATZ, *Groupes de monodromie en géométrie algébrique (SGA 7, II)*, Lecture Notes in Math. no. 340, Springer-Verlag, New York-Heidelberg-Berlin, 1973.

[12] R. M. FOSSUM, *The divisor class group of a Krull domain*, Ergeb. der Math., Bd. 74, Springer-Verlag, New York-Heidelberg-Berlin, 1973.

[13] __________, P. A. GRIFFITH, *A complete local factorial ring of dimension 4 which is not Cohen-Macaulay*, Bull. Amer. Math. Soc. 80 (1974).

[14] ________, ________, Complete local factorial rings which are not Cohen-Macaulay, preprint.

[15] E. FREITAG, R. KIEHL, Algebraische Eigenschaften der lokalen Ringe in den Spitzen der Hilbertschen Modulgruppen, Inventiones Math. 24 (1974), 121-148.

[16] A. GROTHENDIECK, Cohomologie locale des faisceaux cohérents et théorèmes de Lefschetz locaux et globaux (SGA 2), North-Holland, Amsterdam, 1968.

[17] ________, J. DIEUDONNÉ, Élements de géométrie algébrique, IV (quatrième partie), Publ. Math. I.H.E.S. 32 (1967).

[18] R. HARTSHORNE, A. OGUS, On the factoriality of local rings of small embedding codimension, Communications in Algebra 1 (1974), 415-437.

[19] J. HERZOG, E. KUNZ, Der kanonische Modul eines Cohen-Macaulay Rings, Lecture Notes in Math. no. 238, Springer-Verlag, New York-Heidelberg-Berlin, 1971.

[20] M. HOCHSTER, J. L. ROBERTS, Rings of invariants of reductive groups acting on regular rings are Cohen-Macaulay, Advances in Math. 13 (1974), 115-175.

[21] W. KRULL, Beiträge zur Arithmetik kommutativer Integritätsbereiche V, Math. Z. 43 (1938), 768-782.

[22] J. LIPMAN, Rational singularities, with applications to algebraic surfaces and unique factorization, Publ. Math. I.H.E.S. 36 (1969), 195-279.

[23] ________, Picard schemes of formal schemes; application to rings with discrete divisor class group, Classification of Algebraic Varieties and Compact Complex Manifolds, pp. 94-132; Lecture Notes in Math. no. 412, Springer-Verlag, New York-Heidelberg-Berlin, 1974.

[24] ________, Rings with discrete divisor class group: theorem of Danilov-Sanuel, preprint.

[25] D. MUMFORD, The topology of normal singularities of an algebraic surface and a criterion for simplicity, Publ. Math. I.H.E.S. 9 (1961), 5-22.

[26] M. P. MURTHY, A note on factorial rings, Arch. Math. 15 (1964), 418-420.

[27] A. OGUS, Local cohomological dimension of algebraic varieties, Annals of Math. 98 (1973), 327-365.

[28] P. SALMON, Su un problema posto da P. Samuel, Atti. Accad. Naz. Lincei Rend. Cl. Sci. Fis. Mat. Natur. (8) 40 (1966), 801-803.

[29] P. SAMUEL, On unique factorization domains, Illinois J. Math. 5 (1961), 1-17.

[30] ________, Sur les anneaux factoriels, Bull. Soc. Math. France 89 (1961), 155-173.

[31] ________, Sur une conjecture de Grothendieck, C. R. Acad. Sci. Paris, 255 (1962), 3101-3103.

[32] G. SCHEJA, Einige Beispiele faktorieller lokaler Ringe, Math. Annalen 172 (1967), 124-134.

[33] C. S. SESHADRI, Quotient space by an abelian variety, Math. Annalen 152 (1963), 185-194.

[34] U. STORCH, Über das Verhalten der Divisorenklassengruppen normaler algebren bei nicht-ausgearteten Erweiterungen, Habilitationsschrift, Univ. Bochum, 1971.

PURDUE UNIVERSITY
WEST LAFAYETTE, IN. 47907
U.S.A.

Proceedings of Symposia in Pure Mathematics
Volume 29, 1975

DIFFERENTIALS OF THE FIRST, SECOND AND THIRD KINDS

William Messing

The purpose of my lecture was to explain what the various types of differentials are and how they relate to both classical and modern algebraic geometry. The results I want to discuss are all for smooth varieties over the complex numbers. Indeed, De Rham cohomology and differential forms fail, at least in any naive sense, to give a good theory over fields of positive characteristic; and one of the purposes of the seminar was to discuss a possible substitute: namely crystalline cohomology. On the eve of the conference, Ogus, in response to my question, pointed out that, in general, the classical definition of differentials of the second kind does not coincide with the definition of Atiyah and Hodge. This will be discussed fully below. Since all other results cited exist with detailed proofs in the literature, I will usually just give the appropriate reference.

I. DIFFERENTIALS OF THE FIRST KIND

Let X be a projective smooth algebraic variety over $\mathbb{C}$, and let X^{an} be the underlying compact complex manifold. The Poincaré Lemma asserts that $\Omega^{\cdot}_{X^{an}}$ is a resolution of the constant sheaf $\mathbb{C}_{X^{an}}$ and therefore the analytic De Rham cohomology:

$$H^*_{D.R.}(X^{an}/\mathbb{C}) =_{dfn.} H^*(X^{an},\Omega^{\cdot}_{X^{an}})$$

is canonically isomorphic to $H^*(X^{an},\mathbb{C})$. From this together with the isomorphism $H^*(X,\Omega^p_X) \xrightarrow{\sim} H^*(X^{an},\Omega^p_{X^{an}})$ provided us by [GAGA] we obtain

$$(1.1) \qquad H^*_{DR}(X) =_{dfn.} \mathbb{H}^*(X,\Omega^{\cdot}_X) \xrightarrow{\sim} H^*(X^{an},\mathbb{C}) .$$

Thus the cohomology of a projective (smooth) variety can be defined algebraically.

This description of the cohomology of X^{an} identifies it as the abutment of the Hodge to De Rham spectral sequence

$$(1.2) \qquad E_1^{p,q} = H^q(X,\Omega^p_X) \Rightarrow H^*_{DR}(X) \simeq H^*(X^{an},\mathbb{C}) .$$

Hodge theory gives us a description of $H^q(X,\Omega^p_X)$ as the space of <u>harmonic</u> (p,q) forms on X^{an} and similarly a description of $H^*(X^{an},\mathbb{C})$ as the space of all harmonic forms. By using the double complex $\mathcal{E}^{pq}$ of C^∞ (p,q) forms on X^{an} to calculate $H_{DR}(X^{an})$, together with the fundamental indentities between the various operators on (p,q)-forms (Laplacian, Green's operator,...); Hodge showed that the spectral sequence degenerates at E_1. In particular we obtain

(1.3) <u>PROPOSITION</u>: Let $\omega \in \Gamma(X^{an},\Omega^p_{X^{an}})$. Then

1) ω is closed.
2) If $\omega = d\eta$ for some C^∞-form η, then $\omega \equiv 0$.

Thus integrating over p cycles defines an injective map $H^0(X^{an},\Omega^p_{X^{an}}) \hookrightarrow H^p(X^{an},\mathbb{C})$. This result which we can restate as: "A non-trivial differential of the first kind can't have all its periods vanish" was Hodge's important theorem. It was proved by Lefschetz in certain special cases and for a while was regarded as a major open problem in algebraic geometry (c.f. Zariski's review of Hodge's book in Mathematical Reviews).

Let's call differentials of the first kind, the global holomorphic forms on X^{an}. In chapter VII of Zariski's book of surfaces, [17], it is explained how using Lefschetz pencils, normal functions, etc. Picard, Lefschetz, Poincare, ... proved (among other things) that

a) the irregularity q defined classically as $p_g - p_a$ is equal to $\dim H^o(X,\Omega^1)$.

b) Lefschetz's existence theorem: A cycle $j \in H_2(X,\mathbb{Z})$ is algebraic $\Longleftrightarrow \int_j \omega = 0$ for all $\omega \in \Gamma(X,\Omega_X^2)$.

Since such results are now well known consequences of Hodge theory we might say that the theory of differentials of the first kind has been subsumed by Hodge Theory.

II. DIFFERENTIALS OF THE SECOND KIND

The theory in this case has a different flavor for 1-forms as contrasted with higher forms, so I'll discuss them separately. It is convenient to begin by recalling Grothendieck's theorem [Gro], comparing algebraic and analytic De Rham cohomology (which generalizes the fundamental isomorphism (1.1)).

(2.1) THEOREM (Grothendieck): Let U be a smooth algebraic variety over $\mathbb{C}$. Then the canonical map

$$H^*_{DR}(U) \to H^*_{DR}(U^{an}) \cong H^*(U^{an},\mathbb{C})$$

is an isomorphism.

In fact this theorem is a consequence of a more general "local" result which we'll need also. Let's assume $U \overset{j}{\hookrightarrow} X$ is an open immersion of U into a smooth projective variety and that $Y =_{dfn.} X - U$ is a divisor. Denote by $\Omega^{\cdot}_{X^{an}}(*Y)$ the complex of meromorphic forms on X with only polar singularities along Y . Formally if I is the ideal sheaf defining Y , then $\Omega^p_{X^{an}}(*Y) = \varinjlim_n \Omega^p_{X^{an}} \otimes I^n$. Now we can state

(2.2) THEOREM (Grothendieck): With U,X,Y as above

$$\mathcal{H}^q(\Omega^{\cdot}_{X^{an}}(*Y)) \cong R^q j_*(\mathbb{C}_{U^{an}}) .$$

REMARK: For $q = 1$, this result is due to Atiyah and Hodge, who also prove it for any q provided Y is a smooth divisor with normal crossings (a simple divisor in their terminology).

Now let's discuss 1-forms of the second kind. Let X be a projective smooth variety. Then there is no distinction between rational forms on X and global meromorphic forms on X^{an}.

(2.3) PROPOSITION: Let $\omega \in \Gamma(X,\Omega^1_{rat})$ be a closed rational 1-form. The following conditions are equivalent

(1) Locally in the Zariski topology we can write $\omega = \eta + dg$ where η is regular and g is a rational function.

(2) Locally in the classical topology $\omega = dh$ where h is a meromorphic function.

(3) The residues (defined below) of ω are all zero.

(4) If ω is regular in the Zariski open set U, then the cohomology class it defines in $H^1(U^{an},\mathbb{C})$ is in the image of $H^1(X^{an},\mathbb{C})$.

(2.4) DEFINITION: ω is of the second kind if it satisfies the conditions of (2.3).

To explain the notion of residue it is convenient to introduce some notation which will be used later as well. Let $\Omega^p_{X^{an}}(*)$ denote the direct limit of the $\Omega^p_{X^{an}}(*Y)$ taken over all divisors Y on X. (If $Y \subseteq Y'$, there is a natural inclusion $\Omega^p_{X^{an}}(*Y) \subseteq \Omega^p_{X^{an}}(*Y')$). Thus we obtain $\Omega^{\cdot}_{X^{an}}(*)$ the complex of meromorphic forms on X with "polar singularities". We'll be interested in the cohomology sheaves of this complex. For $q = 0,1$, it is easy to compute $\mathcal{H}^q(\Omega^{\cdot}_{X^{an}}(*))$ (c.f. Atiyah-Hodge) and we find

1) $\mathcal{H}^0(\Omega^{\cdot}_{X^{an}}(*)) \cong \mathbb{C}_X$

2) $\mathcal{H}^1(\Omega^{\cdot}_{X^{an}}(*)) \cong \underline{\mathrm{Div}} \otimes \mathbb{C}$, the sheaf of germs of divisors with complex coefficients.

Now if ω is a closed rational 1-form it defines a global section of $\mathcal{H}^1(\Omega^{\cdot}_{X^{an}}(*))$ and hence a divisor $\Sigma c_i Y_i$, $c_i \in \mathbb{C}$.

(2.5) <u>DEFINITION</u>: If ω is a closed rational 1-form and Y is a divisor on X , then the residue of ω along Y is the coefficient of Y in the corresponding global section of <u>Div</u> $\otimes \mathbb{C}$ (i.e. $\mathrm{res}_{Y_i}(\omega) = c_i$).

From the definition of the hypercohomology $\mathbb{H}^1(X,\Omega^{\cdot}_X)$ and condition (1) in (2.3) we obtain the following corollary.

(2.6) $H^1(X^{an},\mathbb{C}) \cong$ the space of 1 forms of the second kind modulo exact 1-forms.

<u>REMARK</u>: For a surface this result is due to Picard. The general result was proved (in 1952 !) by Rosenlicht [Ros] and also appears in Atiyah-Hodge.

We also have the following corollary.

(2.7) <u>COROLLARY</u>: Let ω be a closed 1-form on X , $Y = \bigcup Y_i$ its polar locus (i.e. ω is regular on $X - Y$). Let V be an affine open set in which Y_i has local equation $f_i = 0$. Then there exist $g \in K(X)$, $\eta \in \Gamma(V,\Omega^1_X)$ such that $\omega = \Sigma\, \mathrm{res}_{Y_i}(\omega) \dfrac{df_i}{f_i} + dg + \eta$.

Now let's turn to another classical object, 2-forms of the second kind. We'll adopt the following definition.

(2.8) <u>DEFINITION</u>: A closed rational 2-form ω is said to be of the second kind if there exists an affine open set U such that ω is regular on U and the cohomology class it defines in $H^2(U^{an},\mathbb{C})$ is in the image of $H^2(X^{an},\mathbb{C})$.

REMARKS: 1) Since in general the restriction maps $H^i(U^{an}) \to H^i(U'^{an})$ for $U' \subseteq U$ are not injective one must bear in mind the fact that the above definition refers to a "sufficiently small" U.

2) We've adopted the analogue of condition 4) in (2.3) as our definition. In view of (2.14) below it is essential that we note explicitly that for 2 forms Atiyah and Hodge prove the equivalence of the definition we've adopted with that we'd obtain by taking the analogue of condition 1) in (2.3) as our definition.

We can now easily prove the following result due to Picard.

(2.4) PROPOSITION: $\dfrac{\text{2 forms of the second kind}}{\{d(\text{rational 1-forms})\}} = \dfrac{H^2(X^{an},\mathbb{C})}{\operatorname{im}(\operatorname{Pic}(X)\otimes\mathbb{C})}$

PROOF: For any affine open U whose complement is a divisor look at the sequence of terms of low degree coming from the Leray spectral sequence for $j: U \hookrightarrow X$ $0 \to H^1(X^{an}) \to H^1(U^{an}) \to \bigoplus_{Y_i} \mathbb{C}_{Y_i} \to H^2(X^{an}) \to H^2(U^{an})$. Passing to the direct limit we find

$$H^2(X^{an}) \big/ \operatorname{im}(\operatorname{Pic}(X)\otimes\mathbb{C}) = \varinjlim_{U} \operatorname{im}[H^2(X^{an}) \text{ in } H^2(U^{an})] .$$

But it is clear that the right hand side can be identified with 2-forms of the second kind modulo exact ones.

Now let's turn to the definition of n forms of the second kind (n arbitrary). Using (2.2) we can rewrite the Leray spectral sequence for the inclusion $j: U \hookrightarrow X$ as

$$E_2^{p,q} = H^p(X^{an}, \mathcal{H}^q(\Omega^{\cdot}_{X^{an}}(*Y))) \Rightarrow H^*(U^{an},\mathbb{C})$$

Passing to the direct limit over such U's we obtain

$$E_2^{p,q} = H^p(X^{an}, \mathcal{H}^q(\Omega^{\cdot}_{X^{an}}(*))) \Rightarrow \varinjlim H^*(U^{an}, \mathbb{C})$$

The abutment can be easily "calculated" as follows:

$$\varinjlim_U H^*(U^{an}, \mathbb{C}) = \varinjlim H_{DR}(U) = \varinjlim \mathbb{H}^*(X, j_{U^*}(\Omega^{\cdot}_U))$$

$$= \varinjlim \mathbb{H}^*(X^{an}, \Omega^{\cdot}_{X^{an}}(*(X-U))) = \varinjlim H^*[\Gamma(X^{an}, \Omega^{\cdot}_{X^{an}}(*(X-U)))]$$

(the last equality holding because U affine implies $H^i(X^{an}, \Omega^j_{X^{an}}(*(X-U))) = (0)$ for $i \geq 1$). Since $\varinjlim$ commutes with H^* we see that

$$\varinjlim H^*(U^{an}, \mathbb{C}) = H^*[\Gamma(X^{an}, \Omega^{\cdot}_{X^{an}}(*))] \ .$$

In fact the spectral sequence is the "second" spectral sequence of hypercohomology for the complex $\Omega^{\cdot}_{X^{an}}(*)$, so its abutment $H^*(X, \Omega^{\cdot}_{X^{an}}(*))$ has a natural filtration Fil^p .

We can now define an n-form of the second kind.

(2.10) DEFINITION: A closed rational n-form $\omega \in \Gamma(X, \Omega^n_{rat})$ is of the second kind if its class in $H^n[\Gamma(X, \Omega^{\cdot}_{X^{an}}(*))]$ has smallest possible filtration, i.e. belongs to Fil^n .

(2.11) REMARKS: 1) Since $E_2^{n,o} = H^n(X^{an}, \mathbb{C})$ maps surjectively to $E_\infty^{n,o} = \mathrm{Fil}^n$ the definition means that there exists some open U (whose complement is a divisor) such that ω is regular on U and the cohomology class it defines in $H^n(U^{an}, \mathbb{C})$ is in the image of $H^n(X^{an}, \mathbb{C})$.

2) If for each U as above we let $W_n[H^n(U^{an}, \mathbb{C})]$ denote the image of $H^n(X^{an}, \mathbb{C})$ in $H^n(U^{an}, \mathbb{C})$, then our definition implies that

$$\frac{\text{n-forms of the second kind}}{\text{exact n-forms}} = \varinjlim_{U} W_n[H^n(U^{an})] .$$

Deligne's theory of mixed Hodge structures allows one to interpret $W_n[H^n(U^{an},\mathbb{C})]$ as the smallest step in an increasing filtration, called the weight filtration, on $H^n(U^{an},\mathbb{C})$. Even though we've defined W_n in terms of a fixed $X \supseteq U$, it can be shown [Br.] that the definition is independent of X and that the filtration is "topological" i.e. is defined on $H^n(U^{an},\mathbb{Q})$, [Del.]. Hence we can think of Deligne's mixed Hodge theory as a strengthening and generalization of the theory of differentials of the second kind.

3) Observe that $\varinjlim H^*_{DR}(U) = H^*_{DR}(K(X))$ which is by definition the cohomology of the complex $\Omega^{\cdot}_{K(X)/\mathbb{C}}$. The fact (noted above) that for U open in X, $\text{im}[H^*_{DR}(X) \to H^*_{DR}(U)]$ is independent of the "compactification" X chosen implies that our definition of n-form of the second kind is a birationally invariant one, i.e. it depends only on the field $K(X)$ and not on the particular projective non-singular model chosen.

4) We have defined n-form of the second kind in terms of a filtration on $H^*_{DR}(K(X))$. Grothendieck observed [Gro] that there is a spectral sequence converging to $H^*_{DR}(X)$ with the property that $gr^\circ H^n_{DR}(X) \cong$ n-forms of the second kind (mod exact). The corresponding filtration on H^*_{DR} is defined by $\text{Fil}^p = \bigcup_Z \text{Ker}[H^*_{DR}(X) \to H^*_{DR}(X - Z)]$ where $Z \subseteq X$ is closed of codimension $\geq p$. Bloch and Ogus have shown [B-O] that (from E_2 on) this spectral sequence may be identified with the second spectral sequence for "algebraic" De Rham cohomology.

$$E_2^{pq} = H^p(X, \mathcal{H}^q(\Omega_X^\cdot)) \Rightarrow H_{DR}^*(X) \ .$$

5) In Atiyah-Hodge it is explained how to define the "generalized" residues of closed rational n-forms. Then to say that a differential is of the second kind <u>means</u> that its $0^{\underline{th}}$ through $n - 1^{\underline{st}}$ residues all vanish. The idea is to write down the cohomology sequences corresponding to the short exact sequences:

$$0 \longrightarrow \Omega^n_{X^{an}, d=0}(*) \longrightarrow \Omega^n_{X^{an}}(*) \longrightarrow d\Omega^n_{X^{an}}(*) \longrightarrow 0$$

$$0 \longrightarrow d\Omega^n_{X^{an}}(*) \longrightarrow \Omega^{n+1}_{X^{an}, d=0}(*) \longrightarrow \mathcal{H}^{n+1}(\Omega^\cdot_{X^{an}}(*)) \longrightarrow 0$$

and use the observation that $H^i(\Omega^n_{X^{an}}(*)) = (0)$ for $i \geq 1$ to splice them together and thus explicate the map $H^n(X^{an}, \mathbb{C}) \to H^n[\Gamma(X, \Omega(*))]$.

We have the following simple proposition.

(2.12) <u>PROPOSITION</u>: If ω is an n-form of the second kind, then locally in the Zariski topology we can write $\omega = \eta + d\tau$ where η is regular and τ is a rational n-1 form .

<u>PROOF</u>: Let $x \in X$ and let Y be a very ample divisor such that $x \notin Y$. Set $U = X - Y$. Let $E_r^{p,q}(Y)$ denote the Leray spectral sequence for the inclusion $j: U \hookrightarrow X$, and $E_r^{p,q}(*)$ the direct limit of these spectral sequences over <u>all</u> divisors Y . Then we have a commutative diagram

$$\begin{array}{ccccc} H^n(X^{an}, \mathbb{C}) & \cong & E_2^{n,o}(Y) & \twoheadrightarrow & \mathrm{Fil}^n[H^n(\Gamma(\Omega^\cdot(*Y)))] \\ \| & & \downarrow \wr & & \downarrow \\ H^n(X^{an}, \mathbb{C}) & \cong & E_2^{n,o}(*) & \twoheadrightarrow & \mathrm{Fil}^n[H^n(\Gamma(\Omega^\cdot(*)))] \end{array}$$

Because the right hand vertical arrow is surjective and because U is affine there is a closed n-form, η, regular on U whose cohomology class comes from X and which modulo exact n-forms is equal to ω.

(2.13) REMARK: Picard and Lefschetz took as their definition of differential of the second kind, the condition that be locally the sum of a regular plus an exact rational form. We've noted above that for 1-forms and 2-forms this definition is equivalent with the Atiyah-Hodge definition that we've adopted. Atiyah and Hodge indicate that, despite Lefschetz's assertion that the two definitions are equivalent; (Actually Lefschetz gave another definition in terms of geometric residues which is shown by Atiyah and Hodge to be equivalent to their's.), they cannot prove this. The reason is given by the following result.

(2.14) THEOREM: The Atiyah-Hodge definition is _not_ equivalent to the Picard-Lefschetz definition.

PROOF: We'll rely heavily on Griffiths' counterexample showing that in general a cycle can be homologically equivalent to zero without any multiple being algebraically equivalent to zero [Gri.]. Besides this we'll have to use some of the recent results of Bloch and Ogus [B.-O.], which we now recall. Consider, for the Zariski topology on X, the spectral sequence

$$E_2^{p,q} = H^p(X,\mathcal{H}^q(\Omega_X^{\cdot})) \Longrightarrow H^q(X^{an},\mathbb{C})$$

They prove [B.-O.] that $E_2^{p,q} = (0)$ if $p > q$. This implies the exactness of the following sequence:

(2.15) $$H^3(X^{an},\mathbb{C}) \to \Gamma(X,\mathcal{H}^3) \xrightarrow{d_2} H^2(X,\mathcal{H}^2) \to \mathcal{H}^4(X^{an},\mathbb{C})$$

Furthermore it is proven that $H^2(X,\mathcal{H}^2)$ is canonically isomorphic to $A^2(X) \otimes \mathbb{C}$ where $A^2(X)$ denotes the group of algebraic cycles of codimension 2 modulo algebraic equivalence. Identifying these two groups, the map to $H^4(X^{an},\mathbb{C})$ is identified with the "cycle class map". Thus by Griffiths this map need not be injective and hence the map $H^2(X^{an},\mathbb{C}) \to \Gamma(X,\mathcal{H}^3)$ need not be surjective. But there is a natural map from $\Gamma(X,\mathcal{H}^3)$ to 3 forms of the second kind in the sense of Picard-Lefschetz, modulo exact forms. Namely the definition of $\mathcal{H}^3$ as the sheaf associated to the presheaf $U \longmapsto H^3(U^{an},\mathbb{C})$, makes it clear that a global section of $\mathcal{H}^3$ is given by the following data:

1) An affine open cover $\{U_i\}$
2) closed 3-forms $\omega_i \in \Gamma(U_i,\Omega^3_X)$
3) For each (i,j) a cover $\{U_{i,j,k}\}$ of $U_i \cap U_j$ and 2-forms $\eta_{i,j,k}$ regular on $U_{i,j,k}$

with the compatibility that on $U_{i,j,k}$ $\omega_i - \omega_j = d\eta_{i,j,k}$. If we fix an i, say i_o, then $\omega_{i_o} = \omega_j + d\eta_{i_o,j,k}$ identically because X is irreducible and hence this expression for $\omega_{i_o}|U_j$ shows ω_{i_o} is a differential of the second kind in the sense of Picard-Lefschetz.

The map we seek is given by assigning to the global section corresponding to the above data the class of ω_{i_o}.

But Bloch and Ogus prove also that the map $\Gamma(X,\mathcal{H}^3) \to H^3_{DR}(K(X))$ is injective and we know the image is contained in the Picard-Lefschetz differentials of the second kind (mod exact). Since by Griffiths, $H^3(X^{an},\mathbb{C})$ won't in general "hit" all of $\Gamma(X,\mathcal{H}^3)$ it certainly won't in general "hit" all of the Picard-Lefschetz differentials of the second kind.

(2.16) <u>REMARK</u>: The following reasoning, due to Katz, might make

more plausible the assertion that
$\Gamma(X,\mathcal{H}^n) \longrightarrow H^n_{DR}(K(X))$ is injective. Namely, let's show that the kernel of the map $H^n(X^{an},\mathbb{C}) \to \Gamma(\mathcal{H}^n)$ is the same as the kernel of the composite map $H^n(X^{an},\mathbb{C}) \to H^n_{DR}(K(X))$. In other words we're to show that if a cohomology class, δ in $H^n(X^{an},\mathbb{C})$ vanishes in a neighborhood of one point, then it vanishes in a neighborhood of any point. So say our class dies in $H^n(U^{an},\mathbb{C})$. Set $Y = X - U$. The long exact homology sequence tells us that δ comes from $H_n(Y^{an},\mathbb{C})$. A striking consequence of Deligne's theory of mixed Hodge structures is that if $\tilde{Y}$ is a resolution (of singularities) of Y, then δ is in the image of the Gysin map $f_*:H^{n-2(\text{codim }(Y))}(\tilde{Y},\mathbb{C}) \to H^n(X^{an},\mathbb{C})$. Intuitively we'd like to move Y around so that it avoided every point. There's no reason to think we can do this, but (and this suffices) we can move the graph $\Gamma_f \subsetneq \tilde{Y} \times X$, thought of as a cycle, around in its algebraic equivalence class so that its "projection" to X will miss some neighborhood of any point.

III. DIFFERENTIALS OF THE THIRD KIND

In this case the terminology is not at all standard. The classical definition is a closed 1-form is of the third kind if it is not of the second kind, i.e. if it has non-zero residues. A more restrictive definition is that the form be expressible locally as $\Sigma\, a_i \frac{df_i}{f_i} + \eta$ where η is regular. Using either definition we can (by the discussion following (2.4)) define the residue divisor of such a differential $\Sigma\, \text{res}_Y(\omega)\cdot Y$. We will adopt the more restrictive definition in what follows.

Picard defined the Picard number ρ (which from Severi we know to be the rank of the group of divisors modulo algebraic equivalence) as follows: Given any $\rho + 1$ curves on a surface we can find a closed

1-form whose residue divisor has as its support, the union of these curves, and ρ is minimal with this property. In more modern terminology we see this as a consequence of the following result of Weil; $[W_1]$:

(3.1) PROPOSITION: Let $\Sigma c_i Y_i$ be a divisor on X, $c_i \in \mathbb{C}$. Then this is the residue divisor of a differential of the third kind if and only if its Chern class in $H^2(X^{an},\mathbb{C})$ is zero.

(3.2) Let's discuss briefly (following Serre [Ser]) the relation between differentials of the third kind in the sense of the restrictive definition above and algebraic groups. Fix a variety (projective, smooth) X over $\mathbb{C}$ and a finite number of divisors $D_1,\ldots,D_r$ on X. Let $D = \bigcup D_i$ and set $U = X - D$. Consider the category C of extensions of abelian varieties by algebraic tori. Serre proved that given U, there is an universal map $f:U \to G$ (in the sense of the usual Albanese mapping property) to an object of C. Here's a procedure for construction (G,f).

Let $\mathrm{Div}(X)$ be the group of divisors of X and denote by I the subgroup generated by $\{D_1,\ldots,D_r\}$. Let J stand for the kernel of the composite map.

(3.3) $$I \to \mathrm{Div}(X) \to \mathrm{Pic}(X) \to \mathrm{Pic}(X)/\mathrm{Pic}^o(X)$$

Thus we have a canonical map $J \to \mathrm{Pic}^o$. Let $T = \underline{\mathrm{Hom}}(J,\mathbb{G}_m)$ be the torus whose character group is J. Denote by $\phi:X \to A$ a canonical map of X to its Albanese variety. Because $\underline{\mathrm{Ext}}^1(A,\mathbb{G}_m) \cong \mathrm{Pic}^o$, it follows formally that $\mathrm{Ext}^1(A,T) \cong \mathrm{Hom}(J,\mathrm{Pic}^o)$ and thus we obtain a canonical extension.

(3.4) $$0 \to T \to G \to A \to o$$

corresponding to the above-constructed map $J \to \mathrm{Pic}^o$.

Considering G as a T-tonseur on A we obtain $E = \phi^*(G)$, a T-torseur on X:

$$\begin{array}{ccccccccc} & & & & X & & & & \\ & & & & \downarrow\phi & & & & \\ o & \to & T & \to & G \to A & \to & o & . \end{array}$$

Any $j \in J = \mathrm{Hom}(T, \mathbb{G}_m)$ defines a $\mathbb{G}_m$-torseur = a line bundle, $j_*(E)$, on X. It is a "complete" tautology that the class of $j_*(E)$ in $\mathrm{Pic}(X)$ is the image of j in $\mathrm{Pic}(X)$ under the inclusion $I \hookrightarrow \mathrm{Pic}(X)$. Recall we can identify a divisor D' with a pair consisting of a line bundle $\mathcal{L}$ and a rational section $s: X \dashrightarrow \mathcal{L}$ such that $\mathrm{div}(s) = D'$. Thus there is a rational section $s_j: X \dashrightarrow j_*(E)$ whose divisor is j. Thus there is a rational section $s: X \dashrightarrow E$ such that the composite map $X \dashrightarrow E \to j_*(E)$ is precisely s_j. s is in fact characterized by this property up to translation by an element in T. Our section $s: X \dashrightarrow E$ gives us a rational function $f: X \dashrightarrow G$. Since s is regular on $X - D$ it follows that f gives us a morphism $U \to G$. Thus we've constructed our pair (G,f).

Serre proves the following result.

(3.5) PROPOSITION:

(1) The pair (G,f) satisfies the universal mapping property.

(2) For ω a translation invariant 1-form on G, $f^*\omega$ is a differential of the third kind on X whose residue divisor is supported in D.

(3) Any differential of the third kind whose residue divisor is supported in D is $f^*(\eta)$ for some translation invariant one-form on G.

REMARKS: 1) The preceeding discussion except for 3) in (3.5) is valid over an algebraically closed field of any characteristic.

2) The point of view that differentials of the third (resp. second) kind should be "interpreted" as extensions of the

Albanese by a torus (resp. a vector group) is expressed quite clearly by Weil, in his striking note [W_2].

BIBLIOGRAPHY

[A-.H.] Atiyah, M.F. and Hodge, W.V.D., Integrals of the second kind on an algebraic variety, Annals of Mathematics, Vol. 62, no. 1, pp. 56-91, July 1955.

[B.-O] Bloch, S. and Ogus, A., On Gersten's Conjecture and the Homology of Schemes, to appear in Ann. Ecole Normale Superieure.

[Br.] Grothendieck, A., Le Groupe de Brauer III in Dix Exposés sur la cohomologie des schémas, North Holland, 1968.

[Del] Deligne, P., Théorie de Hodge, II, Publ. Math. IHES 40, pp. 4-57, 1972.

[Gri] Griffiths, P., On Periods of Pational Integrals II, Annals of Math, 90, pp. 496-541, 1969.

[Gro] Grothendieck, A., On the De Rham cohomology of algebraic varieties, Publ. Math IHES (29), pp. 95-103, 1966.

[GAGA] Serre, J.P., Géométrie algébrique et géométrie analytique, Annales de l'Institut Fourier vol. VI, pp. 1-42, 1956.

[Ros] Rosenlicht, M., Simple differentials of second kind on Hodge manifolds, Am. J. Math., 75, pp. 621-626, 1953.

[Ser] Serre, J.P., Morphisms universels et différentielles de troisième espèce, Seminaire Chevalley, 1958-59, exposé 11.

[W_1] Weil, A., Sur la théorie des formes differentielles attachées à une variété analytique complexe, Comment. Math Helvet., vol. 20, pp. 110-116, 1947.

[W_2] Weil, A., Variétiés abélienne in Algèbre et théorie des nombres, pp. 124-127, Paris, Colloques intern. C.N.R.S., 24, 1950.

DEPARTMENT OF MATHEMATICS
MASSACHUSETTS INSTITUTE OF TECHNOLOGY
CAMBRIDGE, MASSACHUSETTS 02139

Proceedings of Symposia in Pure Mathematics
Volume 29, 1975

SHORT SKETCH OF DELIGNE'S PROOF OF THE HARD LEFSCHETZ THEOREM

William Messing

I will try to explain the ideas and the logical structure of Deligne's proof of the hard Lefschetz theorem. Let us begin by recalling the statement.

THEOREM (H.L.): Let X be a projective smooth connected variety over an algebraically closed field k. Fix a prime $\ell \neq \mathrm{char}(k)$ and let $\xi \in H^2(X,\mathbb{Q}_\ell)^*$ denote the cohomology class of a smooth hyperplane section. Let $L_X : H^*(X,\mathbb{Q}_\ell) \to H^{*+2}(X,\mathbb{Q}_\ell)$ be the map "cup product with ξ". Then for $i \leq \dim(X)$, the map $L_X^{\dim(X)-i} : H^i(X,\mathbb{Q}_\ell) \to H^{2\dim(X)-i}(X,\mathbb{Q}_\ell)$ is an isomorphism.

The theorem was first stated in an equivalent form in Lefschetz's fundamental book [11] on the topology of algebraic varieties (over $\mathbb{C}$). There it is the assertion that the "vanishing cycles" do not meet the invariant cycles. We will see below that they are equivalent statements.

In the classical case (over $\mathbb{C}$) one usually formulates the theorem in rational cohomology. But, Artin's comparison theorem [1] implies that the above formulation yields the assertion in rational cohomology.

Whether or not Lefschetz rigorously proved his theorem is a rather

---

* ξ actually is an element of $H^2(X,\mathbb{Q}_\ell(1))$. Since the hard Lefschetz theorem is a geometric assertion, this can be safely ignored by the reader.

delicate question. Hodge using his theory of harmonic integrals (now called Hodge theory) certainly proved the theorem (over $\mathbb{C}$) [9]. This proof is also given in [14]. Since Hodge's proof does not generalize to characteristic p , once "abstract" algebraic geometry became respectable people were very interested in understanding Lefschetz's "geometric proof" in the hope that it would generalize. Even though Deligne has now given an "arithmetic" proof of the theorem (valid in all characteristics) the question of finding a geometric proof, even in the classical case, is still an open and interesting question.

Deligne's proof is basically arithmetic in nature, in that the main result which he utilizes is the Riemann hypothesis on the archimedian absolute values of the zeroes and poles of the zeta function of a variety defined over a finite field [2]. For the purpose of exposition we may divide the proof into three portions:

1) the geometric portion: application of Lefschetz pencils.

2) the arithmetic and/or group-theoretic portion: the main semi-simplicity theorem.

3) the analytic portion: generalization of the Hadamard-de la Vallée Poussin proof that $\zeta(s) \neq 0$ for $\mathrm{Re}(s) = 1$.

The structure of the proof can now be outlined:

a) A preliminary reduction is made to the case where k is an algebraic closure of a finite field.

b) Using Lefschetz pencils and proceeding by induction on $\dim(X)$ we reduce to proving that: If Y is a general member of a Lefschetz pencil on X then the geometric monodromy representation $\overline{\pi}_1$ (open subset of $\mathbb{P}^1) \longrightarrow \mathrm{Aut}(H^{\dim(X)-1}(Y))$ is semi-simple.

c) To verify the assertion of b) a more general result is proven:

THEOREM 1: Let $U/\mathbb{F}_q$ be an open curve, $\overline{U} = U \otimes \overline{\mathbb{F}}_q$. Let $\rho: \pi_1(U) \to GL_n(\mathbb{Q}_\ell)$ be a representation which is pure of some weight β (c.f. definition below). Then $\rho|\pi_1(\overline{U})$ is semi-simple.

It is this theorem which is really the "heart" of the proof of hard Lefschetz. To prove it Deligne shows that (after fixing an isomorphism between $\overline{\mathbb{Q}}_\ell$ and $\mathbb{C}$) the L function corresponding to the representation ρ does not vanish on the line $\mathrm{Re}(s) = \beta/2 + 1$. It is here that the group-theoretic portion of the proof occurs in passing from the ℓ-adic situation to one in which he can show that certain Dirichlet series do not vanish by applying the generalization of Hadamard-de la Vallée Poussin. Granting the non-vanishing of the L function, the above theorem follows easily using Grothendieck's cohomological expression for the L function . Note that the Riemann hypothesis is used precisely to know that the representation $\rho: \pi_1(U) \longrightarrow \mathrm{Aut}(H^{\dim(X)-1}(Y))$ is pure of weight $\dim(X)-1$ and thus that Theorem 1 is applicable.

The rest of this paper will be devoted to amplifying the above summary. Deligne will shortly publish his paper [3] giving the full proofs. Thus I will either omit the proofs entirely or give only very brief indications of the proofs. The reader is also urged to consult Katz's excellent and comprehensive survey, [10], of Deligne's proof of the Riemann hypothesis for a historical discussion (including hard Lefschetz) as well as certain background material.

1) USE OF LEFSCHETZ PENCILS

Let us first explain the preliminary reduction to the case where $k = \overline{\mathbb{F}}_q$. Using the standard techniques of [6], we can find a projective smooth map $f: \mathfrak{X} \to S$, (ℓ invertible on S), with a geometric generic fiber being our given X , and some special fiber X' being defined over a finite field $\mathbb{F}_q ((\ell,q) = 1)$. The proper, smooth base change theorem tells us that each $R^i f_* \mathbb{Q}_\ell$ is a local system on S and hence that the maps $L^{d-i}_{\mathfrak{X}}: R^i f_*(\mathbb{Q}_\ell) \longrightarrow R^{2d-i} f_* \mathbb{Q}_\ell$ are isomorphisms if and only if they induce isomorphisms $H^i(\mathfrak{X}_{\bar{s}}) \longrightarrow H^{2d-i}(\mathfrak{X}_{\bar{s}})$ for one $s \in S$. Hence we may assume $k = \overline{\mathbb{F}}_q$.

In [5, Expose XVIII] it is explained how one utilizes Lefschetz theory to reinterpret hard Lefschetz. The idea is to prove hard Lefschetz by induction on $\dim(X)$ since Lefschetz's techniques (and results) enable one to understand the topology of a variety in terms of the topology of its hyperplane sections. We will briefly review the results.

Let $\tau : Y \hookrightarrow X$ be the inclusion of a smooth hyperplane section. Since we are proceeding inductively we may assume that hard Lefschetz is true for Y. The weak Lefschetz theorem asserts that $\tau^* : H^i(X) \to H^i(Y)$ is an isomorphism for $i \leq \dim(X) - 2$ and is injective for $i = \dim(X) - 1$. Because $\tau_*(1_Y) = \xi$, and L_Y = cup product with $\tau^*(\xi)$, the projection formula $(\tau_*(\tau^*(x) \cdot y) = x \cdot \tau_*(y)$ for $x \in H^*(X)$, $y \in H^*(Y))$ tells us that the following diagram commutes:

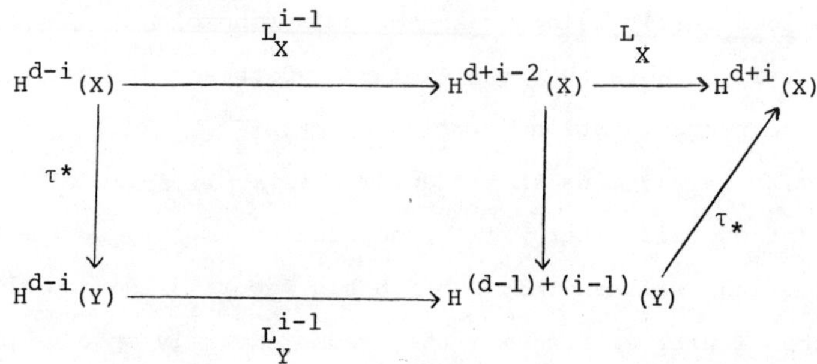

When $i \geq 2$ weak Lefschetz implies τ^* and τ_* (in the diagram) are isomorphisms. By the induction hypothesis L_Y^{i-1} is an isomorphism. Thus the diagram shows that for $i \geq 2$ $L_X^i : H^{d-i}(X) \to H^{d+i}(X)$ is an isomorphism.

Therefore to prove hard Lefschetz for X it suffices to show $L_X : H^{d-1}(X) \to H^{d+1}(X)$ is an isomorphism. By Poincaré duality this is equivalent to showing the pairing on $H^{d-1}(X)$ given by $(a,b) \longmapsto \langle L_X(a) \cdot b \rangle$ is a perfect pairing. By the projection formula and weak Lefschetz this is equivalent to the assertion that the restriction of cup product on $H^{d-1}(Y)$ to $\tau^*(H^{d-1}(X))$ is a perfect pairing.

So far we have only used a single smooth hyperplane section Y. To push our analysis further we will utilize a Lefschetz pencil of hyperplane sections. Recall [5] that a Lefschetz pencil is a family $(Y_t)_{t \in \mathbb{P}^1}$ of hyperplane sections such that:

a) for all t in an open set $\overline{U} \subseteq \mathbb{P}^1$, Y_t is smooth.

b) for $t \in \mathbb{P}^1 - \overline{U}$, Y_t has at most one singular point which is an ordinary double point.

If Y_o and Y_∞ are smooth, then blowing up X along $Y_o \cap Y_\infty$ we obtain a smooth variety $\tilde{X}$ which sits in a diagram $(Y = Y_o)$:

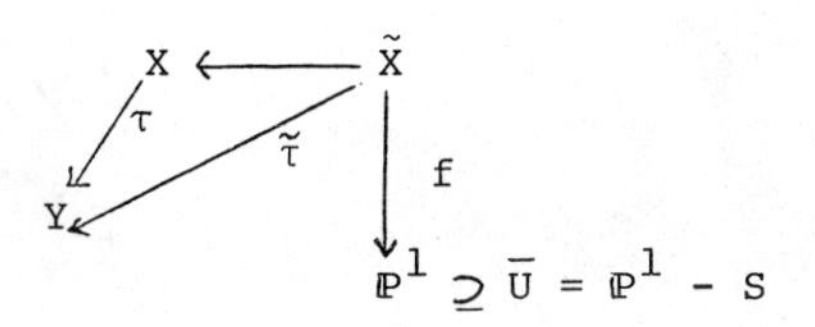

From [5, XVIII, 5.1.6] we know that $\tau^*(H^{d-1}(X)) = \tilde{\tau}^*(H^{d-1}(\tilde{X}))$. Thus we may replace X by $\tilde{X}$, i.e. hard Lefschetz for $\tilde{X}$ implies hard Lefschetz for X. But we are in a better position in that we may assume X is "fibered over $\mathbb{P}^1$". From [5,XVIII, 5.6.10] we know that $\tau^*(H^{d-1}(X)) = H^{d-1}(Y)^{\pi_1(\bar{U})}$, the invariant cohomology. (Recall that since $f_{\bar{U}}: f^{-1}(\bar{U}) \to \bar{U}$ is a proper smooth map $R^{d-1}f_{\bar{U}*}\mathbb{Q}_\ell$ is a local system on $\bar{U}$, i.e. is the representation of $\pi_1(\bar{U})$ on $H^{d-1}(Y)$). This corresponds to Lefschetz's "locus of invariant cycle" construction.

REMARK: To prove the last quoted result one uses the Picard-Lefschetz formula which describes the effect of the local monodromy transformation on $H^{d-1}(Y)$:

For each $s \in \mathbb{P}^1 - \bar{U}$, there is an element $\delta_s \in H^{d-1}(Y)$, the vanishing cycle, such that if $\sigma_s \in \pi_1(\bar{U},o)$ denotes the loop counterclockwise around s we have for $x \in H^{d-1}(Y)$,

$$\sigma_s(x) = x \pm \langle x\cdot\delta_s\rangle\cdot\delta_s .$$

If E is the subspace spanned by the δ_s's, then the Picard-Lefschetz formula tells us that $E^\perp = H^{d-1}(Y)^{\pi_1}$. Thus cup product on $\tau^*(H^{d-1}(X)) = E^\perp$ is a perfect pairing if and only if $E \cap E^\perp = (0)$. This was Lefschetz's original formulation of the hard Lefschetz theorem [11,II,13].

The following proposition shows that hard Lefschetz is a consequence of the semi-simplicity of the monodromy representation $\pi_1(\bar{U}) \to \mathrm{Aut}(H^{d-1}(Y))$ (and hence by the Riemann hypothesis of Theorem 1 above). Note we need only assume that one sequence of π_1-modules splits.

PROPOSITION: Assume the sequence of π_1-modules

$$0 \longrightarrow H^{d-1}(Y)^{\pi_1} \longrightarrow H^{d-1}(Y) \longrightarrow H^{d-1}(Y) \Big/ H^{d-1}(Y)^{\pi_1} \longrightarrow 0$$

splits. Then hard Lefschetz is true (for X).

PROOF: Write $H^{d-1}(Y) = H^{d-1}(Y)^{\pi_1} \oplus M$. Cup product defines a π_1-morphism $\phi : H^{d-1}(Y)^{\pi_1} \longrightarrow \check{M}$ whose image is contained in $(\check{M})^{\pi_1}$. But $(\check{M})^{\pi_1} = (0)$ because $H^{\check{d-1}}(Y) \cong H^{d-1}$ (as π_1-modules) and hence $\dim((\check{H}^{d-1})^{\pi_1}) = \dim((H^{d-1})^{\pi_1})$. This shows ϕ is zero and thus the restriction of cup product to $H^{d-1}(Y)^{\pi_1}$ is a perfect pairing.

2) NON-VANISHING OF L FUNCTION IMPLIES SEMI-SIMPLICITY

Let U be an open subset of a complete nonsingular curve defined over $\mathbb{F}_q$. Let $\overline{U} = U \otimes_{\mathbb{F}_q} \overline{\mathbb{F}}_q$. Let $\pi = \pi_1(U)$, $\overline{\pi} = \pi_1(\overline{U})$ so that we have an exact sequence

$$0 \longrightarrow \overline{\pi} \longrightarrow \pi \xrightarrow{\deg} \mathrm{Gal}(\overline{\mathbb{F}}_q/\mathbb{F}_q) = \hat{\mathbb{Z}} \longrightarrow 0 .$$

For each closed point $x \in U$, there is an element F_x (well defined up to conjugation) in π and satisfying $\deg(F_x) = F^{\deg(x)}$, where $F \in \mathrm{Gal}(\overline{\mathbb{F}}_q/\mathbb{F}_q)$ is the inverse of the Frobenius substitution $\lambda \longmapsto \lambda^q$ and where $\deg(x) = [k(x):\mathbb{F}_q]$. The elements F_x are obtained just as in algebraic number theory [c.f. 12, Chap. I for example].

Let V be a finite dimensional $\mathbb{Q}_\ell$-space and $\rho : \pi \longrightarrow GL(V)$ be a (continuous) representation. The dictionary between galois theory and the étale topology [1] tells us that V, or more precisely ρ, corresponds to a locally constant sheaf $\mathcal{V}$ on U. Recall that the L-function associated to ρ is defined to be the following formal power series in $\mathbb{Q}_\ell[[t]]$:

$$L(\rho,t) = \prod_{x \in U} \frac{1}{\det\left(1 - \rho(F_x)\cdot t^{\deg(x)}\right)}$$

Grothendieck's rationality theorem, [4,8] is the equation:

$$L(\rho,t) = \frac{\det(1 - tF|H_c^1(\overline{U},\mathcal{V}))}{\det(1 - tF|H_c^0(\overline{U},\mathcal{V}))\cdot\det(1 - tF|H_c^2(\overline{U},\mathcal{V}))} .$$

Observe also that $H_c^0 = (0)$ since no section has compact support.

Let us fix an isomorphism $i:\overline{\mathbb{Q}}_\ell \xrightarrow{\sim} \mathbb{C}$. Using this identification we can recall what it mean to say ρ is pure of weight β .

DEFINITION: ρ is pure of weight $\beta \in \mathbb{R}$ if given $x \in U$ and α an eigenvalue of $\rho(F_x)$ we have $|i(\alpha)| = q^{\beta/2\cdot\deg(x)}$ * .

Applying i coefficientwise to $L(\rho,t)$ and making the change of variables $t \longmapsto q^{-s}$ we obtain a Dirichlet series $L_\rho(s)$. Theorem 1, giving the semi-simplicity of $\rho|\overline{\pi}$, is a consequence of:

THEOREM 2): If ρ is pure of weight β , then $L_\rho(s) \neq 0$ for $Re(s) = \beta/2 + 1$.

Let us explain how THEOREM 2 implies THEOREM 1 .
Assume (for the moment):

(*) # of poles of $L(\rho,t) = \dim(V_{\overline{\pi}})$

($V_{\overline{\pi}}$ = coinvariants = largest quotient of V on which $\overline{\pi}$ acts trivially).

Let $\rho^{s.s.}$ denote the semi-simple representation of π obtained by taking the associated graded module, $V^{s.s.}$, corresponding to a Jordan-Hölder series for ρ . Obviously $L(\rho,t) = L(\rho^{s.s.},t)$ and hence

---

* The definition depends on the choice of $i:\overline{\mathbb{Q}}_\ell \xrightarrow{\sim} \mathbb{C}$ and thus differs from that used by Deligne in [2]. Since we make no rationality assumption on the "local factors" $\det(1 - \rho(F_x)t^{\det(x)})$ it would not be natural to make a requirement for all i . The proofs require that we work with lots of representations $\rho:\pi \to GL(V)$ and not just with the monodromy representation $\pi \to Aut(H^{d-1}(Y))$ (which is the representation which we are really interested in).

(by equation (*)) $\dim(V_{\overline{\pi}}) = \dim(V_{\overline{\pi}}^{s.s.})$. Applying this to $\check{V}$, we obtain $\dim(V^{\overline{\pi}}) = \dim(V^{s.s.})^{\overline{\pi}})$. Let $0 \longrightarrow V' \longrightarrow V \longrightarrow V'' \longrightarrow 0$ be a short exact sequence of representations of some fixed weight β . We obtain a corresponding short exact sequence of semi-simplifications

$$0 \longrightarrow V'^{s.s.} \longrightarrow V^{s.s.} \longrightarrow V''^{s.s.} \longrightarrow 0$$

Since taking invariants is exact on this last sequence the equality of dimensions above implies that: $V \longmapsto V^{\overline{\pi}}$ is an exact functor on representations of weight β .

We can now show that any representation $\rho|\overline{\pi}$ is semi-simple (for any ρ of weight β). Let

$$0 \longrightarrow V' \longrightarrow V \longrightarrow V' \longrightarrow 0$$

be exact. Then the map $\mathrm{Hom}_{\overline{\pi}}(V'',V) \longrightarrow \mathrm{Hom}_{\overline{\pi}}(V'',V'')$ is surjective because it is indentified with the map $[\mathrm{Hom}(V'',V)]^{\overline{\pi}} \longrightarrow [\mathrm{Hom}(V'',V'')]^{\overline{\pi}}$ and both $\mathrm{Hom}(V'',V)$, $\mathrm{Hom}(V'',V'')$ are of weight zero.

Now we must verify the equation (*). Since $V_{\overline{\pi}} \cong H_c^2(\overline{U},\mathcal{V})$ (up to a twist), it suffices to show that there is no cancellation between numerator and denominator in

$$L(\rho,t) = \frac{\det(1 - Ft|H_c^1(\overline{U},\mathcal{V}))}{\det(1 - Ft|H_c^2(\overline{U},\mathcal{V}))} .$$

We will verify this statement (using THEOREM 2) in two steps. First let us assume that ρ is an irreducible representation of π . We consider two cases

a) $\rho|\overline{\pi}$ is trivial

In this case $H_c^i(\overline{U},\mathcal{V}) \cong H_c^i(\overline{U},\mathbb{Q}_\ell) \otimes V$. Let $j: U \hookrightarrow Z$ be the smooth compactification of our open curve U . From the exact cohomology sequence

$$\bigoplus_{z\varepsilon Z-U}\mathbb{Q}_\ell \longrightarrow H^1_c(\overline{U},\mathbb{Q}_\ell) \longrightarrow H^1(\overline{Z},\mathbb{Q}_\ell)$$

and the Riemann hypothesis for Z it follows that for α an eigenvalue of F on $H^1_c(\overline{U},\mathcal{V})$ we have (still writing α rather than $i(\alpha)$):

$$|\alpha| = q^{\beta/2} \quad \text{or} \quad q^{\beta/2+1/2}$$

For α an eigenvalue on $H^2_c(\overline{U},\mathcal{V})$ we have $|\alpha| = q^{\beta/2+1}$.

b) $\rho|\overline{\pi}$ is not trivial

Since $V^{\overline{\pi}}$ is π invariant our assumption that ρ is irreducible implies $V^{\overline{\pi}} = (0)$. Because $\overline{\pi}$ is normal in π , $\rho|\overline{\pi}$ is semi-simple. Thus also $V_{\overline{\pi}} = (0)$. Hence $H^2_c(\overline{U},\mathcal{V}) = (0)$ and $L(\rho,t) = \det(1 - Ft|H^1_c)$. We now appeal to THEOREM 2 which implies that for α an eigenvalue on H^1_c , $|\alpha| < q^{\beta/2+1}$. (Note $|\alpha| > q^{\beta/2+1}$ is impossible because $L_\rho(s)$ converges absolutely for $\operatorname{Re}(s) > \beta/2 + 1$).

Combining a) and b) above we conclude that for any irreducible ρ

$$\alpha \text{ on } H^1_c \text{ has } |\alpha| < q^{\beta/2+1}$$

$$\alpha \text{ on } H^2_c \text{ has } |\alpha| = q^{\beta/2+1} .$$

It is now easy to extend this statement to any ρ of weight β . We use induction on the length of the π-module V . Assuming its truth for modules of strictly smaller length, the result follows from the exact sequences

$$H^i_c(\mathcal{V}') \longrightarrow H^i_c(\mathcal{V}) \longrightarrow H^i_c(\mathcal{V}'')$$

corresponding to a short exact sequence

$$0 \longrightarrow V' \longrightarrow V \longrightarrow V'' \longrightarrow 0$$

of π-modules.

3) THE NON-VANISHING OF $L_\rho(s)$ FOR $\mathrm{Re}(s) = \beta/2 + 1$.

If we assume (rather unrealistically) that ρ factors through a finite quotient of π , then $L_\rho(s)$ is an ordinary Artin L function. Then, necessarily $\beta = 0$, and the non-vanishing of $L_\rho(s)$ on $\mathrm{Re}(s) = 1$ is a well known consequence of Brauer's theorem on induced characters [1 bis] which allows the reduction to the case of ordinary abelian L series. Deligne's extension of this result to the situation which interests us is a consequence of a general theorem (Theorem 3 below) which we now prepare to formulate. Consider an extension of groups

$$0 \longrightarrow K^\circ \longrightarrow K \xrightarrow{\deg} \mathbb{Z} \longrightarrow 0$$

where 1) K° is compact and open in K

2) $K/K^\circ \xrightarrow{\sim} \mathbb{Z}$, where $\mathbb{Z}$ has the discrete topology.

3) the image of the center of K , $\deg(Z(K))$, in $\mathbb{Z}$ is not zero.

Fix a non-trivial character $\omega_1 : \mathbb{Z} \longrightarrow \mathbb{R}_+^*$ and for $s \in \mathbb{C}$ write ω_s for the quasi-character $K \longrightarrow \mathbb{C}^*$ given by $k \longmapsto \omega_1(\deg(k))^s$. Let $\{x_\sigma\}_{\sigma\in\Sigma}$ be a countable set of elements of K . We assume that the infinite product

$$\prod_{\sigma\in\Sigma} \frac{1}{1 - \omega_s(x_\sigma)}$$

converges absolutely for $\mathrm{Re}(s) > 1$. Let $\tau : K \to GL(V)$ be a finite dimensional irreducible complex representation. Using the assumption that $\deg(Z(K)) \neq 0$, it is easy to check that there is a unique real number σ such that $\tau \cdot \omega_{-\sigma}$ is a unitary representation. We make that following definition.

DEFINITION: For $\tau : K \to GL(V)$ an irreducible complex representation, the unique $\sigma \in \mathbb{R}$ such that $\tau \cdot \omega_{-\sigma}$ is unitary is called the real part of τ and written $\mathrm{Re}(\tau)$.

Now let $\tilde{K}$ = set of equivalence classes of irreducible representations $\tau : K \to GL(V)$, and let $\hat{K}$ = those which are unitary. We define a complex structure on $\tilde{K}$ by stipulating that for any $\tau \in \tilde{K}$ the set $\{\tau \cdot \omega_s | s \in \mathbb{C}\}$ be open in $\tilde{K}$ and that the map $s \longmapsto \tau \cdot \omega_s$ be holomorphic. This allows us to speak of a function on $\tilde{K}$ as being holomorphic or meromorphic. In particular, the assumption on the above infinite product implies that the function $L(\tau)$ defined by

$$L(\tau) = \prod_{\sigma \in \Sigma} \frac{1}{\det(1 - \tau(x_\sigma))}$$

will converge absolutely in the set $Re(\tau) > 1$.

Now we make the basic assumption concerning the function $L(\tau)$.

<u>ASSUMPTION</u>: The function $L(\tau)$ on $\tilde{K}$ extends to $Re(\tau) \geq 1$, is meromorphic there and is holomorphic in this region except for (at most) a simple pole at ω_1 .

Having completed these preliminaries, we can state Deligne's generalization of the Hadamard-de la Vallée Poussin result.

<u>THEOREM</u> 3): The function $L(\tau)$ is non-zero on $Re(\tau) = 1$ except for at most one representation $\tau_o \cdot \omega_1$. This τ_o is then one-dimensional and defined by a character of order two.

<u>REMARK</u>: The theorem is also true (and the proof is identical) if we assume that $K/K^\circ \simeq \mathbb{R}$. Consider in particular the special case $K = \mathbb{R}$. Then $\tilde{K} = \mathbb{C}$ since a continuous homomorphism $\tau : \mathbb{R} \longrightarrow \mathbb{C}^*$ is or the form $r \longmapsto e^{rt}$ for a uniquely determined $t \in \mathbb{C}$. Letting ω_1 be the exponential and $\{x_\sigma\} = \{-\log p\}$ we find that the L function defined above is precisely the Riemann zeta function. In this case Deligne's proof <u>is</u> the Hadamard - de la Vallée Poussin proof.

We must now explain why THEOREM 2 is a consequence of THEOREM 3 .

Returning to our exact sequence $0 \longrightarrow \overline{\pi} \longrightarrow \pi \longrightarrow \hat{\mathbb{Z}} \longrightarrow 0$, let ρ be any continuous ℓ-adic representation of π . Let $W \subseteq \pi$ be the full inverse image of $\mathbb{Z} \subseteq \hat{\mathbb{Z}}$ so that we obtain the exact sequence

$$0 \longrightarrow \overline{\pi} \longrightarrow W \longrightarrow \mathbb{Z} \longrightarrow 0$$

We retopologize W so that $\overline{\pi}$ with its usually topology is open in W . Then $\rho|W : W \longrightarrow GL(V)$ is still a continuous ℓ-adic representation.

REMARK: W is the quotient of the usual Weil group [6], by the inertia groups corresponding to the points of the complete curve which are not in U . For technical reasons we want W rather than all of π: Basically this is because W is a semi-direct product of $\overline{\pi}$ and $\mathbb{Z}$ and because $\mathrm{Hom}(\hat{\mathbb{Z}}, R_+^*)$ is trivial, (so that $\tilde{K}$ would have been a discrete space had $\hat{\mathbb{Z}}$ been used).

Let G° be the Zariski closure of $\rho(\overline{\pi})$ in $GL(V)$ and $G^{\circ\circ}$ be its connected component. Since W acts (through ρ) on G° we can "push out" the above extension to obtain a diagram, which defines G ,

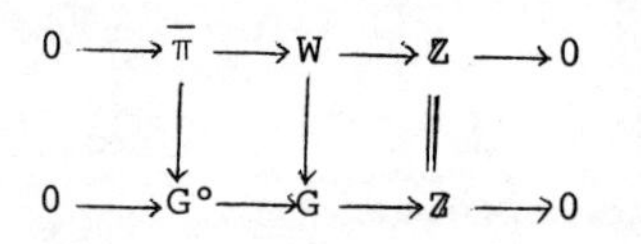

$$\begin{array}{ccccccccc} 0 & \longrightarrow & \overline{\pi} & \longrightarrow & W & \longrightarrow & \mathbb{Z} & \longrightarrow & 0 \\ & & \downarrow & & \downarrow & & \| & & \\ 0 & \longrightarrow & G^\circ & \longrightarrow & G & \longrightarrow & \mathbb{Z} & \longrightarrow & 0 \end{array} .$$

G is an affine group scheme and we still have a representation $\rho : G \longrightarrow GL(V)$. A key result which is used in the proofs of lemmas 1 and 2 below is the following theorem of Grothendieck.

THEOREM (Grothendieck): The unipotent radical of $G^{\circ\circ}$ is equal to the radical of $G^{\circ\circ}$.

REMARK: This result is the global analogue of the key local result [7,13] which Grothendieck uses in his beautiful arithmetic proof of the local monodromy theorem. Since the proof (in a special case) of this local result indicates the ideas which go into the proof of the above theorem without involving us in the group theoretic complications, we will give it.

PROPOSITION: Let $\rho : \mathrm{Gal}\left(\overline{\mathbb{F}_q((T))}/\mathbb{F}_q((T))\right) \longrightarrow GL(V)$ be an ℓ-adic representation. Then the restriction of ρ to $I = \mathrm{Gal}\left(\overline{\mathbb{F}_q((T))}/\overline{\mathbb{F}}_q \cdot \mathbb{F}_q((T))\right)$ is quasi-unipotent, i.e. for $i \in I$, $\rho(i)$ has all its eigenvalues roots of unity.

PROOF: Recall [12] $I \supseteq P$, the pro-ρ group corresponding to wild ramification and that there is a natural isomorphism $t : I/\underline{P} \xrightarrow{\sim} \hat{\mathbb{Z}}(1)(\overline{\mathbb{F}}_q) =_{\text{def.}} \varprojlim \mu_n(\overline{\mathbb{F}}_q)$. Since P is a pro-p group its image $\rho(P) \subseteq GL(V)$ is finite. Thus passing to a subgroup of finite index in I we can assume $\rho(P) = \{1\}$. Let $\bar{x}$ be a topological generator of I/P and $x \in I$ some lifting of $\bar{x}$. For $\psi \in \mathrm{Gal}(\overline{\mathbb{F}_q((T))}/\mathbb{F}_q((T)))$ we know $t(\psi x \psi^{-1}) = \psi(t(x))$. Applying this to some $\sigma \in \mathrm{Gal}$ lifting the Frobenius substitution we obtain $t(\sigma x \sigma^{-1} x^{-q}) = 1 \Longrightarrow \sigma x \sigma^{-1} x^{-q} \in P \Longrightarrow \rho(\sigma x \sigma^{-1}) = \rho(x)^q$. But this obviously implies $\rho(x)$ has eigenvalues roots of 1 since the set of eigenvalues is stable under $\lambda \longmapsto \lambda^q$. This proves the proposition.

Denote by $\overline{G}$, $\overline{G}^\circ$, $\overline{G}^{\circ\circ}$ the quotients by the radical = unipotent radical of $G^{\circ\circ}$.

LEMMA 1: The center, $Z(\overline{G})$, has non-zero image in $\mathbb{Z}$ (under the map $\deg : \overline{G} \longrightarrow \mathbb{Z}$).

The proof of this lemma is easy once we know $\mathrm{Aut}(\overline{G}^{\circ\circ})$ is an extension of a finite group by the group of inner automorphisms. This holds because $\overline{G}^{\circ\circ}$ is connected and semi-simple.

Using Lemma 1, we can define the semi-simple component of an element

$g \in \overline{G}$ as follows: Because we are looking at groups in characteristic zero, there is a unique unipotent element g_u, of $\overline{G}^\circ$ which projects onto the unipotent part of the Jordan decomposition of the image of g in the algebraic group $\overline{G}/Z(\overline{G})$. Define the semi-simple component to be $g_{s.s.} = gg_u^{-1}$.

Now we make use of our identification $i:\overline{Q}_\ell \xrightarrow{\sim} C$ to obtain complex groups $G^{\circ\circ}_{\mathbb{C}}$, $G^\circ_{\mathbb{C}}$, $G_{\mathbb{C}}$ .... If Z' is any subgroup of finite index in $Z(\overline{G}_{\mathbb{C}})$, then $\overline{G}_{\mathbb{C}}/Z'$ is a complex Lie group with only finitely many connected components. But then $\overline{G}_{\mathbb{C}}/Z'$ admits a "polar decomposition" as $K^* \times E$ where K^* is a maximal compact subgroup and E is a Euclidean space. (Think of the decomposition $GL(n,\mathbb{C}) \cong U(n) \times \left\{\begin{matrix}\text{positive definite}\\ \text{Hermetian matrices}\end{matrix}\right\}$). Let K = full inverse image of K^* in $\overline{G}_{\mathbb{C}}$ and $K^\circ = K \cap \overline{G}^\circ_{\mathbb{C}}$. Thus we obtain an exact sequence

$$0 \longrightarrow K^\circ \longrightarrow K \xrightarrow{\deg} \mathbb{Z} \longrightarrow 0$$

which satisfies the assumption that $\deg(Z(K)) \neq 0$. We shall apply THEOREM 3 to this sequence. At this point we must introduce the assumption that our representation $\rho:W \longrightarrow GL(V)$ is pure of weight β. Recall we've chosen a Frobenius element $F_v \in W$ for each closed point v of U. Let F_v still denote the image of our element in G or $\overline{G}$ or $\overline{G}_{\mathbb{C}}$.

LEMMA 2: There are elements x_v in K which are conjugate to the elements $(F_v)_{s.s.}$ of $\overline{G}_{\mathbb{C}}$.

REMARKS: 1) To prove this lemma one passes to the semi-simplification $V^{s.s.}$ of ρ. $\overline{G}$ will then act on this space because the unipotent radical of $G^{\circ\circ}$ will act trivially. Grothendieck's theorem is used precisely to know that

a) $Z(\overline{G})$ is big, which follows because $\overline{G}^\circ$ is semi-simple.

b) $\overline{G}$ acts on $V^{s.s.}$, which follows because we've "only" divided out by the unipotent radical.

2) Recall that K depends upon the choice of a subgroup, Z' of finite index in $Z(\overline{G})$. The passage to such a Z' (rather than just taking $Z(\overline{G})$) is necessitated by the proof of lemma 2.

We'll apply Theorem 3 to the above constructed K, and x_v's. We choose $\omega_1 : \mathbb{Z} \longrightarrow \mathbb{R}_+^*$ to be the map $\omega_1(n) = q^{-n}$. We must now verify that the hypotheses made in the discussion preceding THEOREM 3 are all satisfied. We've noted already that $Z(K)$ has non-zero image in $\mathbb{Z}$ and its clear that

$$\prod \frac{1}{1 - \omega_s(x_v)} = \prod \frac{1}{1 - q^{-\deg(v)s}}$$

converges absolutely for $\mathrm{Re}(s) > 1$. Let $\alpha \in K$ satisfy $\deg(\alpha) = 1$. Then a representation $\tau : K \longrightarrow GL(n,\mathbb{C})$ is simply a representation $\tau_o : K^\circ \longrightarrow GL(n,\mathbb{C})$ plus an automorphism $\tau(\alpha)$ satisfying $\tau_o(\alpha k \alpha^{-1}) = \tau(\alpha)\ \tau_o(k)\ \tau(\alpha)^{-1}$ for all $k \in K^\circ$. There is an identical description of an algebraic representation τ of $\overline{G}_{\mathbb{C}}$ as an algebraic $\tau_o : \overline{G}^\circ_{\mathbb{C}} \longrightarrow GL(n,\mathbb{C})$ plus the giving of an automorphism $\tau(\alpha)$. But the continuous representations of K° are in bijective correspondence with the algebraic representations of $\overline{G}^\circ$ (because this group is semi-simple). By transport of structure we have a bijective correspondence between continuous complex representations of K and the ℓ-adic representations $\tau : \overline{G} \longrightarrow GL(n,\mathbb{Q}_\ell)$. Under this correspondence an irreducible representation τ_ℓ go over into irreducible representation $\tau_{\mathbb{C}}$. It is obvious that we have an identity of formal power series in t:

$$i\left(\prod \frac{1}{\det(1 - \tau_\ell(F_v)t^{\deg v})}\right) = \prod \frac{1}{\det(1 - \tau_{\mathbb{C}}(x_v)t^{\deg v})}\ .$$

Substituting q^{-s} for t in the right hand side gives us $L(\tau_{\mathbb{C}} \cdot \omega_s)$. Hence Grothendieck's rationality theorem [4,8] certainly implies $L(\tau)$ is meromorphic in $\mathrm{Re}(\tau) \geq 1$.

Let $\tau_{\mathbb{C}}$ have real part = 1 . The representation $\tau_\ell : \overline{G} \longrightarrow GL(V)$ when restricted to $\overline{\pi}$ defines an ℓ-adic sheaf $\mathcal{V}$ and the rationality theorem tells us that the denominator of $L(\tau)$ is $i\left(\det(1 - Ft|H_c^2(\overline{U},\mathcal{V}))\right)_{t=1}$. Observe that $\tau_\ell|W$ is irreducible and hence $\tau_\ell|\overline{\pi}$ is semi-simple. If $V^{\tau_\ell(\overline{\pi})} = (0)$, then $H_c^2(\overline{U},\mathcal{V}) \cong V_{\tau_\ell(\overline{\pi})}(-1) = (0)$ and L is holomorphic at $\tau_{\mathbb{C}}$. Otherwise $\tau_\ell|\overline{\pi}$ is trivial. Since our representation τ_ℓ then is simply the giving of the automorphism $\tau_\ell(\alpha)$ (notation as above), it follows that V is necessarily 1-dimensional and $\tau_{\mathbb{C}}$ = multiplication by a scalar λ . The denominator of $L(\tau_{\mathbb{C}})$ is then $i - q\lambda$, because $H_c^2(\overline{U},\mathbb{Q}_\ell) = \mathbb{Q}_\ell(-1)$. Thus if $L(\tau_{\mathbb{C}})$ has a pole it is necessarily a simple pole and we have $\lambda = 1/q$, i.e., $\tau_{\mathbb{C}} = \omega_1$.

This tells us that all the hypotheses of Theorem 3 are satisfied. Hence we know $L(\tau_{\mathbb{C}}) \neq 0$ on $\mathrm{Re}(\tau) = 1$ except for at most one, one dimensional representation $\tau_o \cdot \omega_1$ given by a character τ_o of order two. If τ_o is not the trivial character then it is standard [15] that $L(\tau_o \cdot \omega_1) \neq 0$. If τ_o is the trivial character then the numerator of $L(\tau_o \cdot \omega_1)$ is $i\left(\det\left(1 - \frac{F}{q}|H_c^1(\overline{U},\mathbb{Q}_\ell)\right)\right)$. But the Riemann hypothesis for curves tells us that the eigenvalues of F acting on $H_c^1(\overline{U},\mathbb{Q}_\ell)$ have complex absolute value either 1 or $q^{1/2}$. Thus $L(\tau_o \cdot \omega_1)$ does not vanish in this case either and so we now know $L(\tau) \neq 0$ on $\mathrm{Re}(\tau) = 1$.

Now let us return to our representation $\rho : W \longrightarrow GL(V)$ assumed pure of weight β . Let τ_ℓ be an irreducible representation of $\overline{G}$ occuring in $\rho^{s.s.} : \overline{G} \longrightarrow GL(V)$. Then τ_ℓ has weight β and this means that the real part of the corresponding $\tau_{\mathbb{C}}$ is $-\beta/2$. Hence $L(\tau_C \cdot \omega_s) \neq 0$ on the line $\mathrm{Re}(s) = \beta/2 + 1$. As $L_\rho(s)$, whose definition we recall is $i\left(\prod \frac{1}{\det(1 - \rho(F_v)t)}\right)_{t=q^{-s}}$, is a product of such functions $L(\tau_{\mathbb{C}} \cdot \omega_s)$, it follows that $L_\rho(s) \neq 0$ for $\mathrm{Re}(s) = \beta/2 + 1$.

Thus we've shown THEOREM 3 implies THEOREM 2 and have therefore proven hard Lefschetz.

Bibliography

(1) Artin, M., and Grothendieck, A. and Verdier, J.L., SGA 4 Théorie des topos et cohomologie étale des schemas, Lecture Notes in Mathematics, 269, 270, 305, Springer-Verlag, Berlin, 1972.

(1 bis) Brauer, R., On Artin L series with general group characters, Annals of Math., vol. 48, pp. 502-514, (1947).

(2) Deligne, P., La Conjecture de Weil I., Pub. Math., IHES 43 (1974), pp. 273-307.

(3) Deligne, P., La Conjecture des Weil II, to appear.

(4) Deligne, P., Les Constantes des équations fonctionelles des fonctions L , in Modular Functions of One Variable II, Lecture Notes in Mathematics 349, Springer-Verlag, Berlin, 1973.

(5) Deligne, P. and Katz, N., SGA 7, part II, Groupes de monodromie en géometrie algébrique.

(6) Grothendieck, A., EGA_{IV}, Elements de géométrie algébrique, Pub. Math. IHES 28.

(7) Grothendieck, A., SGA 7, part I., Groupes de monodromie en géométrie algébrique, Lecture Notes in Mathematics 288, Springer-Verlag, Berlin, 1972.

(8) Grothendieck, A., Formule de Lefschetz et rationalité des fonctions L., Seminaire Bourbaki, 1964/65, exposé 279, W.A. Benjamin, New York, 1966.

(9) Hodge, W., The theory and applications of harmonic integrals, Cambridge University Press, 1941.

(10) Katz, N., An overview of Deligne's proof of the Riemann hypothesis for varieties over finite fields, to appear.

(11) Lefschetz, S., L'Analysis situs et la Géométrie Algébrique, Gauthier-Villars, Paris, 1924, (This is reprinted in Selected Papers; Chelsea, New York, 1971).

(12) Serre, J.P., Corps locaux, Hermann, Paris, 1962.

(13) Serre, J.P. and Tate, J., Good reduction of abelian varieties, Ann. of Math 88, (1968), pp. 492-517.

(14) Weil, A., Varietes Kahlériennes, Hermann, Paris, (1958).

(15) Weil, A., Basic Number Theory, Springer-Verlag, Berlin, (1967).

(16) Weil, A., Sur la théorie du corps de classes, J. Math. Soc. Japan, vol. 3, no. 1, (1951), pp. 1-35.

DEPARTMENT OF MATHEMATICS
MASSACHUSETTS INSITUTE OF TECHNOLOGY
CAMBRIDGE, MASSACHUSETTS 02139

Proceedings of Symposia in Pure Mathematics
Volume 29, 1975

P-ADIC INTERPOLATION VIA HILBERT MODULAR FORMS

Kenneth A. Ribet

Katz's article [3] was primarily concerned with the question of constructing p-adic measures $\mu^{(a)}$ on $\mathbb{Z}_p$ whose moments are the values at negative integers of the Riemann zeta function. Here we shall try to generalize one of the approaches discussed by Katz, the technique involving modular forms, to study instead the values at negative integers of the zeta function attached to <u>any</u> number field.

I. Deligne's Integrality Theorem

To begin, let K be a number field and let $\zeta_K(s)$ be its Dedekind zeta function. Since the values ζ_K at negative integers are all zero if K is not totally real, we shall assume that K is a <u>totally real</u> field. The values of ζ_K at negative <u>even</u> integers are in any case zero, and $\zeta_K(0) = 0$ except when $K = \mathbb{Q}$ ($\zeta_{\mathbb{Q}}(0) = -1/2$). Moreover, according to a result of Siegel ([6], p. 136), the numbers

$$\zeta_K(1-k), \quad k \geq 1$$

are rational.

It is therefore natural to ask whether or not the function

$$k \to \zeta_K(1-k)$$

has p-adic properties analogous to those of the Riemann zeta function. Given the situation for $\zeta_{\mathbb{Q}}$, in fact, we might try to construct p-adic measures $\mu^{(a)}$ on $\mathbb{Z}_p$ whose moments satisfy

$$\int x^{k-1}\, d\mu^{(a)} \quad = \quad f(k,\mu^{(a)})\cdot\zeta_K(1-k),$$

where $f(k,\mu^{(a)})$ is a simple "fudge factor" analogous to the factor $1-a^k$ that comes up when $K = \mathbb{Q}$. Again (to continue the analogy) we might look for Eisenstein series g_k which have q-expansions

$$\sum_{n \geq 0} a_{nk} q^n$$

such that $a_{nk} \in \mathbb{Z}$ for $n \geq 1$ and such that a_{0k} is essentially $\zeta_K(1-k)$. These are provided by the following result.

<u>Theorem</u> (Siegel, Serre [5]). For each $k \geq 1$, there exists a modular form g_k of weight $kr = k\cdot[K:\mathbb{Q}]$ whose q-expansion has constant term

$$a_{0k} = 2^{-r}\cdot\zeta_K(1-k)$$

and higher terms

$$a_{nk} = \begin{cases} 0 \text{ if } k \text{ is odd} \\ \displaystyle\sum_{\substack{Tr(x) = n \\ x \in \mathfrak{d}^{-1} \\ x \gg 0}} \sum_{\mathfrak{A} | (x)\mathfrak{d}} (N\mathfrak{A})^{k-1} \quad \text{if } k \text{ is even.} \end{cases}$$

(In the double sum we sum first over a finite set of elements x of the inverse different $\mathfrak{d}^{-1}$ of K, and for each x we then sum over the (finite) set of integral ideals $\mathfrak{A}$ which divide the ideal $(x)\mathfrak{d}$.)

Now let p be a prime. By combining the above theorem with the technique of ([3],§XII) we get measures $\mu^{(a)}$ with the desired property:

<u>Theorem 1</u>. For each $a \in \mathbb{Z}_p^*$ there exists a $\mathbb{V}$-valued measure $\mu^{(a)}$ on $\mathbb{Z}_p$ whose moments are given by the formula

$$\int_{\mathbb{Z}_p} x^{k-1}\, d\mu^{(a)} = (1-a^{kr})g_k.$$

<u>Corollary</u>. For each $a \in \mathbb{Z}_p^*$ there exists a $\mathbb{Z}_p$-valued measure $\mu^{(a)}$ on $\mathbb{Z}_p$ such that

$$\int_{\mathbb{Z}_p} x^{k-1}\, d\mu^{(a)} = (1-a^{kr})\, 2^{-r}\, \zeta_K(1-k)$$

for $k \geq 1$. Consequently, the number $2^{-r}\zeta_K(1-k)$ is p-integral if $kr \not\equiv 0 \pmod{p-1}$.

As Serre pointed out in his Antwerp lectures, this corollary does not give the "best" integrality statement for values of ζ_K at negative integers (cf. [4], p. 164). Indeed, we have the following result of Deligne.

<u>Theorem 2</u>([2]). The quantity $2^{-r}\zeta_K(1-k)$ is p-integral whenever $kd \not\equiv 0$ (mod p-1), where d is the degree over $\mathbb{Q}$ of the intersection of K with the field $\mathbb{Q}(\mu_{p^\infty})$ of p-power roots of unity.

The idea behind this theorem is Serre's suggestion that g_k be viewed not merely as a function on the $\mathcal{M}^{triv}$ of [3] but instead as the restriction to $\mathcal{M}^{triv}$ of a function G_k defined on a larger moduli scheme, the Hilbert-Blumenthal scheme $\mathcal{M}_K^{triv}$ discussed below. This is possible since functions on $\mathcal{M}_K^{triv}$ are p-adic <u>Hilbert</u> modular functions (just as functions on Katz's $\mathcal{M}^{triv}$ are generalized one-variable p-adic modular functions), whereas g_k by its very construction over $\mathbb{C}$ is the restriction to the usual upper half plane of a Hilbert modular form G_k whose (generalized) q-expansion is rational [7]. The point, in other words, is to make algebraic sense out of Siegel's G_k. Doing this allows us to construct a new family of measures μ on $\mathbb{Z}_p$ so that for each k satisfying $kd \not\equiv 0$ (mod p-1) there is a measure μ for which the fudge factor $f(k,\mu)$ is a unit.

II. <u>The Hilbert-Blumenthal Scheme</u> $\mathcal{M}_K^{triv}$

We first need a notion replacing that of an elliptic curve /R. Let O be the integer ring of K. A <u>Hilbert-Blumenthal structure</u> over a ring R is an abelian scheme X/R together with an inclusion

$$m: \mathcal{O} \hookrightarrow \mathrm{End}_R X$$

which makes Lie (X), the tangent space to X at the origin, free of rank 1 over $\mathcal{O} \otimes R$. A trivialization of such a structure over a p-adically complete and separated R is an isomorphism of formal groups

$$\varphi: \mathcal{O} \otimes_{\mathbb{Z}} \hat{\mathbb{G}}_m \xrightarrow{\sim} \hat{X}.$$

If X admits such a trivialization, then X is (fibre-by-fibre) ordinary. Given an ordinary Hilbert-Blumenthal structure over R (i.e., a structure with X ordinary) there will in general be no trivialization over R. However, if φ is one trivialization, then we can get other trivializations $a\varphi$ by "twisting" φ by elements $\underline{a}$ of

$$\mathrm{Aut}(\mathcal{O} \otimes \hat{\mathbb{G}}_m) = (\mathcal{O} \otimes \mathbb{Z}_p)^*.$$

Now we define two stacks on the category of p-adically complete and separated rings:

$$\begin{cases} \mathcal{M}_K^{\mathrm{triv}}(R) = \text{the trivialized Hilbert-Blumenthal structures } /R \\ \mathcal{M}_K^{\mathrm{ord}}(R) = \text{the ordinary Hilbert-Blumenthal structures } /R. \end{cases}$$

These are direct generalizations of the stacks $\mathcal{M}^{\mathrm{triv}}$ and $\mathcal{M}^{\mathrm{ord}}$ of elliptic curves associated to the "one-variable case," and they are connected by a "Galois" covering

$$\begin{array}{c} \mathcal{M}_K^{\mathrm{triv}} \\ \downarrow \\ \mathcal{M}_K^{\mathrm{ord}} \end{array}$$

with structural group $(\mathcal{O} \otimes \mathbb{Z}_p)^*$. As before, the stack $\mathcal{M}_K^{\text{triv}}$ "is" the formal affine scheme over $\mathbb{Z}_p$ which represents the functor

$$R \mapsto \begin{array}{l}\text{isomorphism classes of trivialized}\\ \text{Hilbert-Blumenthal structures} \quad /R\end{array}.$$

In analogy with the elliptic curve case, we call elements of the coordinate ring $\mathbb{V}_K$ of $\mathcal{M}_K^{\text{triv}}$ p-adic Hilbert modular functions (over $\mathbb{Z}_p$). Given any trivialized structure (X,m,φ) over a p-adically complete and separated R, we can evaluate any $f \in \mathbb{V}_K$ at (X,m,φ) to get a number $f(X,m,\varphi)$ in R.

Now let $N:(\mathcal{O} \otimes \mathbb{Z}_p)^* \to \mathbb{Z}_p^*$ be the norm. Also, given $a \in (\mathcal{O} \otimes \mathbb{Z}_p)^*$ and $f \in \mathbb{V}_K$, let $[a]f$ be the function satisfying for each (X,m,φ):

$$([a]f)(X,m,\varphi) \quad = \quad f(X,m,a^{-1}\varphi)$$

This rule defines an operation $[a]$ on $\mathbb{V}_K$, which we extend by linearity to $V_K\left[\frac{1}{p}\right]$.

Definition. A function $f \in \mathbb{V}_K\left[\frac{1}{p}\right]$ has weight k if

$$[a]\, f \quad = \quad (Na)^k f.$$

for every $a \in (\mathcal{O} \otimes \mathbb{Z}_p)^*$.

Theorem (Siegel). For each $k \geq 1$ there exists a function $G_k \in \mathbb{V}_K\left[\frac{1}{p}\right]$ which has weight k and a (generalized) q-expansion whose constant term is $2^{-r}\zeta_K(1-k)$ and whose higher coefficients are all integers.

As mentioned earlier, this is the "key point." To prove this result, one constructs G_k as a classical Hilbert modular form [7] and observes that the "q-expansion" of G_k is rational. By the analogue of the q-expansion principle, G_k is a Hilbert modular form which is "defined over $\mathbb{Q}$." Thus it is a (negative) power of p times a Hilbert modular form over $\mathbb{Z}_p$. But on the other hand, we can view a true modular form as a p-adic modular function just as in the one-variable case; this gives exactly what is desired.

The legitimacy of this chain of reasoning rests on our knowing what the q-expansion of a Hilbert modular form actually is and our knowing that a function is determined by its q-expansion (at least in characteristic 0). We will return soon to the latter point, although we will ignore the former point from now on. Incidentally, the theory of q-expansions for Hilbert modular forms is contained in the (unpublished) work of M. Rapoport on the Hilbert modular scheme.

Assuming a satisfactory theory of q-expansions we will prove for $\mathbb{V}_K$ a

<u>Key Lemma</u>. Let $h \in \mathbb{V}_K\left[\frac{1}{p}\right]$ be a function whose q-expansion is p-integral except perhaps for its constant term. Then for each $a \in (\mathcal{O} \otimes \mathbb{Z}_p)^*$ the difference

$$h - [a]h$$

belongs to $\mathbb{V}_K$.

<u>Proof</u>. Let c be the constant term of the q-expansion of h. Then $h - c$ belongs to $\mathbb{V}_K$ because it is an element of $\mathbb{V}_K\left[\frac{1}{p}\right]$ with integral q-expansion. Since $[a]c = c$, we have

$$h - [a]h = (h-c) - [a](h-c) \in \mathbb{V}_K.$$

<u>Theorem 3</u>. For each $a \in (\mathcal{O} \otimes \mathbb{Z}_p)^*$, the number

$$\{1 - (\mathrm{N}a)^k\}\, 2^{-r}\zeta_K(1-k)$$

is p-integral.

<u>Proof</u>. Since G_K has weight k,

$$G_k - [a]G_k = \{1 - (\mathrm{N}a)^k\}G_k.$$

The former is an element of $\mathbb{V}_K$ by the Key Lemma, so in particular it has an integral q-expansion. Therefore the constant term of $\{1 - (\mathrm{N}a)^k\}G_k$ is integral; this is exactly what we want.

Now if a is in $\mathbb{Z}_p^*$, then $\mathrm{N}a = a^r$. Hence Theorem 3 tells us in particular that

$$(1-a^{rk})2^{-r}\zeta(1-k)$$

is p-integral whenever $a \in \mathbb{Z}_p^*$; this is Theorem 1 (or more precisely its corollary). On the other hand, Theorem 3 is a consequence of Theorem 2. Indeed, let K_p be the largest abelian extension of K which is unramified away from p, and let $G = \text{Gal}(K_p/K)$. Let μ_{p^∞} be the group of p-power roots of unity in $\overline{K}$. Then we have a diagram

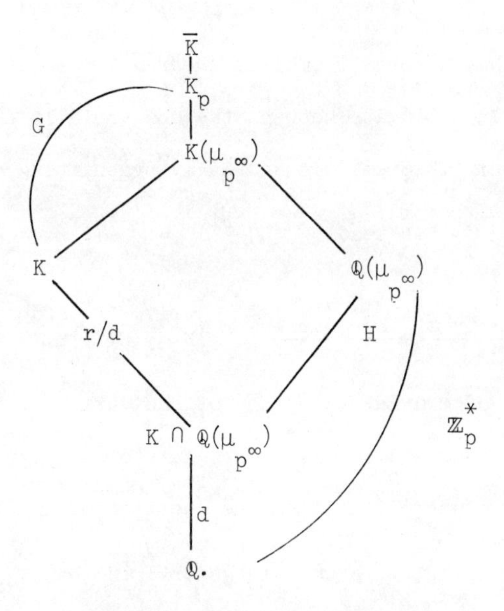

Restriction provides a norm map $N: G \to \mathbb{Z}_p^*$ whose image H is

$$\text{Gal}(\mathbb{Q}(\mu_{p^\infty}):K \cap \mathbb{Q}(\mu_{p^\infty})).$$

Thus the image of G in $\mathbb{Z}_p^*$ has index $d = [K \cap \mathbb{Q}(\mu_{p^\infty}):\mathbb{Q}]$. If j is the composition of the natural inclusion

$$(\mathcal{O} \otimes \mathbb{Z}_p)^* \hookrightarrow (\text{Idèles of } K)$$

with the Artin map

$$(\text{Idèles of } K) \to \text{Gal}(K_p/K),$$

then j is a map $(\mathcal{O} \otimes \mathbb{Z}_p)^* \to G$, and the diagram

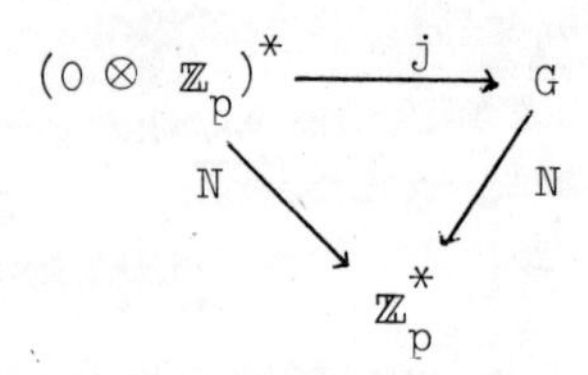

is (anti-) commutative. It follows that

$$(\mathbb{Z}_p^*)^d = H \supseteq N[(\mathcal{O} \otimes \mathbb{Z}_p)^*]$$

and this shows that Theorem 2 implies Theorem 3. In other words, we have obtained a result intermediate between the one-variable theorem (Theorem 1) and the "good" theorem (Theorem 2) essentially by shifting our perspective and working with $\mathcal{M}_K^{\mathrm{triv}}$ instead of $\mathcal{M}^{\mathrm{triv}} = \mathcal{M}_{\mathbb{Q}}^{\mathrm{triv}}$.

III. <u>Irreducibility of the Covering</u> $\mathcal{M}_K^{\mathrm{triv}} \to \mathcal{M}_K^{\mathrm{ord}}$.

In the case of modular functions of one variable, Katz obtained the main facts concerning q-expansions as a corollary of the irreducibility of $\mathcal{M}^{\mathrm{triv}}$ ([3], § XI). Here we will discuss the question of irreducibility for $\mathcal{M}_K^{\mathrm{triv}}$.

The first difficulty is that in general the base $\mathcal{M}_K^{\mathrm{ord}}$ is not connected. Indeed, if X/k is a Hilbert-Blumenthal structure over an algebraically closed field, we define its <u>polarization module</u> $\mathcal{P} = \mathcal{P}(X)$, a certain invertible $\mathcal{O}$-module "with positivity," as follows: $\mathcal{P}$ is the set of $\mathcal{O}$-homomorphisms $f \in \mathrm{Hom}_k(X,\hat{X})$ which are <u>symmetric</u> in the sense that $f = \hat{f}$ (note that $\hat{f}$ is a map $X = \hat{\hat{X}} \to \hat{X}$ just as f is); $f \in \mathcal{P}$ is <u>positive</u> if f is a polarization of X, i.e., an isogeny $X \to \hat{X}$ associated to some ample line bundle on X. Now just as the isomorphism classes of invertible $\mathcal{O}$-modules are the ideal classes of K, so the isomorphism classes of invertible $\mathcal{O}$-modules <u>with positivity</u> are the <u>strict</u> ideal classes of K. Thus, taking $\mathcal{P}(X)$ to its isomorphism class enables us to associate to the Hilbert-Blumenthal structure X a strict ideal class of K. It turns out that this association decomposes $\mathcal{M}_K^{ord}$ into components $\mathcal{M}_{K,\mathfrak{C}}^{ord}$ parameterized by the strict ideal classes of K and that each component $\mathcal{M}_{K,\mathfrak{C}}^{ord}$ is geometrically irreducible.

For each $\mathfrak{C}$, let $\mathcal{M}_{K,\mathfrak{C}}^{triv}$ be the fibre of $\mathcal{M}_K^{triv}$ over $\mathcal{M}_{K,\mathfrak{C}}^{ord}$. We still have for each $\mathfrak{C}$ a covering

$$\begin{array}{c}\mathcal{M}_{K,\mathfrak{C}}^{triv}\\ \downarrow\\ \mathcal{M}_{K,\mathfrak{C}}^{ord}\end{array}$$

with structural group $(\mathcal{O}\otimes\mathbb{Z}_p)^*$

<u>Theorem</u>. The scheme $\mathcal{M}_{K,\mathfrak{C}}^{triv}$ is geometrically irreducible.

As explained by Katz, the covering gives rise to a character

$$\chi:\pi_1(\mathcal{M}_{K,\mathfrak{C}}^{ord})\rightarrow(\mathcal{O}\otimes\mathbb{Z}_p)^*$$

and (given the irreducibility of the base) the theorem is equivalent to the surjectivity of

$$\chi\mid\pi_1(\mathcal{M}_{K,\mathfrak{C}}^{ord}\otimes\bar{\mathbb{F}}_p)$$

For simplicity, we will prove this only when $\mathfrak{C}$ is the class of the inverse different $\mathfrak{d}^{-1}$ of K.

For convenience, let us adopt the following notation:

$$\mathcal{M}^{triv}=\mathcal{M}_{K,\mathfrak{C}}^{triv}\otimes\mathbb{F}_p,$$

$$\mathcal{M}=\mathcal{M}_{K,\mathfrak{C}}^{ord}\otimes\mathbb{F}_p,$$

$$\mathcal{O}_p=\mathcal{O}\otimes\mathbb{Z}_p.$$

Also, let χ now be the character

$$\pi_1(\mathcal{M}) \to \mathcal{O}_p^*$$

arising from the covering

$$\mathcal{M}^{\text{triv}} \to \mathcal{M}.$$

What we want to prove is the surjectivity of

$$\chi \mid \pi_1(\mathcal{M} \otimes \overline{\mathbb{F}}_p).$$

Since π_1 is compact and χ is continuous, it is enough to prove for each positive integer k that the image G of $\pi_1(\mathcal{M} \otimes \overline{\mathbb{F}}_p)$ in $(\mathcal{O}/p^k\mathcal{O})^*$ is all of $(\mathcal{O}/p^k\mathcal{O})^*$. But on the other hand, G is clearly the intersection

$$\bigcap_n G_n,$$

where G_n is the image in $(\mathcal{O}/p^k\mathcal{O})^*$ of $\pi_1(\mathcal{M} \otimes \mathbb{F}_{p^n})$. So it suffices to prove that $G_n = (\mathcal{O}/p^k\mathcal{O})^*$ for all n sufficiently large.

Suppose that $\alpha \in (\mathcal{O}/p^k\mathcal{O})^*$. Choose $a \in \mathcal{O}$ congruent to α mod p^k. Let $n \geq k$ be an integer large enough so that $a^2 - 4p^n$ is totally negative. Let T be the free $\mathbb{Z}$-module

$$\mathcal{O}[x]/(x^2 - ax + p^n)$$

of rank $2\cdot[K:\mathbb{Q}]$ and let F be the endomorphism "multiplication by x" on T. One checks easily that the pair (T,F) satisfies hypotheses (a), (b), and (c) of the main theorem of [1]. Let X be the ordinary abelian variety over $\mathbb{F}_{p^n}$ associated to (T,F) by that main theorem, and let

$$m: \mathcal{O} \hookrightarrow \text{End}_{\mathbb{F}_{p^n}}(X)$$

be the map arising from the F-linear action of $\mathcal{O}$ on T. To check that (X,m) is a Hilbert-Blumenthal structure, we note that $\text{Lie}(x)$ is dual to the kernel of F on $T \otimes_{\mathbb{Z}} \mathbb{F}_{p^n}$ and observe that this kernel is free of rank 1 over $\mathcal{O} \otimes \mathbb{F}_{p^n}$.

Also, I claim that the polarization module attached to X is $\mathfrak{d}^{-1}$. For this, let

$$\hat{T} = T \otimes_{\mathcal{O}} \mathfrak{d}^{-1}$$

and define a pairing $\langle\ ,\ \rangle : T \times \hat{T} \to 2\pi i\, \mathbb{Z}$ by the formula

$$\langle a+bx,\ c+dx\rangle = 2\pi i\cdot \mathrm{tr}_{O/\mathbb{Z}}(ad - bc).$$

Then the pair consisting of $\hat{T}$ and its map "multiplication by x" represents the dual variety $\hat{X}$ to X. Therefore the O-module of O-homomorphisms from $X/\mathbb{F}_{p^n}$ to its dual is given by

$$\mathrm{Hom}_{O[x]}(T,\hat{T}) \xrightarrow{\sim} \hat{T},$$

the isomorphism being the map $f \mapsto f(1)$. Now if f belongs to this module, then its dual is the map $\hat{f} \in \mathrm{Hom}_{O[x]}(T,\hat{T})$ which satisfies

$$\langle f(t),u\rangle^{\wedge} = \langle t,\hat{f}(u)\rangle$$

for all $t, u \in T$, where $\langle\ ,\ \rangle^{\wedge}$ is the pairing on $\hat{T} \times T = \hat{T} \times \hat{\hat{T}}$ analogous to $\langle\ ,\ \rangle$. (Thus $\langle w,z\rangle^{\wedge} = -\langle z,w\rangle$.) Thus f is symmetric (i.e., f is an element of $\mathcal{P}(X)$) if and only if $\langle f(t),u\rangle^{\wedge} = \langle t,f(u)\rangle$. This equation holds exactly when $f(1) \in \mathfrak{d}^{-1}$ as follows immediately from the definition of $\langle\ ,\ \rangle$; thus $\mathcal{P}(X)$ is the O-submodule $\mathfrak{d}^{-1}\cdot 1$ of $\hat{T}$.

Now if T'_p is the Tate module attached to X, viewed as an O_p-module, then Deligne's recipe tells us that the Frobenius endomorphism of X acts on T'_p by multiplication by ξ, where ξ is the unique element of $(O \otimes \mathbb{Z}_p)^*$ which satisfies

$$\xi^2 - a\xi + p^n = 0$$

So if x is the point of $\mathcal{M} \otimes \mathbb{F}_{p^n}$ defined by X and if $F_x \in \pi_1(\mathcal{M} \otimes \mathbb{F}_{p^n})$ is its Frobenius element (well-defined up to conjugation), then

$$\chi(F_x) = \xi.$$

Since

$$\xi^2 \equiv a\,\xi \mod p^n$$

we have

$$\xi \equiv a \mod p^k$$

because $k \leq n$. Thus the image of F_x in G_n is $(a \bmod p^k)$, or in other words α. So $\alpha \in G_n$. Therefore $(O/p^k O)^* = G_n$ provided that n is sufficiently large.

Bibliography

1. P. Deligne, Variétés abéliennes ordinaires sur un corps fini. Inventiones Math. 8 (1969), 238-243.

2. P. Deligne and K. Ribet, Values of abelian L-functions at negative integers. To appear.

3. N. Katz, p-adic L-functions via moduli of elliptic curves. In this volume.

4. J-P. Serre, Cohomologie des groupes discretes. In Prospects in Mathematics, Ann. Math. Studies 70, Princeton University Press (1971).

5. J-P. Serre, Formes modulaires et fonctions zêta p-adiques. Proceedings Antwerp Summer School, 1972. Springer Lecture Notes in Mathematics 350 (1973).

6. C. L. Siegel, Lectures on the analytical theory of quadratic forms (1934-35). Institute for Advanced Study Lecture Notes, Reprinted 1955.

7. C. L. Siegel, Berechnung von Zetafunktionen au ganzzahligen Stellen. Gött. Nach., (1969), 87-102.

PRINCETON UNIVERSITY

Proceedings of Symposia in Pure Mathematics
Volume 29, 1975

INTRODUCTION TO EQUISINGULARITY PROBLEMS

B. TEISSIER

"Better a house without roof than a house without view"
Hunza saying

ABSTRACT

A short journey into the range of equisingularity. In § 1 classical material is presented, in § 2 equisingularity is studied from a numerical view point, by associating to a hypersurface with isolated singularity a generalized multiplicity which has an algebraic definition, but can also be defined topologically with the Milnor numbers of the generic plane sections of X in all dimensions. The results lead to a general conjecture concerning the behaviour of equisingularity with respect to projections and plane sections. In § 3, the general problem of relating the topology of a hypersurface to that of a generic plane section is studied, and new invariants are introduced, which seem to link together many numerical invariants of singularities : e.g. for hypersurfaces with isolated singularities, the Milnor number of the hypersurface and of its generic hyperplane section, the smallest possible exponent θ in the Łojasiewicz inequality $|\mathrm{grad}\, f(z)| \geq C|f(z)|^{\theta}$, and the vanishing rates of certain gradient cells in the nearby non singular fiber.

§ 1

1.1 In these notes I shall present some features of the theory of equisingularity in characteristic zero which is gradually emerging after the fundamental work of Zariski (see $[Z_i]$, $i = 1,2,\ldots$). The basic question is to find algebraic criteria to decide when a space X, algebraic over $\mathbb{C}$, or complex analytic, can be said to be equisingular along a non singular subspace $Y \subset X$ at a point $0 \in Y$, the idea being that the singularities of X at all points of Y near 0 are alike in some sense. Roughly speaking, we would like equisingularity to be a condition as strong as possible, subject to the requirement of openness :

The author was partially supported by NSF G.P. 36269.
AMS Classification : Primary 32C40, 14B05. Secondary : 13B20, 13D10, 14H20, 58C25.

(O.E.) The set of points $y \in Y$ such that X is equisingular along Y at y is the complement of a nowhere dense closed subspace of Y.

This sort of question arises immediately when one tries to give a precise formulation to the intuitive idea that if you take a germ of reduced complex analytic hypersurface $(X_o,0) \subset (\mathbb{C}^{n+1},0)$, or a germ of curve $(\Gamma,0) \subset (\mathbb{C}^{n+1},0)$, then although it will not be true in general that "almost all" sections of $(X_o,0)$ by hyperplanes through the origin in $\mathbb{C}^{n+1}$, or "almost all" images of Γ by linear projections $(\mathbb{C}^{n+1},0) \to (\mathbb{C}^2,0)$ are analytically isomorphic, still, almost all these sections or images must "look alike" in some way, near 0.
As an example, take the cone $x^4 + y^4 + zx^2y = 0$ in $\mathbb{C}^3$, cut it by hyperplanes $z = ax + by$, and see how the cross-ratio of the four lines you get (except for special values of a and b) varies with a and b. Now the cross-ratio is an invariant of the analytic type.
In both cases, we can reduce to the basic question : for hyperplane sections, we choose an open $U \subset \mathbb{C}^{n+1}$ in which an equation $f(z_o,\dots,z_n) = 0$ defining X_o is convergent, homogeneous coordinates $(a_o:\dots:a_n)$ on the $\mathbb{P}^n$ of hyperplanes, and consider the subspace X of $U \times \mathbb{P}^n$ defined by the Ideal $(f(z_o,\dots,z_n), \sum_o^n a_i z_i)$. The fiber of the induced projection $\pi : X \overset{\sigma}{\leftrightarrows} \mathbb{P}^n$ above a hypersurface $(a_o:\dots:a_n)$ is precisely $X_o \cap H$, and π has a section $\sigma : \mathbb{P}^n \to X$ picking the origin in each fiber. So if we set $Y = \sigma(\mathbb{P}^n) = 0 \times \mathbb{P}^n \subset X$ we see that what we really want to do is to study equisingularity of X along Y, i.e. we would like to say that there is a Zariski open dense $V \subset \mathbb{P}^n$ such that X is equisingular along $0 \times \mathbb{P}^n$ at every point of $0 \times V$, and understand what this means for the fibers of π.
There is a similar construction for projections.
Now suppose that we are given in general a situation $\pi : X \overset{\sigma}{\leftrightarrows} Y$ such that the fibers of π, $(X_y,\sigma(y))$ are reduced hypersurfaces (and π is flat).
There is a simple way of saying that two fibers of π "look alike" :

<u>Definition</u> : Two germs of reduced complex analytic hypersurfaces (X_1,x_1) and (X_2,x_2) of the same dimension n are said to be of the same topological type if there exist representatives $(X_i,x_i) \subset (U_i,x_i)$ where U_i is open in $\mathbb{C}^{n+1}$, $i = 1,2$, and a homeomorphism of pairs $(U_1,x_1) \approx (U_2,x_2)$ mapping X_1 to X_2.

In fact, given $\pi : X \overset{\sigma}{\leftrightarrows} Y$ such that Y is reduced and the fibers of π are reduced hypersurfaces one can show with the stratification arguments of Thom-Whitney and Mather, to which I will come back below (see [Ma]), that there exists a nowhere dense closed analytic subset $F \subset Y$ such that for all $y \in Y-F$ the fibers $(X_y, \sigma(y))$ have the same topological type, so that the condition "topological type of the fibers locally constant" almost satisfies (O.E.). It is however a priori rather far from being algebraic !

Now it seems that Zariski's algebraic approach to equisingularity did not quite arise from such problems, but rather from his work on resolution of singularities.
The inductive procedure proposed by Zariski to resolve singularities of a (pure-dimensional) variety X (say, a hypersurface in some non-singular projective variety) was to look at a generic map from X to a non-singular projective variety of the same dimension say $\pi : X \to S$, π a finite map, then look at the branch locus $D \subset S$ of π, and by the inductive assumption, (embedded resolutions ; see Lipman's lectures), obtain a map $h : S' \to S$ with S' non-singular and $h^{-1}(D)$ a divisor with normal crossings in S'. (See Lipman's lectures in this volume). By pull-back we get a finite map $\pi' : X' \to S'$, the branch locus of which has only normal crossings, and therefore one can hope that the singularities of X' are easier to resolve. (On the other hand the map $X' \to X$ is not too hard to analyze). Therefore, the first thing to look at is a finite map such as $\pi' : X' \to S'$, the branch locus of which is non singular (I shall say smooth). Here several beautiful things happen simultaneously, which I shall try to condensate, at least in the special case of hypersurfaces.

In fact, from now on, I shall examine the basic problem only in the case where X is a complex analytic hypersurface (and of course in a neighborhood of a given point $0 \in X$) ; hypersurface because in the general case I know too little, and complex analytic because there will be transcendental avatars of the algebraic problems occasionnally.

<u>Warning</u> : While not writing the structure sheaves,the inverse images will always be meant ideal-theoretically, and Z_{red} will mean the space Z with reduced structure ; furthermore, everything being local, (Z,z) will usually stand for a small enough representative of a germ. I think "small enough" will have a clear meaning in each situation.

1.2 Now for the properties of hypersurfaces with smooth branch locus.

<u>Theorem 1</u> (Zariski $[Z_{1,2}]$) : Let $(X,0) \subset (\mathbb{C}^{N+1},0)$ be a reduced hypersurface, and $\pi : (\mathbb{C}^{N+1},0) \to (C^N,0)$ a holomorphic projection such that $\pi^{-1}(0) \not\subset X$.

By the Weierstrass preparation theorem we can describe this situation by choosing as equation for X :

$$F = z_{N+1}^{\nu} + A_1(z_1,\dots,z_N)z_{N+1}^{\nu+1} + \dots + A_{\nu}(z_1,\dots,z_N) = 0. \quad A_i \in \mathbb{C}\{z_1,\dots,z_N\}.$$

Let $R \in \mathbb{C}\{z_1,\dots,z_N\}$ be the z_{N+1}-discriminant of F. The following conditions are equivalent :

i) The branch locus Y_π of $\pi|X$ is smooth at 0 (i.e., by an analytic change of the coordinates $z_1,\dots,z_N$, we can write in $\mathbb{C}\{z_1,\dots,z_N\}$:

$$R(z_1,\dots,z_N) = z_1^{\Delta}.\varepsilon(z_1,\dots,z_N) \ ; \ \varepsilon(0,\dots,0) \neq 0 \ ; \ \Delta \in \mathbb{N}) .$$

ii) $Y = (\pi^{-1}(Y_\pi) \cap X)_{red}$ is smooth at 0, π induces an isomorphism $(Y,0) \xrightarrow{\sim} (Y_\pi,0)$, and for any retraction $r : (\mathbb{C}^{N+1},0) \to (Y,0)$ there exists a neighborhood U of 0 in Y such that for all $y \in U$, the germs $(r^{-1}(y) \cap X, y) \subset (r^{-1}(y),y)$ are germs of reduced (plane) curves, and have the same topological type. (In particular, if X is singular at 0, Y is the entire singular locus of X.)

iii) For every projection $\pi' : (\mathbb{C}^{N+1},0) \to (\mathbb{C}^N,0)$ which is transversal to X, (i.e. not only $\pi'^{-1}(0) \not\subset X$, but the tangent space $\operatorname{Ker} d\pi' = T_{\pi'^{-1}(0),0} \not\subset C_{X,0}$, the tangent cone of X at 0) the branch locus of $\pi'|X$ is smooth at 0.
(Note that transversality implies $\operatorname{Ker} d\pi \cap T_{Y,0} = (0)$ since $Y \subset X$.)

iv) The normalization $n : \bar{X} \to X$ of X is such that

α) $\bar{X}$ is smooth at every point of $n^{-1}(0)$.

β) The Ideal $\sum_1^{N+1} \frac{\partial F}{\partial z_i} . \mathcal{O}_{\bar{X}}$ is invertible in a neighborhood of $n^{-1}(0)$ in $\bar{X}$). ($n^{-1}(0)$ is a finite set of points, one for each irreducible component of X at 0.)

γ) n is the composition of a finite number of permissible blowing ups [in the sense of Hironaka, i.e. $\bar{X} = X^{(k)} \to X^{(k-1)} \to \dots \to X^{(0)} = X$ and $X^{(i+1)} \xrightarrow{b^{(i)}} X^{(i)}$ is the blowing up of a non singular subspace $Y^{(i)}$ of $X^{(i)}$ such that $X^{(i)}$, which is actually a hypersurface in a non singular space, is locally equimultiple along $Y^{(i)}$], $Y^{(0)}$ is the singular locus of X, the blowing ups $X^{(i+1)} \to X^{(i)}$ induce local isomorphisms $Y^{(i+1)} \to Y^{(i)}$, (i.e. $Y^{(i+1)} \to Y^{(i)}$ is etale) ; and $(b^{(i)^{-1}}(Y^{(i)}))_{red} = Y^{(i+1)}$. (If X is reductible, the $Y^{(i)}$ are not necessarily connected.)

Comments : (I shall give some hints for a proof later.)

1) The first part of condition ii) is the very important nonsplitting phenomenon which was emphasized by Abhyankar in his talk at the Woods Hole conference (see [Ab_1]) (but in characteristic $p>0$, where it is extremely delicate). It says in view of iv), γ) that if the branch locus is smooth, the singular locus Y of X itself is smooth and the multiplicity of X is locally constant on Y, which is the same as saying that the blowing up $X^{(1)} \to X$ of Y is a finite map, and iv) tells us that $X^{(1)}$ has along the reduced inverse image of Y the same properties as X along Y, and we can go on until we reach $\bar{X}$ which is smooth.
I will show below a generalization of the nonsplitting phenomenon.

2) If we view X as a family of plane curves parametrized by Y, with the help of a retraction as in ii), what iv) tells us is that the process of resolution of singularities for all the curves $(r^{-1}(y) \cap X, y)$ is the same for $y \in V$, a neighborhood of 0 in Y (see [Z_1]).

3) The equivalence of i) and iii) is very important for the following reason : in the inductive approach to resolution mentionned above, even if we start with a transversal projection $\pi : X \to S$, after simplification of the branch locus, the map $\pi' : X' \to S'$ obtained will in general no longer be transversal. But often the assumption of transversality makes proofs easier. i) $\Leftrightarrow$ iii) shows that in fact we lose nothing.

1.3 Zariski proposed a general algebraic definition of equisingularity for hypersurfaces, by induction on the codimension, as follows :

Definition (Zariski [Z_4]) :
(Z) Let $(X,0) \subset (\mathbb{C}^{N+1},0)$ be a reduced hypersurface and $(Y,0) \subset (X,0)$ a non singular susbspace of X. X is equisingular along Y at 0 if there exists a projection $\pi : (\mathbb{C}^{N+1},0) \to (\mathbb{C}^N,0)$ such that :

1) $\operatorname{Ker} d\pi \cap T_{Y,0} = (0)$, i.e. π maps isomorphically $(Y,0)$ to a non singular subspace $(Y_\pi,0) \subset (\mathbb{C}^N,0)$, and $\pi^{-1}(0) \not\subset X$.
2) The branch locus $(B_\pi,0) \subset (\mathbb{C}^N,0)$ of $\pi|X$ (which is a reduced hypersurface) is equisingular along Y_π at 0 (in particular $(Y_\pi,0) \subset (B_\pi,0)$).

When Y is of codimension 0 in X, equisingularity is just $Y = X$, and the Theorem 1 above is a list of characteristic properties of equisingularity in codimension 1, some of them independent of the mention of any projection.
Remark also that it follows inductively from this definition that the

condition (O.E.) is satisfied.

Now we have of course :

Problem : Compare (Z) with the other (possibly non-algebraic) definitions of equisingularity, for example :

1) Topological equisingularity (see [Z_5])

(T.E.) : In the same situation as in (Z), X is topologically equisingular along Y at O if for any retraction $r : (\mathbb{C}^{N+1}, 0) \to (Y, 0)$ there exists a neighborhood U of O in Y such that for all $y \in U$, the germs $(r^{-1}(y) \cap X, y) \subset (r^{-1}(y), y)$ are germs of reduced hypersurfaces with the same topological type. In other words, the topological type of a section of X by the non singular subspaces in $\mathbb{C}^{N+1}$ $(r^{-1}(y), y)$ transversal to Y is constant along Y.

We can add the extra condition that the topological type thus obtained is also independent of the choice of the retraction r.

(T.E.) + extra condition will be noted (S.T.E.).

2) Topological triviality

(T.T.) X is topologically trivial along Y at O if there exist a retraction $r : (\mathbb{C}^{N+1}, 0) \to (Y, 0)$ and a germ at O of homeomorphism of pairs

$$(\mathbb{C}^{N+1}, X) \approx (r^{-1}(0) \times Y,\ (r^{-1}(0) \cap X) \times Y))$$

compatible with r i.e. such that

$$\begin{array}{ccc} \mathbb{C}^{N+1} & \approx & r^{-1}(0) \times Y \\ & {\scriptstyle r}\searrow \quad \swarrow{\scriptstyle p_2} & \\ & Y & \end{array}$$

commutes.

3) Whitney conditions

Whitney's stratification theorem (see [Ma], [W_1]) implies the existence of a (locally) finite partition of X into non-singular locally closed subspaces Y_α with the properties that :

i) $\bar{Y}_\alpha$ and $\bar{Y}_\alpha - Y_\alpha$ are closed subspaces of X, $\dim(\bar{Y}_\alpha - Y_\alpha) < \dim Y_\alpha$, $\bar{Y}_\alpha \cap Y_\beta \neq \emptyset \Rightarrow Y_\beta \subset \bar{Y}_\alpha$.

ii) If $Y_\beta \subset \bar{Y}_\alpha$, the following conditions of incidence (Whitney conditions) hold :

a) For any sequence of points $y_i \in Y_\alpha$ converging to a point

$y \in Y_\beta$ and such that the directions of the tangent spaces T_{Y_α, y_i} converge (in the grassmannian of $\dim Y_\alpha$-planes in $\mathbb{C}^{N+1}$) to a plane T, we have : $T_{Y_\beta, y} \subset T$.

b) Given a sequence of points $y_i \in Y_\alpha$ converging to y and a sequence of points $u_i \in Y_\beta$, also converging to y, and such that again T_{Y_α, y_i} converges to a plane T, and also the direction of the secant $\overline{u_i y_i}$ in $\mathbb{C}^{N+1}$ converges (in $\mathbb{P}^n$) to ℓ, we have : $\ell \subset T$.

(If y is fixed we will speak of "Whitney conditions at y".)

iii) The set X^o of smooth points of X is a stratum (we do not ask strata to be connected) and so of course all other strata are in its closure.

Define Whitney equisingularity of X along Y at 0 to be :

(W) Y is a stratum of some Whitney stratification of X (in a neighborhood of 0).

Now the situation in general is this : Varchenko ([V]) gave a topological proof of (Z) ⇒ (T.T.). Thom and Mather ([Ma], [Th_1]) gave differential-geometric proofs of (W) ⇒ (T.T.), and of course (T.T.) ⇒ (T.E.). Speder [Sp] modified Zariski's definition by requiring that the projection π appearing in (Z) should satisfy in addition a certain condition, which is satisfied by "almost all" projections, and implies transversality. With this different definition (Sp), he proved inductively that if X satisfies (Sp) along Y at 0, then the pair (X^o, Y) satisfies Whitney conditions at every point of a neighborhood of 0 in Y.
(Sp) is unfortunately lost when we apply imbedded resolution to the discriminant.
This brings us to a question which is anyway crucial when one tries to compare (Z) with other definitions.
Here, in first analysis, I shall assume for the sake of simplicity that the coordinates are so chosen that Y is a linear subspace in $\mathbb{C}^{N+1}$ and consider only linear projections $\pi : \mathbb{C}^{N+1} \to \mathbb{C}^N$.

<u>Question A</u> : (This is essentially Zariski's question I in [Z_5]).
Assuming that X is equisingular along Y at 0 (in some sense, and in particular of course (Z)), is the set of those linear projections $\pi : \mathbb{C}^{N+1} \to \mathbb{C}^N$ satisfying :

1) $\operatorname{Ker} \pi \cap Y = (0)$

2) The branch locus B_π of $\pi|X$ is equisingular along $Y_\pi = \pi(Y)$ at 0, a dense (Zariski)-constructible subset of the space of linear projections ?

Here I will venture the following :

Conjecture 1 : In the situation of question A, for a linear projection $\pi : \mathbb{C}^{N+1} \to \mathbb{C}^N$ such that $\operatorname{Ker}\pi \cap Y = (0)$ the following conditions are equivalent :

α) B_π is equisingular along Y_π at 0

β) For all i, $\dim Y \leq i \leq N$, there exists a (Zariski) open dense subset $U^{(i)}$ of the Grassmannian of i-planes of $\mathbb{C}^N$ containing Y_π, such that for all $H \in U^{(i)}$, the hypersurface $(\pi^{-1}(H) \cap X)_{red} \subset \pi^{-1}(H)$ is equisingular along Y at 0.

Notice that for $i = \dim Y$, β) is just the nonsplitting of $\pi^{-1}(Y_\pi) \cap X$. This conjecture is proved in the codimension 1 case, where we have :

Theorem 2 (Zariski [Z_4]) : Assume $X \subset \mathbb{C}^{N+1}$ (Z)-equisingular along Y at $0 \in Y$. If $\dim Y = N-1$, then for a linear projection $\pi : \mathbb{C}^{N+1} \to \mathbb{C}^N$ with $\operatorname{Ker} d\pi \cap T_{Y,0} = (0)$ the following are equivalent :

α) The branch locus B_π is equisingular along Y_π at 0 (i.e. $(B_\pi, 0) = (Y_\pi, 0)$).

β') Setting $\ell_y = \pi^{-1}(\pi(y))$, for $y \in Y$, the intersection number $(\ell_y . X)_y$ of the line ℓ_y with X at y, is independent of $y \in Y$ (locally around 0).

Now it is clearly equivalent to say that $(\ell_y, X)_y$ is independent of $y \in Y$, and to say that $(\pi^{-1}(Y_\pi) \cap X)_{red} = Y$, and if $\dim Y = N-1$, the only dimension we have to look at to check β) is $i = \dim Y = N-1$.

Even if one can prove conjecture 1, it does not help much to answer question A unless one can also answer :

Question B : Assume $(X,0) \subset (\mathbb{C}^{N+1},0)$ is equisingular at 0 along $(Y,0) \subset (X,0)$ (here again for simplicity we assume Y linear in $\mathbb{C}^{N+1}$). Is it true that for each i, $\dim Y \leq i \leq N+1$, there exists a dense Zariski-open subset $U^{(i)}$ of the grassmannian of i-planes of $\mathbb{C}^{N+1}$ containing Y, such that $H \in U^{(i)} \Rightarrow X \cap H$ is equisingular along Y at 0 (or, better, a constructible $C^{(i)}$ such that $H \in C^{(i)} \Leftrightarrow X \cap H$ equisingular along Y) ?

A priori, question B looks much simpler that question A, and even rather simple-minded. However, except of course when Y is of codimension

1 in X, there is no answer, for any notion of equisingularity. Zariski proved, however, that (Z) implies equimultiplicity, and Hironaka [H_6] proved that in the most general case, Whitney conditions implie equimultiplicity. Therefore there are affirmative answers to question B for $i = \dim Y+1$. But the point in question B is of course the case $i = N$, the opposite extreme.

Anyway, I think an affirmative answer to question B and conjecture 1 would settle question A.

Now if we want to study equisingularity in codimension > 1, the first thing to look at is the case where Y is the singular locus of X, but not necessarily of codimension 1. Then we can view X as a family of hypersurfaces with isolated singularities, parametrized by Y. So we turn to :

§ 2. ISOLATED SINGULARITIES OF HYPERSURFACES AND THE NUMERICAL APPROACH

2.1 Let me first recall some facts from asymptotic algebra in the sense of Samuel [Sa]. For details I refer to [L.T.], [Se], [P.T.], [T_2]. Let $\mathcal{O}$ be the local analytic algebra of a reduced complex analytic space Z at a point z. Given an ideal I in $\mathcal{O}$ an element $h \in \mathcal{O}$ is said to be integral over I if it satisfies an integral dependence relation

$$h^k + a_1 h^{k-1} + \dots + a_k = 0 \quad \text{with } a_i \in I^i \quad .$$

The set of such elements is an ideal $\bar{I}$ in $\mathcal{O}$, the integral closure of I in $\mathcal{O}$.

This purely algebraic notion due to Prüfer [P] has the following transcendental interpretation, the idea of which is due to Hironaka : If we choose generators $g_1, \dots, g_\ell$ for I, then $h \in \bar{I}$ if and only if there exists a neighborhood U of z in Z and a constant $C \in \mathbb{R}_+$ such that for all $z' \in U$ $|h(z')| \leq C.\mathrm{Sup}|g_i(z')|$.

In the same vein, let me define the Łojasiewicz exponent $\theta_I(h)$ of h with respect to I as the greatest lower bound of those $\theta \in \mathbb{R}_+$ such that there exists a neighborhood U of z in Z and a $C \in \mathbb{R}_+$ (depending on U) such that

$$|h(z')|^\theta \leq C.\mathrm{Sup}|g_i(z')| \quad \text{for all } z' \in U.$$

Then (see [L.T.]) we have the following :

<u>Proposition</u> : $\theta_I(h) = \dfrac{1}{\bar{\nu}_I(h)}$ where $\bar{\nu}_I(h) = \lim\limits_{k\to\infty} \dfrac{\nu_I(h^k)}{k}$

with $\nu_I(g) = \max\{\nu \in \mathbb{N} / g \in I^\nu\}$, (and $\theta_I(h)$ is attained for some U and $C \in \mathbb{R}_+$). If $h \notin \sqrt{I}$, we set $\theta_I(h) = +\infty$. ($\bar{\nu}$ was introduced in [Sa], and Nagata [N] proved $\bar{\nu}_I(h) \in \mathbb{Q}$, so that Łojasiewicz exponents are always rational numbers.) We see that integral dependence will be very useful to study algebraically geometrical incidence (or limit) relations, such as the Whitney conditions, a viewpoint in fact pioneered by Hironaka in [H_1]. Another aspect of integral dependence will be useful for our numerical viewpoint : suppose now that I is primary for the maximal ideal $\mathfrak{M}$ of $\mathcal{O}$. Then the application $H_I : \mathbb{N} \to \mathbb{N}$ defined by $H_I(\nu) = \dim_{\mathbb{C}} \mathcal{O}/I^{\nu+1}$ coincides when ν is large enough with a polynomial in ν of degree $d = \dim \mathcal{O}$ and the highest degree term of this polynomial can be written $e(I)\dfrac{\nu^d}{d!}$, where $e(I) \in \mathbb{N}$ is by definition the multiplicity of the primary ideal I in $\mathcal{O}$. The result I will use is : $e(I) = e(\bar{I})$.

2.2 For example, let $f \in \mathcal{O}_{n+1} = \mathbb{C}\{z_o,\dots,z_n\}$ be such that $f = 0$ defines a germ of hypersurface with isolated singularity $(X,0) \subset (\mathbb{C}^{n+1},0)$. By the nullstellensatz, this means that the ideal in $\mathcal{O}_{n+1}$ generated by $(f, \dfrac{\partial f}{\partial z_o},\dots,\dfrac{\partial f}{\partial z_n})$ is primary for the maximal ideal. Define $\mu(X,0)$ to be the multiplicity of this primary ideal, and remember the very useful

<u>Fact</u> (see [T_2]) : For any choice of the coordinates $(z_o,\dots,z_n)$, f is integral in $\mathcal{O}_{n+1}$ over the ideal generated by $(z_o.\dfrac{\partial f}{\partial z_o},\dots,z_n.\dfrac{\partial f}{\partial z_n})$ (which depends on the choice of coordinates). In particular, f is integral over the product ideal $\mathfrak{M}.j(f)$ where $\mathfrak{M}$ is the maximal ideal of $\mathcal{O}_{n+1}$ and j(f) the ideal generated by $(\dfrac{\partial f}{\partial z_o},\dots,\dfrac{\partial f}{\partial z_n})$. ($\mathfrak{M}.j(f)$ does not depend upon the choice of coordinates).And a fortiori, f is integral over j(f.) (All this is true in general, i.e. for any $f \in \mathfrak{M}$.)

So the multiplicity of (f,j(f)) is equal to that of j(f), which must be also primary, whence $(\dfrac{\partial f}{\partial z_o},\dots,\dfrac{\partial f}{\partial z_n})$ is a regular sequence in $\mathcal{O}_{n+1}$, and then by a theorem of Samuel, $\mu(X,0) = \dim_{\mathbb{C}} \mathcal{O}_{n+1}/j(f)$, the Milnor number of the singularity.

Note that the ideal $(f, \dfrac{\partial f}{\partial z_o},\dots,\dfrac{\partial f}{\partial z_n})$ depends only on the quotient $\mathcal{O}_{n+1}/(f)$, so $\mu(X,0)$ is an analytic invariant, but better still, an attentive reader of Milnor's book [Mi] will remark that :

<u>If $(X_1,0)$ and $(X_2,0)$ are two germs of hypersurfaces with isolated singularity of the same topological type, we have $\mu(X_1,0) = \mu(X_2,0)$.</u>

This follows from the interpretation of $\mu(X,0)$ as "number of vanishing cycles" of the singularity $(X,0)$. See [Mi] and Brieskorn's lectures in this volume. (For a "proof of the remark", see $[T_2]$, and $[L_2]$ for generalization to non-isolated singularities.)

2.3 Now we have a topological invariant defined algebraically, so we should be in good position to study some of the questions of § 1 in the special case considered here. First, a few remarks on the semi-continuity properties of the Milnor number. Given a family of complex hypersurfaces, i.e. a commutative diagram

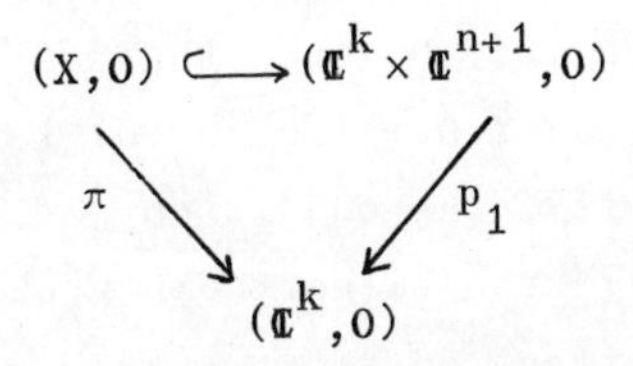

where $(X,0)\hookrightarrow(\mathbb{C}^k\times\mathbb{C}^{n+1},0)$ is defined by $F=0$, $F\in\mathbb{C}\{y_1,\dots,y_k, z_o,\dots,z_n\}$. Assume $F(0,\dots,0, z_o,\dots,z_n)=0$ is a hypersurface with isolated singularity. Then the subspace P of $(\mathbb{C}^{N+1},0)=(\mathbb{C}^{n+k+1},0)$ defined by the ideal J generated by $(\frac{\partial F}{\partial z_o},\dots,\frac{\partial F}{\partial z_n})$ is finite over $(\mathbb{C}^k,0)$ (by the Weierstrass preparation theorem) and $(\frac{\partial F}{\partial z_o},\dots,\frac{\partial F}{\partial z_n})$ is a regular sequence, so that in fact $\mathbb{C}\{y_1,\dots,y_k, z_o,\dots,z_n\}/J$ is a free $\mathbb{C}\{y_1,\dots,y_k\}$-module. This means also that if we take any $y\in\mathbb{C}^k$, $p_1^{-1}(y)\cap P$ is a finite set of points x_i, the local rings $\mathcal{O}_{P,x_i}$ are artinian, i.e. finite dimensional vectorspaces and $\sum_{x_i\in p_1^{-1}(y)\cap P}\dim_{\mathbb{C}}(\mathcal{O}_{P,x_i})$ is independent of $y\in(\mathbb{C}^k,0)$, and is therefore equal to the Milnor number of the hypersurface $(X_o,0)=(\pi^{-1}(0),0)\subset(\mathbb{C}^{n+1},0)$ defined by $F(0,\dots,0, z_o,\dots,z_n)=0$. The singular points of $X_y\in\pi^{-1}(y)$ are the points of $\pi^{-1}(y)\cap P$, which are some of the $x_i\in p_1^{-1}(y)\cap P$, and if $x_i\in\pi^{-1}(y)\cap P$, $\dim_{\mathbb{C}}\mathcal{O}_{P,x_i}$ is just the Milnor number of the isolated singularity of hypersurface $(\pi^{-1}(y),x_i)\subset(p_1^{-1}(y),x_i)$. So we have :

2.3.1 For any $y\in(\mathbb{C}^k,0)$ if the x_i are the singular points of the fiber X_y of π

$$\mu(X_o,0)\geq\sum_{x_i\in X_y}\mu(X_y,x_i)$$

and equality holds for all y, if and only if $P_{red}\subset X$, i.e. : there exists an integer r such that $F^r\in J$.

Assume now that we are given a section $\sigma : (\mathbb{C}^k,0) \to (X,0)$ of π. By a change of the coordinates $(z_o,\dots,z_n)$, we may assume that the image of σ is $(\mathbb{C}^k \times \{0\},0)$ i.e. is defined by the ideal $S = (z_o,\dots,z_n)$. Then, by the semi-continuity property of the fibers of coherent sheaves, we have :

2.3.2 1) For all $y \in (\mathbb{C}^k,0)$, $\mu(X_o,0) \geq \mu(X_y,y)$.

2) In view of the preceding remark, equality holds for all $y \in (\mathbb{C}^k,0)$ if and only if $(P_{red},0) = \sigma(\mathbb{C}^k,0) = (\mathbb{C}^k \times \{0\})$ i.e. iff there exists an integer t such that $S^t \subset J$.

3) Given $(Y,0) \subset (X,0) \subset (\mathbb{C}^{N+1},0)$ with $(Y,0)$ smooth, and a retraction $r : (\mathbb{C}^{N+1},0) \to (Y,0)$ such that $(X_o,0) = (r^{-1}(0) \cap X,0) \subset (r^{-1}(0),0)$ is a hypersurface with isolated singularity, the condition

(μ constant) : the Milnor number of $(r^{-1}(y) \cap X,y) = (X_y,y)$ is locally constant on Y,

satisfies the condition (O.E.) of § 1.

2.4 Applying this last result to the family of sections of a given hypersurface $(X_o,0) \subset (\mathbb{C}^{n+1},0)$, with isolated singularity, we obtain, for each i $(0 \leq i \leq n+1)$ an integer $\mu^{(i)}(X_o,0)$ and a Zariski-open dense $U^{(i)} \subset G^{(i)}$ (= the grassmannian of i-planes) such that $H \in U^{(i)} \Rightarrow \mu(X_o \cap H,0) = \mu^{(i)}(X_o,0)$. $\mu^{(n+1)}(X_o,0)$ is $\mu(X_o,0)$, $\mu^{(1)}(X_o,0) = m(X_o,0) - 1$ where $m(X_o,0)$ is the multiplicity of X_o at 0, and $\mu^{(o)}(X_o,0) = 1$. Set $\mu^*(X_o,0) = (\mu^{(n+1)}(X_o,0),\dots,\mu^{(1)}(X_o,0),\mu^{(o)}(X_o,0))$. Again, for a situation as in 2.3.2, the condition

(μ^* constant) : $\mu^*(X_y,y)$ is locally constant.

satisfies (O.E.).

2.5 In order to state the results which motivate the next questions, I need to introduce still another notion of equisingularity, to add to our collection of § 1. This section owes much to reminiscences of (unpublished) lectures of Hironaka.

Let $(Y,0) \subset (X,0) \subset (\mathbb{C}^{N+1},0)$ be a smooth space in a hypersurface in $\mathbb{C}^{N+1}$ as usual. Choose coordinates $(y_1,\dots,y_k, z_o,\dots,z_n)$ $(N = n+k)$ so that Y is defined by the ideal $S = (z_o,\dots,z_n)\mathcal{O}_{N+1}$ $(\mathcal{O}_{N+1} = \mathbb{C}\{y_1,\dots,y_k, z_o,\dots,z_n\})$. let $F(y_1,\dots,y_k, z_o,\dots,z_n) = 0$ be an equation for X, and J be the ideal $(\frac{\partial F}{\partial z_o},\dots,\frac{\partial F}{\partial z_n})\mathcal{O}_{N+1}$. The new condition of equisingularity is :

(c) $$\frac{\partial F}{\partial y_i} \in \overline{S.J} \quad (\text{in } \mathcal{O}_{N+1}) \quad 1 \leq j \leq k \ .$$

To throw some light on it, let us see that it implies that the smooth part X^o of X satisfies Whitney's conditions a) and b) along Y, near 0. In fact, condition (c) implies that if we take any sequence of points $p_i \to y \in Y$ in $\mathbb{C}^{N+1}$, not necessarily in X, such that the level hypersurface $F(y_1,\dots,y_k, z_o,\dots,z_n) = F(p_i)$ is non singular at p_i, the limit position of the tangent hyperplanes, of homogeneous coordinates $(\frac{\partial F}{\partial y_1}(p_i):\dots:\frac{\partial F}{\partial y_1}(p_i):\frac{\partial F}{\partial z_o}(p_i):\dots:\frac{\partial F}{\partial z_n}(p_i))$ is of the form $(0:\dots:0:a_o:\dots:a_n)$, i.e. contains Y (or $T_{Y,y}$), because condition c) implies : $|\frac{\partial F}{\partial y_j}(p_i)| \leq C \,.\, \mathrm{dist}(p_i,Y)\,.\, \sup_k |\frac{\partial F}{\partial z_k}(p_i)|$, in view of 2.2.
The proof for condition b) is similar, if more delicate. One has to use the important fact, due to Whitney, [W_1] that the pair of strata $(X^o,(0))$ satisfies condition b). Remark finally that in choosing coordinates as we have done, we have in fact chosen a retraction $r:(\mathbb{C}^{N+1},0)\to(Y,0)$, so the following makes sense :

2.6 <u>Theorem 3</u> : Given $(Y,0)\subset(X,0)\subset(\mathbb{C}^{N+1},0)$ as above, and a retraction $r:(\mathbb{C}^{N+1},0)\to(Y,0)$ such that

1) $(X_o,0)=(r^{-1}(0)\cap X,0)$ is a hypersurface with isolated singularity (and Y is the entire singular locus of X)
2) $\dim Y = 1$.

Then : (μ^* constant) $\Leftrightarrow$ condition (c).
(And the μ^* is actually independent of the choice of the retraction.)

The proof of ($\Rightarrow$) goes as follows : first, condition (c) satisfies (O.E.) of § 1. (The proof of the lemma in 2.7 of [T_2] extends easily.) Second, (μ^* constant) is actually a condition of equimultiplicity for the family of ideals induced by S.J in the fibers of r : with the notations of 2.2, given an isolated singularity of hypersurface, we consider the application $K:\mathbb{N}\times\mathbb{N}\to\mathbb{N}$ defined by $K(\nu_1,\nu_2)=\dim_{\mathbb{C}}\mathcal{O}_{n+1}/\mathfrak{M}^{\nu_2}.j(f)^{\nu_1}$. When ν_1 and ν_2 are both sufficiently large, K takes the same values as a polynomial in ν_1, ν_2, of degree n+1, the highest degree part of which is :

$$\overline{K}(\nu_1,\nu_2)=\frac{1}{(n+1)!}\left(\mu^{(n+1)}\nu_1^{n+1}+\dots+\binom{n+1}{i}\mu^{(n+1-i)}.\nu_1^{n+1-i}.\nu_2^{i}+\dots+\mu^{(0)}\nu_2^{n+1}\right),$$

and in particular, $e(\mathfrak{M}.j(f))=\sum_{0}^{n+1}\binom{n+1}{i}\mu^{(i)}$. ($\mu^{(i)}=\mu^{(i)}(X,0)$). So the family of ideals induced by S.J on the fibers of r is equimultiple if (μ^* constant) is realized.
In that case, provided $\dim Y = 1$ (this is the only place where this unnatural assumption is needed) one can prove ([T_2], prop. 3.1) that "$\bar{\nu}$ can

be computed at the generic point of Y". But condition c) is : $\bar{\nu}_{SJ}(\frac{\partial F}{\partial y_i}) \geq 1$ $(1 \leq i \leq k)$ and as mentionned in the beginning, it is realized at the generic point of Y. Equimultiplicity implies that it is also realized at 0. The proof of $(\Leftarrow)$ does not use $\dim Y = 1$. One shows that $(c) \Rightarrow \mu^{(n+1)}(X_y,0)$ constant $(n = N - \dim Y)$: since Y is the singular locus of X, there exists an integer r such that $S^r \subset (\frac{\partial F}{\partial y_1},\ldots,\frac{\partial F}{\partial y_k},\frac{\partial F}{\partial z_o},\ldots,\frac{\partial F}{\partial z_n})\mathcal{O}_{N+1}$. But (c) implies that a power of this last ideal is contained in J. So a power of S in contained in J, which means $\mu^{(n+1)}$ constant (2.3.2). One concludes by proving :

Proposition : For (c)-equisingularity, the answer to question B of § 1 is yes.

I will not give the proof here, but only remark that (c)-equisingularity is in fact an ad-hoc definition for this proposition to hold.

2.7 So if we could prove that $\mu^*(X_o,0)$ is in fact also an invariant of the topological type as $\mu(X_o,0)$ is, we would have, at least in the special case where Y is the singular locus of X, (and $\dim Y = 1$) a completely algebraic interpretation of topological equisingularity (and in fact (S.T.E.)) thus extending Theorem 1. Of course, from the algebraic viewpoint, it is enough to prove : $\mu^{(n+1)}(X_y,y)$ constant $\Rightarrow \mu^{(n)}(X_y,y)$ constant. Then, we can take a hyperplane H in $\mathbb{C}^{N+1}$ containing Y and such that $\mu^{(n)}(X_o \cap H,0) = \mu^{(n)}(X_o,0)$ and by the semi-continuity properties :

$$\begin{array}{ccc} \mu^{(n)}(X_o\cap H,0) & = & \mu^{(n)}(X_o,0) \\ \bigvee\!| & & \| \\ \mu^{(n)}(X_y\cap H,y) & \geq & \mu^{(n)}(X_y,y) \end{array}$$

by our choice of H

by assumption

So that $X \cap H$ again satisfies (μ constant) along Y, and we can go down the staircase of dimensions to prove (μ constant) $\Rightarrow$ (μ^* constant) which is enough to prove the equivalence of all definitions of equisingularity given in § 1 (except Zariski's) with condition (c), in the special case we are considering here. So let me state the :

Conjecture B_μ : $\mu^{(n+1)}$ constant $\Rightarrow \mu^{(n)}$ constant.

which, as we have just seen, is only question B for (μ-constant)-equisingularity.

2.8 It might seem that we have moved very far away from Zariski's discriminant (branch locus) condition.
Let me show that this is not the case . Suppose we want to compare theorem 3 with

Theorem 4 (Lê Dũng Tráng and C.P. Ramanujam [L.R]) : In the situation $(Y,0) \subset (X,0) \subset (\mathbb{C}^{N+1},0)$, with a retraction $r : (\mathbb{C}^{N+1},0) \to (Y,0)$ assume :

1) $(X_o,0)$ is a hypersurface with isolated singularity.
2) The codimension of Y in X is $\neq 2$.
3) X satisfies the condition (μ constant) along Y at 0, with respect to r.

Then, X is topologically equisingular along Y at 0, with respect to r.

Using this result, and a deep theorem of topology, Timourian [Ti] proved :

Theorem 4' (Timourian) : Under the same assumptions, topological triviality holds.

2.8.1 After theorems 3 and 4, we find ourselves in an embarrassing situation. Since we know that the Milnor number is an invariant of the topological type ; theorems 4 and 4' do provide us with an algebraic interpretation of topological equisingularity and topological triviality when Y is smooth and is the entire singular locus of X :
(μ constant) $\Leftrightarrow$ (topological equisingularity) $\Leftrightarrow$ (topological triviality), except when Y is of codimension 2 in X.
On the other hand, if we could prove conjecture B_μ and remove the $\dim Y = 1$ assumption, we would have a complete equivalence
(μ^* constant) $\Leftrightarrow$ (condition c) $\Leftrightarrow$ (Whitney conditions) $\Leftrightarrow$ (topological triviality) $\Leftrightarrow$ (topological equisingularity) $\Leftrightarrow$ (μ constant).

The moment is well chosen to analyze the difference between Whitney conditions and topological equisingularity.
Let us, by a choice of retraction, view (X,0) as a hypersurface in $(Y \times \mathbb{C}^{n+1},0)$. As usual, (X_y,y) will be the intersection of X with $(y \times \mathbb{C}^{n+1}, y \times 0)$.
One of the main points of Whitney conditions (see $[H_1]$, and Lemma 5.1 of [Ma]) is that if they are satisfied by (X^o,Y), then we can find a real number ρ_o (permissible radius) such that for any $y \in (Y,0)$ (i.e. y is

sufficiently close to 0) the spheres $\mathbb{S}_\rho \subset \{y\} \times \mathbb{C}^{n+1}$ of radius ρ (real-analytic manifolds of dimension $2n+1$) are all transversal to the smooth part of X_y, for $0 < \rho \leq \rho_o$. Roughly speaking, this means that if we consider the cylinder $C_\rho = \mathbb{S}_\rho \times Y$, and the tube $T_\rho = \mathbb{B}_\rho \times Y$ in the ambiant space, $X \cap T_\rho$ is "cone like over its boundary $X \cap C_\rho$" (a cone with vertex Y). So theorem 3 implies that if μ^* is constant, we cannot have the situation described by the following picture :

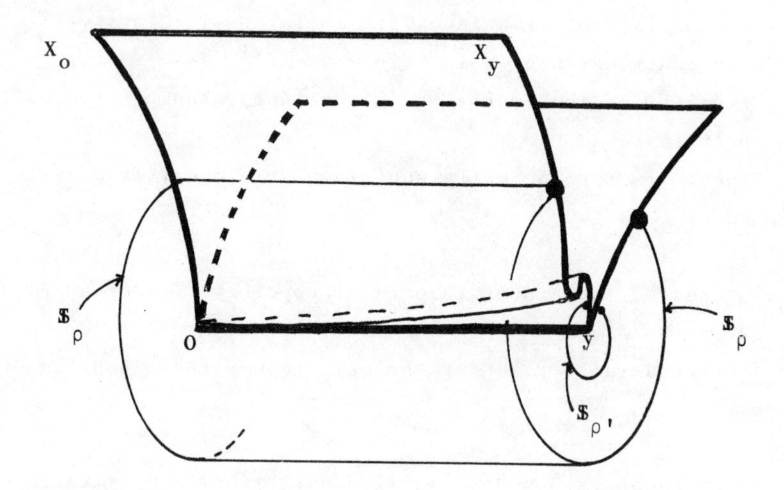

And the gist of theorem 4 is that even if we did have such a situation, if (μ constant) holds, we can, by using the h-cobordism theorem, (this is the reason of the restriction on codimension) and the properties of Milnor fibrations ([Mi] and Brieskorn's lectures), build a homeomorphism $(\mathring{\mathbb{B}}_\rho, X_o) \approx (\mathring{\mathbb{B}}_{\rho'}, X_y)$ where ρ' is a radius permissible for the fiber X_y, and ρ is permissible for X_o.

2.8.2 A few words on discriminants. In what follows we will often consider the following situation :

$$\begin{array}{ccc} (X,0) \subset (\mathbb{C}^{N+1},0) & & \\ \pi \searrow \quad 0 \quad \swarrow p & & \\ (\mathbb{C}^k,0) & & \end{array}$$

where $(X,0)$ is a hypersurface in $(\mathbb{C}^{N+1},0)$, and π is flat. If we choose coordinates $(y_1,\ldots,y_k, z_o,\ldots,z_n)$ on $\mathbb{C}^{N+1}$ $(N = n+k)$ such that p is described by $(y_1,\ldots,y_k, z_o,\ldots,z_n) \mapsto (y_1,\ldots,y_k)$ and an equation

$F(y_1, \ldots, y_k, z_o, \ldots, z_n) = 0$ for X, the critical subspace C_π of π is defined by $(\frac{\partial F}{\partial z_o}, \ldots, \frac{\partial F}{\partial z_n})\mathcal{O}_X$. If C_π is proper over $\mathbb{C}^k$, which in our local situation means it is finite, i.e. $(\pi^{-1}(0), 0) = (X_o, 0)$ has an isolated singularity, we can define the discriminant $(D_\pi, 0) \subset (\mathbb{C}^k, 0)$ of π as the analytic direct image of C_π, the underlying set of which shall be the set of points of $(\mathbb{C}^k, 0)$ such that the fiber has singularities. To be brief, I will just recall the following : (see [T_2] or Brieskorn's lectures). Since $(X_o, 0)$ has an isolated singularity, $\pi : (X,0) \to (\mathbb{C}^k, 0)$ comes by base change from the miniversal deformation of $(X_o, 0)$:

$$\begin{array}{ccc} (X,0) & \longrightarrow & (U,0) \\ \pi \downarrow & & \downarrow G \\ (\mathbb{C}^k,0) & \xrightarrow{h} & (S,0) \end{array}$$

Now G has a discriminant D_G which is a reduced hypersurface in the non-singular space (S,0). The formation of the discriminant commutes with base extension, so the discriminant D_π of π is $(h^{-1}(D_G), 0)$, which is defined by a principal ideal in $\mathcal{O}_k$. The <u>branch locus</u> B_π of π is $D_{\pi,red}$. We will want to speak of the multiplicity at 0 of the discriminant : If $B_\pi \neq \mathbb{C}^k$, it is just the multiplicity at 0 of the hypersurface $(D_\pi, 0)$. By convention, if $B_\pi = \mathbb{C}^k$, "the multiplicity of the discriminant" will be the Milnor number of the fiber $(X_o, 0)$. For example, in this case we will say that "the discriminant is equimultiple along $(\mathbb{C}^k, 0)$" if we are in the equality case of 2.3.1.
Let me also remark that when D_π is a hypersurface, given $(Y,0) \subset (D_\pi, 0)$, it is equivalent to say that D_π is equimultiple along Y, or to say that B_π is so, provided Y is smooth.

2.8.3 The connection between theorems 3 and 4, conjecture B_μ and Zariski's discriminant condition is due to the following :

<u>Proposition</u> ([T_2]) : Let $(X,0) \subset (\mathbb{C}^{n+1}, 0)$ be a hypersurface with isolated singularity, and $p : (\mathbb{C}^{n+1}, 0) \to (\mathbb{C}, 0)$ a projection such that the fiber $(X_o, 0)$ of $\pi = p|X$ again has an isolated singularity. Then, the multiplicity of the discriminant D_π of π is $\Delta = \mu^{(n+1)}(X,0) + \mu^{(n)}(X_o, 0)$. Note that the assumptions imply that $B_\pi = (0)$, i.e. if we take a coordinate z_o in $(\mathbb{C}, 0)$, an equation for D_π is $z_o^\Delta = 0$.

There are generalizations of this formula to complete intersections (see Brieskorn's lectures), and Lê [L_2] generalized the "vanishing cycles" aspect of it to arbitrary singularities of hypersurfaces. We will see more about this proposition in § 3, but what I need here is the

Corollary : Let

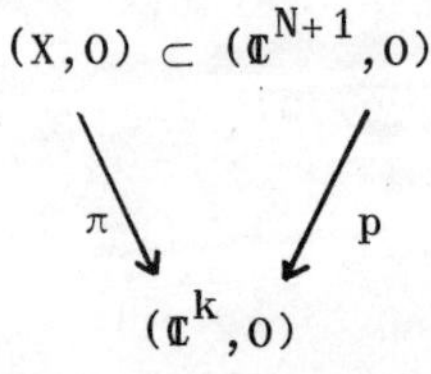

be as in 2.8.2, and assume that $(X_o,0)$ has an isolated singularity and D_π is a hypersurface in $(\mathbb{C}^k,0)$.
Let $(L,0)$ by any smooth curve in $(\mathbb{C}^k,0)$ transversal to $(D_\pi,0)$ (i.e. $(D_\pi.L)_o = m_o(D_\pi)$, the multiplicity of D_π at 0). Then

$$m_o(D_\pi) = \mu(X_o,0) + \mu(X_L,0) \quad \text{where } X_L = p^{-1}(L) \cap X \quad .$$

Proof : The formation of discriminant commutes with base change, so $D_\pi \cap L$ is the discriminant of the induced map $X_L \to L$, to which we can apply the proposition.
Since L is transversal to D_π, the multiplicity of $D_\pi \cap L$ is the multiplicity of D_π at 0.
Notice that when $k = N+1$, the discriminant is X itself, X_o is a point, L a line transversal to X and we recover : $m_o(X) = 1 + \mu^{(1)}(X)$.
At the other extreme, if $k = 0$, there is no L, so we are tempted to set the multiplicity of the discriminant equal to $\mu(X_o,0)$, as I did in 2.8.2. (There are, however, more serious reasons to do that !)

2.8.4 Now, back to B_μ. We take the usual situation, choose a retraction, etc., so we are looking at a family of hypersurfaces $X = \bigcup_{y \in Y} X_y$, $(X,0) \subset (Y \times \mathbb{C}^{n+1},0)$, defined by $F(y_1,\dots,y_n, z_o,\dots,z_n) = 0$, and we assume $\mu^{(n+1)}$ constant along $0 \times Y$. Let us look at the projection $p : (Y \times \mathbb{C}^{n+1},0) \to (Y \times \mathbb{C},0)$ defined by $(z_o,\dots,z_n, \underline{y}) \mapsto (z_o,\underline{y})$.
$p|X = \pi : (X,0) \to (Y \times \mathbb{C},0)$ has a discriminant $(D_\pi,0) \subset (Y \times \mathbb{C},0)$, containing $Y_\pi = 0 \times Y$. By the corollary above, the multiplicity of D_π at a point $y \in Y_\pi$ is $m_y(D_\pi) = \mu^{(n+1)}(X_y) + \mu^{(n)}(X_y \cap H)$, where H is the hyperplane $z_o = 0$.
So to prove B_μ is just the same as to prove that for a sufficiently general hyperplane H containing Y, the discriminant D_π of the corresponding projection is equimultiple along Y_π, i.e. the branch locus B_π coincides with Y_π. And thus we find something very similar to Zariski's discriminant condition ! In fact, if Y is of codimension 1 in X, it is exactly Zariski's discriminant condition.

Remark that if X satisfies (μ constant) along Y, the singular locus of X is Y, and if there are points of D_π outside Y_π, they are the images by π of points of X outside Y where $\frac{\partial F}{\partial z_1} = \dots = \frac{\partial F}{\partial z_n} = 0$, i.e., where the tangent hyperplane to X is parallel to the hyperplane $z_o = 0$.
If $\mu^{(n)}(X_y \cap H, y)$ is not constant, we have in X a curve of such points, which I like to call the curve of vanishing folds of X along Y with respect to H. The picture above can also be deemed to represent vanishing folds. If we agree to say that "X has no vanishing fold along Y" if it has no vanishing folds with respect to a generic hyperplane H containing Y, B_μ becomes the slogan : "(μ constant) $\Rightarrow$ no vanishing folds".

2.8.5 As an application of the numerical viewpoint, let us stop a while to sketch the proof of theorems 1 and 2, and in fact of the equivalence of all definitions of equisingularity, in the case where (Y,0) is (smooth) of codimension 1 in X, and also of dimension 1 since we will use theorems 3 and 4. (The results here are due to Zariski, and although what follows does offer some alternative proofs in the special case $\dim Y = 1$, it is given mostly as an informal introduction to the reading of [Z_2], [Z_4]).
We have therefore, after choice of retraction, a 1-parameter family of reduced plane curves, described by $F(y_1, z_o, z_1) = 0$, $F \in \mathbb{C}\{y_1, z_o, z_1\}$, $F(y_1, 0, 0) \equiv 0$. The first point is that when (Y,0) is of codimension 1 in (X,0), conjecture B_μ is proved, thanks to theorem 4 and to the well known fact that the multiplicity of a reduced plane curve is an invariant of the topological type : (see the appendix at the end of this section). Indeed, here, $n = 1$, so X_y is the curve $F(y, z_o, z_1) = 0$, $(y = (y_1, 0, 0))$ and $\mu^{(1)}(X_y, y) = m_y(X_y) - 1$: by theorem 4, all the (X_y, y) have the same topological type if $\mu^{(2)}(X_y, y)$ is constant, hence $m_y(X_y)$ is constant. (See appendix at the end of this section).
Let us now take a projection $p : (\mathbb{C}^3, 0) \to (\mathbb{C}^2, 0)$ such that $p^{-1}(0) \not\subset X$, and $\operatorname{Ker} dp \cap T_{Y,0} = (0)$. Choose coordinates so that p is written $(y_1, z_o, z_1) \to (y_1, z_o)$. $(Y_\pi, 0)$ is defined by $z_o = 0$. Then, the following are equivalent :

1) $(D_\pi, 0)$ is equimultiple along $(Y_\pi, 0)$, i.e. $(B_\pi, 0) = (Y_\pi, 0)$.
2) $\mu^{(2)}(X_y, y)$ and the intersection number $(\ell_y, X)_y$ are independant of $y \in (Y, 0)$, where ℓ_y is the line $p^{-1}(p(y))$.

Proof : Using 2.8.3, and the fact that if there are several critical points in the same fiber, the multiplicity of the discriminant is the sum of the multiplicities corresponding to each critical point, we find :

$$m_y(D_\pi) = \sum_{x\in\ell_y\cap X} (\mu^{(2)}(X_y,x) + \mu^{(1)}(X_y\cap H,x)) \qquad (*)$$

where H is the plane $z_o = 0$. (Of course, for a small enough representative of $(X,0)$, and in particular we consider only the points in $\ell_y \cap X$ which tend to 0 as $y \to 0$.)
Remark that $\mu^{(1)}(X_y \cap H,x) = \mu^{(1)}(X_y \cap \ell_y,x) = (\ell_y.X)_x - 1$. So by the semi-continuity properties of 2.3, if $m_y(D_\pi) = m_o(D_\pi)$ for all $y \in (Y,0)$, we must have :

$$\mu^{(2)}(X_o,0) = \sum_{x\in\ell_y\cap X} \mu^{(2)}(X_y,x) \qquad \text{for all } y \in (Y,0)$$

$$(\ell_o.X)_o - 1 = \sum_{x\in\ell_y\cap X} ((\ell_y.X)_x - 1) \quad .$$

On the other hand, the basic properties of intersection numbers imply $(\ell_o.X)_o = \sum_{x\in\ell_y\cap X} (\ell_y.X)_x$, so 1) implies $\ell_y \cap X = \{y\}$ for all $y \in (Y,0)$, and therefore 2) by the two equalities above. Conversely, if we have 2) the equality (*) shows that D_π is equimultiple along Y_π at 0.
At this point, we have proved theorem 2, and the fact that (Z)-equisingularity ⇒ (μ constant). But as mentionned above, we know in this case that (μ constant) ⇒ (μ^* constant) and I have already said that (μ^* constant) ⇒ (c) ⇒ (Whitney conditions) ⇒ (Topological triviality) ⇒ (Topological equisingularity) ⇒ (μ constant). So all that remains to prove is that (μ^* constant) ⇒ (Z)-equisingularity, and that (μ^* constant) ⇒ assertion iv) of theorem 1. The proof of the first implication will also prove i) ⇒ iii) of theorem 1 if we do it as follows : (μ^* constant) ⇒ for any projection $\pi : X \to \mathbb{C}^2$ transversal to X, $(D_\pi,0) = (Y_\pi,0)$. But to say that π is transversal means $(\ell_o.X)_o = m_o(X)$ (with the notations introduced above), and if μ^* is constant, $(X,0)$ is equimultiple along $(Y,0)$, and $m_o(X) = m_o(X_o)$, so that $(\ell_y.X)_y \geq m_y(X) = m_o(X) = m_o(X_o) = (\ell_o.X)_o$ which implies $(\ell_y.X)_y = (\ell_o.X)_o$ by semi-continuity, and then the equality (*) implies that $(D_\pi,0)$ is equimultiple along $(Y_\pi,0)$. Q.E.D.
To see that (μ^* constant) ⇒ assertion iv) of theorem 1, we can first reduce to the case where the (X_y,y) are irreducible plane curves : indeed, since we have topological equisingularity,

the number of irreducible components of the fibers (X_y,y) is independent of y, and in fact each irreducible component of X is a (μ^* constant)-family of irreducible plane curves (see the appendix). The proof of the blowing-up part of iv) follows from the following fact : if you blow up the origin in an irreducible curve $(X_o,0)$, there is only

one point lying over 0 in the blown-up curve $(X'_o, 0') \to (X_o, 0)$, and $\mu^{(2)}(X'_o, 0') = \mu(X_o, 0) - m(m-1)$ (see [Z_4], prop. 5.1 or [T_2], lemma 5.16.1) where $m = m_o(X_o)$. So if X is irreductible, μ^* constant, when we blow up Y in X, say $b : X' \to X$, we find that since X is equimultiple along Y and irreducible, $b^{-1}(Y)red = Y'$ is smooth over Y, and $b^{-1}(X_y, y)$ is just the curve obtained by blowing up y in X_y : therefore X' is again a family of irreducible curves with Milnor number constant along Y' and equal to $\mu^{(2)}(X_o) - m(m-1)$. So we can go on, blow up Y' in X', etc. but as the Milnor number strictly decreases at each step, eventually we will reach an $\bar{X}$ which is smooth. By equimultiplicity, each blowing up in the sequence is finite (and bimeromorphic) so $\bar{X} \xrightarrow{n} X$ is finite and bimeromorphic, and since $\bar{X}$ is smooth, n is the normalization of X. Of course this sketch of proof can be read backwards to show that iv) $\Rightarrow$ (μ^* constant).
From the same numerical view point it is possible to prove that if (Y,0) is the entire singular locus, (μ^* constant) $\Leftrightarrow$ (iv,β) of theorem 1 using [T_2], chap. II, prop. 3.1) and the fact that (μ^* constant) is also in this case a condition of equimultiplicity for the ideals generated by $(\frac{\partial F}{\partial z_o}, \frac{\partial F}{\partial z_1})$ in the local rings of the fibers (X_y, y), since we have in general :

<u>Proposition</u> ([T_2], chap. II, cor. 1.5) : Let $(X_o, 0) \subset (\mathbb{C}^{n+1}, 0)$ be a germ of hypersurface with isolated singularity, say with equation $f(z_o, \ldots, z_n) = 0$. Then the multiplicity in $\mathcal{O}_{X_o,0}$ of the jacobian ideal j' $(= j(f).\mathcal{O}_{X_o,0})$ generated by the images of $(\frac{\partial f}{\partial z_i})$ $0 \le i \le n$, is $\mu^{(n+1)}(X_o, 0) + \mu^{(n)}(X_o, 0)$.
In our case this gives precisely $\mu^{(2)}(X_y, y) + \mu^{(1)}(X_y, y)$.
I emphasize however that it is proved in [Z_2] that when Y is of codimension 1 in X, and is the entire singular locus of X, condition (iv), β) of theorem 1 is in itself a necessary and sufficient condition of (Z)-equi singularity.

<u>Appendix</u> : Numerical invariants of reduced plane curves.

Attached to a germ of reduced and irreducible plane curve $(X_o, 0)$, there is a sequence of integers, the characteristic of the curve $(m, \beta_1, \ldots, \beta_g) = C(X_o, 0)$, m is the multiplicity of X_o at 0, and the β_i are the characteristic exponents of the curve (see [Z_3], [P_2]) this characteristic can be quickly, if not very informatively, described by saying that $(X_o, 0)$ has the same topological type as the curve given parametrically by

$$z_o = t^m$$

$$z_1 = t^{\beta_1} + t^{\beta_2} + \dots + t^{\beta_g} \qquad (m < \beta_1 < \dots < \beta_g)$$

and does not have the same topological type as any curve given similarly $z_o = t^{m'}$, $z_1 = t^{\beta'_1} + \dots + t^{\beta'_{g'}}$ with $m' < \beta'_1 < \dots < \beta'_{g'}$, and $g' < g$. So the characteristic is a complete set of topological invariants for irreducible plane curves (see [P_2], [Z_3]).
The situation for reducible plane curves is described by :

Theorem (Zariski) : Two germs of reduced plane curves $(X_1,0)$ and $(X_2,0)$ have the same topological type if and only if there exists a bijection B from the set of irreducible components of X_1 to the set of irreducible components of X_2, which respects topological types and intersection multiplicities, i.e. $B(X_{1,i},0)$ has the same topological type (or characteristic) as $(X_{1,i},0)$ and $(X_{1,i}.X_{1,j})_o = (B(X_{1,i}).B(X_{1,j}))_o$.

Zariski proved the "if" part using his theory of saturation (see [Z_4], theorems 2.1 and 6.1). I cannot, however, find a reference for the "only if" part, except perhaps [H_5]. (The only problem, of course, concerns intersection multiplicities, and they can be defined topologically as linking number of the knots obtained by intersecting $X_{i,o}$ and $X_{j,o}$ with a sufficiently small sphere $\mathbb{S}^3$ in $\mathbb{C}^2$ centered at 0, see Brieskorn's lectures and [H_5].)
Anyhow, the multiplicity of a reduced plane curve, sum of the multiplicities of its irreducible components, depends only on its topological type!

2.9 I want now to explain what looks to me as a set of "axioms" for equisingularity from the view point of plane sections, discriminants, and the nonsplitting phenomenon.
In what follows equisingular will mean : any definition, and when I write discriminants I assume they exist, i.e. the corresponding fiber has an isolated singularity. First a definition : $(X,0) \subset (\mathbb{C}^{N+1},0)$, a hypersurface, is said to be (i)-equisingular along a smooth $(Y,0) \subset (X,0)$ $(i \geq \dim Y)$ if, assuming as usual Y linear in $\mathbb{C}^{N+1}$, there exists a dense Zariski open $U^{(i)}$ in the grassmannian $G^{(i)}$ of i-planes containing Y such that $H \in U^{(i)} \Rightarrow (X \cap H,0)$ is equisingular along $(Y,0)$.
Now, I think a good theory of equisingularity should satisfy the following :

1) Going up : Given a projection (linear, for simplicity)

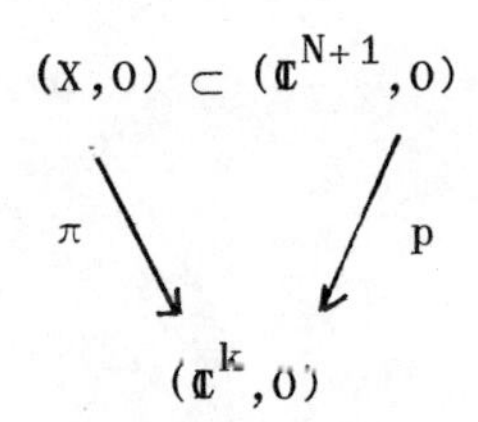

and a $(Y_\pi,0) \subset (B_\pi,0)$ where B_π is the branch locus of π, if $(Y_\pi,0)$ is a linear subspace of $(\mathbb{C}^k,0)$ and $(B_\pi,0)$ is (i)-equisingular along $(Y_\pi,0)$ $(\dim Y_\pi + 1 \leq i \leq k)$, then $Y = (p^{-1}(Y_\pi) \cap C_\pi)_{red}$ is smooth (where C_π is the critical locus of π), p induces an isomorphism $(Y,0) \simeq (Y_\pi,0)$, and (X,0) is (N+1-k+i)-equisingular along (Y,0).

2) Going down : Assume (X,0) is (j)-equisingular along a linear $(Y,0) \subset (X,0)$ $(\dim Y \leq j \leq N+1)$.

Then, for any linear projection $p : (\mathbb{C}^{N+1},0) \to (\mathbb{C}^k,0)$ such that

α) $\operatorname{Ker} p \cap Y = (0)$

β) Setting $(Y_\pi,0) = (p(Y),0)$, for every i-plane H in some dense Zariski open subset of the grassmannian of i-planes in $(\mathbb{C}^k,0)$ containing Y_π, $(i + N + 1 - k \leq j)$, $X \cap p^{-1}(H)$ is equisingular along (Y,0).

we have that B_π is (j+k-N-1)-equisingular along Y_π at 0.

We can alternatively ask :

2') Generic going down : There exists a dense Zariski-open subset of the set of linear projections $p : (\mathbb{C}^{N+1},0) \to (\mathbb{C}^k,0)$ such that this conclusion holds.

Remark : If $i = \dim Y + 1$, X (i)-equisingular along Y means X is equimultiple along Y. (Of course I ask all definitions of equisingularity to give $Y = X$ in codimension 0.) It may happen that $B_\pi = \mathbb{C}^k$. In this case we will say that $(B_\pi,0)$ is (k)-equisingular along $(\mathbb{C}^k,0)$ if we are in the equality case of 2.3.1.

Remark also that any definition of equisingularity which goes up and down has to be equivalent to Zariski's definition. (Indeed, it needs only to go up and down for $k = N$.) To show that this is not just a formality, let me at least prove the nonsplitting part of the going up, assuming D_π is $(\dim Y_\pi + 1)$-equisingular along Y_π, that is, equimultiple. Here the basic result is

Theorem 5 (see [L_5], [La], [T_2] chap. III) : Suppose that we have a projection

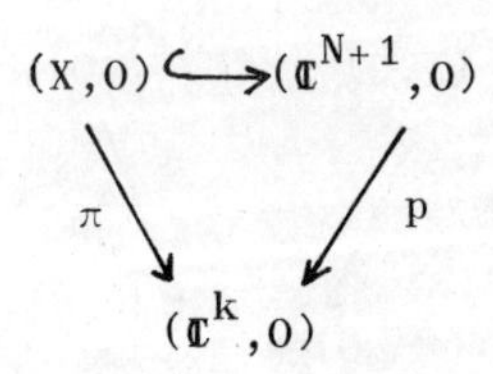

such that $(B_\pi,0) = (\mathbb{C}^k,0)$ and B_π is (k)-equisingular along $(\mathbb{C}^k,0)$, i.e. for any $y \in (\mathbb{C}^k,0)$, $\sum_{x\in\pi^{-1}(y)} \mu(X_y,x) = \mu(X_o,0)$. Then the induced map $\pi:(C_{\pi,red},0) \to (\mathbb{C}^k,0)$ is an isomorphism (and X_y has only one singular point).

This non splitting result has many algebraic translations, for example going back to the notations of 2.3.1 and 2.3.2 :

$\exists\, r : F^r \subset J \Rightarrow \exists t : S^t \subset J$, which is a priori rather odd (see also $[T_2]$ chap. III), but not algebraic proof. I think it is a very good problem to seek an algebraic proof of it. Anyway, it has the

Corollary : Given a projection as in 2.9, 1), suppose there is a linear subspace $(Y_\pi,0) \subset (B_\pi,0)$ such that $(B_\pi,0)$ is equimultiple along $(Y_\pi,0)$. Then if $(C_\pi,0) \subset (X,0)$ is the critical susbpace, $(\pi^{-1}(Y_\pi) \cap C_\pi)_{red}$ is smooth and isomorphic to Y_π by π. (Compare with 2.8.5. This generalizes unpublished results of Zariski, and was also noticed by Lê.)

Proof : As we have already remarked, B_π is equimultiple along Y if and only if D_π is so. Then (this is $\dim Y_\pi + 1$-equisingularity) we can find a $(\dim Y_\pi + 1)$-plane H containing Y_π in $(\mathbb{C}^k,0)$ such that $(D_\pi \cap H)_{red} = Y_\pi$. We make the base change of π by the inclusion $(H,0) \to (\mathbb{C}^k,0)$, i.e. restrict everything over H and thus we obtain a situation

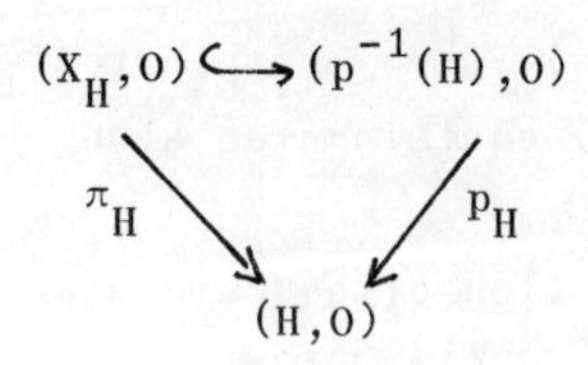

where now $B_{\pi_H} = Y_\pi$, D_{π_H} is equimultiple along Y_π. We can choose a retraction $\rho : (H,0) \to (Y_{\pi_H},0)$ and by 2.8.3 we have

$$m_o(D_{\pi_H}) = \mu(X_o,0) + \mu(\widetilde{X}_o,0) \quad \text{where } \widetilde{X}_o = \pi^{-1}(\rho^{-1}(0))$$

$$= m_y(D_{\pi_H}) = \sum_x \mu(X_y,x) + \mu(\widetilde{X}_y,x) \qquad \widetilde{X}_y = \pi^{-1}(\rho^{-1}(y))$$

the sum being over the points of $\pi_H^{-1}(y) \cap C_{\pi_H}$.

By 2.3.1, we know

$$\mu(X_o,0) \geq \Sigma\, \mu(X_y,x)$$

$$\mu(\widetilde{X}_o,0) \geq \Sigma\, \mu(\widetilde{X}_y,x)$$

so both must be equalities, which implies that if we now restrict over Y_π, $(X_{Y_\pi},0) \to (Y_\pi,0)$ satisfies the assumptions of theorem 5, from which we deduce the desired result.

Notice that we have proved much more, namely that X_H satisfies (μ constant) along $Y = (\pi_H^{-1}(Y_\pi) \cap C_{\pi_H})_{red}$ and even has a hyperplane section containing Y which also satisfies (μ constant), which gives some hope for the going up.

Problem : Take any of the definitions of equisingularity given in § 1 (or (c)) and prove it goes up and down.

All we know in general is that topological triviality goes up (Varchenko).

2.10 With the idea that any good definition of equisingularity has to go up and down, one can give a purely topological equisingularity problem, as follows :

Given a reduced hypersurface $(X,0) \subset (\mathbb{C}^{N+1},0)$, take a non singular subspace $(Y,0) \subset (X,0)$, and for simplicity, I will assume $\dim Y = 1$. Then we can define the "vanishing cycles" of X along Y at 0 with respect to a retraction $r : (\mathbb{C}^{N+1},0) \to (Y,0)$ like this (see [H5]) : If as usual $X_y = r^{-1}(y) \cap X$, one can prove that there exists $\varepsilon_o \in \mathbb{R}^+$, and a monotone increasing function $\lambda : [0,\varepsilon_o] \to \mathbb{R}^+$ (depending on (X,0), (Y,0) and r) such that, if we call y_1 a coordinate on (Y,0), the topological type of $X_y \cap \overset{\circ}{\mathbb{B}}_\varepsilon$ is independent of ε and y provided $0 < \varepsilon < \varepsilon_o$, $0 < |y_1| < \lambda(\varepsilon)$, and independent of the choice of r, provided r is "sufficiently general". The reduced cohomology groups $H^i(X_y \cap \overset{\circ}{B}_\varepsilon, \mathbb{Z})$ will be called the "groups of vanishing (co-) cycles" of X along Y at 0. We will say that X "has no vanishing cycles" along Y at 0 if they are all 0.

Problem : Show that the condition "no vanishing cycle along Y" goes up and down, that is :

1) Given a linear projection $p : (\mathbb{C}^{N+1},0) \to (\mathbb{C}^N,0)$ with $\operatorname{Ker} d\pi \cap T_{Y,0} = (0)$, set $\pi = p|X$, and $Y_\pi = p(Y)$. If the branch locus B_π has no vanishing cycles along Y_π, X has no vanishing cycles along Y.

2) If X has no vanishing cycles along Y, show that there exists a

(Zariski) open dense subset U in the space of linear projections such that if $p \in U$, D_π has no vanishing cycles along Y_π.

This problem is settled by what we have seen above at least in the case where Y is of codimension 1 in X, since we have (using theorem 5) :

Proposition (Hironaka [H_5]) : If Y is the entire singular locus of X, "X has no vanishing cycles along Y" $\Leftrightarrow$ (μ constant) holds, and of course, if Y is of codimension 0 in X, no vanishing cycles means $Y = X$.
By the way, B_μ becomes thus a better slogan : "No vanishing cycles $\Rightarrow$ no vanishing folds". It seems to me that this problem is very attractive for the following reason : its solution in the large requires the construction of what I like to call "the ultimate Morse theory in the complex domain", namely a theory which tells us that under suitable genericity assumptions on π, we can lift (vanishing) cycles from the branch locus to X itself . Of course, we can put a similar problem for projections to $\mathbb{C}^k$ $k \leq N$, and for $k = \dim Y + 1$, it would prove B_μ (see 2.8.4).

2.11 I'll now try to add a little to our experience with the numerical view point.

Certainly, from a differential-geometric view point, the aim of a theory of equisingularity is to obtain a partition of a given complex analytic space X, $X = \cup X_\alpha$ with the following properties :

1) each point $x \in X$ has a neighborhood which meets only finitely many X_α.

2) each X_α is a smooth and locally closed subspace in X, and $\bar{X}_\alpha - X_\alpha$ and $\bar{X}_\alpha$ are closed subspaces of X, and $\dim(\bar{X}_\alpha - X_\alpha) < \dim X_\alpha$.

3) $\bar{X}_\alpha \cap X_\beta \neq \emptyset \Rightarrow X_\beta \subset \bar{X}_\alpha$ and for any $x \in X_\beta$ and any imbedding of a small enough neighborhood U of X in an affine $\mathbb{C}^{N+1}$, $\bar{X}_\alpha \cap U$ is "cone-like over its boundary, with respect to X_β" in the sense of 2.8.1.

These conditions are precisely what was achieved by Whitney [W_1], and what allows one to do some differential geometry on singular varieties. Apart form the motivations given in § 1, the construction of Whitney offers a challenge to algebraic geometers, namely :

Challenge : To describe in many algebraic (i.e. complex analytic !) ways a partition of a singular space which has the properties listed above.

The first thing to do is to describe algebraically what "cone like" can mean. Let me take the case where $X = \bar{X}_\alpha$ is a hypersurface in $\mathbb{C}^{N+1}$, and set

$Y = X_\beta$.

Then we can describe this as usual by $F(y_1,\dots,y_k, z_o,\dots,z_n) = 0$ $(n+k = N)$. Then, an algebraic geometer has a normal cone of X along Y obtained as follows : expand F into homogeneous polynomials in $z_o,\dots,z_n$ with coefficients in $\mathbb{C}\{y_1,\dots,y_n\}$. $F = F_m + F_{m+1} + \dots$, F_i homogeneous of degree i in $(z_o,\dots,z_n)$. Then the normal cone is defined by $F_m(y_1,\dots,y_k ; z_o,\dots,z_n) = 0$, which can be viewed as a family of cones parametrized by Y : $C_{X,Y} \to Y$, (see [H_3]). It turns out that it is hopeless to look only at the normal cone to see if X is cone-like over its boundary with respect to Y, simply because you can find two hypersurfaces with the same normal cone, one of which is cone-like, and the other not. e.g. in $\mathbb{C}^3(y_1,z_o,z_1)$ $z_1^2 - z_o^7 = 0$ (a product of a curve by the y_1-axis) and $z_1^2 - z_o^7 + y_1 z_o^3 = 0$. So to understand what "cone like" might mean algebraically, we have to look further into the expansion of F. Now there are two ways of doing this indirectly.

1) Blow up Y in X and look again (compare theorem 1, iv)).

2) Look at the ideal $J = (\frac{\partial F}{\partial z_o},\dots,\frac{\partial F}{\partial z_n}) \mathbb{C}\{y_1,\dots,y_k, z_o,\dots,z_n\}$.

Now J is the ideal that comes into Whitney conditions and condition (c), and the first point I want to make here is the following : What theorem 3 says, (going back to 2.8.1) is that to ensure that X is "cone-like over its boundary" we can replace the (non-complex analytic) tube $\mathbb{B}_\rho \times Y$ by the (complex analytic) "generic flag of plane sections through Y" which comes into (μ^* constant), for the study of J (or of limit positions of tangent hyperplanes at smooth points of X as they tend to Y, which is the same as the study of J).

The second point is something akin to the problems considered in [H_1] and which throws some new light on what "cone-like" may mean algebraically :

<u>Definition</u> : Given $F(y_1,\dots,y_k, z_o,\dots,z_n)$ as above, let me say that the t-jet of F along Y is strongly sufficient if the following holds : For any $t' \geq t$ and any $G \in S^{t'}$ $(S = (z_o,\dots,z_n)\mathbb{C}\{y_1,\dots,y_k, z_o,\dots,z_n\})$ there exists $\varepsilon_i \in S^{\nu(t')}$ (depending on G) such that

1) $F(y_1,\dots,y_k, z_o,\dots,z_n) + G(y_1,\dots,y_k, z_o,\dots,z_n) \equiv F(y_1,\dots,y_k, z_o+\varepsilon_o,\dots,z_n+\varepsilon_n)$.

2) $\frac{\nu(t')}{t'} > \frac{1}{2}$.

Then

<u>Theorem 6</u> (with J.P.G. Henry, see [H.T]) : For F as above, such that $J \subset S$, the following are equivalent :

1) there exists a t such that the t-jet of F along Y is strongly sufficient.

2) $F = 0$ satisfies (μ constant) along Y, near 0.

So

So the existence of a strongly sufficient jet along Y may be a subtle way to say algebraically in this case, that X is "cone like" with respect to Y (at least if we assume B_μ is true).

§ 3. POLAR VARIETIES

3.1 This is a presentation of some algebraic ways to investigate the relationship between a hypersurface, say, and its generic hyperplane section. (The details will appear in [T_3].) We have many motivations to do this from § 1 and 2, but let me add a few more.

First, remark that the topological type of a generic hyperplane section of a given hypersurface is well defined, as can be seen by applying the Whitney stratification theorem (1.1) to the family of hyperplane sections constructed in § 1, and then using (W) ⇒ (T.E.). (This works for plane sections of any dimension).

Question $B_{top.}$: If two germs of complex analytic hypersurfaces have the same topological type, does it imply that their generic hyperplane sections also have the same topological type ?

(By convention, for a plane curve, the "topological type" of the generic hyperplane section is the multiplicity of the curve. The answer to the question above is known only for curves, see 2.8 App.) An affirmative answer to this question would imply in particular that if two hypersurfaces with isolated singularities have the same topological type, they have the same μ^*, and in general, it would imply that same topological type ⇒ same multiplicity, thus answering a question of Zariski [Z_5] (by descending induction).

I want to relate $B_{top.}$ to the following, inspired by a question of Thom in [Th_3].

Question C : Is it true that any germ of complex analytic hypersurface has the same topological type as a germ of algebraic hypersurface ?

Of course, if (X,0) has an isolated singularity, it is known that (X,0) is even analytically isomorphic to a germ of algebraic hypersurfa-

ce ; in fact (see e.g. [Mi]) given an equation $f(z_o,\dots,z_n)=0$ for a germ of hypersurface with isolated singularity, there exists an integer t depending only on $j(f)$ such that if $g\in\mathfrak{M}^t$, then we can find a change of coordinates $z_i\to u(z_i)=z_i+\varepsilon_i(z_o,\dots,z_n)$ $\varepsilon_i\in\mathfrak{M}^2$, such that $f(\underline{z})+g(\underline{z})=f(\underline{u}(\underline{z}))$, and in practice, if s is the smallest integer such that $j(f)\supset\mathfrak{M}^s$, which exists since $f=0$ has an isolated singularity, we can take $t=2s+1$. Thus $f=0$ is analytically isomorphic to the hypersurface obtained by forgetting all the terms of degree $\geq t$ in the Taylor expansion of f. This method will not work, at least directly, for the non-isolated singularities, even if we want only to preserve the topological type. Look at the Whitney example ($[W_1]$) of a hypersurface in $\mathbb{C}^3$ which is not locally analytically isomorphic to an algebraic hypersurface : it is defined by $F=z_o.z_1.(z_o-z_1)(z_o-(3+y_1)z_1)(z_o-\gamma(y_1)z_1)=0$ where $\gamma(y_1)$ is any transcendental function with $\gamma(0)=4$ e.g. $\gamma=4.e^{y_1}$. This is an equisingular family of curves, each curve consisting of 5 lines. From the fact that equisingularity implies topological triviality, we know that it has the same topological type as an algebraic hypersurface, namely $\tilde{F}=z_o z_1(z_o-z_1)(z_o-3z_1)(z_o-4z_1)=0$. However, for any $t\geq 1$, $F+y_1^t=0$ has an isolated singularity at 0 and therefore does not have the same topological type as $F=0$, as one can check by using for example the first theorem in $[L_2]$. Please compare all this with Theorem 6.
So we set ourselves the following :

<u>Problem</u> : Define algebraically a complete set of invariants for the topological type of a germ of complex hypersurface.

Ideally, this set of invariants should enable me to play the following :

<u>Game</u> : Given a hypersurface $(X,0)\subset(\mathbb{C}^{N+1},0)$, I fill a bag with invariants arranged in layers according to dimensions $1\leq i\leq N+1$, the top layer corresponding to $i=1$. Then I give somebody the first layer of invariants, and he constructs a germ of plane curve, the invariants of which are precisely those in the first layer, and which has the same topological type as the section of my hypersurface by a generic 2-plane. I then give him the second layer, and from this and the curve he constructed before he builds a surface, which has the same topological type as the section of (X,0) by a generic 3-plane, and a generic plane section of which has the same topological type as the curve constructed before, etc. When the bag is empty he has built a hypersurface (X',0) with the same topological type as (X,0). Of course, for sheer economy, at each step he will build

an algebraic hypersurface !
Now the invariants constructed below from polar curves should, I think, be at least part of what one needs to put in the bag.

3.2 Where to look for invariants ? an answer is suggested by the following, which is an improvement on a theorem of Lê and Ramanujam ([L.R.]) :

<u>Theorem 6</u> (see [T_3]) : Let $(X_0,0) \subset (\mathbb{C}^{N+1},0)$ and $(X_1,0) \subset (\mathbb{C}^{N+1},0)$ be two germs of hypersurfaces with isolated singularity, defined respectively by $f(z_0,\ldots,z_n) = 0$, $g(z_0,\ldots,z_n) = 0$. If $\mathcal{O}_{n+1}/\overline{j(f)} \simeq \mathcal{O}_{n+1}/\overline{j(g)}$ are isomorphic as analytic algebras (notations of 2.1 and 2.2), then there exists a one parameter family of complex hypersurfaces $X \overset{\sigma}{\rightleftarrows} \mathbb{D}$ where say $\mathbb{D} = \{y \in \mathbb{C} \,/\, |y| < 2\}$ such that $(X_0,\sigma(0)) = (X_0,0)$, $(X_1,\sigma(1)) = (X_1,0)$ and X is Whitney equisingular (even (c)-equisingular) along $\sigma(\mathbb{D}) = Y$ at every point. In such a case, we will say that $(X_0,0)$ and $(X_1,0)$ are (c)-cosecant. In particular, they have the same topological type. Since $\overline{j(f)}$ is determined by $(f,j(f))$ as we saw in 2.2, and $(f,j(f))$ determines the singular subspace of $(X,0) \subset (\mathbb{C}^{n+1},0)$ we have for isolated singularities of hypersurfaces, the following :
"The analytic type of the singular subspace determines the topological type". (We will see below how false the converse is.)
Therefore, we look for topological invariants in the jacobian ideal $j(f)$, even for non isolated singularities, where a result similar to the one above is not known. In the isolated singularity case, we have already the sequence $\mu^*(X,0)$ of the Milnor numbers of generic plane sections, which can easily be shown, by its algebraic definition as generalized multiplicity (2.6), to depend only on $\overline{j(f)}$.(Perhaps it is time to point out the difference between $j(f)$ and $\overline{j(f)}$. For example the only isolated singularities of hypersurfaces such that $\overline{j(f)}$ is a power of the maximal ideal are those which have the same topological type as their tangent cone ([T_2]). On the other hand, the only isolated singularities of hypersurfaces such that $j(f) = \overline{j(f)}$ are those of type A_k, i.e. which can be defined in suitable coordinates by $z_0^{k+1} + z_1^2 + \ldots + z_N^2 = 0$.) There are however other geometric ways of studying $j(f)$, in the general case :

3.3 <u>Proposition-Definition</u> : Let $(X,0) \subset (\mathbb{C}^{N+1},0)$ be a germ of complex hypersurface, defined by $f(z_0,\ldots,z_N) = 0$. Let $H \in G^{(i)}$ be a direction of i-plane in $\mathbb{C}^{N+1}$. Consider the set P_H of points $p \in (\mathbb{C}^{N+1},0)$ such that

1) $f(z_0,\ldots,z_N) = f(p)$ (X_p, the level hypersurface of f through p) is smooth at p

2) $T_{X_p,p} \supset H$ (i.e. $T_{X_p,p}$ contains a i-plane parallel to H).

The closure Q_H of this set in $(\mathbb{C}^{N+1},0)$ is a closed complex analytic subset of $(\mathbb{C}^{N+1},0)$. In fact, Q_H can be defined as an analytic subspace of $(\mathbb{C}^{N+1},0)$, as follows : choosing coordinates so that H is defined by $z_o = \dots = z_{N-i} = 0$, Q_H is the subspace defined by the ideal $\{(\frac{\partial f}{\partial z_k})\mathcal{O}_{N+1} ; N-i+1 \le k \le N\}$. The union of those irreducible components of $(Q_H,0)$ which are of dimension N+1-i will be called the polar subspace of $f = 0$ with respect to H, and noted S_H.

<u>Theorem 7</u> ([T_3]) : If (X,0) has an isolated singularity, for each $0 < i < N+1$, there exists a Zariski open subset $V^{(i)}$ of the Grassmannian $G^{(i)}$ of i-planes through 0 in $(\mathbb{C}^{N+1},0)$, such that if $H \in V^{(i)}$, S_H is reduced, $m_o(S_H) = \mu^{(i)}$ and the plane (H,0) is transversal to $(S_H,0)$ at 0 in the sense that $\mu^{(i)}$ is also the intersection multiplicity $(S_H.H)_o$. For $i = 0$, we set $S_H = \mathbb{C}^{N+1}$, and for $i = N+1$, S_H is the subspace defined by j(f), which is of multiplicity $\mu^{(N+1)}(X,0)$. Hence $\mu^*(X,0)$ is also the sequence of multiplicities at 0 of the generic polar varieties.

3.4 For $i = N$, S_H is a curve, called the polar curve with respect to the hyperplane direction H, and its consideration has been advocated by Thom for the study of monodromy problems. It also comes into the proof of the proposition in 2.8.3., as follows : (please go back to 2.8.3) : by the projection formula for multiplicities ([Se]), Δ is the multiplicity of the critical subspace of $\pi : (X,0) \to (\mathbb{C},0)$. We can always suppose that the projection $p : (\mathbb{C}^{N+1},0) \to (\mathbb{C},0)$ inducing π is a coordinate function z_o. Then Δ is the multiplicity of the ideal $(\frac{\partial F}{\partial z_1},\dots,\frac{\partial F}{\partial z_n})\mathcal{O}_{X,0}$ if X is defined by $F(z_o,\dots,z_N) = 0$. But that is precisely ([Se]) the intersection multiplicity of the polar curve $S_H : \left(\frac{\partial F}{\partial z_1},\dots,\frac{\partial F}{\partial z_N}\right)$ in $(\mathbb{C}^{N+1},0)$ with our hypersurface (X,0) at 0 . Since we assume (X,0) to be with isolated singularity, $(\frac{\partial F}{\partial z_o},\dots,\frac{\partial F}{\partial z_N})$ is a regular sequence, and hence all the components of $\frac{\partial F}{\partial z_1} = \dots = \frac{\partial F}{\partial z_N} = 0$ are of dimension 1. Let us decompose S_H in irreducible curves (not necessarily reduced, since we do <u>not</u> assume the hyperplane $H : z_o = 0$ is in the open $V^{(N)}$ of theorem 7, but only that $(H \cap X,0)$ has an isolated singularity). Let this decomposition be written $S_H = \cup \Gamma_q$. Now to each Γ_q is attached an integer $n(\Gamma_q)$ such that for any hypersurface $(X',0) \subset (\mathbb{C}^{N+1},0)$ we have $(\Gamma_q.X')_o = n(\Gamma_q).(\tilde{\Gamma}_q.X')_o$ where $\tilde{\Gamma}_q = \Gamma_{q,red}$. Now we remember that $\tilde{\Gamma}_q$ can be parametrized, i.e. the normalization $(\bar{\Gamma}_q,0) \to (\tilde{\Gamma}_q,0) \subset (\mathbb{C}^{N+1},0)$ is such that $(\bar{\Gamma}_q,0) \simeq (\mathbb{C},0)$ since $\tilde{\Gamma}_q$ is reduced and irreducible. Let us choose a local coordinate u_q on $(\bar{\Gamma}_q, 0)$. The composed map above $(\bar{\Gamma}_q, 0) \xrightarrow{h_q} (\mathbb{C}^{N+1}, 0)$ can be described by N+1 holomorphic functions $z_k = z_k(u_q)$. Then the intersection number of (X,0) with $(\tilde{\Gamma}_q, 0)$ is the order in u_q of $f(z_o(u_q), \ldots, z_{N+1}(u_q)) = f \circ h_q$. But we have

$$\frac{d}{du_q} f \circ h = \frac{\partial f}{\partial z_o} \circ h_q \cdot \frac{d(z_o \circ h_q)}{du_q} \quad \text{since on } \tilde{\Gamma}_q \text{ all } \frac{\partial f}{\partial z_i} \ (1 \leq i \leq N+1) \text{ are } 0.$$

Taking orders gives

$$(\tilde{\Gamma}_q . X)_o - 1 = (\tilde{\Gamma}_q . X')_o + (\tilde{\Gamma}_q . H)_o - 1$$

and adding after multiplying by $n(\Gamma_q)$ gives $\Delta = (S_H . X)_o = (S_H . X')_o + (S_H . H)_o$ where X' is $\frac{\partial f}{\partial z_o} = 0$. But it follows from classical results on intersection theory ([Se]) that $(S_H . X')_o = \mu^{(N+1)}(X,0)$ and it is easy to see that $(S_H . H)_o = \mu^{(N)}(X \cap H, 0)$.

3.5 All this was done for a given hyperplane H such that $X \cap H$ has an isolated singularity. Now we have

3.5.1 <u>Proposition</u> ($[T_3]$) : Given a hypersurface $(X,0) \subset (\mathbb{C}^{N+1}, 0)$ with equation $f = 0$, there is a Zariski-open dense $W^{(N)} \subset \mathbb{P}^N$ such that : the polar curve S_H is reduced, the number ℓ of its irreducible components, $\left(S_H = \bigcup_{q=1}^{\ell} \Gamma_q\right)$ and the sets of integers $\{m_q = (\Gamma_q . H)_o\}$, $\{e_q = (\Gamma_q . X)_o - (\Gamma_q . H)_o\}$ are independent of H, provided $H \in W^{(N)}$. Notice $e_q \geq 0$ by the computation in 3.4. If (X,0) has an isolated singularity, we have furthermore :

$$\sum_1^{\ell} e_q = \mu^{(N+1)}(X,0) \quad ; \quad \sum_1^{\ell} m_q = \mu^{(N)}(X,0) \quad \text{by 3.4} .$$

<u>Remark</u> : In the general case, it can happen that $(\Gamma_q, 0) \subset (X,0)$, in which case we set $e_q = +\infty$.
Then we have :

3.5.2 <u>Proposition</u> : The set of quotients $\left\{\frac{e_q}{m_q}\right\}_{red}$, where "red" means that each rationnal number, or $+\infty$, appears only once, depends only upon the integral closure $\overline{j(f)}$ (i.e. can be computed from the ideal $\overline{j(f)}$), and therefore only on the local algebra $\mathcal{O}_{X,0}$. We will denote this set by $F^{(N+1)}(X,0)$.

Let me give an example showing that we have to reduce the sequence if we hope to get topological invariants in this way : consider the family of plane curves : $z_1^4 + z_o^9 + y_1 z_o^5 z_1^2 = 0$. For $y_1 = 0$ the generic polar curve has only one irreducible component, giving $e = 24$, $m = 3$. For $y_1 \neq 0$, it has two irreducible components giving $e_1 = 8$, $m_1 = 1$ and $e_2 = 16$, $m_2 = 2$ respectively ! (This is related to the phenomenon discovered by Pham, that equisingularity does not imply "jacobian equisingularity" see [P_1], [P_3]. In this example, the analytic type of $\mathcal{O}_2/\overline{j(f)}$ varies although the family is equisingular, as can be quickly checked with Zariski's discriminant criterion.)

At least when the singularity is isolated, I can prove that all the invariants $F^{(i)}(X,0)$ of generic i-plane sections of X are well defined and depend only on $\overline{j(f)}$ hence on $\mathcal{O}_{X,0}$ $(0 \leq i \leq N+1)$. ($F^{(1)} = \{\mu^{(1)}\}$, $F^{(o)} = \{1\}$.)

3.5.3 Problem : Show that the $F^{(i)}(X,0)$ are always well defined, and are invariants of the topological type, of $(X,0) \subset (\mathbb{C}^{N+1},0)$.

The answer is now known for irreducible plane curves :

Theorem 8 (M. Merle [Me]) : For an irreducible plane curve (X,0), the datum of $(F^{(1)}(X,0);F^{(2)}(X,0)) = \left(m-1, \frac{e_1}{m_1}, \ldots, \frac{e_g}{m_g}\right)$ is equivalent by a universal algorithm to the datum of the characteristic $(m,\beta_1,\ldots,\beta_g)$ of the curve.

So in this case, $F^{(1)}(X,0)$ and $F^{(2)}(X,0)$ are enough to fill the bag of invariants.

3.6 I want to quote some results in support of the claim that the $F^{(i)}(X,0)$ should go into the bag, the proofs are in [T_3].

3.6.1 Proposition : Given a hypersurface $(X,0) \subset (\mathbb{C}^{N+1},0)$ with equation $f(z_o,\ldots,z_N) = 0$ and isolated singularity. A necessary and sufficient condition for (X,0) to be (c)-cosecant with a hypersurface $(X',0): g(z_1,\ldots,z_N) + z_o^{a+1} = 0$ such that $H : z_o = 0$ is a generic section in the sense that $\mu(X' \cap H,0) = \mu^{(N)}(X',0)$ is that $F^{(N+1)}(X,0) = \{a\}$, $a \in \mathbb{N}$, i.e. all $\frac{e_q}{m_q}$ are equal to the same integer a. In particular this gives inductively a purely algebraic sufficient condition for a hypersurface to have the same topological type as a hypersurface of the Pham-Brieskorn type, i.e. $z_o^{a_o} + \ldots + z_N^{a_N} = 0$.

3.6.2 Proposition : Given a hypersurface $(X,0) \subset (\mathbb{C}^{N+1},0)$. Assume the singular locus Y of X is of dimension 1 and smooth. A necessary and sufficient condition for (X,0) to satisfy "μ constant" along (Y,0) is : $F^{(N+1)}(X,0) = \{+\infty\}$. (This follows from 2.3.2.)

3.6.3 Proposition :

For a hypersurface (X,0) (f = 0) with isolated singularity, set $\eta = \mathrm{Sup}\left(\frac{e_q}{m_q}\right)$. Then the smallest possible exponents for the Łojasiewicz inequalities $|\mathrm{grad}\ f(z)| \geq c_1 |f(z)|^{\theta_1}$, $|\mathrm{grad}\ f(z)| \geq c_2 |z|^{\theta_2}$ to hold in some neighborhood of the origin in the ambiant space are : $\theta_1 = \frac{\eta}{\eta+1}$, $\theta_2 = \eta$. (What one actually shows is $\bar{\nu}_j(f) = \frac{\eta+1}{\eta}$, $\bar{\nu}_j(\mathfrak{m}) = \frac{1}{\eta}$, where $j = j(f)$, see 2.1). So a positive answer to 3.5.3 would in particular show that these Łojasiewicz exponents are in fact topological invariants.

3.6.4 If we go back to the geometric definition of the polar curve, we obtain the following transcendantal interpretation of $\left(\frac{e_q}{m_q}\right)$, which I explain only in the isolated singularity case, for simplicity : let us choose a general hyperplane H (see theorem 7,(3.3)) and look at the polar curve $S_H = \cup \Gamma_q$; by theorem 7,(3.3) we can parametrize Γ_q as follows : $z_o = u_q^{m_q}$, $z_i = a_i u_q^{k_{i,q}} + \ldots$, where $k_{i,q} \geq m_q$, $1 \leq i \leq N+1$, if $H : z_o = 0$. In a sufficiently small neighborhood of 0 in $\mathbb{C}^{N+1}$, the set of points where the tangent space to a level hypersurface $X_t : f(z_o,\ldots,z_N) = t$ ($t \neq 0$, sufficiently small) is parallel to H is precisely $S_H \cap X_t$. (Recall that X_t is smooth for $t \neq 0$.) Furthermore, each Γ_q is smooth outside 0 and meets X_t transversally. The number of these points is $(S_H \cdot X)_o = \mu^{(N+1)}(X,0) + \mu^{(N)}(X,0)$ by Proposition 3.5.1.

Now we can also view these points as the critical points of the function $|z_o|$ ("distance" to H) restricted to X_t, so that we can attach to each of these points a gradient cell of dimension N, which originates at the critical point and ends in $X_t \cap H$. (See [M], [Th_2] and [L_4] where Lê independently introduces rationnal numbers which are actually the $\left(\frac{e_q}{m_q}\right)$, from a topological viewpoint.) In the isolated singularity case the exact sequence of relative homology reduces to : (see Brieskorn's lectures)

$$0 \to H_N(X_t,\mathbb{C}) \to H_N(X_t, X_t \cap H,\mathbb{C}) \to H_{N-1}(X_t \cap H,\mathbb{C}) \to 0$$

and $\dim H_N(X_t,\mathbb{C}) = \mu^{(N+1)}(X,0)$, $\dim H_{N-1}(X_t \cap H,\mathbb{C}) = \mu^{(N)}(X,0)$ since H is general. Hence, our $\mu^{(N+1)}(X,0) + \mu^{(N)}(X,0)$ gradient cells, which generate $H_N(X_t, X_t \cap H)$, are in fact a basis. The idea is in the following picture, (which is misleading because the indices of the critical points are not properly described !).

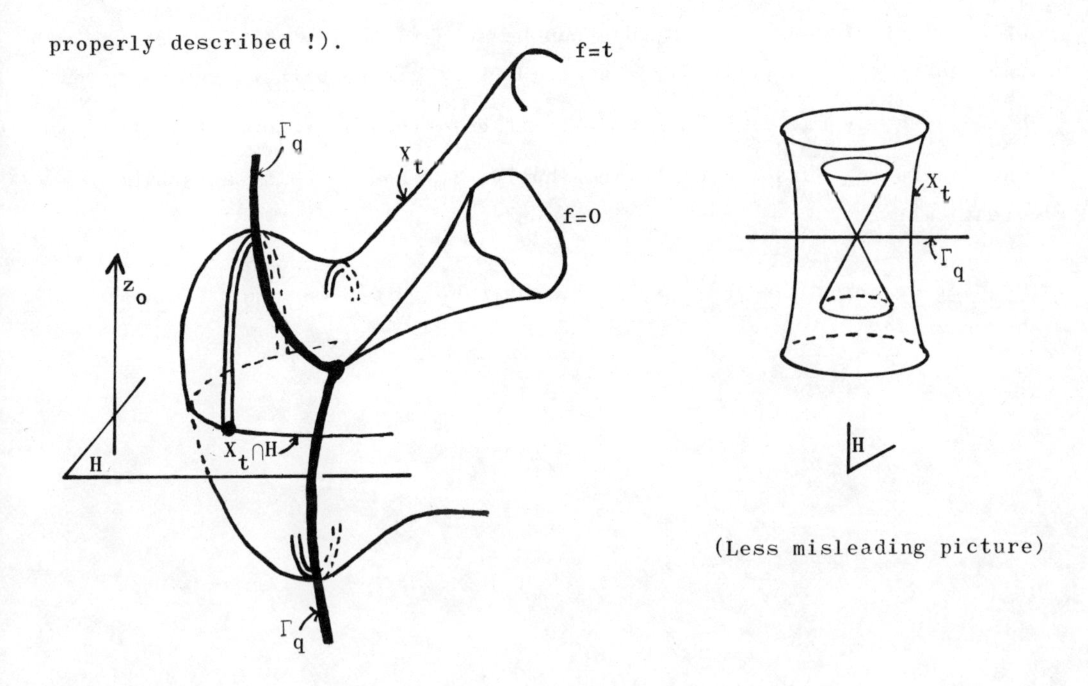

(Less misleading picture)

The double lines picture gradient cells, and we can see them generating the relative homology $H_N(X_t, X_t \cap H)$.

Now these gradient cells do not all vanish at the same rate as $t \to 0$. To see this let us compute the value of f on Γ_q. By the definition of intersection numbers, we have on Γ_q an expansion : $f|\Gamma_q = c_q u_q^{e_q+m_q} + \ldots$ $(c_q \in \mathbb{C}^*)$. Now we want to compute the distance to H of a point of $\Gamma_q \cap X_t$, i.e. the "height" of the corresponding gradient cell, as a function of t, to see how fast it vanishes when $t \to 0$. Since $z_o|\Gamma_q = u_q^{m_q}$, we find a Puiseux expansion

$$z_o|\Gamma_q = \left(\frac{t}{c_q}\right)^{\frac{m_q}{e_q+m_q}} + \ldots .$$

which represents the z_o coordinates of the $e_q + m_q$ intersection points $X_t \cap \Gamma_q$. Thus we have, with the obvious definition of "vanishing rate" :

<u>Proposition</u> : The vanishing rate of the height of a gradient cell attached to a point of $\Gamma_q \cap X_t$ is equal to $\left(\frac{e_q}{m_q} + 1\right)^{-1}$.

So our sequence $\left(\frac{e_q}{m_q}\right)_{red}$ in fact corresponds to all vanishing rates of the gradient cells, or if you prefer, it indexes a certain filtration on the vanishing homology group $H_n(X_t, X_t \cap H, \mathbb{C})$ (by the "rate of vanishing") which a priori depends on H, but can be seen to be in fact independent of the choice of a sufficiently general H. This gives further reasons to put $\left(\frac{e_q}{m_q}\right)_{red}$ in our bag of invariants : it describes an important feature of the "vanishing Morse theory" which builds X_t from $X_t \cap H$ by attaching gradient cells.

Problem : Show that $F^{(i)}(X,0)$ are constant in a μ-constant family of hypersurfaces.

* * *

Post-scriptum : On the real analytic case : it is proved in 2.8.5 that a family of reduced plane complex curves topological equisingularity $\Leftrightarrow$ Whitney conditions. Although Whitney conditions imply topological equisingularity also in the real-analytic (and even differentiable) case, see [Ma], the converse is not true in general. Here is an example : the surface $X : x_o^3 - y_1 x_1^2 - x_1^5 = 0$ in $\mathbb{R}^3$ is easily seen to be topologically trivial along $Y : x_1 = x_o = 0$, but does not even satisfy Whitney's condition a) along Y. Also, after conversations with Merle and Risler, I became convinced that if $f(x_o, \ldots, x_n) \in \mathbb{R}\{x_o, \ldots, x_n\}$, the Łojasiewicz exponents I compute in 3.6.3 for the complexification of f (assuming it has an isolated singularity) are in general larger than the smallest exponents for the corresponding inequalities in the real-analytic case.

Post-sriptum 2 : Important : See after the references for recent developments.

References

[Ab_1] S. Abhyankar , Current status of the resolution problem, Conference at the Woods Hole College on Algebraic Geometry, 1964.

[Ab_2] S. Abhyankar , Ramification and resolution in : Actas del colloquio Internacional sobre Geometria Algebrica, Madrid, Septembre 1965.

[Ab_3] S. Abhyankar , Remarks on equisingularity, Am. J. Math. 90, 1 1968.

[C.H.B.] Cevdet Haş Bey , Application à l'équisingularité, Note aux C. R. Acad. Sc. Paris t. 275, p. 105-107, 1972.

[H_1] H. Hironaka , Equivalences and deformations of isolated singularities, Woods Hole Conference on Algebraic Geometry, 1964.
See also :

[H_2] H. Hironaka , On the equivalence of singularities in : Arithmetical algebraic geometry, ed. by O.F.G. Schilling, Harper and Row editors, 1965.

[H_3] H. Hironaka, Bimeromorphic smoothing of complex analytic spaces, preprint Univ. of Warwick, England, 1971.

[H_4] H. Hironaka, Resolution of singularities of an algebraic variety over a field of characteristic zero, Annals of Math. 79, 1-2, 1964.

[H_5] H. Hironaka , Course on singularities, C.I.M.E., Bressanone, June 1974, to appear Ann. Scuola Norm. Pisa.

[H_6] H. Hironaka , Normal cones in analytic Whitney stratifications, Publ. Math. I.H.E.S., Presses Universitaires de France, No 36, 1969.

[H.T.] J.P.G. Henry et B. Teissier , Suffisance des familles de jets et équisingularité, to appear.

[La] F. Lazzeri , a theorem on the monodromy, Astérisque (Société Mathématique de France) No 7-8, 1973.

[L_1] Lê Dũng Tráng , Topologie des singularités, in "Singularités à Cargèse", Astérisque (Société Mathématique de France) No 7-8, 1973.

[L_2] Lê Dũng Tráng , Calcul du nombre de cycles évanouissants d'une hypersurface complexe, Ann. Institut Fourier t. 23 fasc. 4, 1973.

[L_3] Lê Dũng Tráng , Calcul du nombre de Milnor d'une singularité isolée d'intersection complète, preprint Centre de Mathématiques de l'Ecole Polytechnique, M147.1073. Funkt. An. 2, 1974 (USSR).

[L_4] Lê Dũng Tráng , La monodromie n'a pas de points fixes, to appear.

[L_5] Lê Dũng Tráng , Application d'un théorème d'A'Campo à l'équisingularité, Indagationes Math. 35, 5, 1973.

[Le] M. Lejeune-Jalabert , Sur l'équivalence des singularités des courbes algébroïdes planes : coefficients de Newton, Thèse, Paris VII, 1973, 1ère partie.

[Li_1] J. Lipman , Absolute saturation of one dimensional local rings.

[Li_2] J. Lipman , Relative Lipschitz-saturation, both to appear in the Am. J. Math.

[L.R.] Lê Dung Trang and C.P. Ramanujam , The invariance of Milnor's number implies the invariance of the topological type, to appear in the Am. J. Math. (preprint Centre de Mathématiques de l'Ecole Polytechnique, M120.0573).

[L.T.] M. Lejeune-Jalabert et B. Teissier, Séminaire Ecole Polytechnique 1972-73, to appear.

[M] J. Milnor , Morse theory, Ann. Math. Studies 51, Princeton University Press, 1963.

[Ma] J. Mather , Stratifications and mappings, in : Dynamical systems, Academic Press, 1970; also : Topological stability, Harvard U.P. 1970.

[Me] M. Merle , Idéal jacobien, courbes planes, et équisingularité, thèse 3ème cycle, Paris VII, 1974, to appear (preprint available Centre de Mathématiques de l'Ecole Polytechnique).

[Mi] J. Milnor , Singular points of complex hypersurfaces, Ann. Math. Studies 61, Princeton University Press, 1968.

[N] N. Nagata , Note on a paper of Samuel, Mem. Coll. Sc. Univ. Kyoto t. 30, 1956-57.

[P] H. Prüfer , Untersuchungen über Teilbarkeitseigenschaften in Körpen, J. Reine Angew. Math. 168, 1932.

[P_1] F. Pham , Remarque sur l'équisingularité universelle, Preprint, Université de Nice, 1971.

[P_2] F. Pham , Singularités des courbes planes, cours Paris VII, 1969-70.

[P_3] F. Pham , Déformations équisingulières des idéaux jacobiens ..., in : Liverpool Singularities Symposium II, Springer Lecture Notes No 209, 1971.

[P.T.] F. Pham et B. Teissier , Fractions lipschitziennes d'une algèbre analytique complexe et saturation de Zariski, Preprint Centre de Mathématiques de l'Ecole Polytechnique, M170.0669.

[Re] J. Reeve , Summary of results in the topological classification of plane algebroid singularities, Rendiconti Sem. Mat. Torino 14, 1954-55.

[Ri] J.J. Risler , Sur les déformations équisingulières d'idéaux, Bull. Soc. Math. France 101, fasc. 1, 1973.

[Sa] P. Samuel , Some asymptotic properties of powers of ideals, Ann.

Math. Series 2, t. 56.

[Se] J.P. Serre, Algèbre locale, multiplicités, Springer Lecture Notes No 11, 1965.

[Sp] J.P. Speder, L'équisingularité entraîne les conditions de Whitney, to appear in Am. J. Math., (Preprint, Université de Nice, 1972).

[St] J. Stutz, Equisingularity and equisaturation in codim 1, to appear in Am. J. Math.

[T_1] B. Teissier, Déformations à type topologique constant, in : Séminaire Douady-Verdier 1971-72, Astérisque (Société Mathématique de France) No 16, 1974.

[T_2] B. Teissier, Cycles évanescents, sections planes et conditions de Whitney, Astérisque (Société Mathématique de France) No 7-8, 1973.

[T_3] B. Teissier, Sur les variétés polaires des hypersurfaces, to appear.

[Th_1] R. Thom, Ensembles et morphismes stratifiés, Bulletin A.M.S. 75, 2, 1969.

[Th_2] R. Thom, Sur une partition en cellules associée à une fonctions sur une variété, C. R. Acad. Sc. Paris, Scéance du 14 Mars 1949.

[Th_3] R. Thom, in : Problems concerning manifolds, Manifolds, Amsterdam 1970, Springer Lecture Notes, No 197.

[Ti] Timourian, Preprint I.H.E.S., Bures-sur-Yvette, 1974.

[Tj] G.N. Tjurina, The topological properties of isolated singularities of complex spaces of codimension 1, Izv. Ak. Nauk SSSR, Ser. Mat. 32, 3, 1968 (Math. USSR Iz. 2, 1968).

[V] A.N. Varchenko, Theorems of topological equisingularity ..., Isvestia Ak. Nauk SSSR, Ser. Mat. 36, 1972.

[W_1] H. Whitney, Local properties of analytic varieties, Morse Symposium, Princeton Math. Series No 27, Princeton University Press, 1965.

[Wa] J. Wahl, Equisingular deformations of plane algebroid curves, Preprint, University of Berkeley, 1972.

[Z_1] O. Zariski, Studies in equisingularity I, Am. J. Math. 87, 1965.

[Z_2] - - - - - II, - - - - -

[Z_3] - - - - - III, - - - 90, 1968.

[Z_4] O. Zariski, Contributions to the problem of equisingularity, C.I.M.E. Varenna 1969 (Edizioni Cremonese Roma 1970).

[Z_5] O. Zariski, Some open questions in the theory of singularities, Bull. A.M.S. 77, 4, July 1971.

[Z_6] O. Zariski, General theory of saturation I, II, Am. J. Math. 93,

3-4, 1971.

[Z_7] O. Zariski , Characterization of plane algebroid curves whose module of differentials has maximum torsion, Proc. Nat. Acad. Sciences 56 , 3, 1966.

[Z_8] O. Zariski , General theory of saturation III, to appear in Am. J. Math.

Centre de Mathématiques
de l'Ecole Polytechnique
17, rue Descartes
75230 Paris Cedex 05
(France)

(Added in February, 1975). It has turned out that many problems were too optimistic. Zariski has communicated to me a counterexample to conjecture 1, p. 600 : X is defined in $\mathbb{C}^4$ by $Z^3 - (X_2 - X_1 Z)^2 X_3 = 0$ and Y is the line $X_2 = X_3 = Z$, along which X is even analytically trivial. The branch locus of the projection of X to the (X_1, X_2, X_3) space is not equisingular, although condition β) is satisfied. Briançon and Speder of Nice, Compte rendu notes, Feb. (1975) have given a counterexample to Conjecture B_μ (p. 606) : X is defined in $\mathbb{C}^4$ by $z_o^5 + yz_o z_1^6 + z_1^7 z_2 + z_2^{15} = 0$, and Y is $z_o = z_1 = z_2 = 0$. X is a family of quasi-homogeneous (weights $(z_o, z_1, z_2) = (\frac{3}{15}, \frac{2}{15}, \frac{1}{15})$) hypersurfaces with isolated singularity $(X_y, 0)$, all with the same topological type, and $\mu(X_y, 0)$ is constant (= 364). However, the section of X by a general hyperplane $z_2 = az_o + bz_1$ does not satisfy (μ constant), so X does not satisfy (μ^* constant). Furthermore, (X-Y,Y) does not satisfy Whitney conditions, and worse still, as noted by Henry, X is (Z)-equisingular along Y (the branch locus of the projection to (z_o, z_1, y) is equisingular along Y . This example and [Sp], or other examples of Briançon-Speder, show that the answers to question A (p.599) question I of [Z_5], question B_{top} (p. 620) and the second problem of p. 620 are negative. In the last problem, also, p.628 replace (μ constant) by (μ^* constant).

On the positive side, I have been able to remove the dim Y = 1 assumption in theorem 3 (p. 605), and Briançon-Speder (to appear) have adapted the proof of the proposition p. 606 to show the answer to question B is yes for Whitney-equisingularity when Y the singular locus of X, i.e. in this case (μ^* constant) ⇔ (Whitney equisingularity) ⇔ condition (C). Also I have learnt that Mr. A. Nobile (State U. of Louisiana, Baton Rouge , Louisiana) has given, in his 1970 M.I.T. Thesis, a positive answer to question (C) p. 620 in dimension 2. Zariski communicated to me that the reference needed p. 614 is [Z_6,II, proof of lemme 7.1, page 942-43]

Proceedings of Symposia in Pure Mathematics
Volume 29, 1975

ALGEBRAIC VARIETIES WITH GROUP ACTION

Philip Wagreich

The following is intended to be a brief, unsystematic survey of one part of the algebro-geometric theory of transformation groups and is intended solely to provoke further interest. The author would welcome being informed of any omissions (or errors) he may have made.

For simplicity we shall assume all varieties are reduced, irreducible and finite type over an algebraically closed field k. Let G be an algebraic group. By a G-variety (X,G) we mean a morphism $\sigma: G \times X \longrightarrow X$ satisfying $\sigma(1,x) = x$ and $\sigma(g_1g_2,x) = \sigma(g_1,g_2x)$ for all $g_1,g_2 \in G$, $x \in X$. When no confusion may result $\sigma(g,x)$ is denoted by gx. All actions are assumed effective, i.e., $gx = x$, for all $x \Rightarrow g = 1$.

One approach toward the classification of G-varieties is to find a nice Zariski open subset $X_0 \subset X$ so that X_0/G exists and then to find invariants that, together with X_0/G determine (X,G). In particular we aim to give the invariants explicitly, in terms of the defining equations for X.

§1. Torus Embedding

One extreme case is when $\dim(X_0/G) = 0$. If we assume in addition that $G = (\mathbb{G}_m)^n$ then (X,G) is called a torus embedding [18,26]. If we identify G with $\operatorname{Spec}(k[X_1,X_1^{-1},\ldots,X_n,X_n^{-1}])$ then there is a bijective correspondence between normal affine torus embeddings and finitely generated subsemigroups $S \subset \mathbb{Z}^{(n)}$ which are saturated (i.e., $n\chi \in S$ implies $\chi \in S$). The correspondence is given by $S \longmapsto \operatorname{Spec}(k[S])$ where $k[S] = k[X_1^{i_1}\cdots X_n^{i_n} | (i_1,\ldots,i_n) \in S]$. A subset of $\mathbb{R}^n$ is said to be a convex rational polyhedral cone (c.r.p.c.) if there exist $x_1,\ldots,x_r \in \mathbb{Z}^{(n)}$ so that $\sigma = \mathbb{R}_0x_1 + \cdots + \mathbb{R}_0x_n$ (where $\mathbb{R}_0 = \{x \in \mathbb{R} | x \geq 0\}$). There is a bijective correspondence between c.r.p. cones and semigroups as above given by $\sigma \longmapsto \sigma \cap \mathbb{Z}^{(n)}$. This motivates the following:

Theorem [26; 18, I, Theorem 6]. There is an equivalence of categories between

1) the category of n-dimensional normal torus embeddings
2) the category of finite rational partial polyhedral (f.r.p.p.) decompositions of $\mathbb{R}^n$.

The objects of this second category are defined as follows: A c.r.p. cone is a strongly c.r.p. (s.c.r.p.) cone if $\sigma \cap (-\sigma) = \{0\}$. An (f.r.p.p.) decomposition Σ of $\mathbb{R}^n$ is a collection of s.c.r.p. cones so that

1) if $\sigma \in \Sigma$ and σ' is a face of σ then $\sigma' \in \Sigma$
2) if $\sigma, \tau \in \Sigma$ then $\sigma \cap \tau$ is a face of σ.

The equivalence in the theorem is actually the dual of the one indicated above. Namely, let $(x,y) = \sum_{i=1}^{n} x_i y_i$ and define $\check{\sigma} = \{y \in \mathbb{R}^n | (y,x) \geq 0, \forall x \in \sigma\}$. If $\sigma \in \Sigma$ define $X_\sigma = \mathrm{Spec}(k[\check{\sigma} \cap \mathbb{Z}^n])$. The X_σ patch together to give a separated variety X_Σ with G action. The point of the theorem is that this can all be done canonically, without choosing a basis for $\mathbb{Z}^{(n)}$.

Any property of toroidal embeddings translates into properties of f.r.p.p. decompositions. For example,

1) X is affine $\Leftrightarrow \Sigma$ consists of the faces of one σ
2) X is complete $\Leftrightarrow \mathbb{R}^n = \bigcup_{\sigma \in \Sigma} \sigma$
3) X is non-singular $\Leftrightarrow$ each $\sigma \in \Sigma$ is spanned by a subset of a basis of $\mathbb{Z}^n$.

Torus embeddings have had deep applications in the proof of the "semi-stable reductions theorem" [18, Chapter II] and in construction of compactifications for D/Γ where D is a bounded symmetric domain and Γ is an arithmetic group. Oda and Miyake have applied the theorem above to construct (among others) a 3-dimensional complete non-projective torus embedding (X,G), with an equivariant proper birational morphism $p: X \longrightarrow \mathbb{P}^3$. Note that p cannot be a monoidal transform.

§2. Slice Theorems

A torus embedding will have, in general, orbits of many different dimensions. At the opposite extreme are Seifert actions, i.e., actions which are proper and for which all isotropy groups are finite. (An action is proper if $f: G \times X \longrightarrow X \times X$ defined by $f(g,x) = (gx,x)$ is a proper mapping.) In the category of complex manifolds there is a complete local structure theorem for these actions.

Slice Theorem [17]. Suppose (X,G) is a complex $n+m$ manifold with a

Seifert analytic action of an m-dimensional complex Lie group G. Then for each $x \in X$ there exists

1) a representation σ of the isotropy group G_x in $GL(n,\mathbb{C})$ (called the slice representation)
2) a neighborhood U of $\underline{0}$ in $\mathbb{C}^n$ invariant under σ
3) a neighborhood W of the orbit $G(x)$

such that W is equivariantly isomorphic to $(U \times G)/G_x$ (where the action of G is induced by multiplication on the right hand factor). The image of $U \times \{1\}$ in W is called <u>the slice through</u> x.

It follows from this theorem that locally the orbit space is isomorphic to U/G_x. An algebraic slice theorem has been proved the étale topology by Luna [22] and the classical C^∞ slice theorem appears in [3].

§3. Affine Varieties with Good $\mathbb{C}^*$-action

One situation where the slice theorem has been applied to advantage is in the study of affine varieties with $\mathbb{C}^*$-action. If $q_1,\ldots,q_n$ are integers and $\text{g.c.d.}(q_1,\ldots,q_n) = 1$ then there is an action of $\mathbb{C}^*$ on $\mathbb{C}^n$ defined by $\sigma(t,(z_1,\ldots,z_n)) = (t^{q_1}z_1,\ldots,t^{q_n}z_n)$. If all $q_i > 0$ we say the action is <u>good</u>. In particular, a good action is a proper Seifert action on $\mathbb{C}^n - \{\underline{0}\}$. If V is a closed subvariety invariant under a good $\mathbb{C}^*$-action σ, then V is the zero set of a finite number of quasihomogeneous polynomials, i.e., polynomials so that $f(t^{q_1}z_1,\ldots,t^{q_n}z_n) = t^d f(z_1,\ldots,z_n)$, for all $t \in \mathbb{C}^*$. Some examples of quasihomogeneous polynomials are the Brieskorn polynomials $f(z_1,\ldots,z_n) = \sum_{i=1}^{n} z_i^{a_i}$ $(d = \text{lcm}(a_1,\ldots,a_n)$ and $q_i = d/a_i)$ and polynomials of the form $f(z_1,\ldots,z_n) = z_1^{a_1}z_2 + \cdots + z_n^{a_n}z_1$.

In the above case $V^* = V - \{\underline{0}\}/\mathbb{C}^*$ is a projective algebraic variety. By the slice theorem, if $V - \{\underline{0}\}$ is non-singular then V^* is locally analytically of the form $\mathbb{C}^m/\mathbb{Z}_\alpha$. Letting S denote the unit sphere in $\mathbb{C}^n$ we define $K_V = S \cap V$. This is a smooth manifold if $\underline{0}$ is an isolated singularity. Moreover there is a natural action of S^1 on K_V. Varieties and manifolds of this type were first considered by Brieskorn [4]. He demonstrated precisely which Brieskorn polynomials give rise to a K_V which is a homotopy sphere and then determined the smooth structure on each sphere in terms of the exponents of the polynomial. Affine surfaces with good

$\mathbb{C}^*$-action were studied in [30,31,32]. In particular, the 3-manifold K_V and the resolution of the singularity were determined explicitly in terms of d, q_0, q_1, q_2.

The analytic (= algebraic) classification of certain affine varieties with good $\mathbb{C}^*$-action was obtained in [33]. To give an indication of the nature of that classification we restrict discussion to a special case. Suppose X is a non-singular projective algebraic variety such that $H_1(X,\mathbb{Z}) = 0$ and $H^2(X,\mathbb{Z}) = \mathbb{Z}$.

Theorem. Let W be a variety with $\mathbb{C}^*$-action such that $W/\mathbb{C}^* = X$. Then the set $D \subset X$ corresponding to orbits with non-trivial isotropy is the union of non-singular divisors $D_1, \ldots, D_r$ meeting transversely (this follows from the remark at the end of §2 and the non-singularity of X). If x is a generic point of D_i then the slice representation of the orbit corresponding to x is the sum of a 1-dimensional representation σ_i and a trivial representation. The subgroup A of $\mathbb{C}^*$ generated by the union of all isotropy groups, is finite and W/A is a $\mathbb{C}^*$ bundle over X. Finally W is determined up to equivariant isomorphism by

1) $(D_1,\sigma_1), \ldots, (D_r,\sigma_r)$
2) the element of $\mathrm{Pic}(X)$ determined by W/A.

The relations among the invariants above are described in [33]. This theorem is used there to compute the homology of K_V for certain varieties with $\mathbb{C}^*$-action.

The classification of orbit spaces of quasihomogeneous complete intersections has been studied by Randell [36].

§4. Projective Varieties with $\mathbb{G}_m$-action

A structure theory for complete varieties with $\mathbb{G}_m$-action was initiated by Bialynicki-Birula in [2]. If $\mathbb{G}_m$ acts on X and $x \in X$ is a fixed point then one can find coordinates for the tangent space T_x so that the induced action of $\mathbb{G}_m$ on T_x is of the form $\sigma(t,(v_1,\ldots,v_n)) = (t^{q_1}v_1,\ldots,t^{q_n}v_n)$. Let $(T_x)^+$ be the $\mathbb{G}_m$ submodule spanned by those $v \in T_x$ such that $\sigma(t;v) = t^m v$ for some $m > 0$. Define a morphism $f: U \longrightarrow W$ to be a $\mathbb{G}_m$-fibration if there is a representation $\alpha: \mathbb{G}_m \longrightarrow GL(V)$ and an open cover $\{W_i\}$ of W such that $f^{-1}(W_i) = W_i \times V$ with the action given by α on the second factor.

Theorem [2, 4.3]. Let $k = \bar{k}$ and X a complete k variety. Let

$X^G = \bigcup_{i=1}^{r} (X^G)_i$ be the decomposition of X^G into connected components. Then there exists a unique locally closed G-invariant decomposition of X

$$X = \bigcup_{i=1}^{r} X_i$$

(called the (+) decomposition) and morphisms $\gamma_i : X_i \longrightarrow (X^G)_i$ for $i = 1,\ldots,r$ so that

(i) $(X_i)^G = (X^G)_i$

(ii) γ_i is a G-fibration

(iii) for every closed point $a \in (X^G)_i$

$$T_a(X_i) = T_a((X^G)_i) \oplus (T_a(X))^+.$$

In the same way he defines and proves the existence of a (-) decomposition.

In the case of a non-singular projective surface X with $\mathbb{G}_m$-action this result can be improved. It follows from [2] that if we decompose X^G into irreducible components, $X^G = \bigcup_{i=1}^{r} X_i$, then the X_i are disjoint and at most two of the X_i have dimension 1. Assume that exactly two of the X_i, say X_1 and X_2, are curves (this can always be achieved by blowing up suitable fixed points). An orbit O is said to be ordinary if $\overline{O} \cap X_1 \neq \emptyset$ and $\overline{O} \cap X_2 \neq \emptyset$.[1] Let $O_3,\ldots,O_n$ be the orbits which are not ordinary and define $O_1 = X_1$, $O_2 = X_2$. Define the weighted graph of $(X, \mathbb{G}_m)$ as follows:

1) vertices $e_1,\ldots,e_n$

2) e_i is joined to e_j if $\overline{O}_i \cap \overline{O}_j \neq \emptyset$

3) weight vertex e_i by

$$\underset{[g]}{\textcircled{n}}$$

where $-n$ = self-intersection of X_i and g = genus of X_i (deleted if $g = 0$).

Theorem [34]. (i) X_1 is a non-singular curve isomorphic to X_2.

(ii) The graph of $(X, \mathbb{G}_m)$ is of the form

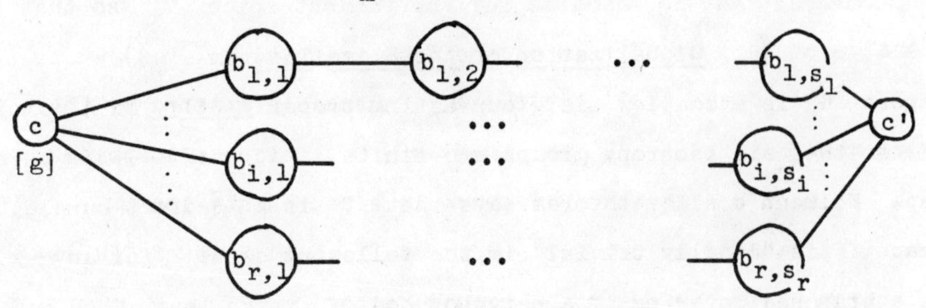

[1] here ‾ denotes closure.

(iii) $b_{ij} \geq 1$ and the continued fraction

$$b_{i,1} - \cfrac{1}{b_{i,2} - \cfrac{1}{\ddots - \cfrac{1}{b_{i,s_i}}}} = 0, \quad \text{for } i = 1,\dots,r$$

(iv) if we define integers c_i by

$$b_{i,1} - \cfrac{1}{b_{i,2} - \cfrac{1}{\ddots - \cfrac{1}{b_{i,s_i-1}}}} = \frac{1}{c_i} \quad \text{for } i = 1,\dots,r$$

then the following equation holds

$$c + c' = \sum_{i=1}^{r} c_i .$$

Conversely any graph as in (ii) satisfying the conditions above arises from some projective surface with $\mathbb{G}_m$ action.

The graph of $(X, \mathbb{G}_m)$ is readily computable using [34], at least for a hypersurface. For example if V is the hypersurface in $\mathbb{P}^3$ defined by

$$Z_0^4 Z_1 Z_3^3 + Z_1^3 Z_2^2 Z_3^3 + Z_0^7 Z_2 = 0$$

with the action

$$\sigma(t,(z_0:z_1:z_2:z_3)) = (t^2 z_0 : t^5 z_1 : t^{-1} z_2 : z_3)$$

and $\tilde{V}$ is the minimal resolution of the singularities of V then σ extends to an action on $\tilde{V}$ and the graph of $(\tilde{V}, \mathbb{C}^*)$ is

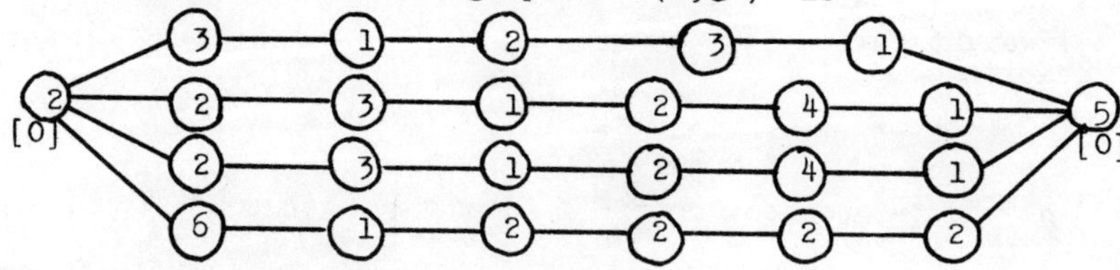

In particular, V is rational.

I understand that Ehlers [13] has obtained interesting results on the resolution of singularities of the form $\mathbb{C}^n/\mathbb{Z}_\alpha$, $n > 2$.

§5. Globalization of the Slice Theorem

Suppose G is a complex Lie group acting properly (§1) on a complex space V so that all isotropy groups are finite. Let $f: V \longrightarrow X$ be the orbit map. Holmann's slice theorem says that f is a "Seifert bundle," i.e., that f is "locally trivial" in the following sense: for any $x \in X$ there is a branched covering of a neighborhood of x so that $V \times_X U \approx G \times U$ over U. One can ask whether this result is true globally. Carrell has shown that this is the case when G is a compact complex torus.

Theorem [5]. Suppose V is a compact Kaehler manifold, G is a complex torus and G acts on V as a group of Kaehler isometries. Then there exists a simply connected complex manifold W, a properly discontinuous holomorphic action of a group N on W and an element $m \in \mathrm{Hom}(N,G)$ such that

1) W/N is compact
2) the action of N on $G \times W$ defined by $n(t,W) = (m(n)t, nw)$ is principal
3) (V,G) is equivalent to $(G \times W/N, G)$ where the second action is obtained by letting G act on the first factor.

It follows from the theorem that $W/N = V/G = X$ and $M \times_X W = G \times W$ so that $f: V \longrightarrow X$ is globally trivialized by the branched cover $W \longrightarrow X$.

In the algebraic case Carrell has obtained the following stronger theorem:

Theorem [5]. If G is an abelian variety acting on a Hodge manifold V then there exists a finite subgroup Δ of G, a connected Hodge manifold F and a holomorphic action (F,Δ) such that (V,G) is isomorphic to $(G \times F/\Delta, G)$. Moreover V is a holomorphic fiber bundle over G/Δ with structure group Δ and fibre F!

The above is probably related to the work of Severi and Kodaira on the classification of elliptic surfaces [19,20].

Seshardri [38] has proven some very general theorems about lifting actions to free actions in the algebraic case.

§6. Modular Forms

It has been observed [7, no. 17] that if Γ is a subgroup of $SL(2,\mathbb{R})$ so that $\Gamma \cap SL(2,\mathbb{Z})$ is of finite index in Γ and $SL(2,\mathbb{Z})$, and A_k denotes the vector space of modular forms of weight k relative to Γ then the graded ring

$$A = \bigoplus_{k=0}^{\infty} A_k$$

is an algebra of finite type over $\mathbb{C}$. The projective variety $\mathrm{Proj}(A)$ is the compactification of H_+/Γ, where H_+ denotes the upper half plane. The affine variety $\mathrm{Spec}(A)$ is a complex surface with a good $\mathbb{C}^*$ action having no singularities other than $\underline{0}$. Dolgacev [11,12] has classified those groups Γ so that H_+/Γ is compact and Spec(A) is a hypersurface in $\mathbb{C}^3$. The group Γ is one of the triangle groups (i.e., the group generated by products of an even number of reflections in the sides of a triangle in H_+ with angles π/n_1, π/n_2, π/n_3). The triples which occur are (2,3,9), (2,4,7), (3,3,6), (2,3,8), (2,4,6), (5,3,3), (2,5,6), (3,4,5), (2,5,5), (3,4,4), (2,3,7), (2,4,5), (3,3,4), and (4,4,4). It is interesting to note that the corresponding surfaces are precisely the unimodal quasihomogeneous singularities of Arnold [1].

Dolgacev's main tools are, I believe, analogues of Carrell's theorems for $G = \mathbb{C}^*$. In particular we have

Theorem [11]. If V is a normal affine surface with good $\mathbb{C}^*$-action then there exists a projective smooth curve X, a finite group of its automorphisms G, and a line bundle $L \in \mathrm{Pic}(X)^G$ such that $V - \{\underline{0}\} \cong L - \{\text{zero section}\}/G$.

This is also closely related to the work of Milnor [24] and the work of Raymond and Vasquez [37].

§7. Vector Fields

Finally, I would like to mention some results on holomorphic vector fields that were communicated to me by Carrell and Lieberman which seem to indicate a strong relationship between varieties admitting vector fields and varieties with group actions. It has been conjectured by Carrell that if V is a non-singular projective variety admitting a vector field $\mathfrak{X}$ having an isolated zero, then V is rational. It follows from [2] and [21] that if V admitted a $\mathbb{C}^*$ action with an isolated fixed point then V would be rational. It is now known that if $\mathfrak{X}$ has a generic zero then V is in fact a torus embedding and if $\mathfrak{X}$ has a simple zero then V is rational. It is also conjectured that if V admits a $\mathbb{G}_a$ action then V is rational. This would imply Carrell's conjecture.

We have the following general theorem.

Theorem [21]. If V is a projective algebraic variety and $\mathfrak{X}$ is a global vector field then $(V,\mathfrak{X})$ is equivariantly birationally equivalent to $(\mathbb{P}^n \times A \times M, \theta \times \tau \times 0)$ where A is an abelian variety, θ is a global vector field on $\mathbb{P}^n$, τ is the vector field generated by a translation and 0 denotes the zero vector field. Moreover, dim M = transcendence degree of the field of functions on V annihilated by $\mathfrak{X}$. If $\mathfrak{X}$ has a zero then A is trivial.

The recent results of Hironaka on resolution of singularities imply that if V and $\mathbb{P}^n \times A \times M$ are birationally equivalent there is a sequence of equivariant blowings up of one of the varieties so that the resultant variety dominates the other. This may be of some use in classifying varieties admitting vector fields.

Bibliography

1. V.I. Arnol'd, Classification of unimodal critical points of functions, Functional Anal. Appl. (3) 7, 230-231 (English translation).

2. A. Bialynicki-Birula, Some theorems on actions of algebraic groups, Ann. of Math. 98(1973), 480-497.

3. G. Bredon, Introduction to Compact Transformation Groups, Academic Press, New York, 1972.

4. E. Brieskorn, Beispiele zur Differentialtopologie von Singularitäten, Invent. Math. 2 (1966), 1-14.

5. J.B. Carrell, Holomorphically-injective complex toral actions, Proceedings of the Second Conference on Compact Transformation Groups, Lecture Notes in Mathematics 299, Springer-Verlag, pp. 205-236.

6. J.B. Carrell and D. Lieberman, Holomorphic vector fields and Kaehler manifolds, Invent. Math. 21 (1973), 303-309.

7. Seminaire Henri Cartan 1957/1958, part 2.

8. P.E. Connor and F. Raymond, Injective operations of the toral groups, Topology 11 (1971), 283-296.

9. P.E. Connor and F. Raymond, Injective operations of the toral groups II, Proceedings of the Second Conference on Compact Transformation Groups, Lecture Notes in Mathematics 299, Springer-Verlag, pp. 109-123.

10. P.E. Connor and F. Raymond, Holomorphic Seifert fibering, ibid., pp. 124-204.

11. I. Dolgacev, letter to P. Wagreich, 1974.

12. I. Dolgacev, Functional Anal. Appl. (Russian) (2) 8 (1974).

13. Ehlers, thesis, University of Bonn (to appear).

14. F. Hirzebruch and K.H. Mayer, O(n)-Mannigfaltigkeiten, exotische Sphären und Singularitäten, Lecture Notes in Mathematics 57, Springer-Verlag, Berlin-Heidelberg-New York.

15. H. Holmann, Quotienten komplexe Räume, Math. Ann. 142 (1961).

16. H. Holmann, Komplexe Räume mit komplexen Transformationsgruppen, Math. Ann. 150 (1963),

17. H. Holmann, Seifertsche Faserräume, Math. Ann. 157 (1964), 138-166.

18. G. Kempf, F. Knudsen, D. Mumford and B. Saint-Donat, Toroidal Embeddings I, Lecture Notes in Mathematics 339, Springer-Verlag, Berlin-Heidelberg-New York.

19. K. Kodaira, On analytic surfaces III, Ann. of Math. 78 (1963), 1-40.

20. K. Kodaira, Compact complex analytic surfaces I, Amer. J. Math. 86 (1964), 751-798.

21. D. Lieberman, forthcoming paper on holomorphic vector fields.

22. D. Luna, Slices étales, Bull. Soc. Math. France. Supplement 1972.

23. J. Milnor, Singular points of complex hypersurfaces, Ann. Math. Stud. 61 (1968).

24. J. Milnor, On the 3-dimensional Brieskorn manifolds $M(p,q,r)$, (preprint).

25. D. Mumford, Geometric Invariant Theory, Academic Press, New York, 1965.

26. T. Oda and K. Miyake, Almost homogeneous algebraic varieties under algebraic torus action (preprint).

27. P. Orlik, Seifert manifolds, Lecture Notes in Mathematics 291, Springer-Verlag, Berlin-Heidelberg-New York.

28. P. Orlik and J. Milnor, Isolated singularities defined by weighted homogeneous polynomials, Topology 9 (1970), 385-393.

29. P. Orlik and R. Randell, The monodromy of a weighted homogeneous variety (preprint).

30.* P. Orlik and P. Wagreich, Isolated singularities of algebraic surfaces with $\mathbb{C}^*$-action, Ann. Math. 93 (1971), 205-227.
*Footnote: Proposition 3.1.2 of the above paper is false. There are actually eight classes of quasihomogeneous polynomials in three variables [29, 34].

31. P. Orlik and P. Wagreich, Singularities of algebraic surfaces with $\mathbb{C}^*$-action, Math. Ann. 193 (1971), 121-135.

32. P. Orlik and P. Wagreich, Equivariant resolution of singularities with $\mathbb{C}^*$-action, Proceedings of the Second Conference on Compact Transformation Groups, Lecture Notes in Mathematics 298, Springer-Verlag, Berlin-Heidelberg-New York.

33. P. Orlik and P. Wagreich, Seifert n-manifolds, Invent. Math. (to appear).

34. P. Orlik and P. Wagreich, Algebraic surfaces with k*-action (preprint).

35. H. Pinkham, Deformations of algebraic varieties with $\mathbb{G}_m$-action, thesis, Harvard University, 1974.

36. R. Randell, The homology of generalized Brieskorn manifolds (preprint).

37. F. Raymond and A. Vasquez, Closed 3-manifolds whose universal cover is a Lie group (to appear).

38. C.S. Seshadri, Quotient spaces modulo reductive algebraic groups, Ann. of Math. 95 (1972), 511-556.

39. H. Sumihiro, Equivariant completion, J. Math., Kyoto Univ. (1) 14 (1974), 1-28.

UNIVERSITY OF ILLINOIS AT CHICAGO CIRCLE